Separation Processes

SECOND EDITION

C. Judson King

Dover Publications, Inc.
Mineola, New York

Bibliographical Note

Separation Processes: Second Edition, first published by Dover Publications, Inc., in 2013, is an unabridged republication of the work originally published by McGraw-Hill Book Company, New York, in 1971 and reprinted in 1980. A new Preface has been provided by the author for the Dover edition.

Library of Congress Cataloging-in-Publication Data

King, C. Judson (Cary Judson), 1934–
    Separation processes / C. Judson King.—Second edition.
        p. cm.
    Originally published: New York : McGraw-Hill, 1971. 2nd ed. Includes a new introduction.
    Includes bibliographical references and index.
    ISBN-13: 978-0-486-49173-8
    ISBN-10: 0-486-49173-0
    1. Separation (Technology) I. Title.

TP156.S45K5 2013
660'.2842—dc23

2013028948

Manufactured in the United States by Courier Corporation
49173001   2013
www.doverpublications.com

CONTENTS

Chapter 3 Additional Factors Influencing Product Purities 103

Chapter 4 Multistage Separation Processes 140

Chapter 5   Binary Multistage Separations: Distillation   206

Chapter 11   Mass-Transfer Rates 508

Chapter **13** Energy Requirements of Separation Processes

PREFACE TO THE DOVER EDITION

It has now been fifty years since I embarked upon the project that led to the publication of the first edition of *Separation Processes,* and thirty-three years since the initial publication of this second edition. What has happened in the field of separation processes over that third to half of a century? It remains one of the most vital components of chemical engineering, with it still often being said that reactions and separations are the two most distinctive features of chemical engineering as a field. We will always have to use our creativity and synthetic and analytic powers in order to separate complex homogeneous solutions into products, recycle material, and treatable wastes. Whereas when I started this book project in the early 1960s, most embodiments of separations were in the petroleum and chemical industries, applications have become much more varied in both nature and scale, ranging now from massive carbon dioxide removal so as to combat global warming to separations on the microscale. Some means of separation have grown substantially in usage, examples being membrane processes and the various recovery processes associated with new biotechnology applications. But the arsenal of methods of separation has remained essentially the same. Only a few new methods of separation have come along, and they are variations on pre-existing themes.

I recently had the opportunity to review my career through an oral history[1]. Despite all conventional advice being to the contrary ("concentrate on your research so as to make tenure"), I chose to undertake this book-writing project early in my faculty career, before tenure came along. In hindsight, I still believe that this was a very good thing to have done. It enabled me to develop and codify my thinking on the subject, gave me the bases for several very fruitful areas of subsequent research, and instilled attitudes and ways of thinking and analyzing

---

1 Regional Oral History Office of the University of California, Berkeley (http://bancroft.berkeley.edu/ROHO/).

that have served me well in a variety of senior university administrative positions during a multi-dimensional career.

I am delighted that Dover Publications has chosen to reissue *Separation Processes*, and I am grateful to Rochelle Kronzek, who generated the idea and shepherded the project. My other thanks and acknowledgments remain the same as for prior editions, but some of them deserve particular emphasis, namely my appreciation and dedication to Jeanne, my wonderful and loving wife of fifty-six years, and my thanks to the University of California and its Berkeley campus for affording me the absolute best of opportunities for professional growth and a most satisfying career.

C. Judson King

Kensington, California
April 2013

PREFACE TO THE SECOND EDITION

My goals for the second edition have been to preserve and build upon the process-oriented approach of the first edition, adding new material that experience has shown should be useful, updating areas where new concepts and information have emerged, and tightening up the presentation in several ways.

The discussion of diffusion, mass transfer, and continuous countercurrent contactors has been expanded and made into a separate chapter. Although this occurs late in the book, it is written to stand on its own and can be taken up at any point in a course or not at all. Substantial developments in computer methods for calculating complex separations have been accommodated by working computer techniques into the initial discussion of single-stage calculations and by a full revision of the presentation of calculation procedures for multicomponent multistage processes. Appendix E discusses block-tridiagonal matrices, which underlie modern computational approaches for complex countercurrent processes, staged or continuous-contact. This includes programs for solving such matrices and for solving distillation problems. Energy consumption and conservation in separations, chromatography and related novel separation techniques, and mixing on distillation plates are rapidly developing fields; the discussions of them have been considerably updated.

At the same time I have endeavored to prune excess verbiage and superfluous examples. Generalized flow bases and composition parameters (B_+, C_-, etc.) have proved to be too complex for many students and have been dropped. The description rule for problem specification remains useful but occupies a less prominent position. Since the analysis of fixed-bed processes and control of separation processes, treated briefly in the first edition, are covered much better and more thoroughly elsewhere, these sections have been largely removed.

In the United States the transition from English to SI (Système International) units is well launched. Although students and practicing engineers must become familiar with SI units and use them, they must continue to be multilingual in units since the transition to SI cannot be instantaneous and the existent literature will not change. In the second edition I have followed the policy of making the units mostly SI, but I have intentionally retained many English units and a few cgs units. Those unfamiliar with SI will find that for analyzing separation processes the activation barrier is low and consists largely of learning that 1 atmosphere is 101.3 kilopascals, 1 Btu is 1055 joules (or 1 calorie is 4.187 joules), and 1 pound is 0.454 kilogram, along with the already familiar conversions between degrees Celsius, degrees Fahrenheit, and Kelvins.

The size of the book has proved awe-inspiring for some students. The first edition has been used as a text for first undergraduate courses in separation processes (or unit operations, or mass-transfer operations), for graduate courses, and as a reference for practicing engineers. It is both impossible and inappropriate to try to use all the text for all purposes, but the book is written so that isolated chapters and sections for the most stand on their own. To help instructors select appropriate sections and content for various types of courses, outlines for a junior-senior course in separation processes and mass transfer as well as a first-year graduate course in separation processes taught recently at Berkeley are given on page xxiv.

Another important addition to the text is at least two new problems at the end of most chapters, chosen to complement the problems retained from the first edition. Since a few problems from the first edition have been dropped, the total number of problems has not changed. A Solutions Manual, available at no charge for faculty-level instructors, can be obtained by writing directly to me.

I have gained many debts of gratitude to many persons for helpful suggestions and other aid. I want to thank especially Frank Lockhart of the University of Southern California, Philip Wankat of Purdue University, and John Bourne of ETH Zürich, for detailed reviews in hindsight of the first edition, which were of immense value in planning this second edition. J. D. Seader, of the University of Utah, and Donald Hanson and several other faculty colleagues at the University of California, Berkeley, have provided numerous helpful discussions. Professor Hanson also kindly gave an independent proofreading to much of the final text. Christopher J. D. Fell of the University of South Wales reviewed Chapter 12 and provided several useful suggestions. George Keller, of Union Carbide Corporation, and Francisco Barnés, of the National University of Mexico, shared important new ideas. Several consulting contacts over the years have furnished breadth and reality, and many students have provided helpful suggestions and insight into what has been obfuscatory. Finally, I am grateful to the University of California for a sabbatical leave during which most of the revision was accomplished, to the University of Utah for providing excellent facilities and stimulating environment during that leave, and to my family and to places such as the Escalante Canyon and the Sierra Nevada for providing occasional invaluable opportunities for battery recharge during the project.

C. Judson King

PREFACE TO THE FIRST EDITION

This book is intended as a college or university level text for chemical engineering courses. It should be suitable for use in any of the various curricular organizations, in courses such as separation processes, mass-transfer operations, unit operations, distillation, etc. A primary aim in the preparation of the book is that it be complementary to a transport phenomena text so that together they can serve effectively the needs of the unit operations or momentum-, heat-, and mass-transfer core of the chemical engineering curriculum.

It should be possible to use the book at various levels of instruction, both undergraduate and postgraduate. Preliminary versions have been used for a junior-senior course and a graduate course at Berkeley, for a sophomore course at Princeton, for a senior course at Rochester, and for a graduate course at the Massachusetts Institute of Technology. A typical undergraduate course would concentrate on Chapters 1 through 7 and on some or all of Chapters 8 through 11. In a graduate course one could cover Chapters 1 through 6 lightly and concentrate on Chapters 7 through 14. There is little that should be considered as an absolute prerequisite for a course based upon the book, although a physical chemistry course emphasizing thermodynamics should probably be taken at least concurrently. The text coverage of phase equilibrium thermodynamics and of basic mass-transfer theory is minimal, and the student should take additional courses treating these areas.

Practicing engineers who are concerned with the selection and evaluation of alternative separation processes or with the development of computational algorithms should also find the book useful; however, it is not intended to serve as a comprehensive guide to the detailed design of specific items of separation equipment.

The book stresses a basic understanding of the concepts underlying the selection, behavior, and computation of separation processes. As a result several chapters are almost completely qualitative. Classically, different separation processes, such as distillation, absorption, extraction, ion exchange, etc., have been treated individually and sequentially. In a departure from that approach, this book considers separations as a general problem and emphasizes the many common aspects of the functioning and analysis of the different separation processes. This generalized development is designed to be more efficient and should create a broader understanding on the part of the student.

The growth of the engineering science aspects of engineering education has created a major need for making process engineering and process design sufficiently prominent in chemical engineering courses. Process thinking should permeate the entire curriculum rather than being reserved for a final design course. An important aim of this textbook is to maintain a flavor of real processes and of process synthesis and selection, in addition to presenting the pertinent calculational methods.

The first three chapters develop some of the common principles of simple separation processes. Following this, the reasons for staging are explored and the McCabe-Thiele graphical approach for binary distillation is developed. This type of plot is brought up again in the discussions of other binary separations and multicomponent separations and serves as a familiar visual representation through which various complicated effects can be more readily understood. Modern computational approaches for single-stage and multistage separations are considered at some length, with emphasis on an understanding of the different conditions which favor different computational approaches. In an effort to promote a fuller appreciation of the common characteristics of different multistage separation processes, a discussion of the shapes of flow, composition, and temperature profiles precedes the discussion of computational approaches for multicomponent separations; this is accomplished in Chapter 7. Other unique chapters are Chapter 13, which deals with the factors governing the energy requirements of separation processes, and Chapter 14, which considers the selection of an appropriate separation process for a given separation task.

Problems are included at the end of each chapter. These have been generated and accumulated by the author over a number of years during courses in separation processes, mass-transfer operations, and the earlier and more qualitative aspects of process selection and design given by him at the University of California and at the Massachusetts Institute of Technology. Many of the problems are of the qualitative discussion type; they are intended to amplify the student's understanding of basic concepts and to increase his ability to interpret and analyze new situations successfully. Calculational time and rote substitution into equations are minimized. Most of the problems are based upon specific real processes or real processing situations.

Donald N. Hanson participated actively in the early planning stages of this book and launched the author onto this project. Substantial portions of Chapters 5, 7, 8, and 9 stem from notes developed by Professor Hanson and used by him for a number of years in an undergraduate course at the University of California. The presentation in Chapter 11 has been considerably influenced by numerous discussions with Edward A. Grens II. The reactions, suggestions, and other contributions of teaching assistants and numerous students over the past few years have been invaluable, particularly those from Romesh Kumar, Roger Thompson, Francisco Barnés, and Raul Acosta. Roger Thompson also assisted ably with the preparation of the index. Thoughtful and highly useful reviews based upon classroom use elsewhere were given by William Schowalter, J. Edward Vivian, and Charles Byers.

Thanks of a different sort go to Edith P. Taylor, who expertly and so willingly prepared the final manuscript, and to her and several other typists who participated in earlier drafts.

Finally, I have three special debts of gratitude: to Charles V. Tompkins, who awakened my interests in science and engineering; to Thomas K. Sherwood, who brought me to a realization of the importance and respectability of process design and synthesis in education; and to the University of California at Berkeley and numerous colleagues there who have furnished encouragement and the best possible surroundings.

C. Judson King

POSSIBLE COURSE OUTLINES

Junior-senior undergraduate course on separation processes and mass transfer

Follows a course on basic transport phenomena (including diffusion), fluid flow, and heat transfer; four units; 10 weeks (quarter system); two 80-min lecture periods per week, plus a 50-min discussion period used for taking up problems, answering questions, etc.

Lecture	Topic(s)	Chapter	Pages
1	Organize course; general features of separation processes	1	1–30
Disc. 1	Review phase equilibrium; bubble and dew points	1, 2	30–43, 61–68
2	Simple equilibration	2	68–80
3	Principles of staging	4	140–164
4	Binary distillation	5	206–223
5	Binary distillation	5	233–237
6	Half-hour quiz; use of efficiencies in binary distillation	5, App. D	237–243 798–801
7	Use of McCabe-Thiele diagram for other processes	6	258–273
8	Dilute systems; absorption and stripping; KSB equations	8	360–370
9	Midterm examination		
10	Continue KSB equations; introduction to multicomponent multistage separations	7	371–376 325–331
11	Total reflux, minimum reflux, and approximate distillation calculations	9	417–432
12	Multicomponent separations: review simple equilibrium and single stage; introduction to multistage calculations	2 10	80–90 446–455
13	Survey of methods for computation of multistage separations	10	466–481
14	Half-hour quiz; mass-transfer coefficients	11	518–528
15	Mass-transfer coefficients; interphase mass transfer	11	536–545

Lecture	Topic(s)	Chapter	Pages
16	Mass transfer; simultaneous heat and mass transfer (introduction only)	11	545–556
17	Transfer units; continuous countercurrent contactors	11	556–566
18	Factors governing equipment capacity	12	591–608
19	Stage efficiency	12	608–617, 621–626
20	Case problem; review	12	641–651
	Final Examination		
Additional or alternate topics:			
	Survey of factors affecting product compositions in equilibration processes	3	103–115
	Ponchon-Savarit diagrams for distillation	6	273–283
	Triangular diagrams for extraction	6	283–293
	Multieffect evaporation	App. B	785–790
	Rayleigh equation	3	115–123
	Convergence methods	App. A	777–784

Graduate course on separation processes

Three units; 10 weeks (quarter system); two 80-min lecture periods per week; substantial class time spent for problem discussion; there is a subsequent graduate course on mass transfer

Class	Topic	Chapter	Pages
1	Organization; common features and classification of separation processes	1	17–48
2	Factors affecting equilibration and selectivity in separation processes, flow-configuration effects; Rayleigh equation	3	103–123
3	Fixed beds; chromatography	3, 4	123–130 175–180, 183–187
4	Staging; graphical analysis of countercurrent staged separation processes	5, quick review, 6	[207–250] 258–273
5	Generalized graphical analysis of countercurrent staged separation processes	6	283–293
6	Patterns of change in countercurrent separation processes; extractive and azeotropic distillation	7	309–349
7	First midterm examination		
8	KSB equations	8	361–376
9	Group methods of calculation: Underwood equations	8	393–406
10	Group methods; limiting conditions	9	414–427

Separation Processes

USES AND CHARACTERISTICS OF SEPARATION PROCESSES

Die Entropie der Welt strebt einem Maximum zu.

CLAUSIUS

When salt is placed in water, it dissolves and tends to form a solution of uniform composition throughout. There is no simple way to separate the salt and the water again. This tendency of substances to mix together intimately and spontaneously is a manifestation of the second law of thermodynamics, which states that all natural processes take place so as to increase the entropy, or randomness, of the universe. In order to separate a mixture of species into products of different composition we must create some sort of device, system, or process which will supply the equivalent of thermodynamic work to the mixture in such a way as to cause the separation to occur.

For example, if we want to separate a solution of salt and water, we can (1) supply heat and boil water off, condensing the water at a lower temperature; (2) supply refrigeration and freeze out pure ice, which we can then melt at a higher temperature; (3) pump the water to a higher pressure and force it through a thin solid membrane that will let water through preferentially to salt. All three of these approaches (and numerous others) have been under active study and development for producing fresh water from the sea.

The fact that naturally occurring processes are inherently mixing processes has been recognized for over a hundred years and makes the reverse procedure of " unmixing " or *separation processes* one of the most challenging categories of engineering problems. We shall define separation processes as *those operations which transform a mixture of substances into two or more products which differ from each other in composition.* The many different kinds of separation process in use and their importance to mankind should become apparent from the following two examples, which concern basic human wants, food and clothing.

1

AN EXAMPLE: CANE SUGAR REFINING

Common white granulated sugar is typically 99.9 percent sucrose and is one of the purest of all substances produced from natural materials in such large quantity. Sugar is obtained from both sugar cane and sugar beets.

$$\begin{array}{cccc}
& & & H_2COH \\
H & & & | \\
\setminus C & \text{---}O\text{---} & C \\
| & & | \\
HCOH & & HOCH \\
| & & | \\
HOCH \quad O & & HCOH \quad O \\
| & & | \\
HCOH & & HC \\
| & & | \\
HC & & H_2COH \\
| & & \\
HOCH_2 & &
\end{array}$$

Sucrose

Cane sugar is normally produced in two major blocks of processing operations (Gerstner, 1969; Shreve and Brink, 1977, pp. 506–523). Preliminary processing takes place near where the sugar cane is grown (Hawaii, Puerto Rico, etc.) and typically consists of the following basic steps, shown in Fig. 1-1.

1. *Washing and milling.* The sugar cane is washed with jets of water to free it from any field debris and is then chopped into short sections. These sections are passed through high-pressure rollers which squeeze sugar-laden juice out of the plant cells. Some water is added toward the end of the milling to leach out the last portions of available sugar. The remaining cane pulp, known as *bagasse*, is used for fuel or for the manufacture of insulating fiberboard.
2. *Clarification.* Milk of lime, $Ca(OH)_2$, is added to the sugar-laden juice, which is then heated. The juice next enters large holding vessels, in which coagulated colloidal material and insoluble calcium salts are settled out. The scum withdrawn from the bottom of the clarifier is filtered to reclaim additional juice, which is recycled.
3. *Evaporation, crystallization, and centrifugation.* The clarified juices are then sent to steam-heated evaporators, which boil off much of the water, leaving a dark solution containing about 65% sucrose by weight. This solution is then boiled in vacuum vessels. Sufficient water is removed through boiling for the solubility limit of sucrose to be surpassed, and as a result sugar crystals form. The sugar crystals are removed from the supernatant liquid by centrifuges. The liquid product, known as *blackstrap molasses*, is used mainly as a component of cattle feed.

The solid sugar product obtained from this operation contains about 97% sucrose, and is often shipped closer to the point of actual consumption for further processing.

Figure 1-2 shows a flow diagram of a large sugar refinery in Crockett, California, which refines over 3 million kilograms per day of raw sugar produced in Hawaii. As a first step, the raw sugar crystals are mixed with recycle syrup in *minglers* so as to soften the film of molasses adhering to the crystals. This syrup is removed in centri-

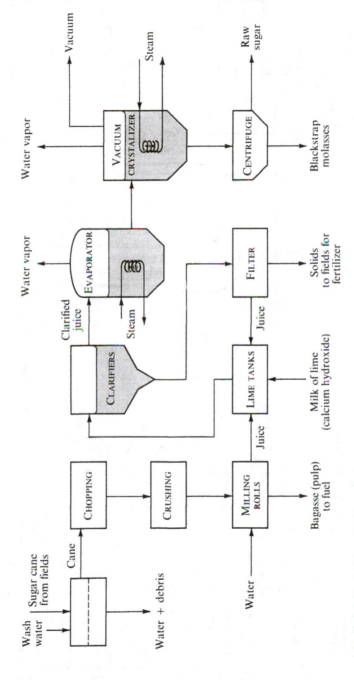

Figure 1-1 Processing steps for producing raw sugar from sugar cane.

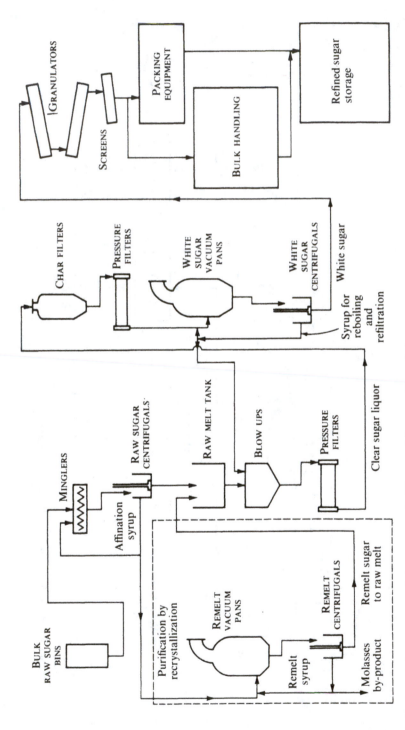

Figure 1-2 Cane-sugar refining process. (*California and Hawaiian Sugar Co.*)

Figure 1-3 Leaf filter used with diatomaceous-earth filter aid to remove insolubles after addition of $Ca(OH)_2$ and H_3PO_4. *(California and Hawaiian Sugar Co.)*

fuges and is both recycled and processed for further sugar recovery. The sugar crystals (now at a sugar content of 99%) are next dissolved in hot water and treated (in vessels called *blowups*) with calcium hydroxide and phosphoric acid to precipitate foreign substances which form insoluble compounds with these chemicals. Diatomaceous earth, a spongy porous material, added as a filter aid, serves to provide an extensive amount of solid surface, which facilitates the removal of insolubles in the *filtration* step that follows. The large leaf-type filter used at this point in the plant under consideration is shown in Fig. 1-3.

The sugar syrup is next passed slowly through large beds of granular animal-derived *charcoal* (boneblack, or char). The purpose of this step (Fig. 1-4) is to adsorb color-causing substances and other remaining impurities onto the surface of the char. (Note the size of the man in Fig. 1-4.) The sugar solution is then freed of excess water by boiling in steam-heated evaporators, or *vacuum pans* (Fig. 1-5). Again, sufficient water is boiled off to cause the solubility limit of sucrose to be exceeded, and crystals of pure sugar form, leaving essentially all remaining impurities behind in the solution, which is recycled to appropriate earlier points in the process. The recycle solution is removed from the pure sugar crystals in large *centrifuges*, 14 of which are shown in Fig. 1-6.

Figure 1-4 Char beds, used for decolorizing sugar syrup. *(California and Hawaiian Sugar Co.)*

Before packaging for sale, the sugar crystals must be *dried*, since they still contain about 1% water. This is accomplished by tumbling the crystals through a stream of warm air of low humidity in large, slightly inclined, horizontal revolving cylinders called *granulators*. Figure 1-7 shows the interior of a granulator, the sugar granules falling off short shelves attached to the rotating wall which serve to distribute the sugar in the warm air. After drying, the sugar is passed through a succession of *screens* of different mesh sizes to segregate crystals of different sizes (fine, coarse, etc.).

There are 11 different classes of separation processes included in the steps shown in Fig. 1-1 for making raw sugar and in the sugar refining processes shown in Fig. 1-2:

1. *Settling (clarifiers)*. A suspension of solids in a liquid is held in a tank until the solids settle to the bottom to form a thick slurry or scum. Solids-free liquid is withdrawn from the top of the vessel. This process requires that the solids be denser than the liquid.
2. *Filtration (scum filter, pressure filters)*. A solid-in-liquid suspension is made to flow through a filter medium such as a fine mesh or a woven cloth. The pores of the filter medium are so small that the liquid can pass through but the solid particles cannot. Sometimes a filter aid (diatomaceous earth) is added to form a still more effective filter medium on top of the mesh or cloth. Filtration is a separation based on *size*, whereas settling is a separation based on *density*.

Figure 1-5 A vacuum pan used for evaporating water and crystallizing pure sugar. *(California and Hawaiian Sugar Co.)*

Figure 1-6 Centrifuges for recovering pure sugar crystals from syrup. *(California and Hawaiian Sugar Co.)*

7

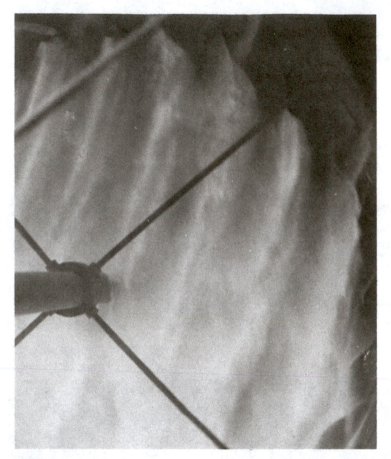

Figure 1-7 Interior view of granulator. *(California and Hawaiian Sugar Co.)*

3. *Centrifugation (raw-sugar centrifuges, white-sugar centrifugals)*. A solid-in-liquid suspension is whirled rapidly. The centrifugal force from the rotation aids the phase separations. Centrifuges may operate on a *settling* principle, wherein the denser phase is brought to the outside by the centrifugal force, or on a *filtration* principle, as in a basket centrifuge, where the mesh of the basket retains solid particles and the centrifugal force causes the liquid to flow through the solids in the basket more readily than in an ordinary filter.
4. *Screening (classification by crystal size)*. Particles are shaken on a screen. The smaller particles pass through the screen, and the larger particles are retained.
5. *Expression (milling rolls)*. Mechanical force is used to squeeze a liquid out of a substance containing both solid and liquid.
6. *Washing and leaching (debris removal, water addition to milling rolls, minglers)*. Soluble material is removed from a mixture of solids by dissolution into a solvent liquid.
7. *Precipitation (lime tanks, blowups)*. A chemical reactant is added to a liquid solution causing some, but not all, of the substances in solution to form new insoluble compounds.
8. *Evaporation (evaporators, vacuum pans)*. Heat is added to a liquid containing nonvolatile solutes in a volatile solvent. The solvent is boiled away, leaving a more concentrated solution. Solvent can be recovered by condensing the vapor.

9. *Crystallization* (*vacuum pans*). A liquid is cooled and/or concentrated so as to cause the formation of an equilibrium solid phase with a composition different from that of the liquid.
10. *Adsorption* (*char filters*). Trace impurities in a fluid phase are retained preferentially on the surface of a solid phase, being held there by van der Waals forces (physical adsorption) or chemical bonds (chemical adsorption).
11. *Drying* (*granulators*). Water is caused to evaporate from a solid substance by the addition of heat and the circulation of an inert-gas stream of low humidity.

ANOTHER EXAMPLE: MANUFACTURE OF *p*-XYLENE

Figure 1-8 presents a simple schematic of an industrial process for the manufacture of *p*-xylene from crude oil. *p*-Xylene is an important petrochemical which is an intermediate for the manufacture of terephthalic acid, $HOOCC_6H_4COOH$, and dimethyl terephthalate, $CH_3OOCC_6H_4COOCH_3$, both of which are raw materials for the manufacture of polyester fibers (Dacron, etc.) (Shreve and Brink, 1977, p. 612). *p*-Xylene is one of the three xylene isomers,

| Ortho | Meta | Para |

all of which have quite similar physical properties. *p*-Xylene was produced to the extent of about 1.4 billion kilograms (3.1 billion pounds) in 1977 at an average sales price of 29 cents per kilogram, for a sales volume of about $400 million (*Chem. Mark. Rep.*, Dec. 26, 1977).

Not shown in Fig. 1-8 is a primary distillation step, which separates crude oil into various streams boiling at different temperatures. The naphtha feed for the xylene manufacture process is typically a stream boiling between 120 and 230 K. The naphtha is charged to a high-temperature high-pressure chemical reactor system called a *reformer*, where reactions convert much of the largely paraffin naphtha into aromatic molecules. Typical reactions include *cyclization*, i.e., conversion of *n*-hexane into cyclohexane, etc.,

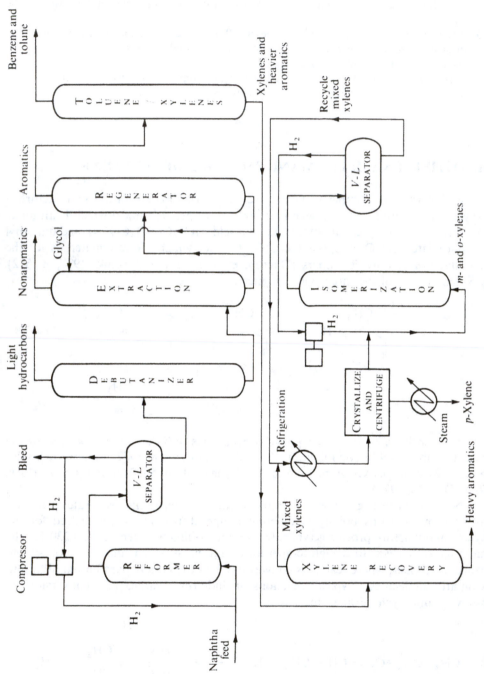

Figure 1-8 Production of *p*-xylene.

10

and *aromatization*, i.e., conversion to cyclohexane into benzene plus hydrogen, etc.,

$$
\begin{array}{c}
H_2 \\
C \\
H_2C \quad CH_2 \\
| \qquad | \\
H_2C \quad CH_2 \\
C \\
H_2
\end{array}
\longrightarrow \quad \bigcirc \quad + \quad 3H_2
$$

The product aromatics are a mixture of benzene, toluene, the xylene isomers, and higher aromatics. The catalyst for the reformer must be protected against deactivation by a hydrogen atmosphere. Since hydrogen is costly, it is recovered in a vapor-liquid separator for recycle. A bleed stream taken off from the recycle gas is necessary in order to remove the net amount of hydrogen formed in the cyclization and aromatization reactions.

In order to obtain an optimal product distribution pattern from the reforming reaction, the reaction is usually carried out in a series of catalyst beds, each operating at a different temperature. The six reforming reactors incorporated into one industrial plant are shown in Fig. 1-9. This plant receives a feed of 40,000 bbl/day of naphtha.

Figure 1-9 Six spheroidal reactors in a catalytic reforming unit at the Texas City, Texas, refinery of the American Oil Company. (*Pullman Kellogg Co.*)

In the units of basic chemistry this turns out to be a flow rate of 74 L/s! (Notice the size of the man in the photograph.)

Most of the output from an oil refinery reforming unit like this becomes high-octane gasoline, but the product stream is also suitable for the production of xylenes and other aromatics. As shown in Fig. 1-8, for *p*-xylene manufacture a portion of the effluent from the separator passes to a distillation tower, which removes butane and lighter molecules. The remaining material passes to a liquid-liquid extraction process, in which the hydrocarbon stream is contacted with an immiscible stream of

Figure 1-10 Rotating-disk contactors used for the extraction of aromatics from other hydrocarbons in the aromatics recovery unit at the Texas City, Texas, refinery of the American Oil Company. The RDC vessels are each 20 m long and 3.4 m in diameter. The solvent in this plant is Sulfolane. (*Pullman Kellogg Co.*)

a solvent, such as diethylene glycol. The aromatics dissolve preferentially in the solvent, while the paraffins and naphthenes (cyclic nonaromatics) do not.

One type of device for carrying out this extraction process is shown in Fig. 1-10. In the foreground are four rotating-disk contactors, which are large vessels containing a number of horizontal disks mounted on a vertical motor-driven stirrer shaft. The four motors mounted on top of the vessels are visible in Fig. 1-10. These rotating disks agitate the two immiscible liquid phases (hydrocarbon and solvent) inside the vessels and thus promote a high rate of dissolution of the aromatics into the solvent phase.

The aromatic-laden solvent stream passes to a distillation tower, which separates the aromatics from the solvent. The solvent is then recycled to the extraction step. The distillation tower for separating the aromatics from the solvent is located behind the rotating-disk contactors in Fig. 1-10. Note the scaffolding around the tower, which is a sign that the photograph was taken during plant construction, and again note the size of the men in the photograph.

Figure 1-11 Distillation towers in the aromatics recovery unit at the Texas City, Texas, refinery of the American Oil Company. (*Pullman Kellogg Co.*)

Figure 1-12 Xylene-feed preparation towers at the Pascagoula, Mississippi, refinery of the Standard Oil Company of California. *(Chevron Research Co.)*

Two more distillation towers follow the extraction step in Fig. 1-8, the first removing benzene and toluene from the xylenes and heavier aromatics and the second removing the heavier aromatics from the xylenes.

The distillation towers used for this purpose are among those shown in the overview of the aromatics recovery unit in Fig. 1-11. A closeup view of two such towers from another plant is shown in Fig. 1-12, where the towers are still under construction, as is evidenced by the crane and scaffolding.

At this point there is a stream of mixed xylene isomers, which is next chilled below the freezing point. The resultant crystals are composed of p-xylene; hence removal of crystals through centrifugation or filtration accomplishes a separation of the para isomer from the ortho and meta isomers. The p-xylene is melted and taken as product, while the supernatant liquid is sent to an isomerization reactor which provides an equilibrium mix of all three xylene isomers. Again there is a solid catalyst in the reactor which must be protected by a high-pressure circulating hydrogen stream. The equilibrium xylene mix is recycled to the crystallizer, and in this way essentially the entire xylene cut is converted into *p*-xylene product.

In practice, the *p*-xylene manufacture process shown in Fig. 1-8 probably would be part of a much larger installation, most likely of a petroleum refinery manufacturing several different gasoline streams, one of which would be a major portion of the total reformer effluent. The aromatics facility might also be larger, including provision for additional products of benzene, toluene, *o*-xylene, etc., all of which are large-volume petrochemicals. There might also be a hydrodealkylation facility for converting toluene into benzene.

Four different classes of separation process are covered in Fig. 1-8:

1. *Vapor-liquid separation (hydrogen recovery)*. A multiphase stream is allowed to separate into vapor and liquid phases of different compositions, each of which is removed individually. In some cases heating (evaporation) or pressure reduction (expansion) may be necessary to cause the formation of vapor.
2. *Distillation (debutanizer, regenerator, toluene-xylene splitter, xylene recovery)*. This is a separation based upon differences in boiling points, obtained by repeated vaporization and condensation steps.
3. *Extraction (preferential dissolution of aromatics into glycol)*. Two immiscible liquid phases are brought into contact, and the substances to be separated dissolve to different extents in the different phases.
4. *Crystallization (p-xylene recovery)*. A solid phase is formed by partial freezing of a liquid, and the two resulting phases have different compositions.

Numerous other approaches have been investigated for separating the xylene isomers (see Chap. 14); some of them are used in large-scale plants.

IMPORTANCE AND VARIETY OF SEPARATIONS

The separation sequence presented in the *p*-xylene manufacture example is representative of processes which are based upon chemical reactions. The reactor effluent is necessarily a mixture of chemical compounds: the desired product, side products, unconverted reactants, and possibly the reaction catalyst. Typically, the desired product must be separated from this mix in relatively pure form, and the unconverted reactants and any catalyst should be recovered for recycle. All the reactants may have to be prepurified. Separation processes of some sort are required for these purposes. Separation equipment accounts for 50 to 90% of the capital investment in large-scale petroleum and petrochemical processes centered around chemical reactions.

Often separation itself can be the main function of an entire process. Such is the case for sugar refining and for such diverse processes as refining natural gas (Medici,

1974); recovery of metals from various mineral resources (Wadsworth and Davis, 1964; Pehlke, 1973); the manufacture of oxygen and nitrogen from air (Latimer, 1967); many aspects of food processing, e.g., dehydration, removal of toxic or objectionable components, and obtaining beverage extracts (Loncin and Merson, 1979); and the separation of fermentation broths and many other details of pharmaceutical manufacture (Shreve and Brink, 1977, pp. 525–544 and 753–788).

Water and air pollution present separation problems of immense social importance. Indeed, water and air pollution are striking examples of the point made earlier that naturally occurring processes are mixing processes. As water is used for various purposes, it picks up undesirable solutes which contaminate it. Similarly, noxious substances emitted into the atmosphere quickly spread throughout the atmosphere and cause smog and related problems. A number of different separation processes may be used for the removal of contaminants in waste air and water effluents. Alternatively, a separation process may be employed at an earlier point, e.g., for the removal of sulfur from fuel oils before combustion. Separations also play an important role in schemes to *reprocess solid wastes*, such as municipal refuse (Mallan, 1976).

The story of the Manhattan Project of World War II is to a major extent a story of efforts to develop a process and construct a plant for the separation of fissionable ^{235}U from the more abundant ^{238}U (Smythe, 1945; Love, 1973). Thermal diffusion, gaseous diffusion, gas centrifuges, and electromagnetic separation by a scheme similar to the mass spectrometer (in a device known as the *calutron*, after the University of California) were all investigated along with other approaches; gaseous diffusion ultimately was the most successful process.

As noted at the beginning of this chapter, a prime example of a separation problem of current importance is the *desalination of seawater* in an economical fashion. Processes considered on research, development, and production scales (Spiegler, 1962) include evaporation through heating, evaporation through pressure reduction, freezing to form ice crystals, formation of solid salt-free clathrate compounds with hydrocarbons, electrodialysis, reverse osmosis, precipitation of the solids content by elevation to a temperature and pressure above the critical point, preferential extraction of the water into phenol or triethylamine, and ion exchange.

ECONOMIC SIGNIFICANCE OF SEPARATION PROCESSES

Often the need for separation processes accounts for most of the cost of a pure substance. Figure 1-13 shows that there is a roughly inverse proportion between the market prices of a number of widely varied commodities and their concentrations in the mixtures in which they are found. This relationship reflects the need for processing a large amount of extra material when the desired substance is available only in low concentration. There is also a thermodynamic basis for such a relationship, since the minimum isothermal work of separation for a pure species is proportional to $-\ln a_i$, where a_i is the activity of the species in the feed mixture. The activity, in turn, is roughly proportional to the concentration.

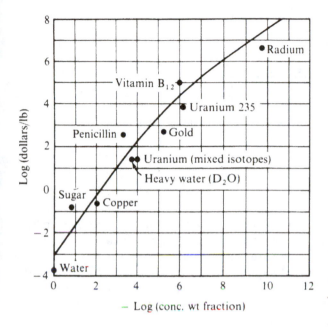

Figure 1-13 Relation between the values of pure substances and their concentrations in the mixtures from which they are obtained. (*Adapted from Sherwood et al., 1975, p. 4; used by permission.)*

It will become apparent that distillation is the most prominent separation process used in petroleum refining and petrochemical manufacture. Zuiderweg (1973) has estimated that the total investment for distillation equipment for such applications over the 20 years from 1950 to 1970 was $2.7 billion, representing a savings of $2.0 billion dollars over what would have been the cost if there had been no improvement in distillation technology over that 20-year period. Clearly there is a large incentive for research directed toward the improvement of separation processes and the development of new ones.

From a consideration of all the various processes which have been mentioned, it is apparent that much careful thought and effort must go into *understanding* various separation processes, into *choosing* a particular type of operation to be employed for a given separation, and into the detailed *design and analysis* of each item of separation equipment. These problems are the main theme of this book.

CHARACTERISTICS OF SEPARATION PROCESSES

Separating Agent

A simple schematic of a separation process is shown in Fig. 1-14. The feed may consist of one stream of matter or of several streams. There must be at least two product streams which differ in composition from each other; this follows from the fundamental nature of a *separation*. The separation is caused by the addition of a *separating agent*, which takes the form of another stream of matter or energy.

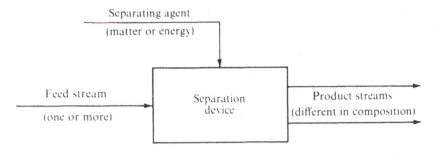

Figure 1-14 General separation process.

Usually the energy input required for the separation is supplied with the separating agent, and generally the separating agent will cause the formation of a second phase of matter. For example, in the evaporation steps in Figs. 1-1 and 1-2 the separating agent is the heat (energy) supplied to the evaporators; this causes the formation of a second (vapor) phase, which the water preferentially enters. In the extraction process in Fig. 1-8 the separating agent is the diethylene glycol solvent (matter); this forms a second phase which the aromatics enter selectively. Energy for that separation is supplied as heat (not shown) to the regenerator, which renews the solvent capacity of the circulating glycol by boiling out the extracted aromatics.

Categorizations of Separation Processes

In some cases a separation device receives a *heterogeneous* feed consisting of more than one phase of matter and simply serves to separate the phases from each other. For example, a filter or a centrifuge serves to separate solid and liquid phases from a feed which may be in the form of a slurry. The vapor-liquid separators in Fig. 1-8 segregate vapor from liquid. A Cottrell precipitator accomplishes the removal of fine solids or a mist from a gas stream by means of an imposed electric field. We shall call such processes *mechanical separation processes*. They are important industrially but are not a primary concern of this book.

Most of the separation processes with which we shall be concerned receive a *homogeneous* feed and involve a diffusional transfer of matter from the feed stream to one of the product streams. Often a mechanical separation process is employed to segregate the product phases in one of these processes. We shall call these *diffusional separation processes*; they are the principal subject matter of this book.

Most diffusional separation processes operate through equilibration of two immiscible phases which have different compositions at equilibrium. Examples are the evaporation, crystallization, distillation, and extraction processes in Figs. 1-1, 1-2, and 1-8. We shall call these *equilibration processes*. On the other hand, some separation processes work by virtue of differences in transport rate through some medium under the impetus of an imposed force, resulting from a gradient in pressure, temperature, composition, electric potential, or the like. We shall call these *rate-governed processes*.

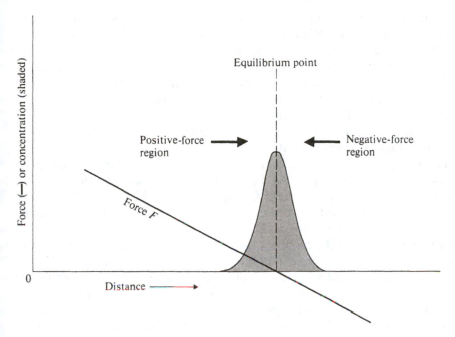

Figure 1-15 Elements of an imposed-gradient equilibration process. *(Adapted from Giddings and Dahlgren, 1971, p. 346; by courtesy of Marcel Dekker, Inc.)*

Usually rate-governed processes give product phases that would be fully miscible if mixed with each other, whereas ordinary equilibration processes necessarily generate products that are immiscible with each other.[†]

Summarizing, we can categorize separation processes in several ways:

Mechanical (heterogeneous-feed) vs. diffusional (homogeneous-feed) processes
Equilibration processes vs. rate-governed processes
Energy-separating-agent vs. mass-separating-agent processes

A relatively new subcategory of equilibration processes is that of *imposed-gradient equilibration* processes, illustrated conceptually in Fig. 1-15. As an example, the process of *isoelectric focusing* is used to separate amphoteric molecules, e.g., proteins, according to their isoelectric pH values. Above a certain pH an amphoteric molecule will carry a net negative charge; in a protein this is attributable to ionized carboxylic acid groups. Below this certain pH the amphoteric molecule will carry a net positive charge; in a protein this is attributable to ionized amino groups which have formed while the carboxylic acid groups become nondissociated. The zero-charge pH, or isoelectric point, varies from substance to substance. In isoelectric

---

[†] Equilibration separation processes have also been referred to as potentially *reversible* and as *partitioning* separation processes, while rate-governed separation processes have also been referred to as inherently *irreversible* or *nonpartitioning* separation processes.

focusing a gradient of pH is imposed over a distance in a complex fashion, using substances called *ampholytes* (Righetti, 1975). If an electric field is imposed in the same direction as the pH gradient, a gradient of force on the molecules of a given substance will result, stemming from the change in the charge per molecule as the pH changes. The force is directed toward the position of the isoelectric point for both higher and lower pH values; at the isoelectric point the force is zero. Therefore the amphoteric molecule will migrate toward the position where the pH equals its isoelectric pH and will stay there. Substances with different isoelectric points migrate to different locations and thus separate.

The imposed-gradient equilibration process creates a force gradient from positive to negative values, through zero, by combining two imposed gradients. In isoelectric focusing these are the gradient from the electric field and the pH gradient. A corresponding rate-governed separation process results from removing the second gradient, in this case the pH gradient. The imposed electric field in a medium of constant pH would cause differently charged species to migrate at different rates toward one electrode or the other, and different species could be isolated at different points by introducing the feed as a pulse to one location and waiting an appropriate length of time. This process, known as *electrophoresis*, is fundamentally different from isoelectric focusing since electrophoresis utilizes differences in rates of migration (charge-to-mass ratio) whereas isoelectric focusing separates according to differences in isoelectric pH.

Imposed-gradient equilibration processes differ from ordinary equilibration processes in that the products are miscible with each other and the separation will not occur without the imposed field.

Some separation processes utilize more than one separating agent, e.g., both an energy separating agent and a mass separating agent. An example is *extractive distillation*, where a mixture of components with close boiling points is separated by adding a solvent (mass separating agent), which serves to volatilize some components to greater extents than others, and then using heat energy (energy separating agent) in a distillation scheme of repeated vaporizations and condensations to generate a more volatile product and a less volatile product.

Table 1-1 lists a large number of separation processes divided into the categories we have just considered. The table shows the phases of matter involved, the separating agent, and the physical or chemical principle on which the separation is based. For mechanical processes the classification is by the principle involved. A practical example is given in each case. References are given to more extensive descriptions of each process. The table is not intended to be complete but to indicate the wide variety of separation methods which have been practiced. Many difficult but highly important separation problems have come to the fore in recent years, and it is safe to predict that challenging separation problems will continue to arise at an accelerated rate. It is possible to devise a separation technique based on almost any known physical mass transport or equilibrium phenomenon; consequently there is a wide latitude of approaches available to the imaginative engineer.

The selection of the appropriate separation process or processes for any given purpose is covered in more detail in Chap. 14.

Name	Feed	Separating agent	Products	Principle of separation	Practical example	References
				Diffusional Separation Processes		
			A. Ordinary equilibration processes, energy separating agent			
1. Evaporation	Liquid	Heat	Liquid + vapor	Difference in volatilities (vapor pressure)	Concentration of fruit juices	Perry and Chilton (1973); Loncin and Merson (1979)
2. Flash expansion	Liquid	Pressure reduction (energy)	Liquid + vapor	Difference in volatilities (vapor pressure)	Flash process for seawater desalination	Spiegler (1962)
3. Distillation	Liquid and/or vapor	Heat	Liquid + vapor	Differences in volatilities (vapor pressure) (repeated internally)	See Fig. 1-8	Hengstebeck (1961), Holland (1963), Robinson and Gilliland (1950), Smith (1963), Van Winkle (1967), Perry and Chilton (1973)
4. Crystallization	Liquid	Cooling or heat causing simultaneous evaporation	Liquid + solids	Difference in freezing tendencies; preferential participation in crystal structure	See Figs. 1-1, 1-2, and 1-8	Findlay and Weedman (1958), Perry and Chilton (1973), Schoen (1962)
5. Drying solids	Moist solid	Heat	Dry solid + humid vapor	Evaporation of water	Food dehydration	Krischer (1956), Perry et al. (1963), Van Arsdel et al. (1972)
6. Freeze drying	Frozen water-containing solid	Heat	Dry solid + water vapor	Sublimation of water	Food dehydration	Van Arsdel et al. (1972), King (1971)
7. Desublimation	Vapor	Cooling	Solid + vapor	Preferential condensation (desublimation); preferential participation in crystal structure	Purification of phthalic anhydride	Perry and Chilton (1973)
8. Dual-temperature exchange reactions	Fluid	Heating and cooling	Two fluids	Difference in reaction equilibrium constant at two different temperatures	Separation of hydrogen and deuterium	Benedict and Pigford (1957)
9. Zone melting	Solid	Heat	Solid of non-uniform composition	Same as crystallization	Ultrapurification of metals	Pfann (1966)

Table 1-1 (continued)

Name	Feed	Separating agent	Products	Principle of separation	Practical example	References
			B. Ordinary equilibration processes, mass separating agent			
1. Stripping	Liquid	Noncondensable gas	Liquid + vapor	Difference in volatilities	Removal of light hydrocarbon from crude oil fractions	Perry and Chilton (1973); Sherwood et al. (1975); Smith (1963)
2. Absorption	Gas	Nonvolatile liquid	Liquid + vapor	Preferential solubility	Removal of CO_2 and H_2S from natural gas by absorption into ethanolamines	Perry and Chilton (1973); Sherwood et al. (1975); Smith (1963); Kohl and Riesenfeld (1979)
3. Extraction	Liquid	Immiscible liquid	Two liquids	Different solubilities of different species in the two liquid phases	See Fig. 1-8	Alders (1959); Perry and Chilton (1973); Smith (1963); Treybal (1963)
4. Leaching or washing	Solids	Solvent	Liquid + solid	Preferential solubility	Leaching of $CuSO_4$ from calcined ore	Perry and Chilton (1973)
5. Precipitation	Liquid	Chemical reactant	Liquid + solid	Formation of insoluble precipitate	Lime-soda water treatment	Shreve and Brink (1977)
6. Adsorption	Gas or liquid	Solid adsorbent	Fluid + solid	Difference in adsorption potentials	Drying gases by solid desiccants	Perry and Chilton (1973); Ross (1963); Schoen (1962); Vermeulen (1963)
7. Ion exchange	Liquid	Solid resin	Liquid + solid resin	Law of mass action applied to available anions or cations	Water softening	Helfferich (1962); Perry and Chilton (1973); Wheaton and Seamster (1966)
8. Ion exclusion	Liquid	Adsorbing solid ion-exchange resin	Liquid + solid resin	Prevention of adsorption of species with same charge as resin	Separation of nucleic acids	Helfferich (1962); Singhal (1975)
9. Paper chromatography	Liquid	Capillarity: paper or gel phase	Regions of moistened paper	Preferential solubilities and adsorption potentials in two phases	Protein separation	Zweig and Whitaker (1967); Weaver (1968)

Process	Phase	Separating agent	Phase added or produced	Basis for separation	Application	References
10. Ligand-specific ("affinity") chromatography	Liquid	Immobilized ligands	Liquid + ligand-bearing solid	Reversible chemical interaction with ligands	Separation of enzymes	May and Zaborsky (1974), Weetall (1974)
11. Bubble fractionation; foam fractionation	Liquid	Rising air bubbles; sometimes also complexing surfactants (see text)	Two liquids	Tendency of surfactant molecules to accumulate at gas-liquid interface and rise with air bubbles	Removal of detergents from laundry wastes; ore flotation	Schoen (1962), Lemlich (1972), Valdes-Krieg et al. (1977)
C. Imposed-gradient equilibration processes						
1. Isoelectric focusing	Liquid	Electric field; pH gradient	Liquids	Movement toward location of isoelectric pH	Protein separation	Righetti (1975), Catsimpoolas (1975)
2. Isopycnic ultra-centrifugation	Liquid	Centrifugal force; density gradient	Liquids	Pressure diffusion	Separation of biological substances	Cline (1971)
D. Equilibration processes with more than one separating agent						
1. Extractive and azeotropic distillation	Liquid and/or vapor	Added liquid and heat	Liquid + vapor	Difference in volatilities	Recovery of butadiene	Perry and Chilton (1973), Smith (1963)
2. Clathration	Liquid	Clathrating molecule and cooling	Liquid + solid	Preferential participation in crystal structure	Hydrate process for seawater desalination	Schoen (1962), Makin (1974)
3. Adductive crystallization	Liquid	Adduct and cooling	Liquid + solid	Preferential participation in crystal structure	Separation of xylene isomers	Findlay and Weedman (1958)
E. Rate-governed processes						
1. Gaseous diffusion	Gas	Pressure gradient (compressor work)	Gases	Difference in rates of Knudsen or surface diffusion through a porous barrier	Concentration of $^{235}UF_6$ from natural UF_6	Benedict and Pigford (1957), Smythe (1945)
2. Sweep diffusion	Gas	Condensable vapor	Gases	Different diffusivities against sweeping motion of cross-flowing vapor	Suggested for isotope separation, helium from methane, etc.	Benedict and Pigford (1957), Smythe (1945)

Table 1-1 (continued)

Name	Feed	Separating agent	Products	Principle of separation	Practical example	References
3. Thermal diffusion	Gas or liquid	Temperature gradient	Gases or liquids	Different rates of thermal diffusion	Isotope separation, etc.	Benedict and Pigford (1957), Schoen (1962), Rutherford (1975)
4. Mass spectrometry	Gas	Magnetic field	Gases	Different charges per unit mass	Isotope separation	Benedict and Pigford (1957), Smythe (1945), Love (1973)
5. Dialysis	Liquid	Selective membrane; solvent	Liquids	Different rates of diffusional transport through membrane (no bulk flow)	Recovery of NaOH in rayon manufacture; artificial kidneys	Rickles (1966), Dedrick et al. (1968)
6. Electrodialysis	Liquid	Anionic and cationic membranes; electric field	Liquids	Tendency of anionic membranes to pass only anions, etc.	Desalination of brackish waters	Perry and Chilton (1973), Spiegler (1962)
7. Gas permeation	Gas	Selective membrane; pressure gradient	Gases	Different solubilities and transport rates through membrane	Purification of hydrogen by means of palladium barriers	Gantzel and Merten (1970), Antonson et al. (1977)
8. Electrophoresis	Liquid containing colloids	Electric field	Liquids	Different ionic mobilities of colloids	Protein separation	Zweig and Whitaker (1967), Everaerts et al. (1977), Bier (1959)
9. Electrolysis plus reaction	Liquid	Electric energy	Liquids	Different rates of discharge of ions at electrode	Concentration of HDO in H_2O^+	Benedict and Pigford (1957)
10. Sedimentation ultracentrifuge	Liquid	Centrifugal force	Two liquids	Pressure diffusion	Separation of large polymeric molecules according to molecular weight	Schachman (1959), Bowen and Rowe (1970), Olander (1972)
11. Reverse osmosis	Liquid solution	Pressure gradient (pumping power) + membrane	Two liquid solutions	Different combined solubilities and diffusivities of species in membrane	Seawater desalination	Merten (1966), Michaels (1968b)
12. Ultrafiltration	Liquid solution containing large molecules or colloids	Pressure gradient (pumping power) + membrane	Two liquid phases	Different permeabilities through membrane (molecular size)	Waste-water treatment; protein concentration; artificial kidney	Michaels (1968a), Dedrick et al. (1968), Michaels (1968b)

Method	Feed	Separating agent	Products	Physical principle	Application	References
13. Molecular distillation	Liquid mixtures	Heat + vacuum	Liquid + vapor	Difference in kinetic-theory maximum rate of vaporization	Separation of vitamin A esters and intermediates	Perry and Chilton (1973), Burrows (1960)
14. Gel filtration	Liquid	Solid gel, e.g., cross-linked dextran	Gel phase + liquid	Difference in molecular size and hence ability to penetrate swollen gel matrix	Purification of pharmaceuticals; separation of proteins	Altgelt (1968), John and Dellweg (1974)
15. Liquid membrane	Liquid	Solvent liquid layer	Liquids	Different rates of permeation through liquid layer	Waste-water processing	Cahn and Li (1974), Cussler and Evans (1975)
16. Nozzle diffusion	Gas	Pressure gradient	Gases	Different rates of outward transport in jet issuing from nozzle	Separation of uranium isotopes	Becker and Schutte (1960)

Mechanical Separation Processes

F. Density-based

Method	Feed	Separating agent	Products	Physical principle	Application	References
1. Settling	Liquid + solid or another immiscible liquid	Gravity	Liquid + solid or another immiscible liquid	Density difference	Clarification of murky solutions	Perry and Chilton (1973)
2. Centrifuge (sedimentation type)	Liquid + solid or another immiscible liquid	Centrifugal force	Liquid + solid or another immiscible liquid	Density difference	Recovery of insoluble reaction products	Perry and Chilton (1973)
3. Cyclone	Gas + solid or liquid	Flow (inertia)	Gas + solid or liquid	Density difference	Recovery of fluidized catalyst fines	Perry and Chilton (1973), Sheng (1977)
4. Sink-float	Solids	Gravity	Two solids	Density difference	Wheat from chaff	Perry and Chilton (1973)
5. Isopycnic centrifugation or settling	Solids	Gravity or centrifugal force; density gradient	Two solids	Density difference	Separation of ash particles	Khalafalla and Reimers (1975)

G. Size-based

Method	Feed	Separating agent	Products	Physical principle	Application	References
1. Filtration	Liquid + solid	Pressure reduction (energy); filter medium	Liquid + solid	Size of solid greater than pore size of filter medium	Recovery of slurried catalysts	Perry and Chilton (1973)

Table 1-1 (continued)

Name	Feed	Separating agent	Products	Principle of separation	Practical example	References
2. Mesh demister	Gas + solid or liquid	Pressure reduction (energy); wire mesh	Gas + solid or liquid	Size of solid greater than pore size of filter medium	Removal of H_2SO_4 mists from stack gases	Perry and Chilton (1973)
3. Centrifuge (filtration type)	Liquid + solid	Centrifugal force	Liquid + solid	Size of solid greater than pore size of filter medium	See Fig. 1-2	Perry and Chilton (1973)
4. Particle chromatography	Solids in liquid	Cooling (freezing)	Solids in frozen liquid	Rejection from frozen crystal if size greater than critical value for freezing rate used	Classification of particles	Kuo and Wilcox (1975)
H. Surface-based						
Flotation	Mixed powdered solids	Added surfactants; rising air bubbles	Two solids	Tendency of surfactants to adsorb preferentially on one solid species	Ore flotation; recovery of ZnS from carbonate gangue	Perry and Chilton (1973); Fuerstenau (1962)
I. Fluidity-based						
Expression	Liquid + solid	Mechanical force	Liquid + solid	Tendency of liquid to flow under applied pressure gradient	See Fig. 1-1	Gerstner (1969); Perry and Chilton (1973)
J. Electrically based						
Electrostatic precipitation	Gas + fine solids	Electric field	Gas + fine solids	Charge on fine solid particles	Dust removal from stack gases	Perry and Chilton (1973)
K. Magnetically based						
Magnetic separation	Mixed powdered solids	Magnetic field	Two solids	Attraction of materials in magnetic field	Concentration of ferrous ores	Perry and Chilton (1973); Mitchell et al. (1975)

The capabilities of the methods indicated in Table 1-1 can be further expanded by *chemical derivatization,* in which the components to be separated are subjected to some form of chemical reaction. Either some components react and other do not, or else all react, such that the products are more readily separable than the initial unreacted mixture. Some examples are the following:

Extraction using chemical complexing agents. The selectively complexed substances have
    greater solubility in the solvent phase, e.g., the use of silver(I) compounds for hydrocar-
    bon separations (Quinn, 1971).
Laser separation of isotopes. Here the idea is to cause one of the isotopes to enter an activated
    state selectively, using a laser of carefully determined and controlled emission frequency.
    If the result is selective ionization, a subsequent ion-deflection device can be used to
    separate the isotope mixture (Letokhov and Moore, 1976; Krass, 1977).
Separation of optical isomers by selective reaction with enzymes (Anon., 1973).
Plasma chromatography. The components of a mixture are ionized to various extents and made
    to undergo different times of flight in an electric field (Cohen and Karasek, 1970; Keller
    and Metro, 1974).

In a number of cases several variations are possible on the methods listed in Table 1-1. As an example, foam and bubble fractionation and flotation methods can be classified according to the scheme shown in Fig. 1-16. All the processes shown effect a separation through differences in surface activity of different components. In *foam fractionation,* foam bubbles rise and liquid drains between the cells of a foam, transporting the surface-active species selectively upward. The species recovered in this way can be either surface-active themselves, e.g., detergents, or substances that

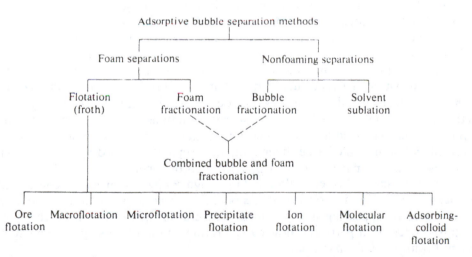

Figure 1-16 Classification of adsorptive bubble separation methods. (*Adapted from Karger et al., 1967, p. 401; by courtesy of Marcel Dekker, Inc.*)

react selectively with a surfactant, e.g., certain heavy metals with anionic surfactants. Alternatively, the process can be operating with a swarm of air bubbles rising through an elongated liquid pool, the surface-active species being drawn off from the top of the pool; this approach is known as *bubble fractionation.* In variants of bubble fractionation, a solvent is used to extract and collect the surface-active species at the top of the pool (*solvent sublation*), or else the surface-active species are collected in a layer of foam at the top of the pool (*combined bubble and foam fractionation*). Flotation processes, on the other hand, recover solid particles or scums from a suspension in liquid and are therefore mechanical separation processes. *Ore flotation* has been a major method of separation for years in the mineral industry. *Macroflotation* and *microflotation* refer to recovery of relatively large and small particles, respectively. In *precipitate flotation* a chemical agent is added to precipitate one or more components selectively from solution as fine particles (chemical derivitization); these particles are then removed by flotation. In *ion* and *molecular flotation* a surfactant forms an insoluble compound with the ion or molecule to be removed, and the substance undergoes flotation and is removed as a scum. Finally, in *adsorbing colloid flotation* a colloidal substance added to a solution adsorbs the substance of interest and is removed by flotation.

Lee, et al. (1977) have proposed classifying separation processes by three vectors, the size of the molecules or particles involved, the nature of the driving force causing transport (concentration, electric, magnetic, etc.), and the flow and/or design configuration of the separator. This approach appears particularly useful for categorising rate-governed separations and indicating promising new methods.

Separations by centrifuging are indicative of the interaction between molecule or particle size and the size of driving force required to achieve separation. Conventional *sedimentation centrifuges,* with speeds in the range of 1000 to 50,000 r/min, are widely used for separating particulate solids from liquids on the basis of density difference. Much higher speeds must be used in *ultracentrifuges,* which can separate biological cell constituents, macromolecules, and even isotopes. The higher speeds are required to create forces large enough to move these much smaller particles or molecules. Alternatively, ultracentrifuges and centrifuges can be made to work on the isopycnic principle as an imposed-gradient equilibration process by creating a density gradient within the centrifuge and withdrawing products at the zones corresponding to their individual densities. The density gradient can be created by such methods as adding stratified layers of sucrose solutions of different strengths or imposing a magnetic field upon a suspension of magnetic material. Yet another form of centrifugation uses the basket filter to make a separation based upon size rather than density. Larger particles cannot pass through the openings provided.

The third vector of Lee, et al., flow and/or design configuration, will be considered in Chaps. 3 and 4. It is appropriate at this point, however, to point out that the term *chromatography,* appearing in the names of several of the separation processes in Table 1-1, refers to a particular flow configuration and not to a single chemical or physical principal for separation.

Categorization of separation processes is also discussed by Strain et al. (1954), Rony (1972), and Giddings (1978).

Separation Factor

The degree of separation which can be obtained with any particular separation process is indicated by the *separation factor*. Since the object of a separation device is to produce products of differing compositions, it is logical to define the separation factor in terms of product compositions†

$$\alpha_{ij}^S = \frac{x_{i1}/x_{j1}}{x_{i2}/x_{j2}} \tag{1-1}$$

The separation factor α_{ij}^S between components i and j is the ratio of the mole fractions of those two components in product 1 divided by the ratio in product 2. The separation factor will remain unchanged if all the mole fractions are replaced by weight fractions, by molar flow rates of the individual components, or by mass flow rates of the individual components.

An effective separation is accomplished to the extent that the separation factor is significantly different from unity. If $\alpha_{ij}^S = 1$, no separation of components i and j has been accomplished. If $\alpha_{ij}^S > 1$, component i tends to concentrate in product 1 more than component j does, and component j tends to concentrate in product 2 more than component i does. On the other hand, if $\alpha_{ij}^S < 1$, component j tends to concentrate preferentially in product 1 and component i tends to concentrate preferentially in product 2. By convention, components i and j are generally selected so that α_{ij}^S, defined by Eq. (1-1), is greater than unity.

The separation factor reflects the differences in equilibrium compositions and transport rates due to the fundamental physical phenomena underlying the separation. It can also reflect the construction and flow configuration of the separation device. For this reason it is convenient to define an *inherent separation factor*, which we shall denote by α_{ij} with no superscript. This inherent separation factor is the separation factor which would be obtained under idealized conditions, as follows:

1. For *equilibration separation processes* the inherent separation factor corresponds to those product compositions which will be obtained when simple equilibrium is attained between the product phases.
2. For *rate-governed separation processes* the inherent separation factor corresponds to those product compositions which will occur in the presence of the underlying physical transport mechanism alone, with no complications from competing transport phenomena, flow configurations, or other extraneous effects.

These definitions are illustrated by examples in the following section.

We shall find that both the inherent separation factor α_{ij} and the actual separation factor α_{ij}^S, based on actual product compositions through Eq. (1-1), can be used for the analysis of separation processes. When α_{ij} can be derived relatively easily, the most common approach is to analyze a separation process on the basis of the inherent separation factor α_{ij} and allow for deviations from ideality through

---

† Notation used in this book is summarized in Appendix G.

efficiencies. This procedure is advantageous since, as we shall see, α_{ij} is frequently *insensitive* to changes in mixture composition, temperature, and pressure. The concept and use of efficiencies are developed in Chaps. 3 and 12.

On the other hand, there are situations where the physical phenomena underlying the separation process are so complex or poorly understood that an inherent separation factor cannot readily be defined. In these instances one must necessarily work with α_{ij}^S derived empirically from experimental data. Such is the case for separations by electrolysis and flotation, for example.

The quantity α_{ij}^S may be closer to, or further from, unity than α_{ij}, but if α_{ij} is unity, it is imperative that α_{ij}^S be unity, no matter what the flow configuration or other added effects. Put another way, no flow configuration can provide a separation if the underlying physical phenomenon necessary to cause the separation is not present.

INHERENT SEPARATION FACTORS: EQUILIBRATION PROCESSES

For separation processes based upon the equilibration of immiscible phases it is helpful to define the quantity

$$K_i = \frac{x_{i1}}{x_{i2}} \qquad \text{at equilibrium} \tag{1-2}$$

K_i is called the *equilibrium ratio* for component i and is the ratio of the mole fraction of i in phase 1 to the mole fraction of i in phase 2 *at equilibrium*. The inherent separation factor is then given by Eq. (1-1) as

$$\alpha_{ij} = \frac{x_{i1}/x_{j1}}{x_{i2}/x_{j2}} = \frac{K_i}{K_j} \tag{1-3}$$

α_{ij} is substituted for α_{ij}^S in Eq. (1-1) in order to obtain Eq. (1-3) since we are considering the special case of complete product equilibrium.

Vapor-Liquid Systems

For processes based on equilibration between gas and liquid phases α_{ij}, K_i, and K_j can be related to vapor pressures and activity coefficients. If the components of the mixture obey Raoult's and Dalton's laws,

$$p_i = P y_i = P_i^0 x_i \tag{1-4}$$

where y_i, x_i = mole fractions of i in gas and liquid phases, respectively
$\qquad$ P = total pressure
$\qquad$ p_i = partial pressure of i in gas
$\qquad$ P_i^0 = vapor pressure of pure liquid i

In such a case

$$K_i = \frac{y_i}{x_i} = \frac{P_i^0}{P} \tag{1-5}$$

and

$$\alpha_{ij} = \frac{P_i^0}{P_j^0} \tag{1-6}$$

The inherent separation factor α_{ij} in a vapor-liquid system is commonly called the *relative volatility*. The reasons for this name are apparent from Eq. (1-6), where for an ideal system α_{ij} is simply the ratio of the vapor pressures of i and j.

From Eq. (1-5) it is apparent that $K_i P$ should be independent of the pressure level. This is true at pressures low enough for the ideal-gas assumption to hold and for the Poynting vapor-pressure correction to be insignificant. When $K_i P$ and $K_j P$ are independent of pressure, α_{ij} will necessarily be independent of pressure. The vapor pressures of i and j depend upon temperature; hence K_i and K_j are functions of temperature. Since α_{ij} is proportional to the ratio of the vapor pressures, and since both vapor pressures increase with increasing temperature, α_{ij} will be less sensitive to temperature than K_i and K_j. Over short ranges of temperature α_{ij} can often be taken to be constant. It is also important to note from Eq. (1-6) that in the case of adherence to Raoult's and Dalton's laws α_{ij} is not a function of liquid or vapor composition. Thus for this situation α_{ij} is independent of pressure and composition and is insensitive to temperature.

Most solutions in vapor-liquid separation processes are nonideal, i.e., do not obey Raoult's law. In such cases Eq. (1-4) is commonly modified to include a liquid-phase activity coefficient γ

$$p_i = Py_i = \gamma_i P_i^0 x_i \tag{1-7}$$

At high pressures a vapor-phase activity coefficient and vapor- and liquid-phase fugacity coefficients also become necessary (Prausnitz, 1969). The liquid-phase activity coefficient is dependent upon composition. The standard state for reference is commonly chosen as $\gamma_i = 1$ for pure component i. If $\gamma_i > 1$, there are said to be *positive* deviations from ideality in the liquid solution, and if $\gamma_i < 1$, there are *negative* deviations from ideality in the liquid solution. Positive deviations are more common and occur when the molecules of the different compounds in solution are dissimilar and have no preferential interactions between different species. Negative deviations occur when there are preferential attractive forces (hydrogen bonds, etc.) between molecules of two different species that do not occur for either species alone.

For nonideal solutions, Eqs. (1-5) and (1-6) become

$$K_i = \frac{\gamma_i P_i^0}{P} \tag{1-8}$$

and

$$\alpha_{ij} = \frac{\gamma_i P_i^0}{\gamma_j P_j^0} \tag{1-9}$$

Both K_i and α_{ij} are now dependent upon composition because of the composition dependence of γ_i and γ_j. It will still be true, however, that α_{ij} is relatively insensitive to temperature, pressure, and composition.

Binary Systems

In a binary mixture containing only i and j, making the substitutions $y_j = 1 - y_i$ and $x_j = 1 - x_i$ in the definition of α_{ij}^S leads to

$$\alpha_{ij}^S = \frac{y_i(1 - x_i)}{x_i(1 - y_i)} \tag{1-10}$$

which can be rearranged as

$$y_i = \frac{\alpha_{ij}^S x_i}{1 + (\alpha_{ij}^S - 1)x_i} \tag{1-11}$$

Equation (1-11) relates y_i to α_{ij}^S and x_i in a binary vapor-liquid system. If the vapor and liquid phases in a binary vapor-liquid system are in equilibrium, we can substitute α_{ij} for α_{ij}^S in Eq. (1-11)

$$y_i = \frac{\alpha_{ij} x_i}{1 + (\alpha_{ij} - 1)x_i} \tag{1-12}$$

The system benzene-toluene adheres closely to Raoult's law. The vapor pressures of benzene and toluene at 121°C are 300 and 133 kPa, respectively. Therefore at 121°C

$$\alpha_{BT} = \frac{300}{133} = 2.25$$

Substituting this value of α_{BT} into Eq. (1-12) gives

$$y_B = \frac{2.25x_B}{1 + 1.25x_B} \tag{1-13}$$

Equation (1-13) provides a relationship between all possible equilibrium product compositions at 121°C. Figure 1-17 shows Eq. (1-13) in graphical form. Two features are characteristic of such plots for constant α: (1) the curve intersects the $y = x$ line only at $x = y = 0$ and $x = y = 1$, and (2) the curve is symmetrical with respect to the line $y = 1 - x$.

In Fig. 1-17 the temperature is held constant for all y_B and x_B, but the pressure necessarily varies as x_B or y_B changes. The *Gibbs phase rule*, developed in most physical chemistry and thermodynamics texts, states that

$$P + F = C + 2 \tag{1-14}$$

where P = number of phases present
 C = number of components
 F = number of *independently specifiable variables*

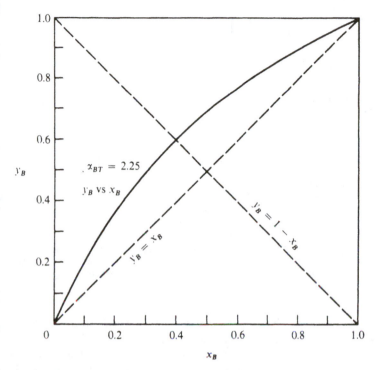

Figure 1-17 Vapor and liquid compositions for a case of constant α_{ij}, corresponding to benzene-toluene at 121°C.

For equilibrium between vapor and liquid phases in a binary system $P = 2$ and $C = 2$; hence $F = 2$. If the composition of one phase is specified, one of the independent variables is consumed, since in a binary system setting x_i for one component necessarily fixes the liquid mole fraction of the other component at $1 - x_i$. If temperature is fixed, as in Fig. 1-17, the second independent variable has been set and all other variables are dependent ones for which we can solve. In Fig. 1-17 the total pressure varies monotonically from 133 kPa (vapor pressure of toluene) at $x_B = 0$ to 300 kPa (vapor pressure of benzene) at $x_B = 1$.

In analyzing separation problems it is frequently more realistic to set the total pressure as a specified variable rather than temperature. If the pressure and x_i (two independent variables) are specified in a vapor-liquid equilibrium binary system, we can then, by the phase rule, solve for temperature and y_i. Figure 1-18 is a plot of the equilibrium temperature vs. x_B for the system benzene-toluene at a constant total pressure of 200 kPa; y_B is also plotted vs. T or x_B. A plot of y_B vs. x_B can be prepared by reading off the values of y_B and x_B which correspond to a given T (dashed line in Fig. 1-18). This yx plot will be different from Fig. 1-17 since the ratio of vapor pressures, and hence α_{BT}, changes with changing temperature. The difference will be slight, however, since α_{BT} changes by only 14 percent as T changes from 103 to 137°C.

The region above the saturated-vapor (i.e., vapor in equilibrium with liquid)

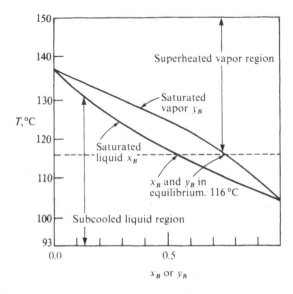

Figure 1-18 Temperature vs. composition for vapor-liquid equilibrium at 200 kPa in the binary system benzene-toluene.

curve in Fig. 1-18 corresponds to the occurrence of superheated vapor with no equilibrium liquid able to coexist. The region below the saturated-liquid curve corresponds to subcooled liquid with no equilibrium vapor able to coexist. The region in between the saturated-vapor and saturated-liquid curves corresponds to a two-phase mixture.

Figure 1-19 shows typical plots of α_{ij} and y_i vs. x_i for binary vapor-liquid systems with positive and negative deviations from ideality. The characteristic trends of α_{ij} vs. x_i for positive and negative deviations follow from the fact that γ_i differs most from unity at *low* x_i whereas γ_j differs most from unity at *high* x_i. Since γ_i/γ_j in a positive system is therefore greatest at low x_i, α_{ij} for a positive system is highest at low x_i. Opposite reasoning holds for a negative system.

Systems where there are large deviations from ideality and/or close boiling points of the pure components involved often produce *azeotropes*, where the y_i-vs.-x_i curve crosses the $y_i = x_i$ line. At the azeotropic composition $y_i = x_i$; therefore $\alpha_{ij} = 1.0$, and no separation is possible. An azeotrope occurs in the chloroform-acetone system.

Liquid-Liquid Systems

For equilibrium between two immiscible liquid phases, such as occurs in liquid-liquid extraction processes, we can use Eq. (1-7) for both phases along with the postulate of a single vapor phase which is necessarily in equilibrium with *both* liquid phases to obtain

$$\frac{x_{i1}}{x_{i2}} = \frac{\gamma_{i2}}{\gamma_{i1}} \tag{1-15}$$

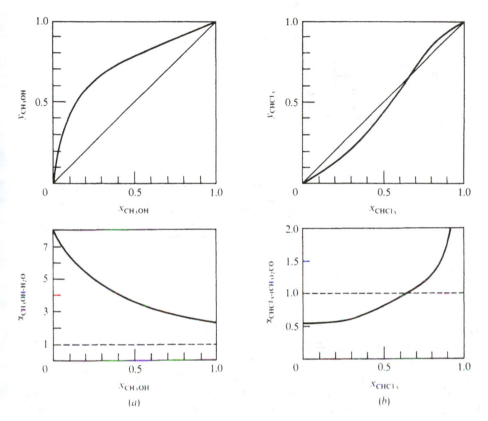

Figure 1-19 Vapor-liquid equilibrium behavior for nonideal binary solutions: (*a*) methanol-water system where $P = 101$ kPa (positive deviations); (*b*) chloroform-acetone system where $P = 101$ kPa (negative deviations).

relating the mole fractions of component i in liquid phases 1 and 2 at equilibrium. For the separation between components i and j in a liquid-liquid process at complete equilibrium we can then write

$$\alpha_{ij} = \frac{x_{i1}/x_{j1}}{x_{i2}/x_{j2}} = \frac{\gamma_{i2}\gamma_{j1}}{\gamma_{i1}\gamma_{j2}} \tag{1-16}$$

Thus we can see that liquid-liquid equilibration processes must necessarily involve nonideal solutions ($\gamma \neq 1$) if α_{ij} is to be different from unity. It follows that α_{ij} will be dependent upon composition in a liquid-liquid system unless the system is so dilute in component i in both phases that the activity coefficients will stay constant at the values for infinite dilution.

When the inherent separation factor varies substantially with composition, it is usually most convenient to present and utilize equilibrium data in graphical form. Consider the process shown in Fig. 1-20 for the removal of acetic acid from a solution of vinyl acetate and acetic acid (feed) by extraction of the acetic acid into water

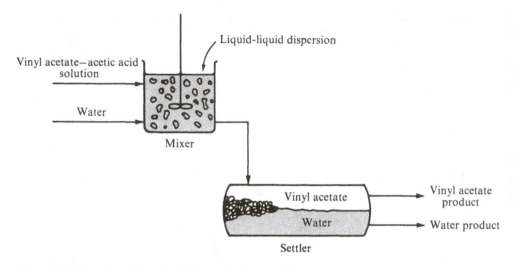

Figure 1-20 Extraction of acetic acid from vinyl acetate by water.

(separating agent) at 25°C. The two partially miscible liquid phases are contacted in an agitated mixture to bring them to equilibrium and are then separated physically in a settler from which products are withdrawn.

A triangular diagram giving miscibility and equilibrium data for this system is presented in Fig. 1-21. An important property of this sort of diagram is that the sum of the lengths of the three lines which can be drawn from any interior point perpendicular to each of the three sides and extending to the three sides ($DE + DF + DG$ in Fig. 1-22) is equal to the altitude of the triangle. Since the altitude of the triangle in Fig. 1-21 or 1-22 is 100 percent of one of the components, any point within the triangle represents a unique composition. The point P represents 37 wt $\%$ vinyl acetate, 36 wt $\%$ acetic acid, and 27 wt $\%$ water. Compositions are expressed as weight percent, but they could as well be mole percent.

Any mixture lying within the phase envelope in Fig. 1-21 corresponds to partial miscibility; two liquid phases are formed, but all three components are present to some extent in both phases. Equilibrium compositions of the two phases are related by the equilibrium tie lines, shown dashed in Fig. 1-21. The composition marked P also corresponds to the *plait point*, the point on the phase envelope when the two phases in equilibrium approach identical compositions. Any higher concentration of acetic acid in the system gives total miscibility, in which case there are no longer two liquid phases, and the separation shown in Fig. 1-20 could not occur.

In this process the separation is between acetic acid and vinyl acetate, the water being present in the capacity of separating agent. Because of this a clearer picture of the separation and of the separation factor can be obtained by displaying the equilibrium data on a water-free basis. Plots of this sort are given in Figs. 1-23 and 1-24. In these diagrams the compositions involve only the two components being separated from each other. In Fig. 1-23 the mass of acetic acid per mass of acetic acid + vinyl

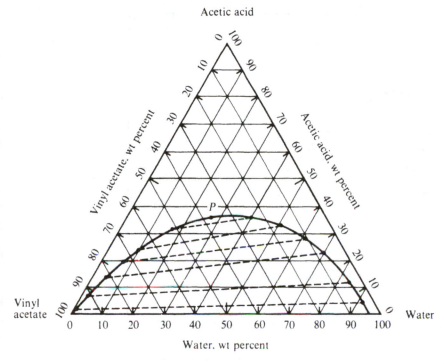

Figure 1-21 Equilibrium data for vinyl acetate–acetic acid–water system at 25°C. (*From Daniels and Alberty, 1961, p. 258: used by permission.*)

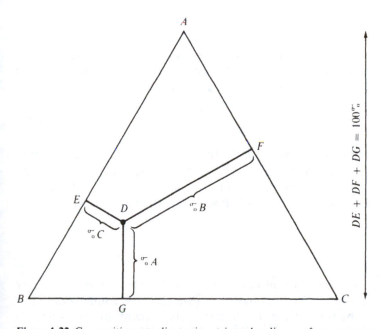

Figure 1-22 Composition coordinates in a triangular diagram for a ternary system.

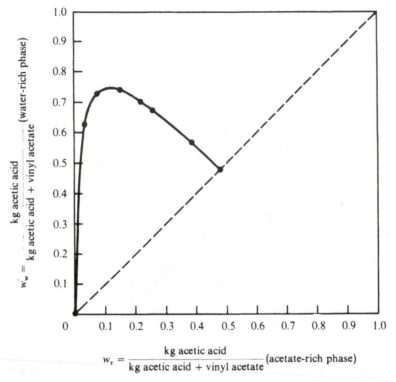

$$w_v = \frac{\text{kg acetic acid}}{\text{kg acetic acid} + \text{vinyl acetate}} \text{(acetate-rich phase)}$$

Figure 1-23 Relation between product compositions for vinyl acetate–acetic acid–water system; weight basis.

acetate in the water-rich phase w_w' is plotted against the same factor in the vinyl acetate-rich phase w_v'. In Fig. 1-24 the equilibrium separation factor is plotted against composition. Here the separation factor is defined as

$$\alpha_{\text{acetic acid – vinyl acetate}} = \frac{w_w'(1 - w_v')}{w_v'(1 - w_w')} \tag{1-17}$$

Figures 1-23 and 1-24 show that α varies markedly with composition and the shape of the $w_w' \, w_v'$ diagram is very different from the shape of the yx diagram (Fig. 1-17) for a constant α. Note again that the separation is independent of the basis of compositions; i.e., the separation factor based on mole fractions is the same as that based on weight fractions.

Liquid-Solid Systems

Figure 1-25 is a phase diagram showing liquid-solid equilibrium conditions for the binary system m-cresol–p-cresol. By the phase rule [Eq. (1-14)] if equilibrium liquid and solid phases are to be present for this binary system, there are two variables which can be specified independently. If these variables are (1) the total pressure and

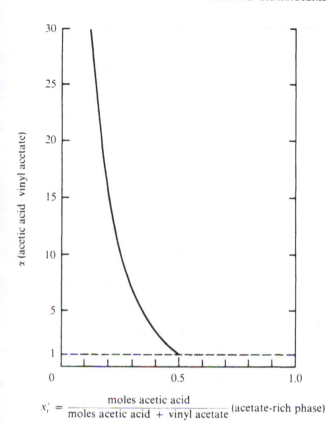

Figure 1-24 Separation factor for acetic acid extraction process.

$$x_i' = \frac{\text{moles acetic acid}}{\text{moles acetic acid + vinyl acetate}} \text{ (acetate-rich phase)}$$

(2) the mole fraction of m-cresol in the liquid phase, the system is then fixed and the temperature and solid-phase composition are dependent variables. The curves marked "solution compositions" in Fig. 1-25 show the temperature at which an equilibrium solid can coexist for any given mole fraction of m-cresol in the liquid. Below 40% m-cresol in the liquid, the equilibrium solid will be pure p-cresol. Between 40 and 90% m-cresol in the liquid, the equilibrium solid is an intermolecular compound containing two molecules of m-cresol for each molecule of p-cresol. Above 90% m-cresol in the liquid, the equilibrium solid is pure m-cresol. The various regions in the phase diagrams have labels indicating the two phases that would coexist at equilibrium if the gross composition of the two phases *combined* were within that region at a particular temperature. For example, if the overall composition were 8% m-cresol at 14°C, the two phases present would be solid p-cresol and a liquid containing 27% m-cresol.

The points marked E_1 and E_2 in Fig. 1-25 are known as *eutectic points*. They represent minima in the plot of freezing point vs. composition (the solution compositions curve). In order to freeze any mixture of m-cresol and p-cresol completely, it is necessary to cool the mixture to a temperature below the appropriate eutectic temperature (1.6°C for an initial mixture in the range $0 < x_L < 0.67$ or 4°C for an initial

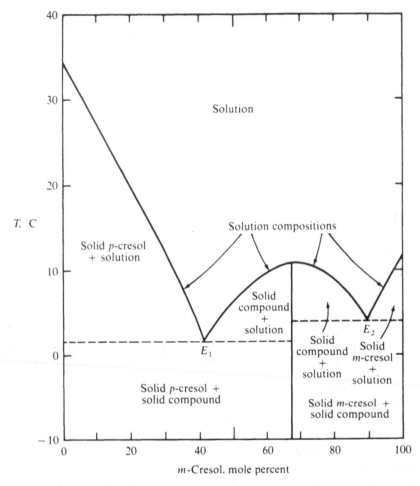

Figure 1-25 Phase diagram for the system m-cresol-p-cresol at 1 atm. (*Adapted from Chivate and Shah, 1956, p. 237; used by permission.*)

mixture in the range $0.67 < x_L < < 1.0$). Consider, for example, the cooling of a liquid mixture containing 60% m-cresol. As this mixture is cooled, it will first begin to form solid at 9.5°C. This solid will be the $x_S = 0.67$ compound. Since the solid formed is richer in m-cresol than the original liquid, the remaining liquid must become more depleted in m-cresol. As the temperature is lowered more, the composition of the remaining liquid will move along the solution composition curve toward the eutectic point E_1. When the residual liquid contains 40% m-cresol and the temperature has therefore reached 1.6°C, any further lowering of temperature will require that *all* the remaining liquid solidify at the eutectic temperature and composition. This analysis will be similar for any other initial liquid composition before freezing.

The x_L-vs.-x_S behavior (liquid mole fraction vs. solid mole fraction) corresponding to Fig. 1-25 is very different from the yx behavior corresponding to

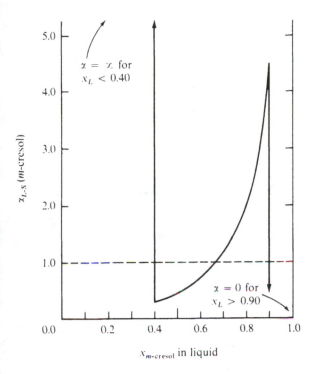

Figure 1-26 Separation factor vs. liquid composition for *m*-cresol–*p*-cresol liquid-solid equilibrium.

Fig. 1-18 for the benzene-toluene vapor-liquid system. For $0 < x_L < 0.40$ (mole fraction *m*-cresol in liquid), $x_S = 0.00$. For $0.40 < x_L < 0.90$, $x_S = 0.67$. For $0.90 < x_L < 1.00$, $x_S = 1.00$. Thus the equilibrium x_S is discontinuous and can assume only three discrete values. A plot of the equilibrium separation factor for this system $\alpha_{m\text{-cresol}-p\text{-cresol}}$ vs. x_L is shown in Fig. 1-26. Readers should establish for themselves the reasons for the form of this plot.

Not all liquid-solid equilibria for mixtures show the behavior of Fig. 1-25 with solids of only a few discrete compositions being formable. Figure 1-27 shows a liquid-solid phase diagram for the gold-platinum system. This diagram is similar to Fig. 1-18, and will lead to an $x_L x_S$ diagram similar to the yx diagram in Fig. 1-17. Equilibrium solid phases of all compositions can be formed, depending upon the composition of the liquid phase from which they are formed. Systems forming solid solutions are rarer than those showing eutectic points and forming solids of only a few discrete compositions. In order to form a solid solution, two substances must form a compatible crystal structure with each other. This occurs for gold and platinum but not for the mixed cresols, even though the cresols are isomers.

Systems With Infinite Separation Factor

For some equilibrium separations the relationship between product compositions is determinable in a far simpler manner. An example is the classical process for the production of fresh water from seawater by evaporation. The desired separation is

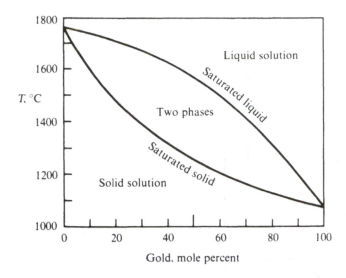

Figure 1-27 Solid-liquid phase diagram for the gold-platinum binary, a system forming a solid solution. (*Adapted from Daniels and Alberty, 1961, p. 252; used by permission.*)

between pure water on the one hand and the dissolved salts on the other hand. In this case, however, the salts are for all intents and purposes entirely nonvolatile. The equilibrium separation factor, or relative volatility, is given by

$$\alpha_{WS} = \frac{y_W}{x_W} \frac{x_S}{y_S} \tag{1-18}$$

where W refers to water and S to salt. Since y_S is necessarily equal to zero and the other three mole fractions are finite, α_{WS} must approach infinity. This infinite α corresponds to a "perfect" separation; no salt is present in the evaporated water.

Solid-liquid equilibrations frequently give an infinite separation factor, too. As we have seen, the m-cresol–p-cresol system shown in Fig. 1-26 gives an infinite equilibrium $\alpha_{m\text{-cresol–}p\text{-cresol}}$ for x_L between 0 and 0.40 and an infinite equilibrium $\alpha_{p\text{-cresol–}m\text{-cresol}}$ for x_L between 0.90 and 1.00. Absorption of a single solute from a gas also involves a nearly infinite separation factor.

Sources of Equilibrium Data

Reid et al. (1977) give an excellent review of sources of data and prediction methods for vapor-liquid and liquid-liquid equilibria, which underlie distillation, extraction, absorption, and stripping processes. In an earlier review, Null (1970) covers the presentation, measurement, analysis, and prediction of vapor-liquid, liquid-liquid and solid-liquid equilibria.

The ultimate source of equilibrium data is reliable experimentation. Compilations of available experimental data for vapor-liquid equilibria are given by Wichterle et al. (1973), Hirata et al. (1975), and in earlier works by Hala et al. (1967, 1968). Solubilities of gases in liquids are compiled and referenced by Seidell and Linke (1958), Hayduk and Laudie (1973), Battino and Clever (1966), Perry et al. (1963), and

Kohl and Riesenfeld (1974). Liquid-liquid equilibrium data have been collected by Francis (1963) and Perry and Chilton (1973). Solid-liquid equilibrium data are available from Seidell and Linke (1958), Timmermans (1959–1960), and Stephen and Stephen (1964). Equilibrium data are also provided in many books on individual separation processes. Many of the references in Table 1-1 are useful in this regard. Equilibrium data for a particular system can often be found by searching the indexes of *Chemical Abstracts.*

Since equilibrium data are subject to error in measurement or interpretation, it is useful to check them by thermodynamic-consistency tests and by comparison to known behavior of similar systems. Methods for doing this are given by Prausnitz (1969), Null (1970), and Reid et al. (1977).

Much progress has been made in the use of theoretical interpretations and models to extend results to different temperatures and pressures and even to different systems of components. These methods can be used in the absence of experimental data if one keeps in mind the degree of uncertainty thereby introduced. Equilibrium ratios in hydrocarbon systems can be obtained from convergence-pressure considerations (NGSPA, 1957) or can be predicted from solubility parameters and other more fundamental information. The latter approach is more appropriate for computer manipulation; Prausnitz and associates have extended it to nonhydrocarbon systems (Prausnitz et al., 1966) and to high-pressure systems (Prausnitz and Chueh, 1968). Methods for the prediction of activity coefficients in fluid-phase systems are comprehensively reviewed by Reid et al. (1977); these methods include various thermodynamically based semitheoretical correlations and two methods (ASOG and UNIFAC) based upon summing contributions of individual functional groups in the molecules concerned. A comprehensive presentation of the UNIFAC method, its applications, and underlying parameters is given by Fredenslund et al. (1977). If the resulting activity coefficients are to be used for the prediction of vapor-liquid equilibria, pure-component vapor-pressure data are needed; they have been compiled by Boublik et al. (1973).

INHERENT SEPARATION FACTORS: RATE-GOVERNED PROCESSES

Gaseous Diffusion

The molecular transport theory of gases is sufficiently well developed to allow us to make reasonable estimations of the inherent separation factor for those separation processes based upon different rates of molecular gas-phase transport. Consider the simple *gaseous-diffusion* process shown in Fig. 1-28. The gas mixture to be separated is located on one side (the left) of a porous *barrier*, e.g., a piece of sintered metal containing open voids between metal particles. A pressure gradient is maintained across the barrier, the pressure on the feed (left) side being much greater than that on the product (right) side. This pressure gradient causes a flux of molecules of the gaseous mixture to be separated across the barrier from left to right.

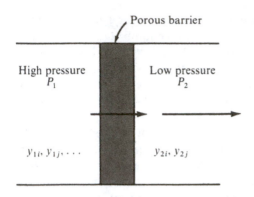

Figure 1-28 Simplified gaseous-diffusion process.

If the barrier has pores sufficiently small, and if the gas pressure is sufficiently low, the mean free path of the gas molecules will be large compared with the pore dimensions. As a result the molecular flux will occur by Knudsen flow, and the flux is describable by the equation

$$N_i = \frac{a(P_1 y_{1i} - P_2 y_{2i})}{\sqrt{M_i T}} \tag{1-19}$$

where N_i = flux of component i across barrier
P_1, P_2 = pressure of the high- and low-pressure sides, respectively
y_{1i}, y_{2i} = mole fractions of component i on high- and low-pressure sides, respectively
T = temperature
M_i = molecular weight of component i
a = geometric factor depending only upon structure of barrier

We shall now presume that the composition of the high-pressure side does not change appreciably through depletion of one of the gas species. The material on the low-pressure side has all arrived through the steady-state transport process described by Eq. (1-19), so that

$$\frac{N_i}{N_j} = \frac{y_{2i}}{y_{2j}} \tag{1-20}$$

We shall also presume for simplicity that $P_2 \ll P_1$. Combining Eqs. (1-19) and (1-20) for that case, we get

$$\alpha_{ij} = \frac{y_{2i} y_{1j}}{y_{2j} y_{1i}} = \sqrt{\frac{M_j}{M_i}} \tag{1-21}$$

which is not dependent upon composition. For the separation of $^{235}UF_6$ from $^{238}UF_6$ by gaseous diffusion as carried out by the United States government, $\alpha_{235-238} = \sqrt{352.15/349.15} = 1.0043$. Thus ^{235}U travels through the barrier preferentially to ^{238}U. Uranium isotopes are separated with the uranium in the form of the hexafluoride, since UF_6 is one of the few *gaseous* uranium compounds.

The inherent separation factor, in principle at least, can also be computed from molecular theory for a number of other rate-governed processes, such as sweep diffusion and thermal diffusion, although the analysis is more complex.

Reverse Osmosis

For rate-governed separation processes where the transport mechanism through the barrier is not well understood, the separation factor can be determined only experimentally. Consider, for example, the *reverse-osmosis* process for making pure water from salt water.

In a reverse-osmosis process the object is to make water flow selectively out of a concentrated salt solution, through a polymeric membrane, and into a solution of low salt concentration. The natural process of osmosis would cause the water to flow in the opposite direction until a pressure imbalance equal to the *osmotic pressure* is built up. The osmotic pressure is proportional to solute activity and hence is approximately proportional to the salt concentration; for natural seawater the osmotic pressure is about 2.5 MPa. This means that in the absence of other forces, water would tend to flow from pure water into seawater—across a membrane permeable only to water—until a static head of 2.5 MPa, or 258 m, of water was built up on the seawater side. This assumes that the seawater is not significantly diluted by the water transferring into it. At this point this system would be at equilibrium, as shown in Fig. 1-29. If the pressure difference across the membrane were less than the osmotic pressure, water would enter the seawater solution by osmosis, but when the pressure difference across the membrane is greater than the osmotic pressure, water will flow from the seawater solution into the pure-water side by reverse osmosis. For the water passing through the membrane to be salt-free in a reverse-osmosis process, the membrane must be permeable to water but relatively impermeable to salt since the salt would pass through the membrane with the water if the membrane were salt-permeable.

Reverse osmosis should be distinguished from ultrafiltration and dialysis, which are also rate-governed separation processes based upon thin polymeric membranes. In *ultrafiltration* relatively large molecules (polymers, proteins, etc.) or colloids are to be concentrated in solution by removing some of the solvent. Since the molarity of solutions of high-molecular-weight materials is quite low, the osmotic pressure is not significant. Pressure is used to drive the solvent (usually water) through membranes in ultrafiltration, but the pressure level can be relatively low (up to 800 kPa) because of the very low osmotic pressure and because a more " open " membrane can be used if salt retention is not required. In *dialysis* the object is to remove low-molecular-weight solutes preferentially from a solution. The process takes advantage of the fact that low-molecular-weight solutes have a higher diffusion coefficient in the membrane material than higher-molecular-weight solutes. Bulk flow of solvent through the membrane is prevented by balancing the osmotic pressure of the feed solution by using a flowing *isotonic* (same osmotic pressure) solution on the other side of the membrane to take up the solutes passing through the membrane.

Figure 1-30 shows a view of the components of an apparatus used to carry out

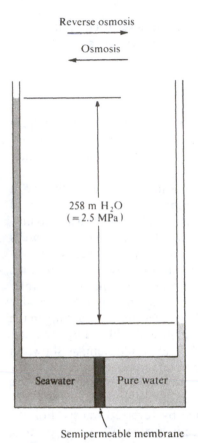

Reverse osmosis

Osmosis

258 m H_2O
(= 2.5 MPa)

Seawater   Pure water

Semipermeable membrane

Figure 1-29 Osmotic pressure.

Figure 1-30 Disassembled apparatus for ultrafiltration. *(Amicon Corp., Lexington, Mass.)*

ultrafiltration. Feed solution enters the bottom of the module and flows along a spiral spacer in rapid laminar flow. Above the spiral spacer is a polymeric membrane, and above that is a porous disk to support the membrane. (Both the membrane and the disk are white in Fig. 1-30.) Another spiral spacer fits on top. The solvent passing through the membrane (*permeate*) travels around the top spiral and out. There is also an exit from the bottom spacer to allow the concentrated solution not passing through the membrane (*retentate*) to exit. If it were made to withstand high pressure, this unit could be used for reverse osmosis as well as ultrafiltration.

A number of devices, many of which are similar to that shown in Fig. 1-30, have been used to evaluate a number of different membrane materials for the desalination of seawater by reverse osmosis. It has been found (Merten, 1966) that the flux of water and salt through a membrane can be described by two equations

$$N_W = k_W(\Delta P - \Delta \pi) \tag{1-22}$$

$$N_S = k_S(C_{S1} - C_{S2}) \tag{1-23}$$

where N_W, N_S = water and salt fluxes, respectively, across membrane, mol/time-area

ΔP = drop in total pressure across membrane

$\Delta \pi$ = drop in osmotic pressure across membrane

C_{S1}, C_{S2} = salt concentrations on two sides of membrane

k_W, k_S = empirical constants depending on membrane structure and nature of salt

If the membrane has a low salt permeability ($C_{S2} \ll C_{S1}$), we can derive the separation factor for the membrane from Eqs. (1-22) and (1-23) as

$$\alpha_{W-S} = \frac{C_{S1}}{C_{S2}} \frac{C_{W2}}{C_{W1}} \approx \frac{N_W}{N_S} \frac{C_{S1}}{\rho_w} = \frac{k_W(\Delta P - \Delta \pi)}{\rho_w k_S} \tag{1-24}$$

if C_{S1} is expressed in moles per liter and ρ_w is the molar density of water in moles per liter. Since the separation factor for a given k_W and k_S should depend upon ΔP and C_{S1} (through $\Delta \pi$), it is appropriate to compare the separation factors of different membranes at the same ΔP and C_{S1}.

Table 1-2 shows how sensitive k_W, k_S, and α_{W-S} are to the nature of the membrane. Since ΔP and C_{S1} are held constant and α_{W-S} is always on the order of 10 or greater, N_W in Table 1-2 is directly proportional to k_W. Data are presented for anisotropic cellulose acetate membranes manufactured from organic casting solutions of various proportions with various processing parameters (Loeb, 1966). Membranes of this sort hold the greatest interest for desalination of salt water. The membranes were originally made in the laboratory by casting the solution onto a glass plate with 0.025-cm-high side runners, evaporating the solution at this temperature, immersing the system in ice water for 1 h, removing the film from the plate and heating it in water for 5 min.

Table 1-2 shows that the membrane permeability and separation factor are quite sensitive to changes in preparation technique. Loeb (1966) and others have also found that α_{W-S} varies over several orders of magnitude when solutions of different

Table 1-2 Properties of cellulose acetate membranes for reverse osmosis (data from Loeb, 1966)

$\Delta P = 4.15$ MPa, $C_{S1} = 5000$ ppm NaCl; casting temperature $= 23°C$

Composition of casting solution, wt %	Evaporation period, s	Heating temperature, °C	N_w, g/m²·s	α_{w-s}
Cellulose acetate 25, formamide 30, acetone 45	60	74.0	14	12.2
Cellulose acetate 25, formamide 25, acetone 50		71.5	7	18.5
Cellulose acetate 14.3, dimethyl formamide 21.4, acetone 64.3	210	60 Unheated	2.4 8.9	16.7 9.1
Cellulose acetate 25, dimethyl formamide 75	480	93	5.1	38
Cellulose acetate 25, dimethyl sulfoxide 37.5, acetone 37.5	480	93	5.1	38

inorganic salts are subjected to reverse osmosis with the same membrane. The membrane properties are reproducible for a given preparation technique to a standard deviation of about 12 percent, so the changes shown in Table 1-2 are significant. From these data it is apparent that separation factors for cellulose acetate membranes can be determined only from experiment, not from first principles.

The data in Table 1-2 were obtained in a device similar to that shown in Fig. 1-30. The function of the rapid flow in the spacers of such a device is to minimize mass-transfer (diffusional) limitations within the fluids on either side of the membrane. In a large-scale practical device it is often difficult to minimize these diffusional resistances within the fluid phases, and the result is that the apparent separation factor α_{w-s}^{S} and the permeability observed are less than those found for the membrane alone. These mass-transfer effects are considered further in Chaps. 3 and 11.

An effective method for increasing the selectivity and throughput capacity of a membrane in a separation process is to incorporate a substance which reacts chemically with the component to be transmitted preferentially (Robb and Ward, 1967; Ward, 1970; Reusch and Cussler, 1973). The reaction increases the concentration of the component within the membrane and thereby increases the concentration-difference driving force for diffusion of the component through the membrane.

REFERENCES

Alders, L. (1959): "Liquid-Liquid Extraction," 2d ed., Elsevier, New York.
Altgelt, K. H. (1968): Theory and Mechanics of Gel Permeation Chromatography, in J. C. Giddings and R. A. Keller (eds.), "Advances in Chromatography," vol. 7, Marcel Dekker, New York.

Anon. (1973): *Chem. Eng.*, Oct. 15, p. 61.

Antonson, C. R., R. J. Gardner, C. F. King, and D. Y. Ko (1977): *Ind. Eng. Chem. Process Des. Dev.*, **16**:463.

Battino, R., and H. L. Clever (1966): *Chem. Rev.*, **66**:395.

Becker, E. W., and R. Schutte (1960): *Z. Naturforsch.*, **15A**:336.

Benedict, M., and A. Boas (1951): *Chem. Eng. Prog.*, **47**: 51, 111.

———— and T. H. Pigford (1957): "Nuclear Chemical Engineering," McGraw-Hill, New York.

Bier, M. (ed.) (1959): "Electrophoresis," vols. 1 and 2, Academic, New York.

Bowen, T. J., and A. J. Rowe (1970): "An Introduction to Ultracentrifugation," Wiley-Interscience, New York.

Boublik, T., V. Fried, and E. Hala (1973): "The Vapour Pressures of Pure Substances," Elsevier, Amsterdam.

Burgess, M., and R. P. Germann (1969): *AIChE J.*, **15**:272.

Burrows, G. (1960): "Molecular Distillation," Oxford University Press, London.

Cahn, R. P., and N. N. Li (1974): *Separ. Sci.*, **9**:505.

Catsimpoolas, N. (1975): *Separ. Sci.*, **10**:55.

Chao, K. C., and J. D. Seader (1961): *AIChE J.*, **7**:598.

Chivate, M. R., and S. M. Shah (1956): *Chem. Eng. Sci.*, **5**:232.

Cichelli, M. T., W. E. Weatherford, Jr., and J. R. Bowman (1951): *Chem. Eng. Prog.*, **47**:63, 123.

Cline, G. B. (1971): Continuous Sample Flow Density Gradient Centrifugation, in E. S. Perry and C. J. van Oss (eds.), "Progress in Separation and Purification," vol. 4, Wiley-Interscience, New York.

Cohen, M. J., and F. W. Karasek (1970): *J. Chromatogr. Sci.*, **8**:330.

Cussler, E. L., and D. F. Evans (1975): *Separ. Purif. Methods*, **3**:399.

Daniels, F., and R. A. Alberty (1961): "Physical Chemistry," 2d ed., chap. 10, Wiley, New York.

Dedrick, R. L., K. B. Bischoff, and E. F. Leonard (eds.) (1968): *Chem. Eng. Prog. Symp. Ser.*, vol. 64, no. 84.

Everaerts, F. M., F. E. P. Mikkers, and T. P. E. M. Verheggen (1977): *Sep. Purif. Methods*, **6**:287.

Findlay, R. A., and J. A. Weedman (1958): Separation and Purification by Crystallization, in K. A. Kobe and J. J. McKetta (eds.), "Advances in Petroleum Chemistry and Refining," vol. 1, Interscience, New York.

Flinn, J. E., and R. H. Price (1966): *Ind. Eng. Chem. Process Des. Dev.*, **5**:75.

Francis, A. W. (1963): "Liquid-Liquid Equilibriums," Interscience, New York.

Fredeslund, A., J. Gmehling, and P. Rasmussen (1977): "Vapor-Liquid Equilibria Using UNIFAC," Elsevier, Amsterdam.

Fuerstenau, D. W. (ed.) (1962): "Froth Flotation," 50th Anniversary Volume, American Institute of Mining, Metallurgical, and Petroleum Engineers, New York.

Gantzel, P. K., and U. Merten (1970): *Ind. Eng. Chem. Process Des. Dev.*, **9**:331.

Gerstner, H. G. (1969): Sugar (Cane Sugar), in R. E. Kirk and D. F. Othmer (eds.), "Encyclopedia of Chemical Technology", 2d ed., McGraw-Hill, New York.

Giddings, J. C. (1978): *Separ. Sci. Technol.*, **13**:3.

———— and K. Dahlgren (1971): *Separ. Sci.*, **6**:345.

Hala, E., J. Pick, V. Fried, and O. Vilim (1967): "Vapor-Liquid Equilibrium," 3d ed., 2d Engl. ed., Pergamon, New York.

————, I. Wichterle, J. Polak, and T. Boublik (1968): "Vapor-Liquid Equilibrium Data at Normal Pressures," Pergamon, Oxford.

Havighorst, C. R. (1963): *Chem. Eng.*, Nov. 11, pp. 228ff.

Hayduk, W., and H. Laudie (1973): *AIChE J.*, **19**:1233.

Helfferich, F. (1962): "Ion Exchange," McGraw-Hill, New York.

Hengstebeck, R. J. (1961): "Distillation: Principles and Design Procedures," Reinhold, New York.

Hirata, M., S. Ohe, and K. Nagahama (1975): "Computer Aided Data Book of Vapor-Liquid Equilibria," Kodansha, Tokyo, and Elsevier, Amsterdam.

Holland, C. D. (1963): "Multicomponent Distillation," Prentice-Hall, Englewood Cliffs, N.J.

John, M., and H. Dellweg (1974): *Separ. Purif. Methods*, **2**:231.

Karger, B. L., R. B. Grieves, R. Lemlich, A. J. Rubin, and F. Sebba (1967): *Separ. Sci.*, **2**:401.

Keller, R. A., and M. M. Metro (1974): *Separ. Purif. Methods*, **3**:207.

Khalafalla, S. E., and G. W. Reimers (1975): *Separ. Sci.*, **10**:161.

King, C. J. (1971): "Freeze Drying of Foods," CRC Press, Cleveland.

Kohl, A. L., and F. C. Riesenfeld (1979): "Gas Purification," 3d ed., Gulf Publishing, Houston.

Krass, A. S. (1977): *Science*, **196**:721.

Krischer, O. (1956): " Die wissenschaftlichen Grundlagen der Trocknungstechnik," Springer-Verlag, Berlin.

Kuo, V. H. S., and W. R. Wilcox (1975): *Separ. Sci.*, **10**:375.

Latimer, R. E. (1967): *Chem. Eng. Prog.*, **63**(2):35.

Lee, H. L., E. N. Lightfoot, J. F. G. Reis and M. D. Waissbluth (1977): The Systematic Description and Development of Separations Processes, in N. N. Li (ed.), "Recent Developments in Separation Science," vol. 3A, CRC Press, Cleveland.

Lemlich, R. (ed.) (1972): "Adsorptive Bubble Separation Techniques," Academic, New York.

Letokhov, V. S., and C. B. Moore (1976): *Sov. J. Quantum Electron.*, **6**:129.

Loeb, S. (1966): Preparation and Performance of High-Flux Cellulose Acetate Desalination Membranes, in U. Merten (ed.), "Desalination by Reverse Osmosis," M.I.T. Press, Cambridge, Mass.

Loncin, M., and R. L. Merson (1979): "Food Engineering," Academic, New York.

Love, L. O. (1973): *Science*, **182**:343.

Makin, E. C. (1974): *Separ. Sci.*, **9**:541.

Mallan, G. M. (1976): *Chem. Eng.*, July 19, p. 90.

May, S. W., and O. R. Zaborsky (1974): *Separ. Purif. Methods*, **3**:1.

Medici, M. (1974): "The Natural Gas Industry," chap. 5, Newnes-Butterworth, London.

Merten, U. (ed.) (1966): "Desalination by Reverse Osmosis," M.I.T. Press, Cambridge, Mass.

Michaels, A. S. (1968): Ultrafiltration, in E. S. Perry (ed.), "Advances in Separations and Purifications," Wiley, New York.

———— (1968b): *Chem. Eng. Prog.*, **64**(12): 31.

Mitchell, R., G. Britton, and J. A. Oberteuffer (1975): *Separ. Purif. Methods*, **4**:267.

NGSPA (1957): "Engineering Data Book," Natural Gasoline Suppliers & Processors Assn., Tulsa, Okla.

Pehlke, R. D. (1973): "Unit Processes of Extractive Metallurgy," American Elsevier, New York.

Perry, R. H., and C. H. Chilton (eds.) (1973): "Chemical Engineers' Handbook," 5th ed., McGraw-Hill, New York.

————, ————, and S. D. Kirkpatrick (eds.) (1963): "Chemical Engineers' Handbook," 4th ed., McGraw-Hill, New York.

Pfann, W. G. (1966): "Zone Melting," 2d ed., Wiley, New York.

Prausnitz, J. M. (1969): "Molecular Thermodynamics of Fluid-Phase Equilibria," Prentice-Hall, Englewood Cliffs, N.J.

————, C. A. Eckert, R. V. Orye, and J. P. O'Connell (1966): "Computer Calculations for Multicomponent Vapor-Liquid Equilibria," Prentice-Hall, Englewood Cliffs, N.J.

———— and P. L. Chueh (1968): "Computer Calculations for High-Pressure Vapor-Liquid Equilibria," Prentice-Hall, Englewood Cliffs, N.J.

Quinn, H. W. (1971): Hydrocarbon Separations with Silver(I) Systems, in E. S. Perry and C. J. van Oss (eds.), "Progress in Separation and Purification," vol. 4, Wiley-Interscience, New York.

Reid, R. C., J. M. Prausnitz and T. K. Sherwood (1977): "Properties of Gases and Liquids," 3d ed., McGraw-Hill, New York.

Reusch, C. F., and E. L. Cussler (1973): *AIChE J.*, **19**:736.

Rickles, R. N. (1966): *Ind. Eng. Chem.*, **58**(6):19.

Righetti, P. G. (1975): *Separ. Purif. Methods*, **4**:23.

Robb, W. L., and W. J. Ward (1967): *Science*, **156**:1481.

Robinson, C. S., and E. R. Gilliland (1950): "Elements of Fractional Distillation," 4th ed., McGraw-Hill, New York.

Rony, P. R. (1972): *Chem. Eng. Prog. Symp. Ser.*, **68**(120):89.

Ross, S. (1963): Adsorption, Theoretical, in R. E. Kirk and D. F. Othmer (eds.), "Encyclopedia of Chemical Technology," 2d ed., vol. 1, Interscience, New York.

Rutherford, W. M. (1975): *Separ. Purif. Methods*, **4**:305.

Schachman, H. K. (1959): "Ultracentrifugation in Biochemistry," Academic, New York.

Schoen, H. M. (ed.) (1962): "New Chemical Engineering Separation Techniques," Interscience, New York.

Seidell, A., and W. F. Linke (1958): "Solubilities of Inorganic and Metal-Organic Compounds," Van Nostrand, New York.

Sheng, H. P. (1977): *Separ. Purif. Methods,* **6:**89.

Sherwood, T. K., R. L. Pigford, and C. R. Wilke (1975): "Mass Transfer," McGraw-Hill, New York.

Shreve, R. N., and J. A. Brink, Jr. (1977): "Chemical Process Industries," 4th ed., McGraw-Hill, New York.

Singhal, R. P. (1975): *Separ. Purif. Methods,* 3:339.

Smith, B. D. (1963): "Design of Equilibrium Stage Processes," McGraw-Hill, New York.

Smythe, H. D. (1945): "Atomic Energy for Military Purposes," Princeton University Press, Princeton, N.J.

Spiegler, K. S. (1962): "Salt Water Purification," Wiley, New York.

Stephen, H., and T. Stephen (eds.) (1964): "Solubilities of Inorganic and Organic Compounds," Pergamon, New York.

Strain, H. H., T. R. Sato, and J. Engelke (1954): *Anal. Chem.,* **26:**90.

Timmermans, J. (1959–1960): "The Physico-chemical Constants of Binary Systems," Interscience, New York.

Treybal, R. E. (1963): "Liquid Extraction," 2d ed., McGraw-Hill, New York.

Valdes-Krieg, E., C. J. King, and H. H. Sephton: (1977): *Separ. Purif. Methods,* **6:**221.

Van Arsdel, W. D., M. J. Copley, and A. I. Morgan, Jr. (eds.) (1972): "Food Dehydration," 2d ed., AVI, Westport, Conn.

Van Winkle, M. (1967): "Distillation," McGraw-Hill, New York.

Vermeulen, T. (1963): Adsorption, Industrial, in R. E. Kirk and D. F. Othmer (eds.), "Encyclopedia of Chemical Technology," 2d ed., vol. 1, Interscience, New York.

Wadsworth, M. E., and F. T. Davis (eds.) (1964): Unit Processes in Hydrometallurgy, *Metallurg. Soc. Conf.,* vol. 24.

Ward, W. J. (1970): *AIChE J.,* **16:**405.

Weaver, V. C. (1968): Review of Current and Future Trends in Paper Chromatography, in J. C. Giddings and R. A. Keller (eds.), "Advances in Chromatography," vol. 7, Marcel Dekker, New York.

Weetall, H. H. (1974): *Separ. Purif. Methods,* **2:**199.

Wheaton, R. M., and A. H. Seamster (1966): Ion Exchange, in R. E. Kirk and D. F. Othmer (eds.), "Encyclopedia of Chemical Technology," 2d ed., vol. 11, Interscience, New York.

Wichterle, I., J. Linek, and E. Hala (1973): "Vapor-Liquid Equilibrium Data Bibliography," Elsevier, Amsterdam.

Zuiderweg, F. J. (1973): *The Chemical Engineer,* no. 277, p. 404.

Zweig, G., and J. R. Whitaker (1967): "Paper Chromatography and Electrophoresis," 2 vols., Springer-Verlag, New York.

PROBLEMS

1-A₁† The forty-niners of California obtained gold from river-bed gravel by *panning* and by such devices as *rockers, long Toms,* and *sluice boxes.* Look up gold mining in a good reference book to see what physical property and principle these devices were based upon.

1-B₂ In general, the degree of miscibility (or similarity of compositions) of the two liquid phases in a liquid-liquid extraction process increases with increasing concentration of the solute being extracted. One manifestation of this behavior is that plait points at some high solute concentration (such as point *P* in Fig. 1-21, where acetic acid is the solute) are common. Suggest a reason why a higher concentration of the solute being extracted tends to increase the miscibility of the two phases.

---

† The number represents the chapter, the letter represents the sequence, and the numerical subscript represents the degree of difficulty, as follows: 1 = problems which are straightforward applications of material presented in the text; 2 = problems which involve more insight but still should be suitable for undergraduate students; and 3 = problems requiring still more insight, appropriate for the most part for graduate students.

1-C$_2$ Assuming that the membrane characteristics are not changed, will the product-water purity in a reverse-osmosis seawater desalination process increase, decrease, or remain the same as the upstream pressure increases? Explain your answer qualitatively in physical terms.

1-D$_2$ As part of the life support system for spacecraft it is necessary to provide a means of continuously removing carbon dioxide from air. The CO_2 must then be reduced to carbon, oxygen, methane, and/or water for reuse or for disposal. For extended space flights it is not possible to rely upon gravity in any way in devising a CO_2-air separation process. Suggest at least two separation schemes which could be suitable for continuous removal of CO_2 from air in spacecraft under zero-gravity conditions. If solvents, etc., are required, name specific substances which should be considered.

1-E$_3$ Gold is present in seawater to a concentration level between 0.03×10^{-10} and 440×10^{-10} weight fraction, depending upon the location. Usually it is present to less than 1×10^{-10} weight fraction. Briefly evaluate the potential for recovering gold economically from seawater.

1-F$_3$ The deuterium-hydrogen isotope-exchange reaction between hydrogen sulfide and water

$$H_2O(l) + HDS(g) \rightleftharpoons HDO(l) + H_2S(g)$$

has the following equilibrium constants at 2.07 MPa and various temperatures:

Temperature, °C	30	80	130
$K = p_{H_2S} C_{HDO}/p_{HDS} C_{H_2O}$ †	2.29	1.96	1.63

SOURCE: Data from Burgess and Germann (1969).

† These equilibrium constants compensate for the solubility of H_2S in water and for the vapor pressure of water by considering liquid-phase H_2S as H_2O and vapor-phase H_2O as H_2S. Hence these effects need not be taken into account any further in this problem.

The reaction occurs rapidly in the liquid phase, without catalysis. The variation of the equilibrium constant with respect to temperature can be used as the basis for a separation process to produce a water stream enriched in HDO and a water stream depleted in HDO from a feed containing both HDO and H_2O (natural water contains 0.0138 atom % D in the total hydrogen). Such a process is called a *dual-temperature isotope-exchange process*.

Figure 1-31 shows three possible simple processes for carrying out this separation. In each case there is a cold reactor operating at 30°C and a hot reactor operating at 130°C. The pressure is the same in both reactors at about 2 MPa. The atom fraction of deuterium in the total hydrogen of a stream will in all cases be very small compared with unity.

(a) For each of the three flow schemes shown in Fig. 1-31 derive an expression for the separation factor provided by the process in terms of (1) the equilibrium constants of the isotope-exchange reaction at the two temperatures; (2) the circulation rate of H_2S, expressed as S mol of H_2S per mole of water feed; and (3) the fraction f of the feed which is taken as enriched water product. Equilibrium is achieved in both reactors.

(b) Over what range of values will the separation factors for these processes vary as S and f are changed?

(c) In practice, f will be quite small compared with unity. Given this fact, what is the relative order of separation factors which would be obtained from each of the three flow schemes of Fig. 1-31 at a fixed value of S; that is, which scheme gives the greatest separation factor and which gives the least? Explain your answer in terms of physical as well as mathematical reasoning.

(d) What factors will place upper and lower limits on the reactor temperatures that can be used for this process in practice?

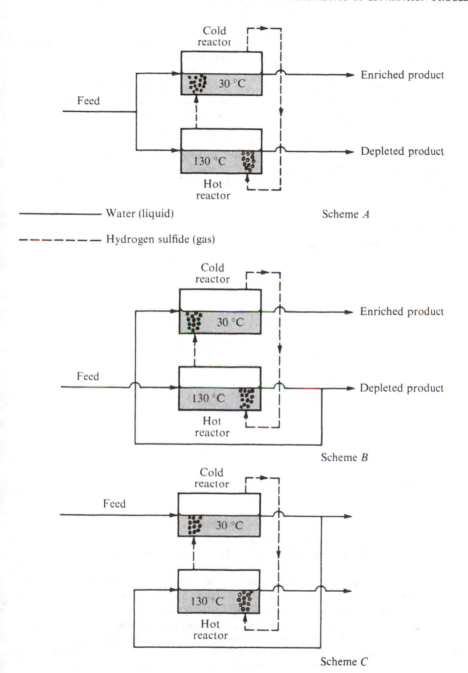

Figure 1-31 Simple dual-temperature isotope-exchange processes for enrichment of deuterium in water.

(e) Why is a relatively high pressure employed?

(f) Where are heaters or coolers required in scheme A? Where might heat exchangers exchanging heat between two process streams be used? Are any compressors or pumps needed?

1-G₃ At Trona, California, in the Mojave Desert, the Kerr McGee Chemical Corp. plant obtains a number of inorganic chemicals from Searles Lake. This is a "dry" lake, composed of salt deposits permeated by concentrated brine solutions. Among the products obtained are potassium chloride, sodium sulfate (salt cake), bromine, borax and boric acid, potassium sulfate, lithium carbonate, phosphoric acid, and soda ash (sodium carbonate). Table 1-3 shows the composition of the upper salt-deposit brine at Searles Lake, which serves as a feed for this operation. Figure 1-32 gives an outline of the Trona processing procedures and Fig. 1-33 a flowsheet for the manufacture of potassium chloride and borax. Furthermore, the flowsheet for the process used at Searles Lake for reclaiming additional boric acid from weak brines and plant end liquors is shown as Fig. 1-34. More detail on these processes can be obtained from Shreve and Brink (1977).

Table 1-3 Composition of upper-deposit brine at Searles Lake, California

Brine composition	Wt %
KCl	4.85
NaCl	16.25
Na_2SO_4	7.20
Na_2CO_3	4.65
$Na_2B_4O_7$	1.50
Na_3PO_4	0.155
NaBr	0.109
Miscellaneous	0.116
Total salts	~ 34.83
H_2O	65.17
Specific gravity	1.303
pH	~ 9.45

SOURCE: Shreve and Brink (1977, p. 266); used by permission.

(a) What are the principal uses of each of the products from these processes?

(b) For each separation step included in Figs. 1-32 and 1-34 indicate (1) the function of the separation step within the overall process, (2) the physical principle upon which the separation is based, (3) the separating agent used, and (4) whether the separation is mechanical, an equilibration process, or a rate-governed process [there is one rate-governed process; consult Shreve and Brink (1977) if necessary].

1-H₃ Oil spills at sea are a major problem. On a number of occasions crude oil from tankers or from offshore drilling operations has been accidentally released in quite large quantities into the ocean near land. The spilled oil forms a thin slick on the ocean surface. The oil spreads out readily, since it is essentially insoluble in water, less dense than water, and lowers the net surface tension. Crude oil from an ocean spill has been washed onto beaches, depositing offensive oil layers which mix with sand and render the beach unattractive. Oil slicks also have a deleterious effect on fish, gulls, seals, and other forms of marine life. Stopping these effects of oil slicks is a separation problem of major proportions. What techniques can you suggest for eliminating an oil slick relatively soon after a spill so as to protect beaches and marine life?†

---

† Reference for consultation after generating suggestions: *Chem. Eng.* (Feb. 10, 1969), pp. 40, 50–54.

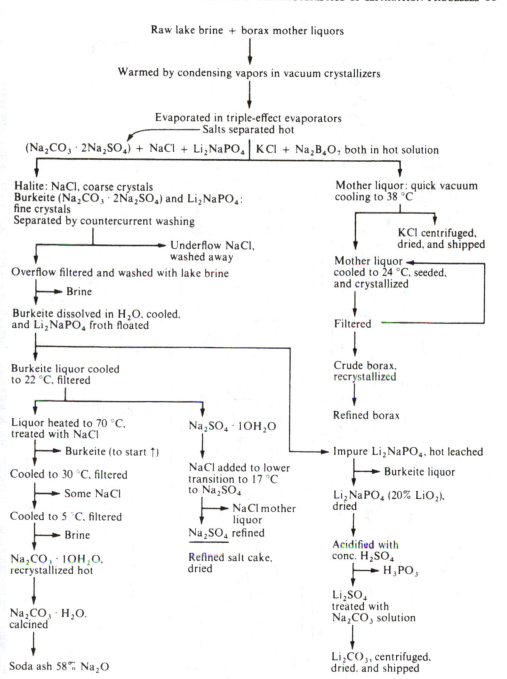

Raw lake brine + borax mother liquors

Warmed by condensing vapors in vacuum crystallizers

Evaporated in triple-effect evaporators
Salts separated hot

$(Na_2CO_3 \cdot 2Na_2SO_4) + NaCl + Li_2NaPO_4 \mid KCl + Na_2B_4O_7$ both in hot solution

Halite: NaCl, coarse crystals
Burkeite $(Na_2CO_3 \cdot 2Na_2SO_4)$ and Li_2NaPO_4:
fine crystals
Separated by countercurrent washing

→ Underflow NaCl,
washed away

Overflow filtered and washed with lake brine

→ Brine

Burkeite dissolved in H_2O, cooled,
and Li_2NaPO_4 froth floated

Burkeite liquor cooled
to 22 °C, filtered

Liquor heated to 70 °C,
treated with NaCl

→ Burkeite (to start ↑)

Cooled to 30 °C, filtered

→ Some NaCl

Cooled to 5 °C, filtered

→ Brine

$Na_2CO_3 \cdot 10H_2O$,
recrystallized hot

$Na_2CO_3 \cdot H_2O$,
calcined

Soda ash 58% Na_2O

$Na_2SO_4 \cdot 10H_2O$

NaCl added to lower
transition to 17 °C
to Na_2SO_4

→ NaCl mother
liquor

Na_2SO_4 refined

Refined salt cake,
dried

Mother liquor: quick vacuum
cooling to 38 °C

KCl centrifuged,
dried, and shipped

Mother liquor ←
cooled to 24 °C, seeded,
and crystallized

Filtered

Crude borax,
recrystallized

Refined borax

→ Impure Li_2NaPO_4, hot leached

→ Burkeite liquor

Li_2NaPO_4 (20% LiO_2),
dried

Acidified with
conc. H_2SO_4

→ H_3PO_3

Li_2SO_4
treated with
Na_2CO_3 solution

Li_2CO_3, centrifuged,
dried, and shipped

Figure 1-32 Processing steps at Searles Lake. *(From Shreve and Brink, 1977, p. 263; used by permission.)*

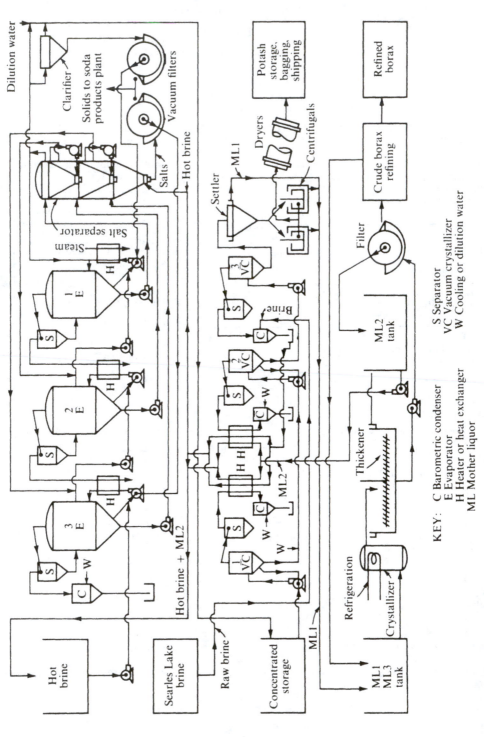

Figure 1-33 Processing steps for manufacture of potassium chloride and borax. (*From Shreve and Brink, 1977, p. 268; used by permission.*)

KEY: C Barometric condenser
E Evaporator
H Heater or heat exchanger
ML Mother liquor

S Separator
VC Vacuum crystallizer
W Cooling or dilution water

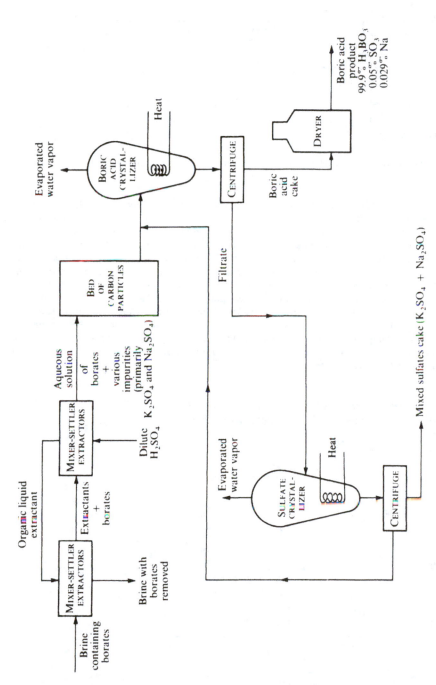

Figure 1-34 Process for obtaining boric acid from weak brines and plant end liquors. (*Adapted from Havighorst, 1963, pp. 228–229; used by permission.*)

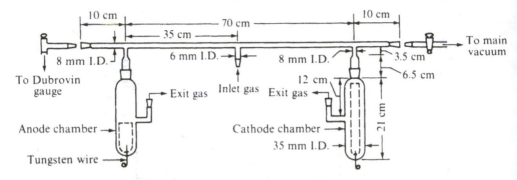

Figure 1-35 Continuous-flow glow-discharge separation device. *(Data from Flinn and Price, 1966.)*

1-I₃ Flinn and Price (1966) investigated the separation of mixtures of argon and helium in a continuous-flow electrical glow-discharge device. The apparatus used is shown in Fig. 1-35. The principle of separation is that the gases will form positive ions within the glow discharge, and the species with the lower ionization potential (argon) should migrate preferentially to the cathode. In the apparatus shown in Fig. 1-35 a luminous glow discharge is formed between a tubular aluminum positively charged anode and a tubular aluminum negatively charged cathode. The mixture of helium and argon enters continuously midway along the discharge path. There are two gas exit streams, one near each electrode. Valves in the exit gas lines are adjusted so that exactly half the feed gas (on a molar basis) leaves in each exit stream. The device is run at low pressures, with the pressure level monitored by a Dubrovin gauge. Table 1-4 shows the separation factors found experimentally for argon-helium mixtures at 0.67 kPa (5 mm Hg) as a function of feed composition, discharge current and flow rate. Suggest physical reasons why the separation factor (a) decreases with increasing feed flow rate, (b) increases with increasing discharge current, and (c) decreases with increasing argon mole fraction in the feed.

Table 1-4 Separation factors α_{Ar-He} found for glow-discharge device, 0.67 kPa (data from Flinn and Price, 1966)

Argon in feed, mol %	Discharge current, mA	Feed flow, mmol/h	α_{Ar-He}
21.30	180	4.9	1.56
21.30	180	9.8	1.30
21.30	180	14.6	1.21
21.30	90	9.8	1.16
21.30	180	9.8	1.30
21.30	270	9.8	1.50
5.43	180	9.8	2.65
21.30	180	9.8	1.30
53.74	180	9.8	1.18

TWO

SIMPLE EQUILIBRIUM PROCESSES

In this chapter we are concerned with calculation of phase compositions, flows, temperatures, etc., in processes where simple equilibrium is achieved between the product phases. Often this requires an iterative calculation scheme, suitable for use with the digital computer. This, in turn, requires selection of appropriate trial variables, check functions, and convergence procedures, which are discussed in texts on numerical analysis and reviewed in Appendix A.

    Procedures for computer calculations and for hand calculations will be discussed interchangeably, since they generally involve the same goals and criteria. The exception is one's ability to monitor the computation as it proceeds in a hand calculation; however, even this distinction is beginning to disappear as interactive digital computing becomes more commonplace.

EQUILIBRIUM CALCULATIONS

If the composition of one of the phases in an equilibrium contacting is known, the composition of the other phase can be obtained by using inherent separation factors, equilibrium ratios, or graphical equilibrium plots. In some instances, notably when few components are present, the determination of the composition of the other phase may be quite easy, but in cases of mixtures of many components an extensive trial-and-error solution may be required.

Binary Vapor-Liquid Systems

For a binary vapor-liquid system we have seen that the composition of the vapor can be determined from the separation factor and the liquid composition by a simple equation

$$y_i = \frac{\alpha_{ij} x_i}{1 + (\alpha_{ij} - 1)x_i} \tag{1-12}$$

The equation can be solved, as well, for x_i in terms of α_{ij} and y_i.

Ternary Liquid Systems

Use of the ternary liquid diagram (such as Fig. 1-21) is also straightforward. Suppose the weight percentage of acetic acid in the water-rich phase of a vinyl acetate-water-acetic acid system is specified to be 25 percent. Since only saturated phases can be in equilibrium with each other, the composition of the water-rich phase must lie on the phase envelope and hence must be 25% acetic acid, 8% vinyl acetate, and 67% water (point A in Fig. 2-1). The composition of the equilibrium acetate-rich phase is obtained by following the appropriate equilibrium tie line, interpolating between those

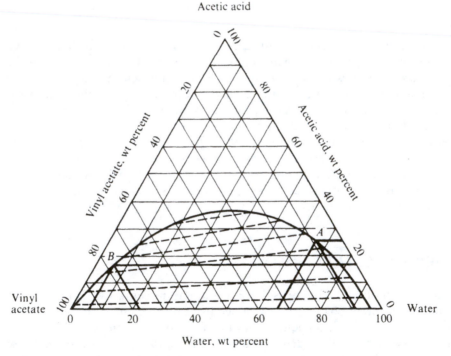

Figure 2-1 Graphical equilibrium calculation in a ternary liquid system. *(Adapted from Daniels and Alberty, 1961, p. 258; used by permission.)*

shown. The indicated equilibrium composition is 6% water, 16% acetic acid, and 78% vinyl acetate (point B in Fig. 2-1).

Multicomponent Systems

A case of multicomponent vapor-liquid equilibrium not requiring trial and error is considered in the following example.

Example 2-1 Find the vapor composition in equilibrium with a liquid mixture containing 20 mol % benzene, 40% toluene, and 40% xylenes at 121°C.

SOLUTION In applying the phase rule

$$P + F = C + 2 \tag{1-14}$$

to this problem we find that $C = 3$ and $P = 2$; hence $F = 3$. All three degrees of freedom have been used in specifying two liquid mole fractions (the third is dependent since $\Sigma x_i = 1$) and the temperature; therefore the pressure is a dependent variable. Note that the problem cannot be solved by use of Eq. (1-12), since that equation is valid only for binary solutions.

This mixture of aromatics is very nearly an ideal solution; hence one method of approach is to calculate the total pressure first from Eq. (1-4) and a knowledge of the pure-component vapor pressures $[P_B^0 = 300, P_T^0 = 133,$ and $P_X^0 = 61$ kPa at 121°C (Maxwell, 1950)]. Since the three xylene isomers have nearly the same vapor pressure, they may be considered as one component.

$$P = (0.2)(300) + (0.4)(133) + (0.4)(61) = 60 + 53.2 + 24.4 = 137.6 \text{ kPa}$$

The vapor composition is then simply derived from Dalton's law:

$$y_B = \frac{(0.2)(300)}{137.6} = 0.436$$

Similarly $y_T = 0.387$ and $y_X = 0.177$. □

Example 2-1 involved calculating the vapor in equilibrium with a known liquid, with the *temperature* known and the *pressure* unknown. The case where *pressure* is known and *temperature* is unknown is usually more difficult to analyze because a knowledge of temperature is necessary in order to define the vapor pressures, the K's, or the α's between any two components, all of which are functions of temperature. Such a calculation of temperature for a completely specified equilibrium must proceed by trial and error in the general case. When the liquid-phase composition is known, the computation is called a *bubble-point calculation*, and when the vapor composition is known, we have a *dew-point calculation*. For a nonideal solution a dew point is even more difficult to compute since liquid-phase activity coefficients are a function of liquid-phase composition, which is also unknown.

Example 2-2 The vapor product from an equilibrium flash separation at 10 atm is 50 mol % n-butane, 20 mol % n-pentane, and 30 mol % n-hexane. Determine the temperature of the separation and the equilibrium liquid composition.

SOLUTION In this case all components are paraffin hydrocarbons. As a result, activity coefficients are near unity and the equilibrium ratios do not depend significantly upon liquid-phase composition. Values of K_i for the three hydrocarbons as a function of temperature are shown in Fig. 2-2 for a pressure of 10 atm.

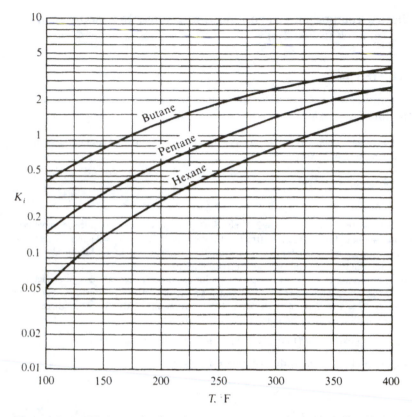

Figure 2-2 Equilibrium ratios for *n*-butane, *n*-pentane, and *n*-hexane; $P = 10$ atm.(*Data from Maxwell, 1950.*)

The problem as stated is a *dew-point* computation. The most common procedure is to assume a temperature and check its validity. Notice that selecting a temperature overspecifies the problem. Since $P + F = C + 2$, where $P = 2$ and $C = 3$, we have but three remaining variables which we can specify, and these have already been set (two vapor mole fractions and the total pressure). Stipulating T overspecifies the system; hence, unless we have happened to select the correct T, we shall find that one of the necessary relationships between variables has been violated. We shall find the K's for the assumed temperature, calculate x_i for each component $(x_i = y_i/K_i)$, and see whether the condition $\Sigma x_i = 1.0$ is violated. The subscripts B, P, and H refer to butane, pentane, and hexane, respectively.

The temperature must obviously lie between that for which $K_B = 1.0$ (176°F) and that for which $K_H = 1.0$ (330°F); otherwise there is no way that $\Sigma(y_i/K_i)$ can equal 1.0. As a first trial assume $T = 250°F$:

	y_i	K_i	$x_i = y_i/K_i$
Butane	0.50	1.87	0.268
Pentane	0.20	0.94	0.213
Hexane	0.30	0.48	0.625
			1.106

The sum of the x's is too high. This is the result of selecting too low a temperature. A higher temperature will increase all the K_i and hence decrease all the x_i.

A close estimate of the correct temperature can now be obtained by making use of the fact that α_{ij} is relatively insensitive to temperature in hydrocarbon systems. This in turn means that the K's can be expressed as

$$K_i = \alpha_{iP} K_P$$

where α_{iP} will be relatively constant with respect to temperature. The $\Sigma x_i = 1.0$ condition can be written as

$$\Sigma x_i = \Sigma \frac{y_i}{K_i} = \frac{1}{K_P} \Sigma \frac{y_i}{\alpha_{iP}} = 1$$

Since $\Sigma(y_i/\alpha_{iP})$ will be insensitive to temperature, we can estimate that the correct K_P is given by

$$K_P = (K_P)_{\text{trial I}} \left(\Sigma \frac{y_i}{K_i} \right)_{\text{trial I}}$$

Thus the indicated K_P is $(0.94)(1.106) = 1.04$. This corresponds to $T = 262°F$, which will be assumed for the second trial:

	y_i	K_i	x_i
Butane	0.50	2.05	0.244
Pentane	0.20	1.04	0.192
Hexane	0.30	0.545	0.551
			0.987

This is close enough. Also, the answer has now been bounded from both sides. One can estimate that

$$T = 262 - \frac{(262 - 250)(1 - 0.987)}{1.106 - 0.987} = 261°F \quad \text{and} \quad x_B = \frac{0.244}{0.987} = 0.247$$

and, similarly, that $x_T = 0.196$ and $x_H = 0.557$. ☐

An analogous procedure holds for bubble-point calculation. One would assume T, calculate K_i, calculate y_i $(= K_i x_i)$, and check whether or not $\Sigma y_i = 1.0$. The indicated T for the next trial would be picked so as to make K_i for a central component equal $(K_i)_{\text{trial I}}/(\Sigma y_i)_{\text{trial I}}$. The logic behind this procedure is analogous to that employed for obtaining a new K_P in Example 2-2.

If the relative volatilities of one component to another within the mixture are totally insensitive to temperature and liquid composition, it is possible to eliminate the trial-and-error aspect of a bubble- or dew-point calculation altogether, as shown in Example 2-3.

Example 2-3 Calculate the bubble point of a liquid mixture containing 20 mol % isobutylene, 30 mol % butadiene, 25 mol % isobutane, and 25 mol % n-butane at a total pressure of 1 MPa.

SOLUTION Since this is a relatively close-boiling mixture, we expect the bubble point to be slightly below 80°C, where K for n-butane (the least volatile component) = 1.0. Between 65 and 80°C the following values of α apply, choosing n-butane as the reference component in each case (Maxwell, 1950):

	α_{iB}
Isobutylene	1.14
Butadiene	1.12
Isobutane	1.25
n-Butane	1.00

We know that

$$\Sigma K_i x_i = 1.0$$

Dividing through by K_B. the equilibrium ratio for n-butane. gives

$$\Sigma \left(\frac{K_i}{K_B} x_i \right) = \Sigma \alpha_i x_i = \frac{1}{K_B}$$

Therefore

$$\frac{1}{K_B} = (1.14)(0.20) + (1.12)(0.30) + (1.25)(0.25) + (1.0)(0.25) = 1.127$$

or

$$K_B = 0.888$$

From Fig. 2-2, K_B is 0.888 at $T = 165°F$ (74°C). which is the bubble point. $\square$

When nonideal liquid solutions or high-pressure vapor phases are involved, the values of K_i become functions of liquid- and/or vapor-phase compositions as well as temperature and pressure. For example, γ_i in Eq. (1-8) will depend upon the mole fraction of each component in the liquid product. These factors make determination of K_i values much more complicated; also going backward in determining the temperature from the K of a reference component, as in the solution to Example 2-2, becomes an iterative solution itself. Furthermore, in these situations the assumption that α_{ij} is constant over moderate ranges of temperature is usually no longer the best convergence procedure.

Let us return to the dew-point problem given in Example 2-2 and consider how this problem should be implemented for a formal computer solution, which can be more easily extended to dew-point problems with more complex phase-equilibrium behavior. We shall ignore the problems associated with nonidealities for the time being and presume that K_j for each component j is given by an Antoine equation with a form such as

$$\ln PK_j = A_j + \frac{B_j}{T} + C_j T \tag{2-1}$$

To apply a convergence method the most obvious procedure is to take

$$f(T) = \Sigma x_j - 1 = \left[\Sigma \frac{y_j}{K_j(T)} \right] - 1 \tag{2-2}$$

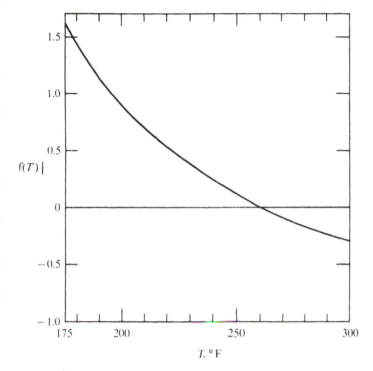

Figure 2-3 Convergence characteristics of $f(T)$ given by Eq. (2-2).

and reduce $f(T)$ to zero. Fig. 2-3 shows a plot of $f(T)$ vs. T based upon calculations similar to those shown in Example 2-2.

Referring to the criteria given in Appendix A for choosing $f(x)$, we find that we can make T a bounded variable. We could build in a check which prevents T from going low enough to give a K for butane less than 1 or high enough to give a K for hexane greater than 1. The function $f(T)$ is monotonic and hence gives no spurious solutions; however, there is a substantial amount of nonlinearity to $f(T)$. If the initial estimate of T were $T_0 = 190°F$, we would find that some five trials would be required to achieve $|f(T)| < 0.005$ by the Newton method described in Appendix A.

The curvature in Fig. 2-3 results from the K_i's not being linear in T. Since the K_j's are related to vapor pressure, we know that $\ln K_j$ will be more nearly linear in T. Because of this behavior it is reasonable to anticipate that a more nearly linear function will be

$$\psi(T) = \ln \Sigma x_j = \ln \sum \frac{y_j}{K_j(T)} \tag{2-3}$$

$\psi(T)$ would then be reduced to zero during the convergence. Note that $\psi(T)$ will be linear in T if the relative volatility of all components with respect to each other is independent of T and if $\ln K$ for any component is linear in T.

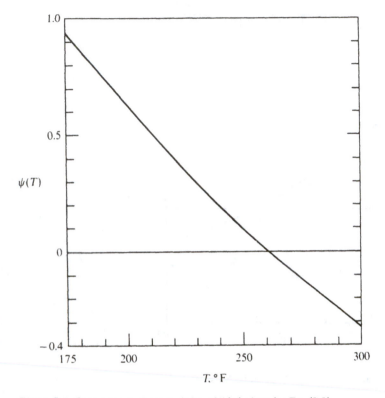

Figure 2-4 Convergence characteristics of $\psi(T)$ given by Eq. (2-3).

Figure 2-4 shows a plot of $\psi(T)$ vs. T for Example 2-2. Note that Fig. 2-4 is considerably more nearly linear than Fig. 2-3 and hence any convergence procedure will reach the required dew-point temperature in a smaller number of iterations. If the initial estimate of T were $T_0 = 190°F$, we would find that three trials would be required to achieve $|\psi(T)| < 0.005$ by the Newton method.

An even more rapid convergence can be achieved if

$$\psi\left(\frac{1}{T}\right) = \ln \sum \frac{y_j}{K_j(1/T)} \tag{2-4}$$

is reduced to zero, since $\ln K_j$ will usually be even more nearly linear in $1/T$. This function is shown in Fig. 2-5. Often a sufficiently accurate dew-point temperature can be obtained by computing $\psi(1/T)$ at two values of temperature T_0 and T_1 and then calculating the dew point by linear interpolation or extrapolation:

$$\frac{1}{T_{DP}} = \frac{1}{T_0} + \left(\frac{1}{T_1} - \frac{1}{T_0}\right) \frac{\psi(1/T_0)}{\psi(1/T_0) - \psi(1/T_1)} \tag{2-5}$$

with no further computation.

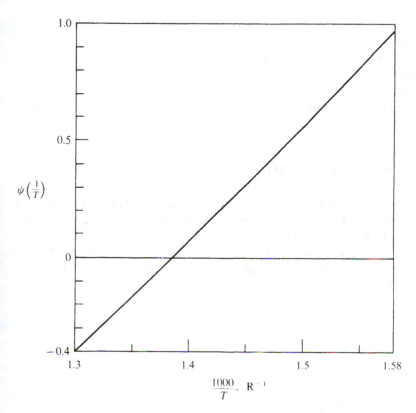

Figure 2-5 Convergence characteristics of $\psi(1/T)$ given by Eq. (2-4).

For computer calculations, the smaller number of iterations using the logarithmic functions must be balanced against the extra time required for computing the logarithm.

Convergence procedures for bubble-point calculations are analogous to those for dew points. The functions corresponding to those given by Eqs. (2-2) to (2-4) are

$$f(T) = [\Sigma x_j K_j(T)] - 1 \tag{2-6}$$

$$\psi(T) = \ln \Sigma x_j K_j(T) \tag{2-7}$$

$$\psi\left(\frac{1}{T}\right) = \ln \Sigma x_j K_j\left(\frac{1}{T}\right) \tag{2-8}$$

Again, Eq. (2-5) can be used in many cases to identify the bubble point after two trials.

When liquid- and/or vapor-phase nonidealities cannot be neglected, each K_j will be a function of *all* the x_j and/or y_j. In many cases the K_j depend rather weakly upon the compositions of the phases, and a successful convergence scheme can be based upon computing the K_j for each trial from the compositions obtained in the last previous trial. When the K_j are affected more strongly by composition, it may be necessary to converge the K_j as an inner loop for each assumed value of T or to converge all composition variables simultaneously in a full multivariate Newton solution (Appendix A).

When two immiscible liquid phases are in equilibrium with a vapor phase, the computation becomes more complex. Henley and Rosen (1969) suggest methods for approaching that problem.

CHECKING PHASE CONDITIONS FOR A MIXTURE

By extending the reasoning involved in dew- and bubble-point calculations it can be seen that a mixture for which $\Sigma K_i x_i < 1$ will be a subcooled liquid, whereas if $\Sigma K_i x_i > 1$, the mixture must contain at least some vapor. Similarly, if $\Sigma(x_i/K_i) < 1$, a mixture will be a superheated vapor, and if $\Sigma(x_i/K_i) > 1$, the mixture must contain at least some liquid. Thus we can set up the following criteria to ascertain the phase condition of a mixture which potentially contains both vapor and liquid:

$\Sigma K_i x_i$	$\Sigma(x_i/K_i)$	Phase condition
< 1	> 1	Subcooled liquid
$= 1$	> 1	Saturated liquid
> 1	> 1	Mixed vapor and liquid
> 1	$= 1$	Saturated vapor
> 1	< 1	Superheated vapor

Similar criteria can be set up for mixtures that are potentially combinations of two immiscible phases, mixtures of a vapor and two liquids, etc.

ANALYSIS OF SIMPLE EQUILIBRIUM SEPARATION PROCESSES

The analysis of equilibrium or ideal separation processes is important for two reasons: (1) It is frequently possible to provide a close approach to product equilibrium in real separation devices. Such is true, for example, for most vapor-liquid separators and for mixer-settler contactors for immiscible liquids. (2) A common practice is to correct for a lack of product equilibrium or ideality by introducing an *efficiency* factor into the calculation procedures used for equilibrium or ideal separations.

In analyzing the performance of a simple separation device one might typically want to calculate the flow rates, compositions, thermal condition, etc., of the products, given the properties and flow rates of the feed and such additional imposed conditions as are necessary to define the separation fully, e.g., the quantity of separating agent employed and the temperature of operation. The quantities to be calculated will vary from situation to situation. For example, one might want to compute the amount of separating agent necessary to give a certain product recovery or the amount of product which can be recovered in a given purity. In any event, the solution will involve:

1. Specifying the requisite number of process variables
2. Developing enthalpy and mass-balance relationships
3. Relating the product compositions through the separation factor or through equilibrium or ideal rate data with corrections for any departure from equilibrium or ideality
4. Solving the resulting equations to obtain values for unknown quantities

Process Specification: The Description Rule

To solve a set of simultaneous equations one must specify the values of a sufficient number of variables for the number of remaining unknowns to be exactly equal to the number of independent equations. If there are five independent equations, there may be no more than five unknown variables for there to be a unique solution. The same reasoning applies to any separation process. If the behavior of the process is to be fully known or is to be uniquely established, there must be a sufficient number of specifications concerning flow rates, temperatures, equipment sizings, etc. The *number* of variables which must be set will depend on the process; on the other hand, the *particular* variables which are set will depend on the problems posed, the answers sought, and the methods of analysis available for calculation. If the problem is overdefined, no answer is possible; if it is underdefined, an infinite number of solutions may exist.

A separation process (or for that matter any process) can always be described by simply writing down all the independent equations which apply to it. Inevitably, the number of unknowns in these equations will be greater than the number of equations. Thus the equations cannot be solved until a sufficient number of the unknowns have had values assigned to them to reduce the remaining number of unknowns to the number of equations. The unknowns to which we assign values are the independent variables of the particular problem under consideration, and the remaining unknowns are the dependent variables.

The procedure of itemizing and counting equations is tedious, however, and is open to error if one misses an equation or counts two equations which are not independent. As we have seen, the phase rule also can be of assistance in determining the number of variables which can be independently specified, but it becomes difficult to apply the phase rule in a helpful way as processes become more complex. A more direct approach is afforded by the *description rule*, originally developed by Hanson et al. (1962), which relies upon one's physical understanding of a process.

Put in its simplest form, the description rule states that *in order to describe a separation process uniquely, the number of independent variables which must be specified is equal to the number which can be set by construction or controlled during operation by independent external means.* In other words, if we build the equipment, turn on all feeds, and set enough valves, etc., to bring the operation to steady state, we have set just enough variables to describe the operation uniquely. In any particular problem where we wish to specify values of any variables which are not set in construction or by external manipulation, we must leave an equal number of construction and external-manipulation variables unspecified. We can replace specified variables on a one-for-one basis.

The description rule is useful for determining the *number* of variables which can be specified. The *particular* variables which are specified will vary considerably from one type of problem to another. If the problem at hand deals with the *operation* of an already existing separation process, the list of specified variables may coincide very nearly or exactly with the list of variables set by construction and controlled during operation. If the problem deals with the *design* of a new piece of equipment, the list of specified variables may be quite different. Typically, variables relating to equipment size will be replaced by variables giving the quality of separation desired.

Often there are upper and lower limits placed upon the values which can be specified for independent variables. For example, amounts of feed must be positive, mole fractions must lie between 0 and 1, etc. For simple equilibration processes identifying these limits is often trivial, but for many more complex processes it is not.

For a new student use of the description rule can be confusing at first because it requires a physical feel for the cause-and-effect relationships occurring. However, it is certainly desirable for an engineer to develop this physical feel, and using the description rule to help specify problems is a direct way of developing it. Furthermore, physical consideration of the degrees of freedom is basic to the selection and understanding of control schemes for separation processes. For that reason control systems are included in diagrams and process descriptions for examples in this book. The number of variables left to be specified after construction of the equipment is the number to be controlled somehow.

A continuous steady-state flash process with preheating of the feed is shown in Fig. 2-6. A liquid mixture (feed) receives heat (separating agent) from a steam heater and then passes through a pressure-reduction, or expansion, valve into a drum in which the phases are separated. Vapor and liquid products are withdrawn from the drum and are close to equilibrium with each other. It is possible to eliminate either the heater or the pressure-reduction valve from the process. In the scheme shown in Fig. 2-6 the drum temperature is held constant by control of the steam rate, the drum pressure controls the vapor-product drawoff, and the liquid product is on level control. Several other control schemes are possible.

Design and construction ensure simple equilibrium between the gas and liquid products, since we assume that there is adequate mixing in the feed line and adequate phase disengagement in the drum. We can apply the description rule further and let the control schemes set the pressure and temperature of the equilibrium. The level control system holds the liquid level in the drum constant, thereby making steady-state operation and satisfactory phase disengagement possible. The process will then be fully specified if the feed composition and flow rate are established before the feed comes to the process. Notice that the temperature and pressure of the feed to the

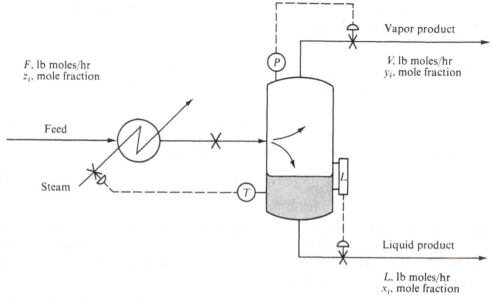

F, lb moles/hr
z_i, mole fraction

Vapor product

V, lb moles/hr
y_i, mole fraction

Feed

Steam

Liquid product

L, lb moles/hr
x_i, mole fraction

Figure 2-6 Continuous equilibrium flash vaporization.

process do not affect the separation since they are both changed to the temperature and pressure desired for the equilibrium.

For the problems to be considered in this chapter, we shall take the feed composition and flow rate to be specified in all cases, postulate simple equilibrium in all cases, and keep the level control loop to hold steady-state operation. Two more variables must be specified in order to define the process. In Fig. 2-6 these are the pressure and temperature of the equilibrium, but in general these variables may be any two of the group:

T = temperature

P = pressure

V/F = fraction vaporization

v_i/f_i = fraction vaporization for component i

H = total product enthalpy (as in adiabatic flash where heater is absent)

We shall consider problems where various pairs of these variables are specified.

Algebraic Approaches

An algebraic approach to the complete analysis of a separation generally involves the use of K's or α's to relate product compositions. We shall develop the appropriate equations and discuss solution procedures in terms of an equilibrium vaporization or flash process; however, the equations will be general to all simple continuous-flow equilibrium separations for which K's and α's can be determined.

Referring to Fig. 2-6, we can write the following mass balance for any component in the continuous equilibrium flash vaporization

$$x_i L + y_i V = z_i F \qquad (2\text{-}9)$$

L, V, and F refer to the total molal flows of liquid product, vapor product, and feed, respectively, and x_i, y_i, and z_i are mole fractions of component i. An overall mass balance gives

$$L + V = F \qquad (2\text{-}10)$$

The product equilibrium expression for any component is

$$y_i = K_i x_i \qquad (2\text{-}11)$$

Binary systems If T and P are specified, K_1 and K_2 are known, providing they are not also functions of composition. Eq. (2-11) can be written once for each component, with $1 - y_1$ substituted for y_2 and with $1 - x_1$ substituted for x_2. These two equations can be solved simultaneously to give

$$x_1 = \frac{1 - K_2}{K_1 - K_2} \qquad (2\text{-}12)$$

and

$$x_2 = \frac{K_1 - 1}{K_1 - K_2} \qquad (2\text{-}13)$$

Values of y_1 and y_2 can then be obtained through Eq. (2-11). The fraction vaporization can be obtained from simultaneous solution of Eqs. (2-9) to (2-11) with the same substitutions of $1 - y_1$ and $1 - x_1$ for y_2 and x_2, respectively, to give, after some algebraic manipulation,

$$\frac{V}{F} = \frac{z_1}{1 - K_2} - \frac{z_2}{K_1 - 1} \qquad (2\text{-}14)$$

(Lockhart and McHenry, 1958).

Multicomponent systems Substituting Eqs. (2-10) and (2-11) into Eq. (2-9), we have

$$x_i L + K_i x_i V = z_i (V + L) \qquad (2\text{-}15)$$

which can be rearranged to give

$$x_i = z_i \frac{1 + V/L}{1 + (K_i V/L)} \qquad (2\text{-}16)$$

Substituting for x_i instead of y_i gives

$$y_i = z_i \frac{1 + L/V}{1 + (L/K_i V)} \qquad (2\text{-}17)$$

If f_i, l_i, and v_i denote the moles of component i in the feed, liquid product, and vapor product ($z_i F$, $x_i L$, and $y_i V$, respectively), we can rearrange Eqs. (2-16) and (2-17) to read

$$l_i = \frac{f_i}{1 + (K_i V/L)} \qquad (2\text{-}18)$$

and

$$v_i = \frac{f_i}{1 + (L/K_i V)} \qquad (2\text{-}19)$$

The factor $K_i V/L$, common in the analysis of vapor-liquid separation processes, is known as the *stripping factor* for component i, since when this factor is large, component i tends to concentrate in the vapor phase and thus be stripped out of the liquid phase. For similar reasons the inverse factor $L/K_i V$ is called the *absorption factor* for component i, since when this factor is large, component i tends to be absorbed more into the liquid phase.

The form of Eqs. (2-16) to (2-19) is such that an iterative solution is needed for most pairs of specified variables if more than two components are present. Criteria for selecting functions and convergence procedures are reviewed in Appendix A and illustrated in the following examples. When there is a choice of trial variables to be used, it generally is desirable to assume trial values for those variables to which the process is not particularly sensitive. For example, one can often make effective use of the fact that α_{ij} in a vapor-liquid system is usually insensitive to changes in pressure and, to a lesser extent, to changes in temperature. Thus the value assumed for P or T will have little effect on α_{ij} in a solution scheme. $K_i P$ is usually insensitive to total pressure in a vapor-liquid process. In a liquid-liquid or liquid-solid process K_i is usually insensitive to pressure. These facts are also useful. As a result, for instance, it is almost always desirable to assume the pressure at an early point in a trial-and-error solution if it has not already been specified in the problem statement.

Although the cases where P and T are specified and where H and P are specified are the most common, we shall also consider cases where other pairs of variables are specified in an equilibrium flash-vaporization process.

In any calculation of a simple-equilibrium separation process it is usually desirable to check first to make sure that two phases are present at equilibrium, using the procedure presented earlier for checking phase conditions.

Case 1: T and v_i/f_i of one component specified This situation corresponds to fixed recovery of a particular component in a flash operating at a temperature that is set, for example, by the maximum steam temperature. One can take advantage of the fact that the relative volatility between any two components in a multicomponent mixture at fixed temperature is frequently highly insensitive to total pressure. Trial and error can usually be avoided altogether in the following way.

Equation (2-19) can be rearranged to give

$$\frac{L}{K_i V} = \frac{f_i}{v_i} - 1 \qquad (2\text{-}20)$$

If we write Eq. (2-20) for components i and j and take the ratio, we get

$$\alpha_{ij} = \frac{K_i}{K_j} = \frac{(f_j/v_j) - 1}{(f_i/v_i) - 1} \tag{2-21}$$

and therefore

$$\frac{f_j}{v_j} = \alpha_{ij}\left(\frac{f_i}{v_i} - 1\right) + 1 \tag{2-22}$$

If T is known, α_{ij} is known for any pair (assuming it to be independent of pressure and composition). Since f_i/v_i is also known, it is possible to compute f_j/v_j for *all* other components by repeated use of Eq. (2-22). This procedure provides a complete solution except for the total pressure, which can be obtained directly from the known liquid composition. If the values of α_{ij} are dependent upon pressure or composition, an iteration loop can be included as before, but it should be rapidly convergent when there is weak dependence of α_{ij} on these variables.

Example 2-4 A hydrocarbon mixture containing 20 mol % n-butane, 50 mol % n-pentane, and 30 mol % n-hexane is fed at a rate of 200 lb mol/h to a continuous steady-state flash vaporization giving product equilibrium at 250°F. Ninety percent of the hexane is to be recovered in the liquid. Calculate the vapor flow rate and composition and the required pressure.

SOLUTION In order to obtain values of α_{ij} it is necessary to assume a pressure, but we shall find that the calculation converges rapidly because of the insensitivity of α_{ij} to pressure. At 250°F and 10 atm from Fig. 2-2, $K_B = 1.87$, $K_P = 0.94$, $K_H = 0.48$. Therefore $\alpha_{BH} = 3.90$, and $\alpha_{PH} = 1.96$. $f_H/v_H = 100\%/10\% = 10$.

	f_j	α_{jH}	$f_j/v_j =$ $(9/\alpha_{jH}) + 1$	v_j	$y_j =$ v_j/V	$x_j =$ $(f_j - v_j)/L$
Butane	40	3.90	3.30	12.1	0.336	0.170
Pentane	100	1.96	5.59	17.9	0.497	0.501
Hexane	60	1.00	10.00	6.0	0.167	0.329
				$V = 36.0$	1.000	1.000
				$L = 164.0$		

Next it is necessary to check the assumed pressure through $\Sigma K_i x_i = 1.000$. Taking $K_i P$ to be independent of pressure, we have

$$\frac{10}{P}[(1.87)(0.170) + (0.94)(0.501) + (0.48)(0.329)] = 1.000$$

$$10(0.947) = P \quad \text{and} \quad P = 9.47 \text{ atm}$$

Using this pressure to determine α_{jH} for each component would give the same values of α; hence no second trial is needed. This procedure would have been effective even if P had turned out to be substantially different from 10 atm, for the relative volatilities at 250°F are essentially constant up to pressures above 30 atm. One must approach the critical pressure before α_{ij} becomes sensitive to pressure. ☐

Case 2: P and T specified This situation corresponds to the process shown in Fig. 2-6; V/L and the product compositions are unknown. The values of K_i for each component are known if they are not functions of composition or if they have been computed using phase compositions from the previous trial.

Various check functions and convergence procedures have been used for solving this problem, but analysis of desirable function properties, by the criteria indicated in Appendix A, has led to a form of function obtained by specifying that $\Sigma y_i - \Sigma x_i$ $(= 1 - 1) = 0$ (Rachford and Rice, 1952). If we use Eq. (2-10) to eliminate L from Eq. (2-16) [or from a combination of Eqs. (2-9) and (2-11)], we obtain

$$x_i = \frac{z_i}{(K_i - 1)(V/F) + 1} \tag{2-23}$$

Applying Eq. (2-11) to Eq. (2-23) yields

$$y_i = \frac{K_i z_i}{(K_i - 1)(V/F) + 1} \tag{2-24}$$

Applying the criterion that $\Sigma y_i - \Sigma x_i = 0$ yields a convergence function

$$f\left(\frac{V}{F}\right) = \sum_i \frac{z_i(K_i - 1)}{(K_i - 1)(V/F) + 1} = 0 \tag{2-25}$$

The iterative calculation involves assuming values of V/F and applying a convergence procedure until a value of V/F is found such that $f(V/F) = 0$. The section on choosing $f(x)$ in Appendix A shows the advantages of the function given in Eq. (2-25) to be that it has no spurious roots, maxima, or minima; that the iteration variable V/F is bounded between 0 and 1; and that the function is relatively linear in V/F.

In calculating any equilibrium separation it is useful to ascertain first that the specifications of the problem do correspond to there being two phases present. This can be done with Eq. (2-25) by checking that $f(V/F)$ is positive at $V/F = 0$ and negative at $V/F = 1$. If $f(V/F)$ is negative at $V/F = 0$, the system is subcooled liquid. If $f(V/F)$ is positive at $V/F = 1$, the system is superheated vapor.

Example 2-5 The feed of Example 2-4 is fed to an equilibrium-flash vaporization yielding products at 10 atm and 270°F. Find the product compositions and flow rates.

SOLUTION First we shall check that two phases are indeed present by calculating $f(V/F)$ at $V/F = 0$ and 1:

	K_i	z_i	$z_i(K_i - 1)$	V/F = 0 Denom.	V/F = 0 Num./denom.	V/F = 1 Denom.	V/F = 1 Num./denom.
Butane	2.13	0.200	0.2260	1	0.2260	2.130	0.1061
Pentane	1.10	0.500	0.0500	1	0.0500	1.100	0.0455
Hexane	0.59	0.300	-0.1230	1	-0.1230	0.590	-0.2085
					+0.1530		-0.0569

The function at $V/F = 0$ is positive ($+0.1530$) and at $V/F = 1$ is negative (-0.0569), so both vapor and liquid are present at equilibrium.

As a next trial we shall assume $V/F = 0.5$, although it would also be defensible to take a linear interpolation between the results at $V/F = 0$ and 1, giving $V/F = 0.1530/(0.1530 + 0.0569) = 0.73$. At $V/F = 0.5$:

	$z_i(K_i - 1)$	Denom.	Num./denom.
Butane	0.2260	1.565	0.1444
Pentane	0.0500	1.050	0.0476
Hexane	-0.1230	0.795	-0.1547
			+0.0373

If we select the regula falsi method for convergence (Appendix A), we take a linear interpolation between the most closely bounding values giving positive and negative values of $f(V/F)$, namely, the results for $V/F = 0.5$ and 1:

$$\frac{V}{F} = \left(\frac{V}{F}\right)_1 + \left[\left(\frac{V}{F}\right)_2 - \left(\frac{V}{F}\right)_1\right]\frac{f(V/F)_1}{f(V/F)_1 - f(V/F)_2}$$

$$= 0.5 + (0.5)\frac{0.0373}{0.0373 + 0.0569} = 0.698$$

(2-26)

	$z_i(K_i - 1)$	Denom.	Num./denom.
Butane	0.2260	1.7887	0.1263
Pentane	0.0500	1.0698	0.0467
Hexane	-0.1230	0.7138	-0.1723
			+0.0007

This is very close to the correct result, which we can estimate to a very high accuracy using the regula falsi method once again:

$$\frac{V}{F} = 0.698 + (1 - 0.698)\frac{0.0007}{0.0007 + 0.0569} = 0.702$$

The equilibrium phase compositions can now be obtained by using this value of V/F in Eqs. (2-23) and (2-24):

	x_i	y_i
Butane	0.112	0.238
Pentane	0.467	0.514
Hexane	0.421	0.248
	1.000	1.000

In Example 2-5 the number of significant figures carried is greater than is warranted by the precision of the K_i values, but this was done to illustrate the convergence properties and speed of convergence better.

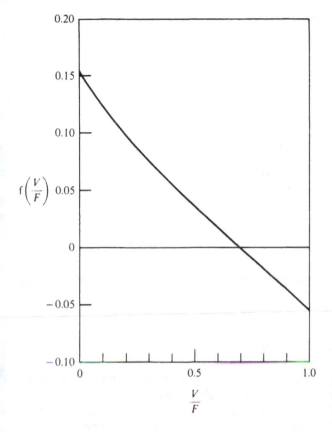

Figure 2-7 Convergence characteristics of Eq. (2-25) for problem of Example 2-5 (Rachford-Rice form).

The regula falsi method used in Example 2-5 is by no means the only convergence method that could be used. A common approach is to use the Newton method [Eq. (A-5)], which requires the derivative of the function, given by

$$f'\left(\frac{V}{F}\right) = -\sum_i \frac{z_i(K_i - 1)^2}{[(K_i - 1)(V/F) + 1]^2} \tag{2-27}$$

The rapid convergence of Eq. (2-25) results from its near linearity, shown in Fig. 2-7; consequently, nearly any convergence method will lead quickly to the answer.

Barnés and Flores (1976) have shown that a logarithmic form of Eq. (2-25) is even more nearly linear and rapidly converging, just as the logarithmic form is more nearly linear for bubble- and dew-point calculations. The resultant function is obtained by combining Eqs. (2-23) and (2-24) to give $\ln(\Sigma y_i / \Sigma x_i) = 0$:

$$G\left(\frac{V}{F}\right) = \ln \frac{\displaystyle\sum_i \frac{K_i z_i}{(K_i - 1)(V/F) + 1}}{\displaystyle\sum_i \frac{z_i}{(K_i - 1)(V/F) + 1}} = 0 \tag{2-28}$$

If we use Eq. (2-28) as a convergence function for Example 2-5, the calculation goes as follows:

	V/F = 0.5		V/F = 1	
	y_i Eq. (2-24)	x_i Eq. (2-23)	y_i Eq. (2-24)	x_i Eq. (2-23)
Butane	0.2722	0.1278	0.2000	0.0939
Pentane	0.5238	0.4762	0.5000	0.4545
Hexane	0.2226	0.3774	0.3000	0.5085
	1.0186	0.9814	1.0000	1.0569

$$G(V/F): \qquad \ln \frac{1.0186}{0.9814} = +0.0372 \qquad \ln \frac{1.0000}{1.0569} = -0.0553$$

If we apply the regula falsi convergence method to find a new value of V/F, we obtain

$$\frac{V}{F} = 0.500 + (1 - 0.500)\frac{0.0372}{0.0372 + 0.0553} = 0.701$$

The computed value of 0.701 is closer to the converged value of 0.702 than is the value of 0.698 obtained using $f(V/F)$ in Example 2-5.

The faster convergence of $G(V/F)$ per iteration is offset by the greater amount of calculation per iteration, since it is necessary to obtain the logarithm.

Equation (2-28) is similar to Eq. (2-25) in that $G(V/F)$ must be positive at $V/F = 0$ and negative at $V/F = 1$ in order for two phases to be present.

It is instructive to compare the convergence properties of Eqs. (2-25) and (2-28) with those of other functionalities which could logically be used. For example, one approach is to assume V/L, compute all l_i from Eq. (2-18), sum the l_i to get L, compute V as $F - L$, and compare the resultant L/V with the assumed value. One problem with this is that L/V is not a bounded trial variable, and the calculation can be sent off to quite large values of V/L. This can be remedied by changing to V/F as the trial variable, since V/F must lie between 0 and 1. Substituting $F - V$ for L in the procedure just described, we obtain

$$\phi\left(\frac{V}{F}\right) = 1 - \frac{1}{F}\sum_j \frac{f_j}{1 + \{K_j(V/F)_i/[1 - (V/F)_i]\}} = \left(\frac{V}{F}\right)_{i+1} \qquad (2\text{-}29)$$

where i refers to the trial number and j to the component. Eq. (2-29) has the form of a direct-substitution convergence method [Eq. (A-2)] and does satisfy the criterion that $d\phi(V/F)/d(V/F)$ at the solution be less than 1 so that the calculation will converge to the desired answer rather than to the spurious roots at $V/F = 0$ and 1. However, convergence is often very slow. $\phi(V/F)$ is plotted in Fig. 2-8 for the problem of Example 2-5. By comparison with Fig. A-1 one can see that the convergence would be extremely slow.

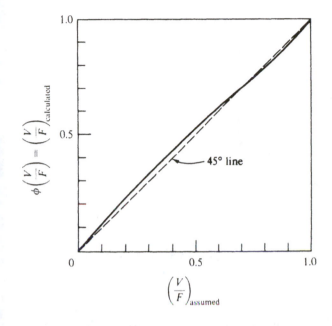

Figure 2-8 Convergence characteristics of direct substitution for Example 2-5 [Eq. (2-29)].

Since the direct-substitution approach is sure but slow for flash calculations with P and T specified, the Wegstein acceleration for direct-substitution convergence (Lapidus, 1962) should be effective and very often is.

For other convergence methods we want an equation of the form

$$f(x) = 0 \qquad (A-1)$$

Equation (2-29) put in this form becomes

$$f\left(\frac{V}{F}\right) = 1 - \frac{V}{F} - \frac{1}{F} \sum_j \frac{f_j}{1 + \{K_j(V/F)/[1 - (V/F)]\}} = 0 \qquad (2-30)$$

Equation (2-30) is shown schematically as the solid curve in Fig. 2-9. Two spurious roots exist, and there are both a maximum and a minimum. These features hamper most convergence procedures; e.g., the Newton method is divergent unless the initial estimate of V/F lies between the maximum and the minimum. The Newton method can also get into a *loop*, shown by the dashed lines in Fig. 2-9, where the computation will cycle without convergence.

Rohl and Sudall (1967) have compared the efficiency of nine different convergence methods for solving equilibrium flash vaporizations of three different feed mixtures. They concluded that the third-order Richmond method and the second-order Newton method [Eq. (A-5)], both applied to the Rachford-Rice form of $f(V/F)$ as given by Eq. (2-25), were most efficient in terms of minimum computation time. The Richmond iteration formula is

$$x_{i+1} = x_i - \frac{2f(x_i)[df(x)/dx]_{x=x_i}}{2[df(x)/dx]_{x=x_i}^2 - f(x_i)[d^2f(x)/dx^2]_{x=x_i}} \qquad (2-31)$$

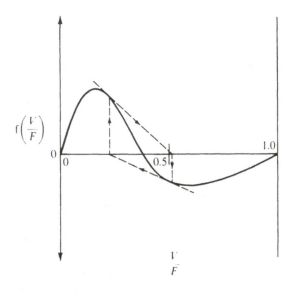

Figure 2-9 Convergence characteristics of Eq. (2-30).

The Wegstein accelerated direct-substitution method applied to Eq. (2-29) is nearly as efficient and can become more efficient than the other methods when the number of components is large. It was found necessary to restrict the movement of V/F between iterations during the early iterations of all three of the methods, in order to prevent the generation of a value of V/F outside the range 0 to 1 during the course of the convergence. Equation (2-28) was not considered in their study.

Case 3: P and V/F specified In this case the values of K_i are not known. Equations (2-25) and (2-28) can be used as functions of T, rather than V/F, with T restricted to the range between the bubble point and the dew point of the feed. Barnés and Flores (1976) and others have found rapid convergence for this case.

Case 4: P and v_i/f_i of one component specified Grens (1967) has shown that an equation similar to Eq. (2-25) but involving v_r/f_r is

$$f(T) = \sum_j \frac{z_j(K_j - 1)}{(K_j - K_r)(v_r/f_r) + K_r} = 0 \tag{2-32}$$

Here r refers to the component for which the split is specified. The values of K_j are unknown and depend on temperature. $f(v_r/f_r)$ has convergence characteristics vs. temperature which are quite good and usually lead to convergence as rapid as that obtained for Eq. (2-25) with V/F specified.

An alternative approach for a hand calculation is equivalent to the inner loop of the hand-calculation method in the next case.

Case 5: P and product enthalpy specified Our presumption so far has been that our equilibrium flash vaporization will be equipped with a steam preheater and expansion valve which will enable us to specify any two of the parameters T, P, V/L, or f_i/v_i without regard to an overall enthalpy balance. The enthalpy difference is made up by heat input from the steam, and, conversely, an enthalpy balance can be used to determine the steam requirement.

On the other hand, flash vaporizations are frequently carried out with no preheater, and the separation is accomplished by forming vapor while throttling the feed through a valve to a lower pressure. This gives an *isenthalpic* flash in which the total product enthalpy must equal the total feed enthalpy. Only one additional variable can now be specified independently (the temperature-control loop has been removed in the application of the description rule). We shall consider the most common case, where P is the additional variable specified.

Two approaches will be presented here. One is suitable for hand calculations in cases where α_{ij} is insensitive to temperature, where equilibrium and enthalpy data are obtained graphically, and where the mixture is sufficiently wide-boiling. The other is more general and suitable for computer implementation for more complex problems.

For simple systems, e.g., hydrocarbons at low to moderate pressures, enthalpies of individual components can be obtained graphically from a source such as Maxwell (1950) and may be considered to be additive. For more general purposes, the algebraic methods based upon correlations and thermodynamic analyses given by Reid et al. (1977) can be used. This includes the enthalpy-departure functions of Yen and Alexander (1965). Holland (1975) presents a method utilizing "virtual" values of partial molal enthalpies to obtain the enthalpy of a mixture.

In general the analysis of an isenthalpic flash with pressure specified requires convergence of *two* trial variables. For α_{ij} insensitive to temperature iteration on the second trial variable can be avoided, however, by incorporating a procedure analogous to that presented already for a flash with T and v_i/f_i of one component specified (Example 2-4). We first assume v_i/f_i for a central reference component, given subscript R. Values of v_i/f_i for all other components are calculated from v_R/f_R by Eq. (2-22). The v_i are then summed to give V. From V/F and v_R/f_R we obtain K_R, which can be converted into T if graphical equilibrium data are available and if K_R is a function of T only. This value of T was determined solely from equilibrium considerations and hence can be checked independently by an overall enthalpy balance. We then iterate upon v_R/f_R until the enthalpy balance converges. Example 2-6 illustrates this method.

Example 2-6 Bubble-point liquid feed containing 30 % *n*-butane, 40 % *n*-pentane, and 30 % *n*-hexane is available at 300°F, 17.5 atm total pressure, and 100 lb mol/h. The mixture is throttled adiabatically to give equilibrium vapor-liquid products at 7.0 atm (88 lb/in² gauge). Find the product temperature and the vapor-liquid component split.

SOLUTION The product temperature will be less than the feed temperature because of the consumption of latent heat in forming vapor. A rough enthalpy balance can be made from the enthalpy data presented in Fig. 2-10. If half the material is vaporized (weight basis),

$$FC_p(300 - T) = \frac{F}{2} \Delta H_V$$

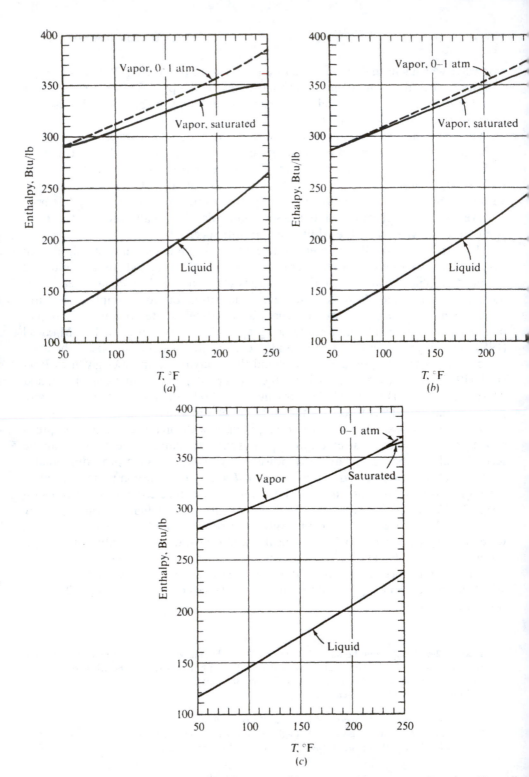

Figure 2-10 Enthalpies of hydrocarbons: (a) n-butane, (b) n-pentane, (c) n-hexane. (*Data from Maxwell, 1950.*)

where C_p is an average heat capacity and ΔH_V is an average latent heat of vaporization; C_p is about 0.65 Btu/lb · °F, and ΔH_V is about 130 Btu/lb. Hence T is roughly

$$300 - \frac{(0.5)(130)}{(0.65)} = 200°F$$

(This is not the only way to generate a first estimate of temperature. Another way is to make bubble- and dew-point calculations for the feed and calculate liquid and vapor enthalpies, respectively, at those two temperatures. One can then make a linear interpolation between these temperatures and enthalpies to obtain the temperature corresponding to the feed enthalpy.) At 200°F, from Fig. 2-2,

$$\alpha_{PB} = \frac{0.57}{1.27} = 0.45 \qquad \alpha_{PH} = \frac{0.57}{0.265} = 2.15$$

The feed enthalpy is computed by extrapolation as follows (M_i = molecular weight of component i):

	M_i	f_i	h_{300}, Btu/lb	$M_i f_i h_i$, Btu/h
Butane	58	30	310	540,000
Pentane	72	40	288	829,000
Hexane	86	30	276	712,000
				$\overline{2,081,000} = h_F F$

Trial 1 Assume $v_P / f_P = 0.500$, $f_P / v_P - 1 = 1.00$.

Component	f_j	f_j / v_j	v_j
Butane	30	1.45	20.7
Pentane	40	2.00	20.0
Hexane	30	3.15	9.5
			$V = \overline{50.2}$

$$K_P = \frac{v_P}{l_P} \frac{L}{V} = \frac{(20.0)(49.8)}{(20.0)(50.2)} = 0.99$$

Following the ideal-gas law, K_P of 0.99 at 7.0 atm corresponds to a temperature where $K_P = 0.99(7/10) = 0.69$ at 10 atm. From Fig. 2-2, K_P of 0.69 at 10 atm corresponds to $T = 217°F$.

The indicated total product enthalpy is found as follows. Specific enthalpies are denoted H_j and h_j for vapor and liquid, respectively.

M_j	Component	v_j	l_j	$H_{j,217}$	$h_{j,217}$	$H_j M_j v_j \times 10^{-3}$	$h_j M_j l_j \times 10^{-3}$
58	Butane	20.7	9.3	359	238	431	128
72	Pentane	20.0	20.0	351	227	505	327
86	Hexane	9.5	20.5	350	218	286	384
						$VH_v = \overline{1222}$	$Lh_L = \overline{839}$

$$VH_v + Lh_L = 2,061,000 \text{ Btu/h}$$

The indicated product enthalpy is less than the feed enthalpy. If a higher v_P/f_P had been assumed, the product enthalpy would have been greater, since the enthalpy increase due to the latent heat of vaporization outweighs any sensible heat effect. The temperature is close enough to 200°F for the same values of α_{ij} to be used.

Trial 2 Assume $v_P/f_P = 0.600$, $f_P/v_P - 1 = 0.667$.

Component	f_j	f_j/v_j	v_j
Butane	30	1.30	23.1
Pentane	40	1.67	24.0
Hexane	30	2.43	12.4
	100		59.5

$$K_P = \frac{v_P}{l_P}\frac{L}{V} = \frac{(24.0)(40.5)}{(16.0)(59.5)} = 1.02$$

K_P of 1.02 at 7 atm corresponds to $K_P = 0.71$ at 10 atm; hence T from Fig. 2-2 = 220°F.

M_j	Component	v_j	l_j	$H_{j,220}$	$h_{j,220}$	$H_j M_j v_j \times 10^{-3}$	$h_j M_j l_j \times 10^{-3}$
58	Butane	23.1	6.9	360	240	482	96
72	Pentane	24.0	16.0	352	229	608	264
86	Hexane	12.4	17.6	351	220	374	333
						$VH_V = \overline{1464}$	$Lh_L = \overline{693}$

$$VH_V + Lh_L = 2{,}157{,}000 \text{ Btu/h}$$

v_P/f_P was increased by too great an amount in trial 2. A better value could have been obtained by a rough prior enthalpy balance. Since the feed enthalpy lies $20/96 = 21\%$ of the way between the product enthalpies from trials 1 and 2, the results can be obtained by linear interpolation. This procedure can be easily accomplished algebraically but is also shown graphically in Fig. 2-11.

The final conditions are:

Component	v_j	l_j
Butane	21.2	8.8
Pentane	20.8	19.2
Hexane	10.2	19.8
	$V = \overline{52.2}$	$L = \overline{47.8}$

and

$$T = 217°F \qquad \square$$

A more general approach, suitable for computer implementation, would involve convergence of T and V/F using check functions based upon mass balances and an

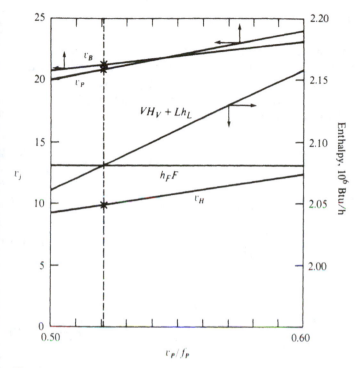

Figure 2-11 Interpolation of results for Example 2-6.

enthalpy balance. A well-behaved function based on mass balances is Eq. (2-25), as used for the flash where T and P are specified:

$$f\left(\frac{V}{F}, T\right) = \sum_i \frac{z_i(K_i - 1)}{(K_i - 1)(V/F) + 1} = 0 \qquad (2\text{-}33)$$

Another well-behaved function for the enthalpy balance can be obtained if the product enthalpies are expressed as $(V/F)\Sigma H_i y_i$ and $[1 - (V/F)]\Sigma h_i x_i$, with y_i and x_i coming from Eqs. (2-24) and (2-23) (Barnés and Flores, 1976; Hanson, 1977):

$$g\left(\frac{V}{F}, T\right) = \left\{ \sum_i \frac{z_i}{(K_i - 1)(V/F) + 1} \left[K_i H_i \frac{V}{F} + h_i \left(1 - \frac{V}{F}\right)\right] \right\} - h_F = 0 \qquad (2\text{-}34)$$

Here the values of H_i and h_i are specific enthalpies of individual components, considered additive, or more generally partial molar enthalpies. Some enthalpy-prediction methods give specific enthalpies of the entire mixture, the properties of individual components having been blended via mixing rules at an early point in the prediction process. Such is the case for vapor-enthalpy prediction methods recommended by Reid et al. (1977). In that case Eq. (2-34) can be modified to

$$G\left(\frac{V}{F}, T\right) = \left\{ \sum_i \frac{z_i}{(K_i - 1)(V/F) + 1} \left[K_i H_V \frac{V}{F} + h_L \left(1 - \frac{V}{F}\right)\right] \right\} - h_F = 0 \qquad (2\text{-}35)$$

where H_V and h_L are the specific enthalpies of the vapor and liquid mixtures, respectively.

When both check functions are substantially influenced by both independent variables, the best convergence procedure is the multivariate Newton method, which for two variables is described by Eqs. (A-6) to (A-10). This requires values of the four partial derivatives, for which analytical expressions can be obtained as follows (Hanson, 1977; Barnés and Flores, 1976):

$$\frac{\partial f}{\partial (V/F)} = -\sum_i \frac{z_i(K_i - 1)^2}{[(K_i - 1)(V/F) + 1]^2} \tag{2-36}$$

$$\frac{\partial f}{\partial T} = \sum_i \frac{z_i}{[(K_i - 1)(V/F) + 1]^2} \frac{dK_i}{dT} \tag{2-37}$$

$$\frac{\partial g}{\partial (V/F)} = \sum_i \frac{z_i K_i(H_i - h_i)}{[(K_i - 1)(V/F) + 1]^2} \tag{2-38}$$

$$\frac{\partial g}{\partial T} = \sum_i \frac{z_i}{(K_i - 1)(V/F) + 1}$$
$$\cdot \left[\frac{(V/F)(1 - V/F)(H_i - h_i)}{(K_i - 1)(V/F) + 1} \frac{dK_i}{dT} + K_i \frac{V}{F} \frac{dH_i}{dT} + \left(1 - \frac{V}{F}\right) \frac{dh_i}{dT} \right] \tag{2-39}$$

Equation (2-36) is, of course, a simple extension of Eq. (2-27). The terms dH_i/dT and dh_i/dT in Eq. (2-39) are equivalent to vapor and liquid heat capacities. For the case where G [Eq. (2-35)] is used as a check function rather than g, the expression for $\partial G/\partial T$ is the same as Eq. (2-39), with H_V and h_L substituted for H_i and h_i, respectively.

Sometimes solutions of equation sets using the multivariate Newton convergence method suffer from stability problems if the initial estimates are well removed from the correct values; however, for isenthalpic flash calculations the near linearity of Eqs. (2-33) and (2-34) or (2-35) appears to make the multivariate Newton method highly stable and rapidly convergent in at least the large majority of cases (Hanson, 1977).

The multivariate Newton convergence method does require the calculation of four partial derivatives per iteration, either analytically by Eqs. (2-36) to (2-39) or by making calculations at incrementally different values of one of the variables. In many cases the physical nature of the problem is such that some of the partial derivatives are necessarily small; i.e., a variable has only a small effect on a check function. As developed further in Appendix A, one can then save computational time by *partitioning* the convergence, or pairing check functions with trial variables on a one-to-one basis. This is equivalent to ignoring the partial derivatives of a function with respect to the variable(s) with which it is not paired. There are two ways of doing this, sequential and paired simultaneous-convergence methods.

The question of which variable to pair with which equation can be viewed in the sense of pairing each variable with that equation which physically has the greater effect in determining the value of that variable. Friday and Smith† give a lucid description of this viewpoint:

> Consider two extreme types of feed mixtures, close boiling and wide boiling, each of which is fed to an adiabatic flash stage which produces vapor and liquid product streams from a completely specified feed. For simplicity let the close boiling feed be the limiting case, a pure component. For such a feed the stage temperature is the boiling point of the particular component at the specified pressure. A change in the feed enthalpy will change the phase [flow] rates but not the stage temperature. Obviously the energy balances should be used to calculate V and L while the [summation-of-mole-fraction] equations are satisfied by a bubble or dew point calculation (trivial in this extreme case) . . . .
>
> Now let the wide boiling material be a mixture of two components, one very volatile and the other quite nonvolatile. For such a feed the amounts of each phase leaving the stage are almost completely determined by the distribution coefficients [i.e., the K_j]. Over a wide temperature range the volatile component will leave predominantly in the vapor, while the heavy component leaves predominantly in the liquid phase. Additional enthalpy in the feed will raise the stage temperature but have little effect on the V and L rates. Obviously in this case the energy balance equation should be used to calculate the stage temperature.

The enthalpy balance is relatively more dependent upon T as opposed to V/F for a wide-boiling flash than for a close-boiling flash. This follows since the enthalpy balance is primarily influenced by latent heats of vaporization (and hence V/F, the degree of vaporization) for a close-boiling flash where the temperature cannot vary greatly. In a wide-boiling flash the V/F cannot vary widely, and the latent heat effect cannot vary much as a result, but the wide range of possible temperatures gives a substantial variable *sensible-heat* effect. These factors again point to pairing f with V/F and g or G with T in a wide-boiling flash and to pairing f with T and g or G with V/F for a close-boiling flash.

Figure 2-12 shows convergence schemes for the pairings associated with a wide-boiling flash. In the *sequential* scheme the inner loop, pairing f with V/F, is converged fully for each assumed value of the outer-loop variable T. The *paired simultaneous* scheme is obtained by introducing the dashed line while removing the one solid line that is labeled "sequential scheme." The paired simultaneous scheme is analogous to the multivariate Newton scheme (also simultaneous), except that the pairing means that not all the partial derivatives are taken into account. In most cases the paired simultaneous scheme will converge more rapidly than the sequential scheme because it is not necessary to converge the inner loop for each value of the outer-loop variable. However, if the paired simultaneous scheme proves unstable, the sequential method should give better stability. For the nearly linear functions considered here, this should be a problem only rarely.

Figure 2-13 shows convergence schemes for the pairing suited to a close-boiling flash. Entirely analogous reasoning applies in comparing the sequential and paired simultaneous versions.

The blocks marked "convergence" in Figs. 2-12 and 2-13 can contain any of th accepted convergence methods, first-order and Newton methods being m

† From Friday and Smith (1964, pp. 701–702); used by permission.

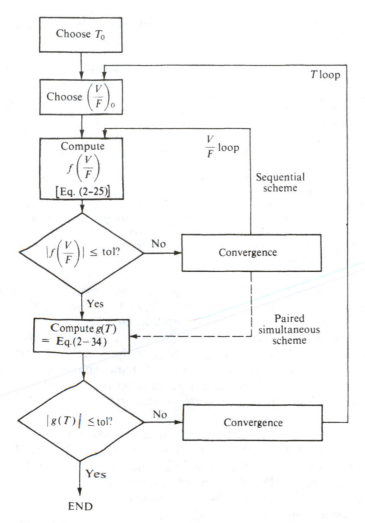

Figure 2-12 Convergence scheme for a wide-boiling adiabatic flash.

common. The ordering of the loops in the sequential schemes is important, however. First, it is desirable to have in the inner loop a trial variable whose converged value is insensitive to the prevailing value of the outer-loop trial variable. This means that the converged value from the previous convergence of the inner loop will be an excellent initial estimate for the next convergence of the inner loop. For a wide-boiling flash the converged V/F is insensitive to the prevailing value of T, following the logic presented by Friday and Smith, above. Similarly, for a close-boiling flash the converged T is insensitive to the prevailing value of V/F. A second reason for having the enthalpy balances in the outer loops is that they can then be based upon values of y_i and x_i which add to unity, giving the enthalpy balances greater physical meaning.

Seader (1978) has found that the paired convergence schemes give rapid and

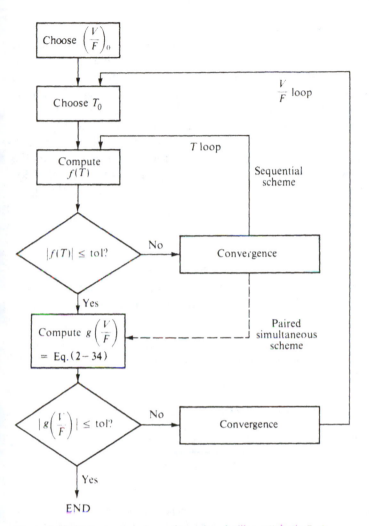

Figure 2-13 Convergence scheme for a close-boiling adiabatic flash.

effective convergence for isenthalpic flashes over a very wide range of conditions. One or the other pairing seems always to converge well, suggesting that the ranges of effective convergence for each pairing overlap.

Case 6: Highly nonideal mixtures The procedures outlined so far for algebraic solution of simple equilibrium processes assume that K_i is not a function of phase compositions or that the dependence of K_i upon phase compositions is weak enough to make it satisfactory to use compositions from the previous iteration to obtain values of K_i for the ensuing iteration. In cases where values of y_i and x_i from the previous iteration do not add to unity, one would normalize them as $x_i/\Sigma x_i$ and $y_i/\Sigma y_i$ before computing activity coefficients.

When K_i values do depend strongly upon composition, it is advisable to converge K_i's and composition variables simultaneously with other variables. This can be accomplished by linearizing the equations and using a general multivariate Newton convergence method (Appendix A). This would be a one-stage version of the calculation method described in Chap. 10 and Appendix E for multistage separations involving highly nonideal mixtures. Another approach is to converge K_i values as an innermost loop in a convergence scheme involving nested loops (see Fig. A-6); Henley and Rosen (1969) give procedures for doing this. The nested-loop approach often requires more computation time, but may give added initial stability to the calculation. Henley and Rosen (1969) also present computational approaches for simple-equilibration situations where a vapor phase and two immiscible liquid phases can be formed.

Graphical Approaches

When phase-equilibrium data are presented graphically, it is possible to employ a graphical solution for analysis of a simple equilibrium process, provided the system is binary or ternary. Example 2-7 illustrates a graphical solution for a binary vapor-liquid equilibrium process:

Example 2-7 100 mol/h of a liquid mixture of 40 mol % acetone and 60 mol % acetic acid is partially vaporized continuously to form one-third vapor and two-thirds liquid on a molar basis Find the resulting phase compositions.

Equilibrium data for acetone–acetic acid at 1 atm (data from Othmer, 1943)

$x_{acetone}$	0.05	0.10	0.20	0.30	0.40	0.50	0.60	0.70	0.80	0.90
$y_{acetone}$	0.162	0.306	0.557	0.725	0.840	0.912	0.947	0.969	0.984	0.993

SOLUTION The equilibrium data are plotted as the curve in Fig. 2-14. Equation (2-9), written as a mass balance for acetone, has the property of being a straight line on a yx plot with a slope equal to $-L/V$ and an intersection at $y_i = x_i = z_i$ with the $y_i = x_i$ line (shown dashed in Fig. 2-14). Therefore the line labeled "mass balance" has a slope of -2 and an intersection with the $y = x$ line at $x_{acetone} = 0.40$. The solution is the intersection of the mass-balance line with the equilibrium curve, giving $y_{acetone}$ in the vapor product $= 0.67$ and $x_{acetone}$ in the liquid product $= 0.27$. ☐

The lever rule If the separation process provides equilibrium between product phases and the equilibrium data are available in graphical form, it is often convenient to employ the *lever rule*. Basically, the application of the rule involves the graphical performance of a mass balance. If the feed (plus separating agent, if it is a stream of matter) contains a mole fraction x_{Fi} of a component and the products contain mole fractions x_{P1i} and x_{P2i} in products P_1 and P_2, respectively, we can write the following mass balance for a continuous steady-state process like that shown in Fig. 2-15:

$$x_{Fi}F = x_{P1i}P_1 + x_{P2i}P_2 \qquad (2\text{-}40)$$

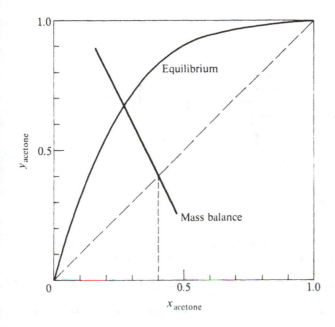

Figure 2-14 Graphical solution of Example 2-7.

F, P_1, and P_2 represent the flow rates of the respective streams (mol/h). Since

$$F = P_1 + P_2 \tag{2-41}$$

we can write

$$\frac{P_1}{P_2} = \frac{x_{P2} - x_F}{x_F - x_{P1}} \tag{2-42}$$

The quantities $x_{P2} - x_F$ and $x_F - x_{P1}$ can often be measured graphically. By Eq. (2-42), *the ratio of the product flows is the inverse of the ratio of the lengths of the lines connecting the feed mole fraction to the mole fractions of each of the products, in order.* This is known as the lever rule.

An example shows the application of this technique.

Example 2-8 Consider the crystallization process shown in Fig. 2-16. A liquid mixture of *m-* and *p*-cresol at 30°C is cooled by refrigeration while flowing inside a pipe long enough to provide equilibrium between solid and liquid in slurry form. The resulting two-phase mixture is then filtered. Assume that it is possible to separate solid and liquid phases completely in the filter.

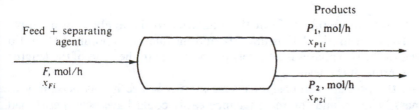

Figure 2-15 Continuous separation process.

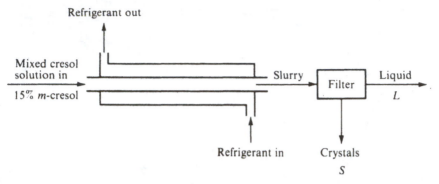

Figure 2-16 Continuous equilibrium crystallization process.

The feed rate is 1000 mol/h, containing 15 mol % m-cresol, and the exit temperature is 6°C. The pressure is 100 kPa. Find the compositions and flows of the product streams.

SOLUTION The phase diagram for the m-cresol–p-cresol system was given in Fig. 1-25 and discussed in Chap. 1. The phase diagram is reproduced in Fig. 2-17.
Applying the description rule to this process gives the variables set during construction and operation:

1. Vessel size and flow configuration (complete product equilibrium)
2. Feed flow rate, temperature, composition, and pressure
3. Total pressure
4. Refrigerant temperature and flow rate

For the problem under consideration we replace the refrigerant temperature and flow by the single variable of product temperature. These two variables can be replaced in this way because they both influence only the product temperature.
The feed condition is shown by point A in Fig. 2-17. Cooling brings us to point B before crystal nucleation can begin. The two equilibrium phases are obtained from the equilibrium isotherm at 6°C as $x_{MS} = 0$ and $x_{ML} = 0.375$ (points C and D, respectively). M refers to m-cresol, S to solid, and L to liquid. Distance CE can be measured as 0.15 and distance DE as 0.225. The lever rule can be applied to give

$$\frac{S}{L} = \frac{x_L - x_F}{x_F - x_S} = \frac{0.225}{0.15} = 1.50$$

where S/L is the molar ratio of solid and liquid product flows. Since $S + L = 1000$ mol/h, we have

$$S = \frac{S}{L + S}(L + S) = \frac{S/L}{1 + (S/L)}(L + S) = \frac{1.5}{2.5}(1000) = 600 \text{ mol/h} \qquad L = 400 \text{ mol/h} \qquad \Box$$

The term "lever rule" follows from the similarity to the analysis of a simple physical lever. The lever in Fig. 2-17 is line CD, and the fulcrum is point E. The ratio of the quantities of the two phases is inversely proportional to the ratio of the lengths of the respective arms of the lever.
In this case the product compositions were known before it was necessary to employ the mass balance, and the mass balance easily could have been performed algebraically. The power of the lever rule is clearer in problems like Examples 2-9 and 2-10, where a simultaneous solution of all relationships is required.

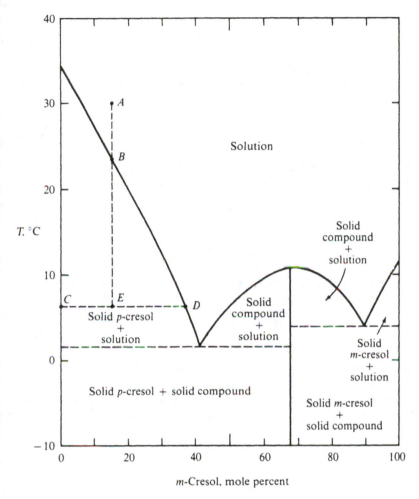

Figure 2-17 Separation of m-cresol and p-cresol by partial freezing. *(Adapted from Chivate and Shah, p. 237; used by permission.)*

Systems with Two Conserved Quantities The solution of an isenthalpic equilibrium flash vaporization of a binary system can be quite simply obtained by a graphical technique. The approach can be generalized to any separation operation involving *two conserved quantities*. In the case of an isenthalpic flash these quantities are mass and enthalpy.

An extension of the lever rule can be made when more than one quantity is conserved. Considering the case of the isenthalpic flash, we can again write the mass balance for component A in the form of the lever rule [Eq. (2-42)]:

$$\frac{y_A - z_A}{z_A - x_A} = \frac{L}{V} \tag{2-43}$$

An enthalpy balance can also be written

$$h_L L + H_V V = h_F F = h_F(L + V) \tag{2-44}$$

where h_L, H_V, and h_F are the specific enthalpies of liquid product, vapor product, and feed, respectively, referred to the same zero enthalpy. Equation (2-44) can readily be rearranged to lever form:

$$\frac{H_V - h_F}{h_F - h_L} = \frac{L}{V} \tag{2-45}$$

Equating the left-hand sides of Eq. (2-43) and (2-45), we have the equation of a straight line, which would relate h_F to z_A if y_A, x_A, h_L, and H_V were fixed:

$$h_F = \frac{H_V - h_L}{y_A - x_A} z_A + \frac{h_L y_A - H_V x_A}{y_A - x_A} \tag{2-46}$$

This line is shown in Fig. 2-18. Note that the line *must* pass through the points (H_V, y_A) and (h_L, x_A). If a value of z_A is now specified in addition to H_V, h_L, y_A, and x_A, the flash can be solved completely; h_F is the point on the line of Fig. 2-18 corresponding to z_A, and L/V is given by either Eq. (2-43) or (2-45) as

$$\frac{L}{V} = \frac{\overline{AB}}{\overline{AC}} = \frac{\overline{AD}}{\overline{AE}} = \frac{\overline{AF}}{\overline{AG}} \tag{2-47}$$

as shown in Fig. 2-19. All these ratios are equal, since triangles ADF and AGE are similar, as are ABF and ACG.

Usually we do not have a situation where the products are specified and the feed is unknown. When the feed is specified and the mass and enthalpy balances must be

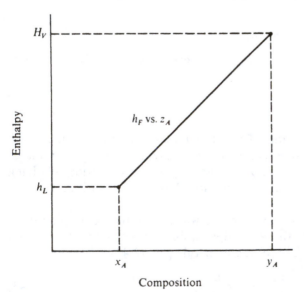

Figure 2-18 Graphical representation of Eq. (2-46).

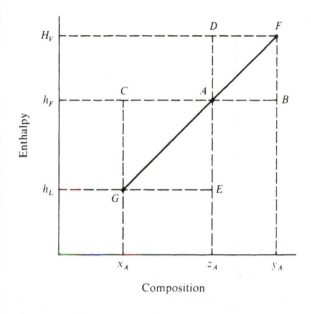

Figure 2-19 Graphical mass and enthalpy balance.

solved in conjunction with the product-composition relationship, we reverse the above procedure, as shown in Example 2-9.

Example 2-9 An enthalpy-vs.-concentration diagram for the system ethanol-water at 1 atm total pressure is given in Fig. 2-20. A mixture containing 60 wt % ethanol and 40 wt % water is received with an enthalpy of 973 kJ/kg at high pressure (referred to the same bases as Fig. 2-20) and is

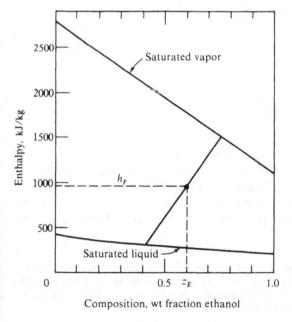

Figure 2-20 Enthalpy-concentration diagram for the ethanol-water system at 101.3 kPa (zero enthalpy = pure liquids at −17.8°C). (*Data from Perry et al., 1963.*)

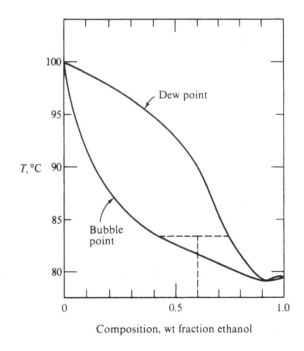

Figure 2-21 Tyx diagram for the ethanol-water system at 1 atm. (*Data from Perry et al., 1963.*)

Composition, wt fraction ethanol

expanded adiabatically to a pressure of 101.3 kPa. Find the product compositions and flow rates and the flash temperature.

SOLUTION Equilibrium data for this system are shown in Fig. 2-21. We know the point representing (h_F, z_E) on Fig. 2-20, and we know that the straight-line connecting (H_V, y_E) and (h_L, x_E) must pass through that point. Since we have an equilibrium flash, we know that the product composition, enthalpies, and temperature must lie on the saturation curves of Figs. 2-20 and 2-21. By trial and error we seek a y and x pair from Fig. 2-21 which will provide a straight line through the known point on Fig. 2-20. The result is $y_E = 0.76$, $x_E = 0.425$, and $T = 83°C$. The product flow rates then come from an application of the lever rule to *either* Fig. 2-20 or 2-21:

$$\frac{L}{V} = \frac{0.76 - 0.60}{0.60 - 0.425} = 0.91 \qquad\qquad \Box$$

The preceding illustration was for conservation of mass of one component and of enthalpy. There are, however, other situations to which this approach can be applied. In a solvent extraction process analyzed on a triangular diagram we can replace the enthalpy restriction with a mass balance on a second component. This is possible since there are two independent composition parameters in a three-component mixture.

Example 2-10 When 40 kg/min of water is added to 20 kg/min of a mixture containing 60 wt % vinyl acetate and 40 wt % acetic acid in a mixer-settler unit like that shown in Fig. 1-20, equilibrium is attained between the products at 25°C. If the operation is steady state and continuous, find the composition and flow rates of the product streams.

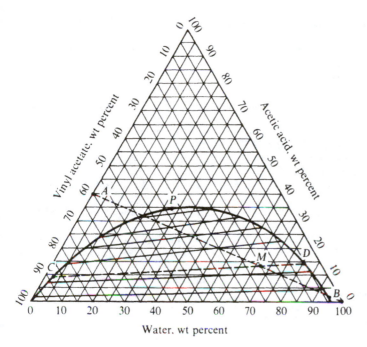

Figure 2-22 Single-stage extraction process. *(Adapted from Daniels and Alberty, 1961, p. 258; used by permission.)*

SOLUTION Points A and B in Fig. 2-22 represent the two feed streams to the extraction process. By the lever rule the point M (20% vinyl acetate, 66.7% water, two-thirds of the way between A and B, since 40 kg of water was added to 20 kg of acetate-acetic acid mixture) represents the gross, combined feed. The point M must also correspond to the gross product if the two products were mixed together. This follows from an overall material balance, which requires no accumulation of mass in a continuous, steady-state process. Thus point M must be related to the product compositions by the lever rule. This implies that M must be collinear with the product compositions and hence that M must lie on an equilibrium tie line. The appropriate tie line (dashed line) is placed by interpolation from the given equilibrium tie lines (solid lines). The indicated product compositions are 3% water, 9% acetic acid, 88% vinyl acetate (point C), and 5.8% vinyl acetate, 79.9% water, 14.3% acetic acid (point D). By the lever rule, the ratio of product flows is

$$\frac{C}{D} = \frac{\overline{MD}}{\overline{MC}} = \frac{20 - 5.8}{88 - 20} = 0.209$$

Hence, the vinyl acetate-rich product flow is $(0.209/1.209)(60) = 10.4$ kg/min, and the water-rich product flow is $60 - 10.4 = 49.6$ kg/min. $\qquad \square$

REFERENCES

Barnés, F. J., and J. L. Flores (1976): *Inst. Mex. Ing. Quim. J.*, **17**:7.
Benham, A. L., and D. L. Katz (1957): *AIChE J.*, **3**: 33.
Friday, J. R. and B. D. Smith (1964): *AIChE J.*, **10**:698.
Grens, E. A., II, Department of Chemical Engineering, University of California, Berkeley (1967): personal communication.

Hanson, D. N., Department of Chemical Engineering, University of California, Berkeley (1977): personal communication.

———, J. H. Duffin, and G. F. Somerville (1962): "Computation of Multistage Separation Processes," chap. 1, Reinhold, New York.

Henley, E. J., and E. M. Rosen (1969): "Material and Energy Balance Computations," chap. 8, Wiley, New York.

Holland, C. D. (1975): "Fundamentals and Modeling of Separation Processes," appendix D, Prentice-Hall, Englewood Cliffs, N.J.

Lapidus, L. (1962): "Digital Computation for Chemical Engineers," McGraw-Hill, New York.

Lockhart, F. J., and R. J. McHenry (1958): *Petrol. Refin.*, **37**:209.

Maxwell, J. B. (1950): "Data Book on Hydrocarbons," Van Nostrand, Princeton, N.J.

Othmer, D. F. (1943): *Ind. Eng. Chem.*, **35**:617.

Perry, R. H., and C. H. Chilton (eds.) (1973): "Chemical Engineers' Handbook," 5th ed., McGraw-Hill, New York.

———, ———, and S. D. Kirkpatrick (eds.) (1963): "Chemical Engineers' Handbook," 4th ed., McGraw-Hill, New York.

Rachford, H. H., Jr. and J. D. Rice (1952): *J. Petrol. Technol.*, vol. **4**, no. 10, sec. 1, p. 19; sec. 2, p. 3.

Reid, R. C., J. M. Prausnitz, and T. K. Sherwood (1977): "The Properties of Gases and Liquids," 3d ed., McGraw-Hill, New York.

Rohl, J. S. and N. Sudall (1967): Convergence Problems Encountered in Flash Equilibrium Calculations Using a Digital Computer, *Midlands Branch Inst. Chem. Eng. Great Britain*, Apr. 19.

Seader, J. D., Department of Chemical Engineering, University of Utah, (1978): personal communication.

Yen, L. C., and R. E. Alexander (1965): *AIChE J.*, **11**:334.

PROBLEMS

2-A$_1$ (a) Find the dew point, at 10 atm total pressure, of a gaseous mixture containing 10 mol % hydrogen, 40 mol % n-butane, 30 mol % n-pentane, and 20 mol % n-hexane. The hydrogen is only very slightly soluble in the liquid phase.

(b) Find the dew point of the above mixture at a total pressure of 8 atm.

2-B$_1$ Find the bubble-point temperature of a mixture containing 35 mol % n-butane, 30 mol % n-pentane, and 35 mol % n-hexane at a total pressure of (a) 10 atm and (b) 8 atm.

2-C$_2$ Find the dew points of the following gas mixtures at 101.3 kPa abs, total pressure. Cite any references you use:

(a) A gas mixture of 60 mol % hydrogen chloride and 40 mol % water vapor.

(b) A gas mixture of 50 mol % hydrogen chloride, 20 mol % nitrogen, and 30 mol % water vapor.

2-D$_3$ From the equilibrium data shown in Fig. 2-23 for binary mixtures of hydrogen and methane one can see that the relative volatility of hydrogen to methane α_{H-M} is relatively high and that it increases with decreasing temperature. One way of effecting a separation of hydrogen-methane mixtures is partial condensation at low temperature. Such an operation must be carried out at a temperature below the dew point of the gas and at a high pressure. Suppose that a gas mixture containing 60 mol % hydrogen and 40 mol % methane is available at 600 lb/in^2 abs and will be cooled so as to form equilibrium vapor and liquid products.

(a) Find the dew-point temperature of this gas mixture.

(b) Apply the description rule to this process. In addition to the feed variables and pressure, how many variables concerning this separation process may be specified independently?

(c) If the gas mixture is cooled to a temperature of $-250°F$, what will be the amounts and compositions of the resulting vapor and liquid phases?

(d) To what temperature must the mixture be cooled so as to provide a recovery v_H/f_H of at least 95 percent of the entering hydrogen in the vapor product at a purity of at least 95 mole percent? As a first step, interpret this portion of the problem in terms of the description rule.

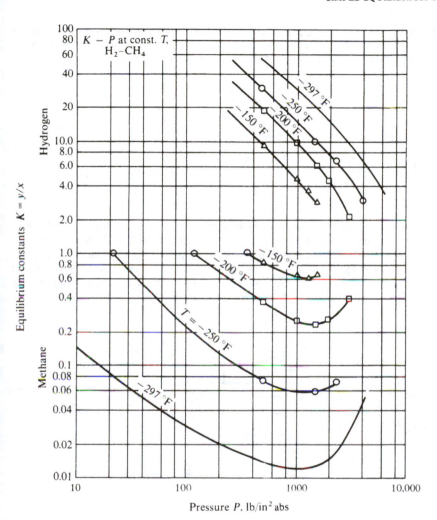

Figure 2-23 Equilibrium ratios for the system hydrogen-methane. (*From Benham and Katz, 1957, p. 33; used by permission.*)

2-E₃ A liquid mixture of 30 mol benzene, 30 mol toluene, and 40 mol water initially at 70°C and 101.3 kPa total pressure is heated slowly at a constant pressure of 101.3 kPa to 90°C. The vapor generated stays in contact with the remaining liquid. Assuming equilibrium between phases at all times, estimate (*a*) the temperature at which vaporization begins, (*b*) the composition of the first vapor, (*c*) the temperature at which vaporization is complete, and (*d*) the composition of the last liquid. *Note:* Water is essentially totally immiscible with benzene and toluene. Each liquid phase contributes to the total vapor pressure. Over the temperature range involved, the vapor pressure of benzene is 2.60 times that of toluene, and the vapor pressure of water is 1.23 times that of toluene. Vapor pressure of water is:

T, °C	70	72	74	76	78	80	82	84	86	88	90
P_w^0, kPa	31.2	34.0	36.9	40.1	43.6	47.3	51.3	55.6	60.1	64.9	70.1

2-F₁† A mixture containing 45.1 mol % propane, 18.3 mol % isobutane, and 36.6 mol % n-butane is flashed in a drum at 367 K and 2.41 MPa. Estimate the mole fraction of the original mixture vaporized at equilibrium, and the compositions of the liquid and vapor phases. Equilibrium vaporization constants at these conditions may be taken to be

	K
Propane	1.42
Isobutane	0.86
n-Butane	0.72

2-G₁ The feed stream of Example 2-4 is fed to an equilibrium flash separation operated at 250°F with $V/L = 1.5$. Find the total pressure, assuming that $K_i P$ is a constant as is predicted by the ideal-gas law.

2-H₂ Consider a continuous flash drum operating at a fixed temperature and pressure, controlled as shown in Fig. 2-24, with the feed a mixture of ethane and hexane in fixed amounts of each. A leak of nitrogen develops into the feed from some source. Will this occurrence cause the percent of the entering hexane lost in the effluent vapor to be more, less, or the same? Explain your answer.

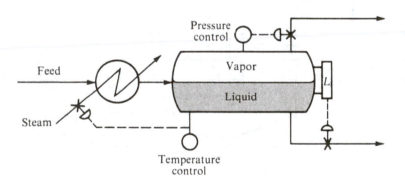

Figure 2-24 Continuous flash drum.

2-I₂ Cumene (isopropylbenzene) is an important chemical intermediate used for the manufacture of acetone and phenol; it is produced by the catalytic alkylation of benzene with propylene. A typical crude yield from the catalytic reactor might be as shown in Table 2-1, first column. The high propane content results from the use of a mixed C_3 feed stream. The products require separation, and in order to hold down utilities consumption, a typical scheme would involve removal of the C_3's first, then separation of the benzene for recycle, and finally removal of the heavies from the cumene product. The first of these steps

† From Board of Registration for Professional Engineers, State of California, Examination in Chemical Engineering, November 1966; used by permission.

Table 2-1 Data for Prob. 2-I

	mol "$_o$	Vapor pressure, MPa		
		37.8°C	65.6°C	93.3°C
Propane	40.0	1.30	2.36	3.95
Propylene	2.0	1.55	2.80	4.66
Benzene	27.0	0.0220	0.0630	0.147
Cumene	30.0	0.0013	0.0050	0.0162
Heavies	1.0			

involves a separation across a relatively wide volatility gap; hence it might be possible to employ a simple equilibrium flash-drum removal of the C_3's rather than bringing in a more expensive fractionator at this point. Neglecting solution nonidealities and gas-law deviations and assuming the reactor effluent is flashed to 241 kPa total pressure, evaluate the flash-drum proposal by finding the percentage loss of benzene and cumene if 90 percent of the propane is to be removed in the flash operation. Vapor pressures are also shown in Table 2-1.

2-J$_3$ Prove that the lever rule for two conserved properties is valid on an equilateral-triangular diagram for three-component liquid-liquid separation processes.

2-K$_2$ (a) Indicate how the lever rule could be employed for graphical computation of a binary equilibrium flash vaporization using a plot of y_A vs. x_A at equilibrium. Consider the case of a specified x_A in the liquid product.

(b) Find the vapor composition and the vapor flow rate if 100 lb mol/h of a solution of 50 mol "$_o$ acetone and 50 mol °$_o$ acetic acid is continuously flashed under conditions to give a liquid product containing 25 mol °$_o$ acetone at 1 atm total pressure. Equilibrium data may be taken from Example 2-7.

2-L$_2$ A mixture of 30 wt °$_o$ acetic acid and 70°$_o$ vinyl acetate is fed at 100 kg/h to a mixer-settler contacting device along with 120 kg/h of water. Use the lever rule and the triangular diagram to ascertain the fraction of the entering acetic acid extracted into the effluent water phase.

2-M$_2$ At what temperature must a mixture of 30 mol °$_o$ gold and 70 mol °$_o$ platinum be equilibrated so as to yield equal molar amounts of liquid and solid products? What is the resulting liquid composition? See Fig. 1-27.

2-N$_1$ Write a digital computer program suitable for solving Prob. 2B by an appropriate algorithm. Supply vapor-liquid equilibrium data as polynomial expressions, curve-fitting the graphical data of Fig. 2-2. Confirm the workability of the program.

2-O$_1$ Write a digital computer program suitable for solving Prob. 2F by an appropriate algorithm. Confirm the workability of the program for Prob. 2F.

2-P$_2$ Write a digital computer program suitable for solving Example 2-6 by an appropriate algorithm. Supply vapor-liquid equilibrium and enthalpy data as polynomial expressions, curve-fitting the graphical data of Fig. 2-2 and 2-10. Confirm the workability of the program.

2-Q$_2$ A liquid mixture of 50 mol °$_o$ ethanol and 50 mol °$_o$ water at elevated pressure and unknown temperature is flashed isenthalpically to 101.3 kPa (1 atm) pressure. The product temperature is measured to be 85°C. Find the specific enthalpy of the feed referred to the same bases as used in Fig. 2-20.

2-R$_2$ A liquid mixture containing 19.2°$_o$ ethanol, 24.7°$_o$ isopropanol, and 56.1°$_o$ n-propanol, molar basis, is flashed isenthalpically from 120°C and high pressure to a final equilibrium pressure of 56.1 kPa. Given the following data, calculate the percentage vaporization, the final temperature and the product phase compositions. Consider the heat capacities to be independent of temperature, as an approximation, and assume ideal solutions. Assume enthalpies to be independent of pressure.

Property	Ethanol	Isopropanol	n-Propanol
Vapor pressure, kPa,			
at 70°C	71.3	60.2	31.7
at 80°C	108.8	93.2	50.1
at 90°C	159.9	139.2	76.5
Normal bp, °C	78.3	82.3	97.2
Latent heat of vaporization at normal bp, kJ/mol	39.4	40.1	41.3
Liquid heat capacity, J/mol·°C	162	219	216
Vapor heat capacity, J/mol·°C	76.5	106	102

2-S₂ Most computer programs in use for solving flashes of feed streams adiabatically to a specified final pressure use the algorithm shown in Fig. 2-12 or a slight modification of it. Assume that K_i is a function of T and P alone in the scheme. It has been found that programs of this sort are incapable of converging upon a solution in the case of an isenthalpic flash of a *single-component* stream.

  (a) Why can't this algorithm handle an isenthalpic flash of a single-component stream?

  (b) How would you modify the algorithm so that it can?

THREE

ADDITIONAL FACTORS INFLUENCING PRODUCT PURITIES

It is difficult to obtain complete thermodynamic equilibrium between products from a continuous separation device or between immiscible contacting phases at any point within a separating device. Similarly, a number of competing effects can influence the product purities from a rate-governed separation process. Many factors can complicate the situation; among them are

1. Incomplete mechanical separation of product phases
2. Flow configuration or mixing effects
3. Mass- and heat-transfer rate limitations

The purpose of this chapter is to develop a qualitative picture of how each of these factors affects the behavior of a separation device and, for the first two, to give quantitative methods which can be used for analysis and design of simple separation devices.

INCOMPLETE MECHANICAL SEPARATION OF THE PRODUCT PHASES

Entrainment

Even though equilibrium between the phases is reached in a separation device, an incomplete mechanical separation of the product phases from each other will result in a seeming lack of equilibrium between the products. This lack of complete separation of the phases is generally known as *entrainment*.

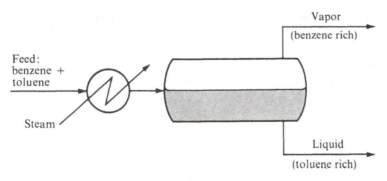

Figure 3-1 Partial vaporization of benzene-toluene mixture.

For example, consider the partial-vaporization process shown in Fig. 3-1. If the separator drum is not oversized, it is possible that the velocity of the exit vapor will be high enough to carry along droplets of liquid from the drum. If the vapor stream is analyzed as a whole (vapor plus entrained liquid), it must of necessity lie closer in composition to the liquid phase than the actual vapor (vapor without entrained liquid) does. If we compare the apparent separation factor with the equilibrium separation factor (relative volatility) α_{BT} for the benzene-toluene separation

$$\alpha_{BT} = \frac{y_B}{x_B}\frac{x_T}{y_T} \qquad \text{at equilibrium} \qquad (3\text{-}1)$$

we find that equilibrium vapor of composition y_B is mixed with liquid of composition x_B to give a stream of net composition y'_B, which must be intermediate between y_B and x_B; thus y'_B/x_B is *less* than y_B/x_B. Similarly, x_T/y'_T is *less* than x_T/y_T. The apparent separation factor α^S_{BT} will be

$$\alpha^S_{BT} = \frac{y'_B}{x_B}\frac{x_T}{y'_T} \qquad (3\text{-}2)$$

Therefore, a general conclusion regarding entrainment is that the separation factor α^S_{ij} based on actual net product-stream compositions will be closer to 1.0 and thus less favorable than the separation factor based upon complete mechanical separation of product streams.

A system possessing a nearly infinite equilibrium separation factor is particularly liable to reductions in apparent separation factor caused by entrainment. In the desalination of seawater by evaporation to make pure water by condensing the vapor, any entrainment of liquid droplets in the escaping vapor will serve to reduce the apparent separation factor between water and salts from infinity to some finite quantity.

Example 3-1 One process that has been considered for the concentration of fruit juices is known as *freeze-concentration*. Selective removal of water from fruit juices is desirable to increase storage stability and to reduce transportation costs. As is discussed further in Chap. 14, evaporation of juices can cause volatile flavor and aroma components to be lost. In freeze-concentration this problem is circumvented by removing water through partial freezing.

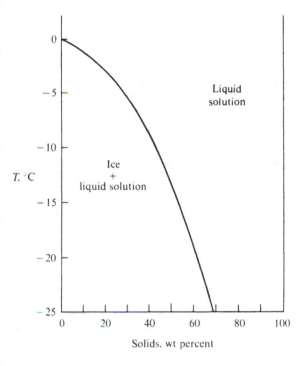

Figure 3-2 Phase diagram for apple juice at different water contents. (*Data from Heiss and Schachinger, 1951.*)

Figure 3-2 shows an estimated solid-liquid phase diagram for apple juice of different water contents. The dissolved solids in the juice are primarily the sugars levulose, sucrose, and dextrose. It has been confirmed (Heiss and Schachinger, 1951) that the ice crystals formed contain negligible impurities. Crystallization of sugars is kinetically impeded even when thermodynamically possible.

The dissolved solids content of natural apple juice is approximately 14 percent. Heiss and Schachinger (1951) measured dissolved solids contents during partial freezing followed by centrifugation to remove as much of the juice concentrate as possible from the ice crystals. Their results for the best freezing and centrifugation conditions indicate that when freeze-concentration is carried out so as to reduce the product juice water content per kilogram of dissolved juice solids by a factor of 2, the residual ice crystal mass will contain about 1.2 weight percent solids. Presumably this solids content is entirely due to entrainment. (*a*) Calculate the weight of entrained concentrate per unit weight of ice crystals, assuming that the entrained concentrate has the same composition as the product concentrate. (*b*) The concentrate product is worth about $1.20 per kilogram to the producer, based on proration of 1978 supermarket prices. Calculate the incremental processing cost caused by the loss of dissolved solids if the residual ice mass is discarded.

SOLUTION (*a*) The fresh juice contains 14 percent dissolved solids, or $14/86 = 0.163$ kg dissolved solids per kilogram of water. The concentrate has double the dissolved solids content, or 0.326 kg dissolved solids per kilogram of water. The residual ice mass is composed of pure ice containing no dissolved solids, along with concentrate. The ice is present in a proportion to make the overall dissolved solids content equal to 1.2 percent by weight. If we take as a basis 1.00 kg H_2O in the *entrained* concentrate, we have

$$\text{Entrained solids} = 0.326 \text{ kg} \quad \text{and} \quad \text{Ice crystals} = w_I \text{ kg}$$

$$\frac{0.326}{1.326 + w_I} = 0.012 = \frac{\text{solids}}{\text{total residual ice mass}}$$

Solving, we have

$$w_l = 25.8 \text{ kg} \quad \text{and} \quad \frac{\text{entrained concentrate}}{\text{ice crystals}} = \frac{1.326}{25.8} = 0.051 \text{ kg/kg}$$

(b) Keep the basis of 1.00 kg H_2O in the entrained concentrate and denote the weight of H_2O in the product concentrate as w_c kg. Then the dissolved solids in the concentrate product are $0.326w_c$ kg. To find the amount of concentrate we use an overall mass balance to satisfy the dissolved solids content of the initial juices

$$\frac{\text{Solids in all products}}{\text{Total weight of all products}} = \frac{\text{solids in feed}}{\text{total weight of feed}}$$

$$\frac{0.326w_c + 0.326}{1.326w_c + 25.8 + 1.326} = 0.14$$

$$w_c = 24.8$$

Hence the amount of feed dissolved solids lost through entrainment is $1/(w_c + 1) = 1/25.8 = 3.9$ percent. The cost of this loss per kilogram of concentrate product is

$$(\$1.20/\text{kg loss}) \frac{0.039 \text{ kg loss}}{0.961 \text{ kg product}} = \$0.049/\text{kg product}$$

Since the total processing cost for concentrating apple juice will be of the order of 4 to 12 cents per kilogram, this loss would be a substantial drawback for this process in comparison with other approaches which entail less loss of dissolved solids.

The large economic detriment caused by entrainment in the case of fruit-juice concentration is the result of an economic structure wherein the processing costs are small compared with the product value. ◻

Washing

A washing or leaching process is an example of a case where entrainment can completely control the separation attainable. Consider a process, as shown in Fig. 3-3, where a soluble substance is to be leached from finely divided solids (feed) by contact with water (separating agent). A commercial example would be the recovery of soluble $CuSO_4$ from an insoluble calcined ore by water washing. The $CuSO_4$ is highly soluble in water, while no other substance is appreciably soluble. Thus the equilibrium separation factor between $CuSO_4$ and ore is essentially infinite. On the other hand, the filter will not be able to provide a complete separation of the liquid and solid phases because liquid will fill the pore spaces between solid particles and will adhere to the particle surfaces. Thus a certain amount of liquid will remain with the solids.

If the mixer has succeeded in bringing all liquid to uniform composition, one can say that the concentration of solute in the exit water is equal to the concentration of solute in the liquid retained by the solids. The percentage removal of solute is then determined solely by the degree of dilution by water. If the entering solids are dry, if the filter retains 50 percent liquid by volume in the cake, and if the amount of water fed is 10 volumes per volume of dry solids, it follows that 10 percent of the solution will remain with the solids and hence 10 percent of the solute will not be removed.

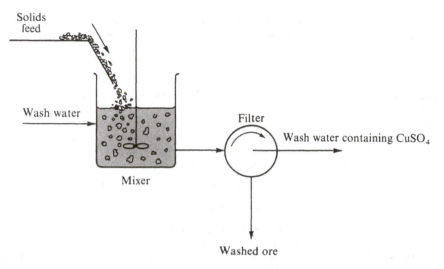

Solids feed

Wash water

Mixer

Filter

Wash water containing CuSO₄

Washed ore

Figure 3-3 Ore-leaching process.

Example 3-2 What incentive is there for washing the residual ice mass after centrifugation in Example 3-1?

SOLUTION In a washing step we could mix the ice bed thoroughly with water at a temperature such that melting would not occur and then recentrifuge. In order to recover the dissolved solids which enter the wash water we should recycle the wash water back to the feed point and reprocess it along with the fresh juice, as shown in Fig. 3-4.

The amount of wash water can be freely set at any value we want. For a first calculation we can set the rate of addition of wash water at a value such that the effluent wash will have the same solids

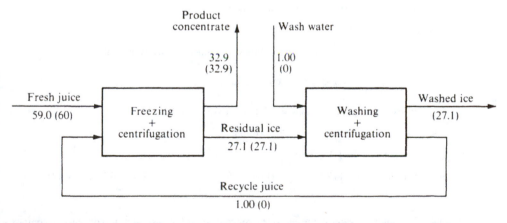

Product concentrate

Wash water

32.9 (32.9)   1.00 (0)

Fresh juice 59.0 (60)

Freezing + centrifugation

Washing + centrifugation

Washed ice (27.1)

Residual ice 27.1 (27.1)

Recycle juice 1.00 (0)

Figure 3-4 Mass balance on freeze-concentration process with washing, where the dissolved-solids content of recycle juice equals the dissolved-solids content of fresh juice. (Parentheses–without washing; no parentheses–with washing.)

content as fresh juice. This will amount to using W kg of wash water, on the same basis of 1.00 kg H_2O in the original entrained concentrate as used in Example 3-1. W is obtained from

$$0.14 = \frac{0.326}{1.326 + W}$$

Solving gives $W = 1.00$ kg, as could have been determined by realizing that the water content of the concentrate had been reduced by half.

At this point we need to estimate how much of this leaner juice will be entrained following the centrifugation after the wash. A convenient and reasonable approximation is that the weight of entrained juice per weight of ice remains the same. Hence 1.326 kg of juice will again be entrained in 25.8 kg of ice crystals. The recycle rate of juice from the wash can be computed as

$$\text{Recycle} = \text{entrained rich concentrate} + \text{wash water} - \text{entrained lean juice}$$

$$= 1.326 + 1.000 - 1.326 = 1.000 \text{ kg}$$

This recycle juice will have to be reprocessed, forming more ice, which entrains more concentrate. Through the mass balance shown in Fig. 3-4, we find that the fresh juice feed to the process, for a basis of 1.00 kg H_2O in the first concentrate, is reduced from 60.0 kg without the wash to $60 \times (60/61) = 59.0$ kg with the wash.[†] The juice loss through entrainment in the washed crystals is thus $1.326/59.0 = 2.2$ percent of the fresh feed juice. The solids loss is proportional to the juice loss, since the entrained juice and feed juice have the same solids content. Hence the loss of dissolved solids is also 2.2 percent of the dissolved solids in the fresh feed.

The solids loss has been reduced from 3.9 percent of the feed solids to 2.2 percent of the feed solids through washing. Hence the washing step is definitely of value, saving

$$4.9\left(1 - \frac{0.022}{0.039}\frac{0.961}{0.978}\right) = \$0.020/\text{kg}$$

of concentrate product.

A wash-water rate of 1.00 kg per kilogram of water in the original entrained concentrate is rather low, in actuality. A calculation for a wash-water flow of 5.00 kg per kilogram of water in the original entrained concentrate follows.

H_2O in entrained concentrate $= 1.00$ kg   Solids in entrained concentrate $= 0.326$ kg

Ice crystals $= 25.8$ kg (assuming same weight ratio of entrainment)

Wash water $= 5.00$ kg

Solids content of recycle juice and of liquid entrained in washed crystals

$$= \frac{0.326}{1.326 + 5.00} = 0.0515 \text{ wt fraction solids}$$

Solids lost with washed crystals $= 1.326(0.0515) = 0.0683$ kg

Recycle juice $= 27.1 + 5.00 - 27.1 = 5.00$ kg

Let the fresh feed flow be F kg. Then

Solids in fresh feed $= 0.14F$ Solids entering freezer $= 0.14F + (0.0515)(5.00) = 0.14F + 0.258$

[†] Because the amount of water in the entrained concentrate is the basis and the concentrate composition is held constant, the amount of ice formed in the freezer will remain constant. Since the recycle is of the same composition as the fresh feed, the ratio of concentrate to ice will be unchanged, and hence the total feed (recycle plus fresh feed) must be unchanged.

Solids in concentrate product = solids in fresh feed − solids lost in the washed crystals

$$= 0.14F - 0.0683$$

The solids content of concentrate must be 0.326 kg per kilogram of water, and so

$$\frac{0.14F - 0.0683}{F + 5.0 - 27.1} = \frac{0.326}{1.326}$$

Solving, we find

$$F = 50.7 \qquad \text{Solids loss} = \frac{0.0683}{(0.14)(50.7)} = 1.0\%$$

The solids loss is cut in half by increasing the wash water by a factor of 5 per kilogram of ice product, or by a factor of 5.8 per kilogram of fresh feed.

An even more efficient use of wash water can be made with a wash column, a fixed-bed process, discussed later in this chapter. □

Leakage

Nonseparating flow in a rate-governed separation process has the same effect as entrainment in an equilibration separation process. For example, consider the gaseous-diffusion process shown schematically in Fig. 1-28. As was established in Chap. 1, this process depends upon the fact that Knudsen flow gives a molecular flux of each component that is inversely proportional to its molecular weight [Eq. 1-19]. In order for Knudsen flow to prevail, the pressure must be low enough and the pore size small enough for the mean free path of the molecules in the gas mixture to be large compared with the pore size of the barrier. Viscous flow will occur through the pore spaces which are large enough to rival the gas mean free path. Viscous flow moves all species together, at a rate dependent upon the mixture viscosity. Since the rate of viscous flow of each species does not depend upon the molecular weight of that species, the viscous-flow contribution will not provide any separation. Avoiding leakage was a major concern in the development of the gaseous-diffusion process for separating uranium isotopes.

For separation processes relying upon differences in rates of flow or diffusion through a thin polymeric membrane (ultrafiltration, reverse osmosis, dialysis, etc.) it is important to guard against any macroscopic holes in the membrane. Just as in the gaseous-diffusion process, any flow through a hole in the membrane will be non-separative and may markedly contaminate the product with feed.

FLOW CONFIGURATION AND MIXING EFFECTS

The product purities from a separation device are often strongly influenced by the flow geometry of the device and by the degree of mixing within the individual phases or product streams. The separation obtained will be different depending upon:

1. The *uniformity* of composition within the bulk of either phase or product stream
2. The *charging sequence* of feed and separating agent
3. The relative *directions of flow* of the phases or product streams

Mixing within Phases

Consider, for example, the ore-leaching process of Fig. 3-3. In addition to the wash-water-to-solids feed ratio, another factor influencing the separation attainable in this process is the degree of *liquid mixing* obtained in the mixer. If mixing is not complete near the surfaces of the ore particles, it is possible that the solution retained by the solids will be richer in $CuSO_4$ than the wash water is. This lack-of-mixing effect will serve to reduce the amount of $CuSO_4$ removed. Similarly, in the freeze-concentration process considered in Examples 3-1 and 3-2 the loss of juice solids will be increased if mixing in the freezer is poor enough for the concentrate retained by the ice to be richer than the concentrate removed as liquid product. Alternatively, a given measured solids loss can correspond to fewer pounds entrained liquid per pound ice.

A lack of complete mixing of the individual phases within a separation device often improves the quality of separation rather than harming it. For example, one popular device for gas-liquid contacting is the cross-flow plate. Figure 3-5 shows such a device as it would be used for a *stripping* operation. A small quantity of ammonia in a water stream (feed) is to be removed by stripping it into air (separating agent). The water flows across the plate while the air passes upward through the liquid in the form of a mass of bubbles, emanating from holes in the plate. The ammonia is at a low enough concentration for the phase equilibrium to be described by Henry's law

$$y_{NH_3} = K_{NH_3} x_{NH_3} \tag{3-3}$$

Because the air has already been saturated with water vapor, there is no evaporation and K_{NH_3} is constant since the operation is isothermal.

On a cross-flow plate it is likely that there will not be sufficient backmixing in the

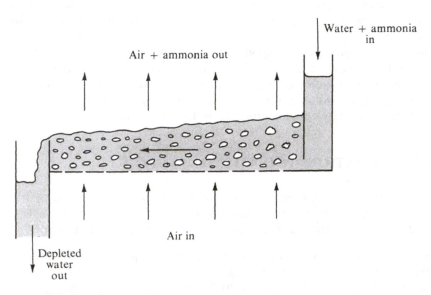

Air + ammonia out

Water + ammonia in

Air in

Depleted water out

Figure 3-5 Cross-flow stripping process.

direction of liquid flow to iron out completely any liquid concentration differences which develop. The water will therefore continually decrease in ammonia concentration as it flows across the plate and ammonia is removed from it. As a result, the exiting air bubbles will contain less ammonia at positions closer to the liquid outlet. If the water on the plate is deep enough and the airflow is low enough, the exiting air bubbles will have nearly achieved equilibrium with the water. At the right-hand (water-inlet) side the concentration of ammonia in the exit air will be nearly in equilibrium with the *inlet* water, and at the left-hand (water-exit) side the concentration of ammonia in the exit air will be nearly in equilibrium with the *exit* water. The exit water has the lowest ammonia concentration of any water on the plate, and the equilibrium concentration of ammonia in the exit air is directly proportional to the ammonia concentration in the liquid, by Eq. (3-3). The exit air at the liquid-outlet side is in near equilibrium with the exit liquid, and all other exit air must have a *higher* ammonia concentration. Therefore the entire exit airstream, taken together, has achieved a *higher* concentration of ammonia than corresponds to equilibrium with the exit water. The cross-flow configuration and lack of liquid backmixing have provided more separation of ammonia from water than would have been obtained by simple equilibration of the two gross exit streams.

Example 3-3 Derive the relationship between the exit-gas and exit-liquid ammonia compositions for the cross-flow stripping process of Fig. 3-5. Assume that the liquid is totally unmixed in the direction of liquid flow and that the gas bubbles achieve equilibrium with the liquid. The inlet air contains no ammonia.

SOLUTION We can write a mass balance for NH_3 upon a differential vertical slice of liquid, as shown in Fig. 3-6:

$$\text{Input} - \text{output} = \text{accumulation}$$

$$y_{in}\, dG + L(x + dx) - y_{eq}\, dG - Lx = 0$$

$$L\, dx = (y_{eq} - y_{in})\, dG \tag{3-4}$$

y_{eq} can be replaced by Kx since the exit gas achieves equilibrium with the liquid at that point. Also we can set $y_{in} = 0$.

$$L\, dx = Kx\, dG \tag{3-5}$$

This expression can be integrated from x_{out} at $G = 0$ to x at any particular point across the plate. The direction of integration comes from the convention of making dx increase from left to right in Fig. 3-6.

$$L \int_{x_{out}}^{x} \frac{dx}{x} = K \int_{0}^{fG} dG \tag{3-6}$$

where f is the fraction of the gas flowing through the plate to the left of the location where the liquid composition is x.

$$\frac{x}{x_{out}} = e^{fKG/L} \tag{3-7}$$

$$y_{eq} = Kx = Kx_{out}e^{fKG/L}$$

$$y_{eq,\,av} = \int_{0}^{1} y_{eq}\, df = Kx_{out} \int_{0}^{1} e^{fKG/L}\, df = \frac{L}{G} x_{out}(e^{KG/L} - 1) \tag{3-8}$$

$$\frac{y_{eq,\,av}}{Kx_{out}} = \frac{L}{KG}(e^{KG/L} - 1) \tag{3-9}$$

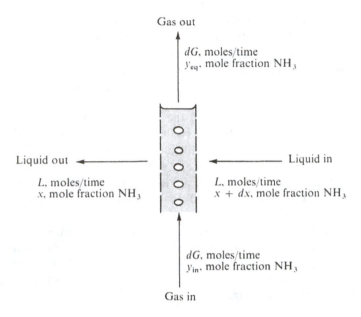

Gas out

dG, moles/time
y_{eq}, mole fraction NH_3

Liquid out

L, moles/time
x, mole fraction NH_3

Liquid in

L, moles/time
$x + dx$, mole fraction NH_3

dG, moles/time
y_{in}, mole fraction NH_3

Gas in

Figure 3-6 Differential mass balance for Example 3-3.

If the combined exit gas were in equilibrium with the exit liquid, $y_{eq, av}/Kx_{out}$ would be unity. Instead $y_{eq, av}/Kx_{out}$ is a function of KG/L, as follows:

KG/L	0	0.2	0.5	1	2	5	∞
$y_{eq, av}/Kx_{out}$	1.00	1.11	1.30	1.72	3.17	29.5	∞

Thus, for any finite gas flow rate, the air picks up *more* ammonia than corresponds to equilibrium with the exit liquid. The separation of ammonia from the water is therefore better than it would be through a simple equilibration of product streams. □

The influence of mixing within the phases upon product compositions is taken up in more detail in Chaps. 11 and 12.

Flow Configurations

Many different flow configurations and feed-charging sequences are employed in separation processes. The feed and/or separating agent may be charged on a *batch*, i.e., at discrete intervals, or a *continuous* basis. If both phases are flowing, there may be *cocurrent flow*, *cross flow*, or *countercurrent flow*. Each phase may be nearly *completely mixed* within itself in the separation device, or it may pass through in a close approach to *plug flow*, i.e., little or no *backmixing*. *Partial mixing* of a phase is also possible, as an intermediate between plug flow and complete mixing. We have already seen in Fig. 3-5 a case of a continuous process with cross flow and with little backmixing of the liquid phase. The extraction and leaching processes of Figs. 1-20 and 3-3 are examples of continuous processes with nearly complete mixing of the continuous phase.

Some further examples of common flow and charging configurations are given in Fig. 3-7. The separatory funnel (Fig. 3-7a) is an example of an entirely *batch* process. For example, a common chemistry laboratory experiment involves adding an aqueous solution containing iodine (feed) to a separatory funnel along with carbon tetrachloride (separating agent). The funnel is then shaken and the iodine is preferentially extracted into the CCl_4 phase, imparting a color to it. This is a batch process for both phases since they are charged to the vessel before the shaking takes place and are withdrawn afterward.

Figure 3-7b shows a simple batch distillation. The liquid feed is charged to the still initially. The steam (separating agent) is then turned on, and vapor is continuously generated. After the desired amount of vapor has been removed, the steam flow is stopped and the remaining liquid is withdrawn. The liquid is a *batch* charge in this process, whereas the vapor flow is *continuous*. The agitation afforded by the boiling makes it likely that both phases in the zone of contact will be relatively well mixed.

Figure 3-7c shows a process in which air (feed) is dried during continuous flow over a bed of solid desiccant (separating agent), such as activated alumina. In this case the solid phase is a batch charge and the gas phase is a continuous flow. The desiccant in the solid phase is unmixed during drying, and there should be little backmixing in the gas.

A packed absorption tower is shown in Fig. 3-7d. Here the liquid absorbent falls downward over the surface of a bed of divided solids (packing). The gas passes upward through the remaining void spaces. Both phases are in *continuous* flow in this process, and the flow is *countercurrent*. There is little backmixing in either phase, provided the tower is tall enough.

The partial condenser shown in Fig. 3-7e is also a *continuous countercurrent* process with little backmixing in either phase. A vapor-phase mixture of components flows upward through tubes, which are cooled by a jacket of refrigerant or cooling water. As liquid condenses on the tube wall, it flows downward, collecting in the bottom of the vessel. As the condensate flows downward, it contacts the vapor stream and has the opportunity to exchange mass with it.

The final process (Fig. 3-7f) is a double-pipe crystallizer, in which a liquid flows through the inner pipe and is cooled by a coolant flowing in the annular outer pipe. A solid phase freezes out and is kept in motion as a slurry within the liquid by a helical scraper. This is a *continuous cocurrent* flow process with little backmixing likely for either phase.

Through reasoning similar to the analysis made of the air stripping process of Fig. 3-5, readers should convince themselves that in Fig. 3-7b to e the actual separation factor can be greater than that corresponding to equilibrium between the two product streams, whereas this cannot be the case for the processes of Figs. 1-21, 3-1, and 3-7a and f. In batch charging of one stream with continuous flow of the other, the product compositions are the average of the effluent collected and the average composition of the material remaining in the device at the end of the run.

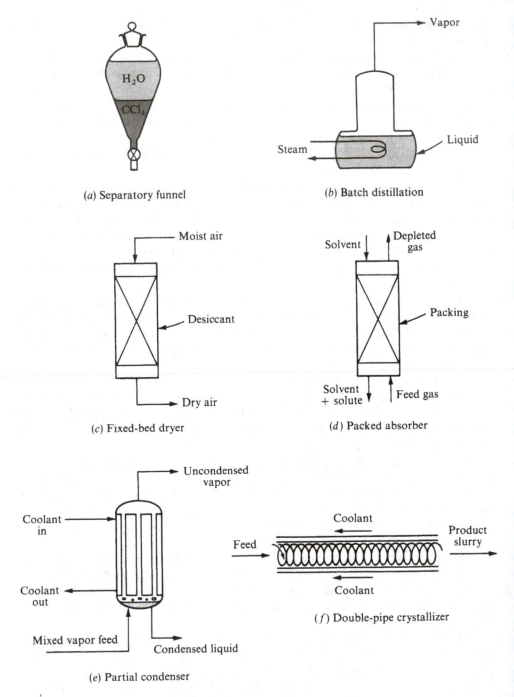

(a) Separatory funnel

(b) Batch distillation

(c) Fixed-bed dryer

(d) Packed absorber

(e) Partial condenser

(f) Double-pipe crystallizer

Figure 3-7 Flow patterns in separation devices.

In general, an actual separation factor higher than the equilibrium separation factor may be obtained:

1. In a process where one phase is batch and the other is continuous (see below)
2. In a continuous cross-flow process where at least one phase is not well mixed in the direction of flow (see Example 3-3)
3. In a continuous countercurrent process where both phases are not well mixed (see Chap. 4).

Countercurrent contacting is even more effective than cross-flow contacting for increasing the apparent separation factor. Since the concepts in countercurrent flow are analogous in many ways to the concepts of *multistage* separations, further discussion of countercurrent-flow systems is deferred until Chap. 4.

BATCH OPERATION

Both Phases Charged Batchwise

Separations in which the feed and both products are charged and withdrawn batchwise almost always involve bringing two phases of matter toward equilibrium. If equilibrium is achieved between the products, the analysis of the separation is entirely analogous to that of the corresponding continuous-flow, steady-state, simple-equilibrium separation. In fact, an identical separation is achieved: the product compositions are related by the equilibrium expression, and the product quantities are related by overall mass balances

$$x_F F' = x_{P1} P'_1 + x_{P2} P'_2 \tag{3-10}$$

$$F' = P'_1 + P'_2 \tag{3-11}$$

F' represents the moles of feed (plus separating agent) charged; P'_1 and P'_2 represent the moles of each of the products removed after the separation is accomplished. Equations (3-10) and (3-11) are identical in form to Eqs. (2-9) and (2-10), with the unprimed flow *rates* (moles per hour) replaced by the primed feed charge and product *quantities* (moles). The same logic as for continuous processes applies in picking effective schemes for solving the equations.

Rayleigh Equation

In many separation processes one stream or phase is charged and withdrawn batchwise and the other stream is fed and removed continuously. Procedures for describing such separations when the batch-charged phase is *well mixed* during operation follow the original developments put forth by Lord Rayleigh in 1902 (Rayleigh, 1902). Let us consider the case of an equilibrium vaporization process in which the feed liquid is initially charged entirely to a still pot and heat is then added continuously. Vapor in equilibrium with the remaining well-mixed liquid is continuously generated and is continuously removed from the vessel. The liquid product is

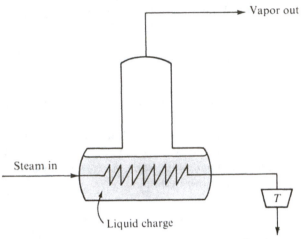

Vapor out

Steam in

Liquid charge

Condensate out   **Figure 3-8** Rayleigh distillation.

removed at the end of the run. This operation, shown in Fig. 3-8, is commonly called a *Rayleigh* or *batch distillation*.

Two differential mass balances can be written for the changes in the still pot as a differential amount of vapor dV' is removed.

$$dV' = -dL' \tag{3-12}$$

$$y_i \, dV' = -d(x_i \, L') \tag{3-13}$$

$$= -L' \, dx_i - x_i \, dL'$$

L' represents the moles of liquid remaining in the still pot, and y_i and x_i represent the mole fractions of component i in the vapor and liquid, respectively. Equation (3-12) is a mass balance for all species, while Eq. (3-13) is a mass balance for component i.

Equation (3-12) can be substituted into Eq. (3-13) to give

$$L' \, dx_i = (y_i - x_i) \, dL' \tag{3-14}$$

which can be integrated between the limits of L'_0 and x_{i0} (initial liquid charge and mole fraction) and L' and x_i (remaining liquid and mole fraction at any subsequent time) to give

$$\int_{L'_0}^{L'} \frac{dL'}{L'} = \int_{x_{i0}}^{x_i} \frac{dx_i}{y_i - x_i} \tag{3-15}$$

$$\ln \frac{L'}{L'_0} = \int_{x_{i0}}^{x_i} \frac{dx_i}{y_i - x_i} \tag{3-16}$$

the *Rayleigh equation*, which relates the composition of the remaining liquid to the *amount* of remaining liquid. In order to proceed further it is necessary to have a relationship between y_i and x_i. If we are dealing with a two-component system where both phases are well mixed and equilibrium is achieved between vapor and liquid, y_i

will be a unique function of x_i. It is then possible to employ a graphical presentation y_i vs. x_i in Eq. (3-16) and obtain the solution of x_i as a function of L' by graphical integration.

In other cases it may be allowable to state that either K_i or α_{ij} is constant during the course of operation. For $K_i = \text{constant}$

$$\ln \frac{L'}{L'_0} = \frac{1}{K_i - 1} \int_{x_{i0}}^{x_i} \frac{dx_i}{x_i} = \frac{1}{K_i - 1} \ln \frac{x_i}{x_{i0}} \qquad (3\text{-}17)$$

If α_{ij} can be assumed constant and there are only two components in the liquid, we can substitute Eq. (1-12) into Eq. (3-16) and obtain

$$\ln \frac{L'}{L'_0} = \int_{x_{i0}}^{x_i} \frac{dx_i}{[\alpha_{ij}/(1/x_i + \alpha_{ij} - 1)] - x_i} = \frac{1}{\alpha_{ij} - 1} \ln \frac{x_i(1 - x_{i0})}{x_{i0}(1 - x_i)} + \ln \frac{1 - x_{i0}}{1 - x_i} \qquad (3\text{-}18)$$

It should be pointed out that in a Rayleigh distillation the changing liquid composition will cause T to change if P is held constant. Since T changes, both K_i and (to a lesser extent) α_{ij} will in general change as the distillation proceeds.

Another point worthy of mention is that Eq. (3-16) applies to any separation in which one phase is charged batchwise and is well mixed and in which the other phase is formed and removed continuously. By following through the derivation readers can convince themselves that y_i refers to the product stream continuously withdrawn and that L' and x_i refer to the other product, which remains in the separation device throughout the separation operation and is well mixed.

Several examples follow to illustrate the applications of these concepts. The first two are simple, but the second two are less obvious and display the power of the Rayleigh equation in different contexts.

Example 3-4 A liquid mixture of 60 mol % benzene and 40 mol % toluene is charged to a still pot, where a Rayleigh distillation is carried out at 121 kPa total pressure. How much of the charge must be boiled away to leave a liquid mixture containing 80 mol % toluene?

SOLUTION The bubble point of a binary mixture containing 60 mol % benzene at 1.20 atm is 89°C and is 103°C for a mixture containing 80 mol % toluene. Over this temperature range α_{BT} varies from 2.30 to 2.52 (Maxwell, 1950). An average α_{BT} of 2.41 should be adequate; therefore Eq. (3-18) can be employed to a good approximation

$$\ln \frac{L'}{L'_0} = \frac{1}{1.41} \ln \frac{(0.20)(0.40)}{(0.60)(0.80)} + \ln \frac{0.40}{0.80} = -1.27 - 0.69 = -1.96$$

$$\frac{L'}{L'_0} = 0.141$$

and 86 mole percent of the initial liquid must be boiled away. □

Example 3-5 Important volatile flavor and aroma compounds can be lost during concentration of liquid foods by evaporation. Methyl anthranilate, an important, characteristic aroma constituent for grape juice, has a relative volatility of 3.5 with respect to water at 100°C. Find the percentage of methyl anthranilate lost if half the water is removed from grape juice by evaporation at 100°C with no inert gases in the vapor and in a batch process.

Methyl anthranilate, like all flavor and aroma compounds, is present at a very low mole fraction. Furthermore, because of the high molecular weight of dissolved solutes (mostly sugars) in grape juice, the mole fraction of water is very nearly unity.

SOLUTION Because of the high mole fraction of water in the liquid and the absence of appreciable quantities of inerts in the vapor, both y and x of water are near 1.00 and K_{H_2O} must be very nearly 1.00. Hence K for methyl anthranilate is constant at 3.5 during the evaporation, and Eq. (3-17) can be used instead of the more complex Eq. (3-18):

$$\ln \frac{0.5}{1.0} = \frac{1}{3.5 - 1} \ln \frac{x_{MA}}{x_{MAO}}$$

$$\ln \frac{x_{MA}}{x_{MAO}} = (\ln 0.5)(2.5) = -1.733$$

$$\frac{x_{MA}}{x_{MAO}} = 0.176$$

$$\% \text{ MA lost} = \frac{(100)(x_{MA}^{\circ} L_0' - x_{MA} L')}{x_{MAO} L_0'} = (100)[1 - (0.5)(0.176)] = 91.2\% \qquad \square$$

Example 3-6 After 0.5 kg of a liquid mixture containing 38 wt % acetic acid and 62 wt % vinyl acetate has been placed in a separatory funnel at 25°C, small amounts of water are added, the funnel is shaken, and the aqueous phase is continually withdrawn from the bottom of the funnel. How much organic phase will remain in the funnel when its acetic acid concentration has fallen to 10 percent by weight on a water-free basis?

SOLUTION The organic phase first saturates with water. This mixing process can be represented by a straight line on the triangular diagram, as shown in Fig. 3-9. This is the same system and same

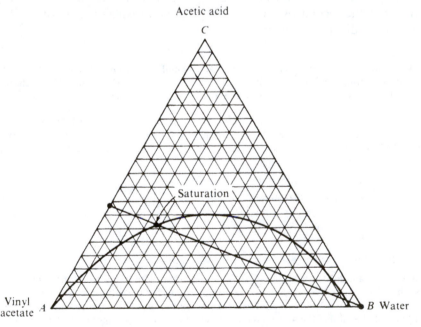

Acetic acid

Figure 3-9 Saturation of organic phase with water in Example 3-6. (*Adapted from Daniels and Alberty, 1961, p. 258; used by permission.*)

equilibrium diagram which appeared in Fig. 1-21 and were discussed in Chap. 1. The saturated composition is 51% acetate, 31% acetic acid, and 18% water. By application of the lever rule, the amount of organic phase at this point is

$$\frac{(0.5)(100)}{100 - 18} = 0.61 \text{ kg}$$

After the organic phase has become saturated, the addition of more water will form a heavier aqueous phase, which is drawn off. In order to be able to say that $dW' = -dV'$, and thereby use the Rayleigh equation, we must let V' and W' represent the weight of solution on a *water-free basis* in the acetate-rich and water-rich phases, respectively. We cannot work on a basis of *total* weight since water is continually added to the system. Adapting Eq. (3-16) to our purposes, we have

$$\ln \frac{V'}{V'_0} = \int_{w'_{V_0}}^{w'_V} \frac{dw'_V}{w'_W - w'_V}$$

where w' is now *weight* fraction of acetic acid on a water-free basis and the subscripts V and W refer to the vinyl acetate rich and water-rich phases, respectively. V' is the weight of acetate-rich phase in the funnel, again on a water-free basis. Figure 1-23 shows the equilibrium relationship between w'_W and w'_V, taking

$$w' = \frac{w_{\text{acid}}}{w_{\text{acid}} + w_{\text{acetate}}}$$

where w is weight fraction. V'_0 is 0.50 kg on the water-free basis, and w'_{V_0} is 0.38; therefore,

$$\ln \frac{V'}{0.50} = \int_{0.38}^{0.10} \frac{dw'_V}{w'_W - w'_V}$$

The graphical integration is shown in Fig. 3-10. The shaded area under the curve is -0.77 (integration is from right to left); therefore

$$\ln \frac{V'}{0.50} = -0.77$$

and $\qquad V' = (0.464)(0.50) = 0.232$ kg of organic phase remaining (water-free)

From Fig. 3-9 this amount of organic phase will contain 3 wt % water, so that the total amount of organic phase is $(0.232)(1.00/0.97) = 0.239$ kg. $\qquad\qquad\qquad\qquad\qquad\qquad\qquad\qquad\qquad\quad\square$

Example 3-7 A common flow configuration used for continuous separation of a gas mixture by gaseous diffusion is shown in Fig. 3-11. We shall assume that

1. The flow of gas through the porous barrier occurs solely by Knudsen flow.
2. The low-pressure side is at a pressure low enough for there to be no appreciable backflow from the low- to the high-pressure side.
3. The gas in the high-pressure side is well mixed in a direction normal to flow but does not undergo appreciable mixing in the direction of flow.

If a stream of UF_6 containing 0.71% $^{235}UF_6$, the rest being $^{238}UF_6$, is passed into the chamber and half of it is passed through the barrier, find the concentration of $^{235}UF_6$ in the low-pressure product. Compare with the result which would have been obtained if the high-pressure stream were assumed to be totally mixed in the direction of flow.

SOLUTION As we have seen in the discussion following Eq. (1-21), the separation factor relating the two streams on either side of the barrier at any point under assumptions 1 and 2, above, is

$$\alpha_{235, 238} = 1.0043$$

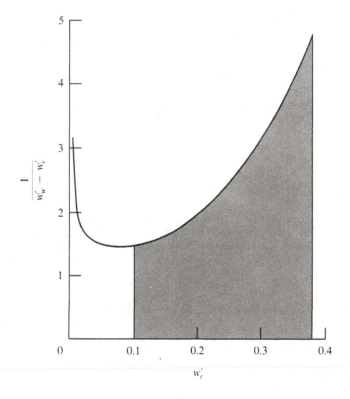

Figure 3-10 Graphical integration for Example 3-6.

Since the ^{235}U is present to only a small amount, we can say that $K_{235U} = 1.0043$ and $K_{238U} = 1.00$, following the same logic used in Example 3-5.

Despite the continuous operation, this process is akin to a Rayleigh distillation: the high-pressure stream is continuously depleted in ^{235}U as it passes through the device, and the low-pressure gas removed through the barrier at any point is related through the α or K expressions to the concentration of the high-pressure stream at that point. If we follow a particular mass of high-pressure gas through the device, we find that it undergoes a Rayleigh distillation, the low-pressure product being continuously removed and the high-pressure product remaining behind.

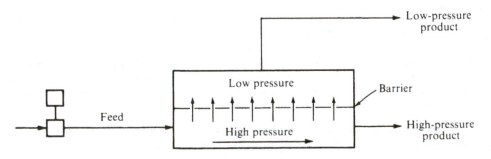

Figure 3-11 Gaseous-diffusion device.

Since K_i is constant, we can apply Eq. (3-17) in the form

$$\ln \frac{H'}{H'_0} = \frac{1}{K_{235U} - 1} \ln \frac{y_H}{y_{H_0}}$$

where y_H, y_{H_0} = mole fractions of ^{235}U
H' = flow rate of high-pressure product
H'_0 = feed flow rate $(H'/H'_0 = 0.50)$

Therefore,

$$\frac{y_H}{y_{H_0}} = (0.50)^{0.0043} = \frac{1}{1.00298}$$

By the lever rule, since the product flows are equal, $y_{H_0} - y_L = y_H - y_{H_0}$, and hence $y_L = y_{H_0}(2 - y_H/y_{H_0}) = (0.007100)(1.00596)/1.00298 = 0.007121$.

If the high-pressure side had been completely mixed in the direction of flow, we would have taken $y_L = 1.0043 y_H$ directly, since the composition of the high-pressure side at all points would have been equal to the exit high-pressure composition due to the mixing. Thus, we would have found by the lever rule for our case of equal molal product flows that

$$y_L = 1.00215 y_{H_0}$$

The lack of mixing has served to increase the enrichment $y_L - y_{H_0}$ by a factor of 0.00298/0.00215, or 39 percent. ☐

Comparison of Yields from Continuous and Batch Operation

An examination of the results of Examples 3-4, 3-5, and 3-7 shows that the operation with one phase batch and the other continuous provided a better separation than would have been obtained in an entirely continuous scheme with product equilibrium and with the same ratio of product quantities. Considering Example 3-4, a continuous equilibrium flash giving 14.1 mol % of the feed as liquid product would have produced a liquid changed from the 60 mol % benzene composition of the feed to only 41 mol % instead of to 20 mol % as found in the Rayleigh distillation. For Example 3-5 a simple continuous vaporization would give a 78 percent loss of methyl anthranilate. In Example 3-7, the product enrichment was increased by 39 percent.

This behavior will be encountered whenever the equilibrium mole fraction in one phase rises as the mole fraction in the other phase rises. Considering a Rayleigh distillation, the combined vapor product is made up of individual bits of vapor which have been removed in equilibrium with *all* liquid compositions, ranging from the initial feed liquid to the final product liquid. Only the last bit of vapor removed is in equilibrium with the final liquid product. All the rest of the vapor removed was in equilibrium with a liquid *richer* in the more volatile component than the final liquid. Thus, the *combined* vapor is richer than the last bit of vapor, and it has a mole fraction of the more volatile component *greater* than that corresponding to equilibrium with the final liquid. Hence the separation is better than can be achieved by simply equilibrating the product phases. The reader should notice that this argument is very similar to that presented for the improved separation obtained in the cross-flow geometry of the air stripping process in Fig. 3-5.

In Example 3-6, one would find that a continuous extraction of the type shown in

Fig. 1-20 would give 0.239 kg of acetate-rich phase containing *less* than the 10 wt %
acetic acid (water-free basis) obtained in the semibatch case. This corresponds to a
better separation in the continuous case than in the semibatch case and follows from
the fact that w'_W at equilibrium (Fig. 1-23) *decreases* with increasing w'_V in the range
of interest. In this case the final bits of water fed produce a better separation than
the initial bits of water do.

In a continuous binary separation it is impossible to obtain a complete separation or even to produce a pure product made up entirely of one component unless α,
the separation factor, is infinite. When one phase is charged batchwise and the other
is removed continuously, it *is* possible to produce a pure phase but only an
infinitesimal amount of it. Considering Eq. (3-16), if all the liquid is boiled away, the
left-hand side, ln (L/L'_0), will approach $-\infty$. This means that the right-hand side
must also approach $-\infty$, which in turn means the integral must become infinite. The
integral will become infinite only if one approaches a point where $y_i = x_i$, in which
case the denominator within the integral will approach zero. This point also can be
seen from Fig. 3-10. As the amount of acetate-rich phase remaining approaches zero,
we must approach $w'_V = 0$, where the curve rises to infinity. At $w'_V = 0$, $w'_W = 0$ and
$w'_V = w'_W$.

For a separation with a constant α, one can see from Fig. 1-17 that $y = x$ only at
$x = 0$ or $x = 1$. Thus the last drop of liquid remaining in a Rayleigh distillation of a
system with relatively constant α will be made up of the less volatile component and
will be pure. At no time will the accumulated vapor be pure. On the other hand, there
are cases where the two equilibrium product compositions become equal to each
other at some intermediate composition. This corresponds to the formation of an
azeotrope in distillation. The last drop in a Rayleigh distillation will therefore either
be pure or have the composition of a *maximum*-boiling azeotrope. It cannot have the
composition of a *minimum*-boiling azeotrope unless that was also exactly the feed
composition, since the liquid composition will move away from a minimum-boiling
azeotrope as the distillation proceeds.

Multicomponent Rayleigh Distillation

Equation (3-16) is usually not suitable for use in the analysis of a Rayleigh distillation
when appreciable amounts of more than two species are present, since the relationship between y_i and x_i is not known a priori. Equation (1-12) is applicable only to a
two-component system. A more convenient expression can be obtained.

Considering components A, B, C, ..., letting l_i equal the number of moles of
component i in the liquid, and letting dv_i equal a differential amount of component i
removed in the vapor, we can put the equilibrium relationship as follows:

$$\frac{dv_A}{dv_B} = \frac{-dl_A}{-dl_B} = \alpha_{AB}\frac{l_A}{l_B} \qquad \frac{dv_A}{dv_C} = \frac{-dl_A}{-dl_C} = \alpha_{AC}\frac{l_A}{l_C} \qquad \cdots \qquad (3\text{-}19)$$

If α_{ij} is constant, we can integrate to get

$$\int_{l_{A0}}^{l_A} \frac{-dl_A}{l_A} = \alpha_{AB} \int_{l_{B0}}^{l_B} \frac{-dl_B}{l_B} = \alpha_{AC} \int_{l_{C0}}^{l_C} \frac{-dl_C}{l_C} = \cdots \qquad (3\text{-}20)$$

and

$$\frac{l_A}{l_{A0}} = \left(\frac{l_B}{l_{B0}}\right)^{\alpha_{AB}} = \left(\frac{l_C}{l_{C0}}\right)^{\alpha_{AC}} = \cdots \qquad (3\text{-}21)$$

The use of Eq. (3-21) to solve a multicomponent Rayleigh distillation is illustrated in Example 3-8.

Example 3-8 A liquid hydrocarbon mixture of 20 mol $\%$ n-butane, 30 mol $\%$ n-pentane, and 50 mol $\%$ n-hexane is subjected to a Rayleigh distillation at a controlled pressure of 10 atm. Find the liquid composition when exactly half of the pentane has been removed.

SOLUTION The bubble point of the feed mixture can be found from the data in Fig. 2-2 to be 263°F. As the vaporization occurs, the bubble-point temperature of the remaining liquid will increase. (Why?) An upper limit on the temperature achieved can be estimated by presuming that 90 percent of the butane and 10 percent of the hexane are removed during the vaporization of half the pentane. This would leave a final liquid with a bubble point of 300°F. Over this temperature range α_{BH}, taken from Fig. 2-2, varies from 3.3 to 3.7 and α_{BP} varies from 1.8 to 1.9. Although the variation of α is appreciable, it is still not excessive and the time saved in using the constant α expressions is often worth a small loss in accuracy. Mean values of α are 3.5 and 1.85.

Basis An initial charge of 100 mol

$$\frac{l_B}{20} = \left(\frac{15}{30}\right)^{1.85} = \left(\frac{l_H}{50}\right)^{3.5}$$

Solving gives $l_B = 5.5$ and $l_H = 34.7$.

	l_i	x_i
Butane	5.5	0.100
Pentane	15.0	0.272
Hexane	34.7	0.628
	55.2	

The liquid is depleted in butane and enriched in hexane. Since the bubble point of the final liquid is 280°F, the original estimates of α_{ij} are satisfactory. □

Simple Fixed-Bed Processes

Separations involving a fluid phase and a solid phase (adsorption, ion exchange, etc.) are usually carried out with the solid charged on a batch basis and with the fluid charged continuously. The solid phase forms a *fixed bed*, through which the fluid flows. This procedure results from the difficulty of providing a continuous feed of a

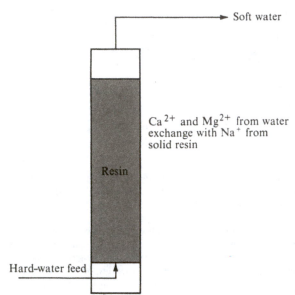

Soft water

Ca^{2+} and Mg^{2+} from water exchange with Na$^+$ from solid resin

Resin

Hard-water feed

Figure 3-12 Fixed-bed ion-exchange process for water softening.

solid substance and of causing a solid phase to move uniformly within a separation device.

As an example, consider the fixed-bed water-softening process shown in Fig. 3-12. Natural waters often contain sufficient trace levels of calcium or magnesium salts for insoluble salts to be formed with household soaps or other substances. As a result scums form in the water, and the cleaning action of the soap is reduced. To avoid the formation of these undesirable precipitates, it is often advisable to remove the calcium and/or magnesium ions from the water. This *softening* of the water is most commonly accomplished by the use of solid ion-exchange resins in the type of process shown in Fig. 3-12. Many homes have this sort of water-softening system incorporated in their water system.

Ion-exchange resins used for water softening are commonly polymeric materials made in the form of small beads. Large organic anions are incorporated in the resin. In the resin fed into the column, these large anions are paired with sodium cations. As the water flows through the column, the sodium ions from the resin *exchange* with the calcium and magnesium ions in solution, by the reactions

$$Ca^{2+} + 2Na^+ \text{ resin}^- \rightarrow Ca^{2+}(\text{resin})_2^{2-} + 2Na^+$$
$$Mg^{2+} + 2Na^+ \text{ resin}^- \rightarrow Mg^{2+}(\text{resin})_2^{2-} + 2Na^+$$

The calcium and magnesium ions are thus removed from the water; they enter the solid phase and are replaced in the water by sodium ions. The calcium ions will continue to be removed as long as the equilibrium constant

$$K_{Ca^{2+} - Na^-} = \frac{(Ca^{2+})_{\text{resin}}(Na^+)^2_{\text{aqueous}}}{(Ca^{2+})_{\text{aqueous}}(Na^+)^2_{\text{resin}}} \tag{3-22}$$

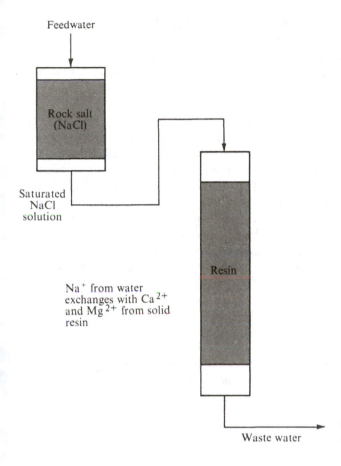

Feedwater

Rock salt
(NaCl)

Saturated
NaCl
solution

Resin

Na^+ from water
exchanges with Ca^{2+}
and Mg^{2+} from solid
resin

Waste water

Figure 3-13 Regeneration of ion-exchange bed.

for the ion-exchange reaction is not reached. A similar criterion holds for magnesium removal.

After a period of time, sufficient Ca^{2+} and Mg^{2+} will have been removed from the water stream passing through for the Ca^{2+} and/or Mg^{2+} content of the resin at all points in the bed to approach the equilibrium value given by Eq. (3-22). Further use of the bed would leave too much Ca^{2+} and Mg^{2+} in the effluent water. At this point the old resin can be removed from the bed and be replaced with fresh Na^+ resin, but a less expensive procedure is to *regenerate* the old resin, accomplished by passing a strong NaCl solution through the bed, as shown in Fig. 3-13. The high Na^+ content in the solution reverses the ion-exchange reaction, and, in accord with Eq. (3-22), Na^+ replaces Ca^{2+} and Mg^{2+} on the resin, making it suitable for reuse in water softening.

The behavior of the fixed-bed processes differs in one very important respect from those batch processes to which the Rayleigh equation can be applied. Although one phase is charged batchwise and the other phase is charged continuously in both cases, the batch-charged phase (the solid) is *not mixed* in the fixed-bed processes, whereas the batch-charged phase is *fully mixed* in the Rayleigh-equation processes. This lack of solid-phase mixing in the fixed-bed processes results in an important separation advantage, for it allows much of the fluid-phase product to be in equilibrium with the initial solid-phase composition, as shown below. This equilibration with the initial solid composition usually gives the purest possible fluid-phase product.

At first the water contacts sufficient calcium- and magnesium-free resin for the effluent water to achieve equilibrium with the initial resin, and hence the water is well softened. As time goes on, more and more of the resin becomes loaded with calcium and magnesium, and the water has less contact time in which to come to equilibrium with calcium- and magnesium-free resin. If the bed is large enough, however, most of the bed can become loaded with Ca^{2+} and Mg^{2+} before the point is reached where the exit water does not achieve equilibrium with the initial resin composition. When the effluent water can no longer reach equilibrium with the initial resin composition, the concentrations of calcium and magnesium in the effluent water begin to rise, as shown in Fig. 3-14, and eventually the feedwater goes through unchanged. At this point the resin has taken up so much calcium and magnesium that the entire resin bed is in equilibrium with the feedwater. The curve shown in Fig. 3-14 is known as a *breakthrough curve*, since it shows when Ca^{2+} begins to break through the bed into the effluent water. Regeneration should be started before or just as breakthrough begins.

Figure 3-15 shows the progress of the Ca^{2+} loading along the resin bed during on-stream service between one regeneration and the next. As long as no appreciable amount of Ca^{2+} has accumulated on the resin at the effluent end of the bed, the exit water will be relatively free of Ca^{2+}.

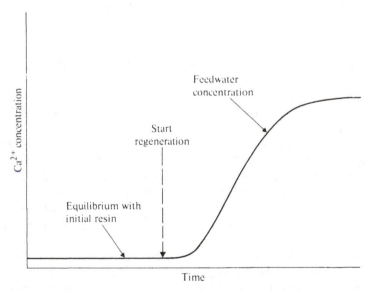

Figure 3-14 Effluent concentrations from fixed-bed water-softening process. Typical breakthrough curve.

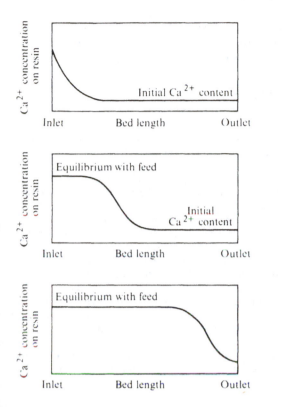

Figure 3-15 Ca^{2+} loadings on resin bed at different times.

The width of the breakthrough, i.e., the time or bed distance elapsed between asymptotic Ca^{2+} concentrations, in Figs. 3-14 and 3-15 reflects a number of factors, including the nature of the resin-water equilibrium expression, rates of mass transfer between the fluid and solid phases, and dispersion (local variations in flow, diffusion, etc.) within the fluid phase. The relationship between these factors and the width of the breakthrough curve is explored briefly in Chap. 11 and in more detail in various references on fixed-bed processes (Helfferich, 1962; Hiester et al., 1973; Vermeulen, 1977a; etc.).

The volume of a fixed bed is set primarily by the solute content of the feed and the desired time between regenerations, as shown in Example 3-9, below. The length-to-diameter ratio of the bed is established as a compromise between a number of factors. High length-to-diameter ratios reduce the effects of axial dispersion and channeling (uneven flow) and can increase rates of mass transfer through higher fluid velocity, thereby reducing the portion of the bed volume occupied by the breakthrough. However, higher length-to-diameter ratios also cause greater fluid-pressure drops, requiring more pumping power. Although the bed geometry can vary widely depending upon the application, a length-to-diameter ratio of the order of 5 is not unusual.

Because of the breakthrough phenomenon and the fact that the initial effluent purity is determined primarily by the condition of the solid at the exit of the bed, it is common for the regeneration flow to be in the reverse direction from the feed flow, e.g., upflow for softening in Fig. 3-12 and downflow for regeneration in Fig. 3-13. In this way the portion of the bed which has been regenerated most thoroughly serves as the final contact for the process stream.

Since intermittent regeneration is required, multiple beds must be supplied if a continuous feed is to be handled over a substantial period of time.

Example 3-9 A process for drying air by adsorption of water vapor onto activated alumina particles is shown in Fig. 3-16. Typical equilibrium data for the adsorption of water vapor onto activated alumina are shown in Fig. 3-17.

Dry activated alumina is placed in a bed and air is passed over it. The alumina adsorbs moisture from the air. Cooling coils are placed in the bed to remove the heat released upon adsorption. When the bed can no longer dry the air sufficiently, it is regenerated by passing heated air through the bed. The heated air removes the adsorbed moisture from the bed, since the equilibrium partial pressure of water over the bed at any moisture content is approximately proportional to the vapor pressure of pure water. At higher temperatures the vapor pressure of water is higher; hence the equilibrium water vapor partial pressure over the bed is higher, and the bed therefore loses moisture to hot air. The regenerated alumina is cooled before being returned to drying service.

Suppose that it is desired to dry 1000 std ft³/min (379 std ft³ = 1.0 lb mol) of air at 30 lb/in² abs and 86°F from an initial 75 percent relative humidity moisture content to a final moisture content corresponding to 3 percent relative humidity or less at 86°F. If the time of drying service between regenerations is to be 3.0 h, calculate the weight of activated alumina required in the bed.

Assume that the shape of the water content breakthrough curve (similar to Figs. 3-14 and 3-15) is such that 70 percent of the bed can come to equilibrium with the inlet water-vapor content before

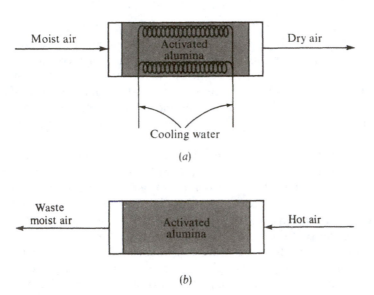

Figure 3-16 Fixed-bed process for air drying: (a) drying air with activated alumina; (b) regeneration of activated alumina.

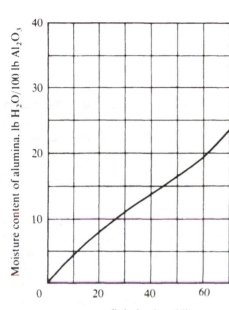

Relative humidity, percent

$$\left(100 \times \frac{\text{partial pressure of water vapor}}{\text{vapor pressure of pure water}}\right)$$

Figure 3-17 Equilibrium moisture content of activated alumina at 86°F vs. partial pressure of water vapor.

regeneration is required and the remaining 30 percent of the bed (the breakthrough zone) can take up an average water loading corresponding to half the moisture loading in equilibrium with the inlet air.

SOLUTION The vapor pressure of pure water at 86°F is 31.8 mmHg or 0.615 lb/in^2 (Perry and Chilton, 1973). Thus the mole fraction of H_2O in the inlet air at 75 percent relative humidity is

$$y_{H_2O, \text{ in}} = \frac{(0.615)(0.75)}{30} = 0.01538$$

For most of the drying cycle, the alumina, if well regenerated beforehand, will provide exit air of substantially less than 3 percent relative humidity. Hence to compute the bed-moisture loading we shall assume that effectively all the H_2O is removed from the air passing through:

Gain in bed moisture = moisture removal from air in 3 h

$$= \left(1000 \frac{\text{std ft}^3}{\text{min}}\right)\frac{1 \text{ lb mol}}{379 \text{ std ft}^3}\left(0.01538 \frac{\text{lb mol } H_2O}{\text{lb mol gas}}\right)\left(18 \frac{\text{lb } H_2O}{\text{lb mol } H_2O}\right)\left(60 \frac{\text{min}}{\text{h}}\right)(3.0 \text{ h})$$

$$= 132 \text{ lb } H_2O$$

Since 70 percent of the bed can equilibrate with 75 percent relative humidity, it can take on 0.26 lb H_2O per pound of Al_2O_3 by Fig. 3-17. The remaining 30 percent of the bed takes on half that moisture loading, or 0.13 lb H_2O per pound of Al_2O_3. The average bed loading is

$$(0.26)(0.70) + (0.13)(0.30) = 0.221 \text{ lb } H_2O/\text{lb } Al_2O_3$$

Hence the alumina required is

$$\frac{132}{0.221} = 599 \text{ lb}$$

or about 0.3 ton to dry this relatively large airflow. □

Numerous other approaches to separations involving beds of solids have been devised, including ones giving more efficient use of the solids and ones capable of separating a multicomponent mixture into individual products. These techniques, including various forms of *chromatography*, are discussed in Chap. 4.

METHODS OF REGENERATION

The degree of solute removal from the fluid product that can be obtained in a fixed-bed separation process is directly related to the degree of regeneration of the solid separating agent. The residual levels of Ca^{2+} and Mg^{2+} in the product water from the water-softening process of Fig. 3-12 are primarily determined by the effectiveness of removal of Ca^{2+} and Mg^{2+} during regeneration with strong salt solution, and the moisture content of the air from the air dryer of Example 3-9 is governed by the degree of water removal from the alumina during hot-air regeneration.

Regeneration requires a change in some thermodynamic variable to make solute removal from the fixed bed become more favorable. If this is not done, the net effect of the process will be to make the adsorbed or extracted solute more dilute in the exit stream from regeneration than it was in the original feed to the process. The thermodynamic variables which can be changed are temperature, pressure, and concentration (or composition). Changes in pressure are effective for gaseous feeds but usually not for liquids, since liquid- and solid-phase chemical potentials tend to be insensitive to pressure. Changes in concentration or composition are effective for liquid feeds but usually not for gases, since chemical potentials of components in gas mixtures tend to be insensitive to the composition of the rest of the mixture.

In the water-softening process of Fig. 3-13 regeneration is accomplished through a change in composition (concentration of Na^+). Ion-exchange processes for water treatment and other purposes can also be regenerated through a change in temperature, the column operating hotter during regeneration than during on-line service. This is the basis of the Sirotherm process and improvements upon it (Vermeulen, 1977*b*).

In the air-drying process of Example 3-9 regeneration is accomplished through an increase in temperature. If the inlet air were at a high enough pressure, regeneration could be accomplished through a reduction in pressure instead (high pressure during drying, low pressure during regeneration). This is the basis of the process known as *heatless adsorption* (Skarstrom, 1972).

These considerations apply to any mass-separating-agent process requiring regeneration. Thus, the absorbent liquid from an absorption process can be regenerated by stripping it at higher temperature and/or lower pressure. The solvent from an extraction process can be regenerated by contacting it with a different immiscible liquid, by distilling it, or in other ways.

MASS- AND HEAT-TRANSFER RATE LIMITATIONS

Equilibration Separation Processes

Phase-equilibration separation processes involve the transfer of material from one phase of matter to another. Often the residence time and intimacy of contact of the two phases in a separation device are not sufficient for the two phases to come to thermodynamic equilibrium with each other. The rate of *mass transfer* across the phase interface will govern the extent to which the phases equilibrate. Mass-transfer rates between phases reflect the phenomenon of diffusion coupled with convective flow, turbulence, and gross mixing. The transferring component(s) must travel from the original phase to the interface and then from the interface to the new phase. Diffusional resistances in either or both phases can be rate-limiting. The subject of interphase mass transfer is complex and is therefore reserved to Chap. 11, where it is considered in moderate detail.

When the feed streams to a separation device have significantly different temperatures, or when there is an appreciable latent heat effect accompanying the transfer from one phase to the other, it is necessary to consider rates of heat transfer in bringing the phases toward the same temperature as each other, as well as mass-transfer effects.

Rate-governed Separation Processes

Since the separation in rate-governed processes is defined by rates of mass transfer through a barrier region, a mass-transfer analysis is essential for defining the separation factor obtained. Mass-transfer limitations in the fluid phases on either side of the barrier can also reduce the rate of product throughput and can alter the separation factor. For example, in a reverse-osmosis process for desalination of seawater salt is rejected at the membrane surface as water passes through. If this salt cannot diffuse back into the main solution fast enough, it will build up to a higher concentration adjacent to the membrane than in the bulk feed solution. This increases the osmotic pressure of the solution adjacent to the membrane, making a higher feed pressure necessary to force the water through at a given rate and at the same time increasing the rate of salt leakage through the membrane. Influences of mass transfer on rate-governed processes are also considered further in Chap. 11.

STAGE EFFICIENCIES

For any real separation device it will be necessary to correct an analysis based upon product equilibrium or ideal separation factors for the effects of entrainment, mixing, charging sequence, flow configuration, and mass- and heat-transfer limitations. One way of correcting for these effects is to use the actual separation factor, which relates actual product compositions, for the analysis. There are, however, often several drawbacks to the use of the actual separation factor as a calculational parameter:

1. The numerical value of the actual separation factor usually has no simple fundamental basis. Whereas for a vapor-liquid process one can show that the equilibrium separation factor $\alpha_{ij} = \gamma_i P_i^0 / \gamma_j P_j^0$ for a nonideal low-pressure system (or P_i^0 / P_j^0 for an ideal system), one in most cases cannot relate α_{ij}^S to fundamental properties in a similar simple way.
2. The value of the actual separation factor is often highly dependent upon composition.
3. When the ideal separation factor is infinite, the actual separation factor does not provide a full enough description of the separation.

To amplify the last point, consider the $CuSO_4$ leaching process mentioned earlier in this chapter (Fig. 3-3). For the leaching process, poor mixing decreases the amount of $CuSO_4$ removed per gallon of added water, but the actual separation factor remains infinite since no solids appear in the aqueous product.[†]

Consequently the actual separation factor is used as a design parameter primarily for complex rate-limited separation processes and for two-phase separation processes where the appropriate equilibrium data have not been determined. When equilibrium or idealized rate data are known, it is more common to employ a *stage efficiency* as a measure of the approach to equilibrium or ideality, instead of using the actual separation factor. The reason for calling this *stage efficiency* rather than simply efficiency is to distinguish it from other efficiencies, i.e., the thermodynamic efficiency (Chap. 13), and because of the usefulness of the concept for analyzing multistage separations (Chaps. 5 et seq.).

Several varieties of stage efficiency have been suggested, but the most common for the description of individual stages is the *Murphree efficiency*, named after its originator (Murphree, 1925). The Murphree efficiency is based upon the compositions of a single phase or stream in the separation

$$E_{M1i} = \frac{x_{1i,\text{out}} - x_{1i,\text{in}}}{x_{1i}^* - x_{1i,\text{in}}} \tag{3-23}$$

The efficiency defined in Eq. (3-23) is for component i and is based upon phase 1 compositions. The subscripts in and out refer to the gross inlet and outlet streams, respectively, of phase 1; x_{1i}^* is the phase 1 composition that would be in equilibrium with the actual outlet composition of phase 2. In this way E_M is a measure of the change in composition occurring in a phase in proportion to the amount of change that could occur if equilibrium with the actual outlet composition of the other phase were reached.

[†] Rony (1968) has suggested analysis of separation processes through a *separation index* ξ, designed to avoid these problems with α_{ij}^S. The separation index is defined for a binary system with two products as $\xi = |(i)_1(j)_2 - (i)_2(j)_1|$, where $(i)_1$ represents the *recovery fraction* of component i in product 1. The recovery fraction is the fraction of the total amount of that component fed which ends up in the particular product; 1 and 2 denote the two products, while i and j denote the two components. The factor ξ must lie between 0 and 1; it is invariant with respect to permutation of component or product indices, and it often gives a more complete description of the separation than α_{ij}^S does. The use of ξ is complicated by the fact that it is complex to evaluate for multicomponent systems. Also, as a measure of the quality of the separation it implicitly gives equal economic value to each of the components, which is not usually the case in practice.

Thus, in a binary vapor-liquid contacting process a Murphree vapor efficiency can be defined as

$$E_{MV} = \frac{y_{out} - y_{in}}{y^* - y_{in}}$$ (3-24)

and a Murphree liquid efficiency can be defined as

$$E_{ML} = \frac{x_{out} - x_{in}}{x^* - x_{in}}$$ (3-25)

Readers should convince themselves (1) that E_{MV} does not in general equal E_{ML} and (2) that in a two-component system the mole fractions of *either* component can be used to compute E_{MV} or E_{ML} without altering the value obtained.

In a system containing n components in a particular phase there will be at most $n - 1$ independent Murphree efficiencies based on different components in that phase. This follows from the requirement that $\Sigma x_i = 1$. Since Murphree efficiencies are applied only to components that transfer between phases, there will be the same number of independent Murphree efficiencies based upon the other phase.

The Murphree definition of efficiency is by no means the only one possible. Alternative definitions which have been used are considered in Chap. 12, where means of predicting and using stage efficiencies are developed in more detail.

If values of the Murphree efficiency are known, it is possible to calculate the composition of one product stream from a stage directly, given the inlet composition of that stream and the outlet composition and temperature of the other product. As we shall see, this make the Murphree efficiency convenient for use in calculations of multistage separation processes. On the other hand, the Murphree efficiency is less convenient in predicting the product compositions from a simple single-stage separation process given the feeds, since one actual outlet stream composition is needed to calculate the other.

Values of Murphree vapor efficiencies vary widely and can typically be in the range of 60 to 90 percent for distillation in a plate column, 3 to 40 percent for absorption of a gas into a heavy oil or for absorption in a chemically reacting system, 85 to 100 percent for extraction in a mixer-settler, and 15 to 50 percent per compartment or plate for column extractors (Perry and Chilton, 1973).

Example 3-10 Suppose that the ammonia-stripping process of Fig. 3-5 is carried out isothermally at 30°C and 101 kPa. The ratio of air to water feeds is 2.0 mol/mol. If the inlet air is free of ammonia and the inlet water contains 0.1 mol $\%$ ammonia, find the exit-water composition. The Murphree vapor efficiency of the plate for ammonia removal is 75 percent, and the Henry's law constant H in the equilibrium expression $p_{NH_3} = Hx_{NH_3}$ is 129 kPa/mole fraction at 30°C.

SOLUTION The equilibrium ratio K_{NH_3} is $129/101 = 1.28$. Hence

$$E_{MV} = 0.75 = \frac{y_{out, NH_3} - y_{in, NH_3}}{1.28x_{out, NH_3} - y_{in, NH_3}}$$

Since $y_{in, NH_3} = 0$.

$$\frac{y_{out, NH_3}}{x_{out, NH_3}} = (0.75)(1.28) = 0.96$$ (3-26)

By material balance

$$L(x_{\text{in, NH}_3} - x_{\text{out, NH}_3}) = G y_{\text{out, NH}_3}$$

Substituting $L/G = 0.50$ mol/mol and $x_{\text{in, NH}_3} = 0.001$ gives

$$y_{\text{out, NH}_3} = 0.50(0.001 - x_{\text{out, NH}_3}) \tag{3-27}$$

Eliminating $y_{\text{out, NH}_3}$ from Eqs. (3-26) and (3-27) gives

$$0.96 x_{\text{out, NH}_3} = 0.0005 - 0.50 x_{\text{out, NH}_3}$$

Solving, we have

$$x_{\text{out, NH}_3} = 0.00034$$

or 66 percent of the ammonia is removed. □

REFERENCES

Heiss, R., and L. Schachinger (1951): *Food Technol.*, **5**:211.

Helfferich, F. (1962): "Ion Exchange," McGraw-Hill, New York.

Hiester, N. K., T. Vermeulen, and G. Klein (1973): Adsorption and Ion Exchange, in R. H. Perry and C. H. Chilton (eds.), "Chemical Engineers' Handbook," 5th ed., sec. 16, McGraw-Hill, New York.

Maxwell, J. B. (1950): "Data Book on Hydrocarbons," Van Nostrand, Princeton, N.J.

Murphree, E. V. (1925): *Ind. Eng. Chem.*, **17**:747.

Perry, R. H., and C. H. Chilton (1973): "Chemical Engineers' Handbook," 5th ed., McGraw-Hill, New York.

Rayleigh, Lord (1902): *Phil. Mag.*, [vi]4(23):521.

Rony, P. R. (1968): *Separ. Sci.*, 3:239.

Skarstrom, C. W. (1972): Heatless Fractionation of Gases over Solid Adsorbents, in N. N. Li (ed.), "Recent Developments in Separation Science," vol. 2, CRC Press, Cleveland.

Vermeulen, T. (1977a): Adsorption, Industrial, in R. E. Kirk and D. F. Othmer (eds.), "Encyclopedia of Chemical Technology," 3d ed., vol. 1, Interscience, New York.

——— (1977b): *Chem. Eng. Prog.*, **73**(10):57.

Wheaton, R. M., and A. H. Seamster (1966): Ion Exchange, in R. E. Kirk and D. F. Othmer (eds.), "Encyclopedia of Chemical Technology," 2nd ed., vol. 11, p. 882, McGraw-Hill, New York.

Weller, S., and W. A. Steiner (1950): *Chem. Eng. Prog.*, **46**:585.

PROBLEMS

3-A$_1$ Consider an evaporation process for separating a dilute solution of salt and water. The salt may be assumed to be totally nonvolatile, so that the equilibrium separation factor α_{W-S} is infinite. Compute the apparent actual separation factor α_{W-S}^S if one-half the water in the feed solution is vaporized and 2 percent of the remaining liquid is entrained in the escaping vapor.

3-B$_1$ (a) Suppose 50 mL of vinyl acetate is added to 50 mL of a solution containing 50 wt % acetic acid and 50 wt % water in a separatory funnel. The separatory funnel is shaken and the phases settle. How many phases will form? What is the composition of the phase(s)?

(b) Repeat part (a) if only 10 mL of vinyl acetate is added to the 50 mL of original acetic acid–water solution.

3-C$_1$ Vapor-liquid equilibrium data are frequently obtained in devices which contact vapor and liquid streams circulating within a closed device. The equilibrium data are obtained by measuring the compositions of the vapor and liquid by means such as gas chromatography. A possible complicating factor in such devices is entrainment. Suppose that a mixture containing 20 mol $\%$ n-butane, 50 mol $\%$ n-pentane, and 30 mol $\%$ n-hexane is brought to equilibrium at 250°F and 9.47 atm to give the vapor and liquid compositions found in Example 2-4. In the product analysis, however, the vapor sample contains 10 mol $\%$ entrained liquid. What would be the apparent values of K_B, K_P, and K_H obtained from the measured compositions of the " vapor " and liquid samples, ignoring the entrainment? By what percentage does each of these K's differ from the actual value?

3-D$_1$ In the derivation of the Rayleigh equation (3-13) the term $L\,dx$ appears, but the term $V'\,dy$ does not appear. Why?

3-E$_2$ Find the liquid composition when 70 percent of the original feed has been removed in the Rayleigh distillation of Example 3-4.

3-F$_2$ Repeat Prob. 2-E for the situation where the vapor is removed from the system as rapidly as it is formed rather than remaining in contact with the liquid phase. Report any additional sources of data you consult.

3-G$_2$ A gas mixture of 0.1 $\%$ ethylene glycol vapor in water vapor at 6.7 kPa is to be purified by passing it in plug flow along a cooled tube which is kept at a temperature below the dew point (60°C). Condensed water is continually removed from the system by passing out through a thin slot in the outer wall in a direction perpendicular to the gas flow.

(a) What fraction of the water vapor must be condensed in order to yield a product water-vapor stream containing only 0.01 $\%$ ethylene glycol?

(b) Compare your answer with the result that would be obtained if the condensate and vapor product were in equilibrium with each other.

Under these conditions the relative volatility of water to glycol is 98. The process is shown schematically in Fig. 3-18.

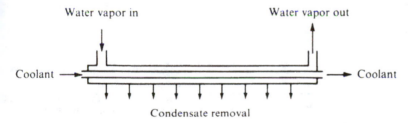

Water vapor in   Water vapor out

Coolant   Coolant

Condensate removal

Figure 3-18 Condensation process with immediate condensate removal at all points.

3-H$_2$† An adsorbent costing $40 per kilogram is being used in a process for selective adsorption of one component from a mixture of several materials. The process consists of a cycle of adsorption followed by reactivation and recharging the unit with the regenerated adsorbent, which after reactivation is only 85 percent as efficient as in the preceding cycle. The cost of reactivation and recharging is $20 per kilogram. What is the optimum number of cycles to use the adsorbent before discarding it?

3-I$_2$ A liquid mixture containing 70 mol % acetone and 30 mol % acetic acid is charged to a batch (Rayleigh) distillation operating at 101.3 kPa (1 atm). What molar fraction of the charge must be removed by the distillation to leave a liquid containing 30 mol % acetone? What will be the composition of the accumulated distillate? Vapor-liquid equilibrium data for this system are given in Example 2-7.

† From Board of Registration for Professional Engineers, State of California, Examination in Chemical Engineering, November 1966; used by permission.

3-J₂ Figure 3-19 shows data for the uptake of water by molecular-sieve desiccants as a function of gas humidity at temperatures near ambient. Compare Fig. 3-19 for molecular sieve with Fig. 3-17 for activated alumina. If the desiccants are to be used for drying a stream of air:

(a) Which desiccant will provide effluent air of the lower water content before breakthrough?

(b) For desiccant beds of equal weight which desiccant will require more frequent regeneration?

(c) Would there ever be an incentive to use two desiccant beds in series for air drying, one bed containing activated alumina and the other containing molecular sieve? If so, which bed would be contacted first by the air?

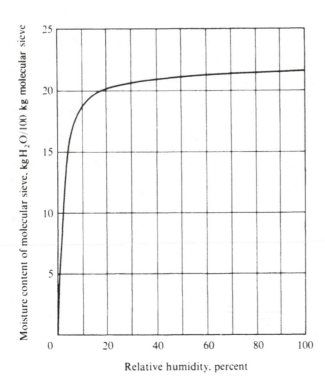

Figure 3-19 Equilibrium moisture content of molecular sieve vs. relative humidity. (*Data from Davison Chemical Division of W. R. Grace & Co.*)

3-K₂ A home water softener will use a bed of sulfonic-resin ion-exchange particles, to be regenerated by a water stream passed through a bed of rock salt, NaCl, particles. The water supply contains 68 ppm (w/w) (= 68 mg per kilogram of water, or 68 mg/L), expressed as Ca^{2+}. Wheaton and Seamster (1966) report that such a resin, regenerated with nearly saturated salt solution, will have an exchange capacity of 27.5 g/L of wet resin bed volume, expressed as weight of Ca^{2+} exchanged.

Assume that it is necessary to supply 35 percent more bed volume than would correspond to the indicated exchange volume of the resin in order to allow for the width of the breakthrough curve and to give some margin for variable conditions. Assume also that it is necessary to pass 2.0 times the stoichiometric amount of salt into the ion-exchange bed to regenerate it fully. Regeneration will be automatically timed to occur once a week, in the middle of the night. For regeneration, the household water supply will be diverted from the bed and salt-bearing solution will be fed into the bed for 1 h, followed by a period of water wash to remove salt from the interstices of the bed. Assume that the regeneration water passing through the bed of rock salt will achieve 90 percent of saturation (solubility of NaCl in water = 36.5 kg/100 kg H_2O). The average daily flow of water to be softened is 1 m³ (1000 L).

(a) Calculate the volume of wet resin needed in the water softener.

(b) What flow rate of water to the salt bed is needed during the regeneration period?

(c) If the bed of rock salt is recharged with 15-kg bags of salt bought at the grocery store, how often will it be necessary to add the contents of a new bag to the bed?

(d) The frequency and duration of regeneration are typically set on a master controller. What factor(s) might warrant resetting these values from time to time? Would a feedback control system for regeneration be more appropriate? If so, how might one be implemented?

3-L₂ A laboratory device is set up to remove a water contaminant from ethylene glycol by stripping the glycol with dry air. The glycol is charged to the vessel, and desiccated air is supplied through a sparger beneath the surface of the glycol. The glycol solution is well mixed by the bubbling action. The air leaves the vessel in equilibrium with the solution. Surprisingly, the water-glycol system obeys Raoult's law closely. Given the following data, find how long air must be passed through the solution to dry the glycol to a water content of 0.1 mol %.

The system is isothermal at 60°C   Vapor pressure of water = 19.9 kPa at 60°C

Relative volatility of water to glycol = 98   Initial charge to vessel = 10.0 mol glycol

Initial H_2O content of glycol = 2.0 mol %   Airflow = 5.0 mol/h

Pressure = 1.00 atm = 101.3 kPa

3-M₃ For the situation of Example 3-2, (a) what is the *maximum operable* wash-water flow rate, expressed as mass per mass of H_2O in the original entrained concentrate?

(b) What is the *optimal* wash-water flow rate for minimum loss of juice solids per unit amount of juice product?

3-N₃ Zone refining is a process which has been used extensively in recent years for the ultrapurification of solid materials. The process operates by passing a heated zone slowly along a thin rod of the solid material which is to be purified, as shown in Fig. 3-20.

Within the heated zone the material is melted, but outside the zone the material is solidified. As the heater passes along the rod it melts the solid. After the heater passes a given location in the rod the material at that point solidifies. The passage of the heated zone along the rod causes the impurity level to vary continuously along the rod, being higher toward one end of the rod and lower toward the other end. The separation can be accentuated by passing the heated zone along the rod in the same direction repeatedly.

Wilcox, in his doctoral dissertation at the University of California, Berkeley, in 1960, examined the removal of β-naphthol impurities from naphthalene by zone refining. Equilibrium measurements for this system have shown that the ratio of the weight fraction of β-naphthol in the solid to the weight fraction in the liquid at equilibrium ($K_{s-l} = w_s/w_l$) is 1.85 as long as the weight fraction of β-naphthol in the solid is less than 0.40. The density of molten naphthalene is 978 kg/m³, and the density of solid naphthalene is 1145 kg/m³.

Consider the zone refining of a 30-cm-long rod of naphthalene which contains an evenly distributed impurity of 0.01 weight fraction β-naphthol w_0. The length L of the heated zone is 1.0 cm. The molten zone is always well mixed.

(a) Will the β-naphthol concentrate toward the end of the rod which is melted first or last? Why?

(b) Obtain an analytical expression for the weight fraction w_s of β-naphthol in the rod as a function of distance after a single pass of the heated zone along the rod. Plot w vs. distance along the rod after the first pass. Make sure to include the last 1 cm at either end.

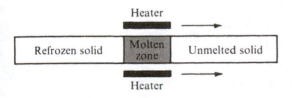

Figure 3-20 Zone-refining process.

(c) What is the level of β-naphthol concentration at the two tip ends of the rod after the *first* pass?

(d) What is the level of β-naphthol concentration at the two tip ends of the rod after the *second* pass?

(e) Assume now that the molten zone is not necessarily totally mixed. Does mixing in the molten zone help or hinder the separation in this process?

(f) Assume once again that the molten zone is totally mixed. A phase diagram for the system m-cresol–p-cresol is shown in Fig. 1-25. The densities of solid and liquid cresols are nearly the same. Suppose that a 30-cm bar of p-cresol containing 1°₀ m-cresol is held in an environment which keeps it at −10°C when it is not heated. The bar is to be purified by passing a molten zone 1.0 cm long slowly along it. What is the composition of the bar as a function of length after the first pass? After the second pass?

3-O₃ The separation of hydrogen and methane is an important industrial problem. One technique which has been suggested is the use of selective polymeric membranes. As shown in Fig. 3-21, one could flow a high-pressure stream of hydrogen and methane into a chamber and allow a portion of that stream to pass through to a low-pressure side of the chamber by means of a diffusion-permeation mechanism. The rate at which component *i* passes through the membrane is given by

$$N_i = K_i(P_H y_{iH} - P_L y_{iL})$$

where N_i = flux of component *i* through membrane, mol/h·(m² membrane area)

K_i = permeability of the membrane to component *i*, mol/h·Pa

P_H, P_L = high- and low-side pressures, respectively, Pa

y_{iH}, y_{iL} = mole fractions of component *i* on high- and low-pressure sides, respectively

For the use of ethyl cellulose membranes to separate hydrogen-methane mixtures, Weller and Steiner (1950) give $\alpha_{H_2-CH_4} = K_{H_2}/K_{CH_4} = 6.6$.

Consider the case where both the high-pressure and the low-pressure sides are well mixed, in which case

$$\frac{y_{iL}}{y_{jL}} = \frac{N_i}{N_j}$$

where *i* and *j* are two components.

(a) Prove that the actual separation factor α_{ij}^S for a binary gas mixture is related to α_{ij}, the pressure ratio, and the high-pressure-side gas composition by

$$\alpha_{ij}^S = \alpha_{ij} \frac{y_{iH}(\alpha_{ij}^S - 1) + 1 - r\alpha_{ij}^S}{y_{iH}(\alpha_{ij}^S - 1) + 1 - r}$$

where $\alpha_{ij} = K_i/K_j$, and $r = P_L/P_H$.

(b) Suppose that an ethyl cellulose membrane process will be used to separate a hydrogen-methane mixture, giving a methane-rich product containing 60 mol °₀ methane and 40 mol °₀ hydrogen. If $P_L/P_H = 0.50$, find the value of $\alpha_{H_2-CH_4}^S$ and the composition of the hydrogen-rich product. Explain physically why $\alpha_{H_2-CH_4}^S$ is so much less than $\alpha_{H_2-CH_4}$.

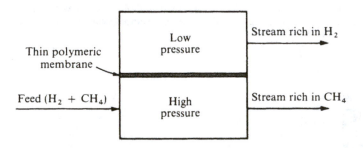

Figure 3-21 Membrane process for separation of hydrogen-methane mixtures.

(c) Show that the minimum value of α_{ij}^S required to provide a given recovery of component j in the retentate (high-pressure product), with fixed compositions of the binary feed and of the permeate (low-pressure product), is given by

$$\alpha_{ij}^S = \frac{\beta - R_j}{1 - R_j}$$

where R_j is moles of component j in the retentate divided by moles of component j in the feed and β is defined by

$$\beta = \frac{y_{jH}}{y_{iH}} \frac{y_{iF}}{y_{jF}}$$

the subscript F referring to the feed.

3-P$_2$ Present the appropriate equations and a computation sequence for finding the maximum percent of the cyclopentane fed which can be recovered in a given purity y_{CP} in the accumulated distillate product from a simple Rayleigh distillation of a liquid mixture of fixed composition containing cyclopentane (CP), cyclohexane (CH), and methylcyclopentane (MCP). Indicate a particular convergence method which you feel would be reliable and effective. Would direct substitution be workable? $\alpha_{CP-CH} = 2.60$ and $\alpha_{MCP-CH} = 1.30$.

3-Q$_2$ Using relative-volatility data from Prob. 3-P, find the mole percentage of cyclopentane in the accumulated distillate from a simple Rayleigh distillation of a mixture containing 70 mol % cyclopentane, 10 mol % cyclohexane and 20 mol % methylcyclopentane, if 20 mol % of the mixture is distilled over.

CHAPTER
FOUR

MULTISTAGE SEPARATION PROCESSES

Separation processes are frequently built in such a way that the same basic separation unit is repeated over and over again. Processes of this sort are called *multistage separations*. This chapter will consider the reasons for devising multistage processes and also explore some of the ways in which multiple staging can be accomplished. The two principal reasons for staging are *to increase product purity* and *to reduce consumption of the separating agent*.

INCREASING PRODUCT PURITY

Often a separation process is called upon to produce one or more relatively pure products. In many cases, however, a sufficiently high degree of product purity cannot be achieved in a simple single-stage contacting device of the sort discussed in Chaps. 2 and 3.

Multistage Distillation

Consider a process in which a benzene-toluene mixture is to be separated into benzene and toluene, both for sale as chemicals. This process might, for example, receive the mixed benzene-toluene effluent stream of Fig. 1-8 as a feedstock. Minimum product purities will be imposed in order for the benzene and toluene to be salable on the market. Typically, these purities might be somewhere in the range of 98 to 99.9 percent or even higher.

As we have seen in Fig. 1-17 and Example 2-1, benzene has a vapor pressure 2.25 times greater than that of toluene at 121°C. This ratio is not very sensitive to temperature. The higher volatility of benzene suggests a process in which the liquid feed is

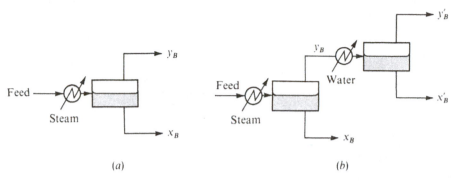

Figure 4-1 (*a*) One- and (*b*) two-stage vaporization processes.

partially vaporized to give a vapor product richer in benzene. This process is shown in Fig. 4-1*a*.

Suppose, however, that the feed consists of 50 mol % benzene and 50 mol % toluene. The first small amount of vapor generated will have a composition given by Eq. (1-12) as

$$y_B = \frac{(2.25)(0.50)}{1 + (2.25 - 1)(0.50)} = 0.691$$

As more vapor is formed, the mole fraction of benzene in the liquid will drop and the equilibrium benzene content of the vapor will also drop. If half the liquid is vaporized, simultaneous solutions of Eqs. (1-12) and (2-9) shows that $y_B = 0.600$ and $x_B = 0.400$. Hence there is no way in which a sufficiently pure benzene product can be made in this single-stage vaporization.

One way to obtain a richer benzene product is to condense a portion of the vapor generated in the first step. Such a process is shown in Fig. 4-1*b*. If the vapor products from both steps are quite small, then

$$y_B' = \frac{(2.25)(0.691)}{1 + (2.25 - 1)(0.691)} = 0.838$$

If the vapor products are both substantial, y_B' will be lower but will still have been increased over the single-step case. For example, if half the feed is vaporized in each step, $y_B' = 0.696$, again by combined use of Eqs. (1-12) and (2-9).

Extending this concept, we can picture a process, shown in Fig. 4-2, in which there are enough successive condensations to give a benzene product of the required purity. Similarly, a sequence of vaporizations of the liquid product from the initial step will serve to create a sufficiently pure toluene product (see Fig. 4-3).

In passing, it should be noted that the successive vaporizations leading to the toluene product in the bottom half of Fig. 4-3 are conceptually quite similar to the Rayleigh distillation discussed in Chap. 3. The two processes give identical products if an infinite number of stages are employed in the bottom half of Fig. 4-3. We have already seen that the last drop of liquid remaining in a Rayleigh distillation will

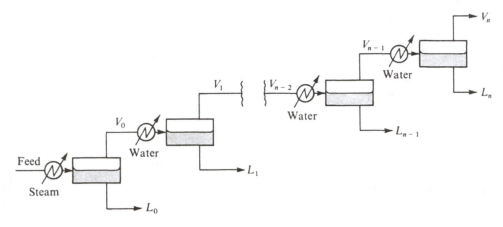

Figure 4-2 Process for production of pure benzene.

consist entirely of the least volatile component. Similarly, an infinite number of stages in the toluene-purification sequence will give a totally pure toluene product.

By analogous reasoning, the benzene-purification train is akin to a batch, or Rayleigh, condensation.

The scheme shown in Fig. 4-3 is capable of giving quite pure products, but the amounts of these products obtained will be quite small. At the same time there will be many products of intermediate composition which have not been brought to the

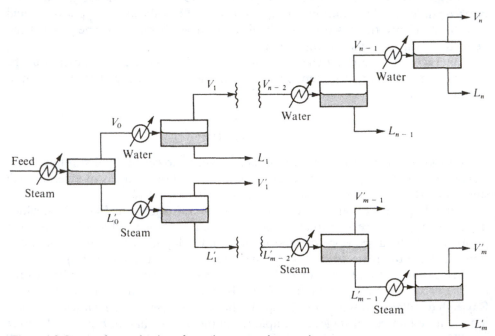

Figure 4-3 Process for production of pure benzene and pure toluene.

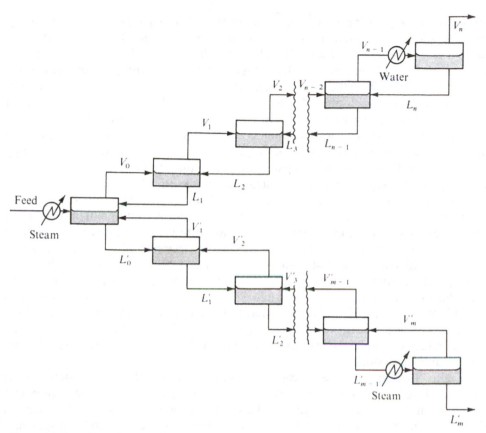

Figure 4-4 Recycling of intermediate products.

required product purity. An obvious improvement is to reintroduce these intermediate products to the process at the point where they correspond to the prevailing
stream composition. Thus the liquid from the second separator of the benzene-
purification train L_1 can be put back into the initial separator. This follows since L_1
has been once enriched in benzene (as V_0) and once depleted; hence L_1 is of approximately the same composition as the main feed. Similarly V_1' has been once depleted
in benzene (as L_0') and once enriched; hence it too has approximately the composition of the main feed and should return to the initial separation stage. By the same
line of reasoning L_{n-1} is enriched $n-1$ times and depleted once; therefore it should
enter the last previous stage along with V_{n-3}, which has been enriched $n-2$ times.
L_n should enter with V_{n-2}, V_{m-1}' should enter with L_{m-3}', and so on.

Figure 4-4 shows the process which results if the intermediate products are all
returned to the appropriate separation stages. It becomes apparent in constructing
this figure that another advantage has accrued from returning the intermediate products to the process. Each stage in the upper train to which a liquid has been
returned no longer requires a water cooler on the vapor feed to that stage. The second

phase for the separation is now formed by the returning liquid. A rough energy balance, allowing for latent heats but ignoring sensible-heat effects, shows that the ratio of product vapor to product liquid from a stage will be the same as the ratio of feed vapor to feed liquid entering that stage. However, the compositions of the two products will differ from the compositions of the two feeds since the two feeds are not in equilibrium with each other. Since L_3 is richer than L_1, etc., the returning liquid necessarily is richer in benzene than corresponds to equilibrium with the vapor fed to a stage. This is what provides the impetus for continual enrichment of the vapors in benzene as we go upward along the sequence of separators. The same reasoning applies to the bottom train of separators. The returning vapors now provide the second phase for separation, and the steam heaters are no longer required.

Note that the heat exchangers on the feeds to the two terminal stages of the purification sequence are still necessary since no returning intermediate product is fed to these stages. The top water cooler and the bottom steam heater are required to generate the second phase in these two terminal separators. Unless the final cooler is used, no available intermediate liquid product has a benzene content higher than corresponds to equilibrium with V_{n-1}; also, unless the final heater is included, no available intermediate vapor product has a benzene content lower than corresponds to equilibrium with L'_{m-1}. Note also that the steam heater on the initial feed to the entire process is not really necessary either, as long as V'_1 is sufficient to generate an adequate vapor phase in the first separator. This will be true unless the original liquid feed is highly subcooled.

Figure 4-4 represents the process of *distillation*, which is the most common of the various staged separation processes carried out industrially. The equipment employed in actual practice is still simpler than that indicated in Fig. 4-4. As long as the vapor and liquid flows can be sent in the proper directions and brought into contact repeatedly to provide the action of the stages, the distillation process can be carried out in a single vessel.

Plate Towers

The most common single-vessel device for carrying out distillation, a *plate tower*, is shown schematically in Fig. 4-5. The tower is a vertical assembly of *plates*, or *trays*, on each of which vapor and liquid are contacted. The liquid flows down the tower under the force of gravity, while the vapor flows upward under the force of a slight pressure drop from plate to plate. The highest pressure is produced by the boiling in the bottom steam heater, called the *reboiler*. The vapor passes through openings in each plate and contacts the liquid flowing across the plate. If the mixing of vapor and liquid on the plates were sufficient to provide equilibrium between the vapor and liquid streams leaving the plate, each plate would provide the action of one of the separator vessels in Fig. 4-4. The portion of the tower above the feed is called the *rectifying* or *enrichment section*. As is clear from the logic leading to Figs. 4-3 and 4-4, this upper section serves primarily to remove the heavier component from the upflowing vapor; it enriches the light product. The portion of the tower below the feed, called the *stripping* section, serves primarily to remove or strip the light

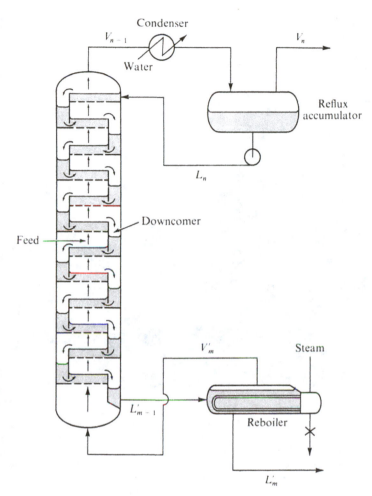

Figure 4-5 Distillation tower.

component from the downflowing liquid. The bottom section thus serves primarily to purify the bottoms product.

The condenser system takes the overhead vapor from the column and liquefies a portion of it to return to the tower as *reflux* (equivalent to L_n). Thus the condenser and reflux accumulator are unchanged in concept from the top stage of Fig. 4-4. The reboiler shown in Fig. 4-5 is a *kettle reboiler*; it combines the heating and phase-separation functions of the bottom stage in Fig. 4-4.

The condenser shown in Fig. 4-5 is a *partial condenser*, so named because it condenses only a fraction of the overhead vapor. Total condensers are also used commonly; with a total condenser the overhead vapor is entirely liquefied and split into two portions, one for use as overhead product and the other for return to the top plate as reflux. Since liquids are more easily stored than vapors, a total reboiler, giving the bottom product as a vapor, is seldom used.

Figure 4-6 Facilities for the primary distillation of crude oil into products of different boiling ranges for further processing. The unit is located at the Naples refinery of Mobil Oil Italiana and processes 82,500 bbl of crude oil per day (1 bbl = 0.159 m³). The distillation is carried out in two towers, atmospheric (narrower) and vacuum (wider), together at the center of the photograph. The tall structure at the left is the flue stack for a furnace preheating the feed to the distillation towers. To the right of the crude distillation towers is a bank of fan-driven air-cooled heat exchangers, which probably serve as the overhead condenser for the atmospheric tower. Four other distillation columns are visible to the right, in the rear. (*The Lummus Co., Bloomfield, N.J.*)

Figures 4-6 and 4-7 give an idea of the importance and scale of distillation towers in petroleum refining and petrochemical manufacture, where they are the principal means of separation.

An enlarged diagram of one type of individual plate is shown in Fig. 4-8. In order to give good mixing between phases and to provide the necessary disengagement of vapor and liquid between stages, the liquid is retained on each plate by a *weir*, over which the effluent liquid flows. To reach the next stage, this effluent liquid flows down through a separate compartment, called a *downcomer*. The downcomer provides sufficient volume and a long enough residence time for the liquid to be freed of entrained vapor before reaching the bottom and entering the next plate.

Many different designs have been proposed and built for the plates themselves. The vapor must enter the plate with a relatively uniform distribution, must contact

Figure 4-7 A giant distillation column, 82 m high and 4.9 m in diameter, being lifted into place at the Imperial Chemical Industries, Ltd., ethylene plant at Wilton, England. This plant produces ethylene, propylene, and butadiene from hydrocarbon feedstocks derived from petroleum refining. Note the scaffolding surrounding the towers under construction in the background. (*The Lummus Co., Bloomfield, N.J.*)

the liquid intimately, and must disengage quickly from the liquid. The simplest type of plate is the *sieve tray*, shown schematically in Fig. 4-8, which consists of a metal plate with 3- to 15-mm holes, spaced in a regular pattern. A photograph of a sieve plate is shown in Fig. 4-9.

Most older distillation towers were built with *bubble-cap* trays, which bring the vapor to a point up in the flowing liquid and then reverse the direction of vapor flow, causing it to jet or bubble out into the liquid through slots. The bubble caps are positioned on a plate in patterns similar to that shown for the sieve-plate holes in Fig. 4-8. Figure 4-10*a* provides a schematic drawing of a bubble cap. Figure 4-11 shows a tray containing a full layout of bubble caps.

Figure 4-10*b* is a schematic drawing of a *valve cap*, another device in widespread

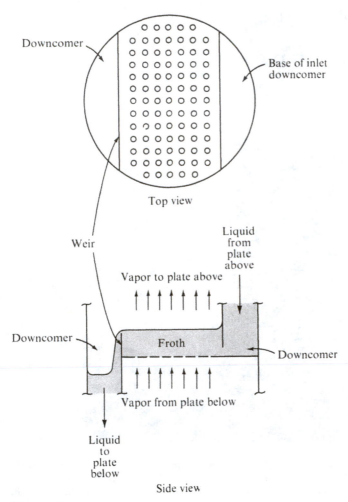

Figure 4-8 Sieve plate.

use. The riser of the valve cap is supported by the momentum of the upflowing vapor. At high vapor velocities the riser is fully open, while at lower vapor velocities the riser is partially or completely lowered. In this way the linear velocity of vapor issuing into the liquid from under the riser is more or less independent of the total amount of vapor being handled by the tower. Thus valve caps provide good vapor-liquid mixing over a wide range of tower flow conditions. Figure 4-12 shows a valve-cap tray with the caps in *down* (low-gas-flow) position.

Further descriptions of these and other types of plates can be found in several distillation texts (Hengstebeck, 1961; Oliver, 1966; Smith, 1963; Van Winkle, 1968; etc.).

As a rule, the individual plates in a distillation column do not provide simple equilibrium between the exiting vapor and liquid. The residence times of the two

Figure 4-9 A 2.1-m-diameter perforated (sieve) tray. The downcomer to the next tray is located on the left. *(Fritz W. Glitsch & Sons, Inc., Dallas, Texas)*

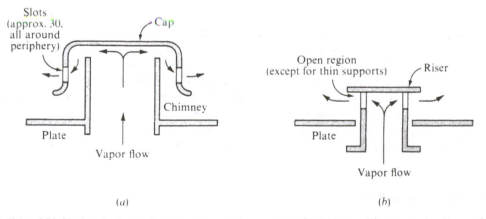

Figure 4-10 Devices for dispersing vapor into liquid on a plate: (*a*) bubble cap; (*b*) valve cap (partly open).

Figure 4-11 Bubble-cap tray 1.2 m in diameter with 7.6-cm cap assemblies. *(Fritz W. Glitsch & Sons, Inc., Dallas, Texas)*

phases and the degree of dispersion of the vapor and liquid represent compromises between effective contacting and reasonable equipment costs; consequently the amount of mass transfer usually falls short of that required to approach equilibrium closely. In the other direction, the cross-flow effect, already analyzed in Example 3-3, can let the change in vapor composition exceed that corresponding to simple equilibrium with the liquid, particularly in towers with large diameters. As a result of these factors, the plates of a distillation tower should not be equated to the equilibrium stages in Fig. 4-4; the difference is usually accounted for through a stage efficiency. Prediction and analysis of stage efficiencies and other plate characteristics are covered in Chap. 12.

Countercurrent Flow

From Figs. 4-4 and 4-5 it should be noted that the distillation is accomplished by *countercurrent flow* of the vapor and liquid phases. The reader is probably already

Figure 4-12 A 1.8-m-diameter valve-cap tray. (*Fritz W. Glitsch & Sons, Inc., Dallas, Texas*)

familiar with the benefits of countercurrent flow in heat exchangers. The important consequence of countercurrent flow in a distillation system is that the overhead product is considerably more enriched in the more volatile component (benzene) than corresponds to equilibrium with the bottom product; this was our reason for *staging* the process in the first place.

It is possible to use several devices other than plate towers to achieve the countercurrent flow required for high-purity products in vapor-liquid systems. One example is a *packed tower*, shown schematically in Fig. 4-13*b*. Here the arrangement of equipment in the packed-tower scheme of Fig. 4-13*b* is the same as for the plate tower of Fig. 4-13*a* except that the internals of the tower itself are different. The packed tower is filled with some form of divided solids, shaped to provide a large particle-surface area. The liquid flows down over the surface of the solids and is exposed to the vapor, which flows upward through the open channels not filled by packing or liquid. Some common types of packing are shown in Figs. 4-14 and 4-15.

Within a packed tower there are no discrete and identifiable stages to provide equilibrium of liquid and vapor. However, it is important to realize that the workability of the plate tower shown in Fig. 4-5 is not predicated upon the attainment of equilibrium between the two product streams leaving a stage. Instead it is only necessary that there be some exchange of material between the phases; this exchange of material will be such as to bring the two phases closer to equilibrium. A large number of stages providing partial equilibrium will perform the same overall separation as a smaller number of stages providing complete equilibrium.

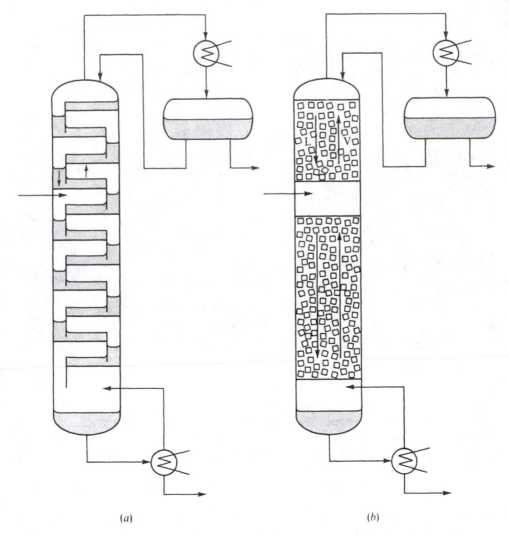

Figure 4-13 Comparison of (a) plate and (b) packed towers.

In the packed tower the vapor and liquid are continuously contacted and are continuously undergoing an exchange of material. As in the plate tower, this exchange of material acts in the direction of bringing the two phases closer to equilibrium. In the benzene-toluene example, if a packed tower were used, the liquid at a given level in the tower would always be richer in benzene than corresponded to equilibrium with the vapor, and there would be a transfer of benzene from the liquid to the vapor at all points in the tower. To preserve the enthalpy balance, there would also be a transfer of toluene from the vapor back to the liquid at all points. Thus the necessary transfer of species between phases also occurs in the packed tower, and the

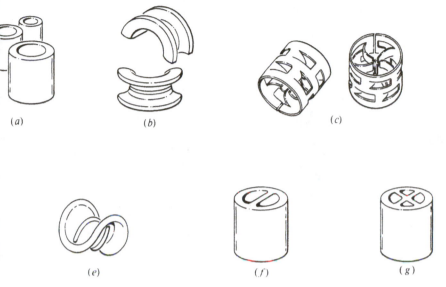

Figure 4-14 Common packing shapes: (*a*) Raschig rings, (*b*) Intalox saddle, (*c*) Pall rings, (*d*) cyclohelix spiral ring, (*e*) Berl saddle, (*f*) Lessing ring, and (*g*) cross-partition ring. (*U.S. Stoneware Corp., Akron, Ohio*)

Figure 4-15 Grid-type packing, 1.8 m in diameter and 2.4 m deep. (*Fritz W. Glitsch & Sons, Inc., Dallas, Texas*)

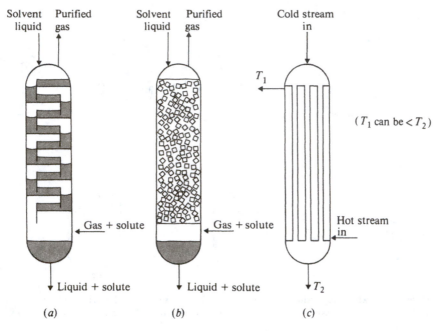

Figure 4-16 Comparison of (a) plate absorber, (b) packed absorber, and (c) shell-and-tube heat exchanger.

packed tower of Fig. 4-13b can give a separation equivalent to that given by the plate tower of Fig. 4-13a.

The analogy between countercurrent mass transfer and countercurrent heat transfer may be even more apparent from Fig. 4-16. In Fig. 4-16a and b plate and packed columns are used as countercurrent absorbers to remove a soluble impurity from a gas into a solvent liquid. In Fig. 4-16c a shell-and-tube heat exchanger is used to remove heat from a hot stream into a cold stream. The driving force for heat transfer is the temperature difference between the two streams at any cross section of the heat exchanger, and the driving force for mass transfer (absorption) is the difference between the partial pressure of solute in the gas and the partial pressure of solute that would be in equilibrium with the liquid at any cross section of the towers. Countercurrency enables the heat exchanger to operate with T_1 becoming less than T_2. Equilibrium between the effluent streams would correspond to $T_1 = T_2$, and so $T_1 < T_2$ corresponds to exceeding the action of simple equilibrium. Similarly, countercurrency in the absorbers enables the partial pressure in the exit gas to be reduced to a value lower than that corresponding to simple equilibrium with the exit liquid.

Since the feeds enter at the ends, the devices in Fig. 4-16 are single-section separators or exchangers and are thereby analogous to either the rectifying section or the stripping section, individually, in a distillation tower.

REDUCING CONSUMPTION OF SEPARATING AGENT

One advantage of staging a separation process is that purer products can be obtained. Another advantage, in many cases, is that the amount of separating agent consumed is less. An example will illustrate this point.

Multieffect Evaporation

Consider the evaporation process shown in Fig. 4-17 for the conversion of seawater into fresh water. Steam condenses in the coils of the evaporator, giving up heat, which serves to evaporate the seawater. The evaporated water is condensed in a water-cooled heat exchanger and forms the freshwater product.

At first glance it may seem strange that condensing water can boil water, which in turn can be condensed by another water stream. A temperature-difference driving force is necessary to cause heat flow across the heat-transfer surfaces. Thus the steam must be at a higher pressure than the pressure in the evaporator chamber, so that the steam-condensation temperature will be higher than the boiling point of the seawater. Similarly, the cooling-water temperature must be less than the condensation temperature of the evaporated water.

Desalination processes for seawater should produce fresh water at a cost on the order of 40 cents or less per cubic meter of fresh water in order to be attractive economically. In the process shown in Fig. 4-17 approximately 1 kg of steam is consumed for 1 kg of fresh water produced since the latent heats of vaporization of steam and seawater are essentially the same. The price of steam is variable, depending upon the pressure, location, etc., between about $1.80 and $9 per 1000 kg. Taking

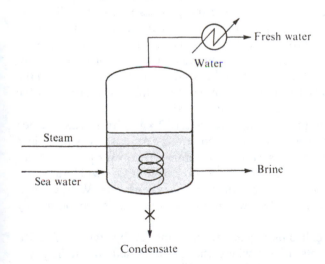

Figure 4-17 Evaporation process for converting seawater into fresh water.

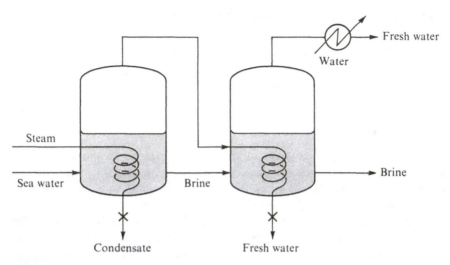

Figure 4-18 Two-stage evaporation process.

a relatively low steam cost of \$2.50 per kilogram, we find that the steam for the process shown in Fig. 4-17 costs

$$\frac{\$2.50}{1000 \text{ kg steam}} \frac{1 \text{ kg steam}}{1 \text{ kg fresh water}} \frac{1000 \text{ kg}}{1 \text{ m}^3} = \$2.50/\text{m}^3 \text{ fresh water}$$

Thus the cost of the separating agent, by itself, is enough to prevent the process from being economically worthwhile. Amortization charges for the plant equipment will raise the cost of water still further.

The separation factor is very large for this process because of the negligible volatility of the dissolved salt contaminants of the seawater; hence there is no advantage in product purity to be gained from staging the process in any way. There is latent heat available, however, in the evaporated water, which can be used to advantage in a second evaporator operating at a lower pressure, as shown in Fig. 4-18. Since the second evaporator is run at a lower pressure than the first, the boiling point of the brine in the second evaporator is less than the condensation temperature of the water vapor in the tubes. Thus the positive-temperature-difference driving force necessary for heat transfer across the coils is established.

In this process 1 kg of steam produces approximately 2 kg of fresh water; hence the steam cost is halved at the expense of increased equipment cost. It is also clear that we can build more evaporators in series, each running at a lower pressure than the previous one. As is evident from Fig. 4-19, the steam consumption will be approximately $1/n$ kg per kilogram of fresh water if there are n evaporators. With enough evaporators the steam costs by themselves will not make the process economically prohibitive.

The process of Fig. 4-19 is called *multieffect* evaporation, each stage being called an *effect*. This process represents a case where the whole benefit of staging is a reduction in the consumption of separating agent for a given amount of product.

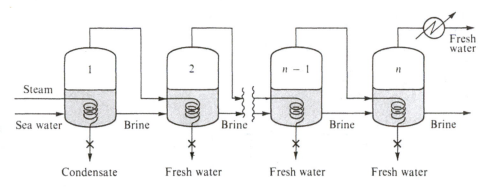

Figure 4-19 Multieffect evaporation.

This saving results from the fact that the separating agent can be reused from one effect to the next. This reuse of separating agent also characterizes the distillation process of Figs. 4-4 and 4-5. Vapor is generated in the reboiler and forms a vapor phase on each plate above, without any need for additional reboilers. Separating agent is introduced in the amount necessary for one stage and then reused in all other stages.

Appendix B gives an analysis of how the design of a multieffect evaporation system depends upon heat-transfer coefficients, thermal driving forces, etc. It also considers how different factors interact to determine the optimal number of effects and presents results of a specific example for seawater desalination.

COCURRENT, CROSS-CURRENT, AND COUNTERCURRENT FLOW

In the discussion thus far, the stages of a continuous-flow multistage process have been shown linked together with countercurrent phase flows. While we have seen that the countercurrent staging arrangement provides improvements over single-stage separations, we have not explored the possibility that other arrangements of stages might provide as good or better a separation. Obviously, many linkages are possible even with a few stages.

Three basic and simple methods of flow arrangement are shown in Fig. 4-20. As an illustration it has been assumed that the products from each stage are in equilibrium, that the phases utilized are vapor and liquid, that the feed is liquid, and that heat is introduced at some point or points in the stage arrangement in order to create a vapor phase. The same conclusions would be reached if the phases were different and produced by different means.

It should be apparent from consideration of Fig 4-20 that the arrangement of parallel, or *cocurrent*, flow results in the separation given by a single stage, no matter how many stages are used. The products from the topmost stage are in equilibrium, and will simply stay in equilibrium upon passing through the next stage. Stages 2 and

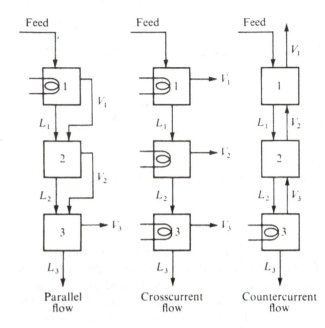

Feed   Feed   Feed

Figure 4-20 Three different flow linkages between stages.

Parallel   Crosscurrent   Countercurrent
flow   flow   flow

3 accomplish nothing. Product purities cannot exceed those attainable in a single equilibrium-stage separation.

The *cross-flow* arrangement possesses the same advantages that a Rayleigh distillation provides, as opposed to a continuous single-stage flash process. Vapors are removed in equilibrium with various liquid compositions, ranging from that of the feed to that of the final liquid product. Since the vapor in equilibrium with the feed is richer in the more volatile component than the vapor in equilibrium with the final liquid product, the cross-flow arrangement gives an improved separation as opposed to the cocurrent arrangement. In the limit of an infinite number of stages, the cross-flow arrangement will give a separation equivalent to the Rayleigh distillation; for any finite number of stages, the separation will be intermediate between a single-stage separation and a Rayleigh distillation.

The countercurrent-flow arrangement can readily exceed the quality of separation attainable in a Rayleigh distillation for a given yield of product. With enough stages, all the vapor product will be in equilibrium with the feed, which is the richest liquid. Hence the countercurrent arrangement is more efficient than the cross-flow arrangement. Similarly, the consumption of separating agent for appropriate comparisons of the three flow arrangements usually increases in the order countercurrent < crosscurrent < cocurrent. These points are illustrated in the following example.

Example 4-1 A process is to be considered utilizing the three-stage and flow arrangements shown in Fig. 4-20. It is a vapor-liquid process using a liquid feed of 0.5 mol A and 0.5 mol B. Neglect the heat capacity of the streams and assume that all heat introduced results in vaporization. Assume also that the latent heats of both A and B are 20 MJ/mol. Each stage provides complete equilibrium. The separation factor between A and B is constant and equal to 4. Calculate with each arrangement the amount of liquid bottom product (purified B stream) produced and the heat requirements if the concentration of A in the bottom product is chosen to be 10 mole percent. Compare also with the results for a one-stage Rayleigh distillation.

SOLUTION *Parallel flow* Since the two products are in equilibrium, we can use the equilibrium expression to obtain

$$y_A = \frac{\alpha_{AB} x_A}{(\alpha_{AB} - 1)x_A + 1} = \frac{4x_A}{3x_A + 1} = \frac{4x_{A,3}}{3x_{A,3} + 1} = \frac{4(0.10)}{3(0.10) + 1} = 0.308$$

The product vapor and product liquid are both depleted in A with respect to the feed. This is an impossible situation, as shown by the mass balances

$$V_3 = 1 - L_3 \quad \text{and} \quad 0.5 = 0.308V_3 + 0.10L_3$$

which yield $V_3 = 1.92$ mol and $L_3 = -0.92$ mol.

The cocurrent-flow arrangement is incapable of producing any bottom product with the mole fraction of A reduced to 0.10.

Crosscurrent flow To solve this case it is necessary to write the material balance for component A at each stage.

Stage 1: $\quad 0.5 = V_1 y_{A,1} + L_1 x_{A,1}$

Stage 2: $\quad L_1 x_{A,1} = V_2 y_{A,2} + L_2 x_{A,2}$

Stage 3: $\quad L_2 x_{A,2} = V_3 y_{A,3} + L_3 x_{A,3}$

If the additional stipulation is made that the amount of heat put into all stages is equal (note that some statement like this is needed to specify the problem completely), then

$$V_1 = V_2 = V_3 = V$$

and the equations can be solved for $x_{A,1}$, $x_{A,2}$, and V, using the equilibrium expression together with the requirement that $x_{A,3} = 0.10$.
The results are

$$V_1 = V_2 = V_3 = 0.303 \text{ mol} \qquad V_1 + V_2 + V_3 = 0.909 \text{ mol total vapor product}$$

$$y_{A,1} = 0.728 \qquad y_{A,2} = 0.584 \qquad y_{A,3} = 0.308 \qquad y_{A,\text{total vapor}} = 0.540$$

$$L_3 = 0.091 \text{ mol liquid bottom product} \qquad \text{Heat requirement} = 18.18 \text{ MJ}$$

Countercurrent flow It is again necessary to write the material balance for component A at each stage.

Stage 1: $\quad 0.5 + V_2 y_{A,2} = V_1 y_{A,1} + L_1 x_{A,1}$

Stage 2: $\quad V_3 y_{A,3} + L_1 x_{A,1} = V_2 y_{A,2} + L_2 x_{A,2}$

Stage 3: $\quad L_2 x_{A,2} = V_3 y_{A,3} + L_3 x_{A,3}$

Since all the heat introduced in the bottom stage results in vaporization and the latent heats of A and B are assumed to be the same,

$$V_1 = V_2 = V_3 \qquad L_1 = L_2$$

Solving the equations under these conditions leads to

$V_1 = 0.646$ mol vapor top product

$y_{A,1} = 0.720$ $y_{A,2} = 0.550$ $y_{A,3} = 0.308$

$L_3 = 0.354$ mol liquid bottom product   Heat required $= 12.92$ MJ

Comparing the two processes that will give the required product, it is apparent that countercurrent flow is markedly better than cross-current flow, producing almost 4 times as much purified bottom product with approximately two-thirds as much heat. In addition, the use of just a few countercurrent stages above the point of feed introduction, provided with liquid flow from condensation of a portion of the vapor, would materially increase the amount of bottom product again.

The amount of bottom product, with $x_A = 0.10$, which could be obtained from the feed by means of a simple Rayleigh distillation can be calculated from Eq. (3-18):

$$\ln \frac{L}{1.0} = \frac{1}{4-1} \ln \frac{(0.1)(0.5)}{(0.5)(0.9)} + \ln \frac{0.5}{0.9} = \tfrac{1}{3} \ln 0.111 + \ln 0.556 = -0.73 - 0.59 = -1.32$$

$L = 0.27$ mol   Heat required $= 14.6$ MJ

Hence we have established that the amount of product obtained lies in the order

Countercurrent > Rayleigh > crosscurrent > cocurrent

The opposite ordering applies to the heat requirement (consumption of separating agent).   □

OTHER SEPARATION PROCESSES

The usefulness of the multistage, or countercurrent-contacting, principle is in no way limited to vaporization-condensation processes such as distillation and evaporation. Any separation process which receives a feed and produces two products of different composition can be staged in exactly the same flow configuration. One result of this generalization is that packed and plate towers can be, and frequently are, used for the other gas-liquid separation processes, such as absorption and stripping.

Liquid-Liquid Extraction

Liquid-liquid extraction can be accomplished by creating a staged arrangement of the mixer-settler devices shown in Fig. 1-20. A three-stage extraction process of this sort is shown in Fig. 4-21. Readers should convince themselves that the operation of this process is entirely analogous to that of the portion of a distillation column lying below the feed (the stripping section), even though the individual items of equipment are quite different. The water (solvent) feed at the right-hand side of the process takes the place of the reboiler vapor in distillation. The water–plus–acetic acid product is equivalent to the vapor leaving the feed stage in distillation. The staged extraction produces purer products than can be achieved in a simple single-stage extraction.

It should also be pointed out that the amount of water (separating agent) required for the recovery of a given amount of acetic acid is less in the three-stage process than in a single-stage process. In a single-stage process the effluent water can contain at most a concentration of acetic acid in equilibrium with the vinyl acetate

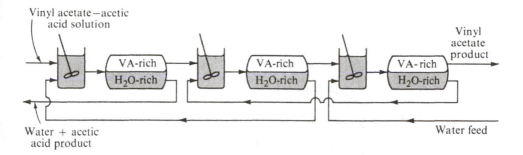

Figure 4-21 Three-stage extraction process for separating vinyl acetate from acetic acid.

product. In the three-stage process the effluent water can contain a higher concentration of acetic acid, corresponding more nearly to equilibrium with the vinyl acetate-acetic acid feed. Since the acetic acid concentration in the water can reach a higher level in the multistage process, less water solvent is required for a given acetic acid recovery. This situation is analogous to the reduction in separating-agent consumption accomplished by staging the evaporation process of Fig. 4-19.

Numerous other devices can be used for carrying out countercurrent, or multistage, liquid-liquid extraction processes. For example, rotating-disk contactors, shown in the aromatics recovery unit of Fig. 1-10, are operated in countercurrent fashion to give purer products than would correspond to simple equilibrium between product streams. As shown in Fig. 4-22, a countercurrent rotating-disk extraction column receives the denser liquid phase at the top of the column, while the other, lighter liquid phase enters the bottom and flows upward, contacting the heavier phase as it goes. Plate and packed towers are also used for multistage liquid-liquid extraction, the less dense liquid taking the place of the vapor in distillation as the fluid which flows upward in the tower.

The process shown in Fig. 4-21 provides the action of the stripping section of a distillation column, but there is no analog to the rectifying section. As a result the vinyl acetate product will be relatively pure, but there still will be a substantial amount of vinyl acetate contaminant in the aqueous product. The vinyl acetate product can equilibrate against the pure water solvent, whereas the aqueous product can equilibrate only against the much less pure feed mixture of vinyl acetate and acetic acid. Similarly, if the rectifying section were left off the distillation column in Fig. 4-5, there would be a sizable toluene contaminant in the benzene product although the toluene product could be quite pure.

Two different approaches can be used to provide the equivalent of rectifying action on the extract and thereby remove vinyl acetate from the acetic acid product in this example. One of these is shown in Fig. 4-23, where a distillation column is used to remove solvent (water, in this case) from a portion of the extract. This converts the overhead from the distillation column into a feed-phase stream, rich in the preferentially extracted component (acetic acid, in this case). Since this *extract-reflux stream* is available for the extract stream to equilibrate against, the main feed

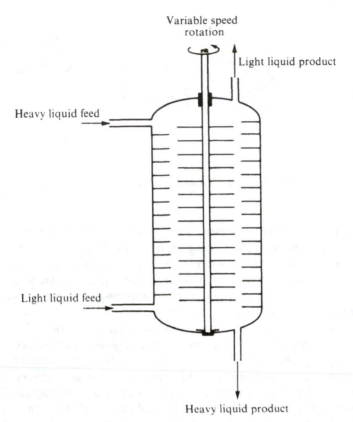

Figure 4-22 Countercurrent rotating-disk column for liquid-liquid extraction.

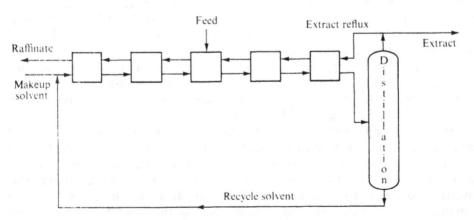

Figure 4-23 Schematic of an extraction process with extract reflux.

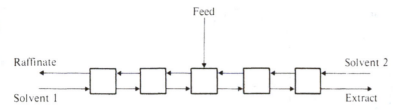

Figure 4-24 Schematic of an extraction process with two counterflowing solvents (fractional extraction).

can be introduced to the middle of the cascade and a two-section extraction process results, capable of giving a low concentration of vinyl acetate in the extract product as well as a low concentration of acetic acid in the raffinate product.

The second approach is shown in Fig. 4-24, where a second solvent, immiscible with the first solvent and the former extract but miscible with the former raffinate, enters at the other end of the cascade. In the present example this second solvent should dissolve vinyl acetate preferentially over acetic acid, thereby purifying the acetic acid product. A heavier ester or an ether might serve as a low-polarity solvent for this purpose.

Two-solvent extraction processes of the sort shown in Fig. 4-24 are sometimes referred to as *fractional extraction*. One commercial example is the Duo-sol process, developed for refining lubricating oils (Hengstebeck, 1959). In that process the counterflowing solvents are liquid propane and Selecto, which is a mixture of 40% phenol and 60% cresylic acids (cresols, etc.). The crude lubricating oil enters midway in the cascade. The desirable noncyclic compounds dissolve preferentially in the propane, while undesirable substances such as asphalts, polycyclic aromatics, and color species are taken up by the phenol-cresylic acid phase. Fractional extraction processes are further discussed by Treybal (1963).

Generation of Reflux

The examples shown so far lead to a generalization of two methods of creating a refluxing stream for a multistage equilibration separation process:

1. Convert a portion of a product stream into the other, counterflowing phase of matter, e.g., by converting vapor into liquid in a condenser or liquid into vapor in a reboiler (distillation) or by converting extract phase into raffinate phase in a distillation column (refluxed extraction). This usually amounts to adding an *energy separating agent.*
2. Add a *mass separating agent,* such as a liquid solvent (simple extraction, absorption) or a carrier gas (stripping).

In two-section multistage processes it is possible to use the first approach at both ends of the cascade of stages, as in distillation, or to use the second approach at both ends, as in the fractional-extraction process of Fig. 4-24 or in a combined absorber-stripper. Alternatively, combination processes can be used, such as the refluxed extraction process of Fig. 4-23 or a reboiled absorber, where solvent is added at the

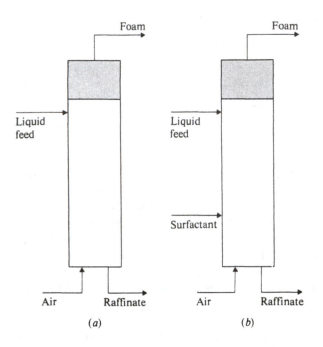

Figure 4-25 Combined bubble and foam fractionation processes: (a) simple configuration; (b) separate surfactant feed.

top, feed enters in the middle, and a reboiler generates counterflowing vapor at the bottom.

Bubble and Foam Fractionation

Figure 4-25a shows the simple flow configuration for combined bubble and foam fractionation. Liquid flows downward through an empty column, and gas rises up from the bottom in the form of fine bubbles. In the configuration shown, foam forms above the feed, being generated by the rising bubbles when they reach the top surface of the liquid pool. The foam is withdrawn as an overhead product. Surface-active species are adsorbed to the bubble surfaces and leave in the foam product.

Figure 4-26 shows measured axial liquid-concentration profiles in the liquid column for a case where Neodol, a commercial anionic surfactant, is removed from water in a bubble column with a length-to-diameter ratio slightly over 20 (Valdes–Krieg, et al., 1977). The fractionation effect from the counterflowing streams (liquid downward and interface rising with the gas upward) is substantial, reducing the Neodol concentration by more than a factor of 20 from column top to bottom.

Since it is an anionic surfactant, Neodol has the property of pairing selectively with certain cations, one of which is copper, Cu^{2+}. However, with the configuration of Fig. 4-25a, where the surfactant would enter in the copper-bearing feed solution, the counterflowing gas and liquid in the bubble column would accomplish relatively little fractionation of copper since not much surfactant is present low in the column (Fig. 4-26). Better fractionation of copper in the bubble column is obtained by introducing the surfactant separately, lower in the bubble column, as shown in Fig. 4-25b.

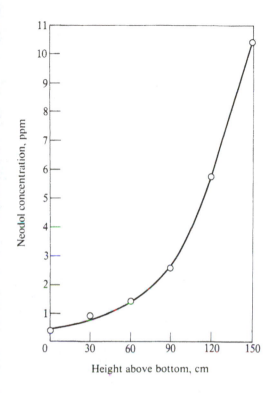

Figure 4-26 Axial concentration profile for Neodol (an anionic surfactant), using configuration of Fig. 4-25a. Column diameter = 6.95 cm. Volumetric gas-to-liquid ratio = 1.88. Neodol concentration in feed = 19.6 ppm. Feed level = 150 cm above bottom. *(Adapted from Valdes-Kreig et al., 1977, p. 274; by courtesy of Marcel Dekker, Inc.)*

An axial copper-concentration profile found with such a configuration is shown in Fig. 4-27, where fractionation reduces the copper concentration by a factor of 3 in the bubble section.

In both Figs. 4-26 and 4-27 the measured concentration at the feed level in the column is less than the concentration in the feedstream (10.6 vs. 19.6 ppm and 0.064 vs. 0.078 mmol/m^3). This is the result of large-scale axial mixing, which causes dilution of the feed by leaner liquid swept up from below. The design of a bubble column like these must take into account both the rate of mass transfer of solute between phases and the amount of axial mixing. Methods for approaching such a problem are outlined in Chap. 11.

Combined bubble and foam fractionation processes are promising for treatment of effluent waters, where environmental regulations necessitate removing solutes from an already quite dilute feed down to still lower concentration levels, e.g., ppm down to ppb. The process works better with very dilute feeds than with more concentrated feeds because of the limited adsorption capacity of the bubble surfaces.

A foam fractionation process can also be run where the column is mostly filled with foam and drainage of liquid in the foam-cell borders gives the counterflow action. To what extent effective fractionation within the foam can be obtained in this way is controversial, however (Goldberg and Rubin, 1972). It is interesting to note that in a combined bubble and foam fractionation process the counterflowing stream at the bottom of the column is created by the second method described above (a mass

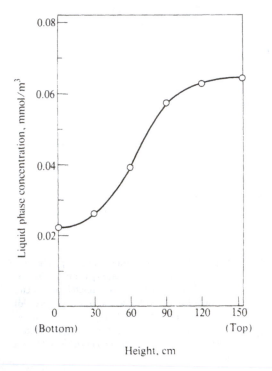

Figure **4-27** Axial concentration profile for copper, using configuration of Fig. 4-25*b*. Column diameter = 6.95 cm. Copper concentration in feed = 0.078 mmol m³. Surfactant feed level = 30 cm above bottom. Main feed level = 156 cm above bottom. (*Adapted from Valdes-Krieg et al., 1977, p. 279, by courtesy of Marcel Dekker, Inc.*)

separating agent, air or bubble surfaces), while at the top of a draining foam the counterflowing reflux is created by changing the phase condition of some of the overhead product from foam surface to draining liquid.

Bubble columns are also used as gas absorbers (Sherwood et al., 1975), but for a large-diameter column the degree of axial mixing is great enough to remove most of the countercurrent action.

Rate-governed Separation Processes

The gaseous-diffusion process was shown in Figs. 1-28 and 3-11 and was discussed at those points. Since single-stage gaseous-diffusion processes are limited in the amount of enrichment they can provide, for the recovery of ^{235}U from natural uranium it is necessary to employ a multistage process. The type of staging employed is indicated by the schematic of a three-stage process in Fig. 4-28. Cooling water in heat exchangers (not shown) is necessary to recool the gas after each compression.

Figure 4-28 is deceptive with regard to the amount of barrier surface area required. In reality, because of the vacuum and fine pore size needed for Knudsen flow, UF_6 permeation rates are very low and the barrier must be very thin and yet strong enough to prevent leaks. At the same time it must be large in expanse for the necessary amount of UF_6 to pass through.

The process shown in Fig. 4-28 provides the action of the rectifying section of a

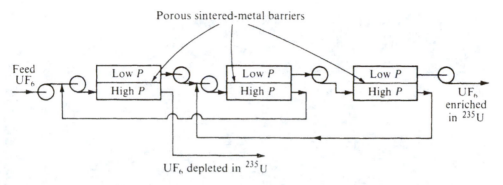

Figure 4-28 Three-stage gaseous-diffusion process for uranium-isotope enrichment.

distillation column and in this way can produce relatively pure $^{235}UF_6$ if the number of stages is adequate. In order to recover more of the ^{235}U out of the rejected stream depleted in ^{235}U, that stream is fed to a series of gaseous diffusion stages to the left of the feed in Fig. 4-28. These stages then produce a more concentrated ^{238}U-rich product and in that way accomplish the same function as the stripping stages of a distillation column.

There is an important conceptual distinction between this process and distillation, however. In the process of Fig. 4-28 the separating agent is energy and takes the form of the various compressors. Note that in this process it is imperative that the compressors be present before *each* stage. Unlike distillation, extraction, etc., gaseous diffusion belongs to a group of separation processes in which the separating agent *must* be added to each stage and cannot be reused. Membrane separation processes (reverse osmosis, ultrafiltration, etc.) are also members of this group. The need for adding separating agent at each stage is a general characteristic of rate-governed separation processes as opposed to equilibration separation processes.

The requirement that separating agent be added to each stage is a definite negative feature because it increases operating costs considerably. Rate-governed separation processes find application when they provide a high enough separation factor for only a single stage or very few stages to be required for a given separation (as in many membrane separation processes) or when they give a separation factor so much higher than those of competitive equilibration separation processes that the inherently higher operating costs are offset (as for gaseous-diffusion separation of uranium isotopes).

Because of the very low ideal separation factor ($\alpha_{235-238} = 1.0043$) for the separation of uranium isotopes, very large numbers of stages and high reflux flows are required to achieve the desired separation. To produce 90% ^{235}U material from the 0.7% ^{235}U found in natural ores requires about 3000 gaseous-diffusion stages in series. Figure 4-29 shows an aerial view of the Oak Ridge gaseous-diffusion plants (K-25, K-27, K-29, K-31, and K-33). These cascade buildings have a ground coverage of over 4×10^5 m^2 and represent a capital investment of some \$840 million. The original World War II separation cascade, the K-25 plant, is the pair of long build-

Figure 4-29 Panorama of the gaseous-diffusion plant at Oak Ridge, Tennessee. *(U.S. Dept. of Energy.)*

ings in the right rear of the photograph. Since the large number of stages and the high interstage flows necessitate numerous extremely large compressors, it was important to locate this plant in a region where electric power is cheap, i.e., that served by the Tennessee Valley Authority (TVA). The Clinch River, a tributary of which runs through the plant, supplies the cooling water for compressor aftercooling.

Other Reasons for Staging

Although increasing product purities and decreasing the consumption of separating agent are the two most common reasons for staging a separation process, occasionally there can be other reasons, e.g., gaining more efficient heat transfer or achieving a more compact geometry. Example 4-2 shows one of these reasons and also illustrates how the description rule (Chap. 2) can be used effectively for process scale-up.

Example 4-2 Electrodialysis is a separation process which has been primarily explored as a means of obtaining fresh water from seawater or from less salty but contaminated brackish water. Zang et al. (1966) describe another use for electrodialysis, in a process for removing excess citric acid from fruit juices.

The tartness of orange and grapefruit juice varies over the course of the growing season. At times the juice is too tart for sale; this has been attributed to the presence of excess citric acid in the juice. Possible ways to circumvent this problem are to blend the juice with less tart juice or to neutralize some of the citric acid by adding a base. The first of these alternatives creates scheduling and storage difficulties, whereas the second affects the taste because of the accumulation of citrate salts.

Electrodialysis affords a means of removing the citric acid instead of neutralizing it. In the process shown schematically in Fig. 4-30, grapefruit juice containing an excess of citric acid flows in alternating chambers between membranes. These membranes are made of a polymeric ion-exchange material, wherein the anions are loosely held and are free to move, while the cations are large organic

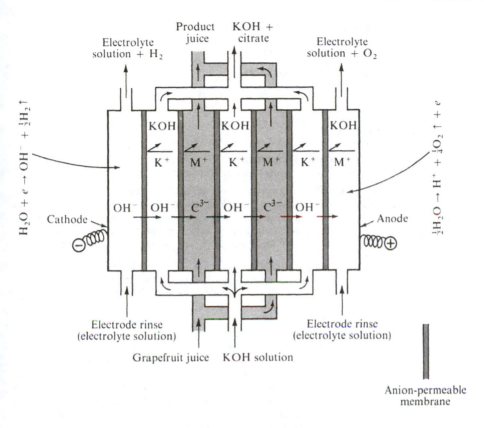

Figure 4-30 Electrodialysis process for removing excess citric acid.

molecules immobilized by the polymeric structure. Potassium hydroxide solution flows through the remaining channels of the device. Passage of an electric current through the device in a direction perpendicular to the flow causes a migration of cations toward the cathode and of anions toward the anode. The anions can be transported through the membranes into the next compartment because of the mobility of the anions in the membrane, but the cations cannot because the cations within the membranes are wholly immobilized. As shown in Fig. 4-30, the result is that citrate ions (C^{3-}) pass from the juice into the KOH, while OH^- ions enter the juice to take their place. The K^+ ions of the KOH and the cations within the juice (M^+) cannot cross the membranes, and hence are not transferred. The net result is that a portion of the citric acid in the juice is converted into water.

Two juice cells with surrounding KOH cells are shown in Fig. 4-30. A typical commercial installation would contain a substantially greater number of each in the alternating array.

Zang et al. (1966) report the results of a pilot electrodialysis run:

No. of cell pairs in stack = 12   Area/membrane = 0.88 m²   Feed temp = 33°C

Feed acidity = 1.52%   Product acidity = 0.90%   Production rate = 0.360 m³/h

Cell velocity = $\begin{cases} 9.2 \text{ cm/s} & \text{juice} \\ 3.1 \text{ cm/s} & \text{KOH} \end{cases}$

Voltage = 167 V   Current = 122 A   Current density = 140 A/m² membrane

Current efficiency = 0.70   dc energy consumption = 209 MJ/m³ juice

The current efficiency is related to the other variables by Faraday's law,

$$\text{Plant capacity (g equiv/h)} = 3.73 \times 10^{-8} E \frac{I}{A} nA \qquad (4\text{-}1)$$

and the power consumption is given by Ohm's law as

$$\text{Power (kW)} = 10^{-3} \left(\frac{I}{A}\right)^2 R_p nA \qquad (4\text{-}2)$$

where E = current efficiency
   I/A = current density, A/m^2
     n = number of cell pairs
     A = cross-sectional area of single membrane
    R_p = resistance of unit area, i.e., product of resistance and area of one cell pair, $\Omega \cdot \text{m}^2$

(a) What is the advantage of the cell geometry shown in Fig. 4-30 as opposed to a system with a single juice channel and a single KOH channel? (b) What size apparatus should be used to process 3.6 m^3/h of grapefruit juice if the feed temperature and acidity and the product acidity are all the same as in the pilot run and the current density, channel widths, and cell velocities are held the same so as to hold the same current efficiency? What will be the voltage and power requirements?

SOLUTION (a) The layout of the electrodialysis stack in Fig 4-30 superficially resembles that of a multistage separation process. Closer inspection reveals that the feedstreams pass through in channels parallel to each other with no sequential flow between channels. There is no purity advantage gained over a simple single-stage single-channel process.

With the arrangement in Fig. 4-30 the electric current passes through all channels in series between electrodes. Thus it may seem that this is an instance where the separating agent (the electric current) is used over and over in each juice channel and that a savings in electric power has been accomplished in a way similar to the saving of steam in a multiple-effect evaporator. Although the current does pass through each channel in series and is reused in that sense, the resistance of the stack increases in direct proportion to the number of cell pairs. Hence, by Ohm's law, the voltage drop necessary to obtain the desired current density increases in direct proportion to the number of cell pairs. Consequently, the wattage requirement $nI^2 R_p /A$ is directly proportional to the number of cell pairs, and there is no apparent saving in electric power gained by using the stack geometry. There is no saving in total membrane area either.

The main advantage of the process of Fig. 4-30 as opposed to a single-channel system is one of structural convenience. With this geometry the separation device can remain reasonably compact, and each flow channel takes the form of flow between flat plates, each of which is a membrane surface. This gives a high membrane area per unit volume and a low electrode area.

(b) As a first step, it is helpful to consider which of the quantities given in the problem statement are truly independent variables. Applying the description rule, we find that the following nine variables can be set by construction or by manipulation during operation:

Number of cell pairs	KOH feed rate
Area/membrane	Membrane spacing in juice channels
Feed temperature	Membrane spacing in KOH channels
Feed acidity	Voltage
Production rate (= feed rate)	

The current, current density, product acidity, current efficiency, and energy consumption are dependent variables in operation.

Considering now the stated problem, we find that the current density, feed acidity, feed temperature, both channel widths, and both cell velocities (seven variables) are fixed at the values for the pilot run. This should hold the current efficiency the same as in the pilot run (dependent variable). The production rate is set, as is the product acidity (a separation variable). This fixes nine variables and hence defines the process. The number of cell pairs, the area per membrane, and the voltage are all dependent variables, having been replaced by the current density, product acidity, and juice channel velocities as independent variables.

The number of cell pairs is determined simply from the production rate, juice velocity, and juice channel spacing. Since the juice velocity and juice channel spacing are to remain at the values for the pilot run, the number of cell pairs must increase in proportion to the juice throughput. Hence the number of cell pairs must be 120.

Since I/A in Eq. (4-6) remains constant and the capacity and n both increase tenfold, A must remain unchanged at 0.88 m² per membrane. Hence the full-scale apparatus must contain 120 cell pairs of the type in the pilot apparatus. One cannot decrease the number of cell pairs and increase the area per membrane to provide the same citric acid removal without increasing either the juice velocity or the juice channel width.

The resistance of the stack increases tenfold due to the greater number of cell pairs with the total current remaining the same. Hence the applied voltage must increase tenfold to 1670 V and the energy requirement also increases tenfold, thereby remaining at 209 MJ/m³. At a power cost of 3 cents per kilowatthour the dc power cost is only \$1.74 per cubic meter of juice; however, the pumping costs and equipment amortization costs are likely to be substantially higher. □

Electrodialysis has been most extensively developed as a process for removing dissolved salt contaminants from seawater or a brackish ground water. Figure 4-31 shows an electrodialysis unit in service for desalting water in Kuwait. The assembly of membranes and flow channels is similar to that shown schematically in Fig. 4-30 except that the membranes and flow channels are horizontal in each unit rather than vertical, as in Fig. 4-30. Also, it is necessary to use membranes that are selectively permeable to cations in the process, as well as others that are selectively permeable to anions (see Prob. 4-F).

Figure 4-31 A 910 m³/day electrodialysis water-desalting plant located in Kuwait. *(Ionics, Inc., Watertown, Mass.)*

FIXED-BED (STATIONARY-PHASE) PROCESSES

Achieving Countercurrency

When a solid phase is involved in a separation process, e.g., adsorption, ion exchange, leaching, and crystallization, it is difficult to design a contacting device which will give continuous countercurrent flow of the phases. Some sort of drive is required to make a bed of solids move continuously, and even then it is very difficult to avoid attrition of the solid particles, to keep the solids in a uniform plug flow, and to avoid channeling the fluid phase through cracks which develop in the bed. In a few cases, e.g., moving-bed ion exchange, these problems have been solved sufficiently to enable continuous countercurrent processes to be built without incorporating a mechanical conveyor for the solids, but such cases are rare.

The more common approach for truly continuous countercurrent contacting with solids uses a helical or screw-type mechanical conveyor to transport the solids in a vertical or sloped device. One large-scale example is the DdS slope diffuser, used for extraction of sugar from sugar beets (McGinnis, 1969). The separation process is actually one of combined *leaching* and *dialysis*. The term "leaching" implies that the valuable component (sugar) is washed away from the solid into a liquid phase. The term "dialysis" refers to the selective action of the cell-wall membranes, which allow sugars to pass while retaining substances of much larger molecular weight and colloids. This device uses a covered trough, sloping at about a 20° angle to the horizontal and typically measuring 4 to 7 m in diameter and 16 to 20 m long. Beet slices (*cossettes*) are conveyed upward by a perforated-scroll motor-driven carrier, while hot water enters at the upper end and flows downward. The countercurrent action makes it possible to reach a sugar content of about 12 percent in the exit solution. So high a concentration would not be possible without the countercurrent design.

Mechanical conveyance of solids is also used in various designs of continuous countercurrent crystallizers, of which the Schildknecht type of column is typical (Betts and Girling, 1971). For crystallizations involving eutectic-forming systems, such as *p*-cresol–*m*-cresol (Fig. 1-25), the column serves primarily to wash concentrate away from otherwise pure crystals of the solid phase. In some cases removal of occluded concentrate (surrounded by crystal structure) is also necessary and requires continual melting along the column. For crystallizations involving a solid solution (as shown for the Au–Pt system in Fig. 1-27) it is necessary to create a temperature gradient along the crystallization column and to provide sufficient residence time to allow for continual melting and recrystallization.

Most large-scale applications of separation processes involving solids either accept the less efficient contacting afforded by simple fixed-bed operation (see Chap. 3) or contrive a closer approach to countercurrency while still keeping the fixed-bed geometry. In this way problems of solids attrition, channeling, and mechanical complexity are minimized.

The longest-established approach for gaining countercurrency while keeping the fixed-bed geometry is a rotating progression of beds into various positions, sometimes

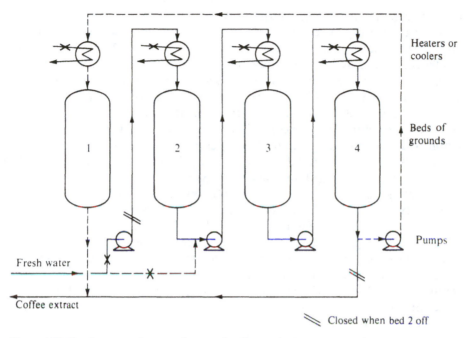

Heaters or coolers

Beds of grounds

Pumps

Fresh water

Coffee extract

Closed when bed 2 off

Figure 4-32 Shanks system for manufacture of coffee extract.

known as the *Shanks system*. A good example of this is the extraction of coffee, as a first step toward the manufacture of instant coffee (see Prob. 14-K). This again is a process of combined leaching and dialysis, where soluble coffee matter is dissolved out of roast and ground coffee beans into hot water (Moores and Stefanucci, 1964). Since the water in this extract is subsequently removed by evaporation or freeze-concentration, followed by spray-drying or freeze-drying, there is a large incentive to obtain as concentrated an extract as possible without impairing flavor through undesirable reactions in concentrated solutions. In order to obtain this high extract concentration, countercurrent contacting of roast and ground coffee and extract is used. As shown in Fig. 4-32, beds of roast and ground coffee are filled in numerical order: bed 1 first, bed 2 second, bed 3 third, bed 4 fourth, then bed 1 again, etc. While bed 1 is off line for emptying and refilling, fresh hot water is passed into bed 2. Solution leaving bed 2 is pumped into bed 3, that leaving bed 3 is pumped into bed 4, and that leaving bed 4 is taken as coffee-extract product. This flow pattern is shown by the full lines in Fig. 4-32. When bed 1 returns to operation and bed 2 is taken off line, the pumping sequence changes to that shown by the dashed lines in Fig. 4-32. The fresh water now enters bed 3 and passes successively to bed 4 and then bed 1. Extract product is now withdrawn from bed 1. In this way water always contacts the most depleted coffee grounds first and contacts the freshest grounds last. The scheme thereby achieves the benefits of countercurrent flow without actual physical movement of the grounds within the beds. The process can obviously be extended to any number of beds.

Extracts containing 30 percent or more coffee solubles are obtained in this way. The percentage recovery of coffee solubles from the grounds also has obvious economic importance for such a comparatively valuable product. Because of that, the fresh water contacting the most depleted grounds is heated to 155°C or higher and the bed is run under the corresponding pressure in order to hydrolyze hemicelluloses into water-soluble materials. About 35 percent of the roast and ground coffee is put into the extract as a result. The heat exchangers before the remaining beds in series give additional degrees of freedom for controlling temperatures at various points in the extraction; their proper manipulation has much to do with product flavor. Other flavor controls are associated with the fineness of the grind, the ratio of coffee to water, and the contact time (Moores and Stefanucci, 1964).

The Shanks system of rotating bed positions has also been used for ion exchange (Vermeulen, 1977) and adsorption. It was used for sugar-beet extraction before the introduction of the slope diffuser. Belter et al. (1973) describe an ion-exchange process for recovering the pharmaceutical novobiocin from fermentation broths. Because the broth contains suspended solid matter that would plug a fixed ion-exchange bed, a fluidized bed of ion-exchange resin is used. Since fluidization causes intense mixing of the resin, the absence of axial mixing typical of a fixed bed is lost, and to compensate for this inefficiency a Shanks rotation of the fluidized ion-exchange beds is used.

A newer but quite successful method for gaining countercurrency with a fixed bed is shown diagrammatically in Fig. 4-33. Here a fluid stream is circulated continuously down through a fixed bed, and the fluid leaving the bottom is pumped back up and into the top. This gives the same effect as if the bed were shaped like a torus, with the bottom connected directly to the top. A rotary valve turns in the direction shown to move the various fluid inlet and outlet points along the bed in a regular progression at predetermined time intervals. At the time shown in Fig. 4-33 the feed mixture is put in at position 8, joins the recirculating flow, and proceeds down the bed. If the separation process is adsorption, the preferentially adsorbed component(s) are held on the bed between points 8 and 11, and hence a raffinate stream rich in the nonadsorbed component(s) can be withdrawn at point 11. Meanwhile, the portion of the bed that was in this adsorption service some time previously is being regenerated by feeding a desorbent stream at point 2. The desorbent is a substance which displaces the adsorbed component(s) from the feed mixture off the bed and back into the circulating fluid stream. Since this displacement occurs between points 2 and 5, an extract stream enriched in the preferentially adsorbed component(s) can be withdrawn at point 5.

Countercurrent flow of the fluid and the solids is achieved by the regular rotation of the valve, which gives the effect of moving the bed upward. For example, at the next turn of the valve, feed enters at point 9, raffinate leaves at point 12, desorbent enters at point 3, and extract leaves at point 6. Then we proceed to points 10, 1, 4, and 7; etc. This is equivalent to moving the bed upward one position per turn of the valve. Although this procedure was developed only recently, it has already seen considerable large-scale application for separations of n-paraffins from branched and aromatic hydrocarbons (manufacture of jet fuel), for separation of p-xylene from mixed

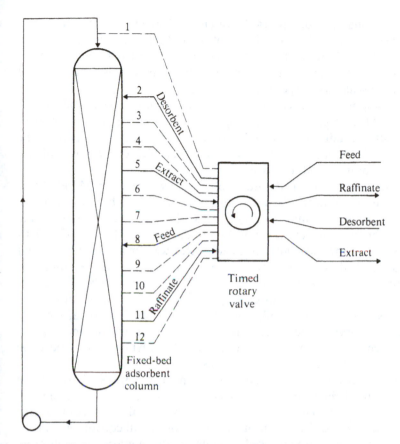

Figure 4-33 Rotating feeds to a fixed bed to achieve the effects of countercurrency. *(Adapted from Broughton, 1977, p. 50; used by permission.)*

xylenes, and for separations of olefins from paraffins in the C_8 to C_{18} range (Broughton, 1977), all carried out by adsorption with molecular sieves.

CHROMATOGRAPHY

The name *chromatography* encompasses a host of different separation techniques with two common features:

1. A *mobile phase* flows along a *stationary phase*. Constituents of the mobile phase enter or attach themselves to the stationary phase to different extents. The greater the fraction of the time that a component spends with the stationary phase, the more it will be retarded behind the average flow velocity of the mobile phase. Different components are retarded to different extents, and this constitutes a separation.
2. At least one of the phases is *very thin*, so that its rates of equilibration with the other phase are quite rapid. As we shall see below and in Chap. 8, this gives the separative action of a large number of stages in series.

Since one phase is stationary, chromatography often involves a fixed bed of particles, which are sometimes coated with a thin layer of separating agent. Because of the thin-layer feature and the dilution that accompanies the most common forms of chromatography, it is more common in laboratory-scale separations, e.g., chemical analyses, than in large-scale plant operation.

Since different components are retarded to different extents in the mobile phase, suitable monitoring of the process can accomplish a separation of a multicomponent mixture into individual products, or signals, for the individual components. The ability to do this in a single operation makes chromatography a very powerful and efficient method of separation and analysis.

The historical development of chromatography has been traced by Mikeš (1961) and Heines (1971), among others. The technique was developed by Tswett in the early 1900's, and was named chromatography because the initial use was to separate plant pigments of different colors. However, it did not develop rapidly until the 1940s and 1950s, when some of the most important advances were made by Martin and Synge, in work that resulted in the 1952 Nobel prize. They developed the concept of partition chromatography and implemented it in the forms that have become countercurrent distribution, paper chromatography, and thin-layer chromatography. Subsequently, James and Martin (1952) introduced gas-liquid chromatography, which has become the workhorse for chemical analysis of gases and organic liquids. Other methods of chromatography are the principal analytical and separative methods for biological and biochemical substances.

Methods of chromatography can be categorized in various ways; Pauschmann (1972) has done this by defining processes as combinations of three different elements: *migrations*, defined by the different rates of travel of different components with the mobile phase; *shifts*, defined as motions transporting all constituents at the same rate; and *gradients*, which are imposed to influence the migration or shift rates. It is also common to classify chromatographic separations in terms of three different process techniques (elution development, displacement development, and frontal analysis), following the early work of Tiselius (1947).

Elution development is shown in Fig. 4-34. A pulse of sample is introduced at the inlet end of the column bearing the stationary phase. A solvent or gas, known as the *carrier* or eluant, flows through the column, conveying the constituents in the sample pulse. The different constituents transfer back and forth between the mobile phase (carrier) and the stationary phase and are retarded according to the fraction of time they spend in the stationary phase. In Fig. 4-34 component A is held in the stationary phase longer than component B. Because of mass-transfer limitations and other factors the peaks for the different components broaden as well as separate as they go along the column. Some form of detector is used to monitor the peaks as they emerge from the column in the carrier stream. Elution development is the most common form of chromatography and is discussed at greater length below.

Displacement development is shown in Fig. 4-35. It is similar to elution development in that a sample pulse is injected at the inlet of the stationary-phase column. However, in this case one or more solvents are used which are more strongly held by the stationary phase than the components of the sample pulse. Hence the solvent(s)

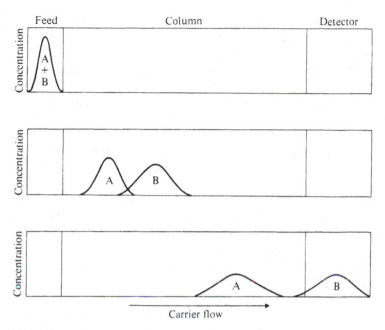

Figure 4-34 Elution-development chromatography. *(From Stewart, 1964, p. 418; used by permission.)*

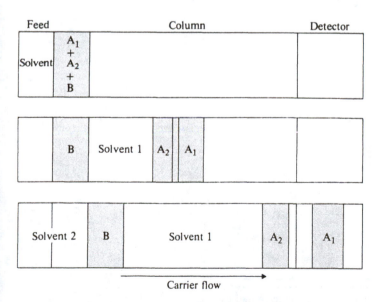

Figure 4-35 Displacement-development chromatography. *(From Stewart, 1964, p. 419; used by permission.)*

displace the sample constituents in the order of the strengths with which they are held into the stationary phase, weakest first. For the example shown in Fig. 4-35, solvent 1 is more strongly held in the stationary phase than components A_1 and A_2 but is less strongly held than component B, which stays behind solvent 1. Subsequent use of a second solvent (solvent 2), more strongly held than component B, will then displace component B and drive it along the stationary phase later. Solvent 1 will still tend to displace and drive component A_2 along the column, and component A_2, in turn, will displace and drive component A_1. To avoid overlap of the peaks for A_1 and A_2, it would be useful to use yet another solvent, held with a strength intermediate between components A_1 and A_2, for a period of time before solvent 1. This would then separate the peaks for A_1 and A_2 more completely.

The process of zone refining (Prob. 3-N) is akin to displacement-development chromatography. In it a molten zone (mobile phase) passed along a solid bar serves to displace forward the component(s) that are least accommodated into the solid phase (stationary phase) as it forms again behind the molten zone.

Frontal analysis, shown in Fig. 4-36, may be regarded as an "integral" form of elution development; i.e., the peak response curves for elution development are the derivatives of the response curve for frontal analysis. Here the mixture to be separated is introduced as a step change, rather than a pulse; the mixture continues to flow in after the change to it as feed has been made. In Fig. 4-36 component B is more strongly held in the stationary phase than component A, so that A proceeds ahead of B. The technique gives a zone of purified A, but because of the sustained feed of the mixture cannot give a zone of B free of A. A B-rich zone would be obtained at the tail if the feed of the mixture were discontinued; such a process would be a hybrid of elution development and frontal analysis.

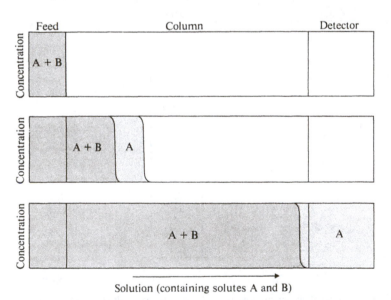

Figure 4-36 Frontal-analysis chromatography. *(From Stewart, 1964, p. 419: used by permission.)*

Although the frontal-analysis technique can isolate only the least strongly held component, it has the advantages of a much greater feed-throughput capacity and less dilution of the products. It is therefore more suitable for scale-up to large capacities. The simple fixed-bed operation described in Chap. 3 may be viewed as a subcase of frontal analysis, as may the rotating-bed and rotating-feed methods, described above. *Electrophoresis* (separation by differential migration of charged particles or macromolecules in an electric field) can be operated effectively in a frontal-analysis mode by utilizing a so-called leading electrolyte. This will isolate a zone of the most rapidly moving component. A further improvement is the use of a tailing electrolyte, which will then work in the mode of displacement development and give a full separation. That process is known as *isotachophoresis* (Everaerts et al., 1977).

Means of Achieving Differential Migration

The mechanism by which the components enter the stationary phase from the mobile phase must be *strong enough* to slow the travel of the components significantly. It must be *selective*, to slow different components to different extents. It must also be *reversible* so that the components readily reenter the mobile phase. For most effective performance, the distribution coefficients between phases for the components of interest should be of order unity rather than very high or very low. The following are some of the mechanisms which have been used:

1. *Partition*, in which a thin layer of liquid is coated onto particles or a solid surface, and the mobile phase is either gaseous (gas-liquid chromatography, GLC) or another immiscible liquid.
2. *Adsorption*, in which the stationary phase is a solid adsorbent and the mobile phase is gas or liquid. Alternatively, adsorption to a gas-liquid interface can be used, either a foam or interstitial liquid being the stationary phase and the other (interstitial liquid or foam) being the mobile phase.
3. *Ion exchange*, in which the stationary phase is particles of ion-exchange resin or of a solid coated with an ion-exchanging liquid and the constituents to be separated are in liquid solution in the mobile phase and bear the opposite charge.
4. *Sieving*, where the stationary phase contains pores of molecular dimensions and effects a separation of substances in the mobile phase on the basis of size. An example is gel-permeation chromatography, where smaller molecules enter the stationary gel phase to a greater extent and are therefore retarded more in the mobile phase.
5. *Reversible chemical reaction*, where the components in the mobile phase to be separated react to different extents (but reversibly) with elements of the stationary phase. An example is affinity chromatography, where certain chemical ligands are immobilized on the stationary phase and react selectively with the species to be separated, typically enzymes.
6. *Membrane permeation*, where substances are separated on the basis of their ability to permeate through a thin membrane material or their ability to react with something encased within the membrane. In one successful implementation of this using the concept of *microencapsulation*, a selective membrane coats small particles whose interiors contain absorbing or reacting solutions (Chang, 1975).

Countercurrent Distribution

The process known as *countercurrent distribution* (CCD) is a form of elution chromatography in which the mobile phase is transferred at discrete times rather than continuously. It has been used extensively in biochemical research. The adjective "counatercurrent" here is something of a misnomer since the second phase remains stationary rather than flowing countercurrent to the mobile phase. The process is illustrated by the following example.

Consider the solvent-extraction process shown schematically in Fig. 4-37. The sequence of four contacting vessels is numbered 1, 2, 3, and 4, from left to right. At the start of the operation 100 mg of substance A and 100 mg of substance B are dissolved in 100 mL of aqueous solution in vessel 1. Then 100 mL of an immiscible organic solvent is added to vessel 1, which is then shaken long enough to bring the two liquid phases to equilibrium. We shall presume that the equilibrium distribution ratio $K'_{i, o-w}$ for substance A is 2.0, while that for substance B is 0.5, where

$$K'_{i, o-w} = \frac{C_{io}}{C_{iw}} \tag{4-3}$$

C_{io} is the concentration (milligrams per milliliter) of component i in the organic phase, and C_{iw} is the concentration of component i in the aqueous phase. Thus substance A is more readily extracted than substance B into the organic phase.

After equilibration of vessel 1, 66.7 mg of substance A and 33.3 mg of substance B are in the organic phase, while 33.3 mg of substance A and 66.7 mg of substance B are in the aqueous phase. These figures are obtained by the simultaneous solution of

$$C_{Ao} = 2.0 C_{Aw} \tag{4-4}$$

$$C_{Bo} = 0.5 C_{Bw} \tag{4-5}$$

$$V_o C_{Ao} + V_w C_{Aw} = 100 \tag{4-6}$$

$$V_o C_{Bo} + V_w C_{Bw} = 100 \tag{4-7}$$

where V_o and V_w are the known volumes in milliliters of the organic and aqueous phases, respectively.

At this point the organic phase from vessel 1 is transferred by decantation into vessel 2. Then 100 mL of fresh aqueous phase (containing no A or B) is added to vessel 2 and 100 mL of fresh organic solvent is added to vessel 1. Both vessels are then shaken and equilibrated. By application to each vessel of Eqs. (4-4) and (4-5) and of Eqs. (4-6) and (4-7) with the right-hand side modified to give the total amount of the solute in that vessel, we find the distribution of the two solutes indicated by the figures given in the row marked "Equilibration 2" in Fig. 4-37.

Next we transfer the organic phase from vessel 2 to vessel 3 and transfer the organic phase from vessel 1 to vessel 2, adding fresh aqueous phase to vessel 3 and fresh organic solvent to vessel 1. Again the vessels are shaken and brought to equilibrium, and the concentration distributions indicated under "Equilibration 3" in Fig. 4-37 result. Following the same procedure, the organic phases are then trans-

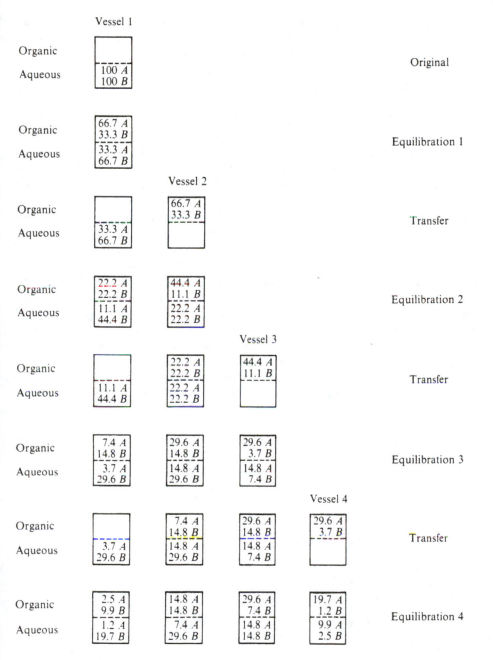

Figure 4-37 Four-vessel countercurrent-distribution process (CCD) for separating substances A and B.

ferred one vessel to the right, fresh aqueous and organic phases are added to the end vessels, and the vessels are once again equilibrated. The final solute concentrations in each vessel are shown in the row for "Equilibration 4."

The net result of this procedure has been to provide a better degree of separation between A and B than would be achieved by a simple one-stage equilibration. As we have seen for equilibration 1 in Fig. 4-37, a single equilibration would have given 66.7 mg of A and 33.3 mg of B in one product and 33.3 mg of A and 66.7 mg of B in the other product. If we compound the contents of vessels 1 and 2 as one product in our four-vessel scheme and compound the contents of vessels 3 and 4 as the other product, we find that one product contains 74 mg of A and 26 mg of B, while the other product contains 26 mg of A and 74 mg of B. A greater resolution of A and B has been achieved than is possible in the one-stage equilibration.

Note also that this degree of separation was achieved even before the final equilibration step. Hence we have effectively a three-stage equilibrate-and-transfer batch separation. A still better separation is achievable if we use more vessels and have more successive equilibrate-and-transfer steps. Figure 4-38 shows the solute concentration as a function of vessel number calculated by Craig and Craig (1956) for a separation of two substances having $K_A' = 0.707$ and $K_B' = 1.414$ when 100 vessels and 100 successive steps are used with equal phase volumes. Here about 96 percent of either component can be recovered in a purity of 95 percent on a binary basis. Another approach for improving the separation achieved is to seek a solvent which gives a higher ratio of distribution ratios for the components to be separated.

Figure 4-39 shows the separation obtained experimentally for a mixture of fatty acids (Craig and Craig, 1956). An aqueous phase 1.0 M in phosphate ion with a pH of

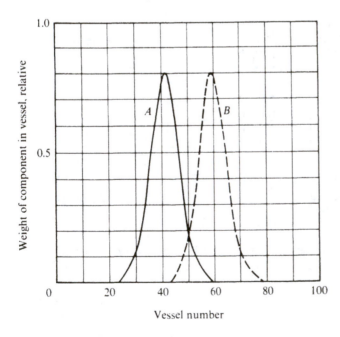

Figure 4-38 Separation of components with $K_A' = 0.707$ and $K_B' = 1.414$ using 100 vessels. *(From Craig and Craig, 1956, p. 194: used by permission.)*

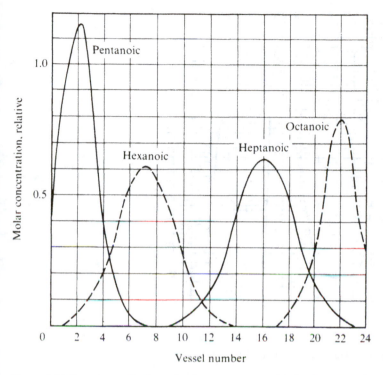

Figure 4-39 Experimentally determined countercurrent distribution of a mixture of fatty acids. (*From Craig and Craig, 1956, p. 267; used by permission.*)

7.88, a solvent of isopropyl ether in the amount of 12 cm^3 to 8 cm^3 of aqueous phase in each vessel, and 24 transfers in 24 vessels were used to separate a multicomponent mixture of pentanoic, hexanoic, heptanoic, and octanoic acids.

Several refinements of the basic pattern of countercurrent distribution presented above (Craig and Craig, 1956; Post and Craig, 1963) involve product removal (the contents of one vessel at a time) at one or both ends of the cascade. Quite elaborate laboratory devices have been developed to allow CCD to be carried out in automatic fashion with a large number of vessels.

Gas Chromatography and Liquid Chromatography

Liquid chromatography is very similar in principle to countercurrent distribution except that the mobile liquid phase flows continuously along a continuous length of stationary phase (the *column*) instead of being transferred at discrete intervals between discrete stages. Analytical liquid-chromatography instruments use quite high pressures to accomplish the flow of the mobile liquid phase along the stationary column, which gives a high pressure drop because of the thin-dimension aspect of chromatography. The thin layer of stationary phase also makes for a much more compact apparatus than the large sequence of batch extractors employed in a CCD

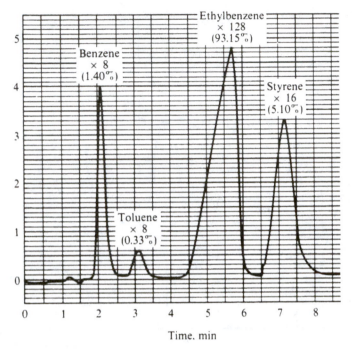

Time, min

Figure 4-40 Typical chromatogram from gas-liquid chromatography.

setup. One result is that liquid chromatography is displacing CCD in routine analytical and research use as further improvements in the design of liquid chromatographs are made.

Gas chromatographs have been in routine analytical and research use since about 1950. For the stationary phase they use a packed bed of solid adsorbent, e.g., molecular sieve, or of liquid coated onto particles of an inert carrier, e.g., firebrick. Alternatively, a capillary column is used, with the stationary liquid phase coated onto the inner wall of the capillary. Capillary columns give an excellent separation but require very small sample volumes and carrier-gas flows.

Figure 4-40 shows a typical chromatogram obtained for a feed mixture containing 1.40% benzene, 0.33% toluene, 93.15% ethylbenzene, and 5.10% styrene. This separation was obtained in a commercial gas chromatograph analytical instrument, using helium as the carrier gas with a 1.5-m-long 6.4-mm-diameter column containing 0.25 wt $\%$ diphenyl ether (liquid) deposited on 170/230-mesh glass beads. The column was maintained at 95°C, and the carrier flow was 16 mL/min. The various substances are detected in the exit carrier gas through measuring the thermal conductivity of the gas. Hence the ordinate is proportional to the thermal conductivity of the exit gas stream and is plotted in Fig. 4-40 against time following the pulse injection of feed. Notice that the components correspond to different *peaks* and appear in the order of ascending solubility (tendency to enter the liquid). Benzene is least soluble and styrene is most soluble. The size (area) of a peak is proportional to the amount of that component in the feed.

The width of a peak corresponds to the degree of nonuniformity of residence time of that component in the column and is related to mass-transfer and axial-dispersion characteristics. The effect of these phenomena on peak spreading is considered in Chap. 8. It should be noted, however, from Fig. 4-40 that chromatography can give a nearly complete separation of the components of a feed.

Gas chromatography and liquid chromatography both require some sort of detector to sense peaks in the effluent carrier. Thermal-conductivity detectors are one kind, used in simple gas chromatographs. Helium is generally used as a carrier with such detectors since it differs considerably in thermal conductivity from other gaseous substances. Other, more sensitive and/or more specific detectors are based on measurement of ionization in a hydrogen flame, electron-capture properties, infrared or ultraviolet absorption, etc.

Retention Volume

A simple expression can be derived for the rate of migration of the peak for a component along the stationary phase in elution chromatography, provided the equilibrium relationship for partitioning solute between the stationary and mobile phases is a simple linear proportionality and diffusional effects within the phases are negligible (Stewart, 1964). This result is independent of the model used to describe peak broadening and applies to CCD with a large number of stages, as well as to elution chromatography. Let K_i = ratio of gas-phase mole fraction to liquid-phase mole fraction at equilibrium, for component i. If M_g represents the moles of carrier gas (void volume) per unit column volume and M_l represents the moles of liquid per unit column volume, the ratio of the effective velocity (u_i) of component i through the column to the actual velocity (u_G) of the carrier gas is given by

$$R_i = \frac{u_i}{u_G} = \frac{M_g}{M_g + (M_l/K_i)} \tag{4-8}$$

Thus a smaller K_i (greater solubility) leads to a smaller R and hence a slower passage of component i through the column. R_i is the ratio of the amount of time spent by component i in the mobile gas phase to the sum of the times spent in the mobile phase and in the stationary phase. Equation (4-8) can also be written in terms of phase volumes and a concentration-based distribution coefficient.

The *retention volume* of component i, the reciprocal of R_i, is the number of column void volumes of carrier that must pass through in order for the peak for component i to appear.

Paper and Thin-Layer Chromatography

Another type of elution chromatography is *paper chromatography*, shown schematically in Fig. 4-41. Here the feed is placed as a spot on a piece of moist, porous paper. The end of the paper is then placed in a solvent (the *developer*), which will rise through the paper by capillary action. Depending upon the partitioning of different feed constituents between the solvent and the aqueous phase held by the paper, the

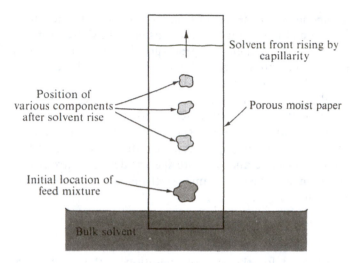

Figure 4-41 Paper chromatography.

constituents will be transported to different extents along the paper by the rising solvent, and different solutes will show up as different spots on the paper, perhaps with characteristic colors. A variant of paper chromatography is *reversed-phase chromatography*, where an organic phase is held in an appropriate paper, and water passes upward.

Thin-layer chromatography (Stahl, 1969) evolved from paper chromatography and is now more commonly used. A thin layer (often of silica gel) on a solid plate takes the place of the impregnated paper. A further improvement is *programmed multiple development* (PMD), which enhances and sharpens the separation of spots or peaks by cycling the solvent up and down along the thin layer. This is accomplished by using a volatile solvent and altering the pressure so that periods of solvent rise are interspersed with periods of solvent evaporation (Perry et al., 1975).

Variable Operating Conditions

The concept of *temperature programming* is sometimes used in elution gas chromatography when the feed mixture contains substances with widely different partition coefficients into the stationary phase. The temperature is changed (usually increased) in a predetermined manner with time after introduction of the feed sample. This makes it possible for there to be good resolution of the more volatile constituents at early times when the temperature is low enough to give values of K_i of order unity for them. The less volatile substances then pass through the column with good resolution at later times when the temperature has risen enough to make values of K_i for them become of order unity.

A related technique is used to separate different molecular-weight fractions of a polymer. The polymer mixture is put onto beads in the column, and then a solvent mixture is passed through, with a composition that varies with respect to time, e.g.,

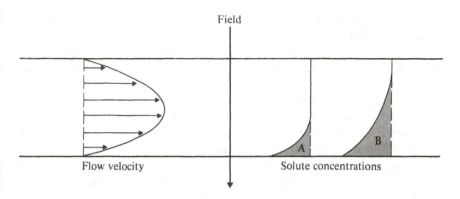

Figure 4-42 Field-flow fractionation (polarization chromatography).

with increasing amounts of methyl ethyl ketone added to ethanol. The solvent power for the polymer thereby becomes better as time goes on, and consequently fractions of different molecular weight elute at different times, the molecular weight of the fraction increasing with time. It is also possible to impose a temperature gradient along the column or to program the temperature over time (Porter and Johnson, 1967). The procedure can be considered to fall in the category of frontal analysis if the entire column is coated with polymer at the start.

Field-Flow Fractionation (Polarization Chromatography)

Different rates of flow of fluid filaments, caused by a velocity gradient, can give an effect analogous to that of the mobile and stationary phases in chromatography. Such a process, shown in Fig. 4-42, has been given the generic names of *field-flow fractionation* by Giddings and coworkers (Grushka et al., 1974; Giddings et al., 1976; etc.) and *polarization chromatography* by Lightfoot and coworkers (Lee et al., 1974; etc.). A fluid moves in laminar flow within a tube or between parallel plates, giving a parabolic velocity profile. A sample of the mixture to be separated is put in the flow near a wall. A field of some sort which promotes transport is imposed transversely across the flow. The components of the sample tend to diffuse outward into the main flow and are also transported back toward the wall by the field. The amount of spreading into the main flow depends upon the ratio of the molecular-diffusion coefficient to the transport coefficient responsive to the imposed field. In Fig. 4-42 component A has a low value of this ratio and concentrates close to the wall; component B has a higher value of this ratio and moves more into the central flow, away from the wall. The concentration level as a function of distance from the wall is depicted as the shaded areas to the right of the curves in the figure.

Since the parabolic velocity profile makes component B move on the average faster than component A in the direction of flow, component B will pass out the end of the flow channel as a peak before the peak for component A. Thus the faster central flow takes the place of the mobile phase, and the slower wall-region flow takes the place of the stationary phase.

Several different transverse fields have been suggested and used, including electric fields (Caldwell et al., 1972; Lee et al., 1974; Reis and Lightfoot, 1976), gravitational fields imposed by centrifugal force (Giddings et al., 1975), temperature gradients (Myers et al., 1974), and transverse fluid flow (ultrafiltration) through permeable walls (Lee et al., 1974; Giddings et al., 1976). They cause a separation based upon differences in the ratio of diffusivity to electrophoretic mobility, the ratio of diffusivity to sedimentation velocity, the ratio of diffusivity to thermal diffusivity, and the diffusivity itself, respectively. It should be noted that these are all rate coefficients and that the "mobile" and "stationary" phases are miscible. Hence these separations fall in the class of *rate-governed* separations, whereas nearly all ordinary chromatographic methods are *equilibration* separations.

Lee et al. (1974) have noted the probable benefits of a tubular geometry over the slit geometry for resolving power, removal of ohmic heat, and control of natural convection. For any geometry it is necessary to preserve stable laminar flow and minimize mixing effects. This complicates scale-up greatly, making the technique most suitable for laboratory analyses and small-scale separations.

Uses

Chromatographic techniques have numerous uses, which can be categorized as follows (Stewart, 1964):

1. *Analysis*
    a. *Identification*, or qualitative analysis, of components of a mixture, on the basis of coincidence of residence times on different columns or other methods
    b. *Quantitative analysis*, by comparing peak sizes with those from known standards of the same substance
    c. *Separation*, as a prelude to other analytical techniques, notably mass spectrometry
    d. *Test of homogeneity*, to see if extraneous peaks are present, in addition to that for the main substance; the term *chromatographically pure* is used to imply the absence of extra peaks
2. Preparation
    a. *Isolation* of a compound from a mixture in small quantities
    b. *Concentration* of substances taken up on the stationary phase and then eluted with a more selective solvent as carrier
3. Research
    a. *Partition coefficients*, determined from retention volumes
    b. *Diffusion coefficients*, determined from peak spreading if proper consideration is given to other, competing broadening mechanisms
    c. *Reaction kinetics*, using a chromatograph for continuous removal of products from one another during reaction in the chromatograph

The use of chromatography for analysis is particularly powerful, as is evidenced by such complicated analyses as determining which crude oil is the source of an oil spill, separation and identification of complex amino acid mixtures, and separation and identification of hundreds of trace volatile flavor and aroma components in the vapor over such foods as orange juice and coffee.

Continuous Chromatography

Elution and displacement chromatography, as described above, are suited for separation of quite small samples on a batch basis. The attempts to turn chromatography into a continuous separation method fall into several categories:

1. *Moving the stationary phase mechanically*, e.g., making the stationary phase a slowly rotating annulus, with different components emerging at differential circumferential positions at the end of the bed (see, for example, Martin, 1949; Sussman, 1976)
2. *Achieving countercurrent flow* of the mobile and "stationary" phases while retaining the thin-layer concept
3. *Utilizing a second separating agent* to accomplish differential migration in a second direction

Expanding upon the second of these, Ito et al. (1974) review methods based upon causing countercurrent flow in liquid chromatography by methods such as a rotating helical coil, which causes phases to flow in different directions depending upon density. Foams have been used for countercurrent chromatographic separations, using either natural drainage (Talmon and Rubin, 1976) or a rotating helix (Ito and Bowman, 1976) to achieve countercurrency. Continuous electrophoresis has been achieved by counterflowing solvent against the direction of transport caused by the electric field (Wagener et al., 1971).

A countercurrent version of the CCD process has been developed (Post and Craig, 1963), in which the upper phase is still moved one vessel to the right in each transfer step but now the lower phase is also moved one vessel to the left at the same time. In this way there is a net flow of the bottom phase to the left and a net flow of the upper phase to the right as time goes on, and the process is entirely analogous to a multistage continuous-flow extraction process in which the immiscible liquid phases flow between stages only at discrete intervals. The feed mixture of components to be separated can be introduced to one or more central vessels in the train at each transfer step, and product can be taken from each end of the cascade at the same time. This modification, known as counter-double-current distribution (CDCD) has a substantially higher throughput capacity for the mixture to be separated than the simple CCD scheme. The operation is shown schematically in Fig. 4-43, where the transfer and equilibration steps alternate during operation.

Liscom et al. (1965) have proposed the counter-double-current contacting scheme for the separation of liquid mixtures by fractional crystallization. In their scheme the liquid in any one of the vessels of the train is partially frozen. The remaining liquid is then decanted into the next vessel to the right while the solid is transferred to the next vessel to the left. The contents of each vessel are then remelted and partially frozen again. Feed is introduced into a central vessel at each step.

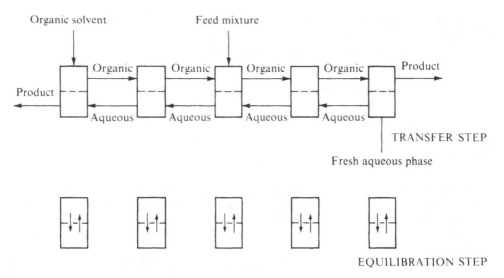

Figure 4-43 Counter-double-current distribution (CDCD) scheme.

Product is withdrawn from either end (solid from the left and liquid from the right). Reflux is achieved by melting a portion of the solid product and returning it as liquid to the end vessel from which it came as solid. Similarly, a portion of the liquid product from the other end of the cascade is frozen and returned to the vessel on that end. In general, reflux is obtained by changing the phase of some of a product stream and returning it to the cascade at the point where that product is withdrawn. Thus for the liquid-liquid extraction form of CDCD, reflux can be obtained (less easily) by removing the solvent from some of the organic-solvent product solution, redissolving the solutes in water, and returning this new solution to the end vessel. Similarly, the water could be removed from the product obtained from the vessel at the other end, the solutes could be redissolved in the organic solvent, and the resulting solution could then be reintroduced at that end.

Figure 4-44 shows two ways chromatography can make use of additional separating agents. As shown in Fig. 4-44a, paper chromatography can be used with two different solvents. After the transport due to one solvent has been accomplished as shown in Fig. 4-41, the paper can be turned 90° crosswise and one of the adjacent edges dipped into another solvent. The result is the displacement of components in two directions, and a pair of components not separated by one solvent can be separated by the other.

Figure 4-44b illustrates the principle of *electrochromatography* (Hybarger et al., 1963; Pucar, 1961), which combines adsorption chromatography and electrophoresis to generate a continuous separation process. Different degrees of adsorption cause the components to move at different net velocities horizontally in the diagram, while different electrical mobilities cause the components to move at different net velocities vertically in response to the transverse electrical field. The net result is a series of

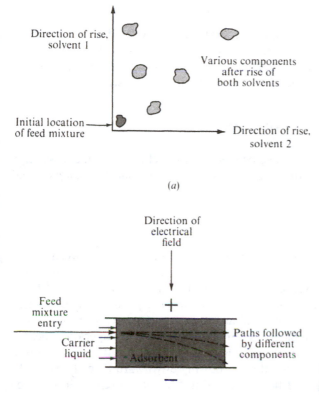

Figure 4-44 Chromatography with additional separating agents: (*a*) paper chromatography with two solvents: (*b*) electrochromatography.

different curved paths for different components, making them appear in the effluent liquids from different cross-sectional locations in the bed. A component must bear a net charge in order to follow a curved path. Such is true, for example, for various amino acids. *Gel electrophoresis* uses gel permeation instead of adsorption in such a process.

Scale-up Problems

Chromatography is inherently difficult to scale up to a production scale for three reasons: (1) the great separating power relies upon one or both phases being thin; (2) elution chromatography and displacement chromatography are inherently batch methods because of the need for handling the feed mixture as a pulse; they also require substantial dilution of the components to be separated with the mobile-phase carrier; and (3) methods relying upon stagnancy or laminar flow, e.g., field-flow fractionation and most electrophoresis configurations, present a problem of stabilizing the system against mixing and/or natural convection upon scale-up. Chromatography with a particulate stationary phase is subject to loss of separating power due to channeling upon scale-up.

Current Developments

Recent developments in chromatographic separations are covered in a number of publications, including the *Journal of Chromatographic Science, Separation Science and Technology*, and *Separation and Purification Methods*, as well as the review series *Advances in Chromatography* and *Methods of Biochemical Analysis*.

CYCLIC OPERATION OF FIXED BEDS

Separations involving fixed beds can also be made using cyclic variation of operating conditions. Two approaches of this type are known as *parametric pumping* and *cycling-zone separation*.

Parametric Pumping

Wilhelm and associates (Wilhelm et al., 1966, 1968; Wilhelm and Sweed, 1968) conceived and demonstrated the possibility of achieving a high degree of separation by means of appropriately phased cycling of fluid flow and temperature within a solid adsorbent bed. This process has been called *parametric pumping*. As illustrated schematically in Fig. 4-45, a single fluid phase is driven alternately up and down within a bed of solid adsorbent particles. In the *direct mode* of operation heat is added to the bed from a jacket while the fluid is flowing upward, and heat is removed from the bed into the jacket while the fluid is flowing downward (Fig. 4-45). In the *recuperative mode* (not shown) fluid flowing up from the bottom reservoir is heated

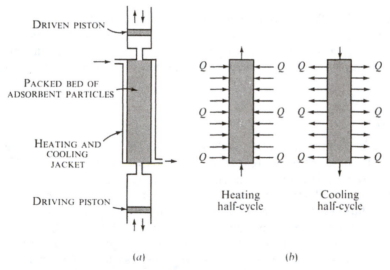

(a)   *(b)*

Figure 4-45 Parametric pumping for separation of a fluid mixture, direct mode. *(Adapted from Wilhelm et al., 1968, p. 340; used by permission.)*

before entering the bed, and fluid flowing down from the top reservoir is cooled before flowing down through the bed. In either case heating causes solutes to enter the fluid, leaving the solid, and cooling causes solutes to enter the solid from the fluid. If heating is coupled with upflow and cooling with downflow, solutes will tend to be "pumped" up to the upper reservoir at rates depending upon their different affinities for the solid phase. Appropriate manipulations of cycle times for flow and temperature can direct some solutes upward and others downward and give quite large separation factors. As in chromatography, the stationary phase can be a liquid coated onto particles rather than an adsorbent. In addition to temperature, the other variable cycled in addition to flow can be pressure (for gases), concentration or pH (for liquids), or an imposed electric field.

Separations obtained during parametric pumping and optimal design have been analyzed by an *equilibrium theory*, formulated by Pigford et al. (1969*a*) and generalized by Aris (1969), assuming local equilibrium between the fluid and stationary phases, a linear equilibrium relationship, and no axial dispersion. This theory can overpredict the ultimate separation obtained after a number of pumping cycles. A better prediction in such cases is given by discrete-stage models (Wankat, 1974) or by models allowing for finite rates of mass transfer and axial mixing.

Parametric pumping, as described, is a batch process. Operation with continuous or semicontinuous feed introduction and product withdrawal is also possible and allows higher throughput capacities; however, the degree of separation is considerably lessened for substantial throughputs. The equilibrium theory has been extended to such systems by Chen and Hill (1971).

Reviews of parametric pumping are given by Sweed (1971) and Wankat (1974).

Cycling-Zone Separations

Simple fixed-bed operation, in which an on-line period is followed by an off-line regeneration period, may be regarded as a cyclic operation. If a fixed bed is kept on line and is subjected to cyclic heating and cooling at a frequency related to its solute capacity, the bed will take up solute from the feedstream during the cold portion of the cycle and release solute back to the fluid during the hot portion of the cycle. Thus an effluent stream can be obtained which alternates between being leaner in solute and richer in solute than the feed. Diverting this effluent stream to different receivers at the appropriate time then gives a semicontinuous flow of enriched product and another of depleted product.

Such a process is named *cycling-zone adsorption* (Pigford et al., 1969*b*) when the fixed bed is an adsorbent. Two methods of implementing it are shown in Fig. 4-46. In the *direct-wave* mode (Fig. 4-46*a*) the bed is heated and cooled by a jacket, while in the *traveling-wave* mode (Fig. 4-46*b*) the fluid passes alternately through a heater or a cooler before entering the bed.

An improved separation results if the fluid flows through a sequence of beds which are cycled in temperature out of phase with each other, as shown in Fig. 4-47. As the solute-enriched portion of the effluent from bed 1 flows through bed 2, bed 2 is heated to release adsorbed solute and further enrich that already once-enriched

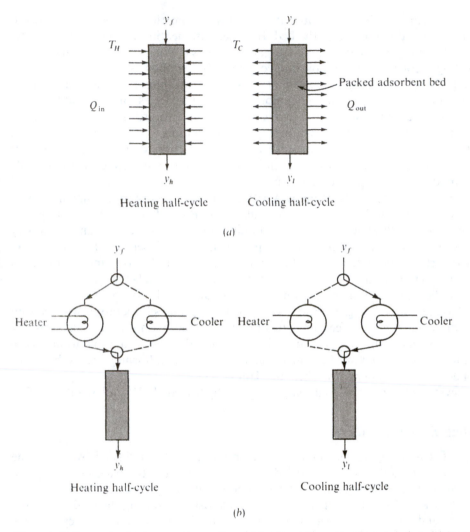

Figure 4-46 Separation by temperature cycling: (a) heating and cooling from a jacket; (b) heater and cooler before bed. (*From Pigford et al., 1969b, p. 849; used by permission.*)

stream. This happens again as the enriched portion of the stream passes through each successive bed. In a sense, the secret of the process is to remove solute from a fluid of low concentration (cold), temporarily store it, and then give it up on command (heating) to a fluid of high concentration (Wankat, 1974).

Cycling-zone separations can be applied to other stationary phases and can be used with other cycled variables (pH, electric field, etc.) besides temperature. Their virtue, compared with parametric pumping, is the ability to handle a substantial feed capacity and to separate a multicomponent feed into more than two products.

Theoretical approaches for analyzing multibed cycling-zone processes, reviewed

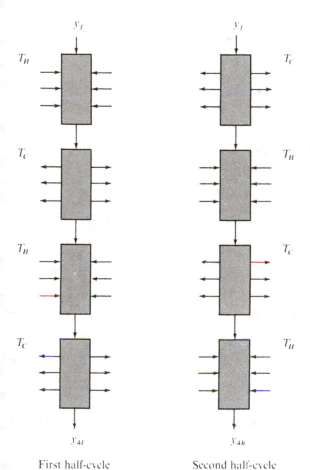

y_f y_f

T_H T_C

T_C T_H

T_H T_C

T_C T_H

y_{4l} y_{4h}

First half-cycle   Second half-cycle

Figure 4-47 Multiple-zone operation of cycling-zone adsorber. (*From Pigford et al., 1969b, p. 849: used by permission.*)

by Wankat et al. (1975), include an extension of the equilibrium theory of Pigford et al. (1969a), as well as staged models either analogous to theories used for counter-current distribution (Chap. 8) or allowing for continuous transfer between phases (Nelson et al., 1978).

TWO-DIMENSIONAL CASCADES

Figure 4-48a depicts a simple cross-flow process for purification by crystallization. A solid feed mixture is contacted with solvent S and is heated and recrystallized upon cooling to form crystals X_1 and supernatant liquid L_1. The crystals are contacted with more solvent, remelted, and recrystallized to form crystals X_2 and liquid L_2. This procedure is repeated until product crystals X_4 are obtained. These are highly purified by virtue of the repeated purification by recrystallization. Liquids L_1 through L_4 contain varying amounts of impurity.

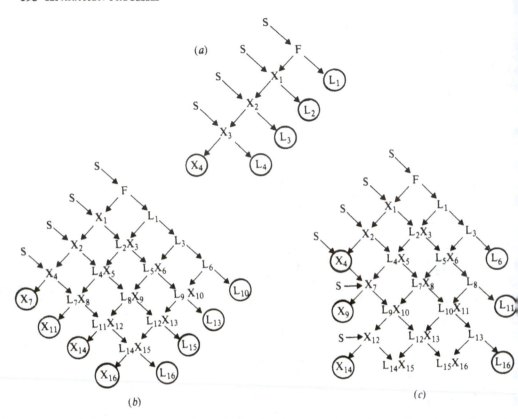

Figure 4-48 Crystallization cascades: (a) simple recrystallization; (b) two-dimensional diamond cascade; (c) two-dimensional double-withdrawal cascade. (*Adapted from Mullin, 1972, p. 237; used by permission.*)

This procedure can be extended to two-dimensional cross-flow cascades (Mullin, 1972), in which F is separated into X_1 and L_1, X_1 is separated into X_2 and L_2, and L_1 is separated by partial crystallization into X_3 and L_3. L_2 and X_3 are then mixed and recrystallized into X_5 and L_5, etc. Figure 4-48b is a *diamond* cascade, which produces purified crystals and impurity-bearing liquids with no intermediate products. Figure 4-48c is a *double-withdrawal cascade*, which produces crystal and liquid products of relatively uniform compositions but also produces some intermediate products (L_{14}, X_{15}, L_{15}, X_{16}).

These processes may be looked upon as two-dimensional versions of equilibrate-and-transfer processes like CCD and CDCD. The approach can readily be extended to other types of separation, e.g., extraction. Separation of stigmasterol from mixed sitosterols by repeated crystallizations is a large-scale application of two-dimensional cascades in the pharmaceutical industry (Poulos et al., 1961).

Wankat (1977) has pointed out the capability of two-dimensional diamond cascades for making multicomponent separations, concentrating different components into different products or sets of adjacent products. There is also a direct analogy between the two-dimensional diamond cascade and some forms of continuous chromatography; e.g., compare the diamond cascade with L_{10}, L_{13}, L_{15}, and L_{16} fed back in at the S points, on the one hand, with the rotating-annulus continuous chromatograph, mentioned above, on the other. There is also an analogy between one-dimensional time-dependent separations, e.g., elution chromatography, and two-dimensional cascades, where the second dimension takes the role of the time variable (Wankat et al., 1977).

Treybal (1963) has pointed out that a double-withdrawal cascade, if deep enough, will simulate a countercurrent continuous-flow equilibrium-stage separation.

REFERENCES

Aris, R. (1969): *Ind. Eng. Chem. Fundam.*, **8**:603.

Belter, P. A., F. L. Cunningham, and J. W. Chen (1973): *Biotechnol. Bioeng.*, **15**:533.

Betts, W. D., and G. W. Girling (1971): *Separ. Purif. Methods*, **4**:31.

Broughton, D. B. (1977): *Chem. Eng. Prog.*, **73**(10):49.

Caldwell, K. D., L. F. Kesner, M. N. Myers, and J. C. Giddings (1972): *Science*, **176**:296.

Chang, T. M. S. (1975): *Separ. Purif. Methods*, **3**:245.

Chen, H. T., and F. B. Hill (1971): *Separ. Sci.*, **6**:411.

Craig, L. C., and D. Craig (1956): Laboratory Extraction and Countercurrent Distribution, in A. Weissberger (ed.), "Technique of Organic Chemistry," vol. 3, Interscience, New York.

Everaerts, F. M., F. E. P. Mikkers, and T. P. E. M. Verheggen (1977): *Separ. Purif. Methods*, **6**:287.

Giddings, J. C., F. J. F. Yang, and M. N. Myers (1975): *Separ. Sci.*, **10**:133.

———, ———, and ——— (1976): *Science*, **193**:1244.

Goldberg, M., and E. Rubin (1972): *Separ. Sci.*, **7**:51.

Grushka, E., K. D. Caldwell, M. N. Myers, and J. C. Giddings (1974): *Separ. Purif. Methods*, **2**:127.

Heines, V. (1971): *CHEMTECH*, **1**:280.

Hengstebeck, R. J. (1959): "Petroleum Processing," chap. 7, McGraw-Hill, New York.

——— (1961): "Distillation: Principles and Design Procedures," Reinhold, New York.

Hybarger, R., C. W. Tobias, and T. Vermeulen (1963): *Ind. Eng. Chem. Process Des. Devel.*, **2**:65.

Ito, Y., and R. L. Bowman (1976): *Separ. Sci.*, **11**:201.

———, R. E. Hurst, R. L. Bowman, and E. K. Achter (1974): *Separ. Purif. Methods*, **3**:133.

James, A. T., and A. J. P. Martin (1952): *Biochem. J.*, **50**:679; *Analyst*, **77**:915.

Lee, H. L., J. F. G. Reis, J. Dohner, and E. N. Lightfoot (1974): *AIChE J.*, **20**:776.

Liscom, P. W., C. B. Weinberger, and J. E. Powers (1965): *AIChE-Inst. Chem. Eng. Symp. Ser.*, no. 1, p. 90.

Martin, A. J. P. (1949): *Discuss. Faraday Soc.*, **7**:332.

McGinnis, R. A. (1969): Sugar (Beet Sugar), in R. E. Kirk and D. F. Othmer (eds.), "Encyclopedia of Chemical Technology," 2d ed., vol. 19, Interscience, New York.

Mikeš, O. (ed.) (1961): "Laboratory Handbook of Chromatographic Methods," trans. by R. A. Chalmers, Van Nostrand, London.

Moores, R. G., and A. Stefanucci (1964): Coffee, in R. E. Kirk and D. F. Othmer (eds.), "Encyclopedia of Chemical Technology," 2d ed., vol. 5, Interscience, New York.

Mullin, J. W. (1972): "Crystallization," chap. 7, Butterworths, London.

Myers, M. N., K. D. Caldwell, and J. C. Giddings (1974): *Separ. Sci.*, **9**:47.

Nelson, W. C., D. F. Silarski, and P. C. Wankat (1978): *Ind. Eng. Chem. Fundam.*, **17**:32.

Oliver, E. D. (1976): "Diffusional Separation Processes: Theory, Design and Evaluation," Wiley, New York.

Pauschmann, H. (1972): *Fresenius Z. Anal. Chem.*, **258**:358.

Perry, J. A., T. H. Jupille, and L. J. Glunz (1975): *Separ. Purif. Methods*, **4**:97.

Pigford, R. L., B. Baker III, and D. E. Blum (1969a): *Ind. Eng. Chem. Fundam.*, **8**:144.

——, ——, and —— (1969b): *Ind. Eng. Chem. Fundam.*, **8**:848.

Porter, R. S., and J. F. Johnson (1967): Chromatographic Fractionation, chap. B3 in M. J. R. Cantrow (ed.), "Polymer Fractionation," Academic, New York.

Post, O., and L. C. Craig (1963): *Anal. Chem.*, **35**:641.

Poulos, A., J. W. Greiner, and G. A. Ferig (1961): *Ind. Eng. Chem.*, **53**:949.

Pucar, Z. (1961): *Chromatogr. Rev.* **3**:38.

Reis, J. F. G., and E. N. Lightfoot (1976): *AIChE J.*, **22**:779.

Rosenzweig, M. D. (1969): *Chem. Eng.*, Apr. 7, pp. 108–110.

Sherwood, T. K., R. L. Pigford, and C. R. Wilke (1975): "Mass Transfer," pp. 651–656, McGraw-Hill, New York.

Smith, B. D. (1963): "Design of Equilibrium Stage Processes," McGraw-Hill, New York.

Stahl, E. (ed.) (1969): "Thin-Layer Chromatography," 2d ed., Springer-Verlag, New York.

Stewart, G. H. (1964): Chromatography, in R. E. Kirk and D. F. Othmer (eds.), "Encyclopedia of Chemical Technology," 2d ed., vol. 5, Interscience, New York.

Sussman, M. V. (1976): *CHEMTECH*, **6**:260.

Sweed, N. H. (1971): Parametric Pumping, in E. S. Perry and C. J. van Oss (eds.), "Progress in Separation and Purification," vol. 4, Wiley (Interscience), New York.

Talmon, Y. and E. Rubin (1976): *Separ. Sci.*, **11**:509.

Tiselius, A. (1947): *Adv. Protein Chem.*, **3**:67.

Treybal, R. E. (1963): "Liquid Extraction," 2d ed., McGraw-Hill, New York.

Valdes-Krieg, E., C. J. King, and H. H. Sephton (1977): *Separ. Purif. Methods*, **6**:221.

Van Winkle, M. (1968): "Distillation," McGraw-Hill, New York.

Vermeulen, T. (1977): *Chem. Eng. Prog.*, **73**(10):57.

Wagener, K., H. D. Freyer, and B. A. Bilal (1971): *Separ. Sci.*, **6**:483.

Wankat, P. C. (1974): *Separ. Sci.*, **9**:85.

——: (1977): *AIChE J.*, **23**:859.

——, J. C. Dore, and W. C. Nelson (1975): *Separ. Purif. Methods*, **4**:215.

——, A. R. Middleton, and B. L. Hudson (1976): *Ind. Eng. Chem. Fundam.*, **15**:309.

Wilhelm, R. H., A. W. Rice, and A. R. Bendelius (1966): *Ind. Eng. Chem. Fundam.*, **5**:141.

——, ——, R. W. Rolke, and N. H. Sweed (1968): *Ind. Eng. Chem. Fundam.*, **7**:337.

—— and N. H. Sweed (1968): *Science*, **159**:522.

Zang, J. A., R. J. Moshy, and R. N. Smith (1966): *Chem. Eng. Progr. Symp. Ser.*, **62**(69):105.

PROBLEMS

4-A₁ (a) When one is rinsing out a drinking glass, is it more efficient to use a given volume of water for a single rinsing or to rinse sequentially with the same volume of water divided into several portions? Explain your answer briefly.

(b) Is there some scheme that is more efficient than either of the schemes mentioned in part (a) for rinsing the glass with the same total amount of water? Explain.

(c) Repeat part (b) for several different glasses to be rinsed at the same time.

4-B₁ Two components A and B are present in a mixture which is 0.05 mole fraction A and 0.95 mole fraction B. Solvent C is to be used to extract A away from B; 100 mol/h of the mixture of A and B are to be treated in this way, and 100 mol/h of solvent C are available for the extraction. B and C are totally insoluble in each other, and it can be assumed that when C is mixed with A and B, two equilibrium liquid phases will result, one containing all the B and the other containing all the C, with A distributed between the phases. From measurements of the equilibrium constant of A in this system it has been found that A will attain the sàme mole fraction in both phases; therefore $K_A = 1$. Calculate the percentage of A in the feed which is extracted into the C phase using three equilibrium stages (a) in parallel, cocurrent flow, (b) in cross-current flow, dividing solvent C equally between the three stages, and, (c) in countercurrent flow.

4-C$_1$ A countercurrent distribution separation of the sort outlined in Fig. 4-37 is carried out using five transfers with five vessels. A feed mixture of amino acids containing 40 mol $^o/_o$ A and 60 mol $^o/_o$ B diluted in a buffered water solution is introduced into the first vessel initially. In each contacting vessel there are 200 mL of the buffered aqueous phase and 300 mL of an organic solvent when an equilibration step is carried out. The aqueous phase is denser and is not transferred from vessel to vessel. The organic phase is transferred to the next adjacent vessel at each transfer step. If $K'_{A, o-w} = 4.0$ and $K'_{B, o-w} = 0.3$ expressed as $(mol/m^3) \cdot (mol/m^3)^{-1}$, find the maximum fraction of A in the feed which can be recovered in such a purity that the molar ratio of A to B in the gross product is at least 4.0.

4-D$_2$ A stream of nearly pure benzene contains a 1.0 mole percent concentration of an impurity which is known to have a relative volatility of 0.20 relative to benzene. It is proposed to purify a portion of the benzene in one of the following ways. In all cases the purified product should contain exactly 50 percent of the benzene charged.

(a) The stream is passed continuously through a heater and into an equilibrium vapor-liquid separator drum. The overhead vapor will be totally condensed and will be the product.

(b) The benzene will be stored and then charged in discrete batches to a simple equilibrium still. After a batch is charged, the still will be heated and vapor will be removed and condensed continuously. The accumulated overhead condensate will form the product.

(c) Operation will be the same as in part (b) except that half of the condensed vapor will be added to the liquid in the still at all times during a run, rather than being taken as product.

(d) Operation will be the same as in part (b) except that the overhead vapor will be continuously passed to a second cooled vessel where half of it will be condensed. The liquid from this second vessel will be returned to mix with the liquid in the still, whereas the vapor from the second vessel will be continuously condensed and taken as product. The holdup in the second vessel may be considered to be very small; a very small amount of liquid is in the second vessel at any one time.

If all vessels can be considered well mixed, what will be the product purity in each of the four above cases? Explain the causes of the differences in product purity.

4-E$_2$ A phase diagram for the system m-cresol–p-cresol is given in Fig. 1-25. Suppose that p-cresol is to be recovered by crystallization from a feed containing 25$^o/_o$ m-cresol and 75$^o/_o$ p-cresol which is available at 25°C. The aims of the process are to achieve as high a recovery of p-cresol as possible and for the p-cresol product to be as pure as possible. The crystallization will be accomplished in one or more refrigerated scraped-surface double-pipe crystallizers (Fig. 3-7f), and the resultant slurry will be separated in a centrifuge. One cause of product impurity will be a small amount of retained liquid in the centrifuged bed of crystals. Even so, no facilities for washing the crystals will be included.

(a) Indicate how this process might be carried out in multiple stages.

(b) What incentive(s), if any, are there for making this a multistage process rather than a single-stage process? The following possibilities are suggested for your evaluation.

1. Greater product purity if equilibrium is achieved in each stage and there is no liquid retained by the centrifuged crystals
2. A greater recovery fraction of p-cresol if equilibrium is achieved in each stage and there is no liquid retained by the centrifuged crystals
3. Greater product purity because the liquid retained by the centrifuged crystals contains less m-cresol
4. Less refrigeration duty (joules per hour) required for a given p-cresol product rate
5. Accomplishing some or all of the refrigeration using a less cold refrigerant

4-F$_2$ If electrodialysis is to be used for desalting either seawater or brackish water, the process must remove both cations and anions of the salts from the water. Typical cations include Na^+, Ca^{2+}, Mg^{2+}, etc., and typical anions include Cl^-, CO_3^{2-}, SO_4^{2-}, etc. The electrodialysis process shown in Fig. 4-30 for grapefruit juice removes only anions from the juice; hence a directly analogous process will not remove the cations from seawater or brackish water.

(a) Create a schematic flow diagram of an electrodialysis process which will remove cations as well as anions from a salty feedwater, producing salt-free water as a product along with a reject stream which is enriched in salt content. Use two different types of membrane in your process, one which is anion-permeable but cation-impermeable and the other which is cation-permeable but anion-impermeable.

(b) For salt contents of the order of 1 percent or higher in the feedwater, the electric energy consumed by this process becomes a very important economic factor. How is the electric energy consumption related to the feed salt content?

4-G₂ Figure 4-49 shows a schematic flow diagram for a process making fresh water by multistage flash evaporation of seawater. In this process some water is flashed off as vapor in each chamber. The flashed vapor is then condensed and taken as freshwater product. The pressure in each successive chamber is less

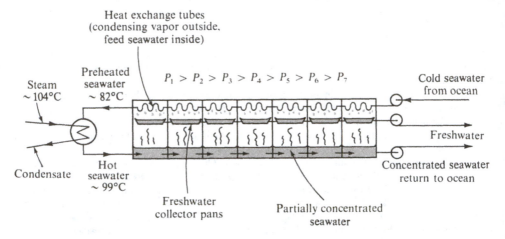

Heat exchange tubes
(condensing vapor outside,
feed seawater inside)

$P_1 > P_2 > P_3 > P_4 > P_5 > P_6 > P_7$

Steam
~104°C

Preheated
seawater
~82°C

Cold seawater
from ocean

Freshwater

Condensate

Hot
seawater
~99°C

Concentrated seawater
return to ocean

Freshwater
collector pans

Partially concentrated
seawater

Figure 4-49 Fresh water produced from seawater by multistage flash evaporation.

than in the one before. Since the saturation temperature of water decreases as pressure decreases, the water will cool and a certain amount of water will boil off in each chamber. The vapor is condensed by heat exchange against the seawater feed, which is preheated by the latent heat released by the condensing water. Typically, the steam used to supply heat to the feedwater before the first stage of flashing has a condensation temperature of about 104°C. Any higher temperature would cause scale to form from the seawater onto the heat-exchanger surfaces in the steam heater. The pressure in the lowest-pressure chamber is given

Figure 4-50 A multistage flash seawater desalination plant with a capacity of 3800 m³/day, originally located at Point Loma, California, and subsequently transferred to the United States Guantanamo Naval Base in Cuba. (*The Fluor Corp., Ltd., Los Angeles, Calif.*)

a lower limit by the need of condensing the vapor generated in that chamber with feedwater at the supply temperature from the ocean.

(a) What is gained by carrying the flashing out in a succession of chambers rather than flashing the feedwater from the same initial temperature and pressure to the same final pressure in one single large chamber?

(b) What would be a typical percentage of the seawater feed recovered as freshwater product in a plant of the design shown in Fig. 4-49? Support your answer by a simple calculation.

(c) Contrast this process with multieffect evaporation (Fig. 4-19).

Figure 4-50 shows a multistage flash plant for seawater conversion into fresh water which was built at Point Loma, California, and went on stream in 1962 with a capacity of 3800 m³/day of fresh water. When Cuba shut off the water supply to the Guantanamo Naval Base, this plant was transferred by ship to Guantanamo and put into service there.

4-H₂ A process is to be devised for leaching a valuable water-soluble substance from an ore. The composition of the ore is 20 wt % desirable water-soluble substance and 80 wt % insoluble residue. The process will follow either scheme I or scheme II, as shown in Fig. 4-51. The solid ore will be ground up, slurried in

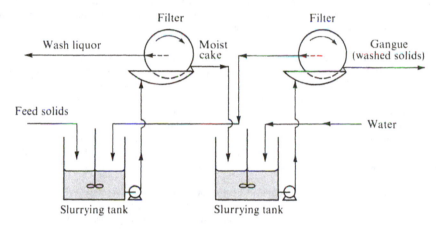

SCHEME I

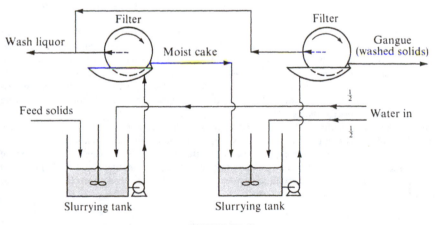

SCHEME II

Figure 4-51 Schemes for leaching ore.

water, and passed to a rotary filter. The filter will leave an amount of *water* in the cake equal to the weight of the remaining insoluble solids. The water retained will contain the prevailing concentration of soluble material. The filter cake will then be removed and reslurried, after which the filtration process will be repeated, as shown. The rotary filters operate with a vacuum inside, drawing water solution through the filter medium. At no time does the concentration of the water-soluble species in the water approach its solubility limit. The soluble material dissolves rapidly. In scheme II the wash water is split into two equal streams.

(a) Which scheme will give the greatest *recovery fraction* of the water-soluble substance in the wash liquor for a given water-to-solids treat ratio?

(b) Which scheme will give the highest *concentration* of the water-soluble substance in the wash liquor for a given water-to-solids treat ratio?

(c) Illustrate the correctness of your answers to parts (a) and (b) by performing the appropriate calculations for the case where the total water consumption is 4 kg per kilogram of total ore fed.

4-I$_2$ Confirm that Eq. (4-8) does correspond to the peak locations shown in Fig. 4-38.

4-J$_2$ Figure 4-52 shows a flow diagram for a hydrometallurgical process used for obtaining high-purity nickel from sulfide ores. The process has been used by Sherritt Gordon Mines, Ltd., at its Fort Saskatchewan, Alberta, refinery in Canada to recover nickel from sulfide-ore concentrates from a mine in Lynn Lake, Manitoba. Rosenzweig gives the following description:†

In the first part of the Sherritt refining process, concentrate is contacted with an ammonia solution. Concentrate from the Lynn Lake mine contains about 10% nickel, 2% copper, 0.4% cobalt, 33% iron and 30% sulfur. The ammonia solution extracts copper, nickel and cobalt.

The hydrometallurgical treatment essentially is a continuous, two-step, countercurrent operation. The sulfide concentrate is leached in two sets of autoclaves. Their optimum operating pressure ranges between 100 and 110 psig; temperature, between 170 and 180°F.

Fresh concentrate goes into the first set of autoclaves where it is treated with leach liquor that has already contacted concentrate in the second set of autoclaves. This intermediate liquor extracts the most easily leached portion of the concentrate—it leaves the vessel with a full-strength solution of dissolved metals.

The partially leached concentrate is then sent to the second set of autoclaves where it contacts fresh leach liquor high in ammonia, and loses more-difficult-to-extract metal values.

Leach residue (iron oxide and other insolubles) is separated from the solution containing dissolved metals by means of thickeners and disk filters. Phases of this liquid-solid separation occur after each of the two hydrometallurgical stages. Following the final leaching and filtering, residue is carefully washed by repulping and filtered to remove all soluble nickel; then it is sent to residue ponds.

A certain quantity of sulfur is also extracted with the metals. Most of this forms ammonium sulfate, but a small portion becomes unsaturated sulfur compounds, such as ammonium thiosulfate.

Following the removal of residue, the pregnant solution from leaching is heated to the boiling point in an enclosed five-stage boiler unit. Most of the uncombined ammonia in the solution is vaporized and, after condensation, is recycled to the hydrometallurgical circuit. The heating also causes reaction between the unsaturated sulfur compounds and the copper in solution—precipitating the copper as black copper sulfide sludge.

Most of the copper sulfide is then removed by passing the solution through a filter press. Then, copper still remaining is stripped by bubbling hydrogen sulfide through the solution.

Copper sulfide produced in the copper boil can be shipped directly to a smelter for recovery of copper. But the copper sulfide formed by using hydrogen sulfide (representing about 15% of the total copper processed) contains considerable quantities of nickel; it is returned to the leach circuit for redissolving.

† From Rosenzweig (1969, pp. 108–110); used by permission.

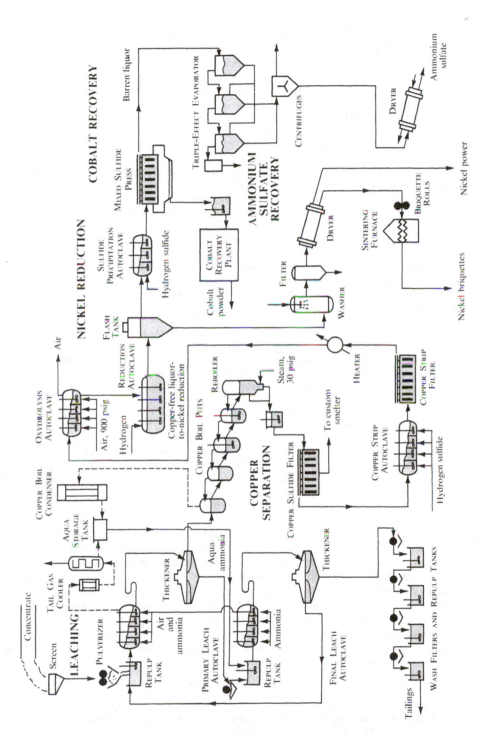

Figure 4-52 Process for obtaining nickel from sulfide ores. *(From Rosenzweig, 1969, pp. 109–110; used by permission.)*

203

Solution passing from the copper-separation circuit contains nickel, cobalt, ammonium sulfamate and small amounts of unsaturated sulfur compounds. The sulfamate and unsaturated sulfur compounds must be removed before the metals can be recovered. This is accomplished by heating the solution under pressure: ammonium sulfamate and unsaturated sulfur compounds are converted into more ammonium sulfate. Nickel recovery then proceeds on a batchwise basis.

Solution is fed into an autoclave containing a small quantity of fine nickel powder. When the autoclave is filled, the powder is brought into suspension by the action of agitators, and hydrogen is passed into the vessel up to a total pressure of 500 psi. The nickel metal in solution precipitates into fine particles of pure nickel that grow on the nickel powder. The process continues until almost all of the nickel has been precipitated. (Very little cobalt will precipitate as long as a small amount of nickel is left in solution.)

At this point, the agitators are stopped, the nickel particles are allowed to settle, depleted solution is drawn off, and fresh solution is added.

After some 40 drawoffs and additions, the nickel particles become so heavy that it is hard to keep them in proper suspension. When this occurs, the solution is drawn off with the agitators running—thus also removing the precipitated nickel.

Then the autoclave reseeds itself via a controlled nucleation reaction to reduce fine nickel powder.

Meanwhile, the entire contents of the autoclave from the completed cycle are discharged to cone-bottomed flash tanks. Mother liquor overflows to a storage tank, and nickel metal settles to the bottom of the cone. Now in slurry form, the nickel is washed, dried and packaged as a powder; or pressed into briquets, sintered and packaged.

The mother liquor sent to storage contains a very small amount of nickel, cobalt, and a high concentration of ammonium sulfate. Nickel and cobalt are extracted from the solution together by treatment with hydrogen sulfide; nickel and cobalt sulfides are formed and precipitate. The precipitate is filtered off and processed elsewhere in the plant for the recovery of pure cobalt metal. The recovery technique is similar to that used for nickel.

The remaining solution contains only ammonium sulfate. It is recovered by evaporation, leaving ammonium sulfate crystals. These are bagged and sold as nitrogenous fertilizer.

All told, the Fort Saskatchewan plant daily produces over 40 tons of nickel, approximately 3,000 lb of cobalt, and 300 tons of ammonium sulfate.

List all separation processes present within this process and indicate, for each, what its *function* is in relation to the overall objectives of the process. Also identify what the *physical phenomenon* is upon which each separation is based. Show which separations are *single-stage* and which are *multistage*. For the multistage separations, indicate which are *cross-flow* and which are *counterflow*. Also, in each instance of a multistage separation, indicate what *economical processing advantage(s)* has been gained by staging the separation.

4-K$_2$ Indicate one or more ways of staging the H$_2$S–H$_2$O process for separating deuterium from hydrogen (both combined into water, H$_2$O or HDO) to produce a deuterium-rich product that is enriched to a substantially greater extent in deuterium. Single-stage versions of this separation process were considered in Prob. 1-F. Recall that the fraction deuterium in the natural water feed is very small. A desirable policy in formulating the process is to make the equipment as compact and simple as possible.

4-L$_2$ Compute the equilibrium ratio $K_i = y_i/x_i$ for each of the four components in Fig. 4-40 if the column of glass beads has a void fraction of 0.40 and the pressure is atmospheric.

4M$_2$ Liquid ion exchangers are organic solvents which can exchange either anions or cations with ions present in a contacting aqueous solution. For example, certain high-molecular-weight organic acids R—H are immiscible with water and will exchange copper according to the reaction

$$2R-H + Cu^{2-} \rightleftharpoons 2H^- + R_2Cu$$

The ionized species exist in the aqueous phase, and the R—H and R$_2$Cu species exist in the organic phase. A typical equilibrium relationship between copper concentration in the organic phase and copper concentration in the aqueous phase is shown in Fig. 4-53.

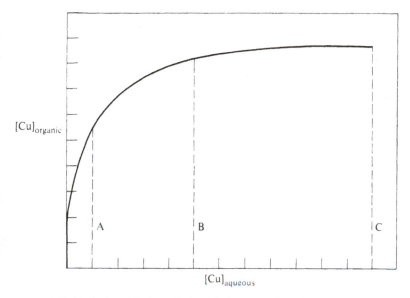

Figure 4-53 Typical equilibrium relationship for extraction of copper ion using a liquid ion exchanger.

(a) Explain why the equilibrium relationship curves in the direction shown.

(b) Would staging the extraction process be more useful for reducing the aqueous copper concentration from the value shown by vertical dashed line C to that shown by vertical dashed line B or for reducing the aqueous copper concentration from the value shown by line A to a very low value? Explain briefly.

(c) What would be a likely method for regenerating the liquid ion exchanger and recovering the copper contained in it?

(d) Suppose that the liquid ion exchanger also has a small capacity for extraction of iron, Fe^{3+}, from aqueous solution. Devise and sketch an extraction process which would recover the copper from solution by liquid ion exchange but which at the same time would reduce the presence of Fe^{3+} in the recovered copper product as much as possible.

4-N$_3$† An aqueous solution is being continuously concentrated in a multieffect system of four evaporators connected in series, with parallel flow of steam and liquor. The following conditions are normal: steam pressure in the coils of the first effect, concentration and temperature of the feed, vacuum in the vapor space of the last effect, and concentration of the product from the last effect. The condensate from each effect is withdrawn from the system.

(a) Assume that the capacity is normal but the steam consumption is abnormally high. State the nature of the trouble and list, in the proper order, the steps to be taken to remedy it.

(b) Now assume that the steam consumed per pound of total evaporation is normal but that the capacity is abnormally low. List, in the proper order, what steps should be taken to locate and remedy the trouble.

† From W. H. Walker, W. K. Lewis, W. H. McAdams, and E. R. Gilliland, "Principles of Chemical Engineering," 4th ed., McGraw-Hill, New York, 1937; used by permission. Review of Appendix B in connection with this problem will be helpful.

FIVE

BINARY MULTISTAGE SEPARATIONS: DISTILLATION

BINARY SYSTEMS

As we have seen, multistage separation operations fall into many categories, depending on which physical phenomenon the separation is based upon. For purposes of calculating the performance of separation devices it is helpful to create a division of a different sort, between *binary* and *multicomponent* systems. This terminology arises from distillation, where a binary system contains only two components and a multicomponent system contains more. For purposes of a general approach we shall adopt a somewhat different definition. *A binary system will be taken as one where both streams flowing between stages have the property that the stream composition is uniquely determined by setting the mole, weight, or volume fraction of one particular component present in the stream.* In a multicomponent system, on the other hand, it is necessary to specify the mole, weight, or volume fraction of more than one species in order to determine the composition of one or both interstage streams.

In most cases this definition of binary systems implies that there are only one or two species present which are capable of appearing to an appreciable extent in both output streams from any stage. An exception occurs for a three-component system forming two saturated phases at a fixed temperature and pressure, as is frequently encountered for liquid-liquid extraction. By the phase rule, the entire composition of a phase is set in this case by fixing the mole fraction of a single component.†

---

† $P + F = C + 2$. Since $P = 2$ (resulting from the specification of *saturation*) and $C = 3$, $F = 3$ corresponding to temperature, pressure, and one mole fraction in the phase under consideration.

Table 5-1 Examples of separations

Separation	Agent added to effect separation	Species appearing appreciably in both output streams from a stage
Binary		
Distillation of mixture of A and B	Heat	A, B
Absorption of A from a noncondensable carrier gas B into heavy non-volatile solvent C	C	A
Absorption of A from a mixture of inert gases (B, C, and D) into a heavy solvent which is a mixture of several nonvolatile components (E, F, G)	E, F, G mixture	A
Washing a soluble constituent A from an insoluble medium	Wash water	A
Gaseous diffusion separation of gas mixture of A and B	Energy (pressure)	A, B
Extraction of A from B in a liquid mixture by adding partially miscible solvent C (fixed temperature and pressure)	C	A, B, C
Fractional crystallization of liquid mixture of A and B	Cooling	A, B
Multicomponent		
Distillation of a mixture containing three or more volatile components	Heat	All volatile components
Absorption of two or more different species (A, B, ...) from a non-condensable carrier gas C into a heavy nonvolatile solvent D	D	A, B, ...
Gaseous diffusion separation of gas mixture of A, B, and C	Energy (pressure)	A, B, C
Separation of A from B by liquid-liquid extraction, using two counter-flowing solvents C and D; all species soluble to some extent in both phases	C and D	A, B, C, D
Fractional crystallization of liquid mixture of A, B, and C, all of which form solid solutions	Cooling	A, B, C
Removal of Ca^{2+} and Mg^{2+} (both) from hard water by ion exchange onto a cation-exchange resin initially loaded with Na^+	Resin	Ca^{2+}, Mg^{2+}, Na^+

In particular this definition of binary systems does not necessitate that only two components be present within each phase; on the other hand it does place restrictions on the nature of any additional components. Table 5-1 gives some examples of binary separations, and multicomponent separations.

Calculational problems involving binary separations can be quite easily handled by several methods. In contrast, the analysis of multicomponent separations is always complex, despite the fact that the basic concepts applying to both types of systems are the same. Treatment of binary systems is facilitated by the fact that we have enough independent variables to stipulate a great deal about the conditions of the separation, whereas in multicomponent systems we can stipulate less about the separation even though there are more total variables.

A second real advantage of dealing with binary systems results from the very definition we have made for them. The determination of a single composition variable serves to set the entire composition of a stream at any point in a separation device.

In this chapter we shall consider only binary distillation, which has classically received the most attention and is probably the most common binary multistage separation process. In Chap. 6 the concepts and procedures developed for binary distillation will be applied to binary multistage separations in general.

EQUILIBRIUM STAGES

We saw in Chap. 3 that there can be several types of complication in the analysis of single-stage separations where two phases tend to equilibrate with each other. Some factors tend to prevent the attainment of complete equilibrium and others provide the wherewithal of exceeding equilibrium between product streams. Once this situation is acknowledged, two general approaches are possible in the analysis of multistage separations: (1) postulate the attainment of complete equilibrium between the product streams from each stage, perform the analysis of the separation, and then correct the computation for the lack of equilibrium as a final step; or (2) allow for all factors occurring within a stage, thereby obtaining the actual relationship between stage product-stream compositions for each stage, and then complete the analysis of the separation. A quantitative allowance for all factors which tend to prevent exact equilibration is well beyond the scope of this book and, indeed, usually beyond present-day capabilities. Therefore, of necessity, we adopt the *first* of these two procedures for use in nearly all cases and defer consideration of quantitative methods for correcting for nonequilibrium until Chap. 12. The exception to this convention will occur for processes, such as a wash, in which the equilibrium-discouraging effects are controlling and are analyzed relatively easily.

McCABE-THIELE DIAGRAM

The pertinent equations for the analysis of continuous-flow binary distillation were developed in the late nineteenth century by Sorel (1893), but the simplest, and most instructive method for analyzing binary distillation columns is the graphical

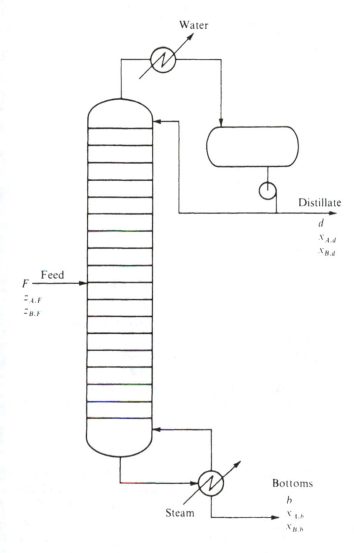

Figure 5-1 Binary distillation.

approach devised by McCabe and Thiele (1925). The method makes use of the fact that the composition at every point is completely described by the mole fraction of only one of the two components.

Consider the equilibrium-stage distillation column shown in Fig. 5-1. The feed is a mixture of two components A and B. All the liquid compositions on the successive stages of the column can be shown as a series of values of x_A and all the vapor compositions as a series of values of y_A. The mole fractions of component B are not independent; they follow by difference from unity.

We can group the values of x_A and y_A into pairs. The pair of compositions, vapor and liquid, leaving each stage is certainly of interest, and the pair of compositions passing each other between stages is of interest. If we can determine the relationships

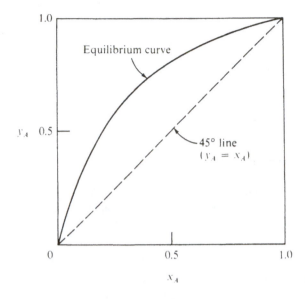

Figure 5-2 Equilibrium curve and 45° line.

for all these pairs, we shall know every composition in the column and have the wherewithal to relate the stages to each other.

A yx diagram can then be set up, where x_A is the abscissa and y_A is the ordinate, both going from 0 to 1, as shown in Fig. 5-2. Any point on this diagram represents a pair of phases, a mole fraction of component A in a vapor phase (and hence the composition of a vapor) together with a mole fraction of component A in a liquid phase (and hence the composition of a liquid).

Equilibrium Curve

A pressure drop from stage to stage upward is necessary to cause the vapor to flow through the column; however, a distillation column is usually considered to be at constant pressure unless the pressure level or height of the column is such that this assumption is clearly in error. In an analysis which considers the distillation as an *equilibrium*-stage process, the liquid and vapor phases leaving any stage are presumed to be in equilibrium with each other at this constant pressure. The phase rule shows that one more degree of freedom aside from the pressure exists. Hence, if some particular liquid composition is chosen for consideration, the vapor composition in equilibrium with this liquid and the temperature at which the two phases can exist are both fixed at unique values. If only one vapor composition can exist in equilibrium with each chosen liquid composition, a single curve plotted on the xy diagram will contain all possible pairs of liquid and vapor compositions in equilibrium with each other at the column pressure and hence all possible pairs of compositions *leaving* stages in a column. This curve is called the *equilibrium curve* and is completely independent of any consideration concerning the column except the total pressure.

A typical equilibrium curve is shown in Fig. 5-2. It should be noted that the equilibrium curve lies above the 45° line (representing $y_A = x_A$) throughout the

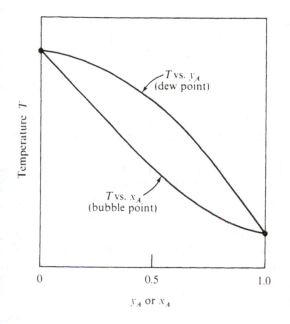

Figure 5-3 Dew- and bubble-point curves for binary distillation.

diagram. Thus y_A is always greater than x_A, indicating that A is the more volatile of the two components since it concentrates in the vapor. The corresponding plots of temperature vs. y_A and x_A are shown in Fig. 5-3. In an azeotropic system the equilibrium curve in Fig. 5-2 will intersect the $y_A = x_A$ line at a point between $x_A = 0$ and $x_A = 1$, and the curves in Fig. 5-3 will show maxima or minima (see, for example, Fig. 1-19b).

Consider a distillation column separating A and B to produce relatively pure B and relatively pure A. The vapor and liquid flows leaving the individual stages of this column would be given by a series of points on the equilibrium curve, progressing upward from the bottom of the curve. Each stage higher in the column is represented by a point higher on the equilibrium curve, since the vapor leaving any stage is enriched in A and flows upward. The temperature of the bottom stage (the reboiler) would be the highest temperature of any stage in the column, corresponding to the highest concentration of component B, and temperature would decrease from stage to stage upward in the column. This fact follows from a consideration of Fig. 5-3. The equilibrium ratios K_A and K_B would be highest at the bottom of the equilibrium curve (or column) and would also decrease upward. The relative volatility, or ratio of equilibrium ratios $(K_A/K_B = \alpha_{AB})$, will in general change much less through the diagram or column than the equilibrium ratios themselves and may in some cases be essentially constant.

Figure 5-2 is drawn for a relative volatility that is constant with respect to composition. Thus the equilibrium curve is described by Eq. (1-12) and is similar to that shown in Fig. 1-17:

$$y_A = \frac{\alpha_{AB} x_A}{1 + (\alpha_{AB} - 1) x_A} \qquad (1\text{-}12)$$

Strictly speaking, the relative volatility will be constant only for an ideal solution where both components have identical molar latent heats of vaporization; however, for many cases of nearly ideal solutions without too wide a boiling range, the assumption of a constant relative volatility is a close approximation. As an exercise the reader could verify that the relative volatility will be constant for a distillation involving an ideal solution with equal molar latent heats of vaporization.

In general, the equilibrium curve must be based upon experimental data or upon thermodynamic extensions of experimental data. Sources of these data are given in Chap. 1.

It might be noted that if we had decided to make an xy plot of concentrations of component B, the less volatile component, rather than component A, the diagram would be inverted diagonally. The equilibrium curve would lie below the 45° line. The bottom stage of the column would have been represented by a point near the upper right-hand corner. The choice of which component to use is purely arbitrary, but custom long has been to use the more volatile component, and this pattern will be followed in the succeeding discussion.

Mass Balances

The equilibrium curve represents all possible pairs of vapor and liquid compositions leaving stages of the column. If we are to relate the stages to each other, we also need a relationship between the pairs of vapor and liquid compositions flowing past each other *between* stages. This relationship is given by a simple mass balance for each component and an energy balance.

Consider a portion of the rectifying section of a simple distillation column, as shown in Fig. 5-4, where stages have been numbered from the bottom of the column. A mass-balance equation can be written to include the vapor and liquid flows passing each other between stages 9 and 10. From the inner mass-balance envelope drawn in Fig. 5-4 it is apparent that the only input of component A is $V_9 y_{A,9}$ and that there are two output streams, $L_{10} x_{A,10}$ and $x_{A,d} d$.† These are molal flows of component A if V and L are taken to be the *total* moles of flow of both components in the vapor and liquid phases, respectively. The mass balance for component A between these two stages is then

$$V_9 y_{A,9} = L_{10} x_{A,10} + x_{A,d} d \qquad (5\text{-}1)$$

since we postulate that the operation occurs at steady state with no buildup of A.

From the outer mass-balance envelope drawn in Fig. 5-4 it is apparent that the mass-balance relation relating flows of component A between stages 9 and 8 is

$$V_8 y_{A,8} = L_9 x_{A,9} + x_{A,d} d \qquad (5\text{-}2)$$

---

† A convention throughout this book is to denote the flow of a vapor product from distillation by a capital letter and the flow of a liquid product by a small letter. The distillate is liquid, hence the flow is denoted by d. If the distillate were vapor it would be denoted by D. Similarly, vapor enthalpies are H and liquid enthalpies are h.

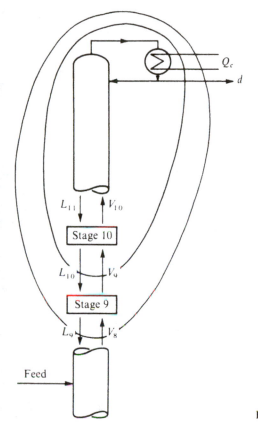

$L_{11}\downarrow$ $\uparrow V_{10}$

Stage 10

$L_{10}\downarrow$ $\uparrow V_9$

Stage 9

$L_9\downarrow$ $\uparrow V_8$

Feed

Figure 5-4 Rectifying-section mass balances.

and the mass-balance relation for an envelope passing between any two adjacent stages, p and $p + 1$, in this section of stages is

$$V_p y_{A, p} = L_{p+1} x_{A, p+1} + x_{A, d} d \qquad (5\text{-}3)$$

Considering the total flows, it is also apparent that

$$V_p = L_{p+1} + d \qquad (5\text{-}4)$$

Consider next the stripping section of a column like that shown in Fig. 5-5. Mass-balance envelopes have been drawn in two ways, both of which include the vapor and liquid flows passing between the general stages p and $p + 1$ in the stripping section. From the mass-balance envelope drawn around the feed and the top product, denoted by the upper solid loop in Fig. 5-5,

$$V'_p y_{A, p} = L'_{p+1} x_{A, p+1} + x_{A, d} d - F z_{A, F} \qquad (5\text{-}5)$$

From the mass-balance envelope drawn around the bottoms product $x_{A, b} b$, denoted by the lower solid loop,

$$V'_p y_{A, p} = L'_{p+1} x_{A, p+1} - x_{A, b} b \qquad (5\text{-}6)$$

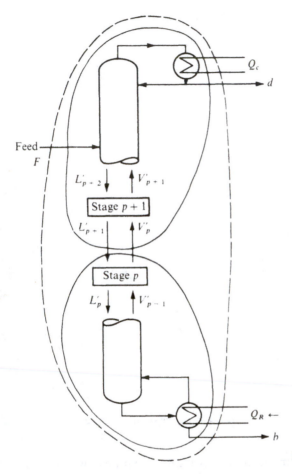

Figure 5-5 Stripping-section mass balances.

Subtracting Eq. (5-5) from Eq. (5-6), one obtains the mass balance around the whole column, denoted by the dashed loop in Fig. 5-5:

$$x_{A,d}d + x_{A,b}b = Fz_{A,F} \qquad (5\text{-}7)$$

In terms of total flows within the stripping section it is readily seen that

$$L'_{p+1} - V'_p = b = F - d \qquad (5\text{-}8)$$

As a general statement, in either section, the mass-balance relation is

$$V_p y_{A,p} = L_{p+1} x_{A,p+1} + \text{net upward product of component A from section} \qquad (5\text{-}9)$$

The net upward product is the algebraic sum of the molar flow of A in all products leaving the system above the two internal flows considered less the amount of A in any feeds above the two internal flows considered. If stages p and $p + 1$ are in the stripping section, the net upward product of A is minus the net downward product, or $-x_{A,b}b$. Considered the other way, the net upward product of A is the top product $x_{A,d}d$ less the feed $Fz_{A,F}$; hence, $x_{A,d}d - Fz_{A,F}$. From this general formulation it is very simple, as we shall see, to write the appropriate mass balances for all sections of any countercurrent staged operation with any number of feeds and products.

Problem Specification

The number of independent variables which can be specified for a distillation calculation is considered in Appendix C, following the description rule. $R + 6$ variables can be specified for a column with a partial condenser, where R is the number of components present. For a column with a total condenser, one additional variable can be set, related to the thermal condition of the reflux.

For a binary distillation with a partial condenser, such as the column shown in Fig. 5-6, there are therefore eight variables which can be set. Nearly always four of these will be the column pressure, the feed flow rate, the feed enthalpy, and the mole

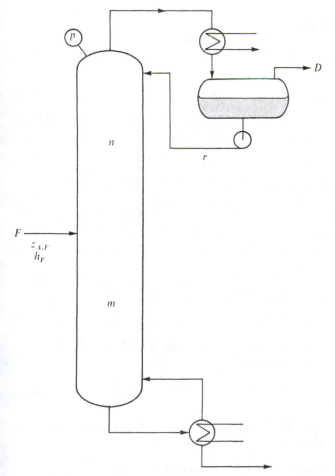

Figure 5-6 A set of specifications for binary distillation.

fraction of a component in the feed. Four variables remain to be set, and for illustration one might set

D = total top product, mol
r = reflux, mol
n = number of stages in column above point of feed introduction
m = number of stages in column below point of feed introduction

Such a combination of specified variables would correspond to the analysis of the operation of an existing column under new conditions.

Numerous other combinations can be set for these other four variables, depending upon the context of a problem. Often *separation variables*, e.g., the mole fraction or the recovery fraction $(/i)_j$, the fraction of component i fed that is recovered in product j, are specified.

Internal Vapor and Liquid Flows

Returning to the envelopes shown in Fig. 5-4 for the rectifying section, we see that for each stage we can write

$$V_p H_p = L_{p+1} h_{p+1} + h_d d + Q_c \tag{5-10}$$

where Q_c is the heat withdrawn in the condenser. Subtracting Eq. (5-10) written for plate $p - 1$ from Eq. (5-10) written for plate p, we have

$$V_p H_p - V_{p-1} H_{p-1} = L_{p+1} h_{p+1} - L_p h_p \tag{5-11}$$

which holds for any section of a column above, below, or in between feeds.

Calculations made employing Eq. (5-11) to ascertain V_p and L_{p+1} at all points are tedious, and experience has shown that a major simplification is at least approximately correct for many problems. This simplification is the assumption of *constant molal overflow*, which corresponds to constant total molar vapor flow rates and total molar liquid flow rates leaving all stages in a given section of the column.

A general expression for the change in vapor rate from stage to stage in the rectifying section can be derived as follows. Equation (5-11) can be rewritten as

$$H_p(V_p - V_{p-1}) - V_{p-1}(H_{p-1} - H_p) = h_{p+1}(L_{p+1} - L_p) - L_p(h_p - h_{p+1}) \tag{5-12}$$

Equation (5-4) indicates that for any two successive stages in the rectifying section $V_p - V_{p-1} = L_{p+1} - L_p$; hence we can rewrite Eq. (5-12) as

$$V_p - V_{p-1} = \frac{V_{p-1}(H_{p-1} - H_p) - L_p(h_p - h_{p+1})}{H_p - h_{p+1}} \tag{5-13}$$

It follows that there are two possible conditions which will cause the total molar vapor flow V to be constant from stage to stage:

Condition 1: $\qquad\qquad H = \text{const} \qquad \text{and} \qquad h = \text{const}$

Condition 2:
$$\frac{H_{p-1} - H_p}{h_p - h_{p+1}} = \frac{L_p}{V_{p-1}}$$

If V is constant, it follows that L is constant. The reader can verify that the same two conditions apply to the stripping section although V' and/or L' will be different from the values in the rectifying section.

From condition 1 it can be seen that constant molal overflow will occur if the molar latent heats of vaporization of A and B are identical, if sensible-heat contributions due to temperature changes from stage to stage are negligible, and if there are no enthalpy-of-mixing effects (ideal liquid and vapor solutions).

Condition 2 indicates that constant molal overflow will also occur if the ratio of the change in vapor molal enthalpy from plate to plate to the change in liquid molal enthalpy is constant and equal to L/V. From Eq. (5-3) it is seen that if L/V is constant from plate to plate, then

$$\frac{L}{V} = \frac{y_{p-1} - y_p}{x_p - x_{p+1}} \tag{5-14}$$

and condition 2 becomes

$$\frac{H_{p-1} - H_p}{h_p - h_{p+1}} = \text{const} = \frac{y_{p-1} - y_p}{x_p - x_{p+1}} \tag{5-15}$$

Thus condition 2 corresponds to the case of $dH_p/dy_p = dh_{p+1}/dx_{p+1}$ under the restriction that the ratio of Eq. (5-15) be a constant equal to L/V. As an application of this result consider the case where there are equal molal latent heats, negligible heats of mixing, and no sensible heat effect but the reference enthalpies of saturated pure liquid A and of saturated pure liquid B are not taken to be equal. Equation (5-15) will hold true in this case, and there will be constant molal overflow even though H and h vary.

A system with a relative volatility that is nearly constant should also exhibit nearly constant molal overflow. This follows since such a system will have activity coefficients close to unity. Therefore the ratio of individual-component vapor pressures will be insensitive to temperature. By the Clausius-Clapeyron equation the percentage change in vapor pressure with respect to temperature is proportional to the latent heat of vaporization; hence constancy of the vapor-pressure ratio implies equal latent heats. Equal latent heats, in turn, imply constant molal overflow.

In the same way one can also reason that constant molal overflow implies constant relative volatility and that this should thereby limit the number of situations in which the constant-molal-overflow assumption is valid. However, as we shall see, it is the constancy of the *ratio* of flows L/V which determines how valid the assumption is for the McCabe-Thiele diagram. Percentage changes in L/V are less than changes in L and V individually because L and V necessarily change in the same direction. Consequently constant molal overflow often turns out to be a good assumption even when the relative volatility varies substantially.

Methods for handling systems with varying molal overflow are presented in Chap. 6.

Subcooled Reflux

In a simple tower with a partial condenser, as illustrated in Fig. 5-6, the reflux is saturated liquid. Hence the assumption of constant molal overflow leads to setting all liquid flows in the rectifying section equal to the overhead reflux rate. The same would be true of a tower equipped with a total condenser which returned saturated reflux. If, on the other hand, a total condenser were used and the *reflux were highly subcooled* below its boiling point, this liquid would in effect have to be heated to its boiling point before leaving the top stage. Such heating would be done through condensation of the vapor rising to the top stage from the stage below, and this condensed vapor would join the reflux flow to produce a larger flow of liquid from the top stage than the reflux flow itself. Thus it would be reasonable to expect the constant liquid flow in the rectifying section to be greater than the rate of reflux return to the tower; the *internal* reflux rate would be greater than the *external* reflux rate. Given the external reflux rate, the internal rate could be determined by a simultaneous solution of Eqs. (5-10) and (5-4), by the procedures developed in Chap. 6, or by the computer calculation methods of Chap. 10.

Operating Lines

Rectifying section Returning to the xy diagram, we see that if the assumption of constant molal flows is made in the rectifying section and a value of L is assigned, the general mass-balance equation for component A, relating the mole fractions of A in the vapor and liquid between stages, follows from Eq. (5-3):

$$V y_{A,p} = L x_{A,p+1} + D y_{A,D} \tag{5-16}$$

Equation (5-16) has been written using D and $y_{A,D}$ so as to reflect the use of a partial condenser. If D has also been set in the problem description, V and L in the equation are then set $(V = L + D)$. If $y_{A,D}$ has been set in the problem description, the equation can be plotted on the xy diagram as a straight line. This line, called the *rectifying-section operating line*, contains all possible pairs of compositions passing countercurrently between stages in the rectifying section. The operating line can be plotted on the diagram from any two of its following properties:

$$\text{Slope} = \frac{L}{V} \qquad \text{Intersection with 45° line} = x_A = y_A = y_{A,D}$$

$$\text{Intercept:} \quad y_A = \begin{cases} \dfrac{L + D y_{A,D}}{V} & \text{at } x_A = 1 \\[2mm] \dfrac{D y_{A,D}}{V} & \text{at } x_A = 0 \end{cases}$$

The intersection with the 45° line is almost always the most useful. It should be noted that the slope must always be *less* than 1 (or equal to 1 for no product). A typical rectifying-section operating line is shown in Fig. 5-7.

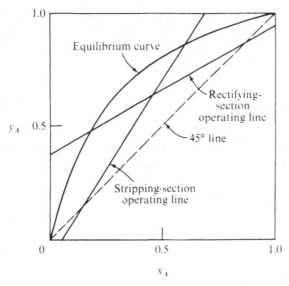

Figure 5-7 Operating lines for simple column with constant molal overflow.

Stripping Section At the point of feed introduction, one or both of the two rectifying-section flows, liquid and vapor, must be changed because of the feed entry. The new flows below the point of feed introduction are labeled L' and V' to distinguish them from the rectifying-section flows. To illustrate the estimation of these new flows, consider a feed which is partially vapor and partially liquid in equilibrium with each other at column pressure. Figure 5-8 shows how such a flashed feed might be introduced between stages.

In Fig. 5-8 the moles of liquid and vapor feed are labeled L_F and V_F, respectively. Following a rough enthalpy balance, a reasonable assumption would be that

$$L' = L + L_F \tag{5-17}$$

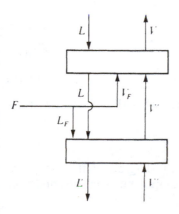

Figure 5-8 Introduction of partially vaporized feed.

V' can next be obtained by the overall mass balance [Eq. (5-8)]. The new flows L' and V' then would be assumed constant for all stages in the stripping section. This includes the reboiler, and hence the vapor flow leaving the reboiler would be V' mol.

Following Eq. (5-6), the operating-line equation for component A in the stripping section is

$$V'y_{A,p} = L'x_{A,p+1} - x_{A,b}b \qquad (5-18)$$

relating the concentrations of all streams passing each other between stages below the feed. If V', L', and $x_{A,b}b$ are fixed, this equation also can be plotted as a straight line on the xy diagram, again from any two of its properties:

$$\text{Slope} = \frac{L'}{V'} \qquad \text{Intersection with } 45° \text{ line} = x_A = y_A = x_{A,b}$$

$$\text{Intercept:} \quad y_A = \begin{cases} \dfrac{L' - x_{A,b}b}{V'} & \text{at } x_A = 1 \\[3mm] -\dfrac{x_{A,b}b}{V'} & \text{at } x_A = 0 \end{cases}$$

Another intersection, that with the operating line of the section above, is more useful and is discussed in the next section. It should be noted that the slope must always be *greater* than 1 (or equal to 1 for no product). A typical stripping-section operating line is shown in Fig. 5-7.

Intersection of Operating Lines

Consider the simple two-section distillation column of Fig. 5-6. The individual section operating lines are given by Eqs. (5-16) and (5-18). The intersection of these lines will be given either by the sum or the difference of the equations. The difference of the equations, however, leads to elimination of the effect of individual values of L, L', V, and V' and expresses the locus of the intersections for all values of these flows. Thus subtracting Eq. (5-18) from Eq. (5-16) and introducing Eq. (5-7) gives

$$Vy_A = Lx_A + x_{A,d}d \qquad (5-16)$$

$$-(V'y_A = L'x_A - x_{A,b}b) \qquad (5-18)$$

$$\overline{(V - V')y_A = (L - L')x_A + Fz_{A,F}} \qquad (5-19)$$

If the assumption is made that $L' - L = L_F$ no matter what the value of L, then Eq. (5-19) is simply

$$V_F\, y_A = -L_F x_A + Fz_{A,F} \qquad (5-20)$$

The locus of intersections is a straight line on the xy diagram with a slope of $-L_F/V_F$. If we substitute $z_{A,F}$, the mole fraction of component A in the total feed

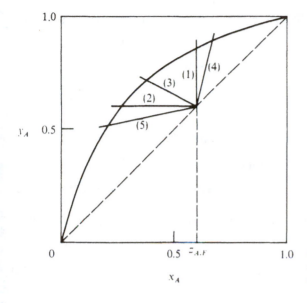

Figure 5-9 Loci of operating-line intersections for different feed phase conditions (see text for key).

(regardless of the condition of the feed or how component A is distributed between the vapor and liquid portions of the feed), for x_A we obtain

$$V_F y_A = (-L_F + F)z_{A,F} = V_F z_{A,F}$$

$$y_A = z_{A,F} = x_A$$

(5-21)

Thus the locus of intersections crosses the 45° line at the point $y_A = x_A = z_{A,F}$.

In order to facilitate calculation it is desirable to estimate the flow changes in Eq. (5-19) from the state of the feed alone, leading to Eq. (5-20). The assumption that $L' - L = L_F$ is not rigorously correct, but it is usually accurate enough for most purposes. Also, this assumption usually will be employed along with the assumption of constant molal flows in the two sections, an assumption which is also in error to about the same extent.

If the composition of the total feed is $z_{A,F}$, all the possible lines representing the locus of intersections of operating lines for the two sections of stages above and below the feed will go through the 45° line at $x_A = y_A = z_{A,F}$ and will have a slope of $-L_F/V_F$, depending on the state of the feed. Figure 5-9 shows typical loci of operating-line intersections for the five possible types of feed. Key numbers refer to the following items.

Saturated liquid feed (**1**) Feed at its bubble point under column pressure. Assume

$$L' - L = L_F = F \qquad \text{and} \qquad V - V' = V_F = 0$$

The slope of the intersection line is thus $-L_F/V_F = \infty$.

Saturated vapor feed (2) Feed at its dew point under column pressure. Assume

$$V - V' = V_F = F \qquad \text{and} \qquad L' - L = L_F = 0$$

The slope of the intersection line is thus $-L_F/V_F = 0$.

Partially vaporized feed (3) Comprises both saturated vapor and liquid portions. Assume

$$L' - L = L_F = \text{mol of liquid in feed}$$

$$V - V' = V_F = \text{mol of vapor in feed}$$

The slope of the intersection line is $-L_F/V_F$, a negative number between 0 and $-\infty$.

Subcooled liquid feed (4) Feed at a temperature below its column-pressure bubble point. Assume

$$L' - L = L_F = F\left(1 + \frac{h^* - h}{H_{eq} - h^*}\right) \tag{5-22}$$

where L_F = change in liquid flow at feed stage
$\qquad F$ = total moles of feed
$\qquad h$ = molal enthalpy of feed as fed
$\qquad h^*$ = molal enthalpy of liquid feed at column-pressure boiling point
$\qquad H_{eq}$ = molal enthalpy of vapor which would exist in equilibrium with feed if feed were at column-pressure boiling point.

With subcooled liquid feed, the increase in moles of liquid flow at the feed stage is greater than the moles of feed. In effect, vapor rising to the feed stage is condensed in order to heat the feed roughly to its boiling point. This condensed vapor adds to the liquid flow leaving the stage. The term $(h^* - h)/(H_{eq} - h^*)$ is an estimate of the quantity of this condensed vapor, the numerator representing the heat necessary per mole of feed and the denominator the heat obtained per mole of vapor condensed. It should be emphasized that this is an approximation, but in the absence of knowledge about the compositions of the streams around the feed stage approximation is necessary. If a correct answer is required, it can be obtained by an enthalpy balance around this stage at an appropriate point in the calculation or by use of the methods of Chap. 6. Following this approximate definition of L_F,

$$V - V' = V_F = F - L_F$$

and it is found that $L_F > F$ and $V_F < 0$. The slope of the intersection line $-L_F/V_F$ is a positive quantity lying between 1 and ∞.

Superheated vapor feed (5) Feed at a temperature above its column-pressure dew point. Assume

$$V - V' = V_F = F\left(1 + \frac{H - H^*}{H^* - h_{eq}}\right) \tag{5-23}$$

where V_F = change in vapor flow at feed stage
$\quad\quad F$ = total moles of feed
$\quad\quad H$ = molal enthalpy of feed as fed
$\quad\quad H^*$ = molal enthalpy of feed at column-pressure dew point
$\quad\quad h_{eq}$ = molal enthalpy of liquid which would be in equilibrium with feed if feed
$\quad\quad\quad$ were at column-pressure dew point

The same reasoning applied to subcooled liquid feeds applies here. Again, following this definition of V_F,

$$L' - L = L_F = F - V_F$$

and, since $V_F > F$, $L_F < 0$. Since $L_F + V_F = F$, the slope of the intersection line, $-L_F/V_F$, is a positive quantity lying between 0 and 1.

Multiple Feeds and Sidestreams

When two streams of different compositions are to be fed to the same tower, it is common to bring them in at different points in the column. In this case the operating-line equations for the rectifying and stripping sections are the same as for the simple column, but a new operating-line equation is required for the section of stages between the two feeds. Such a column is shown in Fig. 5-10, along with a sketch of

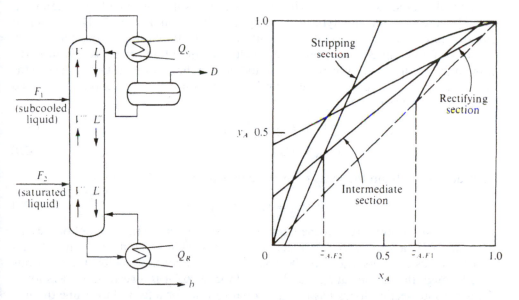

Figure 5-10 Column with two feeds.

typical operating lines for the column on the xy diagram. The equations of the three operating lines are:

Rectifying section: $\qquad V y_A = L x_A + D y_{A,D}$ (5-16)

Intermediate section: $\qquad V'' y_A = L' x_A + D y_{A,D} - F_1 z_{A,F_1}$ (5-24)

Stripping section: $\qquad V' y_A = L' x_A - b x_{A,b}$ (5-18)

The vapor and liquid flows V'' and L' are those estimated for the intermediate section. Thus if L had been estimated for the rectifying section, the liquid flow in the intermediate section normally would be estimated to be

$$L' = L + L_{F_1}$$ (5-25)

and $\qquad V'' = L' + D - F_1$ (5-26)

from the total mass balance. Then, in the same way,

$$L' = L'' + L_{F_2}$$ (5-27)

and $\qquad V' = L' - b$ (5-28)

From Eq. (5-24), estimation of the total vapor and liquid flows in the section, and knowledge of the net upward product for the section, the intermediate-section operating line can be plotted from any two of its properties. It should be noted that the operating line between feeds necessarily has a greater positive slope than the rectifying-section operating line and a lesser positive slope than the stripping-section operating line.

A locus of intersections of the operating lines for the two adjacent sections of stages will exist for any effect which produces changes in flow. As a further example, consider the column and typical xy diagram shown in Fig. 5-11. Here a liquid sidestream of amount L_S and of composition $x_{A,S}$ is drawn from a stage as a third product of different composition. As a result, the liquid flow is changed from L, the flow in the rectifying section, to L', the liquid flow in the section of stages between the side draw and the feed. The operating-line equations for the sections above and below the sidestream are

$$V y_A = L x_A + D y_{A,D}$$ (5-16)

$$V'' y_A = L' x_A + D y_{A,D} + L_S x_{A,S}$$ (5-29)

Subtracting to obtain the locus of intersections of the two lines gives

$$(V - V'') y_A = (L - L') x_A - L_S x_{A,S}$$ (5-30)

Obvious assumptions are that the liquid flow is decreased by the amount of the side draw and the vapor flow is unchanged. Under these changes in flow it will be found that the locus of intersections of the operating lines above and below the side draw goes through the 45° line at $x_{A,S}$ and has a slope of ∞, or is a vertical line as shown. Thus the intersection locus has a slope equal to that for a feed of the same thermal condition as the sidestream; this is a general result. For a side-product withdrawal of

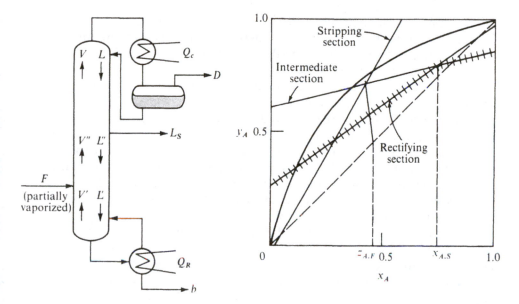

Figure 5-11 Column with a side stream.

this sort, the intermediate-section operating line has a lesser positive slope than the rectifying-section operating line.

The portions of the operating lines denoted by hatched lines in Fig. 5-11 are imaginary in the sense that there is no way they can be used for stage stepping. A side draw must be removed exactly at the intersection of the operating lines above and below the side draw. On the other hand, a feed does not necessarily have to be introduced exactly at the intersection of the operating lines above and below it.

THE DESIGN PROBLEM

Specified Variables

As noted earlier, the number of variables which can be specified independently for a distillation is considered in Appendix C. One common situation, the *design problem*, is easily analyzed and is frequently encountered. In a design problem, the separation desired is specified, a flow at some point is specified (usually the reflux), and the number of stages required in each section of the column is calculated; hence the column to accomplish the chosen separation at a particular reflux is designed. For a binary system in a column of two sections and a partial condenser, like that in Fig. 5-6, the list of variables set is thus

$Fx_{A,F}$ and $Fx_{B,F}$	Second separation variable
h_F	Reflux flow rate
Pressure	Location of feed point
One separation variable	

The composition and amount of the feed are fixed (alternatively, the amount of each component in the feed is fixed). All mass flows and energy flows (the extensive variables) are directly proportional to the amount of feed, while the number of stages required and other intensive variables are independent of the amount of feed. A common practice in distillation calculations is to base the calculation on 1 mol of feed, later multiplying all flows by the actual amount of the feed.

The enthalpy and pressure of the feed once it enters the column are fixed; hence an isenthalpic equilibrium-flash calculation (Chap. 2) can be used to compute the amounts and compositions of the vapor and liquid portions of the feed if both will exist at column pressure. If the feed is wholly liquid or vapor, the changes in flow at the feed point are estimated from the enthalpy of the feed. In either case, for the purposes of the McCabe-Thiele diagram, setting h_F sets values to V_F and L_F. The pressure of the column is also fixed; this in turn fixes the equilibrium curve.

Two separation variables are fixed. One of the properties of a binary distillation system is that if the feed rate and composition and two independent separation variables are fixed, everything about the products from a column producing two products is fixed. For example, if $(/_A)_D$ and $(/_B)_D$ are fixed, $Dy_{A,D}$, D, $y_{A,D}$, $bx_{A,b}$, b, and $x_{A,b}$ are known. Readers should confirm this fact for themselves.

The reflux is fixed. All vapor and liquid flows are assumed constant in both sections. Since a partial condenser is used, the reflux is saturated liquid; therefore $L = r$. Then $V = r + D$, $L' = r + L_F$, $V' = L' - b$. Use of a total condenser would require setting another variable, related to the thermal condition of the reflux.

The last variable to be fixed is the arbitrary location of the feed. This variable will be assigned a value during the calculation.

As an example design problem we shall consider a benzene-toluene distillation, with the pressure set at 1 atm to provide an essentially constant relative volatility of 2.25. Other variables are set as follows:

Feed is 1 mol
Mole fraction benzene in feed is 0.40
Feed enthalpy is such that feed is saturated liquid
Separation variables: 90 percent of the benzene to be recovered in 95 percent purity
Reflux is 1 mole per mole of feed
Location of feed to be determined later

Graphical Stage-to-Stage Calculation

Proceeding to the xy diagram, we can now plot the equilibrium curve and operating lines as shown in Fig. 5-12. The equilibrium curve comes from substituting $\alpha = 2.25$ into Eq. (1-12). The line giving the locus of intersections of the two operating lines is plotted from its slope $-L_F/V_F$ $(= \infty)$ and its intersection with the 45° line at the value of $z_{A,F}$ $(= 0.40)$. The rectifying-section operating line is plotted from $Dy_{A,D}/V$, the intercept at $x_A = 0$, and the intersection with the 45° line at $y_{A,D}$. From the

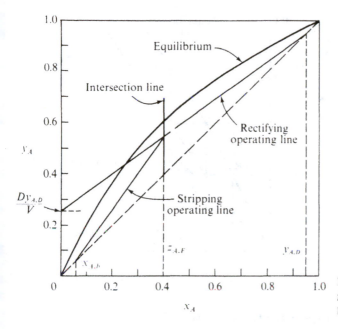

Figure **5-12** McCabe-Thiele diagram for benzene-toluene design problem.

specifications, $Dy_{A,D} = (/_A)_D z_{A,F} F = (0.90)(0.40) = 0.36$. Since $y_{A,D}$ is set at 0.95, $D = 0.36/0.95 = 0.379$, and since $V = L + D = 1.379$,

$$\frac{Dy_{A,D}}{V} = \frac{0.36}{1.379} = 0.261$$

The stripping-section line can now be plotted in several ways, but the easiest is from (1) the triple intersection of the locus of operating-line intersections with the two operating lines (already established by the point where the locus of intersections and the rectifying operating line meet) and (2) the intersection of the stripping-section operating line with the 45° line at $x_{A,b}$. These lines have been drawn in this way in Fig. 5-12, and $x_{A,b}$ is determined as follows: $b = 1 - D = 0.621$; $x_{A,b} b = (/_A)_b z_{A,F} F = (0.10)(0.40) = 0.040$; $x_{A,b} = 0.040/0.621 = 0.064$.

The lower left-hand corner of the xy diagram has been enlarged in Fig. 5-13, and a graphical stage-to-stage calculation is shown, starting at the reboiler. The composition of the bottoms product $x_{A,b}$ is known. This composition is one of a pair of compositions represented by a point on the equilibrium curve, the pair being the composition of the reboiler vapor and the composition of the bottoms product, since these two flows are presumed to be in equilibrium as they leave the reboiler. The composition of the bottoms product is the abscissa of this point on the equilibrium curve, the ordinate is the composition of the reboiler vapor, and the reboiler itself is represented by the point on the equilibrium curve denoted by R.

The composition of reboiler vapor and the composition of the liquid flow from stage 1 of the column are a pair of compositions represented by a point on the operating line of the stripping section, since they pass each other between stages of

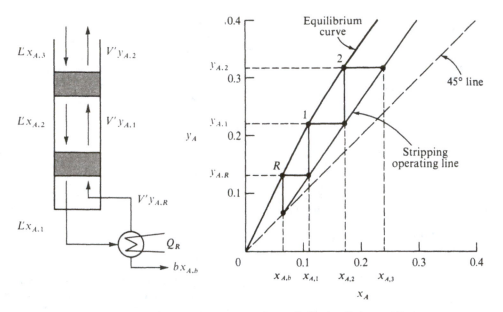

Figure 5-13 Bottom stages of column in benzene-toluene distillation design problem.

the stripping section. The ordinate $y_{A, R}$ of this point is now known, and hence the abscissa $x_{A, 1}$ is easily found. Again, $x_{A, 1}$ is the abscissa of a point on the equilibrium curve representing the pair of equilibrium compositions $x_{A, 1}$ and $y_{A, 1}$. Next, $y_{A, 1}$ is the ordinate of a point on the operating line representing the passing pair of compositions $y_{A, 1}$ and $x_{A, 2}$, etc. The calculation procedure involves the alternating use of equilibrium calculation and mass-balance calculation, proceeding from stage to stage and calculating the next unknown composition by whichever relation applies. The entire computation can thus be performed algebraically rather than graphically if one desires.

Executed graphically on the McCabe-Thiele diagram, the calculation resembles the construction of a staircase leading from the bottom of the column to the top. It would be well to remember, however, that the points of the staircase which fall on the equilibrium curve represent equilibrium stages and the points on the operating lines represent pairs of phases between stages. The lines of the staircase themselves represent nothing physically.

Figure 5-14 shows the complete xy diagram for the same column as Fig. 5-13. It is apparent that a choice of drawing the next horizontal leg of a step to either operating line becomes possible as soon as a stage is reached on the equilibrium curve which is farther to the right than the intersection of the rectifying-section operating line with the equilibrium curve. If the horizontal leg is drawn to the rectifying-section operating line, the arbitrary choice to change flows has been made and the feed has been introduced. This choice of feed stage constitutes the last variable in the list of specified variables. Construction of steps must now continue with horizontal lines drawn to the rectifying-section operating line until a stage is

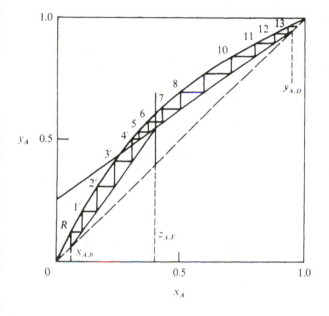

Figure 5-14 Complete McCabe-Thiele diagram for benzene-toluene distillation design problem.

found from which the vapor is equal to, or richer, in component A than the top product. The number of equilibrium stages in each section is then the number of points on the equilibrium curve for that section.

In Fig. 5-14 the feed has been arbitrarily introduced as soon as possible in the construction upward. The stripping-section stages are primed, and rectifying stages are unprimed. Stage 13 produces a vapor richer than the required top product. Obviously all the enrichment produced in stage 13 is not necessary to produce the desired top product, and only a fractional part of the stage is required. The topmost point on the equilibrium curve is the partial condenser, so that the number of equilibrium stages required in the column itself is only 11 plus some fraction.

In a real design problem the stage efficiency would be used for converting from equilibrium stages to the number of actual stages. If the number of actual stages were still fractional (as in all likelihood it would be), the procedure would be to increase the stage requirement to the next highest integral value, since obviously an integral number of stages must be constructed. The separation resulting from this increase to the next higher integral number of stages would necessarily be better than the design specification. This is a comfortable situation.

The construction in the diagram could just as well be started from the top and continued to the bottom or started from any intermediate point and continued in both directions. Starting from the top is illustrated in Fig. 5-15 for a column with a partial condenser and a column with a total condenser. The top stage in the column is labeled t, the next stage $t-1$, etc. The construction for the case of a total condenser is apparent once it is recalled that $y_{A,t}$ must equal $x_{A,d}$ since the vapor from the top stage is totally condensed and then divided into reflux and top product of the same composition.

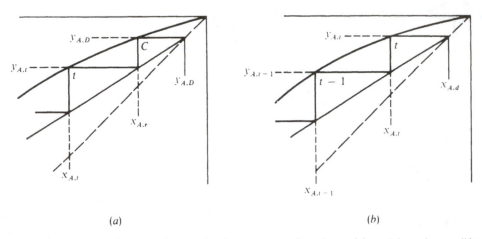

(a)   (b)

Figure 5-15 Graphical construction starting from the top of a column: (a) partial condenser; (b) total condenser.

Feed Stage

Figure 5-16 shows four examples of the choice of the location of the feed point. In Fig. 5-16a construction was started at the bottom and continued as long as possible on the stripping-section operating line. The steps become smaller as construction proceeds toward the intersection of the stripping-section operating line with the equilibrium curve. It is apparent that infinite stages would be required to get to this intersection, and such a section of stages, an infinite stripping section, is a very useful concept even though unbuildable. This intersection is also commonly called a *point of infinitude* in the stripping section or a *pinch point* in the stripping section. The rectifying section in Fig. 5-16a is started at this point and requires relatively few stages. In Fig. 5-16b construction proceeded from the top to a point of infinitude in the rectifying section and then to the stripping section, which required relatively few stages. In Fig. 5-16c and d different choices of feed location have been made.

All these examples have been shown to illustrate the wide choice of feed location available and the column requirements which result. All the columns, if it were possible to build infinite columns and fractional stages, would be operable at the selected reflux to produce the separation specified. These examples do illustrate, however, that an optimum location exists for the feed introduction. It is apparent that drawing steps into either of the constricted areas, following a given operating line beyond the intersection of operating lines, increases the total number of stages required. The optimum point of feed introduction, which yields minimum total stages required at the particular reflux, occurs when steps are always drawn to the operating line that lies *farther* from the equilibrium curve at the particular point under consideration in the column. This policy ensures that all steps are of maximum possible size and thus that the minimum number of stages has been employed.

This point is further illustrated in Fig. 5-17, which qualitatively shows the equilibrium-stage requirement as a function of the liquid composition on the stage above which the feed is introduced. For saturated liquid feed, the optimal feed

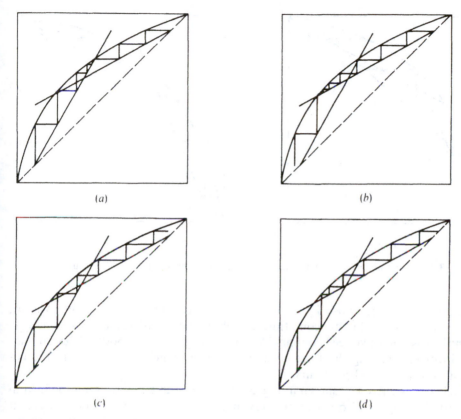

(a) (b)

(c) (d)

Figure 5-16 Alternatives for point of feed introduction.

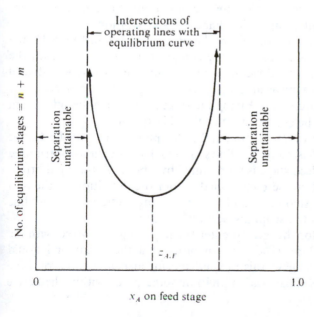

Figure 5-17 Equilibrium-stage requirement vs. feed location (saturated liquid feed).

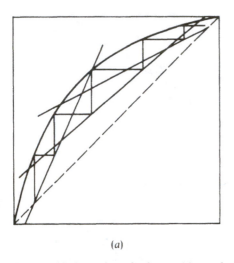
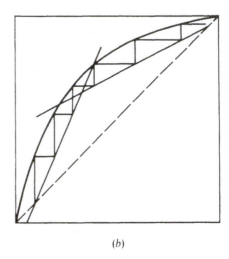

(a)   (b)

Figure 5-18 Operation of column with two feeds: comparison of (a) separate and (b) combined feeds.

location is to the stage whose liquid most closely approximates the feed composition; however, for other feed phase conditions the optimal point of feed entry will change.

In the case of a two-feed column, like that in Fig. 5-10, the optimal design once again involves inserting the feeds so as to keep on whichever operating line lies farthest from the equilibrium curve.

Such a construction is shown in Fig. 5-18a for the same separation as defined in Fig. 5-10. This figure also shows the advantages of introducing different feeds at different points in a column, rather than combining them. More stages are required in Fig. 5-18b, which depicts the same separation problem with the feeds combined. Mixing is the opposite of separation, and for that reason alone it might be apparent that combining feeds of different composition hampers a separation.

In the case of a sidestream withdrawal, on the other hand, as in Fig. 5-11, the composition of the sidestream is necessarily the composition of the liquid on the stage from which it is withdrawn. Readers should convince themselves that if the sidestream is liquid, it must have an x_A corresponding to the intersection of operating lines caused by the sidestream. Fixing the stage from which the sidestream is withdrawn necessarily fixes the composition of the sidestream.

If a column consists of a rectifying section and a stripping section, the feed stage is defined as the stage which has rectifying flows (L and V) above it and stripping flows (L' and V') below it. The feed stage is represented by a point on the equilibrium curve or by a "step" in the staircase constructed. From the definition of the feed stage, the horizontal leg of the step must go to the rectifying-section operating line and the vertical leg must go to the stripping-section operating line.

Physically, it is necessary for the feed to enter the feed stage, be mixed with the other incoming streams, and lose its identity in order that the vapor and liquid leaving be in equilibrium. To achieve this mixing, and for mechanical reasons, the feed is usually added between stages rather than at some point within the active bubbling region composing a stage. Thus, as shown in Fig. 5-8, a liquid feed would be physically introduced above the feed stage. If a feed of both vapor and liquid (partially vaporized) is introduced and a single feed stage is postulated, it will tacitly be assumed that the feed is separated outside the column, the liquid portion then being fed above and the vapor portion below the feed stage. This is not done in practice, the whole feed simply being inserted between two stages, and to be completely correct the feed should then be treated as two feeds, the vapor portion being fed to a feed stage of its own and the liquid portion being fed to a feed stage of its own, which is the next below. The correction for this is minor, however, and can usually be ignored.

Allowable and Optimum Operating Conditions

A binary distillation column can be designed and operated at various combinations of stages and reflux in order to accomplish a given separation. A typical plot of the various solutions for a distillation with two separation variables specified is shown in Fig. 5-19. Higher reflux ratios make the operating lines farther removed from the equilibrium curve and thereby require fewer stages. Near the left of the diagram the stage requirements would be high, but the reflux requirements (and heating and cooling loads, internal flows, and column diameter) would be low. A solution near the right of the diagram would give just the opposite effect. The lowest-cost design will lie at an intermediate condition, since utilities costs become infinite at one extreme and column costs become infinite at the other extreme.

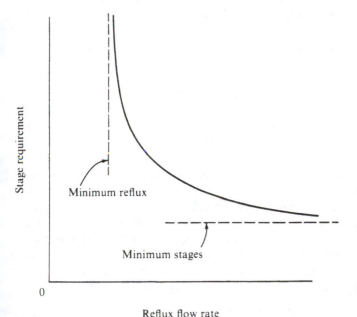

Figure 5-19 Stages vs. reflux.

Appendix *D* explores the economic optimum reflux ratio and gives a worked example. For the late 1970s and the foreseeable future, energy costs are so high relative to fabricated-materials costs that the economic optimum design reflux ratio often falls less than 10 percent above the minimum allowable reflux ratio. However, in such a situation the design can be very sensitive to uncertainties in the vapor-liquid equilibrium data, the stage efficiency, and/or the feed composition. To the extent that such uncertainties exist, it is better to design for a somewhat higher reflux ratio, assuring that the column will be able to meet the design separation and capacity.

Determinations of the optimum column pressure, recovery fractions of components, and the amount of feed preheating are also explored in Appendix D. Feed preheating can reduce the required steam rate to the reboiler, although vapor generated in a feed preheater is useful only in the rectifying section. Other means of reducing the energy consumption in distillation are developed in Chap. 13.

Limiting Conditions

The curve of possible solutions does not come to zero stage requirements as the reflux is increased to infinity, nor does it come to zero reflux as the stages increase to infinity. The limits on Fig. 5-19 are approached asymptotically at each end and are labeled on the curve as "minimum reflux" and "minimum stages." These values are the least amount of reflux and the least number of stages which can possibly give the desired separation. In the usual case, minimum reflux requires infinite stages in *both* sections of the column, and minimum stages require infinite reflux and infinite internal flows. These limits are obviously useful and should be considered. As a matter of fact, estimates of column requirements and reflux flows for separations with finite stages and with finite reflux can be made with reasonable accuracy from a knowledge of the two limits, as shown in Chap. 9.

The limit of minimum reflux is easily shown on the McCabe-Thiele diagram. The operating lines which are obtained for a set separation with a variety of reflux values are shown in Fig. 5-20. As reflux is decreased at a fixed distillate flow, the slope of the rectifying operating line diminishes, since in $L/(L + d)$ the numerator decreases on a percentage basis faster than the denominator. Hence in Fig. 5-20, $r_1 > r_2 > r_3 > r_4$. It is also apparent that at reflux $= r_4$, construction of the stage requirements begun at both ends of the column will never meet and that the separation required cannot be produced by *any* column at this reflux. The lowest value of reflux at which the two constructions will meet, r_3, is the minimum allowable reflux for the separation, and here infinite stages are required in both sections to achieve the separation. The two points of infinitude meet at the feed stage, and the composition of the feed stage is given by the intersection of the feed line and the equilibrium curve. If the coordinates of this point are known, labeled $x_{A, f, r_{min}}$ and $y_{A, f, r_{min}}$ on Fig. 5-20, the minimum reflux is easily obtained from the slope of the rectifying-section operating line

$$\frac{r_{min}}{r_{min} + d} = \frac{x_{A, d} - y_{A, f, r_{min}}}{x_{A, d} - x_{A, f, r_{min}}} \qquad (5\text{-}31)$$

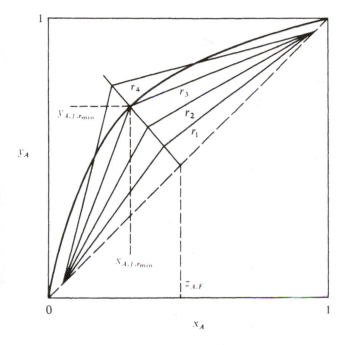

Figure 5-20 Minimum reflux construction; normal case.

It should be noted that the equation is correctly

$$\frac{L_{\min}}{L_{\min} + d} = \frac{x_{A,d} - y_{A,f,r_{\min}}}{x_{A,d} - x_{A,f,r_{\min}}} \tag{5-32}$$

since the minimum liquid flow in the rectifying section is being calculated. If the reflux were saturated, it would be assumed that $L_{\min} = r_{\min}$. If the reflux were subcooled, a correction should be made and $r_{\min}$ would be less than $L_{\min}$. Other expressions useful for calculating minimum reflux are presented in Chap. 9. For minimum reflux the variables specified are

$Fx_{A,F}$, $Fx_{B,F}$	$(/_A)_d$ and $(/_B)_d$ or two other
h_F	separation specifications
Pressure	$n = \infty$
Reflux temperature (if a total condenser)	$m = \infty$

In most cases the point of infinitude occurs at the intersection of the operating lines, and there are infinite plates both above and below the feed. This is not necessarily always the case, however, as shown in Fig. 5-21, which is constructed for a binary mixture showing relatively strong positive deviations from ideality. The minimum reflux condition comes from a *tangent* pinch in the rectifying section. In this case there are infinite plates above the feed but not below, and the $m = \infty$ specification is replaced by a stipulation concerning the point of feed introduction.

The limit of *minimum stages* at infinite reflux (or *total* reflux) is shown in Fig. 5-22. As reflux is increased toward infinity, the slopes of both operating lines tend to unity, since d and b become infinitesimal in comparison with L and L'. Both

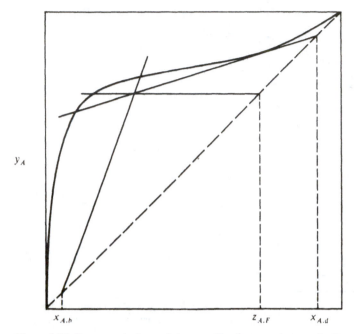

Figure 5-21 Tangent pinch at minimum reflux (saturated-vapor feed).

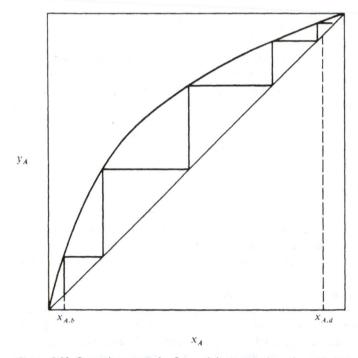

Figure 5-22 Operation at total reflux; minimum number of equilibrium stages.

operating lines lie on the 45° line. Construction of stage requirements is the same as for any reflux except that the position of the feed is immaterial. The problem being solved is described by the following variables

$Fx_{A,F}$, $Fx_{B,F}$	$(/_A)_d$ and $(/_B)_d$ or two other
h_F	separation specifications
Pressure	$r = \infty$
Reflux temperature (if a total condenser is used)	Arbitrary feed-plate location

Of this list of variables, the values of feed-plate location, feed composition and flow rate, h_F, and reflux temperature are immaterial since it is impossible to change the infinite internal flows or the product compositions by their selection.

Allowance for Stage Efficiencies

There are two common approaches to allowing for the influence of nonequilibrium stages on the quality of separation achieved or on the number of stages required for a distillation. The first of these involves the use of an *overall efficiency* E_O defined as

$$E_O = \frac{\text{number of equilibrium stages}}{\text{number of actual stages}} \tag{5-33}$$

The number of equilibrium stages and the number of actual stages are both those required for the specified or measured product purities. In order to use the overall efficiency in a design problem, one carries out an equilibrium-stage analysis and then determines the number of actual stages as the number of equilibrium stages divided by E_O. Thus the overall efficiency concept is simple to use once E_O is known, but it is often not easy to predict reliable values of E_O. In the petroleum industry it has been found that $E_O = 0.6$, or 60 percent, is often a satisfactorily conservative value for analyzing the common distillations; however, E_O can vary widely.

The other commonly used approach involves the concept of the *Murphree vapor efficiency* E_{MV}, defined in Chap. 3,

$$E_{MV} = \frac{y_{\text{out}} - y_{\text{in}}}{y^* - y_{\text{in}}} \tag{3-24}$$

where y^* is the vapor composition which would be in equilibrium with the actual value of x_{out}. There is more theoretical basis for correlating and predicting values of E_{MV} than for E_O, as we shall see in Chap. 12.

If the value of E_{MV} is known for each stage in a binary distillation (or is taken at a single known constant value for the whole column), it can readily be used in the McCabe-Thiele graphical construction. Referring to Fig. 5-23, let us presume that the value of E_{MV} for the distillation under consideration is known to be 0.67. Consider the case of a calculation proceeding up the column. If we know the compositions of the passing streams below the next stage we want to calculate, we then know x_{out} and y_{in} for that stage. This means that we know the point on the operating line marked A in Fig. 5-23. Proceeding upward to the equilibrium curve at that value of $x\ (= x_{\text{out}})$, we find y^*, corresponding to equilibrium with x_{out}. We thus know point B.

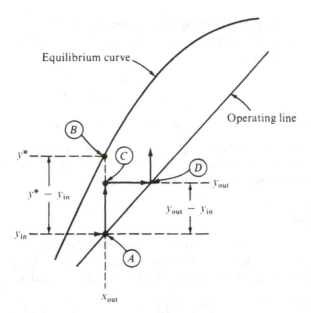

Figure 5-23 Use of Murphree vapor efficiency in McCabe-Thiele construction.

Points A and B fix the value of $y^* - y_{in}$. Now, using Eq. (3-24) and our known $E_{MV} = 0.67$, we are able to fix $y_{out} - y_{in}$ as $0.67(y^* - y_{in})$ and thereby locate point C. This point gives the pair of exit-stream compositions for the stage under consideration. From point C we then step over to point D on the operating line and are ready to go through the same procedure for the next stage.

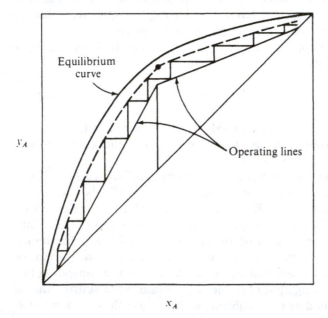

Figure 5-24 Locus of pairs of stage exit compositions for $E_{MV} = 0.67$.

This procedure for using E_{MV} works well for calculations upward in a tower but is more difficult for calculations downward. An expedient for the case where E_{MV} is the same on all stages is shown in Fig. 5-24. A new curve, shown dashed in Fig. 5-24, is drawn in and is so located that it always lies a fraction E_{MV} of the vertical distance to the equilibrium curve from the appropriate operating lines. Actual stages can then be stepped off, as shown, using this curve rather than the equilibrium curve. The dashed curve in Fig. 5-24 corresponds to $E_{MV} = 0.67$, including the reboiler.

Means of predicting and correlating stage efficiencies are covered in Chap. 12.

OTHER PROBLEMS

The solution of a typical design problem has been shown at some length because these problems constitute a large proportion of those encountered and because they can be solved in a straightforward fashion on the McCabe-Thiele diagram. Any other problem which might be encountered in binary distillation also can be solved on the diagram but not necessarily in a straightforward way. Trial-and-error procedures are often required, and the exact way to approach the solution on the diagram must be thought out. In some cases an algebraic approach is as efficient as the graphical approach or more so.

Example 5-1 illustrates the solution of a problem concerning the operation of an existing still.

Example 5-1 A binary mixture is to be separated in a column which contains five *equilibrium* rectifying stages plus a reboiler. The feed is a saturated liquid at column pressure and is introduced into the reboiler. A total condenser is used, and the reflux will be returned to the column at its saturation temperature. The reflux rate is 0.5 mol per unit time. The feed rate is 1 mol per unit time, and $z_{A,F} = 0.5$. The mole fraction of component A in the top product is to be 0.90. Assume that the relative volatility is constant at $\alpha_{AB} = 2$ for the column pressure and the temperature spread of the column. Also assume constant molal overflow. Calculate the mole fraction of component A in the bottom product and calculate the amounts of top and bottom products.

SOLUTION From the foregoing description, the column is as shown in Fig. 5-25. For such a column the number of variables to be set in any problem description is counted as

$F z_{A,F}$ and $F z_{B,F}$	n
h_F	Q_C
Pressure	Q_R
Reflux temperature	

These are the variables which can be set structurally or by external manipulation during operation. In the actual problem description the following variables are set:

$F z_{A,F}$ and $F z_{B,F}$	$F = 1$ mol, $z_{A,F} = 0.5$, $z_{B,F} = 0.5$
h_F	Feed is saturated liquid $L_F = F$
Pressure	$\alpha_{AB} = 2$
Reflux temperature	Reflux is saturated
n	$n = 5$
r (replacing Q_C)	$r = 0.5$ mol
$x_{A,d}$ (replacing Q_R)	$x_{A,d} = 0.90$

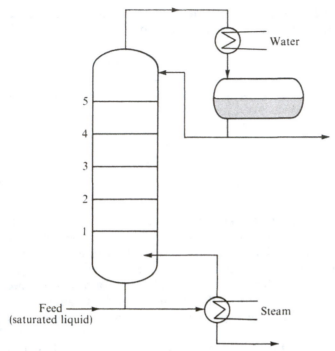

Figure 5-25 Column for Example 5-1.

The equilibrium curve can be drawn on the xy diagram from Eq. (1-12)

$$y_A = \frac{2x_A}{1 + x_A} \tag{5-34}$$

For various values of x_A between 0 and 1, the corresponding values of y_A in equilibrium are calculated and plotted on the xy diagram, as shown in Fig. 5-26.

It can be assumed that a good estimate of the liquid flow in the rectifying section will be $L = r$, since the reflux is saturated. When $x_{A,d}$ is known but d is not known, the rectifying-section operating line cannot be plotted. The locus of operating-line intersections can be plotted and is a vertical line through $z_{A,F} = 0.5$.

Since all independent variables have been set, the total amounts of products and the bottoms product composition are dependent variables whose values are unknown. If a value of d is assumed, the corresponding values of b and $x_{A,b}$ can be calculated by overall mass balance. The assumed value of d will also yield a particular set of operating lines, and consequently five equilibrium stages plus the reboiler can be stepped off on the diagram from the top down. If the assumed value of d is correct, the last vertical leg on the diagram, representing $x_{A,b}$, will be at the same value of $x_{A,b}$ that was obtained from the overall mass balance. If not, another value of d must be assumed until the correct value is found. The solution procedure therefore involves assuming values for one dependent variable d, then calculating another dependent variable $x_{A,b}$ by two different routes until both routes give the same value of $x_{A,b}$. A typical solution follows.

Assume $d = 0.25$ This id sone since the column is inefficient for stripping A out of the bottom product (why?) and hence should not make a large amount of the high-purity A top product.

$$x_{A,d}d = (0.25)(0.9) = 0.225 \qquad V = L + d = r + d = 0.5 + 0.25 = 0.75$$

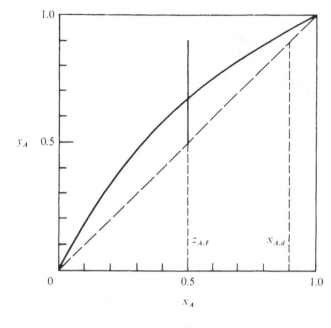

Figure 5-26 Initial operating diagram for Example 5-1.

The intercept of the operating line at $x_A = 0$ is

$$\frac{dx_{A,d}}{V} = \frac{0.225}{0.75} = 0.300$$

$$x_{A,b}b = Fx_{A,F} - x_{A,d}d = 0.5 - 0.225 = 0.275$$

$$b = F - d = 1 - 0.25 = 0.75 \qquad x_{A,b} = \frac{0.275}{0.75} = 0.367$$

Figure 5-27 shows the operating line corresponding to $d = 0.25$ and the construction of steps corresponding to the stages. Note that each time a step is made to an operating line, it is the rectifying-section operating line which is appropriate. From the construction, the vertical leg from the reboiler is at $x_A = 0.47$; this is above the value of $x_{A,b} = 0.367$, which was calculated from the overall mass balance. Hence $d = 0.25$ is wrong. If the next guess of the amount of top product is lower, there will be more A in the bottom product and $x_{A,b}$ calculated from the overall mass balance will be higher on the diagram. Also the slope of the rectifying operating line will be closer to unity and the construction of stages will move further down the diagram. These are opposing effects; hence the calculation will converge readily, and the next assumed value for d should be less than 0.25.

Assume d = 0.18

$$x_{A,d}d = (0.18)(0.9) = 0.162 \qquad V = 0.5 + 0.18 = 0.68$$

The intercept at $x_A = 0$ is

$$\frac{0.162}{0.68} = 0.238$$

$$x_{A,b}b = 0.5 - 0.162 = 0.338$$

$$b = 1 - 0.18 = 0.82 \qquad x_{A,b} = \frac{0.338}{0.83} = 0.412$$

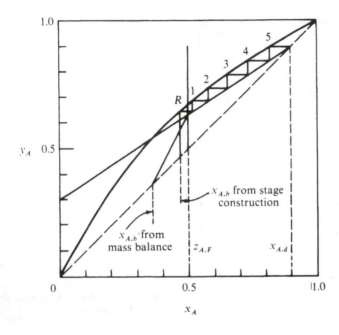

Figure 5-27 First-trial construction for Example 5-1, with $d = 0.25$.

Figure 5-28 shows the new operating lines and new construction. The vertical leg from the reboiler is at $x_A = 0.40$, which is now below the value of $x_{A,b} = 0.412$ obtained from the mass balance. Another assumption of d could be made between $d = 0.25$ and $d = 0.18$, but the inaccuracy in construction does not really justify it. Linear interpolation between the two values is probably the best way to arrive at the answer. On this basis

$$d = 0.18 + (0.25 - 0.18)\frac{0.012}{0.012 + 0.103} = 0.186$$

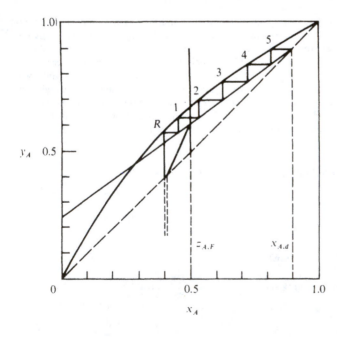

Figure 5-28 Second-trial construction for Example 5-1, with $d = 0.18$.

Thus the answers required are $d = 0.186\,\text{mol}$, $b = 0.814\,\text{mol}$, and $x_{A,b} = 0.408$. It might be noted that the results of the calculation show that a very small fraction of component A in the feed has been recovered as a relatively pure top product. The amount of recovery could be increased by using a feed point higher in the column. □

MULTISTAGE BATCH DISTILLATION

Multistage distillations also can be run on a batch basis. In the column shown in Fig. 5-29 an initial charge of liquid is fed to the still pot. The heating and cooling media are then turned on, and distillation proceeds, continually depleting the liquid in the still pot and building up overhead product in the distillate receiver. The operation is therefore the same as the simple Rayleigh distillation shown in Fig. 3-8 except for the presence of plates above the still pot and for the manufacture of reflux.

Batch stills require considerably more labor and attention than continuous columns. It is also necessary to shut down, drain, and clean the column in between charges, and this can result in a substantial loss of on-stream time. Consequently batch multistage distillation is most often employed when a product is to be manufactured only at certain isolated times and where a number of different mixtures can be handled at different times by the same column. Batch distillations are more common in smaller, multiproduct plants.

In a batch distillation the compositions at all points in the column are continually changing. As a result a steady-state analysis of the type employed for continuous distillation cannot be made of the column behavior. On each plate a mixing process is occurring such that

Input − output = accumulation

$$V_{p-1}y_{A,\,p-1} + L_{p+1}x_{A,\,p+1} - V_p y_{A,\,p} - L_p x_{A,\,p} = \frac{d}{dt}(Mx_{A,\,p}) \qquad (5\text{-}35)$$

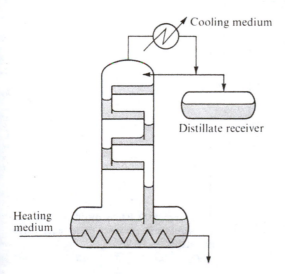

Figure 5-29 Batch multistage distillation.

In Eq. (5-35), M is the liquid holdup, the number of moles of liquid present on plate p. It is assumed that the accumulation of A in the vapor on the plate is negligible because of the low vapor density and that the liquid on the plate is well mixed; otherwise the x_A on the right-hand side should be the average across the plate rather than being the stage exit mole fraction.

The holdup on the plates is often low enough to permit the time-derivative term to be neglected in comparison with the terms on the left-hand side of Eq. (5-35). This situation occurs when the holdup on the plates is a small fraction (5 percent or less) of the charge to the still. One can then employ the steady-state continuous column equations to relate the compositions within the batch column at any time. This, in turn, means that the McCabe-Thiele diagram can be used to relate the compositions, provided the mixture is binary. The holdup in the still pot remains a highly important factor, however.

It is possible to operate a batch distillation column so as to hold the reflux ratio constant throughout the distillation or else the reflux ratio may be allowed to vary in any arbitrary way. Two reflux policies are amenable to relatively simple analysis, i.e., *constant reflux ratio* and *constant distillate composition*.

The McCabe-Thiele analysis for a low-holdup column run at constant reflux ratio is illustrated in Fig. 5-30. The operating lines at different times are a series of parallel lines, the slopes being the same since L/V is constant. The construction for an overhead composition of x_{A,d_1} is shown by solid lines. Since component A is removed preferentially in the distillation, $x_{A,b}$, and hence $x_{A,d}$, will be lower at a later time. The dashed lines give the construction for a later time when the overhead composition is x_{A,d_2}. In each case the operating line of known slope is drawn away from the value of $x_{A,d}$ under consideration. In Fig. 5-30 three equilibrium stages and

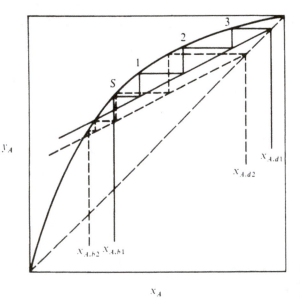

Figure 5-30 Batch distillation at constant reflux ratio; three equilibrium stages plus reboiler.

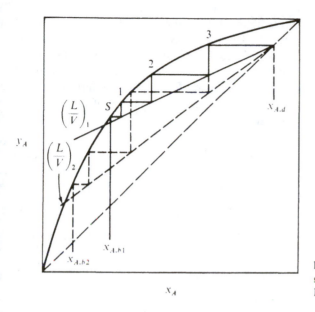

Figure 5-31 Batch distillation at constant distillate composition; three equilibrium stages plus reboiler.

the still pot are stepped off for the two distillate compositions, and values of $x_{A,b}$ are thereby obtained. In this way $x_{A,b}$ can be related to $x_{A,d}$ for all values of $x_{A,d}$.

The Rayleigh equation (3-16) is applicable to a batch distillation at constant reflux ratio. Since $x_{A,d}$ is the composition of the product stream continuously withdrawn and $x_{A,b}$ that of the material left behind in the still pot, the Rayleigh equation takes the form

$$\ln \frac{b'}{F'} = \int_{x_{A,F}}^{x_{A,b'}} \frac{dx_{A,b}}{x_{A,d} - x_{A,b}} \tag{5-36}$$

where F' = amount of initial charge
b' = amount left behind in still pot at end of distillation
$x_{A,F}$ = feed composition
$x_{A,b'}$ = composition of final product b'

A graphical integration is usually required, the relation between $x_{A,d}$ and $x_{A,b}$ at any $x_{A,b}$ being taken from the construction of Fig. 5-30 as previously described. When the integral in Eq. (5-36) has been evaluated, the combined distillate composition can be obtained from an overall mass balance.

Figure 5-31 indicates the analysis for the case of a reflux ratio varying to give constant overhead composition throughout the distillation. The operating lines radiate out from the point representing the constant distillate composition. For each operating line the requisite number of stages can be stepped off to give $x_{A,b}$ as a function of L/d. Subscript 2 in Fig. 5-31 refers to a later time than subscript 1.

The capacity of a distillation column is generally limited by a maximum allowable vapor flow rate, as discussed further in Chap. 12. If the vapor flow is held constant and the reflux ratio continually increases to hold the distillate composition

constant, the distillate flow rate must decrease as time goes on. The total amount of
vapor which must be generated, hence the time required to reach a given bottoms
composition (or a given total amount of collected distillate), can be computed from
the knowledge of L/V as a function of $x_{A,b}$, gained from the construction of
Fig. 5-31. Following a derivation originally given by Bogart (1937) and neglecting
holdup on stages above the still pot, we can relate the amount of the distillate
produced to the amount of vapor generated

$$\frac{dd'}{dV'} = -\frac{db'}{dV'} = \frac{d}{V} = 1 - \frac{L}{V} \tag{5-37}$$

By mass balance

$$b'x_{A,b} = F'x_{A,F} - (F' - b')x_{A,d} \tag{5-38}$$

where $x_{A,d}$ is now a constant. The appropriate form of Eq. (3-14) for a batch distillation is

$$b'dx_{A,b} = (x_{A,d} - x_{A,b})db' \tag{5-39}$$

Substituting Eqs. (5-38) and (5-39) into 5-37 gives

$$dV' = \frac{F'(x_{A,F} - x_{A,d})dx_{A,b}}{(x_{A,b} - x_{A,d})^2[1 - (L/V)]} \tag{5-40}$$

or

$$V_{tot} = F'(x_{A,F} - x_{A,d}) \int_{x_{A,F}}^{x_{A,b}} \frac{dx_{A,b}}{(x_{A,b} - x_{A,d})^2[1 - (L/V)]} \tag{5-41}$$

where V_{tot} is the total amount of vapor which must be generated to produce a
remaining product of composition $x_{A,b}$. The integral is evaluated relating $x_{A,b}$ and
L/V through the construction of Fig. 5-31.

In the case of a constant reflux distillation, the overhead product rate does not
vary, and the time requirement or the total vapor-generation requirement can be
calculated directly from the cumulative amount of distillate and the known reflux
ratio.

The cases of constant reflux and constant distillate composition represent just
two of the infinite number of reflux rate policies that can be followed during a batch
distillation. Converse and Gross (1963) and Coward (1967) have used various optimization techniques to determine the optimal reflux policy which will give a fixed
overall separation with a minimum total amount of vapor generation. In this problem two separation variables are specified. for example, d and $x_{A,d}$, both for the
cumulative product, and r is determined as a function of time to minimize the total
vapor generation. The problem in which V_{tot} and the cumulative $x_{A,d}$ are fixed and
$r(t)$ is determined to maximize the cumulative d has also been explored. For all cases
considered, the optimal reflux policy lies between the cases of constant reflux and
constant distillate composition. For the fixed separation problem, the reduction of
the required total vapor generation was between 1 and 9 percent compared with the
constant-reflux or constant-distillate policy, whichever gave the lower vapor requirement. For the fixed-vapor-generation problem, the distillate recovery increased by up

to 5 percent. Hence it seems safe to say that the optimal reflux policy corresponds to a reflux ratio increasing as time goes on but not as much as would be necessary to hold the distillate composition constant. Conditions are usually sufficiently insensitive for it not to be crucial to hold the optimal reflux policy. Luyben (1971) has considered the more general case of optimizing binary batch distillation with respect to reflux ratio, start-up policy, number of plates, and plate holdup.

Batch vs. Continuous Distillation

It has already been pointed out that a batch distillation provides more operational flexibility than a continuous distillation and is often more suitable for a multiproduct operation. On the other hand, a batch distillation requires considerably more labor and attention. These factors are usually the most important in choosing a type of distillation process; however, it is also instructive to consider the quality of separation afforded by the two types of distillation. The batch distillation has the same advantage in product purity that the single-stage Rayleigh distillation has in comparison to a continuous flash. Referring to Fig. 5-30, for a constant reflux-ratio operation, suppose that the final bottoms composition is to be x_{A, b_2}. In a continuous distillation with three equilibrium stages plus a reboiler, the overhead composition will be x_{A, d_2}. In a batch distillation only the last amount of distillate will have this composition. All the previous portions of distillate will be richer in A, and hence the average $x_{A, d}$ for the cumulative distillate will be greater than x_{A, d_2}.

A disadvantage of batch distillation is that the column shown in Fig. 5-29 provides rectifying action but no stripping action. Consequently it is possible to obtain a distillate of high purity, but the recovery of the more volatile component in the distillate is poor. This follows since $x_{A, b}$ cannot be reduced greatly without reducing $x_{A, d}$ substantially or using a very high reflux ratio. One way of overcoming this difficulty is to take an intermediate cut; the column is first run to collect high-purity distillate. Then the overhead product stream is diverted to an intermediate-product vessel, and distillation proceeds until the bottoms becomes concentrated in the less volatile component. The intermediate product can then be mixed with the charge to the next batch.

Batch distillation columns generally do not contain many stages. From Figs. 5-30 and 5-31 it can be seen that a few equilibrium stages lead rapidly into the region of the pinch where the operating line crosses the equilibrium curve. More stages would be of little or no avail. Greater product purities require more reflux.

Effect of Holdup on the Plates

Allowance for holdup on the plates of a batch distillation column complicates the analysis greatly. Gerster (1963) shows that there are two compensating effects of holdup:

1. After the charge is fed to the still pot, the column must be run at total reflux for a time in order to establish the liquid holdup on the plates. Distillate withdrawal can start only after

the holdup is established. The material on the plates is richer in the more volatile component than was the charge; as a result $x_{A,b}$ at the start of a run is less than $x_{A,F}$. Consequently $x_{A,d}$ is less than would be expected for no holdup, and this effect is detrimental.

2. Holdup on the plates presents an inertia effect, whereby the plate compositions change more slowly than would be expected from the McCabe-Thiele analysis. x_A on any plate decreases as the run proceeds, but the need for depleting component A on each plate as time goes on causes the term on the right-hand side of Eq. (5-35) to be negative, with the result that the downflowing liquid is richer in A than would be predicted when the term is zero. This effect causes the spread between $x_{A,b}$ and $x_{A,d}$ at any time to be greater than given by the McCabe-Thiele analysis and hence improves the separation.

In practice it appears that the second effect is dominant at low holdups on the plates, whereas the first effect becomes dominant at higher holdups.

CHOICE OF COLUMN PRESSURE

The choice of pressure for a distillation column is explored at some length in Appendix D. Factors to be considered include (1) the change in relative volatility with temperature (pressure), (2) greater shell thicknesses at higher pressures, (3) the cost of a vacuum system, (4) the maximum temperature to which the bottoms material can be raised without degradation, (5) the availability and cost of the heating medium to be used in the reboiler, (6) the availability and cost of cooling medium to be used in the condenser, and (7) the increased cost of materials for extreme temperatures.

These factors usually lead to a pressure slightly above atmospheric if the resulting temperatures do not require refrigeration overhead and do not require an unusual heating medium or lead to thermal-degradation problems in the reboiler. Up to about 1.7 MPa (250 lb/in² abs), if cooling water or air can be used overhead, the pressure is usually set to give an average driving force of 5 to 15°C in the overhead condenser. If this would lead to higher pressures, overhead refrigeration becomes likely. An example optimization for such a system (ethylene-ethane) is given in Appendix D.

Steam Distillations

Thermal-degradation problems or the need for very high temperature heating media can lead to vacuum distillation as a means of coping effectively with those problems. For organic mixtures *steam distillation* is sometimes used to solve temperature-related problems. In that process live steam is fed directly into the bottom of the column, serving as the heating medium and the source of a vapor stream. The steam serves, in effect, to lower the pressure of the distillation for the organic mixture, since the steam occupies partial pressure within the vapor phase and thus the partial

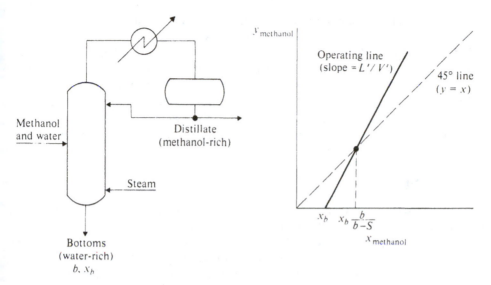

Figure 5-32 Distillation of methanol and water using open steam.

pressures of the organic components add up to less than the total pressure. The lower sum of partial pressures for the organics, in turn, leads to lower temperatures for the distillation. This approach assumes immiscibility of the organics with water.

Steam can also be used as a carrier gas in a Rayleigh distillation, separating an organic mixture. The analysis of such a process is similar to that of the batch air-stripping process in Prob. 3-L; see also Robinson and Gilliland (1950).

An advantage of steam as a vapor-phase diluent in distillations, in addition to the effective pressure reduction, is that, when condensed, the distillate will often break into two liquid phases, the water not diluting the organic distillate product significantly. A disadvantage is that the water effluent contains organic pollutants and requires treatment before discharge or return to a boiler for steam production.

Open steam can also be used in distillations where water is a feed constituent to be recovered in the bottoms product. An example of such a distillation for methanol and water is shown in Fig. 5-32. Instead of Eq. (5-8) the overall mass balance for the stripping section becomes

$$L'_{p+1} - V'_p = b - S \qquad (5\text{-}42)$$

where S is the molar flow rate of steam. The assumption of constant molar overflow leads to $V' = S$ and $L' = b$, and combination with Eq. (5-6) gives

$$V'y_{A,p} = L'(x_{A,p+1} - x_{A,b}) \qquad (5\text{-}43)$$

instead of Eq. (5-18). Whereas Eq. (5-18) crosses the 45° line at $y = x = x_b$, Eq. (5-43) gives $y = 0$ at $x = x_b$, as shown in Fig. 5-32. However, the addition of water in the form of steam means that x_b must be less in the open-steam case by a factor of $b/(b - S)$ for a fair comparison with an ordinary distillation giving the same separation. Since Eq. (5-43) crosses the 45° line at $x = bx_b/(b - S)$, the operating line is effectively unchanged from that in ordinary distillation. The difference is that for the open-steam case additional stages (usually one or a fraction) must be provided to proceed along the operating line below the 45° line to the $(0, x_b)$ point.

The savings with open steam lie in the elimination of the reboiler and the possibility of using somewhat lower pressure steam. Disadvantages are the greater flow of contaminated water effluent and/or the need for reprocessing it before use for the manufacture of additional steam.

AZEOTROPES

Azeotropic mixtures, by definition, give an equilibrium-vapor composition equal to the liquid composition at some point within the range of possible phase compositions (see, for example, Fig. 1-19b). The equilibrium curve crosses the 45° line on the yx diagram at the azeotropic composition and thereby presents a barrier to further enrichment by distillation. Robinson and Gilliland (1950) outline the approaches that can be used to overcome the limitations associated with an azeotrope. They include combining distillation with another type of separation process and altering the relative volatility by combining two distillation columns working at different pressures or by adding a substance which alters the relative volatility (extractive or azeotropic distillation, see Chap. 7).

Heterogeneous azeotropes can form in systems of limited miscibility and have one vapor composition corresponding to equilibrium with a wide range of overall liquid compositions covering the miscibility gap. It has usually been the practice in distillation design to avoid the formation of immiscible liquid phases within a column because of the resulting difficulties in having the two phases flow in the proper proportion between stages. On the other hand, systems with heterogeneous azeotropes do offer the possibility of an extra liquid-liquid separation in the reflux drum, which can be used to advantage; see, for example, Fig. 7-28.

REFERENCES

Bogart, M. J. P. (1937): *Trans. Am. Inst. Chem. Eng.,* **33**:139.

Converse, A. O., and G. D. Gross (1963): *Ind. Eng. Chem. Fundam.,* **2**:217.

Coward, I. (1967): *Chem. Eng. Sci.,* **22**:503.

Davison, J. W., and G. E. Hays (1958): *Chem. Eng. Prog.,* **54**(12):52.

Gerster, J. A. (1963): Distillation, in R. H. Perry, C. H. Chilton, and S. D. Kirkpatrick (eds.), "Chemical Engineers' Handbook," 4th ed., sec. 13, McGraw-Hill, New York.

Luyben, W. L. (1971): *Ind. Eng. Chem. Process Des. Dev.,* **10**:54.

McCabe, W. L., and E. W. Thiele (1925): *Ind. Eng. Chem.,* **17**:605.

Perry, R. H., and C. H. Chilton (1973): "Chemical Engineers' Handbook," 5th ed., McGraw-Hill, New York.

Robinson, C. S., and E. R. Gilliland (1950): "Elements of Fractional Distillation," 4th ed., pp. 196–213, McGraw-Hill, New York.

Sorel, E. (1893): "La rectification de l'alcool," Gauthier-Villars, Paris.

——— (1889): *C. R.,* **58**:1128, 1204, 1317.

——— (1894): *C. R.,* **68**:1213.

PROBLEMS

5-A$_1$ The overhead product from a benzene-toluene distillation column is 95 mol % benzene. The reflux ratio L/d is 3.0. Assuming constant molal overflow, a total condenser, saturated liquid reflux, and a relative volatility of 2.25, calculate *algebraically* the composition of the liquid leaving the second equilibrium stage from the top.

5-B$_1$ A distillation column is to be designed to separate methanol and water continuously. The feed contains 40 mol/s of methanol and 60 mol/s of water and is saturated liquid. The column pressure will be 101.3 kPa (1 atm), for which the following binary equilibrium data are:

Methanol at equilibrium, mol % (data from Perry and Chilton, 1973)

Liquid	2.0	6.0	10.0	20.0	30.0	40.0	50.0	60.0	70.0	80.0	90.0	95.0
Vapor	13.4	30.4	41.8	57.9	66.5	72.9	77.9	82.5	87.0	91.5	95.3	97.9

The feed is to be introduced at the optimal location for minimum stages; 95 percent of the methanol is to be recovered in a liquid distillate containing 98 mol % methanol. The reflux is to be saturated liquid with a flow rate 1.25 times the minimum reflux rate which would correspond to infinite stages. Assuming constant molal overflow, find the number of equilibrium stages required in the column.

5-C$_2$ Find the number of actual plates required for the distillation outlined in Prob. 5-B if E_{MV} is known to be 0.75.

5-D$_2$ An existing tower providing seven equilibrium stages plus a reboiler is being considered for use in the methanol-water distillation described in Prob. 5-B. The feed can be introduced at any point. If the tower will be operated at whatever reflux ratio is required to produce both the purity and the recovery fraction of methanol indicated in Prob. 5-B, what will the necessary rate of vapor production in the reboiler be in moles per second?

5-E$_2$ Suppose that the allowable vapor rate in the tower described in Prob. 5-D is limited by the reboiler capacity to 90 mol/s. For the given 100 mol/s feed, what is the maximum fraction of the methanol fed which can be recovered in a purity of 98 mole percent or higher?

5-F$_2$ One alternative for increasing the capacity of the tower of Probs. 5-D and 5-E is to install a feed preheater which will partially vaporize the feed. If the tower is to have a vapor-generation rate of 90 mol/s in the reboiler, and if the feed can be introduced on any stage, what percentage of the feed must be vaporized in order for 95 percent of the methanol to be recoverable at 98 mole percent purity?

5-G$_2$ Suppose the feed in the tower designed in Prob. 5-B is by oversight introduced to the liquid on the bottom stage of the tower (the stage next above the reboiler) rather than to the stage specified in the design. If the reflux rate, distillate rate, and reboiler vapor-generation rate are all held at the design values, what will be the purity of the methanol product?

5-H$_2$ A batch still is to be used for the separation of a methanol-water mixture. The still consists of an equilibrium still pot surmounted by a number of plates equivalent to two more equilibrium stages. A total condenser is employed, which returns saturated reflux. During operation the overhead reflux ratio L/d is held constant at 1.00. The holdup on the plates and in the condenser system is insignificantly small in comparison with that in the still pot. Suppose that a feed containing 50 mol % methanol and 50 mol % water is charged to the still and distillation is carried out until half the charge (on a molar basis) has been taken as distillate product. What is the composition of the accumulated distillate?

5-I$_2$ Fruit-juice concentrates are prepared commercially by evaporation. One problem is that various volatile components contributing to flavor and aroma tend to be lost in the escaping vapor. The *essence-recovery* process shown in Fig. 5-33 has been developed to recover these volatile substances so that they

PLATE OR PACKED COLUMN

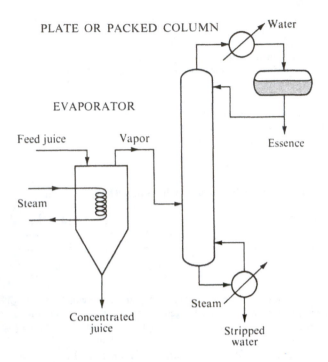

EVAPORATOR

Figure 5-33 Essence-recovery process.

can be reincorporated into the juice concentrate. One of the most important volatile flavor components present in Concord grape juice is methyl anthranilate, which is known to occur at approximately 2×10^{-7} mole fraction in the evaporator vapor. Because of the similarly low concentrations of other volatile flavor components, the methyl anthranilate–water distillation in the essence-recovery tower can be treated as a binary system. The relative volatility of methyl anthranilate to water at high dilution is 3.5 at 100°C. Suppose that the essence-recovery process is to be operated at atmospheric pressure to recover from the vapor at least 90 percent of the methyl anthranilate in an essence which contains only 1.00 percent of the water entering the distillation column. The feed to the column is saturated vapor.

(a) Assuming that the distillation column can be of any size, calculate the minimum steam consumption required in the reboiler of the essence-recovery column, expressed as kilograms per kilogram of entering vapor.

(b) If the reboiler vapor generation is 40 percent greater than the minimum computed in part (a), find the equilibrium-stage requirement for the separation.

5-J$_2$ A continuous distillation column is used to purify n-propanol by stripping a water contaminant from it. The column contains two equilibrium stages plus an equilibrium reboiler and a total condenser. The feed enters as saturated vapor into the vapor space between the two equilibrium stages in the column. Per mole of bottoms product, 1.0 mol of vapor is generated in the reboiler. The feed is dilute water in propanol; hence the equilibrium relationship is $y_W = 2.80x_W$.

(a) Derive an algebraic relationship between the water concentration in the overhead product and the water concentration in the propanol product.

(b) Under what additional specified conditions will the overhead be richest in water? What would the overhead composition be?

5-K$_2$ A plant contains a large distillation tower for the separation of a near equimolal mixture of o-xylene and p-xylene ($\alpha_{p-o} = 1.15$) into relatively pure products. Because of the low relative volatility, a large

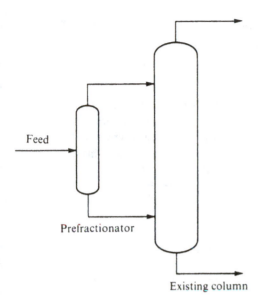

Figure 5-34 Proposed process configuration for Prob. 5K.

number of plates (about 100) are included in the tower and a high reflux ratio (about 18 : 1) is employed. It has been determined that this tower represents the capacity limit to the plant; as a result ways are being sought to increase the capacity, i.e., feed rate, of the tower. One scheme that has been proposed for increasing the xylene separation capacity of the plant is to place a new prefractionator tower before the existing large tower, as shown in Fig. 5-34. The prefractionator will have fewer plates than the existing tower (perhaps 20) and will be much smaller in diameter (perhaps half the diameter). It will therefore operate with a substantially lower reflux ratio. It will provide relatively impure products, one enriched in o-xylene and the other enriched in p-xylene. These streams will be fed to appropriate new feed plates in the existing tower. Will this scheme increase the xylene separation capacity of the plant significantly? Explain your answer qualitatively (no calculations needed).

5-L$_2$ In the design of an acetone production process one of the steps involves a separation of acetone from acetic acid. It is proposed that this be accomplished by batch distillation at atmospheric pressure in a plate column atop a large still. The column will be equipped with a total condenser and will return saturated liquid as reflux. The feed will consist of 65 mol % acetone and 35 mol % acetic acid. Equilibrium data for acetone–acetic acid at atmospheric pressure are given in Example 2-7. For parts (a) to (c) assume that it is necessary to recover 95 mol % of the acetone in a purity of 99.5 mol %. Holdup at points other than the still pot may be neglected, and constant molal overflow from stage to stage may be assumed. The overhead product will be taken at constant purity by varying the reflux ratio.

(a) What is the minimum number of equilibrium stages that must be provided above the still?

(b) What is the maximum reflux ratio that must be provided for, even with an infinite number of plates?

(c) Set the number of equilibrium stages above the still pot at 4. What amount of vapor generation per mole of charge will be necessary to accomplish the separation?

(d) Suppose the operation is carried out at a constant reflux ratio instead of a constant overhead purity. If the same total amount of vapor as found in part (c) is employed to recover the same total amount of distillate, what will be the distillate purity? Explain why this differs from the constant-purity case.

5-M$_2$ The following data were obtained by taking liquid samples from a real 60-plate distillation tower fractionating ethylene and ethane. The tower is equipped with a reboiler and a total condenser. Assume saturated liquid feed.

Bottoms rate = 18,175 lb/h   Distillate rate = 11,110 lb/h   Reflux rate = 96,800 lb/h

Tower pressure = 290 lb/in² abs   Overhead temperature = −20°F

Bottoms temperature = +20°F   Feed plate = no. 31 (from bottom)

Liquid composition

Plate number (from bottom)	Ethylene, mole fraction
Bottoms	0.016
7	0.0885
14	0.260
25	0.599
37	0.783
43	0.9395
Distillate	0.9982

Equilibrium data for the ethylene-ethane system at 290 lb/in² (data interpolated from Davison and Hays, 1958)

$x_{C_2H_4}$	$\alpha_{C_2H_4-C_2H_6}$	$y_{C_2H_4}$
0.000	1.54	0.000
0.100	1.52	0.144
0.200	1.50	0.275
0.300	1.48	0.385
0.400	1.46	0.487
0.500	1.45	0.592
0.600	1.44	0.684
0.700	1.43	0.770
0.800	1.42	0.850
0.900	1.39	0.926
1.000	1.36	1.000

Find the average Murphree vapor efficiencies over (*a*) plates 7 through 14 and (*b*) plates 37 through 43.

5-N₃ In a proposed continuous chemical synthesis process, vapor feed to a reactor is obtained by taking vapor overhead from a partial condenser atop a distillation column. The vapor leaving the partial condenser contains 75 mol °₀ component A and 25 mol °₀ component B, and as it passes through the synthesis section of the process only the A component is consumed. The reaction products (converted A) are removed. The remaining reactor effluent forms a recycle stream containing 40 mol °₀ A and 60 mol °₀ B, which is returned to the distillation column at the appropriate point as saturated vapor. Essentially pure B is withdrawn as bottoms from the reboiler supplying vapor to the column. Makeup A is supplied as saturated liquid containing 90 mol °₀ A and 10 mol °₀ B to the top plate of the column. The volatility of A relative to B is 2.0. In the partial condenser, the liquid leaving may be assumed to be in equilibrium with the vapor leaving. Assume constant molal overflow.

(*a*) Determine the minimum vapor rate required from the reboiler per 100 mol of vapor fed to the reactor. Show the construction corresponding to the minimum reboiler vapor rate on a *yx* diagram.

(*b*) What is the overhead reflux rate corresponding to the minimum reboiler vapor-generation rate?

(*c*) Show the appropriate operating and equilibrium lines on a McCabe-Thiele diagram for a vapor rate from the reboiler equal to 1.5 times the minimum.

(*d*) Find the number of equilibrium stages required above the recycle addition point for a reboiler vapor rate 1.5 times the minimum. Show these steps clearly on the McCabe-Thiele diagram.

5-O₃ An innovative distillation column is built as shown in Fig. 5-35. The liquid from each plate is fed to the second plate below rather than to the next plate below, as in the classical design. The claim is made that the vapor entering a plate will be contacted with a liquid which is richer in the more volatile component and will therefore undergo more enrichment per stage. Hence, for a given boilup rate and given feed conditions, fewer plates should be required to carry out a given separation. Is this claim true? If it holds only for certain types of mixtures or operating conditions, to what types does it apply? Consider binary distillations only.

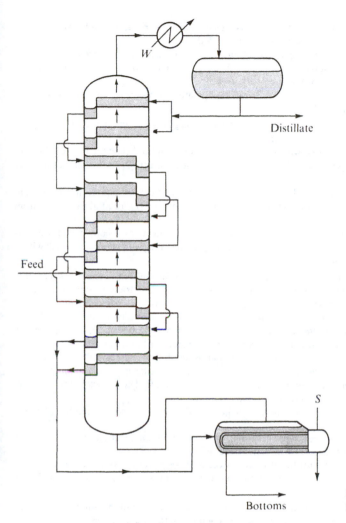

Figure 5-35 An innovative distillation column.

5-P₃ It is desired to design a distillation column to separate methanol from water at a pressure of 1 atm. The following table gives the design requirements for feeds and products.

Stream	Quality	Flow rate, mol/s	MeOH, mole fraction
Feed no. 1	Saturated vapor	400	0.50
Feed no. 2	Saturated liquid	200	0.30
Overhead product	Saturated liquid	150	0.96
Bottoms product	Saturated liquid		0.04
Sidestream product	Saturated liquid		0.70

The column will have a total condenser and a reboiler using steam heating. Constant molal overflow is a satisfactory assumption. Equilibrium data are given in **Prob. 5-B**.

(a) If the liquid reflux rate to the top plate is 400 mol/s, determine (1) the number of equilibrium stages required, (2) the equilibrium stage to which each feed should be added and that from which the sidestream should be withdrawn, and (3) the vapor rate from the reboiler.

(b) What is the minimum possible reflux rate (moles per second) for accomplishing this separation, even with an infinite number of plates?

5-Q₃ For the distillation system of Prob. 5-P determine the operating diagram and the equilibrium-plate requirement if conditions remain the same, *except*:

(a) It is specified that the *vapor* feed will be put in the clear vapor space below the *fifth* equilibrium stage from the bottom and the *liquid* feed will be injected into the liquid on the *seventh* equilibrium stage from the bottom.

(b) It is specified that the *liquid* feed will be injected into the liquid on the *fifth* equilibrium stage from the bottom and the *vapor* feed will be put in the clear vapor space below the *sixth* equilibrium stage from the bottom.

Presume that the sidestream of specified composition will be obtained by mixing liquid drawoffs from two adjacent equilibrium stages. Consider the reboiler to provide an equilibrium stage which will be counted as the "first from the bottom."

5-R₃ A process is required for transferring a heavy polymer from a solution in benzene to a solution in xylene without ever concentrating the polymer or taking it out of solution. This transferral is to be accomplished in a distillation tower which will receive two saturated liquid feeds: a stream containing benzene and the polymer and another stream of pure xylene at a molar flow rate equal to the molar flow of benzene. The benzene product is to be 98 percent pure and should recover 98 percent of the benzene fed. The relative volatility of benzene to xylene may be considered constant at 8.0, and the polymer may be considered to have a very low volatility. A total condenser is used, returning saturated reflux. The overhead reflux ratio L/d is set at a value of 0.79, and the Murphree vapor efficiency E_{MV} for all plates and the reboiler is predicted to be 0.50. Constant molal overflow can be assumed. The presence of the polymer can be ignored in solving this problem since its volatility is effectively zero.

(a) If the two feeds are mixed together and fed at the optimal location, find the number of plates required.

(b) If the two feeds are introduced separately and at their optimum locations for achieving a minimum plate requirement, find the number of plates required.

(c) Suppose that the xylene feed must be introduced exactly three plates above the benzene feed to make sure that the polymer will not come out of solution, even during a tower upset. Find the number of plates required.

5-S₂ Write a digital computer program suitable for carrying out the calculation of the number of equilibrium stages required for a binary distillation where the feed conditions, the pressure, the reflux ratio, and the product recovery fractions of both components are specified and the feed is to be put in on the optimum stage. Equilibrium data will be supplied by giving the relative volatility as a polynomial expression in liquid mole fraction. Assume constant molal overflow. Confirm the workability of your program for an example problem.

5-T₂ Qualitatively sketch an xy diagram for a binary distillation with a vapor sidestream withdrawn midway in the section of the column *below* the feed stage. Indicate the composition of the sidestream.

5-U₂ A distillation column with a partial condenser is built for the separation of benzene and toluene following the design represented in Fig. 5-14. Consider individually the effects of each of the following changes in operation. The variables indicated in the column headed "held constant" remain unchanged; in addition the feed flow rate, the column pressure, the heat duty of the reboiler Q_R, the number of equilibrium stages above, and the number of equilibrium stages below the main feed point remain constant in all cases. For each of the indicated dependent variables, indicate whether it will increase (+), decrease (−), or remain unchanged (0).

Case	Change	Held constant	Dependent variables
(a)	Increase feed preheat h_F	Bottoms flow rate b	Q_c, y_D, x_b
(b)	Increase condenser duty Q_c	Feed enthalpy h_F	D, y_D, x_b
(c)	Increase feed preheat h_F	Condenser duty Q_c	D, y_D, x_b
(d)	Add one-half of feed 2 equilibrium stages higher	h_F, Q_c	D, y_D, x_b
(e)	Withdraw a liquid sidestream from equilibrium stage no. 10	h_F, Q_c	b, y_D, x_b

(*f*) As the flow rate of the sidestream in case (*e*) increases, will the benzene mole fraction x_s in that sidestream increase, decrease, or remain unchanged? Explain.

5-V₁ A methanol-water distillation at atmospheric pressure receives a feed containing 75 mol % methanol as a saturated liquid and produces a distillate containing 98 mol % and a bottoms containing 5 mol % methanol, with an overhead reflux ratio L/d of 1.00. This design utilizes an ordinary reboiler with indirect steam. If the feed, the distillate composition and flow rate, and the reflux ratio remain the same but the reboiler is removed and open-steam heating is substituted, find (*a*) the new bottoms flow rate and composition and (*b*) the number of additional equilibrium stages required in the column. Assume constant molal overflow, and neglect the difference between methanol and water in molar latent heat of vaporization.

SIX

BINARY MULTISTAGE SEPARATIONS: GENERAL GRAPHICAL APPROACH

The McCabe-Thiele, or yx, type of diagram has proved extremely useful for the analysis of binary distillation with constant molal overflow, but its usefulness is by no means limited to that one operation. In this chapter we explore the applications of this diagram to other countercurrent multistage binary separations. Whereas in distillation the separating agent is energy (heat) and the counterflowing streams between stages are vapor and liquid, we now consider the general cases where these streams may be any phase of matter. We also consider processes with a mass separating agent (extraction, absorption, etc.), in addition to those with an energy separating agent.

Following the definition at the beginning of Chap. 5 and the examples in Table 5-1, for binary separations, we can establish the entire composition of either of the phases by setting a single composition parameter (mole fraction, concentration, etc.). For any application of the McCabe-Thiele, or operating, diagram, an appropriate composition parameter for one of the phases is plotted against an appropriate composition parameter for the other phase. Two curves or lines are needed to describe a section of a countercurrent cascade or contactor. One of these is the *equilibrium line* or *curve*, or whatever relationship enables the composition of one outlet stream from a stage to be calculated from the composition of the other outlet stream. The other relationship is the *operating line* or *curve*, which serves to relate compositions of streams passing each other between stages (inlet stream of phase 1 to a stage related to outlet stream of phase 2 from the same stage).

Another useful concept which can be carried onward from the binary distillation analysis is that of the *net upward product* for a section of a countercurrent cascade or contactor. The net upward product can be defined either as a total flow or the flow of an individual component and is a generalization of Eq. (5-9). At steady-state operation, in a section where there is no feed added or product withdrawn at any inter-mediate location, the net upward product must be constant. The net upward product is defined as the difference between the total flow or flow of component i in the upflowing stream and the total flow or flow of component i in the downflowing stream. The net upward product may be positive or negative, depending upon whether or not the amount in the upflowing stream exceeds that in the downflowing stream. It is a useful concept because it stays constant within a given section.

In the McCabe-Thiele diagram for binary distillation, under the assumption of constant molal overflow, the operating lines are straight. The convenience of straight operating lines is apparent to anyone who has used these diagrams. When an operating line is straight, the entire line can be located from a knowledge of two points or of a point and a slope. Such a construction overcomes the necessity of determining each point independently through a succession of calculations.

Operating lines will be straight on a plot of y (mole fraction) vs. x (mole fraction) *if and only if* the total molar flow rates of each of the two streams passing between stages remain constant within the section under consideration. This fact follows from

$$V_p y_{A, p} = L_{p+1} x_{A, p+1} + x_{A, d} d \tag{5-3}$$

which becomes the equation for a straight line if and only if V and, hence, L are constant.

Operating lines will be straight on operating diagrams for any type of counter-current separation process if the mass-balance equation relating streams passing between stages can be expressed in terms of unchanging flows which can serve as coefficients for the composition parameters in the mass-balance equation. These constant coefficients will then lead to straight operating lines.

We shall first consider some situations where total flow rates can be taken to be constant from stage to stage, leading to straight operating lines. Then we shall consider cases where flows of certain nontransferring components can be taken to be constant, leading to straight operating lines if the composition parameters for the counterflowing streams are defined in a different way, as mole, weight, or volume ratios. Next we consider the case where straight operating lines can be achieved by defining hypothetical compositions, taking latent heats of phase change into account. Finally, we shall consider some important cases of binary separations where, in general, neither the operating curve nor the equilibrium curve can be made a straight line.

STRAIGHT OPERATING LINES

Constant Total Flows

Two simple examples of the application of the McCabe-Thiele type of diagram to multistage separation processes other than distillation follow.

Example 6-1 Solutions of tributyl phosphate (TBP) in kerosene serve as a solvent for recovering certain metals selectively from aqueous solution. For example, zirconium nitrate, $Zr(NO_3)_4$, forms a

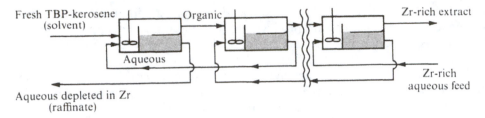

Figure 6-1 Mixer-settler process for selective zirconium extraction.

complex, $Zr(NO_3)_4 \cdot 2TBP$, with TBP, and the complex is readily extracted by TBP solution. Consider a staged mixer-settler extraction process shown schematically in Fig. 6-1. The feed is an aqueous solution of 3.0 M HNO_3 and 3.5 M $NaNO_3$ containing 0.120 mol of zirconium per liter. A solution of 60 vol % TBP in kerosene is employed as the extracting agent. The Murphree efficiency (based on the aqueous phase) of the mixer-settler combinations is 90 percent. The aqueous and organic phases are totally immiscible. Equilibrium data are cited by Benedict and Pigford (1957) as follows:

mol Zr/L		Distribution coefficient
Aqueous phase	Organic phase	mol/L organic / mol/L aqueous
0.012	0.042	3.5
0.039	0.083	2.1
0.074	0.114	1.54
0.104	0.135	1.30
0.123	0.147	1.20

If the entering TBP solution is free of zirconium and 90 percent of the zirconium must be extracted, compute (a) the minimum solvent treat which could be used (liters per liter of feed) to effect the separation, given any number of stages, and (b) the number of mixer-settler units required to accomplish the specified extraction if the solvent treat is 0.90 L of TBP-kerosene per liter of aqueous feed.

SOLUTION From the construction of this sort of extraction unit we can determine the number of process variables to be set in actual operation as follows:

N = number of stages
S/F = ratio of solvent feed rate to aqueous feed rate
x_F = solute concentration in aqueous feed
x_S = solute concentration in solvent feed

T = temp of operation
P = pressure of operation
Mixer stirring speeds

For part (a) we specify N ($= \infty$), x_F, x_S, T, and P and replace S/F by a single separation variable which is the solute concentration in the exit aqueous stream. The stirring speeds are taken into account by the Murphree efficiencies. We then solve for S/F. In part (b) we specify x_F, x_S, T, P, and S/F and replace N with a single separation variable, again the solute concentration in the exit aqueous phase. Once more the Murphree efficiencies replace the stirrer speeds. We then solve for N.

Figure 6-2 shows the appropriate modification of the yx diagram, which in this case is a plot of C_O (moles per liter in the organic phase) vs. C_A (moles per liter in the aqueous phase). The zirconium concentrations are very dilute, and the phases are totally immiscible. The result is that the total flow

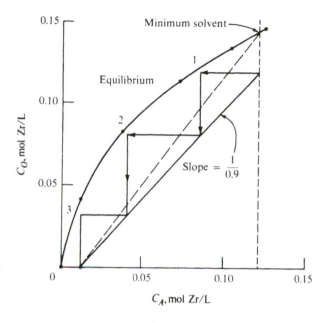

Figure 6-2 Operating diagram for Example 6-1.

rates of both streams are essentially constant, whether stated in molar, mass, or volumetric units. Since the equilibrium data are stated as moles Zr (conserved quantity) per *liter*, we adopt a volumetric flow basis for convenience and anticipate a straight operating line because the flow basis is constant from stage to stage. Since feeds and products occur only at the ends of the cascade, there will be only one operating line.

(a) We know that at the lean (left-hand) end of the cascade $C_O = 0$ and $C_A = (0.12)$ $(1 - 0.9) = 0.012$. This fixes the intersection of the operating line with the horizontal axis. The point corresponding to the rich (right-hand) end of the operating line will lie at $C_A = 0.120$, and the value of C_O will depend upon the solvent-to-feed ratio.

The minimum allowable solvent treat will correspond to the maximum *slope* (liters of feed per liter of solvent) of the operating line, which causes the operating line to touch the equilibrium curve at one point. Observation reveals that this one point will be at the extreme rich end ($C_A = 0.120$), and for this condition $C_O = 0.145$ mol/L. Thus

$$F(C_{A_{in}} - C_{A_{out}}) = S(C_{O_{out}} - C_{O_{in}}) \tag{6-1}$$

and

$$\left(\frac{S}{F}\right)_{min} = \frac{0.120 - 0.012}{0.145 - 0.000} = 0.745 \text{ L solvent/L feed}$$

(b) If the operating S/F is 0.90, we have used a solvent flow 21 percent greater than the minimum. From the known S/F we can obtain the operating line, as shown in Fig. 6-2, from the known intercept and the slope. Since the Murphree efficiency is based on the aqueous phase, it is convenient to determine stages starting at the rich end. Following the definition of the Murphree efficiency [Eq. (3-23)], each horizontal step goes a fraction E_{MA}, or 90 percent, of the way toward the equilibrium curve. The solution stepped off in Fig. 6-2 shows that three mixer-settler units accomplish the separation.

In this simple extraction process there are no stages to the right of the feed in Fig. 6-1. The exit solvent can therefore be no richer in $Zr(NO_3)_4$ than corresponds to equilibrium with the aqueous feed. In order to have a reflux stream and an enriching section to increase the $Zr(NO_3)_4$ concentration in the solvent we would need to provide as input an aqueous solution of $Zr(NO_3)_4$ richer than the aqueous feed. This is not available in any simple fashion.

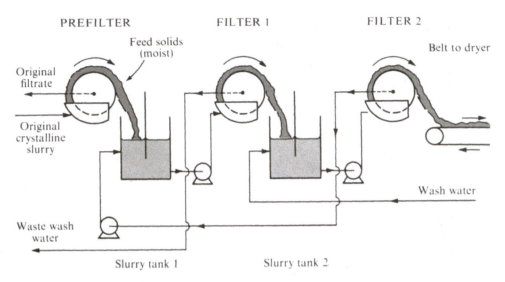

PREFILTER   FILTER 1   FILTER 2

Feed solids (moist)

Belt to dryer

Original filtrate

Original crystalline slurry

Wash water

Waste wash water

Slurry tank 1   Slurry tank 2

Figure 6-3 Two-stage wash process.

Reflux can more readily be obtained and can prove useful in cases of extraction with a partially miscible system (see Example 6-6). □

Example 6-2 A staged countercurrent wash process is to be operated to free an insoluble mass of crystals from supernatant ferric sulfate solution. The slurrying and filtration equipment shown in Fig. 6-3 is currently idle in the plant and may be of use for this process. The equipment consists of two mixing vessels and two filters. The prefilter is already in place for the initial recovery of the crystals.

The liquid portion of the slurry contains 55 wt % $Fe_2(SO_4)_3$ in water. The final solids (sent to a dryer on a conveyor belt) are to contain no more than 0.01 kg $Fe_2(SO_4)_3$ per kilogram of crystalline solid. Each filtration step retains a *volume* of filtrate in the moist solid cake equal to 0.50 m^3 of filtrate per cubic meter of solids. The dry solids have a density of 2880 kg per cubic meter of actual volume occupied by the solid.

Find the necessary wash water input rate (cubic meters per kilogram of final dry solids) required with this equipment. Assume complete mixing in the slurrying tanks.

SOLUTION Readers should convince themselves of the specific boundaries of each stage. Note that the streams passing between stages are (1) the moist cake falling from the filter knife-edge to the next slurry tank and (2) the filtrate before entering the slurry tank. The prefilter is not part of the wash cascade.

Densities of $Fe_2(SO_4)_3$ solutions vary from 1000 kg/m^3 at 0 percent to 1800 kg/m^3 at 60 wt % concentration (Perry et al., 1963). Since the densities are not constant with respect to filtrate composition, we are unable to say that either the total molar flow rate or the total mass flow rate from stage to stage is constant. On the other hand, we can say that the total *volumetric* flow rates of moist cake and filtrate (cubic meters per unit time) are constant from stage to stage since the volume of the solids and the volume of filtrate retained in the cake both remain constant from stage to stage. The flow rates must be volume flows to provide a straight operating line. As a result, the composition parameter must have dimensions of conserved quantity per unit volume; kilograms $Fe_2(SO_4)_3$ per cubic meter is most convenient.

An operating diagram for this example is shown in Fig. 6-4, which plots C_F, kilograms of $Fe_2(SO_4)_3$ per cubic meter of filtrate actually removed, vs. C_S, kg $Fe_2(SO_4)_3$ per cubic meter retained by the cake. As an option we could have taken kg $Fe_2(SO_4)_3$ per cubic meter of total cake as the latter parameter. These are desirable composition parameters because the cubic meters of filtrate

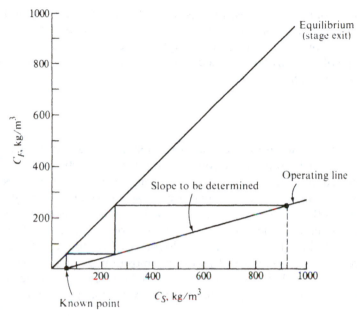

Figure 6-4 Operating diagram for Example 6-2.

actually removed, the cubic meters of filtrate retained by the cake, and the cubic meters of total cake are all constant from stage to stage and give straight operating lines. The equilibrium or exit-stream curve is given simply by the 45° line, since we have assumed sufficient mixing in the slurry tanks to make all the filtrate have uniform composition. Hence the concentration of $Fe_2(SO_4)_3$ in the retained filtrate is the same as the concentration in the filtrate passing through.

At the rich end $C_S = 55$ wt % and $\rho = 1700$ kg/m^3 (Perry et al., 1963); therefore

$$C_S = 0.55(1700) = 935 \text{ kg/m}^3$$

At the lean end C_F is the inlet water concentration and is equal to zero. C_S at the lean end can be obtained from the product solid specification

$$C_S = 0.01 \frac{\text{kg } Fe_2(SO_4)_3}{\text{kg solid}} \times 2880 \frac{\text{kg solid}}{\text{m}^3 \text{ solid}} \times \frac{1 \text{ m}^3 \text{ solid}}{0.5 \text{ m}^3 \text{ filtrate}} = 57.6 \text{ kg/m}^3$$

Thus we know the location of the point at the lean end of the operating lines, as shown in Fig. 6-4.

The slope of the operating line is unknown and is obtained from the criterion that there be two stages in the separation process. The procedure for fixing the slope of the operating line involves trial and error. A series of potential operating lines is drawn, radiating out from the known point on the lean end. The correct operating line, as shown in Fig. 6-4, is that which provides exactly two steps.

The water consumption comes from the slope of this operating line, which is measured as $(251 - 0)/(935 - 57.6) = 0.287$, in units of cubic meters of retained filtrate per cubic meter of filtrate passing through. The rate of filtrate passing through must equal the volumetric water feed rate, and the volume of filtrate is half the dry-solid volumetric rate. Hence

$$\text{Water consumption} = \frac{1 \text{ m}^3 \text{ filtrate passing}}{0.287 \text{ m}^3 \text{ filtrate retained}} \frac{1 \text{ m}^3 \text{ filtrate retained}}{2 \text{ m}^3 \text{ dry solid}}$$

$$= 1.74 \text{ m}^3/\text{m}^3 \text{ dry solid} = \left(1.74 \frac{\text{m}^3}{\text{m}^3}\right) \frac{1 \text{ m}^3}{2880 \text{ kg}}$$

$$= 0.00061 \text{ m}^3/\text{kg dry solid} = 0.61 \text{ L/kg dry solid} \qquad \square$$

Example 6-2 involved only two stages, and as a result an algebraic solution would have been as short or shorter than the graphical one. This would not continue to be the case as the number of stages increased. An algebraic stage-to-stage calculation would begin to involve many simultaneous equations, but the graphical procedure would be no more difficult.

Constant Inert Flows

In many instances when total flows are not constant from stage to stage there may still be inert species present which cannot pass from one counterflowing stream to the other to any appreciable extent. The flow rates of these components by themselves will then be constant from stage to stage. In such a case we can obtain a straight operating line by employing the *flow of inert species* as the flow rate and using mole, weight, or volume *ratios* as composition parameters. If the flow of inert species is expressed in mass per unit time, the mass ratio, kilograms of component A per kilogram of inert species, would be employed as the composition parameter. This procedure usually involves recalculation of the equilibrium data but does provide a straight operating line. Absorption, stripping, and some extraction processes are suitable for this approach.

Our convention will be to denote these ratios of one component to the inert species by capital letters. Thus X_A is the mole ratio of component A, while x_A is the mole fraction of component A.

Example 6-3 A noxious waste-gas stream from your plant containing 70 mol % H_2S, 28 mol % N_2, and 2 mol % other inerts on a dry basis is produced at 1 atm. Upon receiving a complaint from the local pollution-control board, you deem that there is an incentive for removing the H_2S from the gas stream and disposing of the H_2S elsewhere. Distillation is impracticable (why?), and you decide to absorb the H_2S into a suitable solvent in a plate tower. In searching for the suitable solvent you are guided by cheapness and so decide to consider the use of water. If the waste gas must be purified to an H_2S content of only 1.0 mol %, if the temperature is uniform at 21°C, and if the waste gas is initially saturated with water vapor, (a) what is the minimum water flow rate required, expressed as moles per mole of entering waste gas? (b) With a water flow equal to 1.2 times the minimum, how many equilibrium stages are required in the tower?

SOLUTION The tower is shown schematically in Fig. 6-5. In the construction and operation of this tower the following variables are set:

N = number of stages
L_i/G_i = mol water fed/mol gas fed
x_1 = mole fraction H_2S in inlet water
y_2 = mole fraction H_2S in inlet gas

T_{L_i} = temperature of water in
T_{G_i} = temperature of gas in
P = pressure

For part (a) we specify N, x_1, y_2, T_{L_i}, T_{G_i}, and P and replace L_i/G_i with a separation variable y_1, solving for L_i/G_i. In part (b) we replace N by y_1 and solve for N.

We know that the flow rate of N_2 + inerts will be constant from stage to stage, since the solubility of N_2 in water is less than 1 percent of the solubility of H_2S (see Fig. 6-6). The total gas flow rate will lessen considerably as H_2S is absorbed, with the result that some water will be condensed; however, because the water requirement is relatively large, the flow rate of water will remain relatively constant in the liquid from stage to stage. Thus we make moles of N_2 + inerts per unit time the flow rate in the gas phase and make moles of H_2O per unit time the flow rate in the

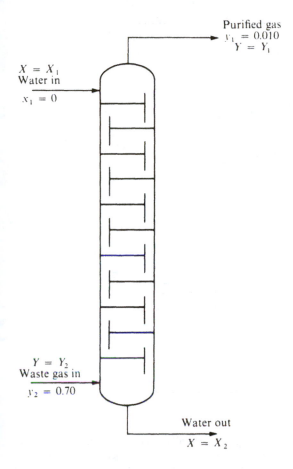

Purified gas
$y_1 = 0.010$
$Y = Y_1$

$X = X_1$
Water in
$x_1 = 0$

$Y = Y_2$
Waste gas in
$y_2 = 0.70$

Water out
$X = X_2$

Figure 6-5 Absorber for Example 6-3.

liquid phase. The composition parameters are Y_{H_2S} (moles H_2S per mole of N_2 + inerts) in the gas phase and X_{H_2S} (moles H_2S per mole of water) in the liquid phase. Because of these definitions the operating line will be straight despite the fact that the *total* flow rates change.

From Fig. 6-6 the solubility of H_2S in water at 21°C is found to be 0.0020 mole fraction (liquid) when the H_2S partial pressure is 1.00 atm. At the low liquid concentrations encountered, Henry's law ($p_{H_2S} = Hx_{H_2S}$) can be invoked. Consequently, the equilibrium expression for a total pressure of 1 atm is

$$y_{H_2S}(0.0020) = x_{H_2S}$$

or

$$y_{H_2S} = 500x_{H_2S} \tag{6-2}$$

We can assume justifiably that the gas phase will always be saturated with water vapor ($P^0_{H_2O} = 2.48$ kPa at 21°C). Therefore y_{H_2O} is constant and equal to $2.48/101.3 = 0.024$.

We wish to work in terms of mole ratios Y and X, where

$$Y = \frac{y_{H_2S}}{1 - y_{H_2S} - y_{H_2O}} \tag{6-3}$$

and

$$X = \frac{x_{H_2S}}{1 - x_{H_2S}} \tag{6-4}$$

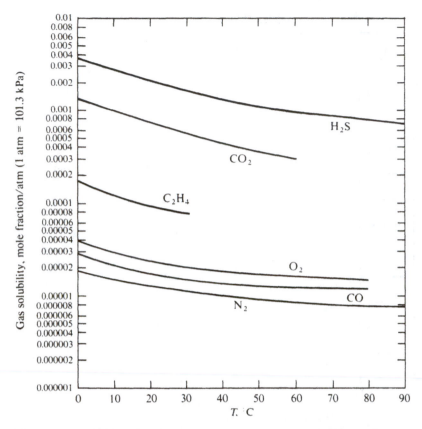

Figure 6-6 Solubilities of various gases in water. Solubility is proportional to partial pressure at 0.5 MPa and less for gases shown. Solubility is nonlinear in partial pressure for gases such as Cl_2 and SO_2 (because of chemical reaction) or for gases with higher solubility, for example, NH_3. *(Data from Perry et al., 1963, and Seidell and Lincke, 1958.)*

Substituting the constant value of y_{H_2O} and Eqs. (6-3) and (6-4) into Eq. (6-2), we have the equilibrium expression in terms of Y and X:

$$Y = \frac{500X}{0.976 - (499 + 0.024)X} \tag{6-5}$$

This expression is plotted on the mole-ratio operating diagram (Fig. 6-7) as the equilibrium curve.

(a) We know that the mole ratio of H_2S to N_2 + inerts entering at the rich end (bottom) of the tower is

$$Y_2 = \frac{0.70}{0.30} = 2.33$$

and the mole ratio at the lean end (top) is

$$Y_1 = \frac{0.01}{0.99} = 0.0101$$

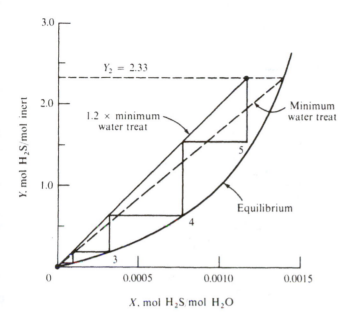

Figure 6-7 Operating diagram for Example 6-3.

X_1 at the lean end is zero, but X_2 at the rich end is unknown. X_1 and Y_1 represent a pair of passing compositions, however, and one point on the operating line is therefore fixed. This point is more clearly shown in the expansion of the lower left-hand corner of the diagram, given in Fig. 6-8.†

The slope of an operating line will now be L/G', where L is the constant water flow and G' is the constant flow of N_2 + inerts (both in moles per unit time). Notice that the operating-line slope is necessarily the ratio of the two constant flow rates. The minimum water flow rate will correspond to the minimum slope. The *pinch*, or touching, point occurs at $Y = 2.33$, where $X_2 = 0.00139$. Hence

$$\left(\frac{L}{G'}\right)_{min} = \frac{2.33 - 0.01}{0.00139 - 0.0000} = 1670 \text{ mol } H_2O/\text{mol } N_2 + \text{inerts}$$

$$= 1670 \times (0.3/1.0) = 500 \text{ mol } H_2O/\text{mol inlet gas (water-free basis)}$$

(*b*) Setting $L/G' = 600$ mol H_2O per mole water-free inlet gas (1.2 times the minimum), we have

$$x_2 = \frac{0.00139}{1.2} = 0.00116$$

and the corresponding operating line is plotted in Figs. 6-7 and 6-8. Starting arbitrarily at the lean end, we find very nearly five equilibrium stages required. □

---

† It is sometimes convenient to use log-log diagrams when a wide range of concentrations is to be considered; one plot is then required instead of two or more. Operating lines usually curve on log-log plots.

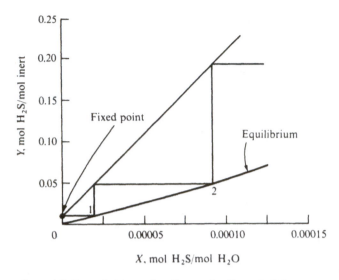

Figure 6-8 Expanded operating diagram for Example 6-3.

Several additional points should be made about the process of Example 6-3.

1. The operating line would still have been straight if x had replaced X as the liquid-composition parameter, since the liquid is highly dilute and its total flow rate is essentially constant. On the other hand, replacing Y with y would have caused the operating line to curve, since the total molar gas flow changes markedly throughout the column.
2. In this case the problem would not have been appreciably more complex if y were plotted vs. x, since the equilibrium line would then have been straight and could have been plotted with less effort, thus compensating for the curvature of the operating line on such a plot. Systems following Henry's law are a special case, however, and one cannot expect equilibrium data to give a straight line on a yx diagram in general.
3. Stage efficiencies for absorption processes are generally quite low, with the result that the actual plate requirement is substantially greater than the equilibrium-stage requirement.
4. The necessary water rate is very large in comparison with the gas rate, costing money and resulting in a lot of unwholesome water to be disposed of somewhere. These facts would lead one to choose an absorbent which could take up more H_2S.

Figure 6-9 shows an operating diagram for a situation where the same gas stream would be contacted with 2.5 N monoethanolamine (MEA) solution in water. The equilibrium data are extrapolated to 21°C from the data of Muhlbauer and Monaghan (1957) [see also Kohl and Riesenfeld (1974)]. The inlet MEA solution to the absorber contains 0.5 moles H_2S per mole MEA. Here the buildup of H_2S in the liquid solution is large enough to make it necessary to use a mole ratio as the composition parameter in the liquid phase in order to obtain a straight operating line. X is defined as moles H_2S per mole MEA; it could as well have been moles H_2S per mole of MEA + water. Here also the equilibrium curve would not have been straight on a yx plot; one can discern this from the fact that the equilibrium curve

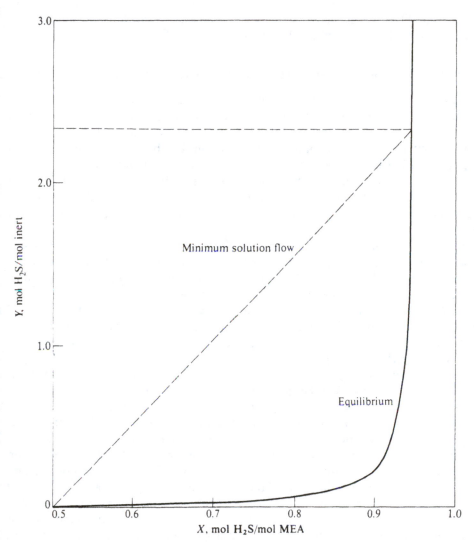

Figure 6-9 Operating diagram for absorption of H_2S from the gas stream of Example 6-3 using 2.5 N MEA in water.

has a very different shape in Fig. 6-9 from that in Fig. 6-7. The necessary circulation rate of MEA solution is much less than that for the water absorbent in Example 6-3. This can be seen from the minimum flow conditions, at which the X change for MEA solution is 0.445 mol H_2S per mole MEA and the X change for water absorbent is much less, 0.00139 mol H_2S per mole water.

Unless the design were modified, the MEA absorber would be complicated by the heat of absorption being large in comparison to the heat capacity of the solution. This would continuously change the temperature, and hence the equilibrium relationship, from one stage to the next. The behavior of such systems is discussed qualitatively in Chap. 7, and design methods for nonisothermal absorbers are discussed by Sherwood et al. (1975) and in Chap. 10.

Accounting for Unequal Latent Heats in Distillation; MLHV Method

The causes of varying molal overflow in distillation can be unequal latent heats of vaporization for the different components, sensible-heat effects due to a wide range of temperatures across the column, and/or heats of mixing. In many cases the influence of unequal latent heats is dominant. These situations can be handled effectively with straight operating lines on a McCabe-Thiele diagram if different composition parameters and flows are used, defined so as to make the latent heats of different components equal. The method was originally developed by McCabe and Thiele (1925) and has also been described by Robinson and Gilliland (1950) and Brian (1972).

The basic idea of the modified-latent-heat-of-vaporization (MLHV) method is to define pseudo molecular weights which serve to make the new molar latent heats of vaporization equal. In a binary system, if the true molecular weights of components i and j are M_i and M_j, the pseudo molecular weights can be taken to be $M_i^* = M_i$ and $M_j^* = M_j \Delta H_i / \Delta H_j$, where ΔH_i and ΔH_j are the true molal latent heats of components i and j, respectively. This will serve to make the latent heats per pseudo mole equal. Alternatively, the pseudo molecular weights could be taken to be $M_i^* = M_i \Delta H_j / \Delta H_i$, and $M_j^* = M_j$.

Mole fractions and molar flows are now defined using these new pseudo molecular weights. These values will be given the superscript *. Thus, $x_i^* = x_i/(x_i + \beta x_j)$, $x_j^* = \beta x_j/(x_i + \beta x_j)$, $d^* = d(x_{id} + \beta x_{jd})$, etc., where $\beta = \Delta H_j / \Delta H_i$.

Two other changes are needed to use the method. If the relative volatility is independent of composition when expressed in terms of true mole fractions, it will be constant and have the same value when defined in terms of the new composition parameters as $y_i^* x_j^* / y_j^* x_i^*$. But if it is a function of composition, the equilibrium data will have to be recalculated in terms of the new composition parameters. The second change involves the line denoting the locus of intersections of the operating lines. Saturated vapor and saturated liquid, respectively, will remain the same, but for other thermal conditions of the feed it is necessary to recompute the slope of this line, converting to the new pseudo-molecular-weight basis. In general, this requires computing the true and then the pseudo mole fractions of the vapor and liquid portions of the feed and then finding the ratio L_F^* / V_F^*. For superheated or subcooled feeds one would need to redefine Eqs. (5-22) and (5-23) in terms of the new basis.

> **Example 6-4** A stream of acetone and water is to be distilled in a plate tower to give an acetone product containing 91.0 mol % acetone; 98 percent of the acetone should be recovered in that product. The feed contains 50 mole percent of either component and has an enthalpy such that the increase in molal vapor rate across the feed tray will be 55 percent of the molal feed rate. There is a total condenser, returning saturated liquid reflux. Use the MLHV method, taking the latent heats of vaporization of acetone and water to be 30.19 and 40.73 kJ/mol, respectively. (a) Determine the minimum allowable overhead reflux ratio r/d. (b) Taking an overhead reflux ratio equal to 1.22 times the minimum, compute the number of equilibrium stages required for the separation.

SOLUTION The construction and operation variables are

P = pressure

Q_C = condenser load

N = number of equilibrium stages

Q_R = reboiler load

T_C = condenser outlet temperature

F = feed rate

$z_{A,F}$ = feed composition

h_F = feed enthalpy

Feed location

In part (a) we replace Q_C, Q_R, and h_F by $x_{A,d}$, $(l/_A)_d$, and $(\Delta V)_f$, defined as the increase in vapor flow at the feed, and solve for the reflux ratio r/d. In parts (b) and (c) in addition we set r/d instead of N and then solve for N taking the feed location to be the optimum.

We know that $x_{A,d} = 0.91$ and that $z_{A,F} = 0.50$. We can solve for $x_{A,b}$, d/F, and b/F from

$$\frac{d}{F} + \frac{b}{F} = 1.0 \qquad 0.91\frac{d}{F} = (0.98)(0.50) \qquad x_{A,b}\frac{b}{F} = (0.02)(0.50) = 0.01$$

Therefore

$$\frac{d}{F} = \frac{(0.98)(0.50)}{(0.91)} = 0.538 \qquad \frac{b}{F} = 1.0 - 0.538 = 0.462 \qquad x_{A,b} = \frac{0.01}{0.462} = 0.0216$$

To apply the MLHV method we first compute the ratio of latent heats (water to acetone) to be $\beta = 40.73/30.19 = 1.349$. Therefore we can treat water as a component having a pseudo molecular weight $M_W^* = 18/1.349 = 13.34$, while keeping the pseudo molecular weight for acetone equal to the true molecular weight of 58.08. Thus

$$x_{A,d}^* = \frac{0.91}{0.91 + (1.349)(0.09)} = 0.882$$

$$z_{A,F}^* = \frac{0.50}{0.50 + (0.50)(1.349)} = 0.426$$

$$x_{A,b}^* = \frac{0.0216}{0.0216 + (0.9784)(1.349)} = 0.0161$$

$$F^* = [0.50 + (1.349)(0.50)]F = 1.174F$$

$$d^* = [0.91 + (1.349)(0.09)]d = 1.031d$$

$$b^* = [0.0216 + (1.349)(0.9784)]b = 1.341b$$

Hence

$$\frac{b^*}{F^*} = \frac{(0.462)(1.341)}{1.174} = 0.528$$

$$\frac{d^*}{F^*} = \frac{(0.538)(1.031)}{1.174} = 0.472 \left[= 1 - \frac{b^*}{F^*} \right]$$

Table 6-1 gives equilibrium data for the acetone-water system at atmospheric pressure. True mole fractions of acetone are given in the first two columns, and computed pseudo mole fractions are given in the last two columns.

To identify the locus of operating-line intersections on the pseudo-mole-fraction diagram, we first calculate a feed equilibrium vaporization giving 55 mole percent vapor. As shown in Chap. 2, this can be done graphically, with the result that the true mole fractions of the equilibrium vapor and liquid in the feed are $x_F = 0.14$ and $y_F = 0.795$. Converting to pseudo mole fractions gives $x_F^* = 0.108$ and $y_F^* = 0.742$. We then have

$$L_F^* = [0.108 + (1.349)(0.892)]L_F = 1.311L_F$$

$$V_F^* = [0.742 + (1.349)(0.258)]V_F = 1.090V_F$$

Table 6-1 Vapor-liquid equilibrium data for acetone-water recomputed to pseudo-mole-fraction basis (data from Treybal, 1968)

x_A	y_A	x_A^*	y_A^*	x_A	y_A	x_A^*	y_A^*
0.00	0.00	0.00	0.00	0.40	0.839	0.331	0.794
0.01	0.253	0.0074	0.201	0.50	0.849	0.426	0.807
0.02	0.425	0.0149	0.354	0.60	0.859	0.527	0.819
0.05	0.624	0.0376	0.552	0.70	0.874	0.634	0.837
0.10	0.755	0.0761	0.696	0.80	0.898	0.748	0.867
0.15	0.798	0.116	0.745	0.90	0.935	0.870	0.914
0.20	0.815	0.156	0.766	0.95	0.963	0.934	0.951
0.30	0.830	0.241	0.784	1.00	1.000	1.000	1.000

and the slope of the locus of intersections is given by

$$-\frac{L_F^*}{V_F^*} = -\frac{(1.311)(0.45)}{(1.090)(0.55)} = -0.984$$

Figure 6-10 shows the equilibrium curve, feed and product compositions, and locus of operating-line intersections in terms of pseudo mole fractions.

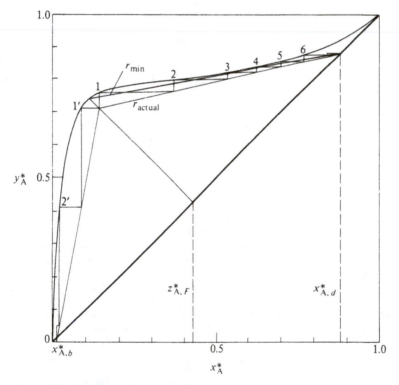

Figure 6-10 Operating diagram for Example 6-4.

(a) To locate minimum-reflux conditions, we take the slope of the line from (x_d^*, x_d^*) through the tangent pinch, which is $0.192 = L_{\min}^*/V_{\min}^*$. This gives $L_{\min}^*/d^* = 0.192/(1 - 0.192) = 0.237$. Converting back to true flows gives $L_{\min}/d = 0.237$ since the compositions of the two streams are the same.

(b) Since $L/d = 1.22(L/d)_{\min}$ for actual operation, $L/d = L^*/d^* = (1.22)(0.237) = 0.288$. Therefore $L^*/V^* = 0.288/1.288 = 0.224$, and the rectifying-section operating line can be located from its crossing of the 45° line at $x_{A,d}^*$, and its slope. The stripping-section operating line is then located from the intersection with the locus of operating-line intersections and its crossing of the 45° line at $x_{A,b}^*$. Stepping off equilibrium stages, we find a total of eight and a fraction equilibrium stages required. It should be noted, however, that the stage requirement is very sensitive to the precision of the equilibrium data in the vicinity of the minimum-reflux tangent pinch.

CURVED OPERATING LINES

Even if it is not possible to find a technique for generating straight operating lines, one can still employ the yx type of diagram for the analysis of a binary staged separation process. The operating line will be curved, however, and it will have to be calculated point by point. The component A mass balance must now be solved in conjunction with some other piece of information which will relate L and V or their equivalents. This "other piece of information" can take the forms of an enthalpy balance, miscibility relationships, or independent specifications. We shall consider cases of each.

Enthalpy Balance: Distillation

In a binary distillation we have seen that the mass balance relating compositions of streams *passing* each other between stages in the rectifying section is

$$V_p y_{A,p} = L_{p+1} x_{A,p+1} + x_{A,d} d \tag{5-3}$$

Similarly, the enthalpy balance for passing streams is

$$V_p H_p = L_{p+1} h_{p+1} + h_d d + Q_C \tag{5-10}$$

Under the assumption of constant molal overflow, V and L are constant throughout the rectifying section, and a straight operating line is obtained from Eq. (5-3) alone. Equation (5-10) is presumed to substantiate the constant-molal-overflow assumption and is not used directly.

To approach the more general case where constant molal overflow is not assumed, let us consider a binary distillation tower with a total condenser, where the specified independent variables correspond to a design problem:

Pressure	Feed rate
Distillate flow rate	Feed composition
Distillate composition	Feed enthalpy
Reflux rate	Arbitrary feed location
Reflux temperature	

In Eqs. (5-3) and (5-10) the quantities $x_{A,d}d$ and $h_d d$ have now been fixed, and Q_C is set through the distillate and reflux rates and temperatures. From the phase rule we know that the temperatures of both the liquid and the vapor streams will be those corresponding to thermodynamic saturation at column pressure; hence H_p will be uniquely related to $y_{A,p}$, and h_{p+1} will be uniquely related to $x_{A,p+1}$ by enthalpy-composition data like those shown in Fig. 2-20. We also know that the difference between V_p and L_{p+1} must be constant and equal to the net upward product flow

$$V_p - L_{p+1} = d \qquad (5\text{-}4)$$

Therefore, we have five equations [(5-3), (5-4), (5-10) and the two enthalpy-composition curves] in six unknown variables (V_p, L_{p+1}, H_p, h_{p+1}, $y_{A,p}$, and $x_{A,p+1}$). By specifying one of these six variables all the others can be obtained.

Algebraic enthalpy balance For example, if we specify $x_{A,p+1}$, we can immediately find h_{p+1} from the enthalpy-composition relationship for saturated liquid. Proceeding onward by trial and error, we can assume a value of $y_{A,p}$ and hence obtain a value of H_p from the enthalpy-composition relationship for saturated vapor. L_{p+1} and V_p can then be obtained from a simultaneous solution of Eqs. (5-4) and (5-10). These values of L_{p+1} and V_p can then be substituted into Eq. (5-3), which can be solved for $y_{A,p}$ to check the assumed value of $y_{A,p}$. A new value of $y_{A,p}$ is then assumed, and the procedure is repeated until the value of $y_{A,p}$ computed from Eq. (5-3) checks with the assumed value. Then a new value of $x_{A,p+1}$ can be specified, and the corresponding $y_{A,p}$ can be found by the same trial-and-error procedure. In this way a curve of $y_{A,p}$ vs. $x_{A,p+1}$ (Fig. 6-11) can be obtained, and this will be the operating curve for the distillation yx diagram.

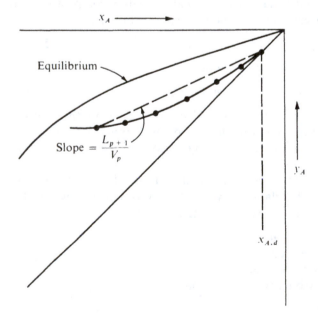

Figure 6-11 Construction of curved operating line in binary distillation.

One important property of this operating curve should be realized. If Eqs. (5-3) and (5-4) are solved simultaneously to eliminate d, we obtain

$$\frac{L_{p+1}}{V_p} = \frac{x_{A, d} - y_{A, p}}{x_{A, d} - x_{A, p+1}} \tag{6-6}$$

Thus the local L/V ratio is the *slope of the chord* connecting the particular point to the $x_{A, d}$ point on the 45° line, as shown on Fig. 6-11. In the case of constant molal overflow the slope of this chord is also the slope of the operating line. For varying L and V this is no longer true, and the local slope of the operating curve itself is not L/V.

In general, L/V will be the slope of a chord connecting a point on the operating line to the 45° line point with the composition of the net product flowing through the particular section of the column. Thus, in the stripping section, L/V is the slope of the chord to the point $(x_{A, b}, x_{A, b})$.

Graphical enthalpy balance At this point it would be useful for the reader to review the discussion surrounding Figs. 2-18 to 2-20. As developed there, a graphical approach can be used to analyze mixing or separation in a process where there are two conserved quantities, e.g., matter and enthalpy. Thus, on a plot of specific enthalpy vs. composition a stream with a given enthalpy and composition is represented by a point. If two streams with different enthalpies and compositions are mixed, the resultant mixture has a composition and enthalpy corresponding to a point lying on a straight line connecting the two points for the initial streams. The location of the mixture point is determined by the lever rule [Eq. (2-47)]. Similarly, if a mixture is separated into two products, the product composition and enthalpy points are collinear with the feed mixture point, the locations again being determined by the lever rule.

The construction of an operating curve when an enthalpy balance must be considered can be simplified by using a graphical construction on an enthalpy-composition diagram.

We can state Eq. (5-3) in its more general form

$$V_p y_{A, p} = L_{p+1} x_{A, p+1} + \text{net upward product of component A from section} \tag{5-9}$$

Within any section of a column the difference between $V_p y_{A, p}$ and $L_{p+1} x_{A, p+1}$ must be a constant amount of component A. Similarly, from Eq. (5-11) or from any equivalent equation for any other section of a column we find that the *difference between $V_p H_p$ and $L_{p+1} h_{p+1}$ must be a constant amount of enthalpy. Thus the differences in enthalpy and mass between any two passing streams in a given section of a column are fixed and can be represented by a single point on an enthalpy-composition diagram.* We shall call that point the *difference point* for the section. By Eqs. (5-3) and (5-10) this difference point must be collinear with the points corresponding to H_p, $y_{A, p}$, and h_{p+1}, $x_{A, p+1}$. Thus if we can establish the position of the difference point on an enthalpy-composition diagram, the compositions and enthalpies of the vapor and liquid streams passing each other between stages at various points in the column

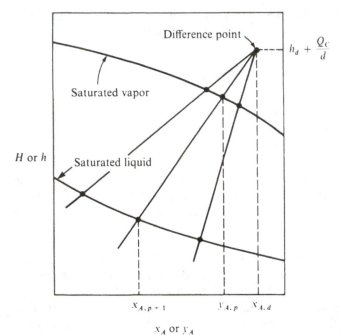

$x_{A,p+1}$ $y_{A,p}$ $x_{A,d}$

x_A or y_A

Figure 6-12 Graphical determination of $y_{A,p}$ and $x_{A,p+1}$; rectifying section.

can be obtained from a series of straight lines radiating out from the difference point. The pairs of vapor and liquid compositions making up the operating curve come from the intersections of these straight lines with the curves for saturated vapor and saturated liquid on the enthalpy-composition diagram, as shown in Fig. 6-12.

In the rectifying section with a total condenser the total net upward flow is d, the net upward product of component A is $x_{A,d}d$, and the constant-enthalpy difference is $h_d d + Q_c$. As a result the difference point for the rectifying section corresponds to a composition (moles of component A per total moles) of

$$x_{A,\text{diff}} = \frac{x_{A,d}d}{d} = x_{A,d} \tag{6-7}$$

and a specific enthalpy (per mole) of

$$h_{\text{diff}} = \frac{h_d d + Q_c}{d} = h_d + \frac{Q_c}{d} \tag{6-8}$$

The coordinates of the difference point in Fig. 6-12 would be $h_d + Q_c/d$ and $x_{A,d}$. Q_c must represent enough cooling to condense all the overhead vapor, which is necessarily a greater quantity than the distillate. Hence Q_c/d must be greater than the latent heat of vaporization of material of composition $x_{A,d}$, and as a result the difference point for the rectifying section necessarily lies above the saturated-vapor curve. Similar reasoning shows that the difference point for a tower with a partial condenser has the coordinates given by Eqs. (6-7) and (6-8) with d replaced by D and that the difference point must still lie above the saturated-vapor curve in Fig. 6-12.

For the stripping section the net upward product is $-b$, the net upward flow of

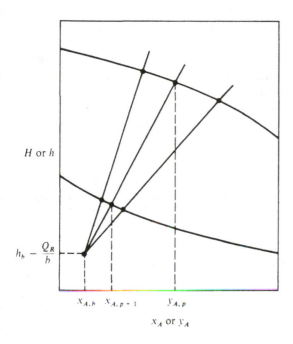

H or h

$h_b - \dfrac{Q_R}{b}$

$x_{A,b}$ $x_{A,p+1}$ $y_{A,p}$

x_A or y_A

Figure 6-13 Graphical determination of $y_{A,p}$ and $x_{A,p+1}$; stripping section.

component A is $-x_{A,b}b$, and the constant difference in enthalpy between the upflowing and downflowing streams is $Q_R - h_b b$. Hence the coordinates of the difference point for the stripping section are

$$x_{A,\text{diff}} = \frac{-x_{A,b}b}{-b} = x_{A,b} \tag{6-9}$$

and

$$h_{\text{diff}} = \frac{Q_R - h_b b}{-b} = h_b - \frac{Q_R}{b} \tag{6-10}$$

Equation (6-10) shows that h_{diff} for the stripping section necessarily lies below the saturated-liquid curve. The construction to find $y_{A,p}$ and $x_{A,p+1}$ in the stripping section is shown in Fig. 6-13.

As can readily be shown from the overall enthalpy and mass balances for the column, the difference points for the rectifying section and the stripping section must be collinear with the point corresponding to the feed enthalpy and composition and the relative distances must be in inverse proportion to the split between distillate and bottoms, by the lever rule.

In order to complete a design problem with the variables listed at the beginning of this section fixed, the bottoms flow and composition are first computed from the overall mass balance. The difference point for the rectifying section is located directly from the information given, and the difference point for the stripping section can be located as the intersection of (1) an extension of the line through the rectifying-section difference point and the point corresponding to the feed enthalpy and composition, and (2) the known $x_{i,b}$. The rectifying-section operating curve is located by extending rays out from the rectifying-section difference point, and the stripping-section operating curve is located by extending rays out from the stripping-section difference point. Stages can then be stepped off on the yx diagram. If a minimum-reflux calculation is desired, it can be made by identifying the point of tangency or first touching of an operating curve with the equilibrium curve on the yx diagram, transferring these saturated-liquid and saturated-vapor conditions to the appropriate points on the enthalpy-composition diagram, drawing a straight line through these two points, and extending this line to its intersections with $x_{i,b}$ and $x_{i,d}$. These procedures are illustrated in Example 6-5.

> **Example 6-5** Consider the same acetone-water distillation described in Example 6-4. Using the data in Tables 6-1 and 6-2 and the enthalpy-composition-diagram approach, (a) determine the minimum allowable overhead reflux ratio r/d and compare the answer with the result obtained using the

Table 6-2 Thermodynamic data for acetone-water system at 1 atm total pressure

			Properties of mixtures				
x acetone in liquid	Integral heat of solution at 59°F, Btu/lb mol of solution	y_{eq} acetone in vapor, equilibrium	Vapor-liquid temperature, °F	Heat capacity at 63°F, Btu/lb sol·°F	h_L, Btu/lb mol†	H_V, Btu/lb mo
---	---	---	---	---	---	---
0.00	0	0.00	212	1.00	0	17,510
0.01	....	0.253	197.1	0.998		
0.02	−81.0	0.425	187.8	0.994	−433	17,210
0.05	−192.3	0.624	168.3	0.985	−776	16,950
0.10	−287.5	0.755	151.9	0.96	−1047	16,650
0.15	−331	0.798	146.2	0.93		
0.20	−338	0.815	143.9	0.91	−984	
0.30	−309	0.830	141.8	0.85	−840	15,690
0.40	−219	0.839	140.8	0.80		
0.50	−150.5	0.849	140.0	0.75	−391	14,975
0.60	−108.6	0.859	139.1	0.70	−262	14,600
0.70	....	0.874	138.1	0.66		
0.80	....	0.898	136.8	0.61	....	13,800
0.90	....	0.935	135.5	0.57		
0.95	....	0.963	134.6	0.55		
1.00	....	1.000	133.7	......	0	12,980

Properties of pure acetone

Temperature, °F	68	100	150	200	212
Heat capacity, Btu/lb·°F	0.53	0.54	0.56	0.58	
Latent heat of vaporization, Btu/lb	242	233	219	206	203

SOURCE: First five columns and acetone data from Treybal (1968, p. 401: used by permission).
† These columns calculated by the author.

MLHV method (Example 6-4) and the result obtained assuming constant molal overflow; (b) taking an overhead reflux ratio r/d of 0.288, as in Example 6-4, part (b), find the equilibrium-stage requirement for the distillation and compare the result with that from the MLHV method and that from assuming constant molal overflow.

SOLUTION Table 6-2 gives data for vapor-liquid equilibrium, heats of solution, temperatures, heat capacities, and latent heats at 1 atm total pressure. The saturated-vapor and saturated-liquid curves on the enthalpy-concentration diagram of Fig. 6-14 have been prepared from these data. The saturated-vapor and saturated-liquid enthalpies are fixed by setting the pressure and the stream composition. The temperature is a dependent variable. The calculated liquid and vapor enthalpies are also given in the last two columns of Table 6-2.

The product flows and compositions are computed at the beginning of the solution to Example 6-4.

The locus of operating-curve intersections on a yx diagram must go through the point $(y_A = 0.50,\ x_A = 0.50)$ with a slope [from Eq. (5-20)] of $-0.45/0.55 = -0.818$. This line and the

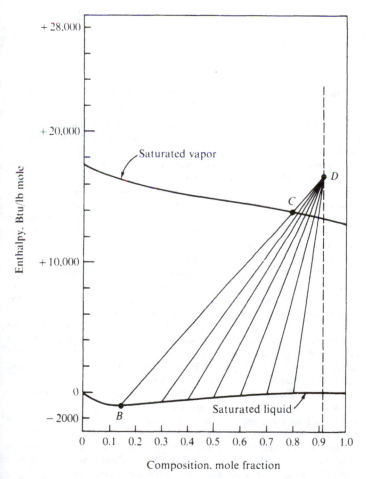

Figure 6-14 Enthalpy-composition diagram for acetone-water system at 1.0 atm total pressure; basis: enthalpies of pure saturated liquids = 0.

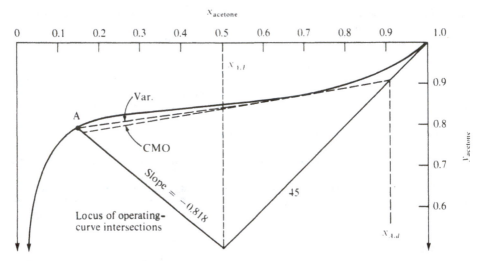

Figure 6-15 Minimum reflux construction for Example 6-5.

known $x_{A,d}$ are shown in Fig. 6-15, which is an expansion of the upper portion of the yx diagram. The equilibrium curve for this diagram is taken from the data of Table 6-1.

(a) If there were constant molal overflow, the minimum reflux ratio would be found from the dashed straight line marked CMO in Fig. 6-15. This would be a case where the pinch does not occur at the feed tray but is a tangent pinch midway in the rectifying section.

Since the more volatile component has the smaller molal latent heat (see Fig. 6-14), we know that the vapor and liquid flows will tend to *decrease* as we pass down the column (see Chap. 7). This point follows from the necessarily constant difference in enthalpies of passing streams. Vapor and liquid flows decreasing downward mean that L/V must become smaller, and the chords reaching back to the $x_{A,d}$ point must progressively decrease in slope as we move down the column. Hence, the rectifying-section operating line will be concave upward. Because of this we may or may not find the minimum-reflux pinch to be at the feed tray in the real case.

If we assume for the moment that the pinch does occur at the feed tray, we can obtain a limiting operating line with the aid of the enthalpy-composition diagram. If point A from Figure 6-15 is on the upper operating curve, it must satisfy both the mass balance [Eq. (5-4)] and the enthalpy balance [Eq. (5-12)]. Therefore, a straight line through points B and C in Fig. 6-14 must represent the combined mass and enthalpy balance. The y and x coordinates of points B and C in Fig. 6-14 correspond to the coordinates of point A in Fig. 6-15. The enthalpy and composition of the *net upward product* must lie on the straight line defined by points B and C if we interpret the line as a graphical subtraction of a liquid from the vapor passing it. We know, however, that the composition of this net upward product is $x_{A,\text{diff}} = 0.91$ [by Eq. (6-7)]. Hence the difference point is point D, and from the graphical construction we find that $h_{\text{diff}} = 16,500$ Btu/lb mol.

Since the net upward product is unchanged throughout the rectifying section, the difference point must be the same between all stages in the rectifying section. The rectifying operating curve is thus obtained from a sequence of lines radiating out from the difference point. The y coordinate will fall on the saturated-vapor curve, and the x coordinate will fall on the saturated-liquid curve. Several such lines are shown in Fig. 6-14, and the resulting operating curve representing the yx pairs is shown in Fig. 6-15, marked "Var." for variable molal overflow. We conclude that this curve does indeed represent minimum reflux since it does not cross the equilibrium curve perceptibly at any point.

An enthalpy h_{diff} represents $h_d + Q_C/d$ [Eq. (6-8)]. Since we have a total condenser giving saturated liquid reflux. Q_C must represent the overhead vapor rate V_o times the enthalpy change between saturated vapor and saturated liquid at $x_A = 0.91$. Hence

$$h_{\text{diff}} - h_d = \frac{V_o(H_d - h_d)}{d} \tag{6-11}$$

or

$$\frac{r}{d} = \frac{V_o - d}{d} = \frac{h_{\text{diff}} - H_d}{H_d - h_d} \tag{6-12}$$

Substituting gives

$$\left(\frac{r}{d}\right)_{\text{min}} = \frac{16,500 - 13,300}{13,300 - 0} = 0.240$$

This result compares with 0.237 obtained by the MLHV method [Example 6-4, part (a)] and essentially the same value obtained by the constant-molal-overflow construction (CMO) shown in

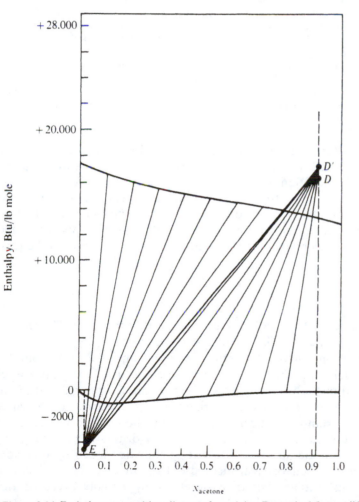

Enthalpy, Btu/lb mole

X_{acetone}

Figure 6-16 Enthalpy-composition diagram for solving Example 6-5, part (b).

Fig. 6-15. The close agreement with the CMO result comes from the curvature of the actual operating curve around the tangent pinch.

(*b*) Since $r/d = 0.288$, we can calculate h_{diff} for the actual operation as

$$h_{\text{diff}} = (0.288)(13,300) + 13,300 = 17,130 \text{ Btu/lb mol}$$

An enthalpy-composition diagram with D' representing this point is shown as Fig. 6-16.

The rectifying-section operating curve for $r = 0.288d$ is obtained as before by radiating a sequence of lines out from point D'. The rectifying-section operating curve of Fig. 6-17 was obtained in this way.

The rectifying-section operating curve joins the locus of operating-curve intersections at $y_A = 0.775$, $x_A = 0.170$. This point must also lie on the lower operating curve. Proceeding as before, we follow a straight line through $y_A = 0.775$ (saturated vapor) and $x_A = 0.170$ (saturated liquid) in Fig. 6-16, and find point E, where $x_A = 0.0216$. This is the difference point for the stripping section, and from the graphical construction $h_{\text{diff}} = -4700$.

The stripping-section operating line on Fig. 6-17 now comes from a sequence of lines radiating out from the stripping-section difference point. The lines are shown in Fig. 6-16.

Equilibrium stages can now be stepped off in Fig. 6-17. Starting at the feed plate we find six and a small fraction stages in the rectifying section and two and a small fraction stages in the stripping section, one of which is the reboiler. The liquid and vapor flows at any point can be determined from the slope of the chord from each step on the operating line back to $x_{A,d}$ or $x_{A,b}$.

The operating lines obtained under the assumption of constant molal overflow with $r = 0.288d$ are shown by dashed lines in Fig. 6-17. The number of equilibrium stages is about $7\frac{1}{2}$, as opposed to the value of about $8\frac{1}{2}$ found allowing for variations in flow rates through the enthalpy-composition diagram or the MLHV method (Example 6-4). The difference is thus some 10 to 15 percent, and the CMO case is *not* conservative. □

We can now reconsider the point made in Chap. 4 and Appendix C concerning the interdependence of reflux flow rate and condenser duty in distillation columns with total condensers. The location of the operating curve in the rectifying section is controlled solely by the location of the difference point on the enthalpy-composition diagram. The enthalpy coordinate of the difference point can be increased either by increasing the condenser duty at fixed reflux and distillate flows (subcooling the reflux) or by increasing the reflux flow at fixed distillate and reflux enthalpies. Considering

$$h_{\text{diff}} = h_d + \frac{Q_C}{d} \tag{6-8}$$

the first alternative increases Q_C/d while decreasing h_d a lesser amount. The second alternative will necessarily increase Q_C. Both changes have equivalent effects.

Notice that the operating-curve intersection line in Example 6-5 came from the statement that the molal vapor flow would increase across the feed plate by 55 percent of the molal feed rate. This is not necessarily the same as saying that the feed was 55 mole percent vapor at 1 atm before entering the column because of the same factors which cause the assumption of constant molal overflow not to be valid.

This approach to the complete analysis of a binary distillation allowing for varying molal overflow is a modification of that originally developed by Ponchon (1921) and Savarit (1922). Their original method did not involve the yx diagram directly. It is interesting to note that their development actually preceded the McCabe-Thiele analysis chronologically.

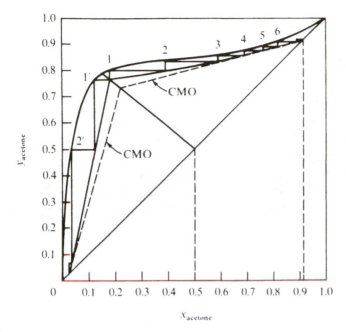

Figure 6-17 A yx diagram for solving Example 6-5, part (b).

Use of the enthalpy-composition diagram to generate operating curves is more complicated than the MLHV method and is probably warranted only where heat-of-mixing effects and sensible-heat effects from the temperature span across a column rival latent-heat effects in importance. The MLHV method has the added advantage of being readily extensible to the analysis of multicomponent distillation (Chaps. 8 and 10). For problems with complex enthalpy effects, binary or multicomponent, it is often just as convenient or more so to use a full computer solution of the mass- and enthalpy-balance equations for each stage.

The use of coupled enthalpy and mass balances to determine operating curves is applicable to several other binary multistage separation processes of the equilibration type where there is an energy separating agent, e.g., crystallization.

Miscibility Relationships: Extraction

In solvent-extraction processes varying total flow rates of the interstage streams arise when there is appreciable miscibility of the two liquid phases. Since the temperature of operation is usually nearly constant, no important restriction is applied by enthalpy balances. The two phases are thermodynamically saturated when passing between stages in an equilibrium-stage extraction process; this imposes a constraint upon allowable stream compositions. Three components are present when two components are being separated in an extraction process, and we therefore are faced with mass-balance restrictions in each of two components (the mass balance on the third

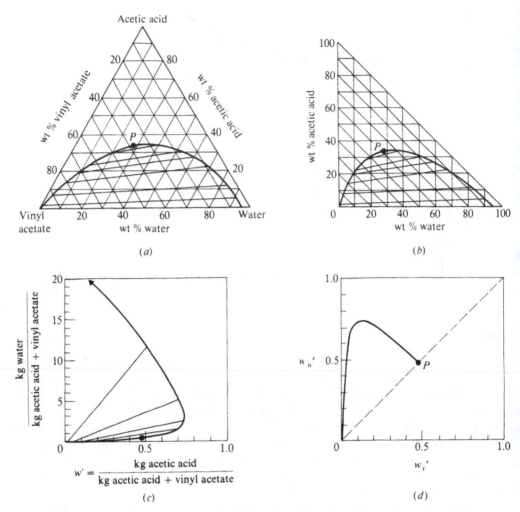

Figure 6-18 Alternative representations of liquid-liquid equilibria for vinyl acetate–acetic acid–water at 25°C: (a) equilateral-triangular diagram (Fig. 1-21); (b) right-triangular diagram; (c) Janecke diagram; and (d) w'_w vs. w'_v (Fig. 1-23).

component is not independent). Thus again we have *two* conserved properties and we can employ a plot of one composition variable vs. another in a way entirely analogous to our use of the enthalpy-composition diagram in Example 6-5.

Figure 6-18 shows four different ways of plotting one composition variable against another for the system vinyl acetate–acetic acid–water. Figure 6-18a shows the equiliateral-triangular diagram, already encountered in Fig. 1-21. Here each apex of the triangle corresponds to a pure component, and each point inside the triangle corresponds uniquely to a composition. The miscibility envelope is shown, and equilibrium tie lines are drawn across the two-phase region. Above the plait point P sufficient acetic acid is present to cause the system to be entirely miscible.

Figure 6-18b is a right-triangular diagram for the same system with the weight percent of acetic acid plotted against the weight percent water. The weight percent of vinyl acetate is obtained by difference. Since the sum of weight percents cannot exceed 100 percent, the hypotenuse of the right triangle represents 0 percent vinyl acetate and is the limit of possible compositions. The right-triangular diagram is a skewed version of the equilateral-triangular diagram. It offers the convenience of being able to use rectilinear coordinates, but the composition variable for the third component (vinyl acetate, in this case) is now measured on a different scale from the other two components.

Figure 6-18c is a plot of the mass ratio of the amount of solvent (water, in this case) to the sum of the amounts of the two components being separated vs. the weight fraction of one of the other two components on a solvent-free basis. This form of plot is known as a *Janecke diagram*. In it the solvent plays a role completely analogous to enthalpy in the enthalpy-composition diagram. The vertical coordinate in the Janecke diagram is the specific solvent content of the mixture being separated, while on the enthalpy-composition diagram it is the specific enthalpy content of the mixture being separated. The curve is the phase envelope, which does not extend all the way across the diagram because the system becomes fully miscible above a certain acetic acid content. Equilibrium tie lines are also shown, transferred from Fig. 6-18a and b. The Janecke diagram is a rather cumbersome representation of this system because the solvent-lean phase compositions are squeezed into the lower boundary of the diagram.

Finally, Fig. 6-18d is a plot of $w'_{w'}$, the weight fraction acetic acid in the water phase on a solvent-free basis, vs. $w'_{v'}$, that in the vinyl acetate phase (shown earlier in Fig. 1-23). Again the equilibrium curve does not extend fully across the diagram because of the existence of the plait point. This diagram is analogous to the yx diagram.

Graphical computations of mixing and separation of streams can be made on the diagrams of Fig. 6-18a to c, in a way fully analogous to the procedure used with the enthalpy-composition diagram. Mixing of two streams with compositions represented by two different points is described by a straight line connecting the points, the mixture composition being located along the line by the lever rule, in the inverse proportion of the amounts of the two feed streams. For a separation, the product compositions are collinear with the feed composition, the lever rule again relating the amounts of the two products. The type of diagram in Fig. 6-18d is not useful for this purpose since it does not define the amount of solvent present in a mixture.

Consider now the single-section countercurrent staged extraction process shown in Fig. 6-19. For purposes of our example, the feed is a mixture of vinyl acetate and acetic acid, and the solvent is water, which preferentially removes acetic acid into the extract, leaving the bulk of the vinyl acetate in the raffinate. Since this is a single section with no intermediate feeds or products, the difference in flows (both total flows and flows of any component) between passing streams must be the same at any interstage location. We can write

$$R - S = F - E = \text{net product flow to left} \qquad (6\text{-}13)$$

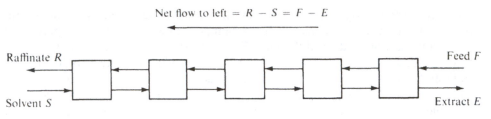

Net flow to left $= R - S = F - E$

Raffinate R

Solvent S

Feed F

Extract E

Figure 6-19 Schematic of extraction process.

where R, S, F, and E are the total flows of raffinate, solvent, feed, and extract, respectively. Similar equations can be written for individual components, e.g.,

$$w_{AR} R - w_{AS} S = w_{AF} F - w_{AE} E = \text{net flow of acetic acid to left} \qquad (6\text{-}14)$$

etc.

Since the differences between total flows and individual-component flows are constant, there will be a unique difference point on the diagrams of Fig. 6-18a to c which will denote the composition of this hypothetical difference stream. This point will then be collinear with the points representing any pair of passing streams at an interstage location. From the nature of the extraction we know that there must be a net flow of the feed components to the left (a large amount of the raffinate component, vinyl acetate, and a small amount of the extracted component, acetic acid). Similarly, we know that there will be a large net flow of solvent (water) to the right. Consequently, if the net total flow is to the left [Eq. (6-13) positive], we know that this net product will contain positive weight fractions of both feed components and a (hypothetical) negative weight fraction of the solvent. This will give a difference point lying outside the diagram, low on the left-hand side, in Fig. 6-18a or b. In Fig. 6-18c the difference point will lie below the diagram, on the left.

If the net total flow is to the right [Eq. (6-13) negative], the difference-point composition contains negative weight fractions of the feed components and a positive weight fraction of the solvent. It lies outside the diagram, to the right and below, on Fig. 6-18a and b, and still lies below the diagram, on the left, in Fig. 6-18c.

The difference point for a section can often be located as the intersection of extrapolated straight lines connecting the feed and extract compositions and the raffinate and solvent compositions. For extractions with an intermediate feed or product stream, there will be different difference points for the different sections.

In extractions, compositions are often considered on a solvent-free w_i' basis. For any composition represented by a point in a triangular diagram (Fig. 6-18a and b), the composition on a solvent-free basis can be determined by a graphical subtraction, in which a straight line from the composition point to the pure-solvent apex is extended to the opposite (solvent-free) side of the diagram.

Minimum solvent flow in an extraction is again found by locating the point of first tangency on the analog of the yx diagram, i.e., Fig. 6-18d, by testing the different difference points resulting as the solvent flow is reduced.

Operating curves can be placed on the yx type of diagram, such as Fig. 6-18d, by radiating a succession of straight lines out from the difference point and taking the compositions corresponding to the intersections of these lines with either side of the phase envelope. For multisection extractions, the operating curves for different sections will intersect on a locus of intersection lines which corresponds to the phase condition of the feed or product between the sections in a way entirely analogous to the result for distillation. Thus, for a raffinate-phase feed or product, the locus of intersections on Fig. 6-18d will be a vertical straight line meeting the 45° line at the feed or product composition, and for a solvent-phase feed or product it will be a horizontal straight line meeting the 45° line at the feed or product composition.

Example 6-6 An existing process in your plant employs a countercurrent cascade of five mixer-settlers of the type shown in Fig. 6-1 for the separation of a feed containing 30 wt "₀ acetone and 70 wt "₀ MIBK, using water as a solvent at 25 to 26°C. During a plant test it has been found that the MIBK product contains 3.0 wt "₀ acetone on a water-free basis when the water feed rate is 1.76 kg per kilogram of ketone feed. It has been found experimentally that this water feed rate corresponds closely to the capacity limit of the plant at the prevailing ketone feed rate. Beyond this point the settlers do not provide sufficient disengagement of the phases.

The process superintendent asks you to explore ways of increasing the plant capacity and the fraction of the MIBK feed that is recovered in the MIBK product. He points out that the present recovery is only about 90 percent and reports that he has heard that the recovery in other extraction units has been increased through the use of *extract reflux.* (a) By how much could the water requirement possibly be reduced if more mixer-settler units were to be added? By how much might the plant capacity increase? (b) Would it be worthwhile to install more powerful motors on the stirrers in the mixing chambers? (c) Is there an incentive for the use of extract reflux in this process? The MIBK purity must be held at 97 percent on a water-free basis.

SOLUTION Figure 6-20 is a plot of the phase-miscibility data on a right-triangular diagram, using the weight fraction of acetone and the weight fraction of MIBK as coordinates. This form of diagram is selected rather than the Janecke diagram since the raffinate-phase compositions would crowd in the lower part of a Janecke diagram, as in Fig. 6-18c. Data underlying Fig. 6-20 are given in the table.

Phase-equilibrium data for water–acetone–methyl isobutyl ketone (MIBK) at 25 to 26°C (data smoothed from Othmer et al., 1941)

Aqueous phase, wt "₀			Ketone phase, wt "₀		
Water	Acetone	MIBK	Water	Acetone	MIBK
98.0	.....	2.0	2.3	.....	97.7
95.2	2.6	2.2	2.7	5.0	92.3
92.2	5.4	2.4	3.0	10.0	87.0
88.9	8.5	2.6	3.2	15.0	81.8
85.3	11.9	2.8	3.7	20.0	76.3
81.5	15.5	3.0	4.3	25.0	70.7
77.2	19.5	3.3	5.3	30.0	64.7
71.8	24.2	4.0	6.8	35.0	58.2
65.7	29.2	5.1	8.8	40.0	51.2
57.5	35.2	7.3	12.6	45.4	42.0
Estimated plait point			34.7	46.5	18.8

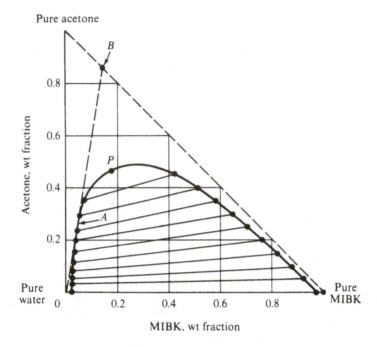

Figure 6-20 Phase-miscibility diagram for Example 6-6.

The problem specification involves compositions on a water-free basis, which amounts to "subtracting" all the water out. The water-free composition must then be on the straight line in Fig. 6-20, which connects pure water with the actual composition under consideration, lying at the point of intersection of that line with the dashed diagonal. This construction is shown in Fig. 6-20 for the ternary composition represented by point A, which corresponds to point B (85.5% acetone) when put on a water-free basis. Point A happens to represent the highest acetone purity available on a water-free basis, since the construction line is tangent to the phase envelope at point A. Figure 6-21 presents the equilibrium data from the table when expressed on a water-free basis (the upper curve). This diagram of Fig. 6-21 will be the equivalent of the yx diagram for our solution.

(*a*) This problem reduces to a determination of the minimum solvent rate required, given any number of stages. The minimum solvent rate will indicate the maximum possible reduction in water consumption that can be achieved by adding stages.

In order to determine the minimum allowable solvent flow rate, we search for an operating curve which touches the equilibrium curve of Fig. 6-21 but does not cross it. Considering the two streams at the left-hand end of the cascade, we know that the difference point must lie on a line connecting pure water S with the specified raffinate R, corresponding to 97% MIBK, water-free basis, denoted by R'. This line is shown on the miscibility-equilibrium diagram in Fig. 6-22. We know that the pinch corresponding to minimum solvent flow does not lie at the left-hand end of the cascade since the solvent enters free of ketones. If the pinch lies at the right-hand end of the cascade, the extract composition E, the feed F, and the ketone stream leaving the right-handmost stage K_1 must be collinear, since the extract composition cannot change across that stage if the stage is in a pinch region. By trial, one can find the equilibrium tie line across the phase envelope which passes through F; this is the line $\overline{FK_1E}$ in Fig. 6-22. The difference point also must be collinear with E and F, the streams passing one another at the right-hand end of the cascade in Fig. 6-19.

The difference point is denoted by D in Fig. 6-22. It is uniquely determined as the point satisfying both lines, $\overline{SRR'}$ and $\overline{FK_1E}$. Points on the operating curve are now found by radiating a

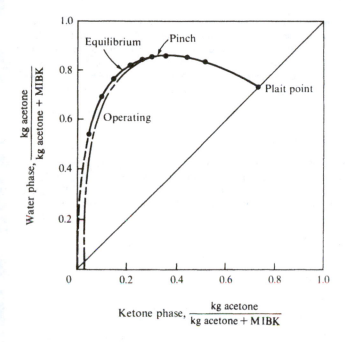

Figure 6-21 Operating diagram for Example 6-6; solvent-free basis, part (a).

sequence of straight lines out from point D in Fig. 6-22, and by taking the coordinates of the two passing streams for the operating diagram from the crossings of the miscibility envelope. These latter compositions can then be converted to a solvent-free basis. An operating curve determined this way is shown in Fig. 6-21. Since the curves touch and do not cross, the pinch does indeed occur at the right-hand end of the cascade and the conditions pictured correspond to the minimum solvent treat.

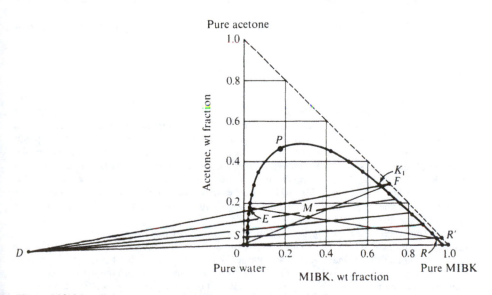

Figure 6-22 Mass-balance construction for Example 6-6, part (a).

By employing a graphical mixing of conserved quantities we know that a point representing the combined products (raffinate + extract) must lie on a straight line between R and E on Fig. 6-22. By an overall mass balance we also know that the combined products must lie on the line connecting points S and F (solvent + ketone feed = combined feeds = combined products, at steady state). Hence point M, the intersection of these two lines, represents both the combined feeds (S mixed with F) and the combined products (R mixed with E). The solvent-to-feed ratio then comes from an application of the lever rule to line $\overline{SMF}$:

$$\left(\frac{S}{F}\right)_{min} = \frac{\overline{MF}}{\overline{SM}} = \frac{0.70 - 0.31}{0.31 - 0.00} = 1.26 \text{ kg water/kg ketone feed}$$

The water rate can be reduced from the present 1.76 kg water per kilogram of ketone feed at most by $(1.76 - 1.26)/(1.76) = 28$ percent. To a crude approximation, the capacity will be limited by the total volumetric flow rate, so as to hold the residence time in the settlers constant. The specific gravity of water is 1.0; that of the ketone feed is 0.81 (Perry and Chilton 1973). Hence the present total volumetric feed flow E_F is

$$V_F = \left(\frac{1.76}{1.0} + \frac{1.00}{0.81}\right)\frac{1}{1000} \text{ m}^3/\text{kg} = 0.0030 \text{ m}^3/\text{kg ketone feed}$$

With a water treat of 1.26 kg per kilogram of ketone feed,

$$V_F = \left(\frac{1.26}{1.00} + \frac{1.00}{0.81}\right)\frac{1}{1000} = 0.0025 \text{ m}^3/\text{kg ketone feed}$$

Thus the plant capacity could be increased by approximately a factor of 0.030/0.025, or only 20 percent, if the number of stages were made infinite. This is not a particularly promising path to pursue.

(b) The question of whether more intense stirrer agitation in each stage would be beneficial reduces to a determination of the equilibrium-stage requirement for the separation now being obtained. If the equilibrium-stage requirement is appreciably less than five, the stage efficiencies are appreciably less than 100 percent. In that case increased stirrer agitation might well be useful since it should bring the effluent streams from each stage closer to equilibrium by virtue of more effective mass transfer within the mixers.

Taking $S/F = 1.76$ kg water per kilogram of ketone feed, we can locate point M' on line $\overline{SF}$ by means of the lever rule.

$$x_{M'} = \frac{0.70}{1 + 1.76} = 0.254$$

M' now represents $S + F$ and is shown in Fig. 6-23. Point E', representing the new extract composition, is then an extension of line $\overline{M'R}$, since M' must also represent E' combined with R (combined products = combined feeds, at steady state). Point D' in Fig. 6-23 represents the new difference point and is the intersection of lines $\overline{E'F}$ and $\overline{SR}$, since the difference point must be collinear with the points representing passing streams at both ends of the cascade. The operating curve shown in Fig. 6-24 is determined from the phase-envelope intersections of a series of straight lines radiating out from point D', as shown in Fig. 6-23. Since the equilibrium and operating curves are now both known, equilibrium stages can be stepped off in Fig. 6-24. Almost exactly, five equilibrium stages are required for the separation. The stage efficiencies are thus close to 100 percent, and it follows that improved agitation in the mixers would be of no use.

(c) The process shown in Fig. 6-19 can at best produce an extract that is in equilibrium with the feed (when saturated with extract). This factor serves to limit the recovery fraction of MIBK which can be obtained in the raffinate, since a substantial amount of MIBK must leave in the extract.

The extraction cascade we have considered serves in a manner similar to the stripping section of a distillation column. It is possible to gain some of the action of the rectifying section of a distillation column by using the scheme developed earlier in Fig. 4-23. The solvent is removed from some of the

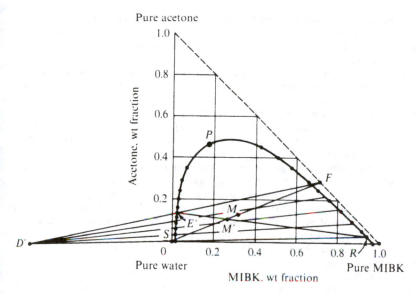

Figure 6-23 Mass-balance construction for Example 6-6, part (*b*).

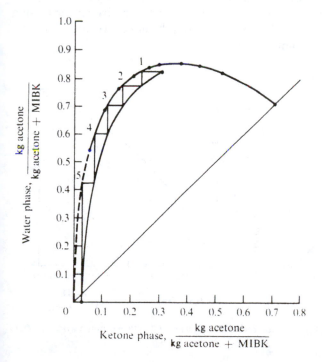

Figure 6-24 Operating diagram for Example 6-6; solvent-free basis, part (*b*).

extract, and this solvent-lean stream is returned to the cascade as extract reflux. This extract reflux should be richer than the feed in terms of the ratio of components being separated; i.e., it should contain more of the preferentially extracted component. This gives the extract material a richer material with which to equilibrate.

In our problem there is little to be gained by extract reflux, for the acetone product is already almost as pure on a solvent-free basis without extract reflux as it can be with extract reflux. We are limited in acetone purity in the nonrefluxed process of Fig. 6-24 by two factors, equilibrium with the feed and total miscibility of the system above 85.5 % acetone (water-free). These two restrictions very nearly coincide. Extract reflux can circumvent the first limitation of equilibrium with the feed, but it cannot circumvent the limitation of total miscibility above a certain acetone content. On the other hand, if our feed had contained only 10 wt % acetone and 90 wt % MIBK, there would have been some incentive for extract reflux. The equilibrium-with-the-feed limit would be more constraining than the total-miscibility limit. By means of reflux in such a case one could raise the acetone product from 69% acetone to 85.5% on a water-free basis and thus increase the MIBK recovery.

The analysis of the extraction cascade when extract reflux is employed would proceed in a fashion analogous to that carried out in parts (a) and (b). The composition of the net product will be different on either side of the feed, however, there being one constant composition of net product to the left of the feed and another to the right. This would provide a discontinuity in slope of the operating curve at the feed stage.

Finally, it should be noted that the separation of acetone from MIBK usually would not be accomplished by extraction. Distillation would be less expensive since the relative volatility is high.

□

Although extract reflux was not helpful in Example 6-6, it can almost always be of significant use in a system where the phase envelope cuts entirely across the triangular diagram or Janecke diagram, i.e., a system with no plait point. The methylcyclohexane–n-heptane–aniline system (Fig. 6-25) is a case in point. Here one can obtain as complete a separation as desired between the hydrocarbons using aniline as a solvent because the equivalent of the yx equilibrium curve does not cross the 45° line at any intermediate point.

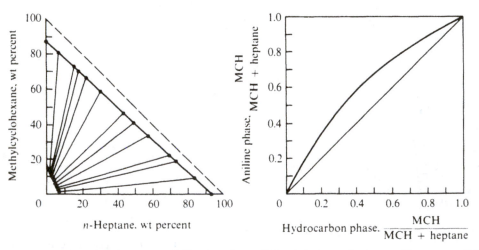

Figure 6-25 Equilibrium and miscibility data for methylcyclohexane–n-heptane–aniline system. (*Data from Varteressian and Fenske, 1937.*)

It is usually possible to select a solvent which will give partial miscibility over the entire composition range, as in Fig. 6-25, but one must compromise the benefit of greater extract fractionation with the fact that the required solvent rate is usually higher for such a system because the extracted component will have less solubility in the solvent.

Notice that the right-triangular diagram in Fig. 6-25 crowds the curve and points for the solvent-rich phase. In systems without a plait point the Janecke diagram is usually more convenient than one of the triangular diagrams because of this feature. Use of the Janecke diagram is described in Perry and Chilton (1973), sec. 15, and graphical analysis of extraction processes in general is covered more extensively by Treybal (1963) and Smith (1963).

For extraction processes with more than three components and where phase-equilibrium data can be represented algebraically, e.g., through activity-coefficient expressions, the computer-based methods described in Chap. 10 are often either necessary or more convenient.

Independent Specifications: Separating Agent Added to Each Stage

In most of the countercurrent staged processes we have discussed so far the separating agent enters at one end of the cascade and flows on through as part of one of the two interstage streams. In distillation heat is put in at the reboiler, and this heat is carried along in the upflowing vapor phase, thereby eliminating the need for adding more separating agent in any of the higher stages. The separating agent is used over and over again and is finally removed in the condenser.

As a result of this situation we have seen that the interstage flows are interdependent. Once the amount of separating agent entering the end stage is specified along with the product leaving the process at that point, all interstage flows within that section of the cascade are fixed and can be found through the various methods described earlier in this chapter.

On the other hand, there is no requirement that we add separating agent only to one end stage and remove it only at the other end stage. In distillation we could equip as many intermediate stages as we wish with heat exchangers, which would either add or remove heat. This would contribute up to another $N - 2$ degrees of freedom (assuming we already have exchangers on both terminal stages), which could be employed to specify the interstage flows at various points. These extra exchangers are ordinarily not added, since the cost of installing them more than offsets savings in operating costs except in certain extreme cases (see Chap. 13). Generally, it is desirable to have the benefit of *all* the separating agent in all the stages. In solvent extraction and gas absorption, liquid-solvent separating agent can be added to intermediate stages at various levels of purity; see, for example, Prob. 6-N, part (*e*), but removal of separating agent from any stage would require some sort of auxiliary separation process, e.g., distillation.

As we have seen, there are some staged separation processes where separating agent cannot flow from stage to stage in any simple fashion. Multistage processes wherein separating agent must be added to each stage generally fall into the category

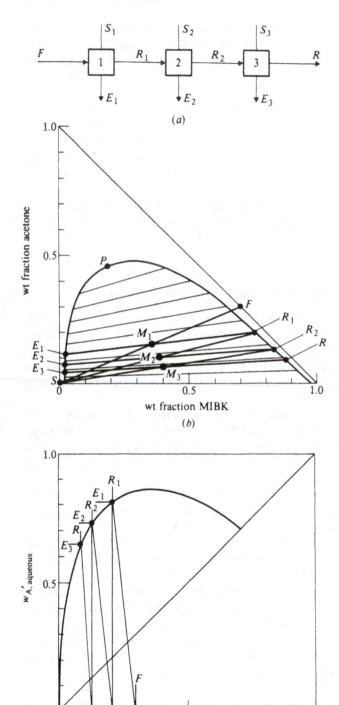

(a)

(b)

(c)

Figure 6-26 Three-stage cross-flow extraction process: (*a*) flow diagram, (*b*) triangular diagram, and (*c*) *yx* analog.

of rate-governed separation processes. Examples are multistage gaseous diffusion, reverse osmosis, ultrafiltration, and sweep diffusion. In gaseous diffusion the gas stream must be recompressed between stages, and the result is an addition of separating agent to each stage independently. There are therefore sufficient degrees of freedom to permit specifying all the interstage flows in one of the directions independently; the interstage flows in the other direction then follow by mass balance. In such processes there are, in effect, separate operating lines for each stage. In general, they must be constructed independently, although in some cases of a regular variation of interstage flows, e.g., the ideal cascade, discussed in Chap. 13, this may not be necessary.

Cross-flow processes Cross-flow staging is another case where separating agent is added individually to each stage, thereby making additional specifications possible. In a typical cross-flow process, however, the separating agent is removed as a product from the stage to which it is added rather than passing onward through the remaining stages.

A three-stage cross-flow version of the extraction analyzed in Example 6-6 is shown in Fig. 6-26a to c. Individual streams of solvent contact the acetone-MIBK feed and the various raffinates (R_1, R_2) successively in each of three stages, as shown in Fig. 6-26a. The process can be analyzed with the triangular diagram (Fig. 6-26b). Mixing F with S_1 in stage 1 gives an overall mixture with composition M_1, which splits along an equilibrium tie line to give raffinate R_1 and extract E_1. The position of M_1 is obtained by applying the lever rule to the flows of F and S_1. Raffinate R_1 is then contacted with S_2 to give overall mixture M_2 (placed by applying the lever rule to the flows of R_1 and S_2), and M_2 splits along a tie line to give R_2 and E_2. Finally R_2 mixes with S_3 to give M_3, which splits into R and E_3. The final raffinate is R, and the extract is a combination of E_1, E_2, and E_3.

The process is shown on the equivalent of a yx diagram in Fig. 6-26c, where the solvent-free weight fraction of acetone in the aqueous (solvent) phase is plotted against the same parameter in the raffinate phase. The successive pairs of raffinate and extract compositions are located by drawing lines from points representing combinations of F, R_1, or R_2 and the solvents ($w'_{A, aq} = 0$). These lines have slopes equal to minus the ratio of the amount of ketones in the organic phase to the amount of ketones in the aqueous phase (a large number). In general, this form of diagram is not particularly useful for cross-flow processes since the ketone flows require the information on the triangular diagram in order to be determined. The triangular diagram is sufficient and simpler to use by itself in this case.

Methods of computation for cross-flow staged processes in general have been reviewed by Treybal (1963). Usually the problem becomes one of computing a succession of single-stage equilibrations, starting at the feed end.

PROCESSES WITHOUT DISCRETE STAGES

As already discussed in Chap. 4, it is not mandatory that we carry out countercurrent multistage separations in equipment that provides a succession of distinct and separately identifiable stages. The only necessity is that there be countercurrent flow of two streams of matter and that these streams be in contact with each other so that they can transfer material back and forth during the countercurrent-flow process. Thus a packed column can provide a separation in distillation equivalent to that achieved in a plate column.

Except for the possible influence of axial mixing, the operating line or curve on a yx diagram for continuous countercurrent contactor, such as a packed column, is the same as for a staged process. Also, the equilibrium curve is necessarily the same. The use of the yx type of operating diagram for analyzing continuous countercurrent processes is considered in Chap. 11.

GENERAL PROPERTIES OF THE yx DIAGRAM

Several properties of the yx diagram were noted in the specific context of binary distillation in Chap. 5 and have been pointed out in the previous discussion in this chapter. Stated in more general form, these properties are the following:

1. The two axes of the diagram should represent *composition parameters* relating to each of the two streams flowing in opposite directions between the stages of the separation cascade.
2. Two distinct types of lines or curves are required in the diagram. One relates the two *exit* compositions from any stage; for an equilibration separation process it relates the compositions that would be obtained if equilibrium were achieved. The other, the *operating curve*, relates the compositions of streams *passing* each other in between stages.
3. There is an advantage to defining composition parameters and flow rates in the mass-balance expressions so that the flow rates will not change from stage to stage. This will give straight operating lines, which can be located from two points or a point and the slope. The use of mole ratios and constant inert flows and the MLHV method are examples of this advantage.
4. In the more general case, the operating curve can be calculated point by point by using an equation involving an additional conserved quantity, e.g., the enthalpy balance in distillation or a second component mass balance in extraction. This calculation can be carried out either graphically or algebraically.
5. A useful concept is that of the *net product* in any section of a countercurrent cascade. Since it remains constant, independent of interstage location, it can be a fixed point in a graphical construction. For mass-separating-agent processes, such as extraction, the net product may have a hypothetical composition, involving negative weight or mole fractions of some components.
6. A condition relating to the *minimum allowable flow* of one or both of the two counterflowing streams and an *infinite number of stages* results when the operating curve or curves are changed so as to touch but not cross the equilibrium or stage-exit-compositions curve at some one point within the range of operation. This condition also corresponds to the *minimum consumption of separating agent*.
7. For energy-separating-agent processes the point of intersection of an operating curve with the 45° line represents the composition of the *net product* leaving from that section of the cascade. In binary distillation these intersections occur at x_d in the rectifying section and at x_b in the stripping section.
8. At an intermediate point in the cascade of stages where a *feed enters* or a *product is withdrawn*, the operating curve must undergo a discontinuity in slope. This discontinuity

will occur on a *locus-of-intersections* line, which crosses the 45° line at a composition equal to that of the feed or product and has a slope relating to the phase condition of the feed or product. If the feed or product is thermodynamically saturated and in the phase condition of the stream denoted by one of the axes, the locus of intersections line will be perpendicular to that axis. The 45° line concept loses its usefulness when different types of composition parameters are used for the two axes.

REFERENCES

Benedict, M., and T. H. Pigford (1957): "Nuclear Chemical Engineering," p. 216, McGraw-Hill, New York.

Benson, H. E., J. H. Field, and W. P. Haynes (1956): *Chem. Eng. Prog.,* **52**:433.

Brian, P. L. T. (1972): "Staged Cascades in Chemical Processing," Prentice-Hall, Englewood Cliffs, N.J.

Ellwood, P. (1968): *Chem. Eng.,* July 1, pp. 56–58.

Kohl, A. L., and F. C. Riesenfeld (1974): "Gas Purification," 2d ed., Gulf Publishing, Houston.

McCabe, W. L., and E. W. Thiele (1925): *Ind. Eng. Chem.,* **17**:605.

Muhlbauer, H. G., and P. R. Monaghan (1957): *Oil Gas J.,* **55**(17):139.

Narasimhan, K. S., C. C. Reddy, and K. S. Chari (1962): *J. Chem. Eng. Data,* **7**:457.

Othmer, D. F., R. E. White, and E. Treuger (1941): *Ind. Eng. Chem.,* **33**:1240.

Perry, R. H., and C. H. Chilton (eds.) (1973): "Chemical Engineers' Handbook," 5th ed., McGraw-Hill, New York.

————, ————, and S. D. Kirkpatrick (eds.) (1963): "Chemical Engineers' Handbook," 4th ed., McGraw-Hill, New York.

Ponchon, M. (1921): *Tech. Mod.,* **13**:20.

Robinson, C. S., and E. R. Gilliland (1950): "Elements of Fractional Distillation," pp. 158–162, McGraw-Hill, New York.

Savarit, R. (1922): *Arts Metiers,* pp. 65, 142, 178, 241, 266, and 307.

Schutt, H. C. (1960): *Chem. Eng. Prog.,* **56**(1): 53.

Seidell, A., and W. F. Linke (1958): "Solubilities of Inorganic and Metal-Organic Compounds," Van Nostrand, Princeton, N. J.

Sherwood, T. K., R. L. Pigford, and C. R. Wilke (1975): "Mass Transfer," McGraw-Hill, New York.

Smith, B. D. (1963): "Design of Equilibrium Stage Processes," chaps. 6 and 7, McGraw-Hill, New York.

Treybal, R. E. (1968): "Mass Transfer Operations," 2d ed., McGraw-Hill, New York.

———— (1963): "Liquid Extraction," 2d ed., McGraw-Hill, New York.

Van Winkle, M. (1967): "Distillation," p. 284, McGraw-Hill, New York.

Varteressian, K. A., and M. R. Fenske (1937): *Ind. Eng. Chem.,* **29**:270.

PROBLEMS

6-A$_1$ An ore containing 15% solubles and 5% moisture by weight is to be leached with 1 kg of water per kilogram of inlet ore in a countercurrent system of three mixer-filter combinations. The solids product from each filter contains 0.25 kg of solution per kilogram inert solids. Determine the percent of the inlet solubles recovered in the product solution from the system if the mixers achieve complete mixing.

6-B$_1$ Repeat Example 5-1 if all specifications are the same except that component A is now acetone and component B is now water. Use the known vapor-liquid equilibrium data and enthalpy-composition data for the acetone-water system (Table 6-2) and allow for changes in molal overflow from stage to stage. Use the enthalpy-composition diagram.

6-C$_1$ A stripping column is used to remove traces of hydrogen sulfide from a water stream. The column uses air at 60.8 kPa abs pressure as a stripping gas and the overhead gases are drawn at that pressure into a vacuum system. The column must remove 98 percent of the hydrogen sulfide from the water stream. The tower is isothermal at 27°C. If the tower provides three equilibrium stages, find the necessary air rate, expressed as moles air per mole water. Figure 6-6 provides equilibrium data.

6-D₂ An acetone-water distillation is to be designed to receive two feeds and provide distillate, sidestream, and bottoms products, operating at atmospheric pressure. The sidestream is withdrawn above the upper feed. The design mass balance is as follows:

	Flow rate, lb mol/h	Mole fraction acetone
Upper feed	40	0.509
Lower feed	60	0.500
Distillate	10	0.995
Sidestream	50	0.800
Bottoms	40	0.010

The sidestream is saturated liquid. The upper feed is saturated liquid, and the lower feed is saturated vapor. The tower utilizes a partial condenser, giving a reflux ratio r/D equal to 7.00. Give the coordinates of the difference points for each section of the column on an enthalpy-composition diagram.

6-E₂ A warm airstream, saturated with water vapor at 80°C and atmospheric pressure, is to be dried by countercurrent contact in a plate column with a feed stream that is a 60 wt % solution of sodium hydroxide in water. The high solubility of sodium hydroxide in water reduces the equilibrium partial pressure of water vapor over the aqueous solution sufficiently for the sodium hydroxide solution to act as an effective liquid desiccant. Vapor-liquid equilibrium data for sodium hydroxide solutions (recalculated from data in Perry et al., 1963):

kg NaOH/100 kg H₂O in solution	Equilibrium partial pressure of H₂O kPa at 80°C	kg NaOH/100 kg H₂O in solution	Equilibrium partial pressure of H₂O, kPa at 80°C
0	47.4	70	12.5
10	43.4	80	9.40
20	38.5	90	7.06
30	32.8	100	5.13
40	26.9	120	2.73
50	21.4	140	1.47
60	16.5	160	0.80

The water-vapor content of the dried air must be no greater than 0.025 mole fraction.

(a) Determine suitable composition parameters which will give a straight operating line for this tower on an operating diagram.

If the tower is cooled internally to maintain the system at 80°C throughout:

(b) Determine the minimum flow rate of sodium hydroxide solution feed required per cubic meter of entering wet air, measured at system pressure and temperature, even for an infinitely high column.

(c) What is the equilibrium-stage requirement if the flow rate of NaOH solution is twice the minimum?

If the internal cooling is not used and the tower operates adiabatically:

(d) Indicate step by step what procedure you would follow in using the vapor-liquid equilibrium data and the enthalpy-concentration diagram (Perry and Chilton, 1973, pp. 3-69 and 3-204, respectively) to determine the answer to part (b). Assume that the heat capacity (flow rate times specific heat) of the gas stream is much less than that of the liquid stream and that the heat of absorption therefore goes entirely to the liquid stream.

(e) Repeat part (d) assuming now that the heat capacity of the liquid stream is much less than that of the gas stream.

(f) Is the assumption of either part (d) or part (e) correct?

6-F₂ The original miscibility and equilibrium data underlying the plots shown in Fig. 6-25 are the following:

Hydrocarbon phase, wt %		Aniline-rich phase, wt %	
Methylcyclohexane	n-Heptane	Methylcyclohexane	n-Heptane
0.0	92.6	0.0	6.2
9.2	83.1	0.8	6.0
18.6	73.4	2.7	5.3
22.0	69.8	3.0	5.1
33.8	57.6	4.6	4.5
40.9	50.4	6.0	4.0
46.0	45.0	7.4	3.6
59.0	30.7	9.2	2.8
67.2	22.8	11.3	2.1
71.6	18.2	12.7	1.6
73.6	16.0	13.1	1.4
83.3	5.4	15.6	0.6
88.1	0.0	16.9	0.0

Source: From Varteressian and Fenske (1937); used by permission.

When analyzed on a solvent-free basis, the equilibrium data correspond to a nearly constant separation factor of 1.90.

(a) Consider a simple extraction process wherein a feed containing 60 wt % methylcyclohexane and 40 wt % n-heptane is contacted in a countercurrent series of equilibrium mixer-settler vessels with a pure aniline solvent. Idle equipment is available, which will provide four equilibrium mixer-settler stages and which, when operated at design capacity, will allow a solvent-feed–hydrocarbon-feed weight ratio of 4.00. If the four equilibrium contractors are used in a simple cascade, with no extract reflux, find the resultant relative product flow rates and product compositions. Compare with the products which would be obtained from a single stage. How useful are the additional stages?

(b) Suppose that the same four equilibrium contractors are now arranged countercurrently in such a way that extract reflux is employed and the hydrocarbon feed is now injected into the hydrocarbon stream flowing between the second and third contractors. The feed compositions remain the same, and the solvent-to-feed weight ratio is reduced to 3.00 in order to stay within the capacity limit. Extract reflux is generated by separating the hydrocarbons completely from the aniline in a distillation column; 80 percent of the hydrocarbon mixture obtained from this distillation column will be returned as reflux, and the other 20 percent will be taken as methylcyclohexane-rich product. Find the relative product flow rates and product compositions. Compare the methylcyclohexane enrichment with that found in part (a) and explain the difference.

(c) Why wouldn't the methylcyclohexane–n-heptane feed stream be separated by distillation rather than by extraction?

6-G₂ Enthalpy-composition data and vapor-liquid equilibrium data for the system ammonia-water are given in Perry and Chilton (1973, pp. 3-67, 3-68, 3-155 to 3-158). Determine the number of equilibrium stages required in a distillation column with a partial condenser which will receive a feed containing 50 wt % ammonia and 50% water and produce an ammonia product containing 99 wt % ammonia which contains 98% of the ammonia fed. The thermal condition of the feed will be such that the mass vapor flow rate does not change across the feed plate. The reboiler vapor will be 1.20 times the minimum. The feed will be introduced at the optimal location. The tower pressure will be such as to make the overhead temperature 40°C and hence make it possible to use water as a coolant in the condenser. Allow for changes in the total stream flow rates from stage to stage by using the enthalpy-composition diagram.

6-H$_2$ Describe step by step the procedure involved in generating saturated-vapor and saturated-liquid enthalpies from the data in Table 6-2.

6-I$_2$ (a) What is the effect of heat loss from a distillation column to the surrounding atmosphere on the quality of separation that can be obtained with a given reboiler heat duty?

(b) If the column operates at subambient temperatures, what is the effect of a heat leak in from the atmosphere at a fixed refrigerant cooling duty in the overhead condenser? Confirm your answers by means of qualitative reasoning using a McCabe-Thiele diagram.

6-J$_2$ Acetylene is frequently an undesirable trace component in ethylene streams, particularly when the ethylene is to be used for the manufacture of polyethylene. Figure 6-27 shows one possible process for removal of acetylene from a low-pressure gaseous ethylene stream. The acetylene is selectively absorbed into circulating liquid dimethylformamide (DMF). Because DMF is expensive, the exit DMF stream is

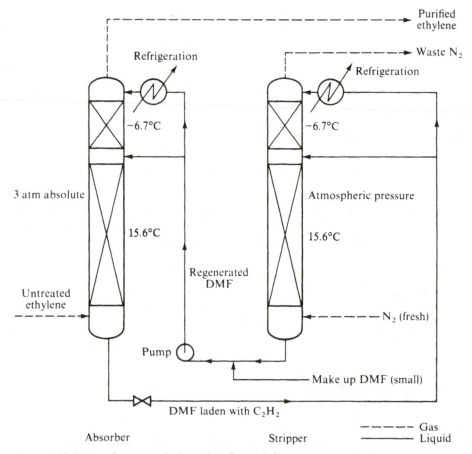

Figure 6-27 Process for removal of acetylene from ethylene.

regenerated in a second tower where the acetylene is stripped into nitrogen. The absorber operates at 304 kPa, and the stripper operates at 101 kPa pressure.

As shown in Fig. 6-27, portions of the DMF feeds to both the absorber and the stripper are passed through a refrigerated cooler and introduced to the top of the columns; the remainder of the DMF is fed to a lower point. The refrigerated solvent feeds are used to reduce the volatility of DMF near the tower tops and thereby reduce losses of DMF in the overhead gases. Assume that the net effect of the two-feed schemes is to make the top sections operate isothermally at $-6.7°C$ and to make the lower sections operate isothermally at $15.6°C$. The following operating conditions are specified:

Ethylene feed rate = 100 mol/s   Acetylene content of raw feed = 1.0 mol %
Acetylene content of purified ethylene = 0.001 mol %

Schutt (1960) gives the following data: solubility of acetylene in DMF at $15.6°C = 1.28 \times 10^{-6}$ mole fraction/Pa and at $-6.7°C = 2.56 \times 10^{-6}$ mole fraction/Pa (estimated). (Both apply at low partial pressures of C_2H_2.) Boiling point of DMF = 153°C; melting point of DMF = $-61°C$. Ethylene is one to two orders of magnitude less soluble in DMF than acetylene. Assume the gas phase is ideal.

(a) Assuming the absorber can be of any size, what is the maximum allowable acetylene content in the regenerated DMF?

(b) Suppose the acetylene content in the regenerated DMF is set at 40 percent of the maximum value computed in part (a). What is the minimum allowable total DMF circulation rate? Does the answer depend on the percentage split of the DMF between the two feed points? Why?

(c) Qualitatively sketch operating diagrams for both the absorber and the stripper under typical conditions. Label the points on the operating and equilibrium curves corresponding to (1) the tower bottom, (2) the point of the lower DMF feed, and (3) the tower top. Distort the scales as necessary to show all points clearly.

6-K₂ A stream of water containing 8.0 wt % phenol is to be purified by means of an extraction process operating at 30°C and employing pure isoamyl acetate as a solvent. The purified water must never contain more than 0.5 wt % phenol. Several different equipment schemes, shown in Fig. 6-28, are being considered for this purpose. In each contacting vessel the exiting streams may be assumed to be in equilibrium with each other.

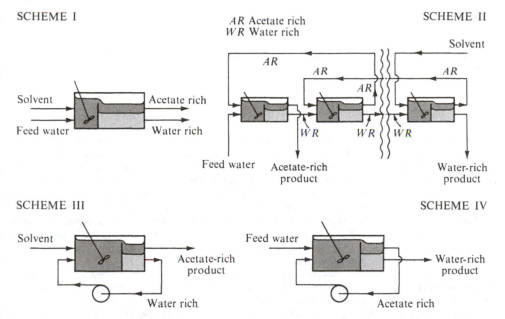

Schemes I and II are continuous processes, while schemes III and IV are batch processes. In scheme III the water feed is initially inside the vessel and acetate is passed through continuously until the water is purified. In scheme IV the acetate is initially inside the vessel and the water is passed through until it can no longer be sufficiently purified. Assume the vessels are well mixed.

(a) What solvent treat per kilogram of feed is required in scheme I?

(b) What solvent treat per kilogram of feed is required in scheme II if a large number of vessels are employed?

(c) What solvent treat per kilogram of feed is required in scheme II if only three stages are employed? Use a yx type of diagram for your analysis.

(d) What solvent treat per kilogram of feed is required in scheme III?

(e) Will the solvent requirement per kilogram of feed in scheme IV differ from that in scheme III? If so, how much solvent per kilogram of feed is required in scheme IV?

Liquid-liquid equilibrium data for the system water–phenol–isoamyl acetate (data from Narasimhan et al., 1962)

Phase I		Phase II	
Phenol, wt fraction	Acetate, wt fraction	Phenol, wt fraction	Acetate, wt fraction
0	0.994	0	0.0022
0.088	0.900	0.0025	0.0020
0.168	0.810	0.005	0.0018
0.230	0.743	0.0085	0.0016
0.320	0.643	0.0165	0.0013
0.415	0.540	0.0252	0.0009
0.490	0.455	0.0308	0.0008
0.565	0.365	0.0425	0.0005
0.660	0.230	0.0570	0.0004
0.680	0.188	0.0595	0.0003
0.718	0.095	0.0685	0.0002
0.696	0	0.0880	0

6-L$_3$ Figure 6-29 shows a photograph of countercurrent dual-temperature H_2S–H_2O isotope-exchange towers at a heavy-water plant. The towers serve to concentrate deuterium in water, first as HDO and then as D_2O. Figure 6-30 shows three conceivable schemes for carrying out dual-temperature hydrogen-deuterium exchange between H_2S and H_2O countercurrently. The three schemes are logical extensions of the three simple equilibration schemes shown in Fig. 1-31 and considered in Prob. 1-F. In each, the single-stage contactors shown in Fig. 1-31 have been made countercurrent plate columns. Necessary heat exchangers and pumps are not shown in Fig. 6-30. In all cases a fraction f of the feedwater is taken as deuterium-enriched product. The top tower is isothermal at 30°C, and the lower tower is isothermal at 130°C in all cases. Recall that the percentage of deuterium in the natural water feed is quite low and that a high degree of enrichment is desired. Equilibrium data are available in Prob. 1-F.

(a) Which of the three schemes shown in Fig. 6-30 do you feel is best for the process purposes and is thus probably the one used at the Savannah River plant? If this is a different scheme from the one that was best in Prob. 1-F, explain what caused the change.

(b) For the scheme which you have picked as best, sketch the conditions of both the hot and the cold tower on one operating yx diagram, labeling points corresponding to various positions within the process.

(c) Suggest a modification to the scheme you have selected as the best one that will maintain high

Figure 6-29 Countercurrent plate columns used for the manufacture of heavy water by hydrogen sulfide–water dual-temperature isotope exchange at the U.S. Department of Energy plant at Savannah River, South Carolina. *(The Lummus Co., Bloomfield, N.J.)*

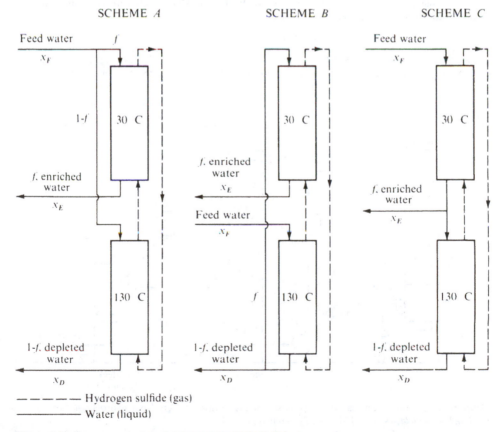

Figure 6-30 Countercurrent dual-temperature H_2S–H_2O isotope-exchange processes.

enriched product purity but will allow a substantially greater recovery fraction of the deuterium fed in the deuterium-enriched product.

6-M₃ Ellwood (1968) describes an isotope-exchange process operated at Mazingarbe in northern France by the French Commissariat à l'Energie Atomique for the manufacture of heavy water. In contrast to the H_2S-H_2O dual-temperature exchange process described in Probs. 1-F and 6-L, this process uses the ammonia-hydrogen exchange reaction

$$NH_3 + HD \rightleftharpoons NDH_2 + H_2$$

and is operated in conjunction with a large plant manufacturing NH_3 from N_2 and H_2. Another striking feature of the process is that the isotope-exchange column is countercurrent, as are those in Fig. 6-30, but operates at only *one* level of temperature rather than the two different levels of temperature required for the dual-temperature process. Thus something aside from the isotope-exchange reaction itself is basically different between the processes.

Consult Ellwood (1968). From his description of the plant, determine why this process can work carrying out the exchange reaction at only a single temperature while still obtaining a high enrichment and recovery fraction of deuterium. Would it be possible to modify the H_2S-H_2O process so it could operate at a single temperature yet provide a high degree of separation?

6-N₃ The hot potassium carbonate process has been developed by the U.S. Bureau of Mines for the removal of CO_2 from high-pressure high-temperature gas streams of substantial CO_2 content. Figure 6-31 shows a flow scheme and operating conditions for a process which removes CO_2 from a gas mixture containing 20% CO_2 and 80% H_2, such as would be encountered in a plant for the manufacture of hydrogen from natural gas (Kohl and Riesenfeld, 1974). The CO_2 reacts with potassium carbonate according to the overall reaction

$$CO_2 + K_2CO_3 + H_2O \rightleftharpoons 2KHCO_3$$

The absorbent is a solution of K_2CO_3 in water (20 wt °₀ equivalent K_2CO_3). The high absorber temperature and pressure, coupled with the low regenerator pressure, make it possible to operate the absorber and

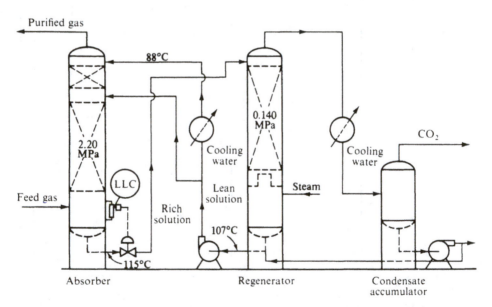

Figure 6-31 Hot potassium carbonate process for absorption of CO_2 from hydrogen. (*Adapted from Kohl and Riesenfeld, 1974, p. 176; used by permission.*)

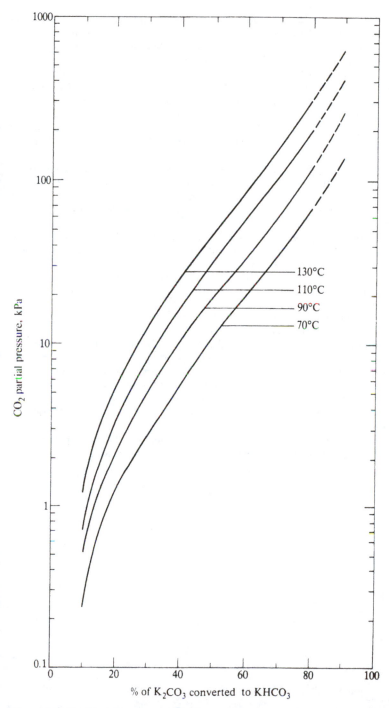

Figure 6-32 Equilibrium pressures of CO_2 over K_2CO_3 solution (20% equivalent K_2CO_3). (*Data from Benson, et al., 1955.*)

regenerator at similar temperatures while still achieving a favorable equilibrium in both towers. This eliminates any need for heat exchange between the rich and lean solution or for the addition of sensible heat to the circulating carbonate solution in the regenerator. Typical Murphree vapor efficiencies for a plate absorber in this process are on the order of 5 percent. This low value probably results from the slow rate of reaction between CO_2 and K_2CO_3. Figure 6-32 shows equilibrium pressures of CO_2 over the K_2CO_3 solution, as measured by Benson, et al. (1956). Although the heat of absorption of CO_2 is somewhat less than the heat of vaporization of water on a molar basis, you may assume they are the same.

(a) Why is the lean carbonate solution split into two portions, one being fed at the top of the absorber and the other being fed part way down? Why is the upper lean carbonate feed cooled while the lower feed is put in at the temperature at which it leaves the regenerator? Why not cool *both* streams or cool *neither* stream?

(b) Draw a qualitative operating diagram for the absorber showing the location and shapes of the equilibrium and operating curves. If possible, choose coordinates so as to give a straight operating line while not requiring an extensive recalculation of the equilibrium data. On the diagram label the points corresponding to the tower top, the tower bottom, and the intermediate carbonate feed. Distort the scale if necessary to show each of these points clearly.

(c) If 99 percent or more of the CO_2 is to be removed from the feed gas and 20 percent of the K_2CO_3 is converted into $KHCO_3$ in the lean carbonate solution, find the *minimum* circulation rate of carbonate solution required per 100 mol inlet gas, even with an infinite number of stages.

(d) Sketch qualitatively a typical operating diagram for the regenerator. Again, concentrate on the shape and location of the operating and equilibrium curves and strive to provide a straight operating line.

(e) A major operating cost of this process is for the regenerator steam. If the carbonate circulation rate is 1.3 times the minimum and the carbonate is to be reduced in the regenerator to a point where only 20 percent of the K_2CO_3 is converted into $KHCO_3$, determine the minimum regenerator steam consumption per 100 mol inlet gas, even with an infinite number of stages. Neglect sensible-heat effects.

(f) Kohl and Riesenfeld also describe a design modification which has been used when more complete removal of CO_2 is needed. The lower carbonate feed to the absorber is withdrawn from a level in the stripping column well above the reboiler, so that only the remaining portion of the carbonate solution passes through the section of the stripping column below that level. The solution emanating from the bottom of the stripping column becomes the top carbonate feed to the absorber. Why can this modification be superior to the process of Fig. 6-31?

6-O₃ Raffinate reflux in a liquid-liquid extraction cascade would correspond to diverting a portion of the raffinate product stream, mixing it with the fresh solvent to convert it to solvent-rich phase, and reintro-

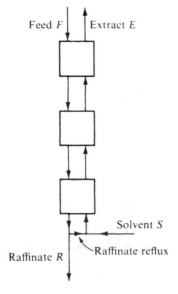

Figure 6-33 Use of raffinate reflux in an extraction cascade.

ducing it as feed to the end stage. Such a process is shown schematically in Fig. 6-33. Consulting the literature, we find one reference which indicates that this can be a useful procedure and another which indicates that it is useless. Can you resolve this controversy?

6-P$_2$ Air is separated into nitrogen and oxygen by distillation in many large plants. The product oxygen is used as the oxidant for rocket fuels, for combustion processes requiring higher temperatures than can be reached using unenriched air, etc. Liquid nitrogen sees use for food freezing, inert-gas blanketing, etc. The small amount of argon in air requires a purge from the distillation system. We shall ignore this complication and also the fact that rather complex column designs are usually employed to minimize energy consumption (see Chap. 13). Instead, we shall assume a simple, one-feed–two-product binary distillation with a total condenser. Data are given in the tables. Saturated-vapor and saturated-liquid enthalpies are both essentially linear in mole fraction.

Vapor-liquid equilibrium and enthalpy data for N_2–O_2 at total pressures of 101.3 kPa (1 atm) and 506.5 kPa (data from Van Winkle, 1967)

Enthalpy data				
	101.3 kPa		506.5 kPa	
	N_2	O_2	N_2	O_2
Boiling point, K	77.3	90.2	94.3	108.9
Latent heat of vaporization, kJ/mol	5.59	6.83	4.89	6.22

Vapor-liquid equilibrium data

	y_{N_2}	
x_{N_2}	101.3 kPa	506.5 kPa
0.00	0.00	0.00
0.05	0.174	0.123
0.10	0.310	0.228
0.20	0.492	0.399
0.30	0.641	0.532
0.40	0.735	0.639
0.50	0.805	0.725
0.60	0.859	0.797
0.70	0.903	0.859
0.80	0.940	0.912
0.90	0.972	0.958
1.00	1.00	1.00

(a) Is the MLHV method likely to be a good approximation for this system? Explain.

(b) Using the MLHV method, calculate the equilibrium-stage requirement for separating a saturated-vapor feed of air into products containing 98 mol % oxygen and 98.5 mol % nitrogen, using a total condenser returning saturated-liquid reflux and an overhead reflux ratio r/d equal to 1.12 times the minimum. The column pressure is 506.5 kPa.

(c) If operation is at 1.12 times the minimum reflux ratio in both cases, compare the overhead-condenser cooling requirement (kilojoules per mole of feed) for 101.3 kPa column pressure with that for 506.5 kPa column pressure.

(d) Using your answer to part (c), discuss the choice between the two column pressures on the basis of minimizing energy consumption.

6-Q₂ Repeat Prob. 6-G using the MLHV method of calculation.

6-R₂ A countercurrent extraction cascade will use water as a solvent to recover acetone selectively from streams containing mixtures of acetone and methyl isobutyl ketone (MIBK). Per unit time, the two ketone feeds to the extraction will contain 11 kg acetone and 100 kg MIBK, and 20 kg acetone and 80 kg MIBK, respectively. The raffinate will be 174 kg MIBK, 4 kg water, and 1 kg acetone. The inlet water flow will be 250 kg. Extract reflux will be employed, created by a distillation column that separates 60 kg acetone and 12 kg MIBK from 246 kg water. The ketone stream from the distillation will be split exactly in half, with one half serving as extract reflux and the other half serving as product. Find the compositions for the difference points for each of the three sections of this extraction cascade and indicate where they would lie with respect to a triangular diagram representing the phase-miscibility data.

SEVEN

PATTERNS OF CHANGE

In Chaps. 5 and 6 we discussed the application of the McCabe-Thiele graphical approach to the solution of a number of problems associated with binary separations. In these cases the graphical technique provided an efficacious solution which was at least as simple as any algebraic computational approach and in most cases simpler. Graphical methods have the unique feature of providing a *visual* representation of the separation process. In terms of ready applicability, however, they are largely limited to situations where the entire composition of either phase is fixed through specification of the concentration of *one* component alone.

Before proceeding to a consideration of various more general plans of computational attack which can be invoked for multicomponent systems, we shall pause and consider multistage separation processes from a more qualitative vantage point, analyzing the general patterns of change in flow rate, composition, and temperature which occur from stage to stage in a separation cascade. To do this, we first reconsider various types of binary multistage processes.

An understanding of the factors at play causing changes throughout a multistage separation cascade helps one select appropriate computational approaches and improve design and operating conditions. On the other hand, the patterns of change for multicomponent systems will be most fully understood after one has gained some experience with multicomponent separations. Hence the reader may find it helpful to review this chapter again after studying Chaps. 8 to 10.

BINARY MULTISTAGE SEPARATIONS

From a standpoint of interpretation the simplest multistage process is one which entails straight equilibrium and operating lines. Figure 7-1 shows an operating diagram for a dilute absorber, such as might be utilized for the removal of a small

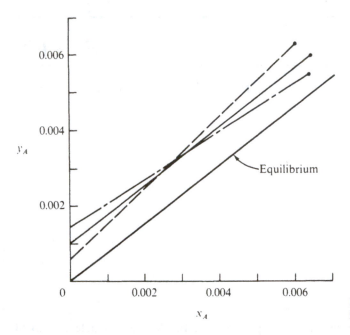

Figure 7-1 Operating diagram for dilute absorber.

amount of a single soluble constituent from a gas stream. The equilibrium line is straight if Henry's law is obeyed by the solute. The operating line is straight if total molar flows are essentially constant, and it lies above the equilibrium line since we have an absorption process where the solute is transferring from gas to liquid. Three possible operating lines are shown corresponding to the ratio of the slope of the operating line to that of the equilibrium line being greater than (dashed), equal to (solid), and less than (dot-dash) 1. Figure 7-2 shows the patterns of change resulting from the three operating lines. The numbers on the abscissa correspond to the various interstage locations (*passing* streams) shown in the accompanying diagram. The total flow rates are nearly constant because of the dilution. The temperatures are constant if the entering temperatures of the two phases are equal and if the system is dilute enough for the heat of absorption to be small compared with the sensible heats of the gas and liquid phases.

If the operating and equilibrium lines are parallel (solid line), the concentration of the solute in the liquid and gas changes at a uniform rate from stage to stage. If the operating line has a *greater* slope (dashed curve) than the equilibrium line, the solute concentrations in liquid and vapor change more rapidly at higher concentrations. On the other hand, if the operating line has a *lesser* slope (dot-dash curve) than the equilibrium line, the solute concentrations in both phases change more rapidly at lower concentrations. Put another way, when the operating line and equilibrium line are closer together, the phase compositions change slowly from stage to stage.

Many factors may arise to complicate the constant-flow constant-temperature straight-line situation of Figs. 7-1 and 7-2. We shall discuss several of them and assess their effects upon the patterns of change.

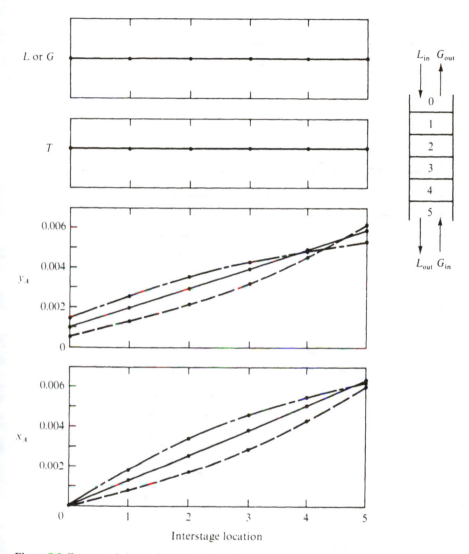

Figure 7-2 Patterns of change for dilute absorber.

Unidirectional Mass Transfer

Multistage separation processes necessarily involve the transfer of material from one counterflowing stream to the other. In a constant-molal-overflow binary distillation the two components which change phase do so in such a way as to leave the total molar flow rates between stages unchanged. The net passage of A from liquid to vapor in a stage is equal in molar rate to the net passage of B from vapor to liquid.

In the simplest absorption process only one component changes phases appreciably. The solute passes from gas to liquid; thus there must be a change in total

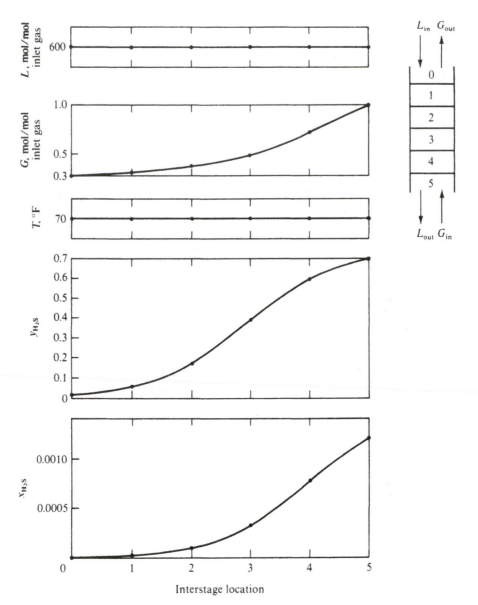

Figure 7-3 Patterns of change for absorber of Example 6-3.

molar flow rates from stage to stage. Unless the solvent has appreciable volatility, nothing returns from liquid to gas to balance the loss of solute from the gas. The absorber of Fig. 7-1 was sufficiently dilute for this change in bulk flow rates to be quite small, less than 1 percent. In Example 6-3, however, the absorber operated such that the change in bulk-gas flow rate was appreciable, as shown in Fig. 7-3. The liquid phase in Example 6-3 was highly dilute; hence the liquid-phase flow rate is nearly constant. The high ratio of liquid to gas flow also means that the liquid sensible heat is large in comparison to the heat of absorption; thus the temperature is constant.

The composition profiles for the absorber of Example 6-3 are also shown in Fig. 7-3. The liquid composition changes most rapidly in the lower stages of the column. This behavior corresponds to the fact that the steps in the x direction shown in Fig. 6-7 are larger toward the rich end. The change in gas composition, expressed as y_{H_2S}, is most rapid toward the middle of the column, and the change at either end is slow. This fact is not immediately obvious from Fig. 6-7. One must recall, however, that Fig. 6-7 is drawn in terms of Y_{H_2S} (mole ratio) instead of y_{H_2S} (mole fraction); y changes more rapidly per unit change in Y at low mole fractions than it does at higher mole fractions. Thus the large steps in Y at higher concentrations in Fig. 6-7 correspond to smaller steps in y.

Constant Relative Volatility

Figure 7-4 shows a McCabe-Thiele diagram for an atmospheric-pressure distillation of a saturated liquid feed containing 50 mol % benzene and 50 mol % toluene. The relative volatility is nearly constant, being 2.38 at $x_B = 0$ and 2.62 at $x_B = 1$ (Maxwell, 1950). The reflux is saturated, and the reflux ratio r/d is 1.57. There are 11 and a fraction equilibrium stages in addition to an equilibrium kettle reboiler and a

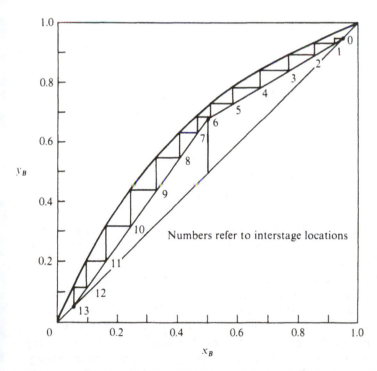

Figure 7-4 McCabe-Thiele diagram for benzene-toluene distillation.

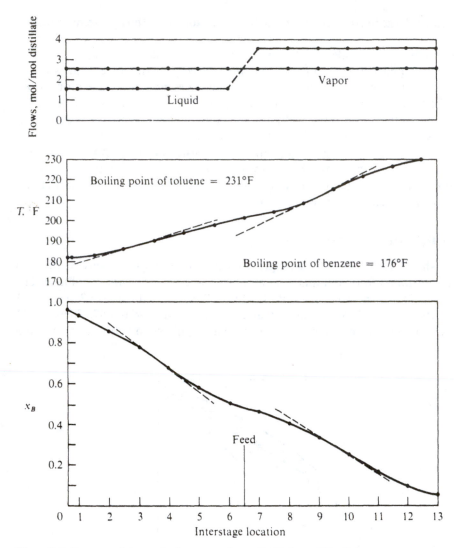

Figure 7-5 Patterns of change for benzene-toluene distillation of Fig. 7-4.

total condenser. The feed is introduced so as to provide maximum separation, which corresponds to a distillate containing 95 mol % benzene and a bottoms product containing 5 mol % benzene.

In Fig. 7-5 the flows, temperatures, and compositions for this distillation are shown as a function of interstage location. The molar flow rates are essentially constant except for the change in liquid flow at the feed plate. This is the case of nearly constant molal overflow. The profile of liquid compositions shows two points of inflection denoted by the dashed tangent lines. The liquid composition changes slowly at the very top of the column, then more rapidly and then more slowly again as the feed stage is approached. The same process is repeated below the feed: slow, then faster, and then slower again. The vapor composition profile is similar to that for the liquid.

Such a composition profile is common in binary distillation. With reference to Fig. 7-4, there will usually be pinches at the very top and very bottom of the column if high product purity is sought. Also, there will be pinches on either side of the feed stage if the operation is not much removed from minimum reflux. Midway in both the stripping and rectifying sections there is more distance between the equilibrium and operating curves and compositions change more rapidly.

The shape of the temperature profile closely follows that of the liquid composition profile, since the two are related through bubble-point considerations. When compositions change rapidly, temperatures also change rapidly; thus in this case temperatures change fastest midway in each of the two column sections. Operating temperatures change from stage to stage in the case of distillation not primarily because of sensible-heat effects but because of the necessity of preserving thermodynamic saturation when compositions change at the constant column pressure.

If the distillation of Fig. 7-4 were carried out at conditions closer to total reflux, the compositions would change slowly at either end and would change fastest near the feed. The pinches above and below the feed would not occur. Another possible situation is that of the acetone-water distillation of Example 6-5. The *tangent pinch* in the rectifying section of Fig. 6-15 causes the compositions and temperatures to change slowly in the middle of the rectifying section and more rapidly at the top and near the feed.

Enthalpy-Balance Restrictions

Another complicating factor in the analysis of multistage separation processes is the necessity of satisfying the first law of thermodynamics. This restriction takes the form of enthalpy balances which determine interstage flow rates and temperatures in processes such as distillation, crystallization, absorption, and stripping, which involve heat effects accompanying phase change. In distillation constant molal overflow is frequently assumed and serves as a sufficiently close approximation in many cases. However, it is important to understand at least qualitatively the factors which determine the change in flows in order to predict the systems for which constant molal flows might be expected to be too much in error. We consider the general case, where more than two components may be present.

Distillation The molecular weight usually changes throughout a distillation column as a result of the fractionation, the average molecular weight generally decreasing upward through the column, since high volatility of a compound generally corresponds to low molecular weight. Since the latent heat of vaporization per mole is usually less for a lower-molecular-weight material, the vapor rising and entering a typical stage when condensed will produce a vapor leaving the stage that has a greater number of moles. Because of this factor the flows usually will tend to increase upward in a column. If the system being fractionated is composed of only two components, the difference in molar latent heats of the pure species is large, or the volatility spread of the feed is small, this effect will tend to predominate. In some cases the higher-boiling component will have the lower heat of vaporization and the latent-heat effect will tend to make flows increase downward.

Second, as the vapor flows upward through a column it must be cooled, since the temperature is decreasing upward. This cooling must be done at the expense of either sensible heat of the liquid or vaporization of the liquid, resulting in flows which increase upward if liquid is vaporized. Third, the liquid flows must be heated as they proceed down through the column, and this heating is done at the expense of either sensible heat or condensation of the vapor, resulting in flows which increase downward if vapor is condensed.

The last two factors can predominate if there are large amounts of components in the feed which are very light or very heavy relative to the components being fractionated or, more generally, if the temperature span from tower top to tower bottom is large.

In order to determine the combined effects of the sensible-heat factors it is necessary to consider a typical plate in each section. Consider first a typical plate in the stripping section. The liquid flow necessarily exceeds the vapor flow, and its heat capacity is greater. Thus heating of the liquid outweighs cooling of the vapor, resulting in condensation of the vapor and increasing flows downward. If the typical stage is in the rectifying section, the vapor flow is larger than the liquid flow, the opposite reasoning holds, and flows tend to increase upward.

It is apparent that the total effect of these three factors is complicated, and no completely general rule can be formulated. However, it is also apparent that the factors are often compensating to a large extent and this is borne out in the usefulness of the assumption of constant molal flows.

Interstage flows are linked together by the overall material balance. Therefore, if the vapor flow increases in a certain direction through the section, the liquid flow will also increase in that direction. Further, since the fractionation is mainly dependent on the ratio L/V, considerable changes can occur in flows without greatly disturbing L/V, and hence the fractionation, from stage to stage. The more nearly equal the two interstage flows, i.e., the closer the operation to total reflux, the less the effect on fractionation caused by changes in the flows.

One example of the effect of varying molar flow rates in a distillation process is the acetone-water separation of Example 6-5. Acetone is the more volatile component and has a lower latent heat of vaporization than water. As a consequence the latent-heat effect is dominant, and the flow rates tend to be higher on the upper stages of the column. This is reflected in the fact that the rectifying-section operating curve of Fig. 6-17 is concave upward and the chord to (x_d, x_d) has a slope that is farther removed from 1.0 on the lower stages.

If molar flows increase upward, the result (as shown in Fig. 6-17) is that the fractionation is poorer (slower changes in temperature and composition) compared with the constant-molal-overflow case at the same overhead reflux ratio r/d. On the other hand, the fractionation is better than that for constant molal overflow when compared at the same bottoms boil-up ratio V'/b.

Absorption and stripping Heat effects are also important in absorption and stripping processes. The temperature will change from stage to stage unless the system is dilute enough for heats of absorption and desorption to be small in comparison with sensible heats of the counterflowing streams. In absorption the unidirectional transfer of solute from gas to liquid brings about a heating effect since the heat of condensation of the solute must be dissipated. This will usually lead to temperatures which increase downward in the column since the liquid generally has a greater sensible-heat consumption than the gas. In absorbers the liquid is sometimes passed through water-cooled heat exchangers, called *intercoolers*, at intermediate points in the column in order to hold the liquid temperature down, preventing absorbent vaporization and loss of favorable equilibrium for absorption.

Conversely, in stripping operations there is a tendency for the liquid to be cooled as it passes downward. The reasons for this are wholly analogous to those developed for absorbers.

If the absorbent liquid has appreciable volatility, it can vaporize partially on the lower stages of the column, so as to bring the inlet gas toward an equilibrium content of vaporized solvent. This phenomenon has been analyzed by Bourne et al. (1974) in the context of absorption of ammonia from air into water at atmospheric pressure. They show that the competing effects of liquid heating from absorption and liquid cooling from solvent evaporation serve to produce a temperature maximum midway along the column.

When the heat capacities (specific heat times flow rate) of the counterflowing streams have roughly equal magnitudes, there is another effect which can cause a temperature maximum. Such a case is shown by Kohl and Riesenfeld (1979) in the form of actual test data for an acid-gas absorber using ethanolamine solution to treat a gas at 3.7 MPa containing 4% CO_2 and 0.8% H_2S. As shown in Fig. 7-6, this absorber operates with inlet and effluent amine temperatures of 40 and 79°C, respectively, while developing an internal maximum temperature of 112°C at a point a few plates above the bottom. Here the hot downflowing liquid loses heat by preheating the incoming gas, and the hot upflowing gas loses heat by preheating the incoming liquid. The preheated gas and preheated liquid both flow away from the ends of the column, serving to reinforce the rise in temperature due to release of the heat of absorption in the middle of the column. This phenomenon has also been noted for countercurrent isotope-exchange towers (Pohl, 1962) and for counterflow heat exchangers where there is generation of heat due to chemical reaction in one of the streams (Grens and McKean, 1963).

It is interesting to observe that in Fig. 7-6 the gas-phase content of H_2S actually undergoes an internal maximum because of the higher equilibrium partial pressures associated with the temperature maximum. The CO_2 content, which is much farther from equilibrium, does not show such behavior.

The high gas pressure in the example of Fig. 7-6 serves to give the counterflowing streams roughly equal heat capacity. In the more usual situation, the liquid heat capacity exceeds that of the gas, tending to make the liquid temperature increase continually down the column. In some ethanolamine absorbers for gases with very

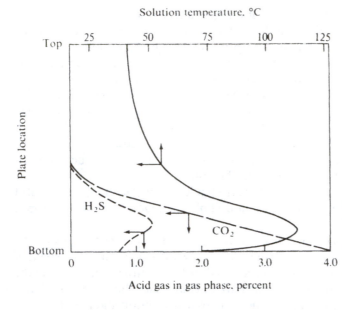

Figure 7-6 Temperature bulge in acid-gas absorber. (*Adapted from A. L. Kohl and F. C. Riesenfeld, Gas Purification, 3d ed., Copyright © 1979 by Gulf Publishing Co., Houston, Texas, p. 62; used by permission. All rights reserved.*)

low CO_2 and H_2S contents the liquid-gas ratio is low enough for the heat of absorption to go primarily to the gas and cause temperatures to increase upward.

The development of an internal temperature maximum complicates the design and analysis of absorbers in two ways: (1) there tends to be an internal pinch which increases the required solvent-to-gas ratio, and (2) there is a strong interaction between the enthalpy balances and the composition changes, which depend upon the equilibrium as influenced by the temperature. Methods of handling such situations are discussed in Chap. 10. Stockar and Wilke (1977) present a method for estimating the temperature profile in packed gas absorbers.

Contrast between distillation and absorber-strippers It is important to note that temperature profiles in ordinary *distillation* columns primarily reflect the compositions of the streams, while total interstage flow profiles primarily reflect enthalpy-balance restrictions. In *absorbers* and *strippers*, on the other hand, the situation is reversed; temperature profiles primarily reflect enthalpy-balance restrictions and interstage flow profiles primarily reflect stream compositions. This distinction will be of considerable use in setting up convergence loops for computer calculations in Chap. 10.

Phase-Miscibility Restrictions; Extraction

In staged liquid-liquid extraction processes there is usually no substantial heat effect accompanying the transfer of solute from one liquid to the other; consequently operation is usually nearly isothermal. In dilute extraction systems the interstage flow rates will remain essentially constant as long as there is no appreciable miscibility of the extract and raffinate phases. When the solvent and the unextracted component are totally immiscible but the solute concentration is high enough, there will be an increase in interstage flows in the direction of extract flow because of unidirectional mass transfer. In still more concentrated extraction systems, however, all components will necessarily become appreciably miscible and the interstage flows will vary to satisfy the phase-equilibrium relationships. This effect again causes flows to increase in the direction of extract flow, as is shown in the following discussion.

Example 6-6 covered a case of extraction involving appreciable miscibility between the phases. The ketones and water in the acetone-MIBK-water system are substantially soluble in each other, and at high enough acetone concentrations total miscibility is reached. Figure 6-24 gives an operating diagram for an acetone-MIBK-water extraction, plotted on a weight-fraction solvent-free basis. The operating curve is not a straight line; hence the mass flow rates of the combined ketones (acetone + MIBK) vary from stage to stage. Similarly, flow rates defined in any other way are not constant from stage to stage.

Figure 7-7 shows the interstage flow rates for the operation, expressed as total mass flow rates of the extract and raffinate phases, and as mass flows of the combined ketones, the two species which are being separated. The interstage flow of combined ketones is greatest at the left-hand, or acetone-rich, end of the cascade. At low acetone concentrations the solubility of *total ketones* in the water-rich (extract) phase is quite small (see Fig. 6-20), but as the acetone concentration increases toward the rich end of the cascade, the solubility of total ketones in the water-rich phase increases. The water-rich and ketone-rich phases become more nearly alike in composition as the acetone content increases. In fact, the compositions of the phases become identical at the plait point. The difference in flows of any of the components between raffinate and extract interstage streams must be constant from stage to stage since the operation is at steady state. As the streams become more alike in composition at higher acetone contents, greater interstage flows of all components—and hence of combined ketones—become necessary in order to preserve the constant difference in flows of those components between streams.

The total flow rates follow suit. The flow of the raffinate phase is nearly equal to the flow of combined ketones in that phase, since the solubility of the water solvent in that phase is always comparatively small. Therefore the raffinate flow is still greatest at the acetone-rich end, being increased somewhat by the higher solubility of water in ketones at that end. Since the difference in total flows of raffinate and extract must remain constant at all interstage positions, the total extract flow must also be higher at the acetone-rich end.

This behavior is characteristic of three-component extraction processes. The main transferring solute (acetone in Example 6-6) will be the component which is relatively soluble in both phases. In many cases the solute will produce complete miscibility when present above some particular concentration. Since presence of the solute promotes miscibility of the phases and similarity of composition of the two phases, the foregoing reasoning leads one to expect higher interstage flow rates at the solute-rich end of the cascade as a general rule.

There is an analogy to be drawn between the governing effect of enthalpy balances on interstage flows in distillation, on the one hand, and the governing effect of miscibility relationships on interstage flows in extraction, on the other. In each case

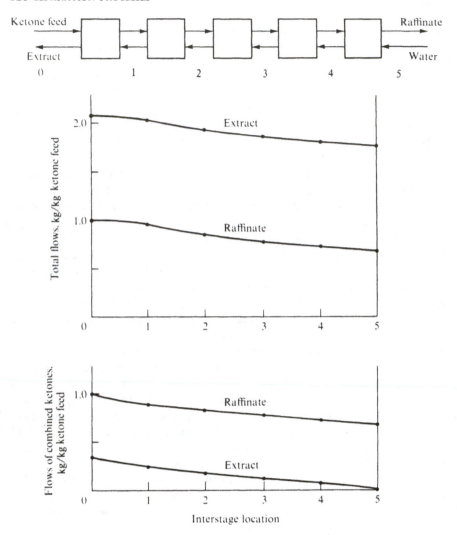

Figure 7-7 Interstage flows in acetone-MIBK-water extraction process of Example 6-6.

the restriction involves conservation of the separating agent—heat or enthalpy in distillation, and solvent in extraction. In distillation, interstage flows are determined from a knowledge of the specific enthalpy content of the appropriate saturated vapor and liquid phases. In extraction, interstage flows are determined from a knowledge of the specific solvent content of the appropriate saturated extract and raffinate phases. In both processes, interstage flows increase when the difference in separating agent content between saturated phases becomes smaller. In distillation, interstage flows increase in the direction of compositions where the difference in enthalpy content per kilogram or mole between vapor and liquid is smaller corresponding to a smaller latent heat. In extraction, interstage flows increase in the direction of compositions where the difference in solvent content per kilogram or mole between the two phases is smaller. This corresponds to the direction of increased miscibility between phases.

MULTICOMPONENT MULTISTAGE SEPARATIONS

The separations we have considered in Chaps. 5 and 6 have been binary, and as a consequence there have been relatively few components present whose properties and behavior had to be considered individually. When more components are present in a separation process, calculational procedures necessarily become more involved because it is not possible to specify as much about the process in a problem description. Graphical computation approaches are of limited usefulness when it is not possible to fix an entire phase composition uniquely by specifying the concentration of a single component on an operating diagram.

In spite of the increased computational difficulties, the qualitative understanding of multicomponent separation processes involves little added complexity beyond an understanding of binary separation processes. The following sections consider published solutions to three different multicomponent separation processes and explore the nature of the patterns of change in temperature, composition, and total flow rates. The computational procedures involved in solving all the various equations describing these processes need not concern the reader at this point; they form much of the subject matter of Chaps. 8 to 10.

Absorption

Horton and Franklin (1940) present a detailed solution to an oil-refinery absorption problem. A schematic of the process is shown in Fig. 7-8. A heavy lean-oil absorbent

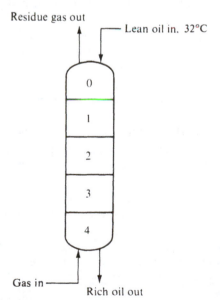

Figure 7-8 Schematic of multicomponent absorption process.

Table 7-1 Composition of residue gas and K_i values (data from Horton and Franklin, 1940)

Component	Composition, mole fraction			K_i
	Lean oil	Wet gas	Residue gas	
Methane (C_1)	....	0.286	0.499	51
Ethane (C_2)	....	0.157	0.250	13
Propane (C_3)	....	0.240	0.214	3.1
n-Butane (C_4)	0.02	0.169	0.025	0.85
n-Pentane (C_5)	0.05	0.148	0.012	0.26
Heavy oil	0.93	......	......	~ 0
	1.00	1.000	1.000	

is employed to recover roughly 60 percent of the propane, and most of the heavier hydrocarbons from a gaseous feed stream. A tower providing four equilibrium stages is used. Operating conditions are

Lean oil inlet temperature = 32°C   Tower pressure = 405 kPa

Lean-oil feed rate = 1.104 mol/mol gas fed

In the four-equilibrium-stage column, 44.8 percent of the gas is absorbed, and the residue-gas composition is shown in Table 7-1. The patterns of change in flows and temperature are shown in Fig. 7-9. The changes in molar flows of the individual components in the gas and liquid phases are also shown ($v_i = y_i V$ and $l_i = x_i L$), along with the gas-phase mole fractions (y_i). The process is similar to the absorption of a single component except that now several individual species are being absorbed.

Some idea of the relative solubilities of the five gas-phase components can be obtained from the values of the equilibrium ratio K_i ($= y_i/x_i$ at equilibrium) at 38°C, also shown in Table 7-1. The total flow rates of both the gas and liquid phases increase downward in the column, in the direction of high-solute contents in the liquid phase (Fig. 7-9a). This increase in flows is the result of unidirectional mass transfer; the components pass from the gas to the liquid without any comparable amount of material passing back the other way.

Temperatures increase downward in the column (Fig. 7-9b); the cause is the heat of absorption released by the phase change of solutes passing from gas to liquid. The heat release serves to increase the sensible heat of the liquid stream, which receives most of the heat released at the interface.

Methane and ethane are sufficiently volatile to remain relatively unabsorbed by the oil. The flow rates of methane and ethane in the vapor are therefore essentially

Figure 7-9 Patterns of change for multicomponent absorption process: (a) total flows; (b) liquid temperature; (c) liquid-component flows; (d) gas-component flows; (e) gas composition. (*Results from Horton and Franklin, 1940.*)

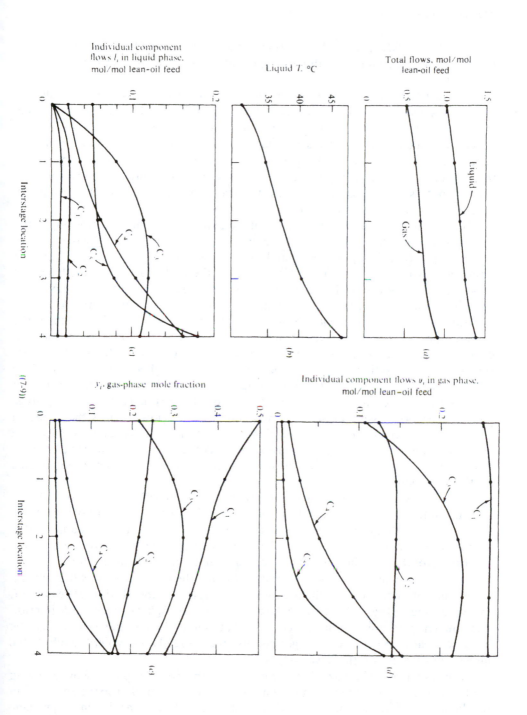

Individual component flows l_i in liquid phase, mol/mol lean-oil feed

Liquid T, °C

Total flows, mol/mol lean-oil feed

Interstage location

y_i, gas-phase mole fraction

Individual component flows v_i in gas phase, mol/mol lean-oil feed

Interstage location

(a)

(b)

(c)

(d)

(e)

(7-9)

constant (Fig. 7-9d). The small depletion in the vapor means that the liquid equilibrates nearly completely with those components in one stage, with little change in l_i thereafter.

Pentane has the least volatility of any component in the gas. As a result it is rapidly absorbed in the lower stages of the column soon after the gas feed enters (Fig. 7-9d). There is little pentane remaining in the gas reaching the upper stages; hence not much pentane enters the liquid on the upper stages. As a result, the pentane flow in the liquid on the upper stages remains constant at the level present in the lean-oil feed until the liquid reaches the lower stages, where rapid absorption of pentane occurs. Butane is the next least volatile component; it also absorbs rapidly on the lower stages but not as rapidly as pentane.

Because of the large absorption of butane and pentane on the lower stages, the mole fractions of these two components fall in the gas phase as the gas passes to higher stages (Fig. 7-9e). Methane and ethane are relatively unabsorbed, and the total gas flow rate decreases upward; as a result the mole fractions of methane and ethane continually rise in the gas as it comes to stages higher in the column.

The amounts of methane and ethane absorbed in the liquid, although small, actually pass through a maximum on an intermediate stage (Fig. 7-9c). This is the result of the changes in temperature and in gas-phase mole fraction. The equilibrium concentration of a solute in the liquid phase is given by $x_i = y_i/K_i$. On the upper stages the mole fractions y_i of methane and ethane are higher in the gas phase. The temperature is also lower on the upper stages, which tends to make K_i lower. As a consequence the equilibrium x_i for methane and ethane is highest on the top stage and becomes progressively lower on lower stages. On the lower stages, methane and ethane tend to absorb to the equilibrium amount, accounting for maxima in the amounts of methane and ethane absorbed. No maxima occur for butane and pentane since they are readily absorbed on the lower stages, reducing y_i for those components on the upper stages.

Propane is a component which is intermediate in volatility. About half the propane in the wet gas is ultimately absorbed (Fig. 7-9d), whereas most of the butane and pentane and very little of the methane and ethane are absorbed. An appreciable amount of propane remains in the gas reaching the upper stages, where it encounters a more favorable equilibrium for absorption in terms of temperature. Thus the maximum of propane in the liquid occurs for much the same reasons as the maxima in amounts of methane and ethane absorbed. Propane is absorbed most readily on the upper stages (Fig. 7-9d) because the combination of high gas-phase mole fraction and low temperature is more effective at that point. Butane and pentane can be absorbed readily on the lower stages because their already low volatility offsets the higher temperature on the lower stages.

The reader should realize that one could not absorb more propane by removing the bottom stage from the column, even though the amount of propane absorbed in the liquid is higher at location 3 than at location 4. The maximum in propane absorption occurs as a direct result of the large absorption of pentane and butane on the bottom stage, which reduces the total gas flow and increases y_{C_3}. This phenomenon will occur no matter what the number of stages.

Table 7-2 Feed and products for depropanizer example (data from Edmister, 1948)

Component	mol %			mol/100 mol feed		
	Feed	Distillate	Bottoms	Distillate	Bottoms	α_i (rel. to C_3)
Methane (C_1)	26	43.5	......	26	.....	10.0
Ethane (C_2)	9	15.0	......	9	.....	2.47
Propane (C_3)	25	41.0	1.0	24.6	0.4	1.0
n-Butane (C_4)	17	0.5	41.7	0.3	16.7	0.49
n-Pentane (C_5)	11	......	27.4	.....	11	0.21
n-Hexane (C_6)	12	......	29.9	.....	12	0.10
	100	100	100	59.9	40.1	

Distillation

Edmister (1948) presents a detailed stage-to-stage solution for a depropanizer distillation column. The column operates at an average total pressure of 2.17 MPa and receives a feed with the composition shown in Table 7-2. The thermal condition of the feed is such that it is 66 mol % vapor at tower pressure. The column is equipped with a kettle-type reboiler and a partial condenser, which allows the manufacture of reflux at 2.17 MPa total pressure while using water for cooling. The product compositions are also given in Table 7-2. The overhead reflux rate r is 0.90 mol per mole of feed.

The example is worked assuming constant molal overflow; 15 equilibrium stages within the tower are required for the separation, the feed being introduced between the ninth and tenth equilibrium stages from the bottom of the tower proper.

The total flow rates are shown as a function of interstage location in Fig. 7-10. The flows are constant above and below the feed, in line with the assumption of constant molal overflow. The changes in total flow of vapor and liquid at the feed point are governed by the fact that the feed is two-thirds vapor.

Figure 7-11 shows the changes of vapor composition from stage to stage for all six of the components present; Fig. 7-12 shows the composition profile in the liquid. In both figures the left- and right-hand sides correspond to the bottoms and distillate, respectively. Values of α_i at 96°C (feed-plate temperature) are also given in Table 7-2.

Key and nonkey components The separation is being achieved primarily between propane and butane, since, as shown in Table 7-2, most of the propane and essentially all the more volatile components appear in the distillate, while most of the butane and essentially all the less volatile components appear in the bottoms. Propane and butane are therefore called the *key components*. The key components, or *keys*, appear to a significant extent in both products, while the other components (called *nonkeys*) are merely "along for the ride," being relegated almost exclusively to

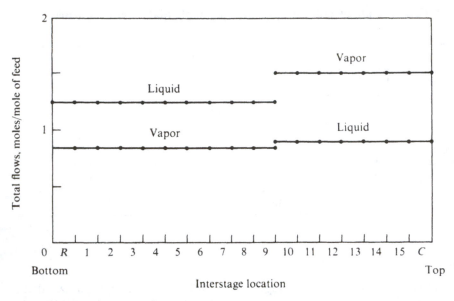

Figure 7-10 Total vapor and liquid flows in depropanizer. *(Results from Edmister, 1948.)*

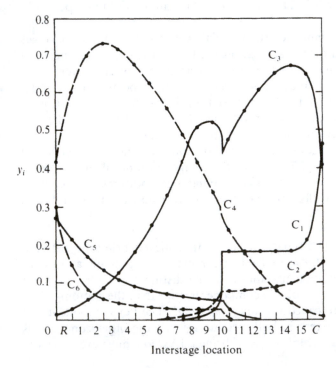

Figure 7-11 Vapor-composition profile in depropanizer. *(Results from Edmister, 1948.)*

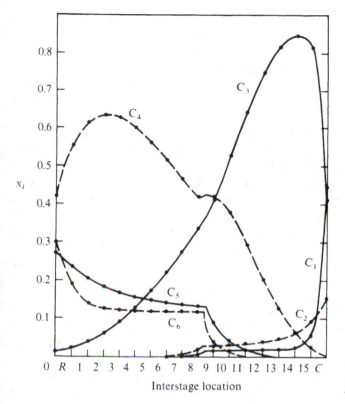

Figure 7-12 Liquid-composition profile in depropanizer. (*Results from Edmister, 1948.*)

one product or the other. Hexane and pentane are called *heavy* nonkeys since they are less volatile than the keys, while methane and ethane are *light* nonkeys since they are more volatile than the keys.

Considering Figs. 7-11 and 7-12, it is apparent that all components are present to a significant amount at the feed stage.[†] This is logical since all components are present in the feed which is introduced at that point. Above the feed the heavy nonkeys (C_5 and C_6) in both liquid and vapor die out rapidly. Because of their low relative volatility with respect to all the other components present, these two components do not enter the upflowing vapor on the stages above the feed to any large extent and thus are not able to pass upward in the column far above the feed. A few stages suffice to reduce the mole fractions of pentane and hexane to a very low value. Since pentane is more volatile than hexane, it persists for a greater number of stages.

Entirely analogous reasoning applies to the light nonkeys, methane and ethane, below the feed point. These components are so volatile that they do not enter the liquid to any great extent and thus are unable to flow down the column in any

---

[†] In Fig. 7-11 the vapor between stages 9 and 10 is arbitrarily taken to be that leaving stage 9 before being mixed with the feed vapor. In Fig. 7-12 the liquid between stages 9 and 10 is that leaving stage 10 before being mixed with the feed liquid.

substantial amount; thus they drop to very low concentrations a few stages below the feed. Ethane persists longer than methane since ethane is less volatile.

Next it should be noted that the heavy nonkeys, pentane and hexane, have relatively constant mole fractions in the liquid and vapor below the feed until a point some three or four stages from the bottom of the column is reached. These two components make up a sizable portion of the bottoms product. The lowest stages of the column are necessarily devoted to a fractionation between these heavy nonkey components and the two keys, propane and butane. The two keys are more volatile than the two heavy nonkeys; hence the keys concentrate in the vapor and increase in mole fraction going upward from the bottom at the expense of the heavy nonkeys.

It is important, however, to realize that the mole fractions of the heavy nonkeys cannot be reduced to zero before the feed point is reached. There must be some certain quantity of these materials in the liquid passing downward from the feed stage, since by mass balance the molar flow of any component in the liquid leaving a stage below the feed must at the very least equal the amount of that component flowing out in the bottoms product. Thus the relatively constant amount of pentane and hexane in the liquid on the stages just below the feed is associated with the necessity of transporting the pentane and hexane downward toward the bottoms product. Proceeding up the tower from the bottom, the mole fractions of the two heavy nonkeys reach values corresponding to these limiting constant flows after the fractionation on the bottom stages has depleted these components as much as possible.

Figure 7-11 reveals that there is also an appreciable, but lesser, constant mole fraction of pentane and hexane in the vapor in the zone where there is a constant liquid mole fraction for these materials. The presence of these components in all vapors below the feed is logical in view of the fact that the downflowing heavy nonkeys in the liquid do have some volatility, and hence the vapor mole fractions of heavy nonkeys correspond to equilibrium with the relatively constant mole fraction in the liquid. In fact, this concept allows us to derive in simple fashion an expression for the limiting mole fraction reached by a heavy nonkey in a zone where it has constant mole fraction below the feed. If mole fractions of heavy nonkeys are constant,

$$x_{HNK,\,lim} L' = x_{HNK,\,b} b + y_{HNK,\,lim} V' \tag{7-1}$$

and

$$x_{HNK,\,lim} = \frac{y_{HNK,\,lim}}{K_{HNK}} \tag{7-2}$$

where the subscript "lim" refers to the heavy nonkey component in the zone where it has constant mole fraction and the subscript b corresponds to the same heavy nonkey in the bottoms product. Hence

$$y_{HNK,\,lim}\left(\frac{L'}{K_{HNK}} - V'\right) = x_{HNK,\,b} b \tag{7-3}$$

Rearranging, we find

$$y_{HNK, \lim} = \frac{x_{HNK, b}\, b/V'}{(L/V'K_{HNK}) - 1} \tag{7-4}$$

To a first approximation K_{LK}, the equilibrium ratio of the light key, is equal to L/V' in the stripping section in this zone of constant heavy nonkey mole fraction.† Since $\alpha_{LK-HNK} = K_{LK}/K_{HNK}$, we have

$$y_{HNK, \lim} \approx \frac{x_{HNK, b}\, b/V'}{\alpha_{LK-HNK} - 1} \tag{7-5}$$

where α_{LK-HNK} is the relative volatility of the light key with respect to the heavy nonkey (a value greater than 1.0), taking $\alpha_{HK} = 0.12$ below the feed (see Table 8-2).

Substituting for hexane in our example,

$$y_{C_6, \lim} \approx \frac{(0.299)(40.1/84)}{(1.0/0.12) - 1} \approx 0.0191$$

and from Eq. (7-2)

$$x_{C_6, \lim} = \frac{0.0191}{K_{C_6}} \approx \frac{0.0191 V' \alpha_{LK-HNK}}{L} \approx \frac{(0.0191)(84)(1.0)}{(124)(0.12)} \approx 0.108$$

Figures 7-11 and 7-12 verify these estimates.

The same reasoning can be applied to the behavior of the light nonkeys, methane and ethane, above the feed. All the light nonkeys in the feed must appear in the overhead product and hence must appear in the upflowing vapor leaving each stage above the feed. Fractionation between the light nonkeys and the keys is effective on the top few stages, which serve to reduce the mole fraction of light nonkeys toward the constant limiting values. A lesser, also nearly constant, amount of light nonkeys must appear in the liquid above the feed because of the equilibrium relationship. A derivation similar to that carried out for the heavy nonkeys shows that

$$x_{LNK, \lim} = \frac{y_{LNK, D}\, D/L}{(VK_{LNK}/L) - 1} \tag{7-6}$$

and

$$y_{LNK, \lim} = K_{LNK} x_{LNK, \lim} \tag{7-7}$$

---

† For a relatively sharp separation $x_{LK, b}$ is relatively small compared with x_{LK} in the zone of constant heavy nonkey mole fraction. Hence the $x_{LK, b} b$ term is small compared with the $x_{LK} L$ and $y_{LK} V'$ terms in a mass-balance equation. If the mole fraction of the light key is changing slowly from stage to stage, an approximate equation is

$$x_{LK} L = y_{LK} V' = K_{LK} x_{LK} V'' \qquad \text{or} \qquad K_{LK} = \frac{L}{V'}$$

In Chap. 8 we see that there is a method for estimating K_{HNK} more accurately from the Underwood equations (8-106) and (8-107).

To a first approximation K_{HK}, the equilibrium ratio of the heavy key, is equal to L/V in the rectifying-section zone of constant light nonkey mole fraction.† Therefore

$$x_{LNK,\,\text{lim}} \approx \frac{y_{LNK,\,D}\, D/L}{\alpha_{LNK-HK} - 1} \tag{7-8}$$

and

$$y_{LNK,\,\text{lim}} \approx \frac{\alpha_{LNK-HK}\, L}{V} x_{LNK,\,\text{lim}} \tag{7-9}$$

The mole-fraction curves for the two keys, propane and butane, in Figs. 7-11 and 7-12 can be understood in the light of the foregoing discussion of the nonkey behavior. The keys must adjust in mole fraction so as to accommodate fractionation against the nonkeys as well as against each other. Thus, the mole fraction of propane does tend to increase upward and the mole fraction of butane tends to increase downward in the column as a simple reflection of the fractionation between the two keys. At the very bottom of the tower the mole fractions of both keys decrease downward. This is the result of fractionation of the keys against the heavy nonkeys. The heavy nonkeys grow rapidly at the bottom at the expense of both the keys, and especially at the expense of the heavy key since it is the more plentiful of the keys. This is the cause of the maximum in butane mole fraction below the feed.

At the very top of the column there is fractionation of the light nonkeys against the keys. The light nonkeys grow on the top few plates at the expense of the keys, especially the more plentiful light key. This is the cause of the maximum in propane mole fraction above the feed in Figs. 7-11 and 7-12.

Just above the feed tray the heavy nonkeys die down from their limiting mole fractions below the feed toward zero. This fractionation is again accomplished against the lighter components, and there is a tendency for mole fractions of the keys and the light nonkeys to rise somewhat more rapidly in the first few stages, proceeding upward away from the feed. Thus we have the slight hump in butane mole fraction in the liquid above the feed (Fig. 7-12). The effect is more marked in the liquid since there is a higher heavy nonkey mole fraction in the liquid. Similarly, the dying out of the light nonkeys in the few stages below the feed is reflected in a slight increase in propane mole fraction in the vapor, as shown in Fig. 7-11.

The temperature profile for the depropanizer column shown in Fig. 7-13 should be compared with the profile shown for a typical binary distillation in Fig. 7-6. In most binary distillations the temperature changes most sharply in the midsections of the rectifying and stripping sections if the operation is near minimum reflux. This reflects the bigger steps in these regions on the McCabe-Thiele diagram, and the corresponding larger changes in mole fraction from stage to stage. As shown in Fig. 7-13, the temperature changes most rapidly at the very top and very bottom of

---

† For a relatively sharp separation, $y_{HK,\,D}$ is small compared with x_{HK} in the zone of constant light nonkey mole fraction. Neglecting the $y_{HK,\,D} D$ term in a mass balance and assuming that x_{HK} changes slowly from stage to stage gives $x_{HK} L = y_{HK} V = K_{HK} x_{HK} V$ or $K_{HK} = L/V$. Again a more accurate value of K_{LNK} is available through the Underwood equations (8-104) and (8-105).

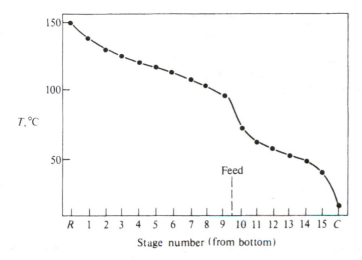

Figure 7-13 Temperature profile for depropanizer. (*Results from Edmister, 1948.*)

the column and in the vicinity of the feed point for the multicomponent distillation. These are the regions where compositions are changing the fastest, but to a large extent it is the nonkey components that are changing. At the top the light nonkey components die out rapidly in the liquid, and the bubble-point temperatures for the individual stage liquids are highly sensitive to the amount of light species present. Below the feed the light nonkeys again die down rapidly in the liquid and again change the bubble-point temperatures markedly. The reduction of heavy nonkeys in the vapor has a similar effect on dew-point temperatures of the vapor at the bottom and just above the feed.

The reader should also note that the presence of nonkey components serves to widen the *span* of temperature across a column.

Equivalent binary analysis Hengstebeck (1961) has suggested that the performance of a multicomponent distillation column be analyzed in terms of an equivalent binary distillation based upon the keys alone. This procedure has the feature of providing a familiar graphical representation of the distillation process which assists in understanding through visualization.

A multicomponent distillation can be treated as a binary involving the keys if the flows and compositions are placed on a basis of the two keys alone. Thus we could use $y'_{C_3} = y_{C_3}/(y_{C_3} + y_{C_4})$ and $x'_{C_3} = x_{C_3}/(x_{C_3} + x_{C_4})$ as effective mole fractions and express the flow as $V(y_{C_3} + y_{C_4})$ for the vapor with similar expressions for liquid, feed, and product flows. The total flows of combined keys in the vapor and the liquid at various interstage locations are shown in Fig. 7-14 for our depropanizer example. The flows are, of course, less than the total flows of vapor and liquid (Fig. 7-10) and show maxima midway along the stripping and rectifying sections. From our previous discussion there is obviously a limit on the flows of combined keys. Below the feed this will correspond to the light nonkeys being absent and the heavy nonkeys being

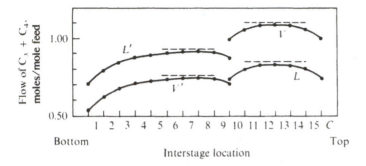

Figure 7-14 Flows of combined keys in depropanizer.

at their limiting mole fractions. Above the feed the situation is reversed, and the limit on the flow of combined keys corresponds to the heavy nonkeys being absent and the light nonkeys being at their limiting mole fractions. These limits on the flows of combined keys can be computed for the depropanizer example and are represented by dashed lines in Fig. 7-14.

Figure 7-15 shows a McCabe-Thiele diagram for the equivalent binary system in the depropanizer. The steps represent the actual changes in propane and butane mole

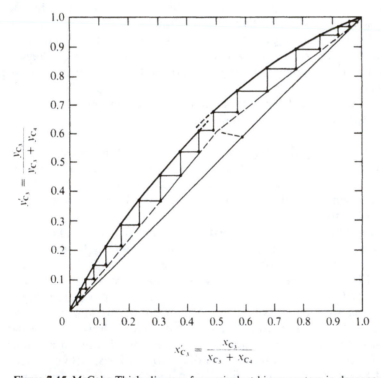

Figure 7-15 McCabe-Thiele diagram for equivalent binary system in depropanizer example.

fractions taken from Figs. 7-11 and 7-12. The equilibrium curve through the stage-exit-stream points forms smooth curves above and below the feed. Readers should verify for themselves that the equilibrium curve can be calculated from Eq. (1-12) using the relative volatility of propane to butane at the temperature of each stage. If this relative volatility were constant, the equilibrium curve would not undergo a sudden shift at the feed; however, in our case $\alpha_{C_3-C_4}$ is a function of temperature, and temperature changes rapidly near the feed because of the rapid changes in nonkey mole fractions. Thus there is a sharp change in $\alpha_{C_3-C_4}$ near the feed.

The operating lines in Fig. 7-15 are drawn for the limiting combined-key flow conditions shown by the dashed lines in Fig. 7-14. The flows of combined keys are always below these upper limits, and the liquid flows and vapor flows of combined keys always differ from each other by a constant amount equal to the amount of keys in the distillate (above the feed) or the bottoms (below the feed). Where flows are less than the upper limits, the point for passing stream compositions necessarily lies *above* the limiting operating line in Fig. 7-15. This follows since lesser total flows and a constant difference between flows necessarily produce an effective value of L/V farther removed from 1.0.

The combined-key flows fall below the limits by the greatest amounts at points where the nonkeys are dying down in mole fraction, namely, at the top, just above the feed, just below the feed, and at the bottom. The operating *points* fall most within the limiting operating lines in these regions. Thus we can conclude that in our depropanizer example the equivalent binary separation resembles that of an ordinary binary distillation, but there are added factors causing *pinches* at the very top, at the very bottom, and near the feed.

The limiting-flow operating lines approximate the separation well in Fig. 7-15. The success of a limiting-flow equivalent binary analysis is not always this good, but it is generally a good first approximation. The equivalent binary analysis, assuming that the nonkeys are at their limiting mole fractions on all stages, often can be used as an effective first analysis of a multistage multicomponent separation. It is also useful as a means of visualizing the stage-to-stage behavior of a separation. It should be noted, though, that the equivalent binary analysis necessarily underestimates the stage requirement for a given degree of separation.

Minimum reflux It is also instructive to consider the behavior of a multicomponent distillation under conditions of minimum reflux. At minimum reflux there must be at least one zone of constant composition of *all* components. Otherwise the addition of more stages must change the separation characteristics, and such a result is contrary to the concept of infinite stages at minimum reflux. If there is one zone of constant composition for all components above the feed, we can convince ourselves that there must be another such zone below the feed unless there is the equivalent of the tangent pinch of binary distillation above the feed or unless the feed is misplaced. If there is not a zone of constant composition below the feed, it must be possible to alter the separation characteristics by shifting some of the stages from the zone of constant composition above the feed to a point below the feed.

There are two possible locations for the zones of constant composition within the rectifying section and within the stripping section. The particular location depends upon the relative volatilities of the nonkey components. It should be recalled that for a binary distillation the zones of constant composition lie adjacent to the feed stage immediately above and immediately below. If there are no heavy nonkey components in a multicomponent distillation, the zone of constant composition above the feed will still be adjacent to the feed. Similarly, if there are no light nonkey components, the zone of constant composition below the feed will still be adjacent to the feed.

When heavy nonkey components are present, the zone of constant composition above the feed may move to a position higher in the rectifying section, partway between the feed and the distillate. Whether or not the zone will move to this new location depends upon whether the heavy nonkeys are *distributing* or *nondistributing* between the products at minimum reflux (Shiras et al., 1950). If one or more of the heavy nonkey components are nondistributing, they will appear at zero mole fraction in the distillate product, and the zone of constant composition above the feed will move away from the feed stage in order to allow the nondistributing heavy nonkey to die down toward zero mole fraction in the stages immediately above the feed. A distributing heavy nonkey will appear to a finite mole fraction in the distillate. If all heavy nonkeys are distributing, the zone of constant composition above the feed will remain immediately adjacent to the feed stage.

Similar reasoning holds for distributing and nondistributing light nonkeys and the location of the zone of constant composition below the feed.

The question of finding whether nonkey components are distributing or nondistributing at minimum reflux is explored further in Chap. 9. By far the most common situation is for the nonkeys to be nondistributing. A nonkey component may be distributing if it has a volatility very close to that of one of the keys or if the specified separation of the keys is not very sharp. A nonkey with a volatility intermediate between the keys also will be distributing.

Figure 7-16 shows a typical vapor-composition profile for a distillation such as our depropanizer example under conditions of minimum reflux. This is a case of nondistributing light and heavy nonkeys. The four zones correspond to those marked on the schematic of the column in Fig. 7-17. If there are nondistributing heavy and light nonkeys, the nature of the distillation dictates that the various nonkeys must necessarily be changing in mole fraction at the top, at the bottom, and on both sides of the feed point. Therefore, the zones of constant mole fraction of all components corresponding to a condition of minimum reflux and infinite stages can only occur midway in the rectifying and stripping sections.

Figure 7-18 displays the minimum reflux condition qualitatively on an equivalent binary McCabe-Thiele diagram. In zone A the heavy nonkeys decrease to their limiting mole fractions, the two keys increase, and there is also effective fractionation between the keys on an equivalent binary basis. Proceeding on upward in the tower past the zone of constant composition to zone B, the nondistributing light nonkeys begin to appear below the feed and the mole fractions of both keys decrease. This turns out to correspond to *reverse* fractionation, for as we proceed upward, the fraction of light key in the combined keys actually decreases. The operating points in

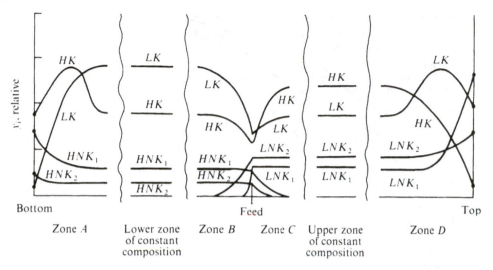

Figure 7-16 Typical vapor-composition profile for multicomponent distillation at minimum reflux.

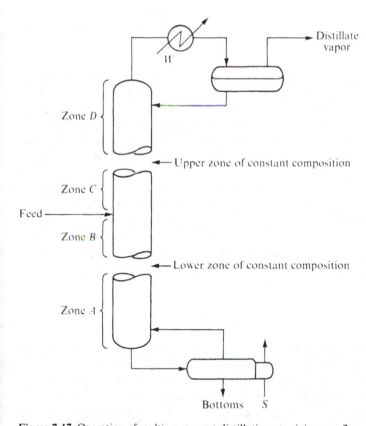

Figure 7-17 Operation of multicomponent distillation at minimum reflux.

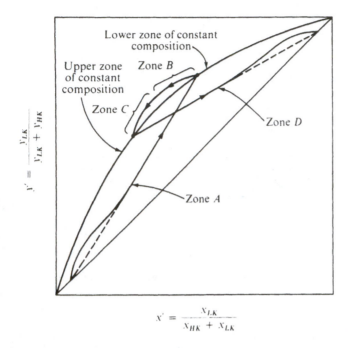

$$x' = \frac{x_{LK}}{x_{HK} + x_{LK}}$$

Figure 7-18 Equivalent binary fractionation at minimum reflux.

zone *B* must lie to the upper left of the limiting operating line for the stripping section. This follows since the flow of combined keys has become less as the feed stage is approached from below, and L/V for the combined keys has therefore become more removed from 1.0. As a result the operating curve for zone *B* necessarily *lies outside the equilibrium curve.* Since the operating curve is above the equilibrium curve, the steps in zone *B* necessarily proceed *downward.* This situation is shown schematically in Fig. 7-19.

Analogous reasoning applies to zone *C*, where the nondistributing heavy nonkeys die out as we proceed upward and the fractionation continues in the reverse direction. We next reach the zone of constant composition above the feed, which corresponds to less light key on a binary basis than the zone of constant composition below the feed. From there we pass to zone *D*, where the light nonkeys increase upward and there is once again effective fractionation in the desired direction.

Extraction

Hanson et al. (1962) present a detailed solution for an isothermal extraction cascade which serves to separate acetone from ethanol by using two different solvents, chloroform and water, as the prime components of the two counterflowing liquid phases. The operation is shown schematically in Fig. 7-20, where it is postulated that equilibrium-staged contactings occur in a plate tower. The two solvents enter at

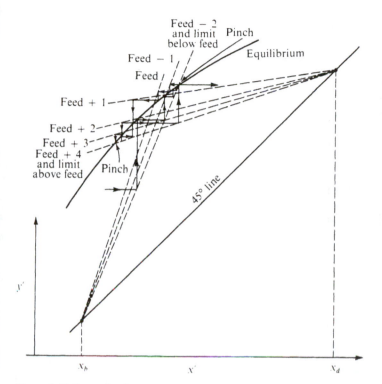

Figure 7-19 Example of "reverse distillation" of keys near feed stage at minimum reflux.

either end of the column, and the feed enters at a point such that there are five equilibrium stages above it and ten below it. The chloroform-rich phase flows downward, since the density of chloroform is greater than that of water. The solvent flow rates are high in comparison with the feed rate of acetone and ethanol to produce a sufficiently high effective reflux ratio of acetone and ethanol on either side of the feed point and also to preserve a high degree of immiscibility between the phases with a consequent high separation factor for acetone and ethanol. As noted in Chap. 4, we can look upon one of the solvents as a substitute for extract reflux in an extraction process of the type more commonly encountered.

The behavior of the fractional-extraction process shown in Fig. 7-20 for separating ethanol from acetone can be understood in terms of a few qualitative facts concerning the phase equilibrium in this four-component system. In binary solutions high activity coefficients correspond to a tendency toward immiscibility and a lack of preference for the two components to dissolve in each other. Table 7-3 gives the activity coefficients at infinite dilution for the various binary systems which can be formed from the four components in the present example.

Several facts are apparent from Table 7-3. First, there is obviously a strong "liking" of acetone and chloroform for each other. Activity coefficients less than 1.0 mean that there are negative deviations from Raoult's law and vapor pressures are

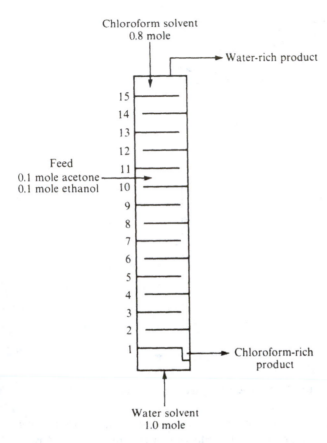

Chloroform solvent
0.8 mole

Water-rich product

15
14
13
12

Feed
0.1 mole acetone
0.1 mole ethanol

11
10
9
8
7
6
5
4
3
2
1

Chloroform-rich
product

Water solvent
1.0 mole

Figure 7-20 Fractional-extraction
example.

Table 7-3 Activity coefficients at infinite dilution for binary solutions represented in fractional-extraction example

Binary solution	Activity coefficient†	Binary solution	Activity coefficient
Acetone in chloroform	0.39	Chloroform in acetone	0.51
Acetone in ethanol	1.72	Ethanol in acetone	1.82
Ethanol in chloroform	5.0	Chloroform in ethanol	1.65
Ethanol in water	4.3	Water in ethanol	2.4
Acetone in water	6.5	Water in acetone	3.8
Chloroform in water	370	Water in chloroform	118

† Referred to Raoult's law.

less than predicted by ideal-solution theory. This is the result of hydrogen bonding between acetone and chloroform

$$\begin{matrix} H_3C \\ \diagdown \\ \diagup \\ H_3C \end{matrix} C{=}O{\cdots}H{-}\underset{\underset{\displaystyle Cl}{|}}{\overset{\overset{\displaystyle Cl}{|}}{C}}{-}Cl$$

while neither molecule hydrogen-bonds appreciably to itself.

The highest activity coefficients are between water and chloroform, indicating almost total immiscibility between those species. Thus water and chloroform serve effectively as prime components of each of the two counterflowing streams, which should be relatively immiscible in order to facilitate the operation of this process.

The next highest activity coefficients belong to the acetone-water binary. Thus acetone will tend to dissolve preferentially in a phase containing ethanol or, especially, chloroform rather than in a phase containing a large amount of water. Ethanol, on the other hand, shows roughly the same activity coefficients in either solvent, water or chloroform. Thus acetone will tend to concentrate in the chloroform phase, and ethanol will be left behind more than acetone in the water phase. Ethanol does show somewhat more preference for acetone than for either of the solvents, which is why there is a separation problem in the first place.

The composition profiles for the extraction column are shown in Fig. 7-21 for the chloroform-rich phase and in Fig. 7-22 for the water-rich phase. The stage numbering corresponds to Fig. 7-20; hence the left-hand sides of Figs. 7-21 and 7-22 refer

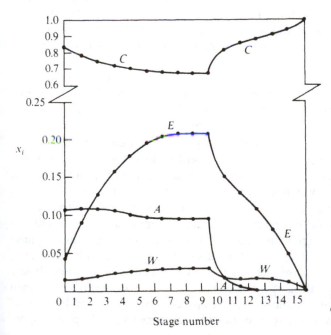

Figure 7-21 Composition profile for chloroform-rich phase. (*Results from Hanson et al., 1962.*)

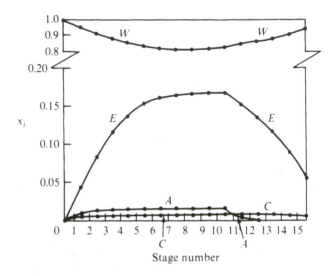

Figure 7-22 Composition profile for water-rich phase. (*Results from Hanson et al., 1962.*)

to the bottom, or acetone-rich, end of the column. The chloroform-rich phase flows from right to left on the composition diagrams, and the water-rich phase flows from left to right. The compositions shown in either diagram refer to points on a multi-dimensional four-component thermodynamic saturation envelope since we have postulated equilibrium between exit streams from a stage. In accord with Table 7-3, the solubility of acetone and chloroform in the water phase is low, while that of water in the chloroform phase is also low. The solubility of ethanol in both phases is about equal, as already noted, although ethanol does show some preference for the chloroform phase when there is a sizable acetone concentration in it. Both solvents constitute 60 percent or more of their respective phases; this is the result of the high solvent-to-feed ratio.

The solutes are carried up the column by the water phase. Acetone does not enter the water phase to any large extent; hence above (or to the right of) the feed, acetone quickly dies down to a very small concentration. This behavior is completely analogous to the dying out of a heavy nonkey above the feed in the previous multicomponent-distillation example. The heavy nonkey does not enter the upflowing vapor appreciably because of its low volatility, as reflected by a K value much less than 1.0; the acetone does not enter the water stream appreciably because of its low solubility in water compared with its solubility in chloroform.

Below (or to the left of) the feed, the acetone behavior is also similar to that of a heavy nonkey. Since more than 99 percent of the acetone must leave in the chloroform product, there must be a significant amount of acetone in the chloroform phase on *all* stages below the feed, and since acetone does have some solubility in water, albeit small, there must also be some acetone in the water phase on all stages below the feed. Because a mass separating agent (water) creates the counterflowing stream at the bottom, the flow rate of the chloroform-rich product is close to the flow rate of that phase within the column. There is no need for the acetone to build up to a higher

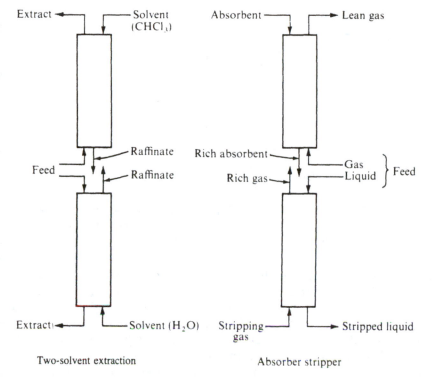

Two-solvent extraction   Absorber stripper

Figure 7-23 Analogy between two-solvent extraction and absorber-stripper.

concentration in the bottom raffinate product, as occurs for a heavy nonkey in distillation where $b < L'$. Thus the acetone fraction does not curve upward on the bottom stages, as a heavy nonkey does in distillation.

Ethanol behaves more like a key component of a multicomponent distillation, but the other key against which it fractionates is the chloroform in one phase and the water in the other.

The ethanol composition profile below the feed is analogous to that for the solute in a single-section extraction column or in a stripping operation, as shown in the lower portion of Fig. 7-23. Stripping agent or solvent (water) is introduced at the bottom and serves to lower continuously the solute concentration in the liquid feed entering the top. The ethanol composition profile above the feed is analogous to the single-section extraction cascade or to the absorber shown in the upper portion of Fig. 7-23. Fresh absorbent or solvent (chloroform) enters the top and serves to lessen the solute concentration in the upflowing feed which enters at the bottom.

The acetone profile can, of course, be interpreted in the same way. The difference between the ethanol and acetone profiles is the result of the different distribution coefficients for these two solutes between the two solvents. Acetone has a greater preference for the chloroform phase and is highly nonvolatile in the absorber-

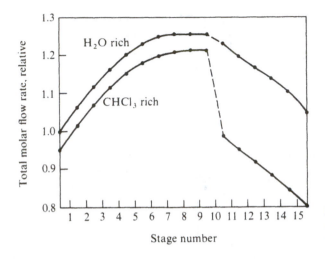

Figure 7-24 Total-flow-rate profile for extraction process. (*Results from Hanson et al., 1962.*)

stripper analogy. Thus acetone is rapidly absorbed or extracted above the feed and dies down to a very low concentration, but below the feed it is not stripped out or extracted to any great extent; therefore the concentration of acetone in the downflowing phase is relatively unaffected. On the other hand, since ethanol has no strong preference for either phase, it is extracted appreciably but less rapidly than acetone above the feed and more than acetone below the feed.

The result of these phenomena is to give a water product out the top which contains about half the ethanol and very little acetone. Thus this acetone-ethanol separation produces a highly pure ethanol product, but there is only 58 percent recovery of ethanol.

Figure 7-24 shows the variation in total flow rates of the two phases with respect to the stage location. There is a trend producing higher flow rates near the feed and lower flows at either end of the column. This result is logical in view of our earlier conclusion regarding the effect of the degree of miscibility on total interstage flows. Ethanol tends to create miscibility in this system; indeed, the ternary system chloroform-ethanol-water probably exhibits a plait point at sufficiently high concentrations of ethanol. The degree of miscibility increases at high ethanol concentrations, the phase compositions become more similar, there is less discrepancy in the amount of any one component per unit amount of any other component in the two phases, and—since the net product flow of any component must be constant in either section of the column—total flows increase toward regions of high ethanol concentration.

One should also note from Fig. 7-24 that most of the feed enters the chloroform phase rather than the water phase. Again, the selective solubility of acetone in chloroform exerts itself and the presence of acetone in the chloroform phase creates a more favorable medium for ethanol in that phase.

This fractional extraction can also be interpreted on an equivalent binary operating diagram. Figure 7-25 is an equivalent binary operating diagram on which the fraction of acetone in the combined acetone + ethanol in the chloroform phase is

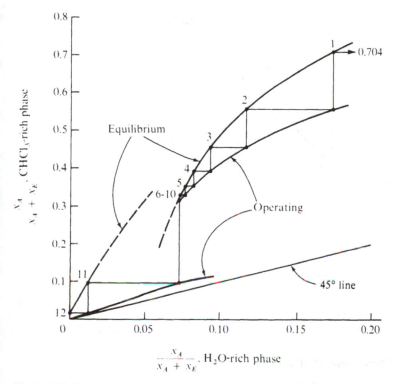

Figure 7-25 Equivalent binary operating diagram for extraction process; rectilinear coordinates.

plotted against the fraction of acetone in the combined acetone + ethanol in the water phase. Figure 7-26 is the same plot on logarithmic coordinates, which serve to expand the low-concentration region.

The equilibrium curves in Figs. 7-25 and 7-26 are smooth above and below the feed, but the abrupt changes in slopes of the composition profiles at the feed point cause a discontinuity in the equilibrium curve. This was also noted for the multicomponent distillation example (Fig. 7-15). The upper operating curve (bottom column section) in Fig. 7-25 is concave downward since the intersection with the 45° line is at the upper end of the plot ($x_A = 0.704$) and combined flows of acetone and ethanol increase toward the feed. The lower operating curve (top column section) is also concave downward since the intersection with the 45° line is at the lower end of the plot, and the combined flows of acetone and ethanol increase toward the feed. The lower operating curve lies closer to the 45° line than the upper operating curve does. This is the result of the higher combined solubility of acetone and ethanol in chloroform than in water which gives a higher reflux ratio [($A + E$ in downflowing chloroform)/($A + E$ in chloroform product)]. There are enough stages in the bottom column section to produce a severe pinch near the feed. Lowering the feed-injection stage would produce a still more acetone-free ethanol product without lessening the ethanol recovery significantly.

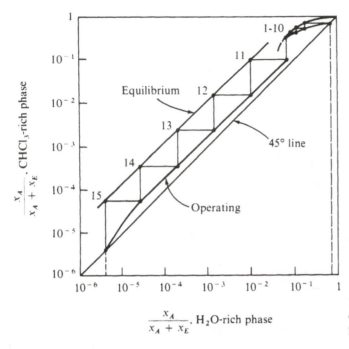

Figure 7-26 Equivalent binary operating diagram for extraction process; logarithmic coordinates.

Extractive and Azeotropic Distillation

Azeotropic and extractive distillations involve the addition of a third component to a binary system to facilitate the separation of the system by distillation. The added component modifies liquid activity coefficients and hence the vapor-liquid equilibria of the other two components in a favorable direction. The third component and the energy input to the reboiler are two different separating agents in these processes.

A typical extractive distillation process is shown in Fig. 7-27. The added component (or *solvent*) is relatively nonvolatile and is present to a high concentration (typically 65 to 90 mole percent) in the liquid within each stage. It is necessary to add the solvent near the top of the column since its lack of volatility will not produce a sufficient solvent concentration to modify the equilibrium in the desired way above the point of introduction. A few stages above the solvent entry point serve to reduce the contaminant level of solvent in the distillate product. The solvent is separated from the bottoms product in a second distillation tower.

The system shown in Fig. 7-27 accomplishes the separation of isobutane from 1-butene using furfural as a solvent (Zdonik and Woodfield, 1950). The relative volatility of isobutane to 1-butene in the presence of 80 mol % furfural is 2.0 at 52°C, as opposed to a relative volatility of 1.16 at the same temperature in the absence of the solvent. Furfural is a polar molecule

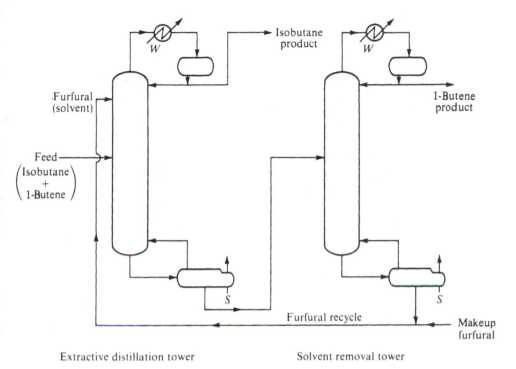

Figure 7-27 Extractive distillation for separation of isobutane from 1-butene using furfural as solvent. *(Adapted from Zdonik and Woodfield, 1950, p. 647; used by permission.)*

the $C-O-C$ and $C=O$ bonds being dipoles. The furfural molecule exerts a selective attraction on 1-butene through dipole-induced dipole interaction with the olefinic bond. In the absence of solvent, the activity coefficients of isobutane and 1-butene are nearly equal to 1.0, and the relative volatility simply represents the ratio of the vapor pressures of these species, which are also not very different from each other. At high dilution in furfural, isobutane at 52°C has an activity coefficient of 12, while 1-butene has an activity coefficient of only 6.2. Thus the addition of a high concentration of furfural increases the volatilities of both hydrocarbons since the polar furfural is a different type of molecule, but it increases the volatility of 1-butene the least because the polar group preferentially polarizes the double bond.

The solvent in extractive distillation is often chosen to be much less volatile than the species being separated in order to facilitate recovery of the solvent in the solvent-removal tower. Furfural, for example, is over two orders of magnitude less volatile than isobutane and 1-butene. As a result, very little furfural appears in the vapor phase, and the furfural molar flow in the liquid is effectively constant at some high value from stage to stage below the point of solvent feed. The other two components take up the difference and would give a composition profile the same as that for a binary distillation (Fig. 7-5), except that the mole fractions add up to $1 - x_{\text{furf}}$. Above the solvent feed, the furfural would die out rapidly.

A typical azeotropic distillation process is shown in Fig. 7-28. The added com-

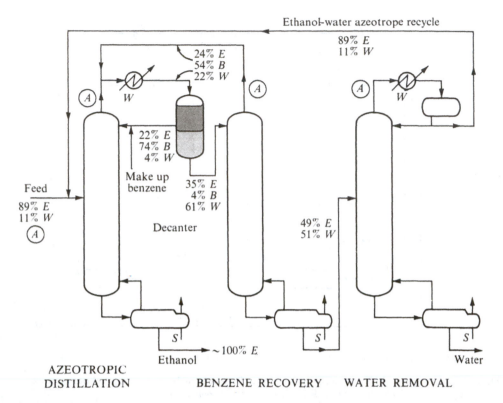

Figure 7-28 Azeotropic distillation for separation of ethanol from water using benzene as entrainer. Compositions are given in mole percent. (*Adapted from Zdonik and Woodfield, 1950, p. 652; used by permission.*)

ponent (or *entrainer*) in this case is relatively volatile and forms an azeotrope with the component to be taken overhead. The entrainer modifies the activity coefficients of the compounds being separated and thereby makes it possible to separate a feed that was originally a close-boiling mixture or a binary azeotrope. The entrainer emerges overhead from the column but must enter the liquid phase sufficiently to affect the equilibria of the other components; hence it must have a volatility comparable to that of the feed mixture. The azeotrope formed by the entrainer is frequently heterogeneous; i.e., it is composed of two immiscible liquid phases when condensed. The heterogeneous nature of the azeotrope facilitates separation of the products from the entrainer.

An azeotropic distillation process (Zdonik and Woodfield, 1950) for the separation of the water from 89 mol % (pure-component basis) ethyl alcohol using benzene as entrainer is shown in Fig. 7-28. The 89 mol % ethanol corresponds to the azeotrope in the ethanol-water binary system and is the highest ethanol enrichment that can be achieved by ordinary distillation. Near-azeotropic compositions are present at points marked A in Fig. 7-28. All towers operate at atmospheric pressure. The presence of the relatively nonpolar benzene entrainer serves to volatilize water (a

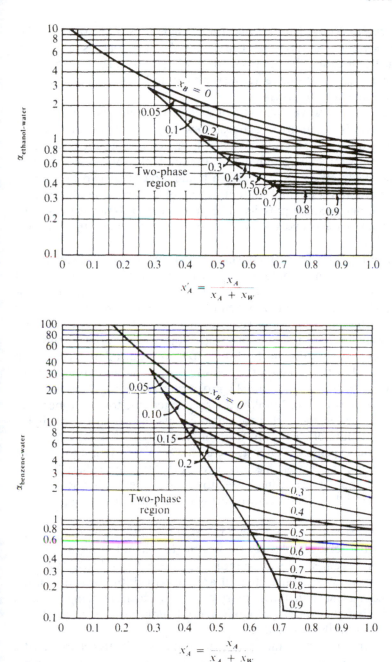

Figure 7-29 Vapor-liquid equilibrium data for ethanol-water-benzene. (*Adapted from Robinson and Gilliland, 1950, pp. 314, 315; used by permission.*)

highly polar molecule) more than it volatilizes ethanol (a moderately polar molecule). Because benzene volatilizes water preferentially, it enables us to obtain a pure ethanol product that cannot be obtained from a binary distillation because of the binary azeotrope. Benzene forms a ternary minimum-boiling azeotrope with water and alcohol at atmospheric pressure.

Figure 7-29 shows the relative volatility of alcohol to water as a function of composition and the relative volatility of benzene to water as a function of composition, as reported by Robinson and Gilliland (1950). The composition parameter is the equivalent binary mole fraction of ethanol in the total ethanol + water. Curves are plotted for different levels of benzene in the liquid. Note that two immiscible liquid phases are formed at low ethanol contents. Ethanol promotes miscibility since it is the component of intermediate polarity. The presence of benzene decreases the ethanol-water relative volatility, and the presence of ethanol reduces the benzene-water relative volatility.

The first tower in Fig. 7-28 forms the ternary azeotrope as an overhead vapor. Nearly pure alcohol issues from the bottom. The ternary azeotrope is condensed and splits into two liquid phases in the decanter. The benzene-rich phase from the decanter serves as reflux, while the water-ethanol-rich phase passes to two towers, one for benzene recovery and the other for water removal. The azeotropic overheads from these succeeding towers are returned to appropriate points of the primary tower.

Figure 7-30 shows a composition profile for the azeotropic distillation column in the process of Fig. 7-28. For the situation they considered, the feed to the azeotropic distillation tower was 89 mol ethanol and 11 mol water per hour, the reflux rate 345 mol/h, and the bottoms rate 82.7 mol/h. Benzene enters the tower by means of the reflux. In the presence of the high concentration of benzene in the rectifying section, the relative volatility of ethanol to water is substantially less than 1, and so the ethanol grows at the expense of water as we go lower in the column toward the feed. Benzene, in the rectifying section, has a relative volatility intermediate between those of water and ethanol. Hence it increases downward where it is fractionating primarily against water and decreases downward where it is fractionating primarily against ethanol.

On the bottom stages of the column there is virtually no water. From Fig. 7-29 (ratio of the two α's) the relative volatility of benzene to ethanol in the absence of water is 1.6 at 40 mol % benzene, 2.8 at 20 mol % benzene, and 4.1 near 0 mol % benzene. Hence, as far as benzene and ethanol are concerned, the behavior in the stripping section is equivalent to that in a binary distillation, benzene being the more volatile component. Thus the benzene dies out and ethanol grows as we go downward toward the bottom of the column.

The behavior of benzene and ethanol below the feed in Fig. 7-30 is characteristic of a binary distillation with a misplaced feed. It appears at first glance that there are many more stages below the feed than are needed, since from stage 10 through stage 21 the benzene and ethanol concentrations change hardly at all. This would correspond to these stages being located in a *pinch* zone at the intersection of the lower operating line and the equilibrium curve in a binary distillation. In the binary distillation we would gain by lowering the feed stage.

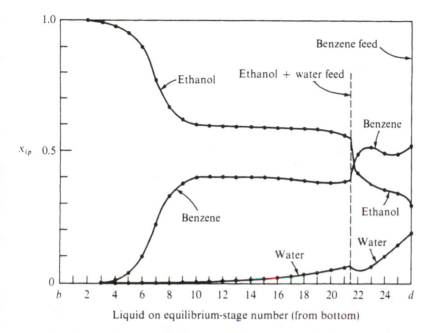

Figure 7-30 Composition profile for azeotropic distillation of ethanol and water with benzene as entrainer. *(Results from Robinson and Gilliland, 1950.)*

In azeotropic distillation having this pinch zone for the ethanol and benzene concentrations is very useful, however, and is in fact necessary for obtaining nearly pure alcohol in the column under consideration. In the pinch zone (stages 10 to 21) the water concentration drops markedly. As can be seen in Fig. 7-29, the relative volatility of ethanol to water is 0.5 at the pinch-zone composition, whereas it is 0.9 or higher (much closer to unity) in the absence of benzene. Hence water can be stripped out of the product ethanol in a reasonable number of stages only in the presence of a high benzene mole fraction. The stages in the ethanol-benzene pinch zone all necessarily have a high benzene concentration. Thus providing the pinch zone is necessary in this tower in order to give an opportunity for stripping water out of the high-purity (99.9 mol %) product ethanol.

REFERENCES

Bourne, J. R., U. von Stockar, and G. C. Coggan (1974): *Ind. Eng. Chem. Process Des. Dev.,* **13**:115, 124.

Edmister, W. C. (1948): *Petrol. Eng.,* **19**(8):128, **19**(9):47; see also "A Source Book of Technical Literature of Fractional Distillation," pt. II, pp. 74ff., Gulf Research & Development Co., n.d.

Grens, E. A., and R. A. McKean (1963): *Chem. Eng. Sci.,* **18**:291.

Hanson, D. N., J. H. Duffin, and G. F. Somerville (1962): "Computation of Multistage Separation Processes," pp. 347ff., Reinhold, New York.

Hengstebeck, R. J. (1959): "Petroleum Processing," pp. 266–268, McGraw-Hill, New York.

——— (1961): "Distillation: Principles and Design Procedures," chap. 7, Reinhold, New York.

Horton, G., and W. B. Franklin (1940): *Ind. Eng. Chem.*, **32**:1384.

Hou, T. P. (1942): "Manufacture of Soda," Reinhold, New York.

Kohl, A. L., and F. C. Riesenfeld (1979): "Gas Purification," 3d ed., p. 72, Gulf Publishing, Houston.

Krevelen, D. W. van, P. J. Hoftijzer, and F. J. Huntjens (1949): *Rec. Trav. Chim. Pays-Bas*, **68**:191.

Maxwell, J. B. (1950): "Data Book on Hydrocarbons," Van Nostrand, Princeton, N.J.

Pohl, H. A. (1962): *Ind. Eng. Chem. Fundam.*, **1**:73.

Robinson, C. S., and E. R. Gilliland (1950): "Elements of Fractional Distillation," 4th ed., chap. 10, McGraw-Hill, New York.

Shiras, R. N., D. N. Hanson, and C. H. Gibson (1950): *Ind. Eng. Chem.*, **42**:871.

Smith, B. D. (1963): "Design of Equilibrium Stage Processes," McGraw-Hill, New York.

Stockar, U. von, and C. R. Wilke (1977): *Ind. Eng. Chem. Fundam.*, **16**:94.

Zdonik, S. B., and F. W. Woodfield, Jr. (1950): Azeotropic and Extractive Distillations, in R. H. Perry et al. (eds.), "Chemical Engineers' Handbook," 3d ed., pp. 629–655, McGraw-Hill, New York.

PROBLEMS

7-A$_1$ In Fig. 7-11 y_{C_1} is greater than y_{C_2} on the stages above the feed, but x_{C_1} is less than x_{C_2} on most of the stages above the feed in Fig. 7-12. Why?

7-B$_1$ Draw qualitatively an equivalent binary McCabe-Thiele diagram, a temperature-profile, and a composition-profile diagram for a multicomponent distillation in which there are no light nonkey components.

7-C$_2$ Sketch the vapor or liquid mole fraction profile of a trace amount of a *sandwich* component, intermediate in volatility between the two main key components in a multicomponent distillation. Explain the shape of the profile. *Note:* One approach to this problem is to derive the K's for the two key components as a function of position from Figs. 7-11 and 7-12 and then use the fact that the K of the sandwich component is intermediate between those of the keys.

7-D$_2$ Often a multicomponent distillation tower is operated to provide one or more sidestream products in addition to the usual overhead distillate and bottoms products. Consider cases where the sidestreams are withdrawn directly from the column, with no sidestream stripper or sidestream rectifier. A sidestream will contain a high fraction of one of the intermediate-boiling components, and it will be desirable to provide for a high purity of that component in the sidestream. For maximum sidestream purity, should the sidestream be withdrawn as liquid or as vapor? Does your answer depend upon whether the sidestream is withdrawn from a plate above or below the feed? Explain briefly.

7-E$_2$ Smith (1963, pp. 424–438) presents a stage-to-stage solution of an extractive distillation process separating methylcyclohexane from toluene using phenol as a solvent. This is interesting as a case of extractive distillation where the solvent has an appreciable volatility, albeit one that is still lower than the volatilities of the keys. Equilibrium data for this system are shown in Fig. 7-31, which gives the relative volatility of methylcyclohexane to toluene and the relative volatility of phenol to toluene as functions of the equivalent binary mole fraction of methylcyclohexane and the mole fraction of the solvent phenol. The solution is derived for an extractive distillation tower of 20 equilibrium stages plus a reboiler and a total condenser, with a feed of 50 mol % methylcyclohexane and 50 mol % toluene entering above the seventh stage from the bottom and a 99 mol % phenol solvent feed entering above the twelfth stage from the bottom; 3.3 mol of phenol is fed per mole of hydrocarbon feed and the overhead reflux ratio r/d is 8.1. The mole fraction of each component in the liquid phase leaving each stage is shown in the composition profile of Fig. 7-32.

(*a*) Which component is preferentially volatilized by the phenol solvent and why?

(*b*) Considering the three different sections of the column—below both feeds, between feeds, and above both feeds—indicate the function of each section in relation to the overall process objectives.

(*c*) Contrast the amount of separation of methylcyclohexane from toluene occurring above the top

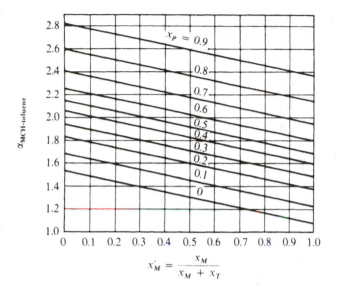

$$x'_M = \frac{x_M}{x_M + x_T}$$

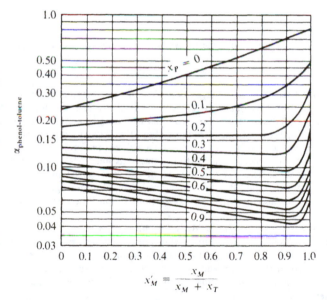

$$x'_M = \frac{x_M}{x_M + x_T}$$

Figure 7-31 Vapor-liquid equilibrium data for methylcyclohexane-toluene-phenol. (*Adapted from Smith, 1963, pp. 428, 429; used by permission.)*

feed with the amount of separation of these components occurring below the top feed. Explain the difference. Why is there a relatively large number of stages above the top feed?

(d) For each of the three column sections indicate whether phenol behaves like a key component, a light nonkey, or a heavy nonkey. Why is there an abrupt change in the mole fraction of phenol in the liquid at the hydrocarbon feed point even though the phenol mole fraction does not change much in the adjoining stages?

(e) What change in column design would you make if you wanted to obtain a higher-purity overhead methylcyclohexane product?

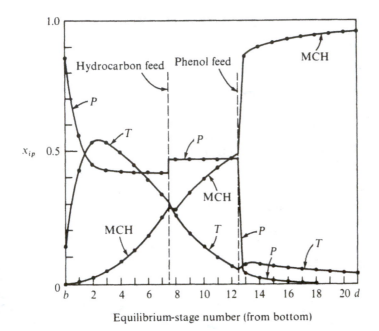

Figure 7-32 Composition profile in extractive distillation of methylcyclohexane and toluene, with phenol as solvent. *(Results from Smith, 1963.)*

7-F₂ Smith (1963, pp. 408–420) presents a detailed stage-to-stage calculation for an azeotropic distillation of *n*-heptane and toluene, using methyl ethyl ketone (MEK) as the entrainer. The entrainer-to-hydrocarbon feed molar ratio is 1.94, half of the entrainer being introduced with the feed above the tenth equilibrium stage from the bottom and the other half being introduced above the sixth equilibrium stage from the bottom. Equilibrium data for this system as presented by Smith are shown in Fig. 7-33 and are plotted as relative volatilities of heptane to toluene and of MEK to toluene as functions of the mole fraction of toluene and the mole fraction of MEK. This system does not form two immiscible liquid phases in the reflux drum, as the water-ethanol-benzene did. The tower contains 16 equilibrium stages, a total condenser, and a reboiler. The overhead reflux ratio r/d is 1.50. The composition profile for this situation is shown in Fig. 7-34.

(*a*) Which component is preferentially volatilized by the MEK entrainer and why?

(*b*) Considering the three different sections of the column—below both feeds, between feeds, and above both feeds—indicate the function of each section in relation to the overall process objectives.

(*c*) In the water-ethanol-benzene system (Figs. 7-28 and 7-30) the benzene entrainer entered the column only through the reflux stream to the azeotropic distillation column. Would that form of adding MEK be suitable in the present system? Explain.

(*d*) Why is a portion of the MEK added to the column as a second feed below the point of the main hydrocarbon feed? Why could the water-ethanol-benzene azeotropic distillation be operated without such a second feed of entrainer?

(*e*) The mole fraction of MEK in the liquid *falls* going from stage to stage upward in the zone between the feeds, while the mole fraction of MEK *rises* going from stage to stage upward in the zone above the feeds. Explain this difference.

7-G₂ The Solvay process, developed to economic fruition by Ernest and Alfred Solvay in 1861 to 1872, has for many years been the source of most of the soda, Na_2CO_3, produced in the world. The process is an excellent example of the recovery and recycle of materials in order to minimize requirements for makeup reactants.

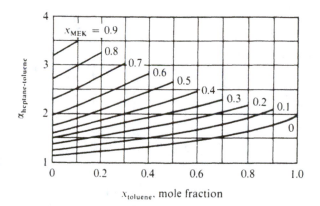

x_{toluene}, mole fraction

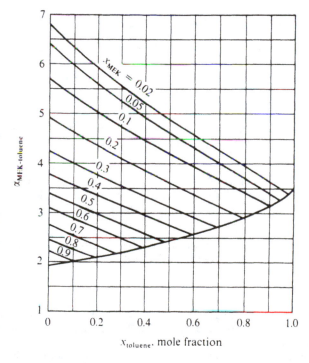

x_{toluene}, mole fraction

Figure 7-33 Vapor-liquid equilibrium data for *n*-heptane–toluene–methyl ethyl ketone. (*Adapted from Smith, 1963, pp. 414, 415; used by permission.*)

The Solvay process uses as feeds (1) a sodium chloride-rich brine (natural brine, dissolved rock salt, or even concentrated seawater) and (2) limestone rock, $CaCO_3$. The process focuses on the reaction of ammonium bicarbonate with the sodium chloride of this brine in concentrated aqueous solution. Of the various compounds which can be formed from the various ions present (sodium, ammonium, chloride, bicarbonate), the least soluble is sodium bicarbonate. The sodium bicarbonate is made to precipitate out of solution, is filtered and washed, and is then calcined (heated) to cause it to decompose into sodium carbonate, with the release of carbon dioxide and water vapor.

The main contribution of the Solvays to the process was to cause the ammonium bicarbonate to be formed in place in a highly concentrated brine solution by the successive absorption of ammonia and then carbon dioxide into the solution. It was also economically necessary to provide for a high degree of

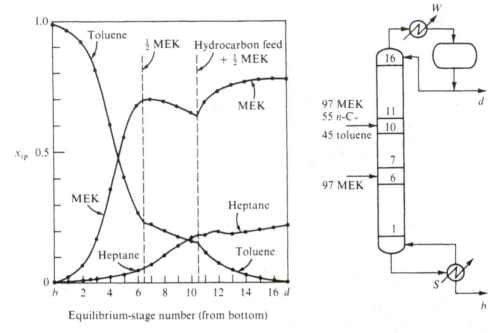

Equilibrium-stage number (from bottom)

Figure 7-34 Composition profile for azeotropic distillation of n-heptane and toluene with MEK as entrainer. (*Results from Smith, 1963.*)

recovery of the ammonia for recycle, to minimize purchases of relatively expensive ammonia as fresh feed. Ammonia recovery is accomplished by calcining the limestone to form lime, CaO, and carbon dioxide:

$$CaCO_3 \xrightarrow{\text{heat}} CaO + CO_2\uparrow$$

The ammonium chloride-rich solution remaining after the precipitation of sodium bicarbonate is treated with the lime to free ammonia, which can then be recovered. The by-product of this step is $CaCl_2$, which is either sold or discarded with unreacted NaCl:

$$2NH_4Cl + CaO \rightarrow 2NH_3 \uparrow + CaCl_2 + H_2O$$

The carbon dioxide from the limestone calcination is used as a portion of the carbon dioxide required as carbonating agent to form ammonium bicarbonate in the ammoniated NaCl brine. Additional carbon dioxide comes from the calcination of sodium bicarbonate.

If there were to be more complete ammonia recovery and pure feeds, the overall stoichiometry of the process would correspond to

$$CaCO_3 + 2NaCl \rightarrow Na_2CO_3 + CaCl_2$$

The heart of the process is the two countercurrent gas-liquid contacting towers shown in Fig. 7-35. The carbonating tower receives as feed an ammoniated sodium chloride-rich brine, known as *green liquor*. This feed typically contains about 5 mol/L NH_3, 4.5 mol/L NaCl, and 1 mol/L CO_2. The CO_2 in the ammoniated brine entered with some of the ammonia-rich gases returned from various places to the ammonia absorber. The carbonating gas typically contains 56 mol % CO_2, the remainder being mostly nitrogen. The carbonating tower is equipped with cooling coils on the bottom stages to remove the heat of the reaction forming ammonium carbonate and bicarbonate. The cooling-water flow and the area of the coil are adjusted to give a temperature profile like that shown in Fig. 7-36. The pressure is maintained at

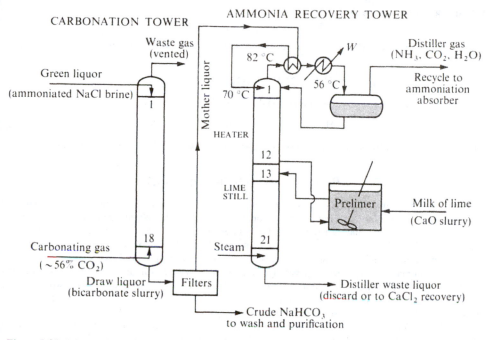

CARBONATION TOWER

AMMONIA RECOVERY TOWER

Waste gas
(vented)

Green liquor
(ammoniated NaCl brine)

Mother liquor

82 °C

W

Distiller gas
(NH₃, CO₂, H₂O)

70 °C

1

56 °C

Recycle to
ammoniation
absorber

1

HEATER

12

13

LIME
STILL

Prelimer

Milk of lime
(CaO slurry)

18

21

Carbonating gas
(∼56% CO₂)

Steam

Draw liquor
(bicarbonate slurry)

Filters

Distiller waste liquor
(discard or to CaCl₂ recovery)

Crude NaHCO₃
to wash and purification

Figure 7-35 Brine-carbonation tower and ammonia-recovery tower in the Solvay process.

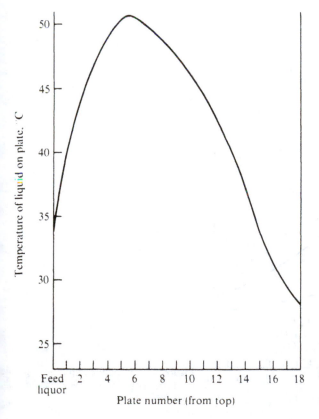

Temperature of liquid on plate, °C

Feed liquor 2   4   6   8   10   12   14   16   18

Plate number (from top)

Figure 7-36 Temperature profile for brine-carbonation tower. (*Data from Hou, 1942.*)

355

about 340 kPa. As the carbon dioxide is absorbed, it first reacts with ammonia to form ammonium carbonate

$$H_2O + 2NH_3 + CO_2 \rightarrow (NH_4)_2CO_3$$

The ammonium carbonate reacts with additional CO_2 to form bicarbonate

$$(NH_4)_2CO_3 + H_2O + CO_2 \rightarrow 2NH_4HCO_3$$

The ammonium bicarbonate then precipitates the less soluble sodium bicarbonate by reaction with the brine

$$NH_4HCO_3 + NaCl \rightarrow NaHCO_3 \downarrow + NH_4Cl$$

The solubility of sodium bicarbonate is about 1.2 mol/L, so an appreciable amount of it remains in solution. The reactions forming ammonium carbonate, ammonium bicarbonate, and sodium bicarbonate all take place in the carbonating tower, and precipitated $NaHCO_3$ emerges as a slurry in the bottoms liquid. Figure 7-37 shows a liquid-phase composition profile for the carbonation tower. The diagram is based upon actual plates, not equilibrium stages, and the data are real plant data.

The ammonia-recovery tower is generally run at atmospheric pressure and consists of two portions called the *heater* and the *lime still*. The feed liquor enters the top heater section and passes down the tower. At a point approximately halfway down the tower (plate 12 in Fig. 7-35) all the downflowing liquid is drawn off to a large agitated vessel, known as the *prelimer*, where it is mixed with milk of lime (an aqueous slurry of the CaO produced in the lime kiln) and is held until the reaction of NH_4Cl with CaO (see above) occurs substantially to completion. The overflow liquid from the prelimer reenters the tower and flows downward through the lime still. Live steam is introduced at the bottom, and the vapors rise through both

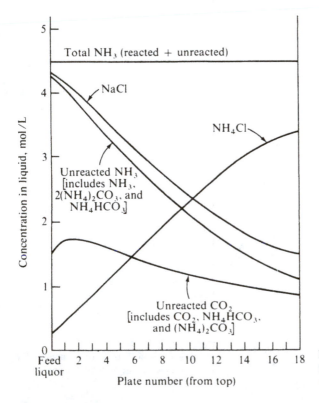

Figure 7-37 Liquid composition profile for brine-carbonation tower. (*Data from Hou, 1942.*)

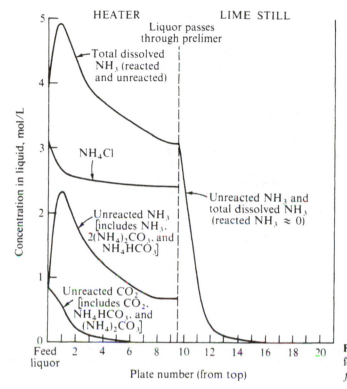

Figure 7-38 Composition profile for ammonia-recovery tower. (*Data from Hou, 1942.*)

sections of the column to a partial condenser overhead, which produces a gas (containing NH_3, CO_2, and water vapor) and a reflux stream. The gaseous product is the main gas stream fed to the absorbers for ammoniating the original brine. Figure 7-38 shows a liquid-phase composition profile for the ammonia recovery tower. Again, the diagram is based upon real plant data.

Figure 7-39 shows vapor-liquid equilibrium data for partial pressures of NH_3 and CO_2 over solutions of these gases in water. The equilibrium would be altered considerably by the high content of other salts in the Solvay process solutions, but the qualitative trends should remain the same.

(*a*) Make a simple schematic flow scheme of the entire Solvay process on the basis of the description given.

(*b*) In the Solvay process it is important that the brine be ammoniated first and then subsequently be carbonated. Why isn't the reverse procedure, absorbing the CO_2 into the brine first and then absorbing the NH_3 second, workable?

(*c*) What is the main function of the heater section of the ammonia-recovery tower (consider your answer carefully)? What is the main function of the lime-still section?

(*d*) What are effectively the key components in the heater section of the ammonia-recovery tower? In the lime-still section?

(*e*) Is the sodium bicarbonate precipitated out primarily in the top or the bottom portion of the carbonating tower, or is it formed to about the same amount on each stage?

(*f*) What *specifically* is the main benefit to the process of the countercurrent operation of the carbonating tower? Of the countercurrent operation of the ammonia-recovery tower?

(*g*) Why is the total NH_3 (reacted + unreacted) concentration in the carbonating tower essentially constant from plate to plate? Suggest a specific cause for the decrease in NH_4Cl concentration from plate to plate downward in the heater section of the ammonia-recovery tower.

(*h*) In most countercurrent absorbers the concentration of the transferring solute in the liquid phase

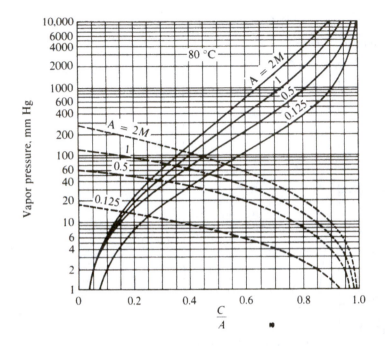

Figure 7-39 Vapor-liquid equilibrium data for solutions of CO_2 and NH_3 dissolved in water at 80°C; solid curves = CO_2 partial pressure; dashed curves = NH_3 partial pressure; A = moles per liter of NH_3 in solution; and C = moles per liter of CO_2 in solution. (*Adapted from Krevelen, 1949, p. 205; used by permission.*)

decreases from plate to plate going upward in the column. This is not the case for the carbonating tower. Explain why the CO_2 concentration *increases* from plate to plate going upward in this column.

(*i*) The unreacted NH_3 concentration in the liquid of the ammonia recovery tower is higher in the liquid leaving the top plate than it is in the feed liquor. How can this be?

(*j*) The unreacted ammonia concentration in the liquid seems to drop asymptotically to a finite lower limit in the lower part of the heater section. What factor would set this lower limit?

(*k*) Sketch operating diagrams for each of the towers.

(*l*) A serious water-pollution problem results from Solvay plants. This, plus the increased availability of mined natural trona, has led to the closing of several Solvay-process plants in recent years. What is the source of the water-pollution problem from the Solvay process?

7-H₃ The trays and downcomers of a distillation column are usually metallic and are therefore good conductors of heat. Suppose that heat transfer occurs across the metal wall of the downcomers of a column between the vapor rising up toward a plate and the liquid leaving that plate. Neglecting any effects on tower capacity (tendencies toward flooding, etc.), will this heat transfer have any effect on the degree of separation of a given feed provided by a fixed number of plates at a fixed reflux ratio? If so, will it serve to improve the degree of separation or to lessen it? Why? Will the same conclusions apply to heat transfer across the metal of a plate itself between the vapor rising to the plate and the liquid on the plate? Why?

7-I₂ Multicomponent distillation columns from which sidestreams are withdrawn frequently use sidestream strippers, of the sort shown in Fig. 7-40. This is common practice, for example, in the primary fractionation of crude oil. What is the purpose of the sidestream strippers? Why can't this purpose be accomplished by the main column?

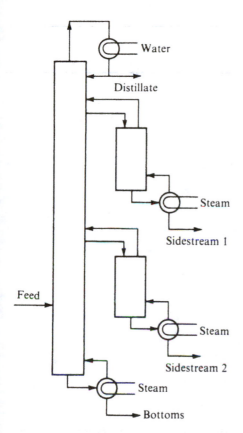

Water

Distillate

Steam

Sidestream 1

Steam

Sidestream 2

Feed

Steam

Bottoms

Figure 7-40 Multicomponent distillation column with sidestream strippers.

7-J₂ Draw a qualitative yx diagram for a single-solute absorber developing an internal temperature maximum. Assume that the isothermal equilibrium relationship is relatively linear. What factor sets the minimum flow of absorbent liquid to accomplish a given separation?

CHAPTER
EIGHT

GROUP METHODS

Group methods of calculation serve to relate the feed and product compositions in a separation process to the number of stages employed, without considering composition at intermediate points in the cascade. This approach therefore requires the development of *algebraic* equations which represent the combined effects of many stages on product compositions. As a result, group methods can be used only in situations where both the equilibrium and operating curves can be approximated satisfactorily by simple algebraic expressions.

The group methods most commonly employed apply to one of two situations: (1) *flow rates* that are *constant* from stage to stage, coupled with simple *linear* stage-exit-composition relationships, and (2) flow rates that are *constant* from stage to stage, coupled with *constant separation factors* α_{ij} between all components present. The first case corresponds to straight equilibrium and operating lines for binary phase-equilibration separations. Examples would therefore include absorption, extraction, and stripping in dilute systems, including chromatography; binary distillations with one of the components present at low concentration; and washing. The second case applies primarily to binary and multicomponent distillations under the assumptions of constant molal overflow and constant relative volatility.

We consider each of the two situations at some length in this chapter, along with the approach presented by Martin (1963) for cases in which the flow rates and the stage-exit relationship can each be hyperbolic functions.

LINEAR STAGE-EXIT RELATIONSHIPS AND CONSTANT FLOW RATES

Countercurrent Separations

The development of an analytical equation covering the effects of a group of stages for the case of constant flow rates and linear stage-exit relationships is based upon the combination of two families of equations. In terms of a vapor-liquid separation process with constant molar flows and with mole fractions as composition parameters, the equations are

$$y_p = mx_p + b \tag{8-1}$$

and

$$y_p V = x_{p+1} L + x_d d \tag{5-3}$$

for the rectifying section of a distillation column or

$$y_p V = x_{p+1} L + y_{\text{out}} V - x_{\text{in}} L \tag{8-2}$$

for a general countercurrent gas-liquid separation process. y and x may refer to any component in a binary or multicomponent mixture which obeys these particular equations and assumptions.

A plot of Eqs. (8-1) and (8-2) is given in Fig. 8-1 for component A of a mixture. The assumptions we have made in this analysis dictate straight operating and stage-exit lines, although these lines are not necessarily parallel. A section of the staged countercurrent separation process is shown in Fig. 8-2. The relative positions of the operating and stage-exit lines correspond to those for an absorber.

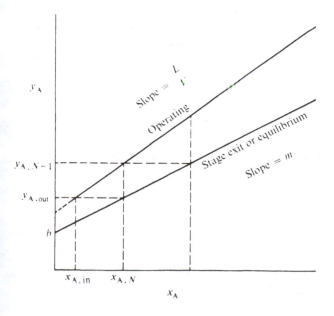

Figure 8-1 Operating diagram.

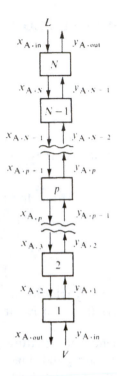

Figure 8-2 Separation cascade.

To obtain a group-method equation covering the action of a number of stages, our approach will be to obtain expressions for the increase in y_A from stage to stage, starting with the top stage in Fig. 8-2. We shall relate these changes to the distance between the two lines at the upper end of the cascade, i.e., to $y_{A,\text{out}} - y^*_{A,\text{out}} = y_{A,\text{out}} - mx_{A,\text{in}} - b$, where y^*_A denotes the value of y_A in equilibrium with the prevailing value of x_A. Rearranging Eq. (8-2) for $p = N - 1$, we have

$$y_{A,N-1} = y_{A,\text{out}} + \frac{L}{V}(x_{A,N} - x_{A,\text{in}}) \tag{8-3}$$

If we make an equilibrium-stage analysis, $y_{A,N}$ must be in equilibrium with $x_{A,N}$, and so we have from Eq. (8-1)

$$x_{A,N} = \frac{y_{A,\text{out}} - b}{m} \tag{8-4}$$

Combining Eqs. (8-3) and (8-4) gives

$$\frac{y_{A,N-1} - y_{A,\text{out}}}{y_{A,\text{out}} - mx_{A,\text{in}} - b} = \frac{L}{mV} \tag{8-5}$$

Combining Eq. (8-2) written for $p = N - 1$ and $p = N - 2$ with Eq. (8-1) for $p = N - 1$ yields

$$\frac{y_{A,N-2} - y_{A,N-1}}{y_{A,N-1} - y_{A,\text{out}}} = \frac{L}{mV} \tag{8-6}$$

or

$$\frac{y_{A,N-2} - y_{A,N-1}}{y_{A,\text{out}} - mx_{A,\text{in}} - b} = \left(\frac{L}{mV}\right)^2 \tag{8-7}$$

Thus a general expression for any number of equilibrium stages is

$$\frac{y_{A,p} - y_{A,p+1}}{y_{A,\text{out}} - mx_{A,\text{in}} - b} = \left(\frac{L}{mV}\right)^{N-p} \tag{8-8}$$

The aim of this derivation is to relate terminal concentrations; hence we add Eq. (8-8) repeatedly to itself for p ranging from $N - 1$ to zero (bottom end of cascade)

$$\frac{y_{A,\text{in}} - y_{A,\text{out}}}{y_{A,\text{out}} - mx_{A,\text{in}} - b} = \sum_{p=N-1}^{0} \left(\frac{L}{mV}\right)^{N-p} = \sum_{n=1}^{N} \left(\frac{L}{mV}\right)^{n} \tag{8-9}$$

substituting n as a dummy variable for $N - p$.

Usually one will want to determine an exit composition $y_{A,\text{out}}$ from the separation process from a knowledge of the two inlet compositions, N and L/mV; therefore it is desirable to modify the left-hand side of Eq. (8-9) so that it contains $y_{A,\text{out}}$ only once. Eq. (8-9) converts directly into

$$\frac{y_{A,\text{in}} - y_{A,\text{out}}}{y_{A,\text{in}} - mx_{A,\text{in}} - b} = \frac{\displaystyle\sum_{n=1}^{N} \left(\frac{L}{mV}\right)^{n}}{1 + \displaystyle\sum_{n=1}^{N} \left(\frac{L}{mV}\right)^{n}} = \frac{\left[\displaystyle\sum_{n=0}^{N} \left(\frac{L}{mV}\right)^{n}\right] - 1}{\displaystyle\sum_{n=0}^{N} \left(\frac{L}{mV}\right)^{n}} \tag{8-10}$$

including the term $(= 1)$ for $n = 0$ in the summation. The sum of a power series is given by

$$\sum_{i=0}^{k} ar^i = \frac{a(1 - r^{k+1})}{1 - r} \tag{8-11}$$

for $|r| < 1$. Hence for $L/mV < 1$ we have

$$\frac{y_{A,\text{in}} - y_{A,\text{out}}}{y_{A,\text{in}} - mx_{A,\text{in}} - b} = \frac{(L/mV) - (L/mV)^{N+1}}{1 - (L/mV)^{N+1}} \tag{8-12}$$

Replacing $mx_{A,\text{in}} + b$ by $y^*_{A,\text{out}}$, we have

$$\frac{y_{A,\text{in}} - y_{A,\text{out}}}{y_{A,\text{in}} - y^*_{A,\text{out}}} = \frac{(L/mV) - (L/mV)^{N+1}}{1 - (L/mV)^{N+1}} \tag{8-13}$$

$y^*_{A,\text{out}}$ is that value of y_A which would be in equilibrium with the prevailing value of $x_{A,\text{in}}$. For $L/mV > 1$, we can divide the numerator and denominator of Eq. (8-10) by

$(L/mV)^N$ and proceed in analogous fashion, obtaining an equation identical to Eq. (8-13). The right-hand side of Eq. (8-13) reduces to $N/(N + 1)$ for $L/mV = 1$.

Equation (8-13) was first developed by Kremser (1930) and by Souders and Brown (1932). As presented, it is useful for solving problems where N is fixed and the quality of separation is to be determined. When equilibrium is closely approached, the following form of Eq. (8-13) is more convenient:

$$\frac{y_{A,\,out} - y^*_{A,\,out}}{y_{A,\,in} - y^*_{A,\,out}} = \frac{1 - (L/mV)}{1 - (L/mV)^{N+1}} \tag{8-14}$$

For a design problem where the separation is specified but N is unknown, the equation can be converted into a form explicit in N

$$N = \frac{\ln\left\{[1 - (mV/L)][(y_{A,\,in} - y^*_{A,\,out})/(y_{A,\,out} - y^*_{A,\,out})] + (mV/L)\right\}}{\ln\,(L/mV)} \tag{8-15}$$

Figure 8-3 presents Eqs. (8-14) and (8-15) in graphical form.

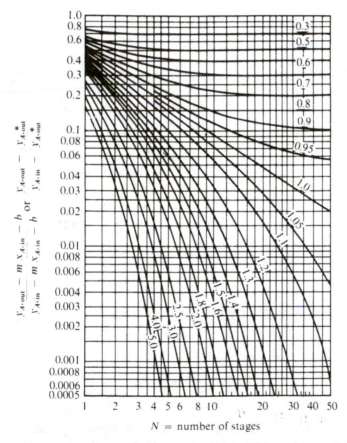

Figure 8-3 Plot of Eqs. (8-14) and (8-15) for equilibrium-stage contactor. Parameter is L/mV.

For a constant Murphree efficiency E_{MV} based on the phase flowing in the positive direction, it can be shown that the actual number of stages is given by

$$N_{act} = \frac{\ln\{[1-(mV/L)][(y_{A,in}-y^*_{A,out})/(y_{A,out}-y^*_{A,out})]+(mV/L)\}}{-\ln\{1+E_{MV}[(mV/L)-1]\}} \quad (8\text{-}16)$$

By solving Eq. (12-33) for E_{MV} and substituting the result into the denominator of Eq. (8-16) it can be shown that

$$-\ln\left[1+E_{MV}\left(\frac{mV}{L}-1\right)\right]=\ln\left[1+E_{ML}\left(\frac{L}{mV}-1\right)\right] \quad (8\text{-}17)$$

Hence the right-hand side of Eq. (8-17) can be substituted for the denominator in Eq. (8-16) if E_{ML} is used instead of E_{MV}.

The entire derivation can be carried out turning the cascade of Fig. 8-2 upside down, i.e., interchanging y and x, L and V, m and $1/m$, and E_{MV} and E_{ML} throughout the previous equations and Fig. 8-3. In that way, Eqs. (8-14) to (8-16) become

$$\frac{x_{A,out}-x^*_{A,out}}{x_{A,in}-x^*_{A,out}}=\frac{1-(mV/L)}{1-(mV/L)^{N+1}} \quad (8\text{-}18)$$

$$N=\frac{\ln\{[1-(L/mV)][(x_{A,in}-x^*_{A,out})/(x_{A,out}-x^*_{A,out})]+(L/mV)\}}{\ln(mV/L)} \quad (8\text{-}19)$$

and

$$N_{act}=\frac{\ln\{[1-(L/mV)][(x_{A,in}-x^*_{A,out})/(x_{A,out}-x^*_{A,out})]+(L/mV)\}}{-\ln\{1+E_{ML}[(L/mV)-1]\}} \quad (8\text{-}20)$$

Again, Eq. (8-17) gives a way of expressing the denominator of Eq. (8-20) in terms of E_{MV} rather than E_{ML}. Figure 8-3 represents the solution to Eqs. (8-18) and (8-19) if the vertical axis is changed to $(x_{A,out}-x^*_{A,out})/(x_{A,in}-x^*_{A,out})$ and the parameter is changed from L/mV to mV/L.

It should also be pointed out that the Kremser-Souders-Brown (KSB) equations are valid independent of the direction of transfer. That is, although Eqs. (8-14) to (8-16) were derived in the context of an absorber, where $y_{A,out}-y^*_{A,out}$ and $y_{A,in}-y^*_{A,out}$ are both positive, they are valid as well for stripping, where both those quantities are negative. The same logic applies, in reverse, to Eqs. (8-18) to (8-20).

Smaller values of the vertical coordinate in Fig. 8-3 correspond to high degrees of removal of a gaseous solute in an absorber or to a high degree of equilibration of $y_{A,out}$ with the inlet liquid, i.e., low $y_{A,out}-y^*_{A,out}$. If the inlet liquid is free of solute ($x_{A,in}=y^*_{A,out}=0$), the vertical coordinate is $y_{A,out}/y_{A,in}$, or the fraction of the solute in the entering gas which remains in the leaving gas. It is apparent that values of L/mV greater than 1 are effective for achieving a high degree of solute removal in an absorber. However, for L/mV less than 1, both Eq. (8-14) and Fig. 8-3 show that the vertical coordinate reaches an asymptotic value at a large number of stages. This asymptotic value is

$$\frac{y_{A,out}-y^*_{A,out}}{y_{A,in}-y^*_{A,out}}\to 1-\frac{L}{mV}, \text{ as } N\to\infty \qquad \frac{L}{mV}<1 \quad (8\text{-}21)$$

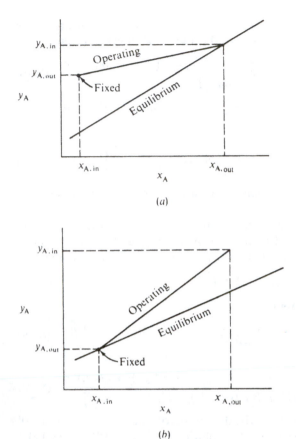

(a)

(b)

Figure 8-4 Operating diagrams for absorbers with infinite stages and (a) $L/mV < 1$ and (b) $L/mV > 1$.

This asymptote can limit removal severely. For example, for $L/mV = 0.2$, 80 percent of the solute would remain in the gaseous product even if there were no solute in the entering liquid and there were infinite stages; there would be only 20 percent removal.

The cause of the asymptotic removals at low L/mV can be understood from Fig. 8-4, which shows operating diagrams for $L/mV < 1$ and $L/mV > 1$. For $L/mV > 1$ (Fig. 8-4b) the pinch with infinite stages occurs at the bottom of the diagram (or top of an absorber), and $y_{A, \text{out}}$ can achieve equilibrium with $x_{A, \text{in}}$, giving $y_{A, \text{out}} - y_{A, \text{out}}^*$ and the vertical coordinate of Fig. 8-3 equal to zero. However, for $L/mV < 1$ (Fig. 8-4a) the pinch with infinite stages must occur at the other end of the diagram, giving $x_{A, \text{out}}$ in equilibrium with $y_{A, \text{in}}$. The solvent capacity has been reached, and there is no way to reduce $y_{A, \text{out}}$ further since it is impossible to transfer more solute to the liquid and increase $x_{A, \text{out}}$ further. This leads to the asymptotic removal shown in Fig. 8.3.

Minimum flows and selection of actual flows The analysis surrounding Fig. 8-4 and the asymptotes in Fig. 8-3 leads to the conclusion that L/mV must be greater than 1 for a high degree of solute removal to be obtained in an absorber; otherwise the removal will be limited to a low value by solvent capacity. Although the exact value will be somewhat less, depending upon the separation specified, $L = mV$ will correspond rather closely to the minimum absorbent flow required by an absorber, even with infinite stages.

If we now consider Fig. 8-3 expressed in terms of x variables with mV/L as the parameter, following Eqs. (8-18) and (8-19), we can consider stripping a volatile solute from a liquid, where we wish to reduce $x_{A, \text{out}} - x^*_{A, \text{out}}$ to some low value. The same logic leads to the fact that mV/L must be greater than 1 (or L/mV less than 1) for a high removal of solute from the liquid. Although the exact value will be somewhat less, depending upon the separation specified, $V = L/m$ will correspond rather closely to the minimum flow of stripping gas, even with infinite stages.

Returning to Fig. 8-3 for the absorber, we can see that beyond L/mV equal to about 3 there is less and less additional benefit from each unit increase in L to make L/mV still higher. Because of the diminishing returns of higher L/mV, the design economic optimum will typically be in the range $1.2 < L/mV < 2.0$, often around 1.4. Similarly, for a stripping column the design economic optimum will typically be in the range $1.2 < mV/L < 2.0$, often around 1.4. A simplified analysis of rules of thumb for optimizing L/V, recoveries, etc., in dilute absorbers and strippers has been given by Douglas (1977).

Limiting components For multicomponent absorption, stripping, and extraction processes, the concept of *limiting components* is useful. For an absorption where several components are to be taken from the gas phase into the liquid, consideration of the KSB equations and Fig. 8-3 shows that the magnitude of the necessary L/V will be established by that component to be absorbed which has the highest value of K_i. It will be necessary to have L/V large enough so that $L/K_i V$ (L/mV) is greater than 1 for this component, which will then mean that $L/K_i V$ is still greater for other components to be absorbed. The extreme sensitivity of the fraction removal to $L/K_i V$ in the vicinity of $L/K_i V = 1$ (see Fig. 8-3) indicates that the fraction removal for the components with lower K_i will be substantially closer to unity than that for the absorbed component with the largest K_i. Therefore the component to be absorbed which has the greatest K_i usually sets the lower limit on the necessary circulation rate of absorbent liquid and is thus called the *limiting component*.

A similar analysis can be made for multicomponent stripping or extraction. For a stripping process, the component to be vaporized that has the least value of K_i is the limiting component and usually sets the lower limit on the necessary flow of stripping gas V/L.

Using the KSB equations The form of the solution shown in Fig. 8-3 leads to some general conclusions regarding the most effective ways of using the KSB equations. Suppose, for example, that we want to compute the stage requirement for a certain removal of a solute with $L/mV < 1$ in an absorber. Since the curves in Fig. 8-3 reach horizontal asymptotes at moderate to high values of N, it will be difficult to determine the resultant value of N with any precision. It would be preferable to invert the equations to the x form [Eqs. (8-18) to (8-20)], specify $x_{A,\,out}$ by means of an overall mass balance, and solve for N using Fig. 8-3 or Eq. (8-19) or (8-20), where mV/L will now be greater than 1 (since L/mV was less than 1). We can now solve for N with precision since we are away from the horizontal-asymptote region of Fig. 8-3.

Next, consider the case of an absorber with a given number of stages and a specified L/mV, which is greater than 1 so as to give a high degree of solute removal. One approach to calculating the separation would be to use Eq. (8-18) or the x form of Fig. 8-3 to solve for $x_{A,\,out}$ and then use an overall mass balance to find $y_{A,\,out}$. However, since $y_{A,\,out}$ will be very close to $y^*_{A,\,out}$, the difference between $y_{A,\,out}$ and $y^*_{A,\,out}$ will not be known with much precision unless a large number of significant figures is carried in the calculation. It is preferable from the standpoint of precision to solve for $y_{A,\,out} - y^*_{A,\,out}$ directly, using Eq. (8-14) or the y form of Fig. 8-3.

Both the examples above lead to the conclusion that it is better to stay away from the horizontal-asymptote region of Fig. 8-3 for calculations. Greater precision can be obtained if Eqs (8-14) to (8-16) and the y form of Fig. 8-3 are used when $L/mV > 1$ and if Eqs. (8-18) to (8-20) and the x form of Fig. 8-3 are used when $L/mV < 1$ $(mV/L > 1)$. Since L/mV is usually greater than 1 for the principal solute in an absorber, Eqs. (8-14) to (8-16) and the y form of Fig. 8-3 are sometimes known as the *absorber form* of the equations. Conversely, Eqs. (8-18) to (8-20) and the x form of Fig. 8-3 are then known as the *stripper form*.

Another way of stating the above conclusions is that it is best to use the KSB equations in a form which solves for (or involves) the concentration difference at the more pinched end of the cascade.

Although the KSB equations have been derived and considered so far in the context of absorbers and strippers, with appropriate changes in notation they are applicable to any staged single-section countercurrent process for which there are straight operating lines and straight equilibrium lines. In general, this means any dilute-solute system with a constant equilibrium ratio. For transfer of a solute with constant K_i in an extraction with constant interstage flows, x and L would apply to one phase and y and V (or appropriately changed symbols) would apply to the other phase. m would be K_i, expressed appropriately as mole, weight, or volume fraction or concentration in the second phase divided by that in the first phase. Similar reasoning would apply to any other type of process obeying the assumptions of straight operating line and straight equilibrium line.

It is often convenient to handle problems involving a large number of stages but without straight operating lines and/or equilibrium lines by dividing the cascade into sections of stages over which straight operating and equilibrium lines are a good assumption, i.e., using successive linearizations. An example of such a problem would be a binary distillation with α_{ij} close to 1 but with $\alpha_{ij} - 1$ varying appreciably. Such a problem cannot be handled with good precision by the constant-α Underwood equations developed later in this chapter but can be handled well by successive applications of the KSB equations over short ranges of composition.

Example 8-1 Use a group-method approach to solve Example 6-2.

SOLUTION The reader should first review the original statement and solution of Example 6-2. The problem concerns a two-stage washing process which removes $Fe_2(SO_4)_3$ solution from insoluble solid crystals. From Fig. 6-4 the stage exit compositions are given by $C_S = C_F$. The operating line is linear when we work on a volumetric flow basis; $C_{S, in}$ is specified as 935 kg/m³, $C_{S, out}$ is specified as 57.6 kg/m³, and $C_{F, in}$ is zero. Hence

$$\frac{C_{S, out} - C^*_{S, out}}{C_{S, in} - C^*_{S, out}} = \frac{57.6}{935} = 0.0616$$

Since the stage-exit-composition relationship and the operating-line relationship are both linear in terms of these flow rates and composition parameters, we can use Eqs. (8-13) to (8-15) or Fig. 8-3 for the solution with the substitution of C_S for y_A, C_F for x_A, cubic meters filtrate retained for V, cubic meters filtrate passing through for L, and $m = 1$.

Since $N = 2$ is specified, we can use Fig. 8-3, with interpolation, to obtain

$$\frac{m^3 \text{ filtrate passing through}}{(m)\,(m^3 \text{ filtrate retained})} = 3.5$$

Since $m = 1$, we have (filtrate retained)/(filtrate passing through) = 0.29, which compares with the value of 0.287 obtained in Example 6-2 by means of the graphical construction. □

Example 8-2 A plate tower providing six equilibrium stages is employed for removing ammonia from a waste-water stream by means of countercurrent stripping at atmospheric pressure and 27°C into a recycle airstream. (a) Calculate the concentration of ammonia in the exit water if the inlet liquid concentration is 0.1 mol% ammonia in water, the inlet air is free of ammonia, and 2.30 kg of air is fed to the tower per kilogram of waste water. Treybal (1968) gives $K = y/x$ at equilibrium = 1.41 at 27°C and atmospheric pressure, at high dilution. (b) Repeat part (a) if the inlet air now contains 1.0×10^{-5} mole fraction (10 ppm v/v) ammonia. (c) Repeat part (a) if the tower provides 10 actual stages with $E_{MV} = 0.45$.

SOLUTION (a) The following variables are fixed:

$$N = 6 \qquad y_{A, in} = 0 \qquad x_{A, in} = 0.001 \qquad b = 0$$

$$m = \left(\frac{y_A}{x_A}\right)_{eq} = 1.41$$

$$\frac{V}{L} = \frac{(2.30 \text{ kg air/kg } H_2O)(18 \text{ kg } H_2O/kmol)}{29 \text{ kg air/kmol}} = 1.43 \text{ mol air/mol } H_2O$$

$$\frac{mV}{L} = (1.43)(1.41) = 2.02$$

Since V, L, and m are constant, the resultant operating-line and equilibrium expressions will both be linear.

If we use Eq. (8-14) we solve for $y_{A, out}$ directly. We can do this since mV/L, N, $y_{A, in}$, $x_{A, in}$, and hence $y^*_{A, out}$ are all known. The problem calls for us to find $x_{A, out}$, however. In principle, we can find $x_{A, out}$ once we know $y_{A, out}$ by using

$$L(x_{A, in} - x_{A, out}) = V(y_{A, out} - y_{A, in})$$

rearranged in the form

$$x_{A, out} = x_{A, in} - \frac{V}{L}(y_{A, out} - y_{A, in})$$

However, since $x_{A,\text{out}}$ is much smaller than $x_{A,\text{in}}$, it is not possible to obtain adequate precision this way. It is more effective to use

$$\frac{x_{A,\text{out}} - x^*_{A,\text{out}}}{x_{A,\text{in}} - x^*_{A,\text{out}}} = \frac{1 - (mV/L)}{1 - (mV/L)^{N+1}} \tag{8-18}$$

which enables us to solve for $x_{A,\text{out}}$ directly. Precision is gained since $x_{A,\text{out}}$ is close to $x^*_{A,\text{out}}$, and it is therefore important to use a form of the equation which will give us the difference between $x_{A,\text{out}}$ and $x^*_{A,\text{out}}$.

Substituting into Eq. (8-18) we have, since $x^*_{A,\text{out}} = 0$,

$$\frac{x_{A,\text{out}}}{0.001} = \frac{1 - 2.02}{1 - (2.02)^7} = \frac{1.02}{136} = 0.0075 \quad \text{and} \quad x_{A,\text{out}} = 7.5 \times 10^{-6}$$

Figure 8-3 could also have been used in the x form. For a parameter mV/L of 2.02 and six equilibrium stages, the vertical coordinate is 0.0075, as found above.

(b) $y_{A,\text{in}}$ is now changed to 1.0×10^{-5}, and $x^*_{A,\text{out}}$ thereby becomes $(1.0 \times 10^{-5})/1.41$, or 0.71×10^{-5}. Since the right-hand side of Eq. (8-18) from part (a) is unchanged,

$$\frac{x_{A,\text{out}} - x^*_{A,\text{out}}}{x_{A,\text{in}} - x^*_{A,\text{out}}} = \frac{x_{A,\text{out}} - (0.71 \times 10^{-5})}{0.001 - (0.71 \times 10^{-5})} = 0.0075$$

$$x_{A,\text{out}} = (0.0075)(0.000993) + (0.71 \times 10^{-5})$$

$$= (0.74 \times 10^{-5}) + (0.71 \times 10^{-5}) = 1.45 \times 10^{-5}$$

This small amount of ammonia in the inlet air very nearly doubles the residual ammonia content of the effluent water.

(c) We want the x form of the KSB equation, with E_{MV} included. This leads to Eq. (8-20) with the denominator changed through Eq. (8-17). Denoting $(x_{A,\text{out}} - x^*_{A,\text{out}})/(x_{A,\text{in}} - x^*_{A,\text{out}})$ by R, we have

$$N_{\text{act}} = \frac{\ln\{[1 - (L/mV)](1/R) + (L/mV)\}}{\ln\{1 + E_{MV}[(mV/L) - 1]\}}$$

$$10 = \frac{\ln\{[1 - (1/2.02)](1/R) + (1/2.02)\}}{\ln[1 + 0.45(2.02 - 1)]}$$

$$= \frac{\ln[0.505(1/R) + 0.495]}{\ln 1.459}$$

$$\ln\left(\frac{0.505}{R} + 0.495\right) = (10)(0.3778) = 3.778$$

$$\frac{0.505}{R} + 0.495 = 43.71$$

$$\frac{0.505}{R} = 43.2$$

$$R = \frac{0.505}{43.2} = 0.0117 = x_{A,\text{out}}/x_{A,\text{in}}$$

since $x^*_{A,\text{out}} = 0$. Hence

$$x_{A,\text{out}} = (0.001)(0.0117) = 1.17 \times 10^{-5} \qquad \square$$

Multiple-section cascades As was discussed in the text surrounding Figs. 4-23, 4-24, and 7-23, two-section cascades, e.g., fractional extraction or absorber-strippers, are used when we want to fractionate two solutes with high recovery fractions of each in two different products. In a single-section cascade the product leaving at the end where the feed enters must usually contain substantial amounts of all solutes present. When the solutes are dilute and equilibrium distribution ratios and flow rates are constant within a section, multiple-section countercurrent cascades can be analyzed by algebraic combinations of various forms of the KSB equations.

Brian (1972) has explored forms and applications of the KSB equations for multiple-section countercurrent cascades and presents equations for two useful sub-cases, both in terms of equilibrium stages, as follows.

Case 1 Refer to the two-section separation described in vapor-liquid nomenclature in Fig. 8-5 and suppose that the amount of feed is significant compared with the

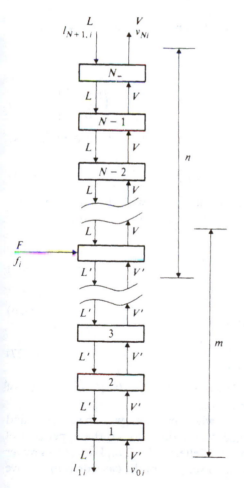

Figure 8-5 Two-section countercurrent staged cascade.

vapor and liquid flows. Therefore either or both the vapor and liquid flow rates change at the feed point. L and L' represent the liquid flows above and below the feed, respectively, and V and V' represent the vapor flows above and below the feed, respectively. It is assumed that the gas and liquid entering the end stages do not contain any of the solutes being separated. n is the number of equilibrium stages above the feed (counting the feed stage), m is the number of equilibrium stages below the feed (counting the feed stage), and $N = n + m - 1$. The ratio of the amount of a solute issuing in the vapor product overhead v_{Ni} to the amount of that solute issuing in the liquid product below l_{1i} is then given by

$$\frac{v_{Ni}}{l_{1i}} = \frac{(m_i V/L)[(m_i V'/L')^m - 1][1 - (L/m_i V)]}{[(m_i V'/L') - 1][1 - (L/m_i V)^n]} \tag{8-22}$$

Case 2 Suppose now that the feed is small compared with the total vapor and liquid flows, and hence $V = V'$ and $L = L'$. However, now f_i mol of solute i enters in the main feed, while v_{0i} and $l_{N+1,i}$ mol of i can also enter in the gas and liquid, respectively, entering the end stages. The moles of solute i issuing in the vapor-product overhead v_{Ni} are now related to the other feeds of component i as follows:

$$v_{Ni}\left[\frac{L}{m_i V} - \left(\frac{m_i V}{L}\right)^N\right] = f_i\left[\left(\frac{m_i V}{L}\right)^{n-1} - \left(\frac{m_i V}{L}\right)^N\right] + \cdots$$

$$+ l_{N+1,i}\left[\left(\frac{m_i V}{L}\right)^{N-1} - \left(\frac{m_i V}{L}\right)^N\right] + v_{0i}\left[1 - \left(\frac{m_i V}{L}\right)^N\right] \tag{8-23}$$

The amount of i leaving in the bottom liquid product comes from an overall mass balance

$$l_{1i} = f_i + l_{N+1,i} + v_{0i} - v_{Ni} \tag{8-24}$$

A subcase of both these cases occurs where no solute enters with the inlet vapor and liquid at either end of the cascade and the feed flow is small compared with both stream flows. In that case

$$v_{Ni} = \frac{(m_i V/L)^{n-1} - (m_i V/L)^N}{(L/m_i V) - (m_i V/L)^N} f_i \tag{8-25}$$

$$l_{1i} = \frac{(L/m_i V)^{m-1} - (L/m_i V)^N}{(m_i V/L) - (L/m_i V)^N} f_i \tag{8-26}$$

and

$$\frac{v_{Ni}}{l_{1i}} = \left(\frac{m_i V}{L}\right)^n \frac{(m_i V/L)^m - 1}{(m_i V/L)^n - 1} \tag{8-27}$$

Equation (8-27) is a combination of Eqs. (8-25) and (8-26); Eq. (8-25) is a subcase of Eq. (8-23); and Eq. (8-27) is a subcase of Eq. (8-22).

If components i and j are to be fractionated, where $m_i > m_j$, we can set upper and lower limits on the V/L or V'/L' ratio for effective fractionation. If the upper part of the cascade is to remove j from the upper product effectively, $L/m_j V$ must be greater than 1. This sets an upper bound on V/L. If the lower part of the cascade is to remove

i from the bottom product effectively, $m_i V'/L'$ must be greater than 1. This sets a lower limit on V'/L'. Furthermore, if the feed flow is significant, either V must be greater than V' and/or L' must be greater than L; hence V/L must be greater than V'/L'. We therefore have $1/m_j > V/L \geq V'/L' > 1/m_i$.

The optimum value of V/L will lie somewhere between these extremes. We can determine the optimum for the fully symmetrical case where there are the same number of stages above the feed as below, where the recovery fraction of i overhead equals that of j at the bottom, and where the feed flow is small compared with the counterflowing stream flows $(V/L \approx V'/L')$, as follows. For $n = m = (N + 1)/2$, Eq. (8-27) becomes

$$\frac{v_{Ni}}{l_{1i}} = \left(\frac{m_i V}{L}\right)^{(N+1)/2} \tag{8-28}$$

For the case of equal recovery fractions of i and j above and below, Eq. (8-28) written for i times Eq. (8-28) written for j must equal unity:

$$\left[m_i m_j \left(\frac{V}{L}\right)^2\right]^{(N+1)/2} = 1 \tag{8-29}$$

This leads to

$$\frac{L}{V} = (m_i m_j)^{1/2} \tag{8-30}$$

which makes L/V the geometric mean between the two limits and makes $m_i V/L = L/m_j V$.

Both components i and j are being stripped from the liquid below the feed and are being absorbed from the vapor above the feed in the process of Fig. 8-5. The fractionation comes from the fact that j is being absorbed much more effectively $(L/m_j V > 1 > L/m_i V)$ above the feed and i is being stripped much more effectively $(m_i V'/L' > 1 > m_j V'/L')$ below the feed. However, because of the simultaneous stripping and absorption of both components, there will be a buildup of solute concentrations at the feed stage to values greater than those present at either end of the cascade. The amount of this buildup can be determined by applying the single-section KSB equations to each of the sections separately. The degree of solute buildup near the feed is larger when m_i/m_j is nearer unity (and hence V/L and V'/L' are nearer unity) and when the number of stages is large.

Example 8-3[†] Streptomycin is a pharmaceutical product used for the treatment of tuberculosis, meningitis, urinary-tract infections, and other diseases. It is manufactured by aerobic fermentation from soybean meal, glucose, and other nutrients. Separation of the products from the fermentation is complex (Perlman, 1969). As a final step it is necessary to isolate streptomycin A (the active product) from streptomycin B (mannosidostreptomycin), the molecular structures of which are shown in Fig. 8-6. (a) A mixture of streptomycin A and streptomycin B is to be fractionated by dual-solvent extraction in a series of centrifuges, as shown in Fig. 8-7. Each centrifuge gives equilibrium between

---

† Adapted from Belter (1977), courtesy of Mr. Belter.

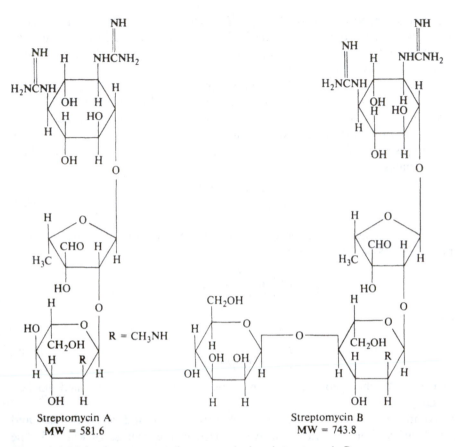

Streptomycin A
MW = 581.6

Streptomycin B
MW = 743.8

Figure 8-6 Molecular structures of streptomycin A and streptomycin B.

the product streams. The entering amyl acetate and aqueous buffer solution are both free of the streptomycins. The aqueous feed volume is negligible. The solvent flows are such that SK_A/W and SK_B/W are 1.50 and 0.45 for streptomycins A and B, respectively. S is the mass flow of amyl acetate, W the mass flow of aqueous buffer solution, and K_A and K_B equilibrium distribution coefficients, expressed as weight fraction in the amyl acetate phase per weight fraction in the aqueous phase. Calculate the recovery fractions of streptomycins A and B in the two product streams. (b) Calculate the solute buildups of the two streptomycins, compared with the concentration in the product where each principally appears. (c) Compare the recovery fractions with those which would have been

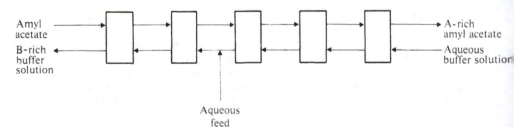

Amyl acetate

A-rich amyl acetate

B-rich buffer solution

Aqueous buffer solution

Aqueous feed

Figure 8-7 Fractional-extraction system for streptomycins.

obtained if the mixed-streptomycin feed had entered with the inlet aqueous buffer solution. (*d*) How would the results of part (*a*) change if 20 percent of the original aqueous buffer solution were used to solvate the feed and enter with it and if the end feed rate of aqueous buffer solution were decreased by only 10 percent to compensate for this change?

SOLUTION (*a*) We make the amyl acetate stream analogous to the vapor in previous examples and make the aqueous stream analogous to the liquid. Furthermore, weight fractions and mass flow rates will be used in the KSB equations. Because the feed enters the second stage from the left, $m = 2$, $n = 4$, and $N = 5$.

For streptomycin A the applicable expression is

$$(/A)_S = \frac{(1.50)^{4-1} - (1.50)^5}{(1/1.50) - (1.50)^5} = 0.609 \tag{8-25}$$

Equations (8-26) or (8-27) would give somewhat more precision, since the smaller driving force is at the acetate-inlet end.

For streptomycin B Eqs. (8-25) or (8-27) should be used because the recovery fraction will be high and the smaller driving force is at the aqueous-inlet end. Using Eq. (8-27), we have

$$\frac{(/B)_S}{1 - (/B)_S} = (0.45)^4 \frac{(0.45)^2 - 1}{(0.45)^4 - 1} = 0.0341$$

$$(/B)_S = \frac{0.0341}{1 + 0.0341} = 0.0330$$

$$(/B)_W = 1 - (/B)_S = 1 - 0.0330 = 0.970$$

The recovery fraction of streptomycin B is large because the design of the process has been pushed toward more complete removal of B from A than of A from B by making the number of stages to the right of the feed greater than to the left and by making $W/K_B S$ greater than $K_A S/W$.

(*b*) For streptomycin A we can use Eq. (8-14) in the form

$$\frac{w_{AS,\,out}}{w_{AS,\,f}} = \frac{1 - (W/K_A S)}{1 - (W/K_A S)^n}$$

where $w_{AS,\,f}$ is the weight fraction A in the solvent stream leaving the feed stage, to give

$$\frac{w_{AS,\,out}}{w_{AS,\,f}} = \frac{1 - (1/1.5)}{1 - (1/1.5)^4} = \frac{1}{2.41}$$

Therefore streptomycin A builds up to a concentration 2.41 times its concentration in the acetate product.

For streptomycin B, adaptation of Eq. (8-18) gives

$$\frac{w_{BW,\,out}}{w_{BW,\,f}} = \frac{1 - (K_B S/W)}{1 - (K_B S/W)^2} = \frac{1 - 0.45}{1 - (0.45)^2} = \frac{1}{1.45}$$

so that streptomycin B builds up to 1.45 times its concentration in the aqueous product.

(*c*) The process is now a five-stage single-section cascade. The operation is analogous to a stripping column, and so we use the *x* form of Fig. 8-3, converting the parameters so that the vertical coordinate is $w_{iW,\,out}/w_{iW,\,F}$ and the parameter is $K_i S/W$. For streptomycin A, $K_A S/W = 1.50$; combining this with $N = 5$, we get $w_{AW,\,out}/w_{AW,\,F} = 0.049$, so that $(/A)_W = 0.049$. For streptomycin B the operation approaches the asymptote; $K_B S/W = 0.45$, which for $N = 5$ gives $w_{BW,\,out}/w_{BW,\,F} = (/B)_W = 0.56$. The shift in feed location considerably improves the recovery fraction of streptomycin A into the acetate product, but now streptomycin B is poorly removed from that product.

(d) We now use an adaptation of Eq. (8-22). For the given changes we have W less by 10 percent and W' greater by 10 percent, so that

Streptomycin	$K_i S/W$	$K_i S'/W'$	$K_i S/W'$
A	1.65	1.35	1.35
B	0.495	0.405	0.405

$$\frac{(/A)_S}{1 - (/A)_S} = \frac{(1.35)[(1.35)^2 - 1][1 - (1/1.65)]}{(1.35 - 1)[1 - (1/1.65)^4]} = 1.44$$

Hence

$$(/A)_S = \frac{1.44}{2.44} = 0.591$$

$$\frac{(/B)_S}{1 - (/B)_S} = \frac{(0.405)[(0.405)^2 - 1][1 - (1/0.495)]}{(0.405 - 1)[1 - (1/0.495)^4]} = 0.0371$$

$$(/B)_S = \frac{0.0371}{1.0371} = 0.0357$$

$\square$

Chromatographic Separations

In this section we develop group methods for the analysis of peak shape and degree of separation between products in both Craig countercurrent distribution (CCD) and elution chromatography. The development is closely related to that given by Keulemans (1959).

Intermittent carrier flow Figure 8-8 recalls the process of countercurrent distribution, examples of which were presented in Figs. 4-37 to 4-39 and which was discussed at that point. At discrete intervals transfers of the upper phase take place from one

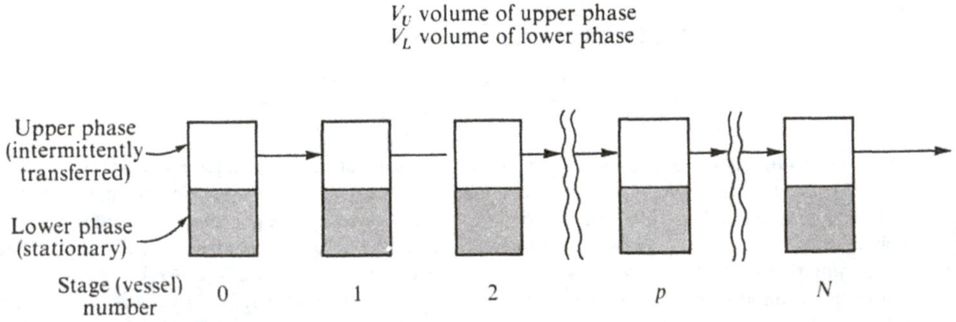

V_U volume of upper phase
V_L volume of lower phase

Figure 8-8 Countercurrent distribution (CCD).

vessel to the next, going from left to right in Fig. 8-8. Between these transfer steps the upper phase then present in each vessel is equilibrated with the lower phase in that vessel. A small amount of feed mixture is initially present in the left-hand stage and is carried along from vessel to vessel in the distribution process.

Let us consider the case where the solutes being separated have constant equilibrium ratios: $K_i' =$ (concentration of component i in upper phase)/(concentration of i in lower phase). The volume of the upper phase in each vessel will be V_U, and the volume of the lower phase in each vessel will be V_L. As a result, we know that the ratio of the amount of component i in the upper phase to the amount in the lower phase of that vessel at equilibrium is given by

$$\frac{\text{Moles } i \text{ in upper phase}}{\text{Moles } i \text{ in lower phase}} = \frac{C_{iU} V_U}{C_{iL} V_L} = \frac{K_i' V_U}{V_L} \tag{8-31}$$

Therefore, the fraction f_i of the total amount of i in any vessel that is present in the upper phase of that vessel at equilibrium is given by

$$f_i = \frac{K' V_U / V_L}{1 + (K' V_U / V_L)} \tag{8-32}$$

and the fraction present in the lower phase is, by difference,

$$1 - f_i = \frac{1}{1 + (K' V_U / V_L)} \tag{8-33}$$

Note that f_i is independent of the vessel number but is different for components with different K_i'.

The moles of component i within vessel or stage p at any time will be designated M_{ip}. During any transfer step s we shall transfer into stage p an amount of component i equal to the amount in the upper phase of stage $p - 1$ after the last equilibration. At the same time an amount of i equal to the amount in the lower phase of stage p after the last equilibration remains behind in p. Hence we can write the following recursion expression relating the amount of i in stage p after transfer s (M_{ips}) to the amounts of i in stages p and $p - 1$ after transfer $s - 1$ ($M_{ip, s-1}$ and $M_{i, p-1, s-1}$):

$$M_{ips} = f_i M_{i, p-1, s-1} + (1 - f_i) M_{ip, s-1} \tag{8-34}$$

Furthermore, as a starting condition for applying Eq. (8-35) successively, we know that

$$M_{i00} = M_{iF} \tag{8-35}$$

and

$$M_{ip0} = 0 \qquad \text{for } p > 0 \tag{8-36}$$

Here stage 0 ($p = 0$) is the left-handmost stage, where an amount of i given by M_{iF} is initially put as part of the feed. The subscript 0 refers to the time before the first transfer step.

In another way of looking at the problem, we can consider the distribution of component i among stages after s transfer steps and equilibrations through a probability analysis. In order to have reached stage p after s transfers, a solute molecule must have been taken to the succeeding stage in the upper phase on exactly p of the transfer steps and must have been left behind in the lower phase on $s - p$ of the transfer steps. On the other hand, we know that the probability of a solute molecule being in the upper phase and hence transferring in the next step is simply f_i, the fraction of total solute in a stage that is in the upper phase at equilibrium. The ratio M_{ips}/M_{iF} giving the fraction of the amount of component i fed which appears in stage p after s transfers is simply the probability $[P_i(p)]_s$ that a molecule of i has made p jumps during s transfer steps. Such a probability is described by the *binomial distribution* (Wadsworth and Bryan, 1960; etc.):

$$\frac{M_{ips}}{M_{iF}} = [P_i(p)]_s = C(s, p) f_i^p (1 - f_i)^{s-p} \qquad (8\text{-}37)$$

Here f_i, the probability of a jump in any one transfer, is given by Eq. (8-32) in our case. The *binomial coefficient* $C(s, p)$ represents the number of different combinations which can be made from s distinct objects taken p at a time; $C(s, p)$ is given by

$$C(s, p) = \frac{s!}{p!\,(s - p)!} \qquad (8\text{-}38)$$

Hence Eq. (8-37) becomes

$$\frac{M_{ips}}{M_{iF}} = \frac{s!}{p!\,(s - p)!} f_i^p (1 - f_i)^{s-p} \qquad (8\text{-}39)$$

Equation (8-39) evaluated for all values of p from 0 to s will give the distribution of component i after s transfers in CCD. Equation (8-39) can be evaluated for other components (j, k, etc.) which have different values of K and hence different values of f, and from the resulting distributions the degree of separation between components can be judged.

Example 8-4 Use Eq. (8-39) to verify the distribution of components A and B shown after three transfers in Fig. 4-37. Recall that $K'_A = 2.0$ and $K'_B = 0.5$, and that the upper and lower phases have equal volumes.

SOLUTION This problem involves three CCD transfers; hence $s = 3$. $M_{iF} = 100$ mg/L for both A and B. Hence Eq. (8-39) becomes

$$M_{ip3} = \frac{100(3!)}{p!\,(3 - p!)} f_i^p (1 - f_i)^{3-p} \qquad (8\text{-}40)$$

For component A, $K'_A = 2.0$. Since $V_U = V_L$, Eq. (8-32) gives

$$f_A = \frac{2}{1 + 2} = \frac{2}{3}$$

Hence for component A, Eq. (8-40) becomes

$$M_{Ap3} = \frac{100(6)}{p!\,(3 - p!)} \left(\frac{2}{3}\right)^p \left(\frac{1}{3}\right)^{3-p} \qquad (8\text{-}41)$$

Evaluating Eqs. (8-41) for $p = 0$, 1, 2, and 3, we have

$$M_{A0,3} = \frac{100(6)}{1(6)}\left(\frac{2}{3}\right)^0\left(\frac{1}{3}\right)^3 = \frac{100}{27} = 3.7 \text{ mg/L in vessel 1} \qquad p = 0$$

$$M_{A1,3} = \frac{100(6)}{1(2)}\left(\frac{2}{3}\right)^1\left(\frac{1}{3}\right)^2 = \frac{200}{9} = 22.2 \text{ mg/L in vessel 2} \qquad p = 1$$

$$M_{A2,3} = \frac{100(6)}{2(1)}\left(\frac{2}{3}\right)^2\left(\frac{1}{3}\right)^1 = \frac{400}{9} = 44.4 \text{ mg/L in vessel 3} \qquad p = 2$$

$$M_{A3,3} = \frac{100(6)}{6(1)}\left(\frac{2}{3}\right)^3\left(\frac{1}{3}\right)^0 = \frac{800}{27} = 29.6 \text{ mg/L in vessel 4} \qquad p = 3$$

These results agree with those shown in the last two rows of Fig. 4-37.
Similarly for component B, $K_B' = 0.50$, $f_B = \frac{1}{3}$:

$$M_{B0,3} = \frac{100(6)}{1(6)}\left(\frac{1}{3}\right)^0\left(\frac{2}{3}\right)^3 = \frac{800}{27} = 29.6 \qquad M_{B1,3} = \frac{100(6)}{1(2)}\left(\frac{1}{3}\right)^1\left(\frac{2}{3}\right)^2 = \frac{400}{9} = 44.4$$

$$M_{B2,3} = \frac{100(6)}{2(1)}\left(\frac{1}{3}\right)^2\left(\frac{2}{3}\right)^1 = \frac{200}{9} = 22.2 \qquad M_{B3,3} = \frac{100(6)}{6(1)}\left(\frac{1}{3}\right)^3\left(\frac{2}{3}\right)^0 = \frac{100}{27} = 3.7 \qquad \square$$

Continuous carrier flow The first effective picture of zone spreading and quality of separation in chromatographic processes came through the use of an idealized equilibrium-stage model (Martin and Synge, 1941). As pictured conceptually in Fig. 8-9, this model depicts a gas-liquid chromatographic process as a succession of well-mixed equilibrium stages. The nonvolatile liquid absorbent (stationary phase) is contained within each stage, and the carrier gas passes continuously through the stages in series, carrying the volatile solutes being separated from each other along from stage to stage. Usually gas-liquid chromatography involves the liquid absorbent being distributed evenly along a continuous length of a solid support; there are no discrete stages. In that sense the equilibrium-stage model is a poorer physical representation than other models based upon diffusional analysis, random-walk analysis, etc., as applied to a continuous length of stationary phase (Giddings, 1965, etc.). Nonetheless, the equilibrium-stage model is widely employed and leads to a useful conceptual picture of chromatography. It is presented briefly here.

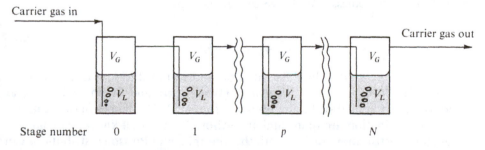

Figure 8-9 Equilibrium-stage model of chromatography.

It is interesting to note that the difference between the equilibrium-stage model for chromatography and the CCD model lies in the continuous flow of carrier phase in the chromatography model as contrasted to the transfers of the carrier phase at discrete intervals in CCD. In other respects the models (and the processes) are alike.

Considering the equilibrium-stage model of Fig. 8-9, let us denote the volume of the gas (carrier) phase within each vessel by V_G and the volume of each liquid phase by V_L. Once again the equilibrium ratio of solute component i between phases will be K_i'. Using C_{iG} and C_{iL} for the concentrations (moles per volume) of component i in the gas and liquid, respectively, we have for any stage p

$$C_{iG,\,p} = K_i' C_{iL,\,p} \tag{8-42}$$

Let us next consider the transfer of an amount of carrier gas dV from stage $p - 1$ to stage p, and the simultaneous transfer of the same amount of carrier from stage p to $p + 1$, $p + 1$ to $p + 2$, etc. We can write a differential mass balance for input − output = accumulation for stage p as

$$\underbrace{C_{iG,\,p-1}\,dV}_{\text{Input}} - \underbrace{C_{iG,\,p}\,dV}_{\text{Output}} = \underbrace{V_G\,dC_{iG,\,p} + V_L\,dC_{iL,\,p}}_{\text{Accumulation}} \tag{8-43}$$

Differentiating Eq. (8-42) and substituting into Eq. (8-43) gives

$$\frac{dC_{iG,\,p}}{dV} = \frac{C_{iG,\,p-1} - C_{iG,\,p}}{V_G + (V_L/K_i')} \tag{8-44}$$

The initial conditions for the various stages corresponding to the injection of a pulse of feed containing F_i mol of i into stage 0 at a cumulative carrier gas volumetric flow of zero are

$$C_{iG,\,0} V_G + C_{iL,\,0} V_L = C_{iG,\,0}\left(V_G + \frac{V_L}{K_i'}\right) = F_i \tag{8-45}$$
$$\text{at } V = 0$$

$$C_{iG,\,p} = 0 \quad \text{for } p > 0$$

The solution of Eq. (8-44) with the initial condition given by Eq. (8-45) is

$$C_{iG,\,p} = \frac{F_i}{V_G + (V_L/K_i')} \frac{e^{-v_i} v_i^p}{p!} \tag{8-46}$$

where v_i is a dimensionless cumulative carrier gas flow,

$$v_i = \frac{V}{V_G + (V_L/K_i')} \tag{8-47}$$

This solution can be verified by differentiation and resubstitution into Eq. (8-44).

Equation (8-46) is a *Poisson distribution* (Wadsworth and Bryan, 1960; etc.), as opposed to the binomial distribution obtained for CCD. The shapes of the peaks for the Poisson distribution are quite similar to those for the well-known gaussian, or error-function, distribution. In fact, both the binomial and Poisson distributions can be well approximated by the gaussian distribution for large values of p. For the

Poisson distribution at large values of p, the error-function approximation (Keulemans, 1959) is

$$C_{iG, p} = \frac{F_i}{V_G + (V_L/K_i')} \frac{1}{\sqrt{2\pi p}} e^{-(v_i - p)^2/2p} \qquad (8\text{-}48)$$

Equations (8-46) and (8-48) can be looked upon as discrete distributions of C_{iG} at various values of p for a fixed value of v_i (corresponding to positions along the column at a fixed point in time). They can also be looked upon as continuous functions relating C_{iG} to v_i for a fixed value of $p = N$ (corresponding to a record of the effluent-carrier-gas composition as a function of time issuing from a column of N equivalent equilibrium stages):

$$C_{iG} = \frac{F_i}{V_G + (V_L/K_i')} \frac{e^{-v_i}v_i^N}{N!} \qquad (8\text{-}49)$$

$$C_{iG} = \frac{F_i}{V_G + (V_L/K_i')} \frac{1}{\sqrt{2\pi N}} e^{-(v_i - N)^2/2N} \qquad (8\text{-}50)$$

When one wants to obtain the area up to a certain point under a peak measured in the carrier gas issuing from a chromatography column, the error-function integral

$$\text{erf}(x) = \sqrt{\frac{2}{\pi}} \int_0^x e^{-x^2/2} \, dx \qquad (8\text{-}51)$$

is required. This integral is tabulated in several references, e.g., Carslaw and Jaeger (1959). A related integral, the *normal-probability integral*, is also tabulated in a number of mathematical handbooks.

Several properties of chromatography peaks predicted by Eqs. (8-46) and (8-49) are of interest. As an example, Fig. 8-10 shows a plot of Eq. (8-49) for $N = 25$. The peak maximum is located at $v_i = N$ or, in this case, at $v_i = 25$. The inflection points of the curves, designated by the dashed tangent, are located at $v_i = N - \sqrt{N}$ and $v_i = N + \sqrt{N}$ or, in this case, at $v_i = 20$ and $v_i = 30$. The lines tangent to these inflection points intersect the base line (v axis) at $v_i = N + 1 - 2\sqrt{N}$ and $v_i = N + 1 + 2\sqrt{N}$, or 16 and 36 in this case. The *peak width w* often has been defined as the distance between the two points where these tangents strike the v_i axis; thus $w = 4\sqrt{N}$.

The prediction that the peaks tend to spread as the square root of the number of stages ($w \approx \sqrt{N}$) or as the square root of the length of flow path along the stationary phase has often been confirmed experimentally. However, this prediction is common to all theories for zone spreading in chromatography (Giddings, 1965).

It is also interesting to note that the carrier-gas throughput required to transport the peak maximum for component i a distance equal to one equilibrium stage along a chromatography column can be obtained by starting with Eq. (8-47) for $v_i = p$

$$p = \frac{V}{V_G + (V_L/K_i')} \qquad (8\text{-}52)$$

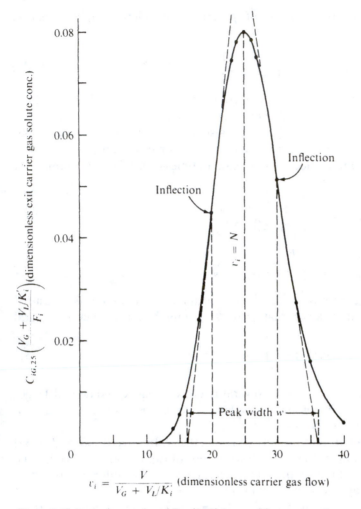

Figure 8-10 Peak shape; plot of Eq. (8-49) for $p = 25$.

where p is now the stage location of the peak maximum. Hence

$$\frac{\Delta p}{\Delta V} = \frac{1}{V_G + (V_L/K_i')} \tag{8-53}$$

is the rate of travel of the peak maximum as the cumulative carrier-gas flow increases. Since the gas-phase volume per equilibrium stage is equal to V_G, the ratio of the velocity of the peak maximum along the column u_i to the superficial carrier-gas velocity u_G is

$$\frac{u_i}{u_G} = \frac{V_G}{V_G + (V_L/K_i')} \tag{8-54}$$

This equation is identical to Eq. (4-8) for the relative peak velocity, except that K_i', based upon concentrations (moles per liter), is used instead of K_i, based upon mole fractions. As a result, Eq. (8-54) involves V_G and V_L, rather than M_G and M_L, which were used in Eq. (4-8). Thus we see that peak broadening does not alter the basic conclusion about peak velocity.

The number of equilibrium stages provided for each component by a chromatography column can be inferred from various properties of the effluent peaks, as shown in Fig. 8-10. For the various peak detectors used in commercial gas-chromatography and liquid-chromatography instruments, the amplitude of the response curve (the recorder output) is directly proportional to the solute concentration in the effluent gas or liquid, if it is dilute. Similarly, if the carrier flow rate is constant, the time elapsed since the sample injection is directly proportional to V. Hence the constructions to find peak width and such properties can be made directly on the recorder-output chart. When measured on the recorder output, the time t_i elapsed between the sample injection pulse and the emergence of the peak maximum of component i is $k_i N$, where k_i is the proportionality constant between time and v_i. Similarly the time lapse w_i' corresponding to the peak width is $4k_i\sqrt{N}$. If we take the ratio of the peak-width time difference to the elapsed time after sample injection, k_i will drop out

$$\frac{w_i'}{t_i} = \frac{4}{\sqrt{N}} \tag{8-55}$$

Equation (8-55) can be solved for the indicated number of equilibrium stages in the column

$$N = \left(\frac{4t_i}{w_i'}\right)^2 \tag{8-56}$$

For different components N may be different, because of the discrete-stage approximation. If the length of the chromatograph column is h, the length equivalent to an equilibrium stage H_S is equal to h/N. The number of equivalent equilibrium stages should be directly proportional to h; hence the H_S value should be a constant reflecting the column geometry, the flow conditions, and, to a lesser extent, the nature of individual components.

There have been numerous studies of the factors influencing values of H_S; Deemter et al. (1956) related H_S to the gas velocity through the equation

$$H_S = A + \frac{B}{u} + Cu \tag{8-57}$$

where A is an axial-dispersion term related to the geometry of the supporting bed of solids and directly proportional to the particle size, B is proportional to the solute diffusivity in the mobile phase and nearly equal to it, and C is proportional to the mass-transfer coefficient for the solute between phases. More refined analyses of this sort are covered by Giddings (1965).

One of the great advantages of gas-liquid chromatography is that H_S is typically on the order of 1 cm or even less. This is a result of the use of a thin stationary phase and reduced axial mixing. H_S is much less than values of equivalent stage height in continuous countercurrent packed absorption-columns which employ Raschig rings, etc., as a solid phase to disperse the downflowing liquid. Values of H_S in chromatography are also substantially less than the plate spacing required in plate columns. On the other hand, as Keulemans (1959) and others point out, the number of equilibrium stages required for a given degree of separation with a given separation factor is greater for chromatography than for countercurrent multistage contacting because only the stages in the vicinity of the peaks at any time are operative in improving the separation.

When the equilibrium relationship for solute partitioning is not a simple linear proportionality, the peak shape becomes nongaussian and the expression for peak retention volume or residence time changes. Approaches for describing such cases are outlined by De Vault (1943) and others.

Peak resolution In chromatographic separations it is desirable to obtain a good resolution between peaks for different components, i.e., to avoid peak overlap. The resolution between two adjacent peaks is related to two factors, the difference in average peak residence time and the amount of spreading of the individual peaks. By Eqs. (4-8) and (8-54) the difference in average residence time reflects the difference in equilibrium distribution ratios K_i' and K_j', it being necessary for the values of $K_i' V_G / V_L$ of at least one of the components to be of order unity or less in order to give appreciable slowing of the peak compared with the carrier velocity. By Eq. (8-56) the amount of spreading reflects the equivalent number of equilibrium stages, more stages giving sharper peaks.

In gas and liquid chromatography several different approaches can be used to increase the resolution between adjacent peaks:

1. *Use a longer column.* The difference in average residence times is directly proportional to N, since the residence times themselves are directly proportional to N. On the other hand, the peak spreads increase only as $N^{1/2}$, giving a ratio of peak spread to residence-time difference varying with $N^{-1/2}$ [Eq. (8-55)].
2. *Change the temperature.* If the peaks are not much delayed by the column, a lower temperature will delay the peaks more and tend to separate them. Also, K_i'/K_j' can change with temperature; for gas-liquid systems K_i'/K_j', written so as to be greater than 1, usually increases with decreasing temperature. For peaks delayed enough by the column for the second term in the denominator of Eq. (8-54) to control, a lower temperature can then increase resolution. On the other hand, the temperature should be high enough in gas chromatography for the peak to come through in a reasonable length of time. In liquid chromatography a change in the nature of the carrier liquid can have the same effect as a change in temperature does in gas chromatography.
3. *Program the temperature or the carrier composition.* In gas chromatography a low temperature early will serve to delay sufficiently peaks that would have come through without adequate delay at higher temperatures. Higher temperatures later in the run can then bring through peaks that would not have issued in a reasonable time at the lower temperature. In this way a mixture of components with a wide range of volatilities can be analyzed with good resolution over the entire range. In liquid-chromatography systems changing the carrier-liquid composition with respect to time can have the same effect.

4. *Change the column material.* The values of K_i' will depend upon the composition of the stationary phase. In gas chromatography with a liquid stationary phase, a relatively nonpolar stationary phase will tend to separate roughly in the sequence of boiling points of pure components, but a relatively polar stationary phase will serve to delay the more polar solute components to a greater extent than the less polar ones.

The resolution between peaks is usually only weakly affected by the carrier velocity, although Eq. (8-57) indicates that there should be an optimum intermediate value of velocity which will give minimum H_S and hence maximum N.

Example 8-5 Suppose that gas-liquid chromatography is to be used to separate a feed mixture of benzene and cyclohexane. The stationary phase will be polyethylene oxide 400 diricinoleate deposited to the extent of 25 wt % on 10/20-mesh firebrick. The column temperature will be 100°C, at which temperature $K_B' = 0.0133$ for benzene and $K_C' = 0.0270$ for cyclohexane (Barker and Critcher, 1960). Assume that the bed void fraction is 0.4 and that the solvent density and firebrick density are 100 and 2300 kg/m³, respectively. If the carrier-gas superficial velocity provides $H_S = 1.0$ cm, find the column length required to separate these components into two distinct peaks if the criterion of a good separation is (a) that the inflection tangents for the trailing edge of the first peak to emerge and the leading edge of the second peak meet at a common point on the v axis or (b) that the distance between the peak maxima should be at least equal to the *sum* of the peak widths for the individual peaks.

SOLUTION (a) Since K_i' is lower for benzene, it will be the peak to emerge last. Figure 8-11 shows two $N = 25$ peaks placed together by the criterion that the inflection-point tangents for the leading side of the benzene peak and the trailing side of the cyclohexane peak intersect at a common point of the v axis. There is some peak overlap, but two peaks are clearly observable.

The v-axis intersection point of the trailing inflection tangent of the cyclohexane peak emerges at

$$v_C = \frac{V}{V_G + (V_L/K_C')} = N + 1 + 2\sqrt{N} \tag{8-58}$$

Similarly, the v-axis intersection point of the leading inflection tangent of the benzene peak emerges at

$$v_B = \frac{V}{V_G + (V_L/K_B')} = N + 1 - 2\sqrt{N} \tag{8-59}$$

Since these two intersection points are to be coincident, V (the cumulative carrier-gas flow) in Eqs. (8-58) and (8-59) must be the same. Eliminating V from these equations gives

$$\frac{N + 1 + 2\sqrt{N}}{N + 1 - 2\sqrt{N}} = \frac{V_G + (V_L/K_B')}{V_G + (V_L/K_C')} \tag{8-60}$$

Dividing the numerator and denominator of the right-hand side of Eq. (8-60) by V_G gives

$$\frac{N + 1 + 2\sqrt{N}}{N + 1 - 2\sqrt{N}} = \frac{1 + (V_L/V_G K_B')}{1 + (V_L/V_G K_C')} \tag{8-61}$$

The groups on the right-hand side can be evaluated from the data given

$$\frac{V_L}{V_G} = \frac{1 - \text{void fraction}}{\text{void fraction}} \frac{0.25/1.0}{(0.25/1.0) + (0.75/2.3)} \frac{\text{m}^3 \text{ solvent}}{\text{m}^3 \text{ firebrick} + \text{solvent}}$$

$$= \frac{0.60}{0.40} \frac{0.25}{0.25 + 0.326}$$

$$= 0.65$$

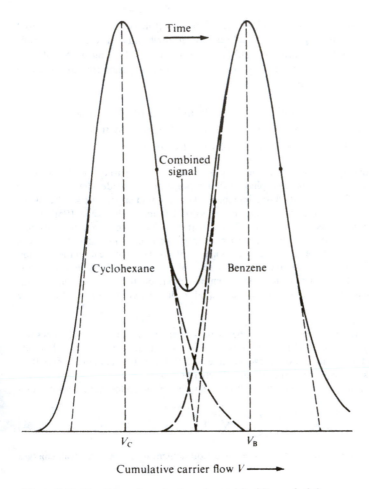

Cyclohexane

Benzene

Combined
signal

Time

V_C

V_B

Cumulative carrier flow V ⟶

Figure 8-11 Conditions for criterion of part (a) of Example 8-5.

Since the values of K'_i are so low, the right-hand side of Eq. (8-60) is nearly equal to $K'_C/K'_B = 0.0270/0.0133 = 2.03$. Hence

$$\frac{N + 1 + 2\sqrt{N}}{N + 1 - 2\sqrt{N}} = 2.03$$

or

$$1.03N - 6.06\sqrt{N} + 1.03 = 0$$

This equation can be solved by the quadratic formula to yield

$$\sqrt{N} = \frac{6.06 \pm \sqrt{36.7236 - 4.2436}}{2.06} = \frac{6.06 \pm 5.71}{2.06}$$

The plus sign corresponds to the practical answer (why?); hence

$$\sqrt{N} = \frac{6.06 + 5.71}{2.06} = 5.71 \quad \text{and} \quad N = 32.6$$

Since $H_S = 1.0$ cm, the column length required is $(1.0 \text{ cm})(32.6) = 32.6$ cm. This would be a relatively short length for a chromatography column; however, the separation is by no means complete, either.

(*b*) If the distance between peak maxima is the *sum* of the peak widths, the separation will be essentially complete. The peak maxima in Fig. 8-11 would be roughly twice as far apart, although the particular shapes shown correspond only to the case of $N = 25$.

The point on the v axis for the location of the cyclohexane peak maximum *plus* the peak width is

$$v_C = \frac{V}{V_G + (V_L/K_C)} = N + 4\sqrt{N} \qquad (8\text{-}62)$$

The point for the location of the benzene peak maximum *minus* the peak width is

$$v_B = \frac{V}{V_G + (V_L/K_B)} = N - 4\sqrt{N} \qquad (8\text{-}63)$$

Again the V's in Eqs. (8-62) and (8-63) refer to a common point because of the criterion stated for part (*b*). Hence we have

$$\frac{N + 4\sqrt{N}}{N - 4\sqrt{N}} = 2.03$$

Solving, we find

$$1.03N - 12.12\sqrt{N} = 0$$

$$\sqrt{N} = \frac{12.12}{1.03} = 11.76$$

$$N = 138.5$$

Since $H_S = 1.0$ cm, the column length is 138.5 cm. This is a more typical column length for a laboratory chromatograph. $\qquad \square$

NONLINEAR STAGE-EXIT RELATIONSHIPS AND VARYING FLOW RATES

Binary Countercurrent Separations: Discrete Stages

When flow rates vary from stage to stage and/or the stage-exit relationships are nonlinear, equations like (8-14) to (8-20) can still be employed in many cases over shorter ranges within a countercurrent separation cascade. Within each range of stages the flow bases would then be assumed constant and the stage-exit relationships would be approximated by linear equations.

In some cases the range of stages over which a group-method equation will apply can be increased or the entire separation can be represented by a single equation if the approach of Martin (1963) is employed.

Martin considers a succession of countercurrent discrete stages and suggests that the stage-exit relationship be represented in gas-liquid contacting notation by

$$y_{A,p} = a_1 x_{A,p} + a_2 x_{A,p} y_{A,p} + a_3 \qquad (8\text{-}64)$$

and that the operating-line relationship be given by

$$y_{A, p} = a_4 x_{A, p+1} + a_5 x_{A, p+1} y_{A, p} + a_6 \tag{8-65}$$

where a_6 is the net upward product of $A(P_{+A})$.

Although these equations appear to be somewhat unusual at first, they are realistic approximations in a number of instances. For example,

$$y_{A, p} = \frac{\alpha_{AB} x_{A, p}}{1 + (\alpha_{AB} - 1)x_{A, p}} \tag{1-12}$$

for a binary separation with a constant separation factor, reduces to Eq. (8-64) with

$$a_1 = \alpha_{AB} \qquad a_2 = 1 - \alpha_{AB} \qquad a_3 = 0$$

Also, the operating curves on a yx diagram for binary distillation with unequal latent heats and negligible sensible-heat and heat-of-mixing effects can be described by Eq. (8-65), as shown by Singh (1972) and others.

Equations (8-64) and (8-65) can be combined to form a Riccati nonlinear difference equation, which can be solved to relate $y_{A, in}$, $y_{A, out}$, and N (the number of intervening equilibrium stages). The solution is one of the following five equations, depending upon the characteristics of the problem.

$$N = \frac{\ln \dfrac{(y_{A, out} + A - E_2)(y_{A, in} + A - E_1)}{(y_{A, out} + A - E_1)(y_{A, in} + A - E_2)}}{\ln (E_1/E_2)} \tag{8-66}$$

$$N = \frac{E_1(y_{A, in} - y_{A, out})}{(y_{A, out} + A - E_1)(y_{A, in} + A - E_1)} \tag{8-67}$$

$$N = \frac{\theta_2 - \theta_1}{\theta} \tag{8-68}$$

$$N = \frac{\ln \dfrac{y_{A, out}(a_4 + a_2 a_6 - a_1 - a_3 a_5) - a_3 a_4 + a_1 a_6}{y_{A, in}(a_4 + a_2 a_6 - a_1 - a_3 a_5) - a_3 a_4 + a_1 a_6}}{\ln [(a_1 + a_3 a_5)/(a_4 + a_2 a_6)]} \tag{8-69}$$

$$N = \frac{y_{A, out} - y_{A, in}}{a_3 - a_6} \tag{8-70}$$

A, B, and C are new constants, defined as

$$A = - \frac{a_4 + a_2 a_6}{a_2 - a_5} \tag{8-71}$$

$$B = \frac{a_1 + a_3 a_5}{a_2 - a_5} \tag{8-72}$$

$$C = \frac{a_3 a_4 - a_1 a_6}{a_2 - a_5} \tag{8-73}$$

E_1 and E_2 are the two roots of the equation

$$E_{1,2} = \frac{A - B}{2} \pm \sqrt{\left(\frac{A + B}{2}\right)^2 - C} \qquad (8\text{-}74)$$

When $a_2 \neq a_5$, A, B, and C are finite. If the roots of Eq. (8-74) are real and unequal, i.e., if $(A + B)^2 > 4C$, the number of stages required for the separation is given by Eq. (8-66). If the roots of Eq. (8-64) are equal, i.e., if $(A + B)^2 = 4C$, the number of stages is given by Eq. (8-67). If the roots of Eq. (8-74) are complex, the number of stages is given by Eq. (8-68) with

$$\theta_1 = \arctan \frac{\beta}{y_{A,\text{in}} + (A + B)/2} \qquad (8\text{-}75)$$

$$\theta_2 = \arctan \frac{\beta}{y_{A,\text{out}} + (A + B)/2} \qquad (8\text{-}76)$$

$$\theta = \arctan \frac{2\beta}{A - B} \qquad (8\text{-}77)$$

$$i\beta = + \sqrt{\left(\frac{A + B}{2}\right)^2 - C} \qquad (8\text{-}78)$$

where i is $\sqrt{-1}$.

If $y_{A,\text{in}} + (A + B)/2$ is greater than 0, the angle θ_1 lies in the first quadrant; that is, $0 < \theta_1 < 90°$. If $y_{A,\text{in}} + (A + B)/2$ is negative, the angle θ_1 lies in the second quadrant; that is, $90° < \theta_1 < 180°$. Similarly, if $y_{A,\text{out}} + (A + B)/2$ is positive, the angle θ_2 lies in the first quadrant. If $y_{A,\text{out}} + (A + B)/2$ is negative, θ_2 is in the second quadrant. The sign (and quadrant) of θ is taken so as to make the number of stages positive and reasonable.

If $a_2 = a_5$ the number of stages required is given by Eq. (8-69), with Eq. (8-20) applying instead for the special case of $a_2 = a_5 = 0$ and $a_1 = a_4$.

These equations can be used with other composition parameters; also, y_A and x_A can be interchanged.

Some illustrations of the application of these equations are given in the following examples. Martin (1963) gives examples of the application of the method to distillation with varying relative volatility and varying molal overflow and to extraction with reflux.

An integral method for determining the stage requirement with arbitrary equilibrium and operating expressions is given by Pohjola (1975). The method requires an integration and uses the assumption that N is a continuous variable, which is valid for large numbers of stages where the method might be used.

Example 8-6 Determine the number of equilibrium stages for the atmospheric-pressure benzene-toluene distillation shown in Fig. 7-4. Use the equations of Martin.

SOLUTION The benzene-toluene distillation is one exhibiting a nearly constant relative volatility. Constant molal overflow can also be assumed with little error. The following conditions were specified in connection with Fig. 7-4:

$$\alpha_{BT} = \begin{cases} 2.38 & \text{at } x_B = 0 \\ 2.62 & \text{at } x_B = 1 \end{cases}$$

$$x_{B.F} = 0.50 \qquad \text{feed is saturated liquid}$$

$$x_{B.b} = 0.05 \qquad x_{B.d} = 0.95 \qquad r/d = 1.57 \qquad \text{feed at optimum location}$$

As a result $d/F = b/F = 0.50$.

Using Eqs. (8-64) and (8-65), we can obtain an exact representation of the operating and equilibrium curves provided we consider the rectifying and stripping sections separately. Relating Eq. (5-9) for a straight operating line in the stripping section to Eq. (8-65), we have

$$a_4 = \frac{L'}{V'} \qquad a_5 = 0 \qquad a_6 = -\frac{b}{V'} x_{B.b}$$

For the rectifying section

$$a_4 = \frac{L}{V} \qquad a_5 = 0 \qquad a_6 = \frac{d}{V} x_{B.d}$$

Taking $r/d = 1.57$, as specified, we have

$$\frac{L}{V} = \frac{r}{r+d} = \frac{1.57}{2.57} = 0.611 \qquad \frac{d}{V} = 0.389 \qquad \frac{d}{V} x_{B.d} = 0.3695$$

$$\frac{L'}{V'} = \frac{3.57}{2.57} = 1.389 \qquad \frac{b}{V'} = 0.389 \qquad \frac{b}{V'} x_{B.b} = 0.0195$$

As is developed during the discussion below of the Underwood equations, it is reasonable to take α_{BT} at the feed tray to be the geometric mean of the extreme values $= \sqrt{(2.62)(2.38)} = 2.50$. Hence $y_{B.out}$ from the stripping section will be

$$y_{B.out} = \frac{2.50}{1.50 + (1/0.5)} = 0.714$$

It is also logical to take $\alpha_{BT} = \sqrt{(2.50)(2.38)} = 2.44$ in the stripping section and $\alpha_{BT} = \sqrt{(2.62)(2.50)} = 2.56$ in the rectifying section. We can relate a_1, a_2, and a_3 to these α_{BT} values by means of the representation of Eq. (1-12) developed above. On this basis we have:

	Rectifying	Stripping
a_1	2.56	2.44
a_2	-1.56	-1.44
a_3	0	0
a_4	0.611	1.389
a_5	0	0
a_6	0.3695	-0.0195
$y_{A.in}$	0.714	0.050
$y_{A.out}$	0.950	0.714

The choice of $y_{A, in}$ for the stripping section means that the reboiler will be counted as a stage; the total condenser, of course, will not.

Starting with the stripping section, we find that $a_2 \neq a_s$. Hence we evaluate A, B, C, E_1, and E_2:

$$A = -\frac{1.389 + (-1.44)(-0.0195)}{-1.44 - 0} = 0.984 \qquad B = \frac{2.44 + 0}{-1.44 - 0} = -1.69$$

$$C = \frac{0 - (2.44)(-0.0195)}{-1.44 - 0} = -0.0330 \qquad \left(\frac{A + B}{2}\right)^2 - C = \left(\frac{0.984 - 1.69}{2}\right)^2 + 0.033 = 0.158$$

$$E_{1, 2} = \frac{0.984 + 1.69}{2} \pm 0.398 \qquad E_1 = 1.735 \qquad E_2 = 0.939$$

E_1 and E_2 are real and unequal; Eq. (8-66) applies:

$$N_S = \frac{\ln \dfrac{(0.714 + 0.984 - 0.939)(0.050 + 0.984 - 1.735)}{(0.714 + 0.984 - 1.735)(0.050 + 0.984 - 0.939)}}{\ln (1.735/0.939)}$$

$$= \frac{\ln [(0.759)(-0.701)/(-0.037)(0.095)]}{\ln 1.85} = 8.2$$

For the rectifying section, again $a_2 \neq a_s$, and we must evaluate A, B, C, E_1, and E_2:

$$A = -\frac{0.611 + (-1.56)(0.3695)}{-1.56 - 0} = 0.022 \qquad B = \frac{2.56 + 0}{-1.56 - 0} = -1.641$$

$$C = \frac{0 - (2.56)(0.3695)}{-1.56 - 0} = 0.608 \qquad \left(\frac{A + B}{2}\right)^2 - C = \left(\frac{0.022 - 1.641}{2}\right)^2 - 0.608 = 0.048$$

$$E_{1, 2} = \frac{0.022 + 1.641}{2} \pm 0.220 \qquad E_1 = 1.052 \qquad E_2 = 0.612$$

Again E_1 and E_2 are real and unequal and Eq. (8-66) is used:

$$N_R = \frac{\ln \dfrac{(0.950 + 0.022 - 0.612)(0.714 + 0.022 - 1.052)}{(0.950 + 0.022 - 1.052)(0.714 + 0.022 - 0.612)}}{\ln (1.052/0.612)}$$

$$= \frac{\ln [(0.360)(-0.316)/(-0.080)(0.124)]}{\ln 1.71} = 4.5$$

Thus $N_R + N_S = 12.7$ in good agreement with the result of 12 and a fraction stages (counting the reboiler) obtained graphically in Fig. 7-4. $\qquad\qquad\square$

Example 8-7 Rework the liquid-liquid extraction problem of Example 6-1 using the equations of Martin.

SOLUTION For this problem

C_o (mol Zr/L organic phase) replaces y_A
C_A (mol Zr/L aqueous phase) replaces x_A
S (flow of organic phase, L/h) replaces V
F (flow of aqueous phase, L/h) replaces L

Since the operating line is straight, the application of Eq. (8-65) is straightforward. The equilibrium curve is not straight and does not present a constant separation factor; thus we shall obtain the constants in Eq. (8-64) by curve fitting

$$C_{o,p} = a_1 C_{A,p} + a_2 C_{o,p} C_{A,p} + a_3 \tag{8-79}$$

$C_o = 0$ when $C_A = 0$. Hence $a_3 = 0$; a_1 and a_2 will be determined from two equilibrium points ($C_o = 0.147$, $C_A = 0.123$; $C_o = 0.083$, $C_A = 0.039$):

$$0.147 = 0.123 a_1 + (0.147)(0.123) a_2$$
$$0.083 = 0.039 a_1 + (0.083)(0.039) a_2$$
$$a_1 = 3.34 \qquad a_2 = -14.6$$

Checking another equilibrium point $C_o = 0.114$, $C_A = 0.074$, we have

$$C_o = (3.34)(0.074) - (14.6)(0.074)$$

$$C_o = \frac{0.247}{2.08} = 0.119$$

Thus the fit appears to be satisfactory but not perfect. a_4, a_5, and a_6 come from the mass-balance expression

$$a_4 = \frac{F}{S} = 1.111 \qquad a_5 = 0$$

$$a_6 = \frac{P_{+A}}{S} = C_{o,\,in} - \frac{F}{S} C_{A,\,out} = -(1.111)(0.012) = -0.0133$$

$$C_{o,\,in} = 0 \qquad C_{o,\,out} = 0.120$$

Solving for A, B, C, E_1, and E_2, we have

$$A = -\frac{1.111 + (-14.6)(-0.0133)}{-14.6} = 0.0894 \qquad B = \frac{3.34}{-14.6} = -0.229$$

$$C = -\frac{(3.34)(-0.0133)}{-14.6} = -0.00304$$

$$\left(\frac{A+B}{2}\right)^2 - C = \left(\frac{0.089 - 0.229}{2}\right)^2 + 0.00304 = 0.00794$$

$$E_{1,2} = \frac{0.0894 + 0.229}{2} \pm 0.089 \qquad E_1 = 0.248 \qquad E_2 = 0.070$$

The roots are real and unequal; hence once again we use Eq. (8-66)

$$N = \frac{\ln \dfrac{(0.120 + 0.089 - 0.070)(0 + 0.089 - 0.248)}{(0.120 + 0.089 - 0.248)(0 + 0.089 - 0.070)}}{\ln (0.248/0.070)} = \frac{\ln \left[(0.139)(-0.159)/(-0.039)(0.019) \right]}{\ln 3.54} = 2.7$$

This result is in good agreement with Fig. 6-2. □

From these two examples it is apparent that the equations of Martin are workable; however, it is not obvious that they effect any savings in time. The advantage of this approach becomes more marked when the separation is more difficult and a large number of stages is involved or when a large number of different operating conditions are to be studied and the effects of various independent variables are to be determined quantitatively. The benefits of the approach must be balanced against any approximations involved in fitting the operating and equilibrium curves. In separations with many stages it is often imperative to carry a large number of significant figures in these and other group-method equations.

CONSTANT SEPARATION FACTOR AND CONSTANT FLOW RATES

Separation processes involving unchanging flow rates and constant separation factors are typified by, and for the most part limited to, simple binary or multicomponent distillations. For this reason the following development and discussion is carried out in the context of distillation.

Binary Countercurrent Separations: Discrete Stages

For calculation of equilibrium-stage requirements in binary distillation processes the use of the group method developed in the following discussion requires two assumptions: the molar flows in the section considered must be constant from stage to stage, and the relative volatility must be constant through the section. These assumptions are more or less wrong depending on the system. For highly nonideal systems they are prohibitively in error. For systems which are relatively ideal but in which the components are quite different in volatility, the assumptions may again be in considerable error, but since such systems are easily separated and require few stages, the error is not serious from a practical standpoint. For systems which are difficult to separate and reasonably ideal the assumptions are generally good and the percentage error is relatively small. Hence, the equations can be usefully applied in many process calculations, particularly if caution is exercised in accepting the results as final design calculations.

Several different equations or sets of equations have been proposed as group methods for the calculation of equilibrium stages under these assumptions (Lewis, 1922; Ramalho and Tiller, 1962; Robinson and Gilliland, 1950; Smoker, 1938; Underwood, 1944, 1945, 1946, 1948; Murdoch, 1948; etc.). As has been pointed out, a special case of the Martin equations (Martin, 1963) can also be used for this purpose. Since all these equations necessarily reduce to each other, we need only consider one. The equations of Underwood (1944, 1945, 1946, 1948) will be developed here because (1) they are relatively simple to use, (2) they are useful for minimum reflux considerations to be taken up in Chap. 9, and (3) there is a logical extension of the equations to multicomponent systems.

Mass balances and equilibrium expressions can be written for the two components A and B about a plate in a rectifying section as follows:

$$\frac{L}{V} x_{A,n+1} + \frac{d}{V} x_{A,d} = y_{A,n} = \frac{\alpha_A x_{A,n}}{\alpha_A x_{A,n} + \alpha_B x_{B,n}} \tag{8-80}$$

$$\frac{L}{V} x_{B,n+1} + \frac{d}{V} x_{B,d} = y_{B,n} = \frac{\alpha_B x_{B,n}}{\alpha_A x_{A,n} + \alpha_B x_{B,n}} \tag{8-81}$$

α_A and α_B are relative volatilities with respect to some arbitrary basis. If the volatilities are referred to component B, then $\alpha_A = \alpha_{AB}$ and $\alpha_B = 1$. If Eq. (8-80) is multiplied by $\alpha_A/(\alpha_A - \phi)$ and Eq. (8-81) is multiplied by $\alpha_B/(\alpha_B - \phi)$, where ϕ is an as yet undefined parameter, and the resulting equations are added, there results

$$\frac{L}{V}\left(\frac{\alpha_A x_{A,n+1}}{\alpha_A - \phi} + \frac{\alpha_B x_{B,n+1}}{\alpha_B - \phi}\right) + \frac{d}{V}\left(\frac{\alpha_A x_{A,d}}{\alpha_A - \phi} + \frac{\alpha_B x_{B,d}}{\alpha_B - \phi}\right)$$
$$= \frac{[\alpha_A/(\alpha_A - \phi)]\alpha_A x_{A,n} + [\alpha_B/(\alpha_B - \phi)]\alpha_B x_{B,n}}{\alpha_A x_{A,n} + \alpha_B x_{B,n}} \quad (8\text{-}82)$$

We now define ϕ through the following implicit equation so as to make the second major term on the left reduce to unity:

$$V = \frac{\alpha_A d x_{A,d}}{\alpha_A - \phi} + \frac{\alpha_B d x_{B,d}}{\alpha_B - \phi} \quad (8\text{-}83)$$

Algebraic manipulation of Eq. (8-82) then results in

$$\frac{L}{V}\left(\frac{\alpha_A x_{A,n+1}}{\alpha_A - \phi} + \frac{\alpha_B x_{B,n+1}}{\alpha_B - \phi}\right) = \frac{\phi\{[\alpha_A x_{A,n}/(\alpha_A - \phi)] + [\alpha_B x_{B,n}/(\alpha_B - \phi)]\}}{\alpha_A x_{A,n} + \alpha_B x_{B,n}} \quad (8\text{-}84)$$

There are two values of ϕ which will satisfy Eq. (8-83), noted as ϕ_1 and ϕ_2 in decreasing order of magnitude, and they possess numerical values such that

$$\alpha_A > \phi_1 > \alpha_B \qquad \alpha_B > \phi_2 > 0$$

This conclusion can be reached as follows. All terms other than ϕ in Eq. (8-83) are necessarily positive. Since $\alpha_A > \alpha_B$ by convention, the right-hand side of the equation must be negative for $\phi > \alpha_A$ and no solution is possible. For $\phi = \alpha_A$ the right-hand side is infinite. As ϕ decreases from α_A to α_B, the right-hand side decreases from $+\infty$ to $-\infty$, and a solution ϕ_1 necessarily occurs in this range. As ϕ decreases from α_B to 0, the right-hand side decreases from $+\infty$ to d, which is necessarily less than V; hence again a solution ϕ_2 necessarily occurs.

If Eq. (8-84) is written with each of the two values of ϕ and the resulting equations are divided into each other, it is found that

$$\frac{[\alpha_A x_{A,n+1}/(\alpha_A - \phi_1)] + [\alpha_B x_{B,n+1}/(\alpha_B - \phi_1)]}{[\alpha_A x_{A,n+1}/(\alpha_A - \phi_2)] + [\alpha_B x_{B,n+1}/(\alpha_B - \phi_2)]}$$
$$= \frac{\phi_1}{\phi_2}\left|\frac{[\alpha_A x_{A,n}/(\alpha_A - \phi_1)] + [\alpha_B x_{B,n}/(\alpha_B - \phi_1)]}{[\alpha_A x_{A,n}/(\alpha_A - \phi_2)] + [\alpha_B x_{B,n}/(\alpha_B - \phi_2)]}\right| \quad (8\text{-}85)$$

The major terms on the left- and right-hand sides are identical except for the plate-number subscript. Thus the equation relates an expression for plate $n + 1$ to the same expression for plate n. Since ϕ_1 and ϕ_2 are constants unaffected by plate number,

Eq. (8-85) can be made into a geometric progression relating the composition above the top stage of the column to the liquid composition leaving the feed stage

$$\left| \frac{[\alpha_A x_A/(\alpha_A - \phi_1)] + [\alpha_B x_B/(\alpha_B - \phi_1)]}{[\alpha_A x_A/(\alpha_A - \phi_2)] + [\alpha_B x_B/(\alpha_B - \phi_2)]} \right|_d$$

$$= \left(\frac{\phi_1}{\phi_2}\right)^{N_R} \left| \frac{[\alpha_A x_A/(\alpha_A - \phi_1)] + [\alpha_B x_B/(\alpha_B - \phi_1)]}{[\alpha_A x_A/(\alpha_A - \phi_2)] + [\alpha_B x_B/(\alpha_B - \phi_2)]} \right|_f \quad (8\text{-}86)$$

The subscripts on the braced terms indicate the location at which x_A and x_B are evaluated. When $x_{A,d}$ and $x_{B,d}$ are substituted, the left-hand side becomes unity; therefore

$$\left(\frac{\phi_2}{\phi_1}\right)^{N_R} = \frac{[\alpha_A x_{A,f}/(\alpha_A - \phi_1)] + [\alpha_B x_{B,f}/(\alpha_B - \phi_1)]}{[\alpha_A x_{A,f}/(\alpha_A - \phi_2)] + [\alpha_B x_{B,f}/(\alpha_B - \phi_2)]} \quad (8\text{-}87)$$

To solve for N_R, the number of equilibrium stages in the rectifying section, both values of ϕ must be calculated from Eq. (8-83) and the composition of liquid on the feed stage must be known. Notice that $x_{A,f}$ and $x_{B,f}$ in Eq. (8-87) refer to the composition of the liquid on the feed stage and not to the feed composition itself. The two compositions are, in general, different.

A similar development for the stripping section leads to analogous equations

$$-V' = \frac{\alpha_A b x_{A,b}}{\alpha_A - \phi'} + \frac{\alpha_B b x_{B,b}}{\alpha_B - \phi'} \quad (8\text{-}88)$$

$$\left(\frac{\phi'_2}{\phi'_1}\right)^{N_S} = \frac{[\alpha_A x_{A,f}/(\alpha_A - \phi'_2)] + [\alpha_B x_{B,f}/(\alpha_B - \phi'_2)]}{[\alpha_A x_{A,f}/(\alpha_A - \phi'_1)] + [\alpha_B x_{B,f}/(\alpha_B - \phi'_1)]} \quad (8\text{-}89)$$

Again, two roots of Eq. (8-88) exist such that

$$\phi'_1 > \alpha_A \quad \text{and} \quad \alpha_A > \phi'_2 > \alpha_B$$

and to solve Eq. (8-89) for N_S, the number of equilibrium stages in the stripping section, both values of ϕ' must be calculated from Eq. (8-88), and the composition of the liquid on the feed stage must be known.

From the equations it can also be noted that N_R includes the feed stage and that N_S includes the reboiler but not the feed stage. If a partial condenser is used, it is also included in N_R as another equilibrium stage.

The Underwood equations are most useful for design problems in which (1) two separation variables have been set, thereby describing the products; (2) reflux or another flow has been set, thereby describing the flows; and (3) a fourth variable remains to be set. Equations (8-87) and (8-89) provide two equations in three unknowns: $x_{A,f}$, N_R, and N_S. Since $x_{B,f}$ follows as $1 - x_{A,f}$, it is not an independent variable. The fourth specified variable is usually the composition on the feed stage, and usually it is desirable to set the composition to correspond to the optimum feed-stage location in order to ensure that the equivalent graphical construction will always step to the operating line farthest removed from the equilibrium curve.†

The operating-line equations are

$$y_A = \frac{L}{V} x_A + \frac{x_{A,d} d}{V} \tag{5-5}$$

and

$$y_A = \frac{L'}{V'} x_A - \frac{x_{A,b} b}{V'} \tag{5-8}$$

Equating the two equations and solving for x_A gives

$$x_A = \frac{(x_{A,d} d/V) + (x_{A,b} b/V')}{(L'/V') - (L/V)} \tag{8-90}$$

If x_A calculated from Eq. (8-90) is taken as $x_{A,f}$, the calculated numbers of stages from Eqs. (8-87) and (8-89) will be effectively the minimum possible to produce the desired separation under the set flows. For the case of saturated liquid feed, Eq. (8-90) reduces to the very simple $x_{A,f} = x_{A,F}$, and the desired feed-plate composition is equal to the composition of the feed.

The group methods of calculation presented by Underwood's equations can easily be extended to binary separations in columns producing more than two products and to columns with multiple feeds. Equation (8-85) can be written to define the number of stages required between any two sets of compositions, and ϕ values for any net upward product from the section can be obtained from Eq. (8-83). By using Eq. (8-85) over small ranges of composition, molar flow changes and changes in relative volatility can also be partially corrected for. Analogous reasoning applies to the equations for the stripping section.

The restriction of constant molar overflow can also be alleviated through use of the modified-latent-heat-of-vaporization (MLHV) method, outlined in Chap. 6, according to which one of the components is given a pseudo molecular weight in order to produce components of equal modified latent heats. Recall that a system with constant relative volatility will retain it when transformed to these new composition parameters. Hence use of the MLHV method with the Underwood equations should not greatly affect the degree of validity of the constant-α assumption.

A disadvantage of the Underwood equations is that they are derived in terms of

---

† Hanson and Newman (1977) observe that setting the composition of the liquid on the feed stage in a binary distillation equal to the x coordinate of the intersection of the operating lines does not correspond to the exact minimum in $N_R + N_S$ as a function of $x_{A,f}$ for the Underwood equations with a specified overall separation. Apparently, by displacing the feed-stage composition slightly one can achieve a lower indicated value of $N_R + N_S$ by trading a fraction of a stage in one section for a smaller fraction of a stage in the other section. However, one can still show mathematically that stepping to the operating line which is farthest from the equilibrium curve must still give a smaller stage requirement than would be obtained by violating that policy. Hence the difference between the equilibrium-stage requirement and the indicated optimum feed location obtained with the feed-stage composition corresponding to the intersection of the operating lines, on the one hand, and the "true" minimum, on the other, must be only a small fraction of an equilibrium stage.

equilibrium stages. Although one can employ an overall stage efficiency [Eq. (12-34)] in straightforward fashion, there is no way to use a Murphree stage efficiency with the equations as they stand. This is in contrast to the KSB equations, for which there are forms [Eqs. (8-16) and (8-20)] employing Murphree efficiencies.

Selection of average values of α When relative volatilities vary somewhat with composition and temperature, it is necessary to use average values in connection with the Underwood equations. The appropriate average could be viewed as that corresponding to the average temperature of the section. The average temperature will be the arithmetic average if the temperature profile in the section is assumed to be linear. Logarithms of vapor pressures, and hence relative volatilities, tend to be nearly linear in $1/T$, following the Clausius-Clapeyron equation. Over short ranges of temperature they may be assumed to be linear in T. If the relative volatilities (α_1 and α_2) at either end of a section are known, the relative volatility corresponding to the average temperature, by these assumptions, would be given by

$$\ln \alpha_{av} = \tfrac{1}{2}(\ln \alpha_1 + \ln \alpha_2) \tag{8-91}$$

which leads to

$$\alpha_{av} = (\alpha_1 \alpha_2)^{1/2} \tag{8-92}$$

or to the *geometric-mean* average relative volatility.

Similarly, the feed-stage relative volatility can be approximated as the geometric mean of the relative volatilities at distillate and bottoms conditions if the column is relatively balanced in conditions between the rectifying and stripping sections.

Looking ahead to the Fenske equation for total reflux [Eqs. (9-17) to (9-24)], we see that the appropriate average relative volatility for a section at total reflux would be $(\alpha_1 \alpha_2 \cdots \alpha_{N-1} \alpha_N)^{1/N}$. This again leads to the geometric mean of the terminal values as a good approximation.

Example 8-8 Determine the number of equilibrium stages for the atmospheric-pressure benzene-toluene distillation shown in Fig. 7-4 and treated in Example 8-6. Use the equations of Underwood.

SOLUTION The specified conditions for this problem were listed in the solution to Example 8-6 and are repeated here:

$$\alpha_{BT} = \begin{cases} 2.50 & \text{at feed stage} \\ \cdot\ 2.56 & \text{average above feed} \\ 2.44 & \text{average below feed} \end{cases}$$

$x_{B,F} = 0.50$, saturated liquid $x_{B,b} = 0.05$ $x_{B,d} = 0.95$
$r/d = 1.57$ $d/F = 0.50$ $V/d = 2.57$ $V'/d = 2.57$ $L/d = 3.57$

Considering the rectifying section first we have, substituting into Eq. (8-83),

$$\frac{V}{d} = 2.57 = \frac{(2.56)(0.95)}{2.56 - \phi} + \frac{(1.00)(0.05)}{1.00 - \phi}$$

x_T has been taken to be unity. By trial and error we have

$$\phi_1 = 1.64 \quad \text{and} \quad \phi_2 = 0.953$$

These results could also have been obtained through use of the general algebraic solution to quadratic equations.

Substituting into Eq. (8-87), we have

$$\left(\frac{0.953}{1.64}\right)^{N_R} = \frac{[(2.56)(0.5)/(2.56 - 1.64)] + [(1.0)(0.5)/(1.0 - 1.64)]}{[(2.56)(0.5)/(2.56 - 0.953)] + [(1.0)(0.5)/(1.0 - 0.953)]}$$

$$(0.581)^{N_R} = \frac{1.391 - 0.781}{0.80 + 10.64} = 0.0533$$

$$N_R = \frac{\ln 0.0533}{\ln 0.581} = 5.4$$

Substituting into Eq. (8-88) for the stripping section, we have

$$-\frac{V'}{b} = -2.57 = \frac{(2.44)(0.05)}{2.44 - \phi'} + \frac{(1.0)(0.95)}{1.0 - \phi'}$$

$$\phi'_1 = 2.503 \qquad \phi'_2 = 1.354$$

$$\left(\frac{1.354}{2.503}\right)^{N_S} = \frac{[(2.44)(0.5)/(2.44 - 1.354)] + [(1.0)(0.5)/(1.0 - 1.354)]}{[(2.44)(0.5)/(2.44 - 2.503)] + [(1.0)(0.5)/(1.0 - 2.503)]}$$

$$(0.540)^{N_S} = \frac{1.123 - 1.412}{-19.36 - 0.33} = 0.01468$$

$$N_S = 6.9$$

$N_R + N_S = 12.3$, in good agreement with the results obtained by the McCabe-Thiele method (Fig. 7-4) and by the Martin equations (Example 8-6). ☐

Multicomponent Countercurrent Separations: Discrete Stages

The equations of Underwood developed for solution of binary distillation systems can readily be extended to multicomponent systems by the simple addition of terms for the added components (Underwood, 1948). Thus Eq. (8-83) becomes

$$V = \frac{\alpha_A dx_{A,d}}{\alpha_A - \phi} + \frac{\alpha_B dx_{B,d}}{\alpha_B - \phi} + \frac{\alpha_C dx_{C,d}}{\alpha_C - \phi} + \cdots \tag{8-93}$$

written more concisely as

$$V = \sum_{i=1}^{R} \frac{\alpha_i x_{i,d} d}{\alpha_i - \phi} \tag{8-94}$$

where R is the number of components. The other pertinent equations are

$$-V' = \sum_{i=1}^{R} \frac{\alpha_i x_{i,b} b}{\alpha_i - \phi'} \tag{8-95}$$

$$\left(\frac{\phi_k}{\phi_j}\right)^{N_R} = \frac{\displaystyle\sum_{i=1}^{R} \frac{\alpha_i x_{i,f}}{\alpha_i - \phi_j}}{\displaystyle\sum_{i=1}^{R} \frac{\alpha_i x_{i,f}}{\alpha_i - \phi_k}} \tag{8-96}$$

$$\left(\frac{\phi'_k}{\phi'_j}\right)^{N_S} = \frac{\displaystyle\sum_{i=1}^{R} \frac{\alpha_i x_{i,f}}{\alpha_i - \phi'_k}}{\displaystyle\sum_{i=1}^{R} \frac{\alpha_i x_{i,f}}{\alpha_i - \phi'_j}} \tag{8-97}$$

where i = any component
R = total number of components
j = *particular* ϕ or ϕ' value
k = second *particular* ϕ or ϕ' value

Again, $x_{i,f}$ refers to the *liquid leaving the feed stage*, not to the composition of either the feed itself or the liquid portion of the feed.

It is apparent that there are as many values of ϕ and ϕ' obtainable from Eq. (8-94) and (8-95) as there are components. Reasoning similar to that employed for binary systems shows that in the rectifying section $\alpha_A > \phi_1 > \alpha_B > \phi_2 > \cdots > \alpha_R > \phi_R > 0$ if A, B, ..., R is the order of decreasing volatility of components. Also, in the stripping section, $\phi'_1 > \alpha_A > \phi'_2 > \alpha_B > \cdots > \phi'_R > \alpha_R$. In the rectifying section each value of ϕ can be associated with the component whose α_i lies next above it, and in the stripping section each value of ϕ' can be associated with the component whose α_i lies next below it.

Both Eqs. (8-96) and (8-97) can be written $R - 1$ times in different independent combinations of the individual values of ϕ and ϕ'. For a multicomponent distillation problem which has been specified through setting the feed flow, composition and enthalpy, the pressure, optimal feed location, the reflux rate, and two separation variables, these $2R - 2$ independent equations can be solved to give N_S, N_R, $x_{i,f}$, and the remaining product composition and flow-rate variables. In a multicomponent problem, specifying two separation variables does not fix the distillate and bottoms products; there are $R - 2$ additional degrees of freedom involving product flows and compositions. These $R - 2$ additional degrees of freedom combine with the $R - 1$ independent feed-stage liquid mole fractions and with N_R and N_S to give $2R - 1$ unknowns, which can be obtained using the $2R - 2$ different versions of Eqs. (8-96) and (8-97) and the criterion of optimal feed-stage location. Note that Eqs. (8-94) and (8-95) also must be included in any such solution scheme since the values of ϕ and ϕ' are themselves functions of the product flows and compositions.

The equations are complex, but Hanson and Newman (1977) give an efficient method for obtaining the solution for a multicomponent distillation specified through all feed conditions, column pressure, reflux rate and thermal condition, two separation variables involving the two key components, and the ratio of the mole fractions of the key components in the liquid leaving the feed stage. The method makes use of the insensitivity of various of the equations to the assumed values of certain of the unknown variables and is an improvement over earlier methods outlined by Alder and Hanson (1950) and Klein and Hanson (1955). Even though it is an iterative method requiring a computer, it is rapidly convergent. Furthermore, it has the attractive feature that the result of the first iteration is itself usually quite sufficient for the purposes of a calculation.

To begin with, it is clear that if the feed-stage liquid composition and the product compositions could be estimated, any version of Eqs. (8-96) and (8-97) could be solved for N_R and N_S, respectively, in the same way that the analogous equations were used for binary systems. If the versions of Eqs. (8-96) and (8-97) involving the ϕ and ϕ' values associated with the key-component relative volatilities are used for obtaining N_R and N_S, it is found that these ϕ and ϕ' values as obtained from Eqs. (8-94) and (8-95) are quite insensitive to the actual distributions of nonkey components between the distillate and bottoms. All that is needed is to say that the nonkeys will appear almost entirely in one or the other of the products. If this appears to be a poor assumption, one can use the nonkey splits predicted for total reflux (Chap. 9); the assumption is not critical. The ϕ values associated with the heavy nonkeys and the ϕ' values associated with light nonkeys are not well known if the splits of those components are not established, but they are not needed yet if the values of ϕ and ϕ' associated with the key components are used to solve for the number of stages. Since the necessary pairs of ϕ and ϕ' values can be closely estimated in this way, the problem of finding a close approximation to N_R and N_S becomes one of finding a close approximation to the composition of the liquid leaving the feed stage.

To approximate the feed-plate composition, we can make use of the concept of limiting flows of the nonkey components in the sections above and below the feed. Referring back to Figs. 7-11 and 7-12 and the discussion concerning them, we recall that the light nonkeys tend to approach limiting mole fractions in the rectifying section whereas the heavy nonkeys tend to approach limiting mole fractions in the stripping section. Equations (7-4) and (7-6) were derived as first approximations for prediction of these limiting mole fractions of the nonkey components.

In Chap. 7 it was assumed that $K_{HK} = L/V$ just above the feed and $K_{LK} = L'/V'$ just below the feed in order to provide K's for estimating the limiting mole fractions of the nonkey components. A better estimation of the K's of the nonkey components in their zones of limiting mole fraction can be made if we realize that these same limiting nonkey mole fractions would occur if the column sections above and below the feed both contained infinite stages. Thus if we postulate a hypothetical rectifying section of infinite stages with the same total molar interstage vapor and liquid flows and with the same distillate flow and composition, there will be some point (see Figs. 7-16 to 7-18) below which no component changes in mole fraction.[†] At this point we can write the equivalents of Eqs. (7-6) and (7-7) for all components:

$$x_{i,\infty} = \frac{x_{i,d}\,d/L}{(VK_{i,\infty}/L) - 1} \tag{8-98}$$

$$y_{i,\infty} = K_{i,\infty}\,x_{i,\infty} \tag{8-99}$$

[†] Hypothesizing an infinite number of stages in the rectifying section constitutes an overspecification of the original problem and would require the feed location to be nonoptimal. Since this development is concerned exclusively with the zone of constant composition which would occur *above* the feed, this change is not important as long as d, V, L, and $x_{i,d}$ are kept the same.

As in Chap. 7, the successive stage subscripts are dropped from Eqs. (8-98) and (8-99) since we are considering a zone of constant composition. Combining these equations and rearranging gives

$$V y_{i,\,\infty} = \frac{x_{i,\,d} d}{1 - (L/V K_{i,\,\infty})} \tag{8-100}$$

If we multiply both the numerator and denominator of the right-hand side of Eq. (8-100) by α_i, we obtain

$$V y_{i,\,\infty} = \frac{\alpha_i x_{i,\,d} d}{\alpha_i - (\alpha_i L/V K_{i,\,\infty})} \tag{8-101}$$

Adding Eq. (8-101) for all components gives

$$V = \sum_{i=1}^{R} \frac{\alpha_i x_{i,\,d} d}{\alpha_i - (\alpha_i L/V K_{i,\,\infty})} \tag{8-102}$$

The group $\alpha_i L/V K_{i,\,\infty}$ is the same for all components as a consequence of the definition of α

$$\frac{\alpha_A}{K_{A,\,\infty}} = \frac{\alpha_B}{K_{B,\,\infty}} = \cdots = \frac{1}{K_{r,\,\infty}} \tag{8-103}$$

where r denotes that reference component whose α has arbitrarily been set equal to unity. Since $\alpha_i L/V K_{i,\,\infty}$ is the same for all components, a comparison of Eq. (8-103) with Eq. (8-94) shows that $\alpha_i L/V K_{i,\,\infty}$ must be identical to one of the values of ϕ. The appropriate value of ϕ is the one associated with the heavy key (next below α_{HK}), which we shall denote ϕ_{HK-}. Thus values of $K_{i,\,\infty}$ for any component in a zone of constant composition above the feed can be obtained from

$$K_{i,\,\infty} = \frac{L}{V} \frac{\alpha_i}{\phi_{HK-}} \tag{8-104}$$

The reader should note that this analysis has led to a physical interpretation of one of the values of ϕ. ϕ_{HK-} is the value of the group $L/V K_r$ for the reference component ($\alpha_r = 1$) in the zone of constant composition in a rectifying section with infinite stages.

For the purposes of our approximate solution of the Underwood equations we assume that $K_{i,\,\infty}$ for light nonkey components is the same in a zone where only the light nonkeys are at constant mole fraction as in a zone where all components are at constant mole fraction. Hence values of $K_{i,\,\infty}$ determined for the light nonkeys from Eq. (8-104) can be substituted into Eq. (8-98) to give the limiting mole fractions of these components in the liquid above the feed stage:

$$x_{LNK,\,\infty} = \frac{x_{LNK,\,d} d/L}{(\alpha_i/\phi_{HK-}) - 1} \tag{8-105}$$

Similar reasoning for the stripping section leads to

$$K_{i,\,\infty} = \frac{L'}{V'} \frac{\alpha_i}{\phi'_{LK+}} \tag{8-106}$$

for a zone of constant composition below the feed with infinite stages in the stripping section, with the same interstage flows in the stripping section, and with the same bottoms flow and composition. Here ϕ'_{LK+} is the value of ϕ associated with the light key component (next above α_{LK}). This value of ϕ' can be interpreted physically as the value of the group $L'/V'K_r$ for the reference component in the zone of constant composition in a stripping section with infinite stages.

A combination of Eqs. (7-2) and (7-4) gives

$$x_{HNK,\infty} = \frac{x_{HNK,b}b/L'}{1 - (K_{HNK,\infty}V'/L')} \tag{8-107}$$

Assuming that $K_{i,\infty}$ for heavy nonkey components is the same in a zone where only the heavy nonkeys are at constant mole fraction as in a zone where all components are at constant mole fraction, we can substitute Eq. (8-106) into Eq. (8-107) to obtain

$$x_{HNK,\infty} = \frac{x_{HNK,b}b/L'}{1 - (\alpha_i/\phi'_{LK+})} \tag{8-108}$$

Returning to our approximate method for determination of N_R and N_S, we can now estimate the composition of the liquid leaving the feed stage for use in Eqs. (8-96) and (8-97). We can set the mole fractions of the heavy nonkey components in this liquid equal to the mole fractions calculated from Eq. (8-108), following the assumption that these components are at their limiting stripping-section mole fractions. Similarly, we can set the light nonkey mole fractions in the liquid leaving the feed stage as approximately equal to the limiting light nonkey mole fractions in the liquid above the feed $x_{LNK,\infty}$ obtained from Eq. (8-105).

The ratio of $x_{LK,f}$ to $x_{HK,f}$ is the remaining unused specified variable for this solution. As a first approximation to an optimum value it is reasonable to extend the logic which led to setting the liquid composition leaving the feed stage in a binary distillation equal to the x value of the intersection of the operating lines. If we write operating-line expressions based upon total flows and ignoring nonkeys, Eq. (8-90) yields

$$\left(\frac{x_{LK}}{x_{HK}}\right)_f = \frac{(x_{LK,d}d/V) + (x_{LK,b}b/V')}{(x_{HK,d}d/V) + (x_{HK,b}b/V')} \tag{8-109}$$

as the ratio in the feed-stage liquid. Equation (8-109), together with the fact that

$$x_{LK,f} + x_{HK,f} = 1 - \Sigma x_{NK,f} \tag{8-110}$$

makes calculation of an approximate feed-stage composition possible.

We can now complete the approximate solution (and the first iteration of the Hanson-Newman full solution) by using this feed-stage composition along with values of ϕ_{LK-}, ϕ_{HK-}, ϕ'_{LK+}, and ϕ'_{HK+} from Eqs. (8-94) and (8-95) in Eqs. (8-96) and (8-97) to find N_R and N_S. For most purposes for which the Underwood equations would be used, stopping at this point is sufficient.

To proceed further with an exact solution, Hanson and Newman (1977) first converge the values of $x_{NK, f}$ by using versions of Eq. (8-96) written in terms of ϕ_{LNK-} and ϕ_{LK-} for each light nonkey to calculate $x_{LNK, f}$. Similarly, values of $x_{HNK, f}$ are calculated using versions of Eq. (8-97) written in terms of ϕ'_{HNK+} and ϕ'_{HK+} for each heavy nonkey. The new feed-stage composition is then obtained using the specified $x_{LK, f}/x_{HK, f}$ and Eq. (8-110). The calculation of N_R and N_S is repeated, new values of $x_{NK, f}$ are obtained, etc., convergence usually being quite rapid. This gives a solution for the assumed nonkey splits between overhead and bottoms. To correct the $x_{HNK, d}$ and $x_{LNK, b}$ values, Eqs. (8-96) written in terms of ϕ_{HNK-} and ϕ_{LK-} for each heavy nonkey are solved for the dominant $\alpha_{HNK} - \phi_{HNK-}$ term in the denominator of the right-hand side, using the fact that ϕ_{HNK-} is close to α_{HNK} typically for the left-hand side. The values of $\alpha_{HNK} - \phi_{HNK-}$ can then be used in Eqs. (8-94) to solve for $x_{HNK, d} d$. Similarly, Eqs. (8-97) written in terms of ϕ'_{LNK+} and ϕ'_{HK+} for each light nonkey are solved for the dominant $\alpha_{LNK} - \phi'_{LNK+}$ term in the denominator of the right-hand side, using the fact that ϕ'_{LNK+} is close to α_{LNK} typically for the left-hand side. The values of $\alpha_{LNK} - \phi'_{LNK+}$ are then used in Eqs. (8-95) to solve for $x_{LNK, b} b$. With the new diluent compositions, the whole procedure is repeated as an inner loop, until both the outer and inner loops have converged.

Hanson and Newman (1977) show how the specification of $x_{LK, f}/x_{HK, f}$ can be relaxed and the problem can be solved instead for the optimum feed location to give minimum stages. Here it turns out to be convenient to solve for the optimum feed location before correcting the splits of the nonkeys, since the splits of the nonkeys depend upon the number of stages present above and below the feed. Results cited in their study show that the total equilibrium-stage requirement is rather insensitive to the feed location but that the optimum feed location can be substantially different from that corresponding to the ratio given by Eq. (8-109), especially when the nonkey mole fractions are large and/or unbalanced between light nonkeys and heavy nonkeys. Light nonkeys serve to raise the optimum $x_{LK, f}/x_{HK, f}$, and heavy nonkeys tend to lower it.

As for binary distillations the MLHV method can be used effectively with the multicomponent Underwood equations to compensate for unequal molal overflow. Also, as for binary distillations, the method cannot handle specified Murphree efficiencies in any simple manner; overall stage efficiencies must be used instead. Since in general there will be different overall stage efficiencies for different pairs of components, this feature can cause the Underwood equations to predict the splits of nonkeys incorrectly, even for a fully converged solution.

Solving for ϕ and ϕ' The functions on the right-hand sides of Eqs. (8-94) and (8-95) are so structured that they approach $+\infty$ or $-\infty$ for ϕ approaching one of the values of α_i. Even though the function becomes infinite for ϕ exactly equal to one of the α_i values, the desired root will often fall very close to one of the values of α_i (see, for instance, Examples 8-9 and 9-1). A convergence method for solving the Underwood equations for specific values of ϕ must be capable of locating the root in a highly nonlinear region close to an asymptote toward infinity without crossing the bounding value of α_i and entering a region of the function corresponding to a different ϕ root.

When it is apparent that the desired value of ϕ will be quite close to a particular value of $\alpha_i (= \alpha_k)$, as is often true for ϕ_{HK-} and ϕ'_{LK+}, an effective approach is to substitute α_k for ϕ in all terms except for the one involving $\alpha_k - \phi$ in the denominator. The equation can then be solved explicitly for ϕ. That indicated value of ϕ can then be substituted for the variable ϕ into all terms, except again for that term involving $\alpha_k - \phi$. Once again the equation can be solved explicitly for ϕ, and the procedure repeated. This direct-substitution technique will converge rapidly for finding a value of ϕ close to a value of α_i.

Ripps (1968) presents an efficient method for solving the Underwood equations, using a Newton method to accelerate the above procedure.

Example 8-9 Compute the number of equilibrium stages required for the depropanizer distillation column outlined in Table 7-2 and considered in Chap. 7. The ratio of reflux to feed flows for this example is specified as 0.90 mol/mol by Edmister (1948). The feed is specified to be 66 mol % vapor at tower conditions. Use the Underwood equations, solved by the approximate method. The split of the key components (propane and butane), is specified in Table 7-2, repeated as Table 8-1. For purposes of the approximate solution, assume that the nonkeys go entirely to one product or the other, as shown in Table 8-1.

Table 8-1 Feed and products for depropanizer example (data from Edmister, 1948)

Component	mol %			mol/100 mol of feed	
	Feed	Distillate	Bottoms	Distillate	Bottoms
Methane (C_1)	26	43.5	......	26	
Ethane (C_2)	9	15.0	......	9	
Propane (C_3)	25	41.0	1.0	24.6	0.4
n-Butane (C_4)	17	0.5	41.7	0.3	16.7
n-Pentane (C_5)	11	......	27.4	.....	11
n-Hexane (C_6)	12	......	29.9	.....	12
	100	100	100	59.9	40.1

SOLUTION The first step is to solve for values of ϕ in the rectifying section. For this purpose we need to determine average values of α_i for the various components in that section. We can prepare Table 8-2, which gives an idea of the extent to which α_i is constant. The values of α_i are those used in Edmister's paper and would be changed somewhat if current sources of thermodynamic data were employed. Temperatures change most rapidly at either end and near the feed (see Fig. 7-13). Hence, both for calculations based on infinite sections and for the determination of the equilibrium-stage requirement, it is desirable to use values of α_i corresponding to conditions midway along either section of the column since we want to solve for only one set of ϕ values and one set of ϕ' values. As discussed above, geometric mean values of α_i in either section are appropriate (Table 8-2).

Table 8-2 α_i relative to propane (data from Edmister, 1948)

| | Temperature | | | Geometric mean | |
	Top (17°C)	Feed (79°C)	Bottoms (149°C)	Rectifying section	Stripping section
Methane	15.4	10.0	8.0	12.4	8.9
Ethane	3.00	2.47	2.4	2.72	2.43
Propane	1.0	1.0	1.0	1.0	1.0
Butane	0.32	0.49	0.54	0.396	0.514
Pentane	0.108	0.21	0.30	0.151	0.25
Hexane	0.029	0.10	0.18	0.054	0.13

From Table 8-1, $D = 59.9$ mol and $b = 40.1$ moles on the basis of 100 mol of feed. Since r is given as 0.90 mol per mole of feed and the feed is 66 mole percent vapor at tower conditions, we can solve for the interstage flows:

$$L = 90.0 \qquad V = 90.0 + 59.9 = 149.9$$

$$V' = 149.9 - 66.0 = 83.9 \qquad L' = 90.0 + 34.0 = 124.0$$

Equation (8-94) becomes

$$V = 149.9 = \frac{12.4(26)}{12.4 - \phi} + \frac{2.72(9)}{2.72 - \phi} + \frac{1.0(24.6)}{1.0 - \phi} + \frac{0.396(0.3)}{0.396 - \phi}$$

For use in Eq. (8-96) we need ϕ_3 and ϕ_4 if the different values of ϕ are identified through $\alpha_{C_1} > \phi_1 > \alpha_{C_2} > \phi_2 > \alpha_{C_3} > \phi_3 > \alpha_{C_4} > \phi_4 > 0$. We also need ϕ_4 in order to use Eq. (8-105).
Solving for ϕ_3 and ϕ_4 by trial and error gives

$$\phi_3 = 0.776 \quad \text{and} \quad \phi_4 = 0.39435$$

The reader should note that the solution for ϕ_4 does not really require trial and error. Since ϕ_4 is close to α_{C_4} in value, α_{C_4} can be substituted for ϕ_4 in all terms but the last one on the right with little loss in accuracy, and we can then solve for ϕ_4 directly. Equation (8-95) becomes

$$-83.9 = \frac{1.0(0.4)}{1.0 - \phi'} + \frac{0.514(16.7)}{0.514 - \phi'} + \frac{0.25(11)}{0.25 - \phi'} + \frac{0.13(12)}{0.13 - \phi'}$$

If the different ϕ' values are denoted through $\phi'_1 > \alpha_{C_3} > \phi'_2 > \alpha_{C_4} > \phi'_3 > \alpha_{C_5} > \phi'_4 > \alpha_{C_6}$, we need ϕ'_1 and ϕ'_2 for use in Eq. (8-97), and we need ϕ'_1 again for use in Eq. (8-108).
Solving for ϕ'_1 and ϕ'_2 trial and error gives

$$\phi'_1 = 1.00655 \qquad \phi'_2 = 0.629$$

Again, since ϕ'_1 is so close to α_{C_3}, we can substitute α_{C_3} for ϕ'_1 in all terms but the first and then solve directly for ϕ'_1.
The next step is to determine the approximate feed-stage liquid composition. Using Eq. (8-108) for $x_{HNK, f} = x_{HNK, x}$, we have

$$x_{C_5, f} = \frac{x_{C_5, b} b/L'}{1 - (\alpha_{C_5}/\phi'_1)} = \frac{11/124.0}{1 - (0.25/1.007)} = 0.118$$

$$x_{C_6, f} = \frac{12/124.0}{1 - (0.13/1.007)} = 0.111$$

From Eq. (8-105) for $x_{LNK,f} = x_{LNK,x}$

$$x_{C_1,f} = \frac{26/90}{(12.4/0.394) - 1} = 0.010$$

$$x_{C_2,f} = \frac{9/90}{(2.72/0.394) - 1} = 0.0169$$

Hence, from Eq. (8-110)

$$x_{C_3,f} + x_{C_4,f} = 1 - (0.010 + 0.017 + 0.118 + 0.111) = 0.744$$

From Eq. (8-109) we have

$$\frac{x_{C_3,f}}{x_{C_4,f}} = \frac{(24.6/149.9) + (0.4/83.9)}{(0.3/149.9) + (16.7/83.9)} = 0.840$$

Therefore $\qquad x_{C_3,f} = 0.340 \qquad$ and $\qquad x_{C_4,f} = 0.404$

These values compare well with the mole fractions obtained from the full stage-to-stage solution and shown in Fig. 7-12.

The number of equilibrium stages in the rectifying section can now be obtained by means of Equation 8-96

$$\left(\frac{0.776}{0.394}\right)^{N_R}$$

$$= \frac{\dfrac{12.4(0.010)}{12.4 - 0.4} + \dfrac{2.72(0.017)}{2.72 - 0.39} + \dfrac{1.0(0.340)}{1.0 - 0.394} + \dfrac{0.396(0.404)}{0.396 - 0.39435} + \dfrac{0.15(0.118)}{0.15 - 0.394} + \dfrac{0.05(0.111)}{0.05 - 0.394}}{\dfrac{12.4(0.010)}{12.4 - 0.8} + \dfrac{2.72(0.017)}{2.72 - 0.78} + \dfrac{1.0(0.340)}{1.0 - 0.776} + \dfrac{0.396(0.404)}{0.396 - 0.776} + \dfrac{0.15(0.118)}{0.15 - 0.776} + \dfrac{0.05(0.111)}{0.05 - 0.776}}$$

$$(1.97)^{N_R} = \frac{97.5}{1.09} = 89.4$$

$$N_R = 6.6$$

The terms involving the keys are controlling. Substituting into Eq. (8-97) gives

$$\left(\frac{1.007}{0.629}\right)^{N_S}$$

$$= \frac{\dfrac{8.9(0.010)}{8.9 - 1.0} + \dfrac{2.43(0.017)}{2.43 - 1.01} + \dfrac{1.0(0.340)}{1.0 - 1.00655} + \dfrac{0.514(0.404)}{0.514 - 1.007} + \dfrac{0.25(0.118)}{0.25 - 1.01} + \dfrac{0.13(0.111)}{0.13 - 1.01}}{\dfrac{8.9(0.010)}{8.9 - 0.63} + \dfrac{2.43(0.017)}{2.43 - 0.63} + \dfrac{1.0(0.340)}{1.0 - 0.63} + \dfrac{0.514(0.404)}{0.514 - 0.629} + \dfrac{0.25(0.118)}{0.25 - 0.63} + \dfrac{0.13(0.111)}{0.13 - 0.63}}$$

$$(1.600)^{N_S} = \frac{-52.2}{-0.97} = 53.8$$

$$N_S = 8.5$$

Thus $N_R + N_S = 15.1$, compared with the conservative value of 17, found by Edmister by stage-to-stage calculation. $\qquad\qquad\qquad\qquad\qquad\qquad\qquad\qquad\qquad$ □

REFERENCES

Alder, B. J., and D. N. Hanson (1950): *Chem. Eng. Prog.*, **46**:48.

Barker, P. E., and D. Critcher (1960): *Chem. Eng. Sci.*, **13**:82.

Belter, P. A. (1977): "Biochemical Engineering: Separation Processes," pres. at Summer Sch. Chem. Eng. Fac., Am. Soc. Eng. Educ., Snowmass, Colo.

Brian, P. L. T. (1972): "Staged Cascades in Chemical Processing," pp. 54–96, Prentice-Hall, Englewood Cliffs, N.J.

Carslaw, H. S., and J. C. Jaeger (1959): "Conduction of Heat in Solids," 2d ed., Oxford University Press, New York.

Deemter, J. J. van, F. J. Zuiderweg, and A. Klinkenberg (1956): *Chem. Eng. Sci.*, **5**:271.

De Vault, D. (1943): *J. Am. Chem. Soc.*, **65**:532.

Douglas, J. M. (1977): *Ind. Eng. Chem. Fundam.*, **16**:131.

Edmister, W. C. (1948): *Petrol. Eng.*, **19**(9):47; see also "A Source Book of Technical Literature on Fractional Distillation," pt. II, pp. 74ff., Gulf Research & Development Co., n.d.

Giddings, J. C. (1965): "Dynamics of Chromatography," pt. I, Dekker, New York.

Guenther, E. (1949): "The Essential Oils," vol. 3, pp. 260–281, Van Nostrand, New York.

Hanson, D. N., and J. S. Newman (1977): *Ind. Eng. Chem. Process. Des. Dev.*, **16**:223.

Hill, A. B., R. H. McCormick, P. Barton, and M. R. Fenske (1962): *AIChE J.*, **8**:681.

Keulemans, A. I. M. (1959): "Gas Chromatography," 2d ed., Reinhold, New York.

Klein, G., and D. N. Hanson (1955): *Chem. Eng. Sci.*, **4**:229.

Kremser, A. (1930): *Natl. Petrol. News*, **22**(21):42 (May 21).

Lewis, W. K. (1922): *Ind. Eng. Chem.*, **14**:492.

Martin, A. J. P., and R. L. M. Synge (1941): *Biochem. J.*, **35**:1358.

Martin, J. J. (1963): *AIChE J.*, **9**:646.

Massaldi, H. A., and C. J. King (1973): *J. Chem. Eng. Data*, **18**:393.

Murdoch, P. G. (1948): *Chem. Eng. Prog.*, **44**(11):855.

Perlman, D. (1969): Streptomycin, in R. E. Kirk and D. F. Othmer (eds.), "Encyclopedia of Chemical Technology," vol. 19, Interscience, New York.

Pohjola, V. J. (1975): *Chem. Eng. Sci.*, **30**:1527.

Ramalho, R. S., and F. M. Tiller (1962): *AIChE J.*, **8**:559.

Reamer, H. H., and B. H. Sage (1951): *Ind. Eng. Chem.*, **43**:1628.

Ripps, D. L. (1968): *Hydrocarbon Process.*, **47**(12):84.

Robinson, C. S., and E. R. Gilliland (1950): "Elements of Fractional Distillation," 4th ed., McGraw-Hill, New York.

Shiras, R. N., D. N. Hanson, and C. H. Gibson (1950): *Ind. Eng. Chem.*, **42**:871.

Singh, R. (1972): *Chem. Eng. Sci.*, **27**:677.

Smoker, E. H. (1938): *Trans. Am. Inst. Chem. Eng.*, **34**:165.

Stoll, M. (1967): Oils, Essential, in R. E. Kirk and D. F. Othmer (eds.), "Encyclopedia of Chemical Technology," vol. 14, Interscience, New York.

Souders, M., and G. G. Brown (1932): *Ind. Eng. Chem.*, **24**:519.

Treybal, R. E. (1968): "Mass Transfer Operations," 2d ed., McGraw-Hill, New York.

Underwood, A. J. V. (1944): *J. Inst. Petrol.*, **30**:225.

———(1945): *J. Inst. Petrol.*, **31**:111.

———(1946): *J. Inst. Petrol.*, **32**:598, 614.

———(1948): *Chem. Eng. Prog.*, **44**:603.

Wadsworth, G. P., and J. G. Bryan (1960): "Introduction to Probability and Random Variables," McGraw-Hill, New York.

Weast, R. C. (ed.) (1968): "Handbook of Chemistry and Physics," 49th ed., CRC Press, Cleveland.

Won, K. W., and J. M. Prausnitz (1974): *AIChE J.*, **20**:1187.

———and———(1975): *J. Chem. Thermodynam.*, **7**:661.

PROBLEMS

8-A₁ Rework Prob. 6-C using a group method for the calculation.

8-B₁ Rework Example 6-3 using a group method for the calculation.

8-C₁ Derive the value H_S of the column length equivalent to an equilibrium stage for each component in the gas-liquid chromatography recorder output shown in Fig. 4-40 and discussed in the surrounding text. Suggest reasons for the differences found between components.

8-D₁ In the manufacture of higher alcohols from carbon monoxide and hydrogen, a mixture of alcohols is obtained and must be separated into the desired products. As an example consider a feed mixture containing

Component	mol %
Ethanol	25
Isopropanol	15
n-Propanol	35
Isobutanol	10
n-Butanol	15

This mixture has been isolated from methanol and from heavier alcohols by prior distillation steps. The mixture is to be split into three products:

1. A stream containing at least 98 percent of the ethanol at a purity of 98.0 mole percent
2. A stream containing isopropanol along with essentially all the remaining ethanol and no more than 5 percent of the n-propanol in the total feed mixture
3. A stream containing no more than 2 percent of the isopropanol in the total feed mixture and containing most of the n-propanol, isobutanol, and n-butanol

The separation will be accomplished in two distillation towers. Column I will receive the entire mixture as feed. The distillate from column I will be the feed to column II, which will produce as products the ethanol-rich stream and the isopropanol-rich stream. Both columns will be at atmospheric pressure, with total condensers, saturated-liquid refluxes, and saturated-liquid feeds. The overhead reflux ratio r/d will be 2.70 in column I and 14.0 in column II. The overall stage efficiency E_0 [= (equilibrium-stage requirement)/(actual stage requirement)] will be 0.70 in both columns. Use group methods (approximate, where necessary or where warranted) to find the number of actual stages and the optimal feed location in each column. Assume constant molal overflow.

Equilibrium data (derived from vapor-pressure data, assuming Raoult's law)

Parameter	Component	70°C	80°C	90°C	100°C	110°C
K	n-Propanol	0.313	0.495	0.755	1.11	1.59
α	Ethanol	2.25	2.17	2.09	2.02	1.95
	Isopropanol	1.90	1.86	1.82	1.78	1.75
	n-Propanol	1.00	1.00	1.00	1.00	1.00
	Isobutanol	0.662	0.669	0.677	0.687	0.697
	n-Butanol	0.399	0.412	0.428	0.444	0.459

8-E₂[†] A plastic coating is applied to cardboard containers by spraying them with resin containing hexane as a solvent. The containers are passed through a curing oven where the resin sets, and hexane is removed by a current of warm air. To minimize explosion hazard, the hexane content of the air is maintained at a maximum of 1 vol %. Under normal processing conditions, 6000 ft³/min of air at 140°F and 1 atm containing 1% hexane is discharged from the oven.

[†] Adapted from Board of Registration for Professional Engineers, State of California, Examination in Chemical Engineering, November 1966; used by permission.

To permit recirculation of the air to the oven it is necessary to remove hexane from the air. One method to be investigated is absorption of the hexane in a nonvolatile hydrocarbon oil followed by recovery in a stripping column. As part of a study of the process, it is desired to estimate the number of equilibrium stages required for absorption and stripping, cooling-water requirements, and steam requirements.

The absorber accomplishes 90 percent removal of hexane from the air, and the stripper accomplishes 95 percent removal of hexane from the oil.

Air enters the absorber at 140°F, 1 atm. Lean oil enters the absorber at a rate corresponding to $KV/L = 0.7$ at 140°F, 1 atm, at the top of the column, where K = equilibrium ratio, V = molar vapor rate, and L = molar liquid rate. The oil actually enters the absorber at a temperature such that the heat of absorption of the hexane will bring the temperature of the oil leaving the absorber to 140°F. Air may be assumed to leave the absorber at the same temperature as the oil entering. To simplify the calculations absorption may be assumed to occur at 140°F, 1 atm.

Stripping is accomplished with pure steam available at 230°F, 1 atm. The stripping column is operated at flow rates corresponding to $KV/L = 1.5$ at 230°F, 1 atm, at flows prevailing at the bottom of the column. The oil feed to the stripper actually enters at a temperature sufficiently above 230°F for the heat requirement for stripping to be met by cooling of the liquid stream. Steam plus hexane can be assumed to leave the stripper at the same temperature as the entering oil. To simplify the calculations stripping can be assumed to occur at 230°F, 1 atm. Hexane is recovered by condensation of the stripper vapors.

Heat is exchanged countercurrently between the hot stripped oil and the cool oil from the absorber with a 10°F minimum approach in the exchanger. A supplementary steam heater is used for the stripper feed, and a supplementary water cooler is used for the absorber lean oil.

(a) Draw a schematic flow sheet of the process showing the necessary vessels, heat exchangers, fluid streams, flow rates and process conditions, and principal control instruments.

(b) Calculate the number of equilibrium stages for the absorber.

(c) Calculate the number of equilibrium stages for the stripper.

(d) Calculate the cooling and heating requirements of the process in Btu per pound of hexane recovered.

(e) Briefly describe two alternate separation processes which might be employed for hexane recovery.

(f) On the basis of the results of your analysis of the process, discuss the merits of recovering the hexane as opposed to merely venting the air from the oven to the atmosphere without recirculation.

Basic data

Absorption oil		Hexane	
Boiling point	500°F (mean average)	Average heat capacity,	
		liquid	0.6 Btu/lb. °F
Molecular weight	200	vapor	0.46 Btu/lb. °F
Gravity	40° API	Average heat of	
		vaporization	140 Btu/lb
Heat capacity,		Equilibrium constant K,†	
At 140°F	0.547 Btu/lb. °F	At 140°F, 1 atm	0.79
At 230°F	0.592 Btu/lb. °F	At 230°F, 1 atm	3.0

$$† K = \frac{y}{x} = \frac{\text{mole fraction in vapor}}{\text{mole fraction in liquid}}$$

8-F$_2$ Figure 4-39 shows an experimentally measured concentration distribution pattern for the separation of pentanoic, hexanoic, heptanoic, and octanoic acids by simple countercurrent distribution, using 24 vessels and 24 transfers. For the aqueous phase, 1.0 M in phosphate with a pH of 7.88, 8 cm^3 was used in each vessel along with 12 cm^3 of isopropyl ether solvent.

(a) By means of an appropriate analysis of Fig. 4-39, derive the values of K_i for each of these acids under the experimental conditions.

(b) Compare the *shape* of the peak for heptanoic acid in Fig. 4-39 with an appropriate theoretical prediction.

8-G₂ The primary tower for deuterium enrichment by dual-temperature H$_2$S–water isotope exchange at Savannah River has the configuration shown as scheme C in Fig. 6-30. It has been reported that the cold tower contains 70 trays and runs at about 30°C, whereas the hot tower contains 60 trays and runs at about 130°C. Both towers exhibit a Murphree vapor stage efficiency 100E_{MV} of about 65 percent. The feed water will be natural water containing 0.000138 atom fraction deuterium. Equilibrium data for the H$_2$S–H$_2$O isotope-exchange reaction are given in Prob. 1-F.

(a) Between what limits must S (the number of moles of H$_2$S circulated per mole of natural water feed) lie for this scheme to be operable to produce water substantially enriched in deuterium? Explain your answer by means of a qualitative operating diagram.

(b) Assume that a value of S equal to the geometric mean $\sqrt{S_1 S_2}$ of the two limiting values calculated in part (a) will be near optimal. Calculate the atom fraction deuterium $D/(H + D)$ in the enriched water if S is at this near optimal value.

(c) Calculate the atom fraction deuterium in the enriched water if S is 5 percent higher than the value used in part (b). Explain what has caused the change in the degree of enrichment of the product water.

8-H₂ Suppose that a sample of mixed xylene isomers is to be analyzed in the laboratory by means of a gas-liquid chromatography apparatus which can handle a column up to 6 m long and which will provide a length of an equivalent equilibrium stage equal to 0.5 cm. What is the minimum separation factor $\alpha_{12} = K'_1/K'_2$ that must be provided by the liquid solvent in order to give two separate and distinguishable peaks on the recorder output?

8-I₂ A solvent-extraction process uses *n*-butyl acetate as solvent to remove pyrocatechol (*o*-dihydroxybenzene) from a waste-water stream where it is present at an average concentration of 800 ppm w/w (0.08 weight percent). Won and Prausnitz (1975) report K_D = (weight fraction in solvent phase)/(weight fraction in aqueous phase) at equilibrium = 13.2 for this system at 298 K. If the regenerated butyl acetate solvent contains 100 ppm pyrocatechol, if the solvent-to-water ratio is 1.70 times the minimum which could give complete recovery with infinite stages, and if there are six actual stages in the extractor, each with a Murphree efficiency of 0.75, based on the water phase, calculate the percentage removal of pyrocatechol achievable.

8-J₂ Propylene, one of the leading petrochemicals manufactured industrially, is used for a variety of purposes including the production of polypropylene, isopropanol, and propylene oxide. The final purification step required for propylene production is almost always separation from propane by distillation. A field test of an existing propylene-propane splitter distillation column gave the following data:

Average column pressure = 1.86 MPa   Overhead temperature = 44.4°C

Bottoms temperature = 55°C   Reflux ratio L/d = 21.5

Propylene product purity = 96.2 mol °₀ (contaminant = propane)

Propane product purity = 91.1 mol °₀ (contaminant = propylene)

Propylene in feed = 50.45°₀   feed rate = 84.2 m^3/day (saturated liquid)

The tower is 1.21 m ID, with 90 sieve trays, the feed being introduced to the forty-fifth from the top. The tray spacing is 46 cm.

(a) What factor(s) probably led to the selection of 1.86 MPa as the column operating pressure when it was designed?

(b) Determine the overall stage efficiency (equilibrium-stage requirement for observed separation divided by the number of actual trays) exhibited by the column.

Vapor-liquid equilibrium data at 1.86 MPa (data interpolated from Reamer and Sage, 1951)

$x_{C_3H_6}$	0.0	0.2	0.4	0.6	0.8	1.0
$\alpha_{C_3H_6-C_3H_8}$	1.166	1.153	1.141	1.130	1.119	1.109

8-K$_3$ The existing propylene-propane distillation facilities described in Prob. 8-J are to be extended to produce polymerization grade propylene (99.7 mole percent). Possibilities under consideration are (1) relocation of the feed in the existing tower and (2) construction of a second tower to act as additional rectifying section for the existing tower, as shown in Fig. 8-12. The percentage loss of propylene in the bottoms product and the reflux ratio are to remain essentially the same.

(a) Is the feed injected to the present tower in the optimal location? If not, what propylene product purity can be achieved by the simple expedient of relocating the feed point and not building a new tower?

(b) How many plates are required in the new tower if the feed is put in at the optimal point?

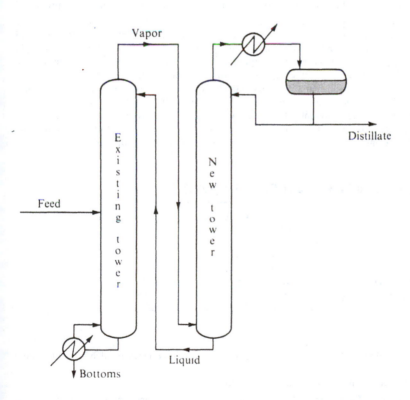

8-L$_3$ A considerable amount of capacity is being installed in the United States and other countries for the manufacture of propylene through thermal cracking of saturated hydrocarbons. Small amounts of propyne (methylacetylene) and propadiene are formed in the cracking process and would be highly deleterious if they were to appear in the propylene product. Hill et al. (1962) report vapor-liquid equilibrium data for small amounts of these two compounds in propylene-propane mixtures. Interpolation of their data in the light of the Reamer and Sage data (Prob. 8.J) shows that propyne is the species more likely to enter the propylene product and that its relative volatility at high dilution in saturated propylene-propane mixtures at 1.86 MPa is as follows:

$x_{C_3H_6}$	0.0	0.2	0.4	0.6	0.8	1.0
$\alpha_{C_3H_4-C_3H_8}$	1.17	1.12	1.07	1.02	0.97	0.92

SOURCE: Data interpolated from Hill et al. (1962).

Propyne concentrations in the mixed C_3's fed to a propylene recovery column might typically be on the order of 0.4 percent. Polymer-grade propylene may contain at most 20 ppm propyne. Can the distillation towers of Prob. 8-K accomplish this amount of propyne removal, or must some other means be provided?

8-M₂ A waste water from a petrochemical operation is to be treated by single-section extraction, using liquid isobutylene (under pressure) as the solvent. The critical components present in the water are listed in Table 8-3, along with their molecular weights and their equilibrium distribution ratios [K_D = (weight fraction in isobutylene)/(weight fraction in water), at equilibrium and at high dilution]. Crotonaldehyde is a bactericide and must be removed to less than 4 ppm before the water can be sent on to the plant biological treatment unit. The three phenolics present difficulties in the biological unit and have chemical value when recovered. They should be reduced to a point where no more than 500 ppm w/w of chemical oxygen demand (COD) due to phenolics is present. COD is defined as the weight of oxygen required to react with the compounds completely to form CO_2 and H_2O. A mixer-settler cascade providing five equilibrium stages is to be used. The entering isobutylene contains no significant levels of pollutants.

Table 8-3 Data reported by Won and Prausnitz (1974)

	MW	Concentration, ppm w/w	K_D
Phenol, C_6H_5OH	94.1	300	0.70
o-Cresol, $CH_3C_6H_4OH$	108.1	1730	4.8
m-Cresol, $CH_3C_6H_4OH$	108.1	2320	2.7
Crotonaldehyde, $CH_3CH{=}CHCHO$	70.1	800	2.48

(a) Does the restriction of COD due to phenolics or the restriction on crotonaldehyde set the more severe limit?

(b) What flow ratio of isobutylene to water is required?

8-N₂ Bergamot oil is one of the principal *essential oils* used for perfumes, colognes, cosmetics, etc. (Guenther, 1949). The bergamot tree is a member of the citrus family and is cultivated extensively in Calabria, southern Italy. The average content of the four most plentiful components is shown in Table 8-4. The most active ingredient is linalyl acetate. The terpenes (d-limonene, β-pinene) have much lower water solubilities than the esters, alcohols, and aldehydes present in the oil. Because of this, bergamot oil is frequently subjected to a terpene-removal process in order to make it more suitable for use in dilute aqueous alcohol solution.

Table 8-4 Average composition of bergamot oil and vapor pressures of components

	wt % in oil (Stoll, 1967)	Vapor pressure at 50°C, Pa (Weast, 1968)
Linalyl acetate, $CH_3CO_2C_{10}H_{17}$	29.8	103.5
d-Limonene, $C_{10}H_{16}$	28.4	1078
Linalool, $C_{10}H_{17}OH$	16.5	259
β-Pinene, $C_{10}H_{16}$	7.7	1890

One possible terpene-removal process would involve combined steam stripping and absorption. Steam would be fed to the bottom of a column and flow upward through a series of stages. An inert nonvolatile liquid, free of bergamot-oil constituents, would enter the top of the column at steam temperature and would flow downward, leaving at the bottom. Bergamot oil would be introduced partway up at a flow rate equal to approximately 1 percent of the steam rate, by weight. Steam leaving the column overhead would be condensed, separating into organic and aqueous phases. The organic phase would be the terpene by-product, and the aqueous phase would be sent to a water-treatment system. The bottom product would be subjected to a steam distillation to recover the deterpenized bergamot oil from the inert liquid.

One problem in processing an essential oil is the need for keeping temperatures low to avoid thermal degradation. For this reason the column will be operated under vacuum, at an absolute pressure of 10 kPa and a temperature of 50°C. This corresponds to superheated steam, which is used to avoid any chance of water condensation within the column. The inert nonvolatile liquid is used both as a carrier and to avoid having high concentrations of oil components in the liquid at column temperatures, which would accelerate degradation reactions.

Vapor pressures of the four principal oil components at 50°C are shown in Table 8-4. Assume that the activity coefficients for the various constituents in the inert liquid are essentially unity. The solubility of d-limonene in water is 13.8 mg/L at 25°C (Massaldi and King, 1973).

(a) It is desired to recover the linalyl acetate and linalool primarily in the bottom product and to have the d-limonene and β-pinene primarily in the overhead product. Choose an appropriate molar steam–inert-liquid ratio.

(b) Using the steam–inert-liquid ratio calculated in part (a), calculate the recovery fractions that would be obtained for the various oil components in a column providing six equilibrium stages, with the feed introduced to the third stage from the bottom.

(c) What fraction of the distilled d-limonene will be lost with the effluent water stream?

(d) Stoll (1967) indicates that the three principal processes in use for removing terpenes from bergamot oil are (1) *chromatography* with a silica-gel column, in which the oil feed is placed initially on the column, the terpene fraction is eluted with petroleum ether, and the deterpenized fraction is then eluted with alcohol; (2) *fractional extraction*, using counterflowing solvents of pentane and aqueous alcohol; and (3) *vacuum distillation*. Assess each of these approaches, along with the process evaluated in this problem, from the standpoints of throughput capacity and ability to recover desirable light-aldehyde components into the deterpenized bergamot oil.

NINE

LIMITING FLOWS AND STAGE REQUIREMENTS; EMPIRICAL CORRELATIONS

A countercurrent multistage separation process being designed to provide a specified degree of separation between feed components is characterized by certain minimum interstage flows and by a certain minimum number of stages required. With lower flows or stages than these minimum values the separation cannot be accomplished with the desired sharpness. It is helpful to identify these limiting conditions in order to orient oneself toward desirable operating conditions for the design of a multistage separation device. Furthermore, it has been found that the limiting flows and stage requirements can often serve as the basis for empirical correlations which predict the actual stage and reflux requirements under design conditions and which predict the split of nonkey components between products.

This chapter is concerned with the prediction of minimum flows and stage requirements and with those empirical correlations which utilize these limiting conditions. Emphasis is on distillation and other processes where reflux is generated by phase change.

MINIMUM FLOWS

It is useful to distinguish two major categories of multistage separation processes for the analysis of minimum-flow conditions.

For *mass-separating-agent processes* (absorption, stripping, washing, extraction without reflux, etc.) it is generally possible to change the flow of the stream containing the mass separating agent without changing the flow of the other, counter-flowing stream. The recovery fraction of one solute can be kept the same by selecting appropriate combinations of stages and mass-separating-agent flow, but it is usually not possible to keep the recovery fractions of each of two solutes unchanged in that way. The minimum flows can be determined for a binary system by reducing the input of mass separating agent until an operating line or curve first touches the equilibrium curve (Chap. 6), or for dilute systems they can be determined as L/mV or $mV/L = 1$, or less, as described in Chap. 8.

In the other class of processes *reflux* is *generated through phase change*, as in distillation and other energy-separating-agent processes, and in refluxed extraction. Here it is possible to keep the same recovery fractions of each of two components as stages and reflux are changed. In distillation, if the distillate flow remains unchanged, the flow rates of the counterflowing streams are necessarily linked and must change together since V must be the sum of L and d. The same applies whenever the flow of net upward product remains unchanged as internal flows are changed. Minimum flows to achieve a given separation in this class of process have been the subject of quite a bit of analysis and will be considered in more detail here. We shall use the notation of distillation, which is the principal case of this sort.

As noted in Chap. 7, the various components in a distillation can be either *distributing* or *nondistributing* at minimum reflux. A nondistributing component reaches a concentration of zero in one of the products, while a distributing component may reach a very low concentration in a product but not zero (Shiras et al., 1950). The key components in a multi-component distillation and both components in a binary distillation must necessarily be distributing at minimum reflux. The nonkey components in a multicomponent distillation can be either nondistributing or distributing at minimum reflux but are usually nondistributing. For a simple, two-section distillation nonkey components will be nondistributing unless the specified separation of the keys is relatively poor and/or the nonkey in question has a relative volatility very close to that of one of the keys or between those of the keys.

Different approaches are appropriate for calculating minimum flows when all components distribute and for the more general case where some of the components are nondistributing. We shall first consider the case where all components are distributing, because it necessarily applies to binary distillation and also places an upper limit on the minimum flows when some components are nondistributing.

All Components Distributing

When all components distribute at minimum reflux, the zones of constant composition above and below the feed will both be immediately adjacent to the feed stage. This is shown by the discussion surrounding Fig. 7-16, and by Shiras et al. (1950). The exception to this behavior occurs when the system is sufficiently nonideal to give the equivalent of the tangent pinch shown in Fig. 5-21. Such cases are rare, however, and we shall not consider them further.

Suppose that recoveries in the distillate are specified for each of two components i and j. These would typically (but not necessarily) be the two key components of a multicomponent distillation, or they would be the two components of a binary distillation. If we write Eq. (8-98) [Eqs. (7-6) and (7-7) combined] for each component in the zone of constant composition just above the feed, the resulting equations can be solved for K_i and K_j at the point of infinite stages and then divided, yielding

$$\alpha_{ij,\,\infty} = \frac{K_{i,\,\infty}}{K_{j,\,\infty}} = \frac{(x_{i,\,d}d/L_\infty x_{i,\,\infty}) + 1}{(x_{j,\,d}d/L_\infty x_{j,\,\infty}) + 1} \tag{9-1}$$

The subscript ∞ denotes conditions in the zone of infinite stages. Equation (9-1) can be rearranged to read

$$L_\infty = \frac{(x_{i,\,d}d/x_{i,\,\infty}) - \alpha_{ij,\,\infty} x_{j,\,d}d/x_{j,\,\infty}}{\alpha_{ij,\,\infty} - 1} \tag{9-2}$$

Since the stage below the feed $f-1$, the feed stage f, and the stage above the feed $f+1$ are all in the zone of constant composition, we have

$$x_{i,\,f-1} = x_{i,\,f} = x_{i,\,f+1} \tag{9-3}$$

and

$$y_{i,\,f-1} = y_{i,\,f} = y_{i,\,f+1} \tag{9-4}$$

A material balance around the feed stage then gives

$$L_{f+1}x_{i,f} + Fx_{i,F} + V_{f-1}y_{i,f} = L_f x_{i,f} + V_f y_{i,f}$$

$$Fz_{i,F} = (V_f - V_{f-1})y_{i,f} + (L_f - L_{f+1})x_{i,f}$$

and

$$Fz_{i,F} = (\Delta V)_f y_{i,f} + (\Delta L)_f x_{i,f} \tag{9-5}$$

where $(\Delta V)_f$ and $(\Delta L)_f$ represent the changes in flows at the feed tray, $V - V'$ and $L' - L$.

Since $y_{i,f}$ and $x_{i,f}$ are in equilibrium, Eq. (9-5) represents an equilibrium flash equation, where $(\Delta V)_f$ is the portion of the flashed feed which is vapor and $(\Delta L)_f$ is the portion of the flashed feed which is liquid. If the composition $x_{i,f}$ of the liquid is substituted for $x_{i,\infty}$ in Eq. (9-2), the general equation for $L_{min}(=L_\infty)$ results

$$L_{min} = \frac{(\Delta L)_f[x_{i,d}d/(\Delta L)_f x_{i,f} - \alpha_{ij}x_{j,d}d/(\Delta L_f)x_{j,f}]}{\alpha_{ij} - 1} \tag{9-6}$$

Note that $(\Delta L)_f$ can be removed altogether from the numerator of Eq. (9-6) if desired.

If $(\Delta L)_f$ is equal to F (saturated liquid feed), a very simple equation results:

$$L_{min} = \frac{(l_i)_d - \alpha_{ij}(l_j)_d}{\alpha_{ij} - 1} F \tag{9-7}$$

Similarly, for saturated vapor feed it can be shown that

$$V_{min} = \frac{\alpha_{ij}(l_i)_d - (l_j)_d}{\alpha_{ij} - 1} F \tag{9-8}$$

For feeds which are not saturated liquid or saturated vapor Eq. (9-6) must be used, the values of $(\Delta L)_f \, x_{i,f}$ being obtained from the standard flash equations. For subcooled liquid or superheated vapor feeds where either $(\Delta L)_f$ or $(\Delta V)_f$ is greater than F, the solution of the flash equation has no physical counterpart but does yield the mole fractions at the feed plate under the conditions of minimum reflux.

An interesting point can be noted from Eqs. (9-7) and (9-8). The minimum flow requirements for a perfect separation of i and j for saturated-liquid feed and saturated-vapor feed, respectively, are simply

$$L_{\min} = \frac{1}{\alpha_{ij} - 1} F \tag{9-9}$$

and

$$V_{\min} = \frac{\alpha_{ij}}{\alpha_{ij} - 1} F \tag{9-10}$$

These minimum flows apply to any distillation where $(/_i)_d$ is very high and $(/_j)_d$ is very low. Thus the reflux requirements do not increase rapidly as the separation required becomes increasingly difficult but become relatively constant. The stage requirements mount rapidly instead of reflux and if the separation is to be made more stringent, it can best be accomplished through an increase in the number of stages.

It should be stressed that no assumptions have been made stipulating constant flows or constant α. The proper value of α is that corresponding to the temperature given by an isenthalpic flash of the feed, and the value of $L_{\min}$ calculated is that flowing from the plate above the feed plate. The minimum overhead reflux flow or minimum condenser duty then can be obtained by means of combined enthalpy and mass balances relating the flows just above the feed plate to those at the overhead of the column.

For binary distillation, Eqs. (9-6) to (9-10) give simple expressions for obtaining minimum reflux, even for nonideal systems. For multicomponent distillations, they give an exact solution for the unusual case where all components distribute, and they provide an upper limit to the value of minimum reflux when one or more nonkeys are nondistributing. For processes other than distillation where reflux is generated through phase change, e.g., refluxed extraction, the equations apply with appropriate changes in notation, subject to the same restrictions. Now $x_{i,d} d$ becomes the net upflowing product of component i, etc.

General Case

In the more general case of a multicomponent distillation where some or all of the nonkey components are (or may be) nondistributing at minimum reflux, it cannot be presumed that the zones of constant composition are immediately adjacent to the feed stage. Nonetheless, relatively simple equations can be derived for the limiting flows in a single section, and they can be combined for a two-section column. We shall consider how to ascertain which nonkeys are distributing and which are nondistributing and shall explore ways of solving the equations. The most common multicomponent situation is that where all nonkeys are nondistributing. The solution for that case is simple and is often a good approximation even when some of the nonkeys distribute. We shall also consider allowance for varying molal overflow, varying relative volatilities, and multiple-section columns.

Single section For a single-section separation cascade the mole fraction of any component and the total flow in a zone of constant composition are described by Eqs. (8-101), (8-102), and (8-104) combined

$$V_\infty y_{i,\,\infty} = \frac{\alpha_i x_{i,\,d} d}{\alpha_i - \phi_{HK-}} \qquad (9\text{-}11)$$

$$V_\infty = \sum_{i=1}^{R} \frac{\alpha_i x_{i,\,d} d}{\alpha_i - \phi_{HK-}} \qquad (9\text{-}12)$$

These equations are written for the rectifying section of a distillation column but can be made to apply to single-section multicomponent separation processes in general if $x_{i,\,d} d$ is identified as the net flow of component i in the positive direction along the cascade, if V is the interstage flow in the positive direction, and if y_i is the mole fraction of component i in the stream flowing in the positive direction. Thus, for example, the appropriate equations for the stripping section of a distillation column can be obtained by selecting the direction of liquid flow as the positive direction. Following that definition, we substitute x_i for y_i, L' for V, and $1/\alpha$ for α, since α_{ij} is still defined as $y_i x_j / y_j x_i$ at equilibrium. Since the ϕ's are related to the α's, we also substitute $1/\phi'_{LK+}$ for ϕ_{HK-}, and thereby convert Eqs. (9-11) and (9-12) into

$$L'_\infty x_{i,\,\infty} = \frac{x_{i,\,b} b}{1 - (\alpha_i / \phi'_{LK+})} \qquad (9\text{-}13)$$

and

$$L'_\infty = \sum_{i=1}^{R} \frac{x_{i,\,b} b}{1 - (\alpha_i / \phi'_{LK+})} \qquad (9\text{-}14)$$

Two sections As illustrated in Figs. 7-16 to 7-19, if nondistributing components are present, one or both of the zones of constant composition will not end at the feed stage and the equations developed for all components distributing are then no longer valid. One or more nondistributing heavy nonkeys serve to displace the zone of constant composition in the rectifying section away from the feed stage, and one or more nondistributing light nonkey components serve to displace the zone of constant composition in the stripping section away from the feed stage. We then cannot substitute $x_{i,\,f}$ for $x_{i,\,\infty}$ in Eq. (9-1) or its counterpart written for the stripping section. In between the two zones of constant composition the nondistributing nonkeys change in concentration, and the keys undergo reverse fractionation (Figs. 7-16 to 7-19).

If constant relative volatilities and constant molal overflow within either section are assumed in the region between the zones of constant composition, the Underwood equations can be used to calculate the minimum flows directly, without having in some other way to solve for the composition changes between the zones of constant composition (Underwood, 1946, 1948). Consider the equations defining ϕ and ϕ'

$$V = \sum_{i=1}^{R} \frac{\alpha_i x_{i,d} d}{\alpha_i - \phi} \tag{8-94}$$

$$-V' = \sum_{i=1}^{R} \frac{\alpha_i x_{i,b} b}{\alpha_i - \phi'} \tag{8-95}$$

Under the conditions of infinite plates in both sections, certain of the ϕ and ϕ' roots of Eqs. (8-94) and (8-95) are identical, as shown by Underwood (1946; 1948). Since these roots are solutions of both equations, they are roots of the sum

$$(\Delta V)_f = V - V' = \sum_{i=1}^{R} \frac{\alpha_i x_{i,d} d}{\alpha_i - \phi} + \sum_{i=1}^{R} \frac{\alpha_i x_{i,b} b}{\alpha_i - \phi}$$

or, since $x_{i,d} d + x_{i,b} b = F z_{i,F}$,

$$(\Delta V)_f = \sum_{i=1}^{R} \frac{\alpha_i F z_{i,F}}{\alpha_i - \phi} \tag{9-15}$$

Equation (9-15) requires only a knowledge of the state and composition of the feed, making it possible to calculate values of ϕ common to Eqs. (8-94) and (8-95) without full knowledge of the product compositions. If any one of these common values of ϕ obtained from Eq. (9-15) is used in Eq. (8-94), the value of V calculated is $V_{\min}$. The values of ϕ which are common to both Eqs. (8-94) and (8-95) are those which lie *between the α values of distributing components* (Underwood, 1948). Also, values of $x_{i,d} d$ are needed for use in Eq. (8-97). Thus it is necessary to know or be able to calculate which components are distributing. This knowledge can be obtained formally by a method outlined by Shiras et al. (1950), where all possible values of ϕ are computed from Eq. (9-15) as follows.

We shall presume that we have a problem in which the specified separation variables serve to set the recovery fractions of the two key components in the two products. From Eq. (9-15) we can obtain $R - 1$ values of ϕ lying between the α values of the different components, where R is the number of components in the feed. Writing Eq. (8-94) $R - 1$ times leads to a set of equations in which the unknowns are V and the $R - 2$ values of $x_{i,d} d$ for the various nonkey components. If these equations are solved simultaneously, components which do not distribute when there are infinite stages will be shown by calculated values of $x_{i,d} d$ for those components which either are negative or are greater than the corresponding values of $F z_{i,F}$. The next step is to eliminate the ϕ's bounded by these nondistributing components and set recovery fractions for those components at 0 or 1, whichever is appropriate. The equations for the remaining values of ϕ are then solved again to see if other components now become nondistributing. Since some of the equations written the first time were not correctly applied, all the calculated values of $x_{i,d} d$ from the first solution are wrong to some extent, and reduction of the set of equations should be done cautiously. When the set of equations has been reduced to a number of equations correctly 1 less than the number of distributed components, the calculated values of

$x_{i,d}d$ for the various unspecified components will be correct, and the value of V calculated from Eq. (8-94) written as often as necessary is equal to V_{min} for the separation required.

A small degree of experience leads quickly to the ability to judge whether or not a particular component will distribute, but this can also be readily ascertained without experience. In order for a nonkey component to be distributing it must either have a volatility intermediate between the keys or be close to one of the keys in volatility. The less sharp the specified separation of the keys, the farther a nonkey may be from the keys in volatility and still be distributing. Barnés (1970) has shown that ϕ_{HK-} and ϕ'_{LK+} are identical to the α_i values of hypothetical trace components which are just on the border line of being distributing or nondistributing.

The value of V_{min} calculated from Eq. (8-94) is relatively insensitive to the actual values of $x_{i,d}$ for nonkey components. Often the nonkey recovery fractions at minimum reflux can be taken to be 0 or 1 without the introduction of serious error into Eq. (8-94) even though not all the nonkeys are nondistributing. As a still better approximation, it may be noted that Eqs. (9-6) to (9-8), used previously for the case where all components distribute, give relatively accurate values of the recovery of fractions of the distributing nonkey components even when some of the other nonkeys are nondistributing. If these distributing equations are used to determine values of $x_{i,d}d$ for the nonkey components, only one value of ϕ from Eq. (9-15) need be calculated and one form of Eq. (8-94) can be solved for V_{min}. This one necessary value of ϕ from Eq. (9-15) is that lying between the α values of the key components if no other component is intermediate in volatility between the keys.

To summarize, the minimum vapor rate and hence the minimum reflux in multi-component distillation can be obtained by alternate use of two equations

$$(\Delta V)_f = \sum_{i=1}^{R} \frac{\alpha_i F z_{i,F}}{\alpha_i - \phi} \tag{9-16}$$

$$V_{min} = \sum_{i=1}^{R} \frac{\alpha_i x_{i,d} d}{\alpha_i - \phi} \tag{8-94}$$

If the degree of vaporization of the feed $(\Delta V)_f$ and the distillate composition can be estimated with some precision, a single value of ϕ (that lying between the relative volatilities of the keys) can be obtained by iterative solution of Eq. (9-16). This value of ϕ, substituted into Eq. (8-94), will give V_{min}. When the distillate composition cannot be estimated with sufficient precision, we can obtain several values of ϕ from Eq. (9-16) and then write Eq. (8-94) as many times as there are values of ϕ, solving the set for V_{min} and the unknown $x_{i,d}d$. If any of the $x_{i,d}d$ come out to be unreal, those components are branded nondistributing and the procedure is repeated with those components put entirely into the distillate or into the bottoms. Usually the remaining values of $x_{i,d}d$ obtained from the first solution are accurate enough, and the entire solution need not be repeated.

The comments immediately before Example 8-9 concerning methods of solving for ϕ apply as well to the iterative solution of Eq. (9-16) for the ϕ values between the distributing components.

The assumptions of constant molal overflow and constant relative volatility in the Underwood equations, as used for minimum flows, are less restrictive than may at first appear. Since relative volatilities and molar flows are those applying at the two zones of constant composition, unless there is the equivalent of a tangent pinch, the assumptions are actually that constant relative volatilities and constant molal overflow prevail in the region between the zones of constant composition, which is a much narrower range of compositions than occurs over the entire column.

When relative volatilities are variable, the estimated average values between the two zones of constant composition should be used. Usually these will be at the estimated feed-stage temperature, obtained either as the equilibrium temperature of the feed upon entry or as the geometric mean of the overhead and bottoms temperatures.

When constant molal overflow does not occur, the Underwood equations still give a good estimate of the minimum flows at the zones of constant composition, subject only to the assumption of constant molal overflow between these zones. The minimum overhead-reflux or reboiler-vapor flow can then be obtained by solving combined enthalpy and mass balances written for an envelope cut by the terminal flows and the flows in the zone of constant composition. A still more accurate approach for varying molal overflow is the modified-latent-heat-of-vaporization (MLHV) method, converting the Underwood equations to pseudo-mole-fraction bases, as described in Chap. 6. The same appropriate average values of relative volatility apply. Sensible-heat effects can be allowed for, as well, by taking latent heats to correspond to the difference between vapor enthalpies at the temperature of the rectifying-section zone of constant composition and liquid enthalpies at the temperature of the stripping-section zone of constant composition.

For highly nonideal systems and/or systems with complex stream-enthalpy effects, Tavana and Hanson (1979a) give an exact method using Eqs. (8-98), (8-99), and (8-107) to derive the flows and compositions in the two zones of constant composition, the method of Ricker and Grens (see Chap. 10) being used to solve for composition and flow changes between these two zones.

Several other approaches developed for determining minimum reflux in multicomponent distillations (Bachelor, 1957; Gilliland, 1940a; Shiras et al., 1950; Erbar and Maddox, 1962; Chien, 1978; etc.) involve either a more complex hand-calculation method than the approach using the Underwood equations presented above or do not appear to present advantages over the Tavana-Hanson approach for an exact solution.

Finally, it should be stressed that the value of L_{min} or V_{min} calculated by applying the equations for all components distributing to the key components alone is always equal to or larger than the true minimum flow for a multicomponent system and hence is conservative. If the system consists mostly of the two key components with high recovery fractions, V_{min} computed by means of these equations usually will be sufficiently close to the correct value to be used directly in setting an operating value for the column reflux rate.

Example 9-1 A multicomponent feed to a distillation is described as shown. The feed is saturated liquid. For the set of separation specifications $(/_C)_d = 0.9$ and $(/_D)_d = 0.1$ calculate the minimum reflux.

$$L_s = L_F + L_r$$
$$V_r = V_F + V_S$$

Component	$F(x)_F$	α
A	0.05	3
B	0.10	2.1
C	0.30	2
D	0.50	1
E	0.05	0.8
	1.00	

SOLUTION It is rather clear from the relative volatilities of the various components that A will be a nondistributing component, appearing exclusively in the distillate at minimum reflux. The real question is whether B and E will distribute. To illustrate the method, however, we shall test all three nonkey components to see whether they are distributing or not.

Equations for all components distributing If all components distribute, the equations based on the two infinite sections meeting at the feed plate apply. Using Eq. (9-7) for the saturated liquid feed gives

$$L_{min} = \frac{(/_C)_d - \alpha_{CD}(/_D)_d}{\alpha_{CD} - 1} F = \frac{0.9 - 2(0.1)}{2 - 1} = 0.7 \text{ mol/mol of feed}$$

This value of L_{min} is the first (conservatively high) approximation to the minimum reflux flow.

Assumed distributions of nonkeys Solving for values of $(/_i)_d$ for the nonkey components by means of Eq. (9-7) gives

$$0.7 = \frac{(/_A)_d - \alpha_{AD}(/_D)_d}{\alpha_{AD} - 1} \qquad (/_A)_d = 1.7$$

$$0.7 = \frac{(/_B)_d - \alpha_{BD}(/_D)_d}{\alpha_{BD} - 1} \qquad (/_B)_d = 0.98$$

$$0.7 = \frac{(/_E)_d - \alpha_{ED}(/_D)_d}{\alpha_{ED} - 1} \qquad (/_E)_d = -0.06$$

These recovery fractions obtained by the equations for all components distributing indicate that components A and E are probably nondistributing. The approximate values of $x_{i,d}d$ to be used in the calculation of V_{min} are thus

	$x_{i,d}d$
A	0.05
B	0.098
C	0.27
D	0.05
E	0
	0.469

From Eq. (9-16)

$$(\Delta V)_f = 0 = \frac{3(0.05)}{3 - \phi} + \frac{2.1(0.10)}{2.1 - \phi} + \frac{2(0.3)}{2 - \phi} + \frac{1(0.50)}{1 - \phi} + \frac{0.8(0.05)}{0.8 - \phi}$$

The value of ϕ lying between the α values of the key components is $\phi = 1.386$. Substituting this value of ϕ into Eq. (8-94) gives

$$V_{min} = \sum \frac{\alpha_i x_{i,d} d}{\alpha_i - \phi}$$

$$= \frac{3(0.05)}{1.614} + \frac{2.1(0.098)}{0.714} + \frac{2(0.27)}{0.614} - \frac{1(0.05)}{0.386} = 1.131 \text{ mol/mol feed}$$

$$L_{min} = V_{min} - d = 1.131 - 0.469 = 0.662 \text{ mol/mol feed}$$

Rigorous solution The values of L_{min} and $x_{i,d} d$ can be calculated even more closely by use of equations for V_{min} [Eq. (8-94)] written in the four possible ϕ values. Using these values,

$$\phi_1 = 2.8788 \qquad \phi_2 = 2.07485 \qquad \phi_3 = 1.3857 \qquad \phi_4 = 0.81179$$

and solving the four simultaneous equations, dropping first the equation written in ϕ_1 as it becomes clear that component A is nondistributing, then solving the three remaining equations and dropping the equation written in ϕ_4 as it becomes clear that component E is nondistributing, and finally solving the two remaining valid equations gives

$$V_{min} = \frac{3(0.05)}{3 - 2.07485} + \frac{2.1 x_{B,d} d}{2.1 - 2.07485} + \frac{2(0.27)}{2 - 2.07485} + \frac{1(0.05)}{1 - 2.07485}$$

$$V_{min} = \frac{3(0.05)}{3 - 1.3857} + \frac{2.1 x_{B,d} d}{2.1 - 1.3857} + \frac{2(0.27)}{2 - 1.3857} + \frac{1(0.05)}{1 - 1.3857}$$

It is found that

$$V_{min} = 1.132 \text{ mol/mol of feed} \qquad x_{B,d} d = 0.0986$$

$$(/_B)_d = 0.986 \qquad L_{min} = 0.663 \text{ mol/mol of feed}$$

Thus the assumption that all components distribute gave L_{min} 6 percent too high, and the application of the equations for all components distributing to the nonkeys followed by the use of Eqs. (9-16) and (8-94), each written once, gave L_{min} indistinguishable from the exact answer. The recovery fraction obtained for component B from the equations for all components distributing was also very nearly correct. For many purposes the most approximate solution was probably sufficiently accurate. In general, the use of the approximate values of $x_{i,d} d$ and a single value of ϕ gives a reliable and easily obtained answer. ☐

Multiple Sections

If a separation process has several intermediate feeds and/or products, or if there is intermediate heating or cooling, the analysis of limiting interstage flows becomes more complex. This complexity is only a matter of locating the appropriate pinch point(s) or zone(s) of infinite stages, however, and simply involves keeping one's eyes open to all prospective pinch points. For example, for a two-feed binary distillation column like that shown in Fig. 5-10, it is not immediately clear whether the operating lines will first touch the equilibrium curve at the upper feed point or at the lower feed point as the overhead reflux ratio is decreased. The best approach is to assume that the pinch occurs at one of the points and then check to make sure that the other point has not then crossed the equilibrium curve. Each prospective pinch point in a multifeed binary distillation can be checked by considering $x_{i,d}d$ for that point to be the sum of component i in products leaving the cascade above that point minus the amount of component i in feeds entering *above* that point and by considering the feed to be the feed entering *at* that point.

Barnés et al. (1972) have shown that the same basic strategy can be used for multicomponent, multifeed towers, using the Underwood equations written to allow for the multiple feed points. The procedure is more subtle, however. Among the complications is a change from single-feed experience regarding which nonkeys will be distributing and which will be nondistributing. For example, an intermediate nonkey missing from a feed altogether can become nondistributing, and a light nonkey present to a substantial amount in the lowest feed can be distributing.

Tavana and Hanson (1979b) found that the Underwood equations can be used to determine minimum reflux for a column with a sidestream. If the sidestream is a liquid withdrawn above the feed, the basic change is to replace the equation for the rectifying section with an equation for the section between the sidestream and the feed; i.e., in Eq. (8-94) $x_{i,d}d$ is replaced by the net upward product flow of i between the feed and the sidestream, which is $x_{i,d}d + Sx_{i,s}$. Similarly, for a sidestream withdrawn as a vapor below the feed $x_{i,b}b$ in Eq. (8-95) becomes $x_{i,b}b + S'x_{i,s}$, which is the net downward product of i in the section below the feed and above the sidestream.

MINIMUM STAGE REQUIREMENTS

Energy Separating Agent vs. Mass Separating Agent

The minimum equilibrium-stage requirement for a given quality of separation in a countercurrent multistage process occurs when one or both of the interstage flows are infinite or the product and feed flows are zero. In a process such as absorption, where the flows are not linked by a difference equation, it is conceptually possible for one flow to become infinite while the other flow remains fixed at some finite value. Thus to achieve a given separation in an absorber the absorbent liquid flow could be increased without limit while the gas feed rate to the absorber remained fixed. On the other hand, for a process such as distillation, where reflux is generated through phase change and the two interstage flows are linked to each other by a difference relationship, it is impossible for one flow to become infinite unless the other flow also becomes infinite. Thus infinite vapor flow in distillation also implies infinite liquid flow.

In a process with a mass separating agent where the flows are not linked, the condition of infinite flow of only one of the streams implies that the ratio V/L or its equivalent has become either zero or infinite. The situation can be pictured for a binary separation on an operating diagram where the operating line is completely horizontal or completely vertical. The stream with infinite flow rate does not change in composition within the separation process; hence one equilibrium stage necessarily provides equilibrium of the finite stream with the inlet composition of the infinite stream. Any real separation which does not provide complete equilibrium takes less than one equilibrium stage. Thus for a process with unlinked interstage flows the minimum requirement is necessarily one stage or less.

When the two interstage flows are linked, on the other hand, it is possible to conceive of and determine the minimum stage requirement for a given separation more clearly. We have already seen a graphical construction of the minimum stage requirement for a binary distillation in Fig. 5-22. On the McCabe-Thiele diagram the case of minimum stages occurs when b and d, the amounts of products, are infinitely small in comparison with L and V. Because of this L/V necessarily equals 1.0 in both sections. In general, for processes with an energy separating agent or for mass separating agent processes where the stream flows are written on a separating-agent-free basis, the minimum stage condition corresponds to both counterflowing streams having the same flow rate and the same composition. This is the case of more interest and is developed further here.

Binary Separations

The graphical construction for the minimum equilibrium stages condition for a specified binary separation is simple, as shown in Fig. 5-22. It is also possible to develop a simple analytical relationship which applies to both binary and multicomponent systems where the separation factor is constant or can itself be related to composition or temperature through a simple function. We consider the case of a distillation.

If a distillation column is considered equilibrium stage by equilibrium stage starting at the bottom, we can write the following equation from the definition of the separation factor (equilibrium reboiler = Stage 1):

$$\left(\frac{y_A}{y_B}\right)_1 = \alpha_1 \left(\frac{x_A}{x_B}\right)_b \tag{9-17}$$

where α is understood to be α_{AB} evaluated at the stage temperature. $x_{A,2}$ is related to $y_{A,1}$ by material balance,

$$V_1 y_{A,1} = L_2 x_{A,2} - x_{A,b} b \tag{9-18}$$

but for infinite flows or zero feed and product flows $V \gg b$, so that

$$y_{A,1} = x_{A,2} \tag{9-19}$$

Thus, combining Eqs. (9-17) and (9-19) gives

$$\left(\frac{x_A}{x_B}\right)_2 = \alpha_1 \left(\frac{x_A}{x_B}\right)_b \tag{9-20}$$

Similarly, to relate equilibrium stage 3 to equilibrium stage 2 and the bottoms we can write

$$\left(\frac{x_A}{x_B}\right)_3 = \alpha_2 \left(\frac{x_A}{x_B}\right)_2 = \alpha_2 \alpha_1 \left(\frac{x_A}{x_B}\right)_b \tag{9-21}$$

If the development is continued to the top of the column, the final equation results:

$$\left(\frac{y_A}{y_B}\right)_t = \alpha_t \alpha_{t-1} \cdots \alpha_1 \left(\frac{x_A}{x_B}\right)_b \tag{9-22}$$

or

$$\left(\frac{x_A}{x_B}\right)_d = (\alpha_{AB})^{N_{min}} \left(\frac{x_A}{x_B}\right)_b \tag{9-23}$$

if α_{AB} is constant across the stages. The minimum number of equilibrium stages N_{min} includes an equilibrium reboiler and/or an equilibrium partial condenser, if used. However, a total condenser and/or a reboiler that generates its product by stream splitting rather than equilibration would not contribute to N_{min}.

Solving for N_{min} gives

$$N_{min} = \frac{\log \left[(x_A/x_B)_d/(x_A/x_B)_b\right]}{\log \alpha_{AB}} \tag{9-24}$$

If α_{AB} varies from stage to stage, the value at the feed stage or the geometric mean of the values at the top and bottom of the column can again be used as an approximation. The logarithms may be natural or base 10.

Another form of the equation may be more convenient to remember:

$$N_{min} = \frac{\log \left[(DR)_A/(DR)_B\right]}{\log \alpha_{AB}} \tag{9-25}$$

where $(DR)_i$ is defined as the *distribution ratio* of component i and is equal to the ratio of the recovery fractions of i between the top and bottom products

$$(DR)_i = \frac{(f_i)_d}{(f_i)_b} \tag{9-26}$$

Equation (9-24), generally known as the *Fenske equation*, was derived independently by Fenske (1932) and Underwood (1932). It should be noted that the minimum number of equilibrium stages to accomplish a certain separation mounts as the separation required becomes more difficult [more extreme $(DR)_i$, or α_{AB} closer to unity] and that the stage requirements are independent of the feed and depend only on the separation requirements. From this result and from the fact that minimum reflux requirements become insensitive to product purities for difficult separations, as noted earlier in this chapter, it is apparent that increased stages are generally more effective than increased flow for increasing product purity for difficult separations.

A revised version of the Fenske equation for use in cases where the relative volatility is not constant has been suggested by Winn (1958). If

$$\alpha_{ij} = \beta K_j^{\theta-1} \tag{9-27}$$

where β and θ are empirical constants, a derivation similar to that above gives

$$N_{min} = \frac{\log \left[(DR)_i/(DR)_j^{\theta}\right]}{\log \beta} \tag{9-28}$$

Multicomponent Separations

The computation of minimum stages for multicomponent systems uses the same equations as those developed under the assumption of constant α for binary systems. The development of the equations contains no assumption limiting the number of components in the system. For multicomponent systems the minimum number of equilibrium stages is calculated from the set separation on the two key components. If desired, the separation on all nonkey components at total reflux can then be calculated by resubstitution into Eq. (9-24) or (9-25) utilizing the known minimum number of equilibrium stages and the specified separation of either of the keys.

For binary separations where the equilibrium relationships have an irregular behavior or the minimum number of actual stages with a known Murphree vapor efficiency is to be determined, a graphical construction is best.

Douglas (1977) has shown that an estimate of N_{min} can be made from the ratio of the sum of the absolute boiling points to the difference in boiling points of the two key components. If 97 percent recovery fractions are specified, the multiplying factor is about 0.33; if 99 percent recoveries are specified, it is about 0.43, etc.

Example 9-2 Determine the minimum stage requirement for the distillation described in Example 9-1. Compare the recovery fractions of the nonkeys in the overhead product at total reflux with those determined in Example 9-1 for minimum reflux.

SOLUTION Applying the Fenske equation to components C and D, we have

$$N_{min} = \frac{\log\left[(DR)_C/(DR)_D\right]}{\log \alpha_{CD}} = \frac{\log\left[(0.9/0.1)(0.9/0.1)\right]}{\log 2} = \frac{4.4}{0.693} = 6.34$$

The distribution ratios and overhead recoveries of the nonkeys at total reflux can also be calculated from the Fenske equation:

$$6.34 = \frac{\log\left[(DR)_A/(DR)_D\right]}{\log \alpha_{AD}} = \frac{\log\left[9(DR)_A\right]}{\log 3}$$

$$(DR)_A = \frac{3^{6.34}}{9} = 120 \qquad (\mathnormal{f}_A)_d = \frac{(DR)_A}{1 + (DR)_A} = \frac{120}{121} = 0.992$$

Similarly, for the other nonkeys,

$$(DR)_B = \frac{(2.1)^{6.34}}{9} = 12.5 \qquad (\mathnormal{f}_B)_d = \frac{12.5}{13.5} = 0.926$$

$$(DR)_E = \frac{(0.8)^{6.34}}{9} = 0.0269 \qquad (\mathnormal{f}_E)_d = \frac{0.0269}{1.0269} = 0.026$$

Comparing, we find

	Total reflux	Minimum reflux
$(\mathnormal{f}_A)_d$	0.992	1.000
$(\mathnormal{f}_B)_d$	0.926	0.986
$(\mathnormal{f}_E)_d$	0.026	0.000

□

EMPIRICAL CORRELATIONS FOR ACTUAL DESIGN AND OPERATING CONDITIONS

Stages vs. Reflux

The determinations of minimum flows and minimum stages for a given separation are helpful for fixing the allowable ranges of flow conditions and stages. They are also useful guidelines for picking particular operating conditions for a subsequent design calculation. Since the procedures for obtaining an exact relationship between flow requirements and number of stages are relatively complex, there have been several attempts to establish empirical correlations for the flow-stage relationship. These correlations almost invariably have been based upon prior knowledge of the minimum flows and minimum stage requirements. They correspond to cases where both flows must become infinite together (linked flows) and where relative volatilities do not vary greatly through the separation process.

The most widely used correlation of this sort is that developed by Gilliland (1940b) on the basis of stage-to-stage calculations for over 50 binary and multicomponent distillations. The correlation is shown in Fig. 9-1 on arithmetic and logarithmic scales. The points in Fig. 9-1 give some idea of the scatter. Notice that even though the logarithmic form of the plot is extended to low values of the ordinate, there are few points below an ordinate value of 0.05.

Use of the correlation should be based upon the following qualifications. †

> It can be shown theoretically that a single line cannot represent all cases exactly, and the correlation can be improved by using more than one line. For example, the position of the line is a function of the fraction of the feed that is vapor. The best line drawn through the all-vapor feed cases on [this] plot ... is lower than the corresponding line for all-liquid feeds. It is also possible to improve the correlation by changing the variable groups, but it is doubtful whether the increased accuracy justifies the added complications. The accuracy of such a correlation will always be limited by the errors in N_{min} and $(L/d)_{min}$. It is believed that it is of real value when it is applied as (1) a rapid but approximate method for preliminary design calculation or (2) a guide for interpolating and extrapolating plate-to-plate calculations. In this latter case, if only one plate-to-plate result is available at a reflux ratio from 1.1 to 2.0 times $(L/d)_{min}$, this point can be plotted on the diagram and a curve of similar shape to the correlation curve fitted to it. Such a method should give good results for other reflux ratios, assuming the values of N_{min} and $(L/d)_{min}$ are reasonably accurate.

When the Gilliland correlation is being used repeatedly, as for approximating stage requirements during computer calculations for an entire process, it is helpful to have an analytical expression for the correlation. Eduljee (1975) has found that the following simple equation represents the correlation well:

$$Y = 0.75 - 0.75X^{0.5668} \qquad (9\text{-}29)$$

where $Y = (N - N_{min})/(N + 1)$ and $X = [(L/d) - (L/d)_{min}]/[(L/d) + 1]$. This expression gives $Y = 0$ at $X = 1$, as it should, but does not extrapolate to $Y = 1$ as X

† From Robinson and Gilliland (1950, p. 348); used by permission.

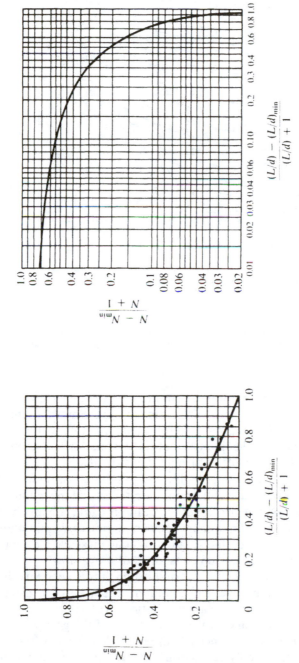

Figure 9-1 Gilliland correlation. *(Adapted from Robinson and Gilliland, 1950, pp. 348 and 349; used by permission.)*

429

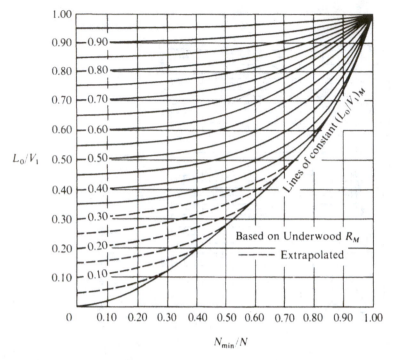

Figure 9-2 Erbar and Maddox correlation. (L_0/V_1 = ratio of overhead reflux to vapor flow off top plate. Subscript M denotes condition of minimum reflux.) (*From Erbar and Maddox, 1961, p. 185; used by permission.*)

approaches zero. When a good representation of realistic behavior is needed, including very low values of X, it may work better to use the more complex representation of the Gilliland correlation given by Molokanov et al. (1972).

Erbar and Maddox (1961) found that the correlation shown in Fig. 9-2 provides a better fit to a large set of multicomponent distillation results, which they had obtained by stage-to-stage computations using a digital computer. Their correlation is also based upon knowledge of the minimum reflux and the minimum stage requirement. Notice that flows at the top of the column L_0/V_1 are involved in the correlation. Since the Underwood minimum-reflux equations give minimum flows near the feed stage, solution of a combined enthalpy and mass balance may be required to obtain $(L_0/V_1)_M$ when there is no constant molal overflow.

A more complex correlation of equilibrium stages vs. reflux, based upon prior calculations of minimum stages and minimum reflux, has been given by Strangio and Treybal (1974). It is exact for binary distillations with constant relative volatility and constant molal overflow since it is based on one of the group-method solutions for such binary distillations. Another advantage is that it yields the number of equilibrium stages for the rectifying section and the number for the stripping section separately, thereby giving the feed location, which the Gilliland and Erbar-Maddox correlations do not provide. Since the amount of fractionation from stage to stage is generally different in the stripping section and in the rectifying section, it should be advantageous to divide the sections for purposes of such correlations. Thus the Strangio-Treybal approach appears to give better results than the other correlations when the stripping-section stages are preponderant over those in the rectifying section. On the other hand, the Strangio-Treybal relationships are substantially more complicated to use than either the Gilliland or Erbar-Maddox correlations, and the extra effort, with uncertainty still remaining in the result, should be balanced against the additional effort of obtaining a full solution by the methods described in Chap. 10.

Example 9-3 Use the Gilliland and Erbar-Maddox correlations to estimate the equilibrium-stage requirement for the depropanizer example considered in Example 8-9 and outlined in Table 7-2. Use the key component splits as separation specifications.

SOLUTION Values of α_i at the feed temperature will be employed for the estimation of $(L/d)_{min}$ via the Underwood equations and for the estimation of N_{min} from the Fenske equations:

	α_i
Methane	10.0
Ethane	2.47
Propane	1.0
Butane	0.49
Pentane	0.21
Hexane	0.10

It is apparent that only the keys will distribute at minimum reflux (this conclusion can be checked by the appropriate calculation, if desired). Substituting the feed composition from Table 7-2 into Eq. (9-16) and working on the basis of 1 mol of feed gives

$$0.67 = \frac{10.0(0.26)}{10.0 - \phi} + \frac{2.47(0.09)}{2.47 - \phi} + \frac{1.0(0.25)}{1.0 - \phi} + \frac{0.49(0.17)}{0.49 - \phi} + \frac{0.21(0.11)}{0.21 - \phi} + \frac{0.10(0.12)}{0.10 - \phi}$$

since the feed is 67 percent vapor. Solving for $1.0 > \phi > 0.49$ gives

$$\phi = 0.678$$

and substitution into Eq. (8-94) leads to

$$V_{min} = \frac{10.0(0.26)}{10.0 - 0.678} + \frac{2.47(0.09)}{2.47 - 0.678} + \frac{1.0(0.246)}{1.0 - 0.678} + \frac{0.49(0.003)}{0.49 - 0.678}$$

$$= 0.279 + 0.124 + 0.764 - 0.008 = 1.159$$

$$\left(\frac{V}{D}\right)_{min} = \frac{1.159}{0.599} = 1.935$$

$$\left(\frac{L}{D}\right)_{min} = 1.935 - 1 = 0.935$$

$$\left(\frac{L}{V}\right)_{min} = \frac{0.935}{1.935} = 0.483$$

The specified L/D of 90/59.9 is 1.60 times minimum reflux.

Using the Fenske equation for the minimum equilibrium-stage requirement gives

$$N_{min} = \frac{\ln\left[(24.6/0.4)/(0.3/16.7)\right]}{-\ln 0.49} = 11.4$$

It is also possible to allow for variations in the relative volatility of propane to butane by using the equation of Winn (1958). Taking equilibrium ratios from Edmister (1948) for the overhead and bottoms temperatures, we have

	K_{C_4}	$\alpha_{C_3-C_4}$
Top	0.16	3.13
Bottom	0.45	1.85

Next we fit the constants β and θ of the Winn equation:

$$\ln 3.13 = \ln \beta + (\theta - 1)\ln 0.16 \qquad \ln 1.85 = \ln \beta + (\theta - 1)\ln 1.45$$

$$1.14 = \ln \beta - 1.83(\theta - 1) \qquad 0.62 = \ln \beta + 0.37(\theta - 1)$$

$$\theta = 0.766 \qquad \beta = 2.04$$

Substituting into Eq. (9-28) gives

$$N_{min} = \frac{\ln\left[(24.6/0.4)/(0.3/16.7)^{0.766}\right]}{\ln 2.04} = 8.7$$

(a) Using the Gilliland correlation, we get

$$\frac{(L/D) - (L/D)_{min}}{(L/D) + 1} = \frac{1.50 - 0.935}{1.50 + 1} = 0.226$$

From Fig. 9-1

$$\frac{N - N_{min}}{N + 1} = 0.4$$

Using $N_{min} = 8.7$ from the Winn equation yields

$$N - 8.7 = 0.4N + 0.4 \qquad N_{Gill} = \frac{9.1}{0.6} = 15.2$$

(b) Using the Erbar-Maddox correlation, we get

$$\frac{L}{V} = \frac{90}{149.9} = 0.60 \qquad \left(\frac{L}{V}\right)_{min} = 0.483$$

From Fig. 9-2

$$\frac{N_{min}}{N} = 0.64 \qquad N_{Erb-Madd} = \frac{8.7}{0.64} = 13.6$$

Larger values of N (19.7 and 17.8) would have been obtained from the two correlations by using the value of $N_{min} = 11.4$ from the Fenske equation with the average value of α. These results compare with $N = 15.1$ from the group method in Example 8-9 and Edmister's (1948) conservative value of $N = 17.0$ obtained by stage-to-stage calculations. □

Distribution of Nonkey Components

Because the number of separation variables which can be independently specified in a multicomponent distillation problem is limited, recovery fractions of nonkey components are very rarely specified in a problem description. Some means of estimating the distribution of nonkey components between products is useful in several ways:

1. If the splits of the nonkey components can be estimated with some reliability in situations where they appear to a significant degree in both products, one can use the Underwood minimum reflux equations directly, without having to solve simultaneously several different versions of Eq. (8-94).
2. If the stages-vs-reflux correlations of the preceding section have been used in a design problem, the distributions of the nonkey components must be obtained in some other way.
3. In order to start a stage-to-stage calculation (Chap. 10) when it is not possible to specify the composition of either product completely, it is necessary to obtain the best possible initial estimate of the splits of the nonkey components. This is especially important for a nonkey component appearing at a very low mole fraction in the product stream from which the calculation is started.

As we have seen earlier in this chapter, it is a relatively simple matter to solve for the distribution of all components at total reflux. In many instances it will not be necessary to know the splits of the nonkey components with any greater precision than is afforded by equating the split at finite reflux ratio to that at infinite reflux. It is somewhat more difficult to find the distributions of nonkey components at minimum reflux; however, it is possible to get these values without too much work, and in that way the distribution of nonkeys at the other limiting operating condition can be obtained. In the solution to Example 9-2 the distributions of nonkey components for a given distribution of key components were compared at minimum reflux and total reflux.

Geddes fractionation index Geddes (1958) and Hengstebeck (1961) have noted that for two of the limiting extremes of distillation of multicomponent mixtures log-log plots of $x_{i,d}/x_{i,b}$ vs α_i are straight lines. As shown in Fig. 9-3, one of these extremes is the case of total reflux, for which Eq. (9-24) can be written as

$$\log \frac{x_{A,d}}{x_{A,b}} - \log \frac{x_{B,d}}{x_{B,b}} = N_{min} \log \alpha_{AB} \tag{9-30}$$

When Eq. (9-30) is written for all independent combinations of components, it indicates that $\log(x_{i,d}/x_{i,b})$ will be a straight-line function of $\log \alpha_i$, the slope being N_{min}. The other extreme is the case of a single-stage equilibration of vapor and liquid, equivalent to a single-stage distillation. In such a simple equilibration

$$\log \frac{y_i}{x_i} = \log \alpha_i + \log K_r \tag{9-31}$$

The equilibrium ratio for the component whose α has been set equal to 1 is K_r. Equation (9-31) defines a straight line on the coordinates of Fig. 9-3, with the slope

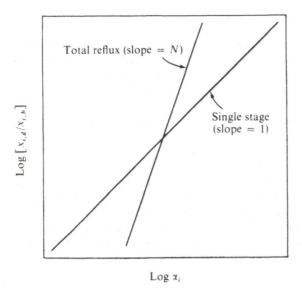

Total reflux (slope $= N$)

$\mathrm{Log}\,[x_{i,\,d}/x_{i,\,b}]$

Single stage
(slope $= 1$)

Log α_i

Figure 9-3 Distribution ratio vs. relative volatility for multicomponent distillation. (*Adapted from Stupin and Lockhart, 1968.*)

equal to unity. Geddes (1958) found that the light key and light nonkey components in a distillation at finite reflux tend to lie on a straight line when plotted as log $(x_{i,\,d}/x_{i,\,b})$ vs α_i on a plot like Fig. 9-3, with the slope between 1 and $N_{\min}$. He found the same to be true of the heavy key and the heavy nonkeys, although the straight line for these components does not necessarily have the same slope as the slope for the light key and light nonkeys. He proposed the name *fractionation index* for the slope of these lines and suggested that they be used to predict distributions of nonkey components from a minimum of information. The fractionation index for the light key and light nonkeys is primarily related to the number of stages in the stripping section, since the light nonkeys die down in mole fraction over these stages from their limiting flow rates arriving at the feed stage from the rectifying section. Similarly the fractionation index for the heavy key and heavy nonkeys reflects primarily the number of stages in the rectifying section. Hengstebeck (1961) presents approximate equations for use in predicting these straight lines and their slopes from problem specifications.

Effect of reflux ratio Stupin and Lockhart (1968) examined a number of different multicomponent distillations and came to the conclusion that the situation is more complex. They point out that the plot of component distributions at minimum reflux must have a nonlinear shape on a plot like Fig. 9-3. The necessary form of the curve for minimum reflux is shown in Fig. 9-4, where it is also compared with the situation for total reflux. In Fig. 9-4 the distribution ratios of the light and heavy keys are presumed to have been set in the problem description. The total reflux curve is shown as a solid straight line. The component distributions at minimum reflux follow the solid curve marked 4 in Fig. 9-4. Above a critical α_i, somewhat above that of the light key, the distribution ratios for light nonkeys become infinite, corresponding to a total

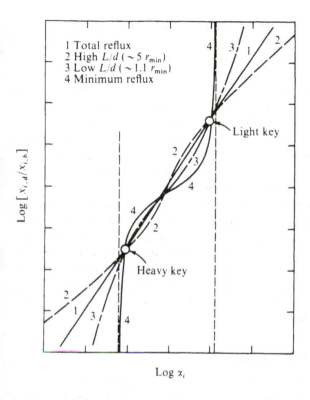

Figure 9-4 Distribution of components at various reflux ratios. (*Adapted from Stupin and Lockhart, 1968.*)

Log α_i

recovery of the component in the distillate. Thus the curve for component distributions must become asymptotic to the vertical dashed line shown in Fig. 9-4. A similar behavior is shown by heavy nonkeys below another critical α_i; they accumulate entirely in the bottoms. Components of intermediate volatility, however, have a distribution ratio at total reflux which is more removed from unity than that at minimum reflux; these components are separated to a better degree at total reflux.

An interesting behavior observed by Stupin and Lockhart (1968) is shown by the family of four curves in Fig. 9-4. One might expect that the component distributions at total reflux and at minimum reflux would bound the component distributions at intermediate reflux ratios. Instead, as the reflux ratio is reduced downward from the infinite L/d corresponding to total reflux, the component distribution curve first moves *away* from the ultimate position of the minimum reflux curve. At an L/d approximately 5 times the minimum, the curve has moved from position 1 at total reflux to position 2. As the reflux ratio is further lowered, the component distribution curve moves back toward the total-reflux curves and approximates the total-reflux distribution again at L/d in the range of 1.2 to 1.5 times the minimum. Still lower reflux ratios bring the distribution curve through position 3 and ultimately to position 4 for minimum reflux. Tsubaki and Hiraiwa (1972) have also explored these trends and methods for analyzing them quantitatively.

A knowledge of these trends along with the calculable component distributions for total reflux and minimum reflux should enable one to estimate component distributions at various operating reflux ratios without making detailed calculations. As shown by Barnés (1970), the two critical α_i values in Fig. 9-4 are ϕ_{HK-} and ϕ'_{LK+} if the critical α_i values refer to components present to small extents in the feed.

As noted in Chap. 5 and Appendix D, economic optimum design reflux ratios tend to be 1.10 times the minimum or less, higher multiples of the minimum being used for particularly difficult separations and/or when there is uncertainty in the vapor-liquid equilibrium data or stage efficiencies. From Fig. 9-4, the nonkey distributions calculated for total reflux should be a good approximation to the actual distributions in the range of reflux ratios 1.15 to 1.25 times the minimum. Also, in this range the nonkey distributions are relatively linear on a plot like Fig. 9-4, lending support to the use of the Geddes fractionation index.

It should be stressed, however, that a multicomponent distillation column has $R - 1$ independent stage efficiencies for the various components on each stage. Consequently, the number of equilibrium stages presented by a tower for one component may not be the same as the number presented for another component. The analyses of component distributions presented here for total, minimum, and intermediate refluxes have been based on the idealized assumption that the distillation column will provide the same number of equilibrium stages for the nonkey-component separations as for the key-component separation.

Distillation of Mixtures with Many Components

Sometimes the feed to a distillation contains so many different components that it is essentially impossible to allow for each component separately in a computation of the distillation performance. The most important example of such a situation is the primary distillation of crude oil. The distillation shown in Fig. 9-5 converts crude oil into seven fractions (vapor and liquid distillates, four sidestreams, and a bottoms, which is sent to a vacuum fractionator for further separation).

Crude oil contains so many different components that the only efficient way to analyze it is through boiling-point curves. So-called true-boiling-point (TBP) curves for a typical feed and the resultant products from the process of Fig. 9-5 are shown in Fig. 9-6. TBP analyses are made by carrying out a slow batch distillation in a column with many stages at high reflux ratio and measuring the overhead temperature as a function of the percentage of the charge that has been converted into distillate. Less efficient analyses of boiling-point curves are the ASTM and EFV methods; Van Winkle (1967) and Nelson (1958) give procedures for converting between different types of boiling curves.

The sidestreams are all withdrawn above the main feed, with the result that a certain amount of very light components is present in each of the streams where they are withdrawn from the main fractionator. The sidestream strippers serve to remove these light components; without the strippers the boiling curves for the sidestream products would show much more downward curvature on the left-hand side of Fig. 9-6.

Temperature and liquid-flow-rate profiles, obtained by a computer simulation of the process of Figs. 9-5 and 9-6, are shown in Fig. 9-7. Because of the extremely wide

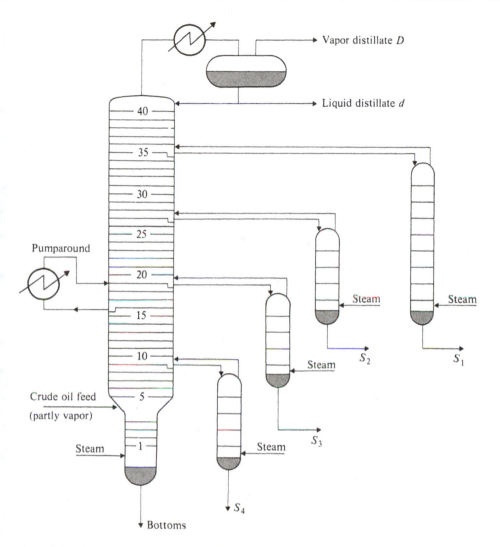

Figure 9-5 Primary fractionation system for crude oil. (*Adapted from Cecchetti et al., 1963, p. 161; used by permission.*)

boiling range of the feed constituents (Fig. 9-6), the temperature changes greatly across the column—by more than 250°C. The liquid-flow profile is complicated, in part because of the withdrawal of liquid sidestreams. However, even apart from the effects of sidestream withdrawal, there is a marked tendency for the liquid flow to decrease downward between sidestream withdrawal points. This is largely the result of a sensible-heat effect. The vapor upflow exceeds the liquid downflow, and the very large temperature reduction of the vapor from stage to stage upward requires vaporization of the liquid, resulting in flows which decrease downward. Another synergistic

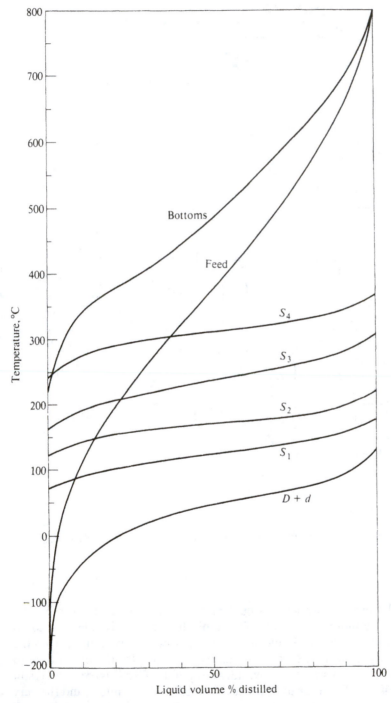

Figure 9-6 True-boiling-point (TBP) curves for feed and products from crude-oil distillation of Fig. 9-5. (*Adapted from Cecchetti et al., 1963, p. 163; used by permission.*)

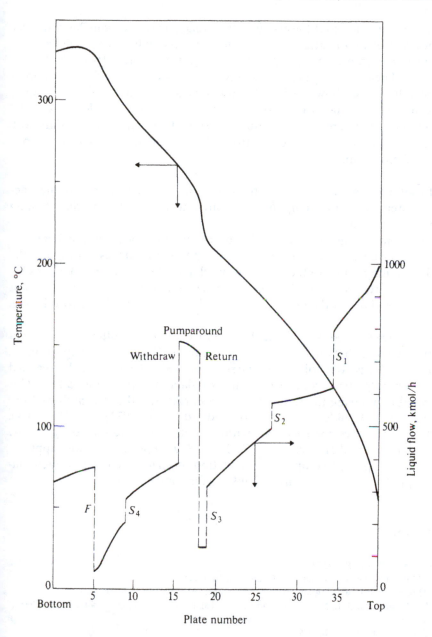

Figure 9-7 Temperature and liquid-flow profiles from simulation of crude-oil distillation of Fig. 9-5. (*Data from Cecchetti, et al. 1963.*)

effect comes from the fact that lower-molecular-weight hydrocarbons have smaller molar latent heats of vaporization than high-molecular-weight hydrocarbons.

The *pumparound* shown in Fig. 9-5 removes liquid from plate 16, cools it in an elaborate external heat-exchange system, and returns it as subcooled liquid at a point higher in the column. As shown in Fig. 9-7, this serves to make much larger liquid flows on the intermediate stages and produces a net increase in liquid flow, due to the cooling, from above the pumparound return to below the pumparound withdrawal. The pumparound serves as a point of control to keep the plates immediately above the feed point from running dry.

Methods of computation There are two basic approaches for determining the fractionation in systems of many components in order to predict the product boiling curves shown in Fig. 9-6.

In the older, much simpler, but highly approximate method empirical correlations, e.g., those given by Packie (1941) and Nelson (1958), are used to relate the number of intervening stages to the degree of overlap of the boiling-point curves for adjacent product streams. These methods work satisfactorily if conditions are not very different from those in the experimental distillations on which the correlations are based.

The more exact method is to divide the feed mixture into a sufficient number of pseudo components, each with a particular boiling point, latent heat of vaporization, heat capacity, etc. A rigorous solution of the sort outlined in Chap. 10 is then carried out, including enthalpy balances as well as mass balances in the computation. This method is capable of giving the resultant boiling curves and the stream flows from different stages. Van Winkle (1967), Taylor and Edmister (1971), and Hess et al. (1977) outline procedures for such calculations, and typical results are given by Cecchetti et al. (1963) and Hess et al. (1977). Jakob (1971) presents a more approximate approach to a pseudo-component calculation, utilizing the Geddes fractionation index to predict the separations of nonkey components.

REFERENCES

Bachelor, J. B. (1957): *Petrol. Refin.*, **36**(6):161.

Barnes, F. J. (1970): M.S. thesis, University of California, Berkeley.

———, D. N. Hanson, and C. J. King (1972): *Ind. Eng. Chem. Process Des. Dev.*, **11**:136.

Cecchetti, R., R. H. Johnston, J. L. Niedzwiecki, and C. D. Holland (1963): *Hydrocarbon Process. Petrol. Refin.*, **42**(9):159.

Chien, H. H. Y. (1978): *AIChE J.*, **24**:606.

Douglas, J. M. (1977): *Hydrocarbon Process.*, **56**(11):291.

Edmister, W. C. (1948): *Petrol. Eng.*, **19**(8):128, **19**(9):47; see also "A Source Book of Technical Literature on Fractional Distillation," pt. II, pp. 74ff., Gulf Research & Development Co., n.d.

Eduljee, H. E. (1975): *Hydrocarbon Process.*, **54**(9):120.

Erbar, J. H., and R. N. Maddox (1961): *Petrol. Refin.*, **40**(5):183.

Erbar, R. C. and R. N. Maddox (1962): *Can. J. Chem. Eng.*, **40**:25.

Fenske, M. R. (1932): *Ind. Eng. Chem.*, **24**:482.

Geddes, R. L. (1958): *AIChE J.*, **4**:389.

Gilliland, E. R. (1940a): *Ind. Eng. Chem.*, **32**:1101.

———— (1940b): *Ind. Eng. Chem.*, **32**:1220.

Hengstebeck, R. J. (1961): "Distillation: Principles and Design Procedures," chap. 8, Reinhold, New York.

Hess, F. E., C. D. Holland, R. McDaniel, and N. J. Tetlow (1977): *Hydrocarbon Process.*, **56**(5):241.

Jakob, R. R. (1971): *Hydrocarbon Process.*, **50**(5):149.

Molokanov, Y. K., T. P. Korablina, N. I. Mazurina, and G. A. Nififorov (1972): *Int. Chem. Eng.*, **12**:209.

Nelson, W. L. (1958): "Petroleum Refinery Engineering," 4th ed., McGraw-Hill, New York.

Packie, J. W. (1941): *Trans. AIChE*, **37**:51.

Robinson, C. S., and E. R. Gilliland (1950): "Elements of Fractional Distillation," 4th ed., pp. 347–350. McGraw-Hill, New York.

Shiras, R. N., D. N. Hanson, and C. H. Gibson (1950): *Ind. Eng. Chem.*, **42**:871.

Strangio, V. A., and R. E. Treybal (1974): *Ind. Eng. Chem. Process Des. Dev.*, **13**:279.

Stupin, W. J., and F. J. Lockhart (1968): *AIChE Annu. Meet.*, *Los Angeles*, December.

Tavana, M., and D. N. Hanson (1979a): *Ind. Eng. Chem. Process Des. Dev.*, **18**:154.

———— and ———— (1979b): Personal Communication.

Taylor, D. L., and W. C. Edmister (1971): *AIChE J.*, **17**:1324.

Tsubaki, M., and H. Hiraiwa (1972): *Kagaku Kogaku*, **36**:880; *Trans. Int. Chem. Eng.*, **13**:183 (1973).

Underwood, A. J. V. (1932): *Trans. Inst. Chem. Eng.*, **10**:112.

———— (1946): *J. Inst. Petrol.*, **32**:598, 614.

———— (1948): *Chem. Eng. Prog.*, **44**:603.

Van Winkle, M. (1967): "Distillation," McGraw-Hill, New York.

———— and W. G. Todd (1971): *Chem. Eng.*, Sept. 20, p. 136.

Winn, F. W. (1958): *Petrol. Refin.*, **37**(5):216.

PROBLEMS

9-A$_1$ Find the minimum air rate (moles air per mole water) required for the separation specified in Prob. 6-C, if the tower may now contain any number of stages.

9-B$_1$ For the two-column separation of alcohols by distillation specified in Prob. 8-D:

(a) Find the minimum overhead reflux ratio in each column, given infinite stages in both.

(b) Find the minimum number of actual stages in each column, given infinite reflux in each and E_O still equal to 0.70.

(c) Use the Gilliland correlation to predict the number of actual stages required in each column at the reflux ratio specified in Prob. 8-D. If you have worked Prob. 8-D, compare your result from the correlation with the result obtained by group methods.

(d) Repeat part (c) using the Erbar-Maddox correlation.

9-C$_1$ Find the distributions of nonkey components in the columns of Prob. 8-D at total reflux and minimum reflux.

9-D$_1$ Repeat part (b) of Prob. 8-K using a stages-vs-reflux correlation. Use an overall stage efficiency of 94 percent.

9-E$_2$ A distillation tower receives a feed containing 30 mol % benzene, 40 mol % toluene, and 30 mol % xylenes. Over the temperature range of interest the equilibrium data can be satisfactorily represented by constant relative volatilities, equal to 2.5 for benzene, 1.0 for toluene, and 0.40 for xylenes. In an effort to avoid using multiple towers, the toluene product will be withdrawn as a liquid sidestream from a point in the tower well above the feed. Benzene and xylene will be obtained at high recovery fractions in the top and bottom products, respectively.

(a) Find the minimum vapor flow if the desired product purities are high and the feed is saturated liquid.

(b) If the overhead ratio of V to d is 5.0 and the feed and reflux are saturated liquids, find the maximum toluene purity that can be achieved, even with an infinite number of plates. The ratio of sidestream to distillate flow can be varied.

9-F₂ A distillation column of three equilibrium stages, a reboiler, and a total condenser is charged with a mixture of A and B which is 0.1 mole fraction A. The column is started and run under total reflux for a long period of time. Assume that the holdup on the plates is negligible but that the molal holdup in the reflux drum is one-third the molal holdup in the reboiler. The relative volatility α_{AB} is 2. Calculate the compositions of the material in the reboiler and the condenser after steady state has been reached.

9-G₂ Using only the Fenske equation, the Underwood minimum reflux equations, and the Gilliland correlation, demonstrate whether or not the presence of a single heavy nonkey component in a saturated liquid feed to a distillation column *must* increase the equilibrium-stage requirement at a given overhead reflux ratio, as compared to the equivalent simple binary distillation of the keys alone. Compare for given percentage recoveries of the two keys. Consider the case of equal molal overflow above and below the feed, and constant relative volatilities, independent of composition and temperature. To what extent is your answer or the effect in general dependent upon the relative volatility of the heavy nonkey?

9-H₂ If 95 percent of the cyclopentane is to be recovered by batch rectification from a liquid mixture containing 20 mol % cyclopentane, 30 mol % methylcyclopentane, and 50 mol % cyclohexane, what will the percentage purity of the cyclopentane product be? The rectification will be carried out at essentially total reflux in a tower of four equilibrium stages above the still pot with insignificant liquid holdup above the still pot.

	α
Cyclopentane	2.60
Methylcyclopentane	1.30
Cyclohexane	1.00

9-I₃ Explain physically the reasons for the directions of the trends in light and heavy nonkey component distributions indicated by the sequence of curves 1, 2, 3, 4 in Fig. 9-4 as the overhead reflux ratio is progressively reduced for a given split of the key components.

9-J₃ A mixture of aromatics is to be separated in a column which will have a sidestream stripper equipped with a reboiler, as shown in Fig. 9-8. The following feed and product specifications are provided.

		kmol/h			
Component	Relative volatility	Benzene product	Toluene product	Heavy product	Feed
Benzene	2.25	34.5	0.5	Small	35
Toluene	1.00	0.5	29.0	0.5	30
Xylenes	0.330	Small	0.5	19.5	20
Cumene	0.210	Small	Small	~15	15

Constant molal overflow can be assumed in each section of the column and the stripper. The boilup ratio in the stripper reboiler is set at 1.33, that is, 40 kmol/h of vapor generated. The feed is saturated liquid.

  (a) Calculate the minimum overhead reflux ratio which would be required with an infinite number of stages in any section. Be sure to consider all prospective pinch points.

  Assume that the overhead reflux ratio is fixed at 3.50 for design purposes:

  (b) If the sidestream stripper were not employed, what would be the highest toluene purity attainable with any possible number of stages?

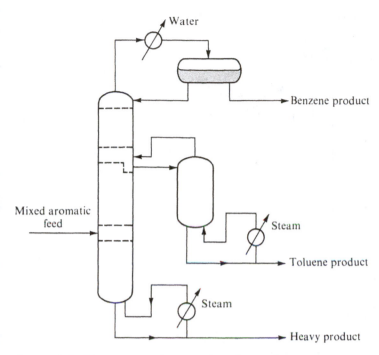

Figure 9-8 Distillation scheme for separation of aromatics.

(*c*) Once again, assuming that the sidestream stripper is to be installed, use the Underwood equations to determine the number of equilibrium stages required in the sections of the column above the sidestream draw and below the main feed.

(*d*) Compare your answers to part (*c*) with the results which would be obtained using equivalent binary McCabe-Thiele analyses for the benzene-toluene and toluene-xylene separations. For the equivalent binary analyses assume that all nonkey components are at their limiting flow rates throughout each section.

Equilibrium points for binary systems:

$\alpha = 2.25$		$\alpha = 2.25$		$\alpha = 3.0$		$\alpha = 3.0$	
x	y	x	y	x	y	x	y
0.10	0.20	0.50	0.69	0.10	0.25	0.57	0.80
0.20	0.36	0.64	0.80	0.20	0.43	0.75	0.90
0.30	0.49	0.80	0.90	0.30	0.56		
0.40	0.60			0.44	0.70		

9-K$_2$ A distillation column is to be built, as shown in Fig. 9-9, to transfer a polymer product from a light solvent to a heavy solvent. The elaborate scheme is necessary to avoid taking the polymer out of solution and bringing polymer into contact with hot heat-transfer surfaces, where fouling might occur. The feed rates of the two streams are fixed, the distillate rate is on level control from the reflux drum, the bottoms is on level control from the liquid in the tower bottom, and the tower pressure controls the cooling water rate to a total condenser. The temperature of an intermediate plate governs the steam pressure, and the

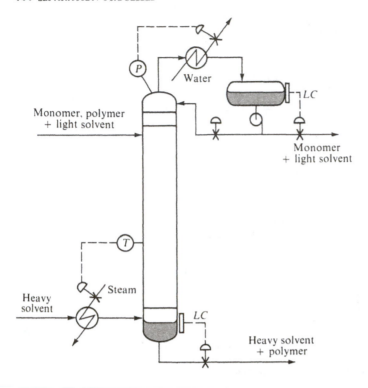

Figure 9-9 Distillation column for Prob. 9-K.

reflux rate is set manually. Only a trace of unreacted monomer is present in the light feed, which is a 10 mole percent solution of polymer in the light solvent. The product polymer stream is to be a 15 mole percent solution in the heavy solvent. There is to be essentially complete separation of the light solvent from the heavy solvent. Relative volatilities are as follows:

	Relative volatility
Monomer	1.5
Light solvent	1.0
Heavy solvent	0.2
Polymer	0.01

Constant molal overflow may be assumed. The light solvent feed enters as saturated liquid.

(a) With an overhead reflux ratio L/d equal to 0.2, what will be the phase condition of the heavy solvent feed?

(b) For the conditions of part (a) what number of equilibrium stages is required if there is to be a recovery of 99.0 percent for both the light solvent and the heavy solvent? The number of equilibrium stages above the top feed is fixed as 2.

(c) After the tower is built and in operation, an upset occurs and monomer is found in the bottoms product, even though the steam valve is wide open. You are aware that there must be no detectable amount of monomer in the product polymer solution. Being a sound engineer you reach immediately for the reflux valve. Do you increase or decrease the flow of reflux? Why?

9-L$_2$ Draw a block diagram for a computer calculation scheme for finding the minimum reflux ratio in a single-feed two-product multicomponent distillation column. The input data include the composition and degree of vaporization of the feed, the predicted distillate composition, and the relative volatilities of all components (assumed constant). The key components are identified in the input. There may be both distributing and nondistributing nonkey components. Constant molal overflow may be assumed. Indicate appropriate equations to be used in your block diagram. Also indicate specific convergence methods and any constraints necessary to insure convergence to the right answer.

9-M$_2$ Consider a separation of methylcyclohexane from *n*-heptane by extraction, using aniline as the solvent and using extract reflux obtained through distillation. Phase-equilibrium data are given in Prob. 6-F and Fig. 6-25.

(*a*) If the feed contains 60 wt % methylcyclohexane and 40 wt % *n*-heptane, and if there is to be 95.0 percent recovery of the feed constituents into their respective products, find the split between product and recycle required for the hydrocarbon product from the distillation column, assuming that an infinite number of stages can be used in the extraction.

(*b*) Find the split between product and recycle required if the extraction provides 16 equilibrium stages. Use the Gilliland correlation.

EXACT METHODS FOR COMPUTING MULTICOMPONENT MULTISTAGE SEPARATIONS

The widespread availability of digital computers and programmable calculators has made it possible to solve routinely the simultaneous equations which describe multi-stage multicomponent separations, even though these equations are not simple and may be highly nonlinear. In this chapter we focus upon computation methods for distillation (including extractive and azeotropic distillation), absorption and stripping, and solvent extraction. The methods for these processes have been developed most extensively, and these processes are also the most common separations in the organic-chemical, petroleum, plastics, and fuels industries. Furthermore, the methods for these processes have important common features which allow important generalizations to be made.

The approach of this chapter will be to consider first several special cases in which simpler approaches can be used. We shall then build to more complex approaches, which also have more general applicability to wide classes of processes.

UNDERLYING EQUATIONS

If we use the nomenclature of vapor-liquid contacting processes, there are four main classes of equations which describe an equilibrium-stage process. We shall write these in terms of individual component flows in the vapor and liquid (v_j and l_j), equilibrium ratios ($K_j = y_j/x_j$, at equilibrium), total stream flows (V and L), vapor-

and liquid-stream enthalpies (H and h), and both individual-component and total feed flows (f_j and F):

1. Equilibrium relationships **E**: $N \times R$ equations:[†]

$$v_{j, p} = \frac{K_{j, p} V_p}{L_p} l_{j, p} \qquad (10\text{-}1)$$

2. Component mass balances **M**: $N \times R$ equations:[‡]

$$l_{j, p} + v_{j, p} - l_{j, p+1} - v_{j, p-1} - f_{j, p} = 0 \qquad (10\text{-}2)$$

3. Enthalpy balances **H**: N equations:[‡]

$$L_p h_p + V_p H_p - L_{p+1} h_{p+1} - V_{p-1} H_{p-1} - F_p h_{F_p} - q_p = 0 \qquad (10\text{-}3)$$

4. Summation of flows **S**: $2N$ equations:

$$\sum_j l_{j, p} = L_p \qquad (10\text{-}4)$$

and

$$\sum_j v_{j, p} = V_p \qquad (10\text{-}5)$$

In addition $K_{j, p}$, h_p, and H_p are in general themselves functions of temperature and of the entire composition of either phase. Therefore these four types of equation comprise $N(2R + 3)$ simultaneous nonlinear equations, containing $N(2R + 3)$ unknowns, i.e., the values of $v_{j, p}$; $l_{j, p}$; L_p; V_p, and (implicitly) T_p, assuming that pressure, the number of stages, all feed and intermediate product flows, all q_p, and the enthalpies and compositions of all feeds are fixed.

If we work in terms of actual stages rather than equilibrium stages and use the Murphree vapor efficiency, the $N \times R$ equilibrium relationships, **E** [Eqs. (10-1)] are replaced by $N \times R$ equations of the form

$$E_{MV j, p} = \frac{y_{j, p} - y_{j, p-1}}{y_{j, p}^* - y_{j, p-1}} \qquad (10\text{-}6)$$

or

$$E_{MV j, p} = \frac{v_{j, p} - v_{j, p-1} V_p / V_{p-1}}{(K_{j, p} V_p / L_p) l_{j, p} - v_{j, p-1} V_p / V_{p-1}} \qquad (10\text{-}7)$$

This introduces another $N \times R$ Murphree efficiencies as variables. Another complication is then that only $R - 1$ of the Murphree efficiencies for any stage are independent of each other; one of the E_{MV} values must be dependent, so that the values of $y_{j, p}$ in Eqs. (10-6) can add up to unity. If the $R - 1$ independent E_{MV} are taken to be

---

[†] Subscript j refers to component number $1 \le j \le R$; subscript p refers to the stage number, counting upward from the bottom $1 \le p \le N$. Subscript i will be used later to refer to iteration number. $x_j = l_j / L$, and $y_j = v_j / V$.

[‡] The term $f_{j, p}$ will be positive for j entering in a feed and negative for j leaving in a product. Similar reasoning holds for $F_p h_{F_p}$; q_p is the amount of heat added in a heater or is the negative of the heat lost in a cooler.

equal, the remaining E_{MV} will have the same value; however, in general, E_{MVj} should be different for different components (see Chap. 12). The $N \times (R - 1)$ independent Murphree efficiencies can then be specified variables or can be determined through an additional class of equations describing the mass-transfer processes on individual stages.

GENERAL STRATEGY AND CLASSES OF PROBLEMS

The general strategy for solving Eqs. (10-1) to (10-5) has been well analyzed by Friday and Smith (1964). The most important questions involve the choice of iteration variables and convergence procedures. Some of the basic criteria for solving equations and families of equations in general are developed in Appendix A.

At one extreme is the possibility of treating all $N(2R + 3)$ equations as a convergence function and employing a multivariate Newton convergence scheme. This involves calculating the $[N(2R + 3)]^2$ partial derivatives in Eq. (A-11) and then inverting that matrix during each iteration. Even allowing for partial derivatives which would be zero, this is a large task, consuming much computer storage and computing time. It is therefore desirable to seek ways of reducing the number of equations by algebraic manipulation, pairing certain convergence functions with certain unknown variables (and possibly nesting the resultant convergence loops), and/or making simplifying assumptions which will make it possible for the equations to be solved more efficiently.

The following distinctions permit classification of multistage multicomponent separation problems into categories which allow different choices of effective calculation procedures:

1. *Design problems vs. operating problems.* Assuming that feed variables, pressure, and a reflux flow (for refluxed separations) are specified, a design problem is one where two separation variables (typically recovery fractions of the two key components) are specified in addition and the number of stages is a dependent variable. Criteria for locating feed and sidestream points are based upon optimization methods and/or composition specifications. On the other hand, an operating problem is one where the number of stages, the feed and product locations, and another flow are specified without setting separation variables. A design problem is typical of what is encountered in the design of a new process to accomplish a certain separation, while an operating problem is typical of a calculation analyzing the performance of an existing separator under fixed operating conditions.
2. *Whether or not either all light nonkeys or all heavy nonkeys are absent.*
3. *Whether the major components have values of K_j close to 1.0 or well removed from 1.0.* Most ordinary distillations have major components with K_j of the order of unity. Distillations with "dumbbell" feeds (much light material, much heavy material, and little in between), most extractions, and simple absorbers and strippers tend to have K_j values well removed from unity for the major components. Reboiled absorbers, wide-boiling distillations with a continual distribution of components, and azeotropic and extractive distillation can have mixed features.
4. *Whether or not the K_j values are strong functions of composition* (nearly ideal vs. highly nonideal solutions).

5. *Whether equilibrium or actual stages are considered.* In the former case an overall efficiency would typically be applied later. In the latter case the E_{MV} or E_{ML} equations would be brought into the calculation.

We shall consider different solution techniques which are effective for different classes of problems, and at the end of the chapter the recommended procedures for different types of problems will be summarized.

STAGE-TO-STAGE METHODS

Stage-to-stage methods are effective for design problems when there are either no light nonkey components or no heavy nonkey components. These methods involve the calculation of conditions in a separation cascade one stage at a time. The computation is usually started at one end of the cascade, where the terminal flows, concentrations, flows, etc., are known or have been assumed. The computation moves away from that point obtaining results for each stage, by trial and error if necessary, before attacking the next stage.

Stage-to-stage methods are ideal for the analysis of those binary-separation design problems where the terminal composition conditions and flows are set in the problem description. The graphical procedures carried out in Chaps. 5 and 6 for binary-separation problems are, in fact, stage-to-stage calculations. Stage-to-stage methods for multicomponent systems are a direct logical extension of the methods used for binary separations.

In order to start a stage-to-stage calculation it is necessary to have set or assumed enough variables to specify completely the performance of the first stage considered. It is nearly always a terminal stage for which this can be done if it can be done at all. Since a product stream will leave from a terminal stage, it is necessary to specify separation variables in order to establish the composition of the product stream. This is the reason for the restriction to design problems.

Furthermore, if nonkey components are present in the product stream, they will not have been set in the problem description. It will be necessary to assume their concentrations in the product stream to start the calculation. Such an assumption can be made with percentwise high accuracy for heavy nonkeys in a bottom product or for light nonkeys in a top product because nearly all of those components go to those products. But it is not possible to estimate the concentrations of light nonkeys in the bottom product or of heavy nonkeys in the top product with much percentwise accuracy. Since the nonkey components build up greatly in concentration as the calculation moves away from these products where they do not primarily appear, small errors (on an absolute basis) in the estimation of these components in those products will propagate to become very large. This is the reason for the restriction to problems where either light nonkeys or heavy nonkeys are completely absent. If light nonkeys are not present, one can estimate the concentrations of all components in the bottom product with percentwise high precision. If heavy nonkeys are absent, one can estimate all concentrations in the top product with percentwise high precision.

Multicomponent Distillation

The stage-to-stage approach for solving multicomponent distillation problems is often called the *Lewis-Matheson method*, after the original developers of the approach (Lewis and Matheson, 1932).

As an example for stage-to-stage calculations, consider a mixture of four components, A, B, C, and D, in order of volatility (highest first). Stage-to-stage methods can be used if the two keys are either A and B or C and D. If the keys are A and B, the calculation would start at the bottom stage (no light nonkeys), and it is necessary to estimate concentrations or recovery fractions of C and D in the bottom product. A first estimate might be that $(/C)_b$ and $(/D)_b$ are both unity. Following the discussion in Chap. 9, a better assumption might be that the recovery fractions of these components are equal to the recovery fractions at total reflux.

If all $x_{j, b}$ are known from the problem specification for A and B and from the assumptions for C and D, it is next possible to calculate the composition of the vapor leaving the bottom equilibrium stage, since it is in equilibrium with the bottom product. For each component $y_{j, p} = K_{j, p} x_{j, p}$, or $v_{j, p} = (K_{j, p} V_p / L_p) l_p$. If the α_j, relative to a reference component, are independent of temperature and composition, the need for solving for the stage temperature and the $K_{j, p}$ values can be avoided by recognizing that the $v_{j, p}$ values are in proportion to $\alpha_{j, p} x_{j, p}$ $[v_{j, p} / v_{k, p} = (\alpha_{j, p} x_{j, p} / \alpha_{k, p} x_{k, p})$, etc.]. Therefore,

$$v_{j, p} = \frac{V_p}{\Sigma \alpha_{j, p} l_{j, p}} \alpha_{j, p} l_{j, p} \tag{10-8}$$

If the vapor composition leaving the bottom stage is found from Eq. (10-8), the liquid composition from the next-to-bottom stage, which passes this vapor between stages, can be found from the component mass balances between this level and the bottom of the column

$$l_{j, p+1} = v_{j, p} + b_j \tag{10-9}$$

Alternating use of Eqs. (10-8) and (10-9) would then give successive vapors and liquids going up the column, until a feed or sidestream point is reached, when the appropriate mass balance would change; i.e., above the feed in a simple distillation column

$$l_{j, p+1} = v_{j, p} - d_j \tag{10-10}$$

Example 10-1 A four-component mixture is to be separated by distillation, as shown below. Find the equilibrium-stage requirement and the optimum feed location for the separation. Assume constant molar sectional flows and constant α_{ij} throughout the column for the calculation.

Conditions		Feed	z_j
Feed	Saturated liquid	C_3H_8	0.40
Pressure	1.38 MPa	C_4H_{10}	0.40
Condenser	Total	$i\text{-}C_5H_{12}$	0.10
Separation	98% recovery and	$n\text{-}C_5H_{12}$	0.10
	purity of C_3H_8		
Reflux	2 mol/mol feed		

SOLUTION To define the products, assume that i-C_5H_{12} and n-C_5H_{12} go completely into the bottom product.

	Top product		Bottom product	
	d_j	$x_{j.d}$	b_j	$x_{j.b}$
C_3	0.392	0.980	0.008	0.013
C_4	0.008	0.020	0.392	0.653
i-C_5	......	......	0.100	0.167
n-C_5	......	......	0.100	0.167
	0.400	1.000	0.600	1.000

Since there are no light nonkeys, stage-to-stage calculations are appropriate and should begin from the bottom of the column.

K values for the four components are taken from Maxwell (1950). In order to obtain the proper average α's we need the temperatures of the overhead and bottoms. For the temperature at the top plate find the dew temperature of top product.

	Assume $T_t = 43°C$	
	K_j	$\dfrac{y_j}{K_j} = \dfrac{x_{j.d}}{K_j}$
C_3	1.043	0.940
C_4	0.372	0.054
		0.994

A top-plate temperature of 43°C is close enough for determination of α values. Equilibrium ratios K_j of i-C_5 and n-C_5 at 43°C are 0.155 and 0.128, respectively. $\alpha_{j.t} = K_{j.t}/K_{r.t}$. The reference component $\alpha_r = 1$ is arbitrarily taken to be butane.

	$K_{j.t}$	$\alpha_{j.t}$
C_3	1.043	2.80
C_4	0.372	1.00
i-C_5	0.155	0.42
n-C_5	0.128	0.34

The temperature at the reboiler is the bubble-point temperature of the bottom product:

	Assume $T_R = 104°C$		Assume $T_R = 110°C$	
	K_j	$K_j x_{j,b}$	K_j	$K_j x_{j,b}$
C_3	2.35	0.031	2.42	0.031
C_4	1.082	0.707	1.154	0.754
$i\text{-}C_5$	0.581	0.097	0.638	0.107
$n\text{-}C_5$	0.499	0.083	0.550	0.092
		0.918		0.984

A reboiler temperature of 110°C is close enough for determination of α values. Since the values of α do not change greatly, the arithmetic mean is a satisfactory approximation to the geometric mean.

	$\alpha_{j,R}$	$\alpha_{j,av}$
C_3	2.10	2.45
C_4	1.00	1.00
$i\text{-}C_5$	0.55	0.49
$n\text{-}C_5$	0.48	0.41

We next need the interstage flows based on $F = 1$. Since the reflux is specified as 2 mol/mol feed, and the feed is saturated liquid, we have

$$L = 2F = 2 \qquad L' = L + F = 3 \qquad V = L + d = 2.4 \qquad V' = V = 2.4$$

We can now use Eq. (10-8) with the known bottoms composition to obtain $v_{j,R}$, where subscript R refers to the reboiler. Equation (10-9) then gives $l_{j,1}$, where subscript 1 refers to the bottom stage in the column itself. Alternating use of Eqs. (10-8) and (10-9) then proceeds up the column. The results are shown in Table 10-1.

The ratio of x_{C_3}/x_{C_4} on stage 6 is approximately the same as the ratio in the liquid portion of the feed (here, the same as the whole feed), and stage 6 should therefore be investigated to see if it is the optimum feed stage. This is done by examination of the liquid on stage 7 under the two possible assumptions (1) that stage 6 is the feed stage and rectifying flows exist above it and (2) that stage 6 is not the feed stage and stripping flows exist above it. One criterion is to have the ratio x_{C_3}/x_{C_4} increase as fast as possible; that will be the one used here. General criteria for feed-stage location will be discussed later in this chapter.

Stage 6 $\neq$ feed stage

$$\frac{x_{C_3}}{x_{C_4}} = \frac{1.756 + 0.008}{0.587 + 0.392} = 1.80$$

Stage 6 = feed stage Since $V = V'$ in this example, $l_{j,7} = v_{j,6} - d_j$.

$$\frac{x_{C_3}}{x_{C_4}} = \frac{1.756 - 0.392}{0.587 - 0.008} = 2.36$$

Table 10-1 Stage-to-stage calculations for Example 10-1 below feed stage

Component	Reboiler			Stage 1			Stage 2			Stage 3		Stage 4		Stage 5		Stage 6	
	h_j	$x_j b_j$	$r_{j,R}$†	$l_{j,1}$	$x_j l_{j,1}$	$v_{j,1}$	$l_{j,2}$	$\alpha_j l_{j,2}$	$v_{j,2}$	$l_{j,3}$	$v_{j,3}$	$l_{j,4}$	$v_{j,4}$	$l_{j,5}$	$r_{j,5}$	$l_{j,6}$	$v_{j,6}$
C_3	0.008	0.020	0.096	0.104	0.255	0.218	0.226	0.554	0.431	0.439	0.750	0.758	1.135	1.143	1.493	1.501	1.756
C_4	0.392	0.392	1.874	2.266	2.266	1.938	2.330	2.330	1.812	2.204	1.537	1.929	1.179	1.571	0.838	1.230	0.587
$i\text{-}C_5$	0.100	0.0490	0.234	0.334	0.164	0.140	0.240	0.118	0.0918	0.192	0.0656	0.166	0.0497	0.150	0.0392	0.139	0.0325
$n\text{-}C_5$	0.100	0.0410	0.196	0.296	0.121	0.104	0.204	0.0836	0.0650	0.165	0.0472	0.147	0.0368	0.137	0.0300	0.130	0.0254
	0.600	0.502	2.400	3.000	2.806	2.400	3.000	3.086	2.400	3.000	2.400	3.000	2.401	3.001	2.400	3.000	2.401

$$\dfrac{V'}{0.502} = 4.78 \qquad \dfrac{V''}{2.806} = 0.855 \qquad \dfrac{V'}{3.086} = 0.778$$

† $v_{j,R} = 4.78\, x_j b_j$.

Table 10-2 Stage-to-stage calculations for Example 10-1 above feed stage

Component	d_j	Stage 7			Stage 8		Stage 9		Stage 10	
		$l_{j,7}$	$x_j l_{j,7}$	$v_{j,7}$	$l_{j,8}$	$r_{j,8}$	$l_{j,9}$	$v_{j,9}$	$l_{j,10}$	$v_{j,10}$
C_3	0.392	1.364	3.34	2.031	1.639	2.207	1.815	2.305	1.913	2.354
C_4	0.008	0.579	0.579	0.352	0.344	0.189	0.181	0.0938	0.0858	0.0431
$i\text{-}C_5$	......	0.0325	0.0159	0.0097	0.0097	0.0026	0.0026	0.00067	0.00067	0.00017
$n\text{-}C_5$	......	0.0254	0.0104	0.0063	0.0063	0.0014	0.0014	0.00030	0.00030	0.00006
		2.001	3.95	2.399	1.999	2.400	2.000	2.400	2.000	2.398

$$\dfrac{V}{3.95} = 0.608 \qquad (v_{C_3,9} = 0.960) \qquad (v_{C_3,10} = 0.981)$$

453

Since the ratio of x_{C_3}/x_{C_4} on stage 7 is higher with stage 6 as the feed stage, the feed should be introduced on stage 6. A similar calculation for stage 5 shows a slight further gain for introducing the feed there; however we shall proceed with stage 6, since the difference is very small.

The material balance is now changed to Eq. (10-10), and the calculation is continued through the rectifying section, as shown in Table 10-2.

The separation requirement of $x_{C_{3,d}} = 0.98$ is obtained in slightly less than 10 equilibrium stages beyond the reboiler.

It is interesting to examine the indicated overhead recovery of the two pentanes:

$$x_{i-C_5}d = 2.8 \times 10^{-5} \qquad x_{n-C_5}d = 1.0 \times 10^{-5}$$

Any correction applied to $x_{i-C_{5,b}}b$ or $x_{n-C_{5,b}}b$ would be percentwise very small and would be insignificant in a recalculation. $\square$

In the preceding example it was presumed that the relative volatilities were constant throughout the column and that the molar vapor and liquid flows were constant within each section. It is not necessary to make these assumptions in order to use a stage-to-stage calculation method. If the relative volatilities were not constant throughout the column, it would be necessary to determine the temperature of each stage in order to ascertain the appropriate values of α_i for use on that stage. The temperature can be determined from a bubble-point calculation on the known liquids in a computation which starts from the bottom or from a dew-point calculation on the known vapors when the computation begins at the top.

If the relative volatilities depend upon the liquid composition as well as temperature, a bottom-up calculation remains straightforward, since the liquid composition is known when Eq. (10-8) is used. For a top-down calculation an iterative loop would be necessary for each stage, wherein values of γ_j, would be assumed, a liquid composition would be calculated, new values of γ_j would be obtained, etc., until convergence. If α_j is a function of both vapor and liquid compositions, such a procedure would be needed for bottom-up calculations as well. The loop converging the activity coefficients would be part of the bubble- or dew-point calculation.

Varying molar flows can be taken into account by means of enthalpy balances on each stage. In a bottom-up calculation, the reboiler temperature and vapor composition can still be computed as in Example 10-1. The reboiler vapor flow would typically be specified instead of the overhead reflux rate. The flow and composition of the liquid from the bottom stage of the tower proper (stage 1) then come from mass balances, and the temperature of stage 1 comes from a bubble-point calculation. The vapor composition from stage 1 is determined by the bubble-point calculation, but the total vapor flow from stage 1 is unknown and must come from a trial-and-error computation. One can assume V_1', calculate L_2' and $x_{i,2}$ by mass balance, and then check the assumed V_1' by means of an enthalpy balance involving V_1', L_2', and b. For each stage in turn as the calculation proceeds it will be necessary to obtain the vapor rate by such an iteration. Analogous, but reverse, logic applies for a top-down calculation.

Example 10-1 also could have been solved for the actual plate requirement if Murphree vapor efficiencies were available. $R - 1$ independent forms of Eqs. (10-6) or (10-7) are required to obtain values of $v_{j,p}$ once the values of $l_{j,p}$ and $v_{j,p-1}$ are known in a bottom-up stage-to-stage calculation. As long as $K_{j,p}$ is independent of

$y_{j,p}$, no trial and error is required since T_p is known from a bubble-point calculation involving $l_{j,p}$. One usually assumes that interstage vapors and liquids are at their respective dew- and bubble-point temperatures even though the E_{MV} values are different from unity. As shown in Chap. 12, there is experimental support for this assumption.

In a top-down calculation, E_{ML} can be used without causing an iterative loop, but use of E_{MV} would require iteration since the values of $y_{j,p-1}$ are not known when the E_{MV} equation is needed.

Another point concerning Example 10-1 is worth stressing again. The concentrations of the nonkeys C and D on each stage are reliable to several significant figures even though the concentrations of these components in the bottoms product were initially guessed. As a result the effect of the nonkey components is clearly established, and a recalculation using improved values of $b_{i\text{-}C_5}$ and $b_{n\text{-}C_5}$ would yield a miniscule percentwise effect on the stage requirement or on the concentration of the two keys on the various stages. On the other hand, the values of $d_{i\text{-}C_5}$ and $d_{n\text{-}C_5}$ indicated by the calculation are not percentwise highly accurate. This fact follows from the neglect of the terms d_j in the mass-balance relationships for these components on the top stages. In this example inclusion of d_j in the mass balances for $i\text{-}C_5$ and $n\text{-}C_5$ when computing $l_{j,10}$ from $v_{j,9}$ would change the d_j values for these components by less than 10 percent. (Readers should confirm this statement for themselves.) The effect of neglecting d_j will be greatest for nonkeys which have volatilities closest to the keys and in distillations with a relatively low r/d or V'/b.

In any event it is likely that the recovery fractions obtained for nonkey components by means of an equilibrium-stage analysis will not be highly accurate percentwise even when a converged solution is reached, because, as already noted, it is quite possible that there may be different Murphree vapor efficiencies for different components.

Extractive and Azeotropic Distillation

As discussed in Chap. 7, extractive and azeotropic distillations involve the addition of another component, the *solvent* or the *entrainer*, to modify the equilibrium and thereby facilitate separation of a feed mixture by distillation. When the feed mixture to be separated contains only two components, extractive and azeotropic distillations involve three component systems. As a result they are amenable to stage-to-stage calculations. In extractive distillation, stage-to-stage calculations can start at the bottom since the solvent is a heavy nonkey and the feed species are keys. In the azeotropic distillation of ethanol and water with benzene as entrainer, shown in Fig. 7-28, the water is an effective light nonkey and the other two components may be considered keys; hence stage-to-stage calculations could start at the top. Examples of stage-to-stage calculations for extractive and azeotropic distillation are given by Robinson and Gilliland (1950), Hoffman (1964), and Smith (1963).

Absorption and Stripping

Stage-to-stage methods often have particular advantages when applied to cases where the equilibrium behavior is complex and yet it is possible to specify all com-

Figure 10-1 Portions of a plant for hydrogen manufacture. incorporated into the nitrogen fertilizer plant of Apple River Chemical Company at East Dubuque, Illinois, and built by the Lummus Company. The large object in the center is the steam reforming furnace, where natural gas, CH_4, is converted into hydrogen. Acid-gas, CO_2, removal is accomplished in the absorber and stripper towers to the right. (*The Lummus Co., Bloomfield, N.J.*)

ponents of one of the products with percentwise high accuracy. One such case is absorption with simultaneous chemical reaction in the liquid phase, as long as there are no components which enter the liquid phase significantly but do not have high recovery fractions in the bottom product.

An important example of such a process occurs in the manufacture of hydrogen from methane (natural gas) by a steam reforming reaction in a heated furnace followed by reaction of the remaining CO to CO_2 in a shift converter reactor:

$$CH_4 + H_2O \rightarrow CO + 3H_2$$

$$CO + H_2O \rightarrow CO_2 + H_2$$

The carbon dioxide is removed from the hydrogen product by multistage absorption into a basic solution such as monoethanolamine, $HO—CH_2—CH_2—NH_2$, or hot potassium carbonate (see Prob. 6-N). Figure 10-1 shows the reforming furnace, the CO_2 absorption tower, and the absorbent regeneration tower for a plant manufacturing hydrogen for conversion into ammonia, which is used directly as a fertilizer or turned into other nitrogenous fertilizers.

In oil refineries and natural gas plants it is often necessary to remove both H_2S and CO_2 from gas streams where they form undesirable impurities. Again, this is commonly accomplished by absorption into a basic solution. The equilibria are even more complex because each solute affects the solubility of the other as they compete for the available base. Example 10-2 illustrates the application of stage-to-stage methods to such a process.

Example 10-2 Figures 10-2 to 10-5 show data for the partial pressures p_j^* of CO_2 and H_2S over aqueous monoethanolamine (MEA) solutions when both solutes are present simultaneously. As is shown by Kohl and Riesenfeld (1979), ln p^* can be assumed linear in the reciprocal of absolute temperature for interpolation between temperatures. Consider the design of a plate-absorber column which will employ a 2.5 N (15.3 wt %) solution of MEA in water to remove CO_2 and H_2S from an otherwise inert gas stream. A schematic of the apparatus is shown in Fig. 10-6. The following specifications are made:

Absorber pressure = 1.38 MPa   Inlet gas temperature = 25°C

Inlet amine temperature = 38°C   Inlet gas composition = $\begin{cases} 10\% \; CO_2 \\ 6.0\% \; H_2S \end{cases}$

Inlet amine loading = $\begin{cases} 0.150 \text{ mol } CO_2/\text{mol MEA} \\ 0.030 \text{ mol } H_2S/\text{mol MEA} \end{cases}$ Effluent gas purity < $\begin{cases} 0.05\% \; CO_2 \\ 0.02\% \; H_2S \end{cases}$

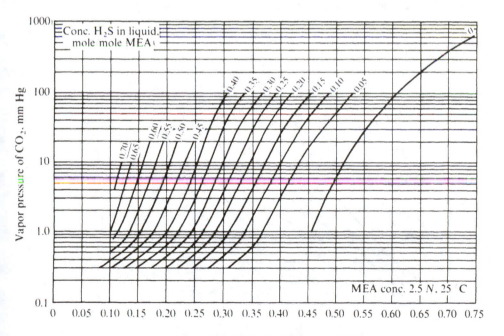

Figure 10-2 Effect of dissolved hydrogen sulfide on vapor pressure of carbon dioxide over 2.5 N MEA solution at 25°C. (*From Kohl and Riesenfeld, 1979, p. 47; used by permission.*)

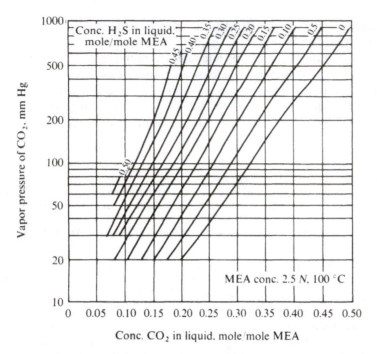

Figure 10-3 Effect of dissolved hydrogen sulfide on vapor pressure of carbon dioxide over 2.5 N MEA solution at 100°C. *(From Kohl and Riesenfeld, 1979, p. 47; used by permission.)*

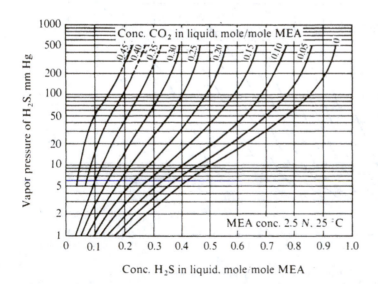

Figure 10-4 Effect of dissolved carbon dioxide on vapor pressure of hydrogen sulfide over 2.5 N MEA solution at 25°C. *(From Kohl and Riesenfeld, 1979, p. 51; used by permission.)*

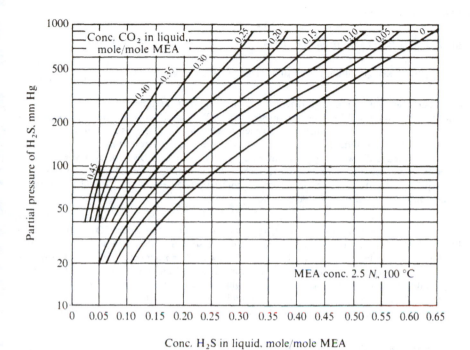

Conc. H_2S in liquid, mole/mole MEA

Figure 10-5 Effect of dissolved carbon dioxide on vapor pressure of hydrogen sulfide over 2.5 N MEA solution at 100°C. *(From Kohl and Riesenfeld, 1979, p. 51; used by permission.)*

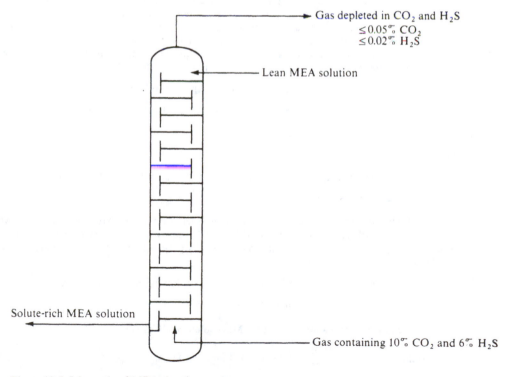

Figure 10-6 Schematic of MEA absorber.

459

Additional data (Kohl and Riesenfeld, 1979):

Heat capacity of MEA solution = 3.98 kJ/K · kg   Heat of absorption = $\begin{cases} 1.92 \text{ MJ/kg } CO_2 \\ 1.91 \text{ MJ/kg } H_2S \end{cases}$

(a) Compute the equilibrium-stage requirement, assuming that the number of equilibrium stages provided by the tower will be the same for each solute. Take an amine circulation rate 1.20 times greater than the minimum. (b) Murphree efficiency data for the simultaneous absorption of H_2S and CO_2 in ethanolamine solutions are not generally available, and it is known that efficiencies in such absorbers vary markedly with stage location, by as much as a factor of 10 or more. For purposes of illustration of incorporating efficiencies into such a calculation, however, compute the number of plates required if the Murphree vapor efficiency is constant at 15 percent for CO_2 and 40 percent for H_2S, and the thermal efficiency is 100 percent.

SOLUTION From Figs. 10-2 to 10-5 it is apparent that the solubility of either CO_2 or H_2S in MEA solution is a strong function of the concentration of both solutes in the MEA solution. We cannot define the absorption behavior of one solute without knowing the amount of the other solute absorbed at any particular stage under consideration. As a consequence, the graphical methods of Chap. 6 are not applicable, even though the gas phase is not highly concentrated in CO_2 and H_2S. (For cases of dilute gas and liquid phases and independent solubility relationships for all species it is possible to perform a binary analysis of the absorption of each solute without considering the others.)

(a) *Determination of minimum MEA circulation rate* The minimum possible MEA circulation rate (infinite stages) can be determined by postulating equilibrium between the phases at the bottom (rich end) of the column. The limiting pinch must occur at that end of the column since the inlet MEA solute loadings are specified and since the equilibrium CO_2 and H_2S pressures rise rapidly with increasing solution concentrations. (A preliminary calculation might check that the specification of gas effluent and MEA inlet do not exceed equilibrium at the top of the column.)

In order to compute equilibrium conditions at the rich end of the column we need to determine the effluent amine solution temperature. A rough calculation will confirm that the sensible heat of the liquid is much larger than that of the gas; hence we can equate the entire heat of absorption to the rise in liquid-phase enthalpy between the tower top and tower bottom. Per mole of inlet gas,

$$\Delta H_{abs} = (0.10 \text{ mol } CO_2)(1.92 \text{ MJ/kg})(44 \times 10^{-3} \text{ kg/mol})$$

$$+ (0.06 \text{ mol } H_2S)(1.91 \text{ MJ/kg})(34 \times 10^{-3} \text{ kg/mol})$$

$$= 12.3 \text{ kJ/mol gas}$$

The heat capacity of the liquid is

$$C_p = (3.98 \text{ kJ/K} \cdot \text{kg soln})(1 \text{ kg soln}/0.153 \text{ kg MEA})(61 \times 10^{-3} \text{ kg MEA/mol})$$

$$= 1.59 \text{ kJ/mol MEA} \cdot °C$$

The equilibrium computation involves trial and error in temperature and MEA loading. We know that in the effluent MEA the solute loadings will be the inlet values plus the total pickup from the gas. Assuming $T_{MEA, out} = 60°C$, we can use an overall enthalpy balance to find the corresponding MEA circulation rate.

$$12.3 \text{ kJ/mol gas} = (1.59 \text{ kJ/mol MEA} \cdot °C)(60 - 38°C)(r \text{ mol MEA/mol gas})$$

$$r = 0.352 \text{ mol MEA/mol gas in}$$

For this rate of MEA circulation, the effluent MEA solute loadings are

CO_2 loading: $\dfrac{0.10}{0.352} + 0.150 = 0.434 \text{ mol/mol MEA}$

H_2S loading:
$$\frac{0.06}{0.352} + 0.030 = 0.200 \text{ mol/mol MEA}$$

The assumed MEA outlet temperature of 60°C can now be checked by seeing if these solute loadings are in equilibrium with the inlet gas composition. This equilibrium is presumed to occur at an interfacial temperature equal to the bulk-liquid temperature. At equilibrium, from Figs. 10-2 to 10-5,

	Partial pressure, mm Hg	
	At 25°C	At 100°C (extrap)
$p^*_{CO_2}$	140	4000
$p^*_{H_2S}$	47	1700

Interpolating, taking $\ln p^*_j$ to be linear in $1/T$ as was indicated, we have at 60°C

$$\ln p^*_{CO_2} = \ln 140 + \left(\frac{1/298 - 1/333}{1/298 - 1/373} \ln \frac{4000}{140}\right) = 4.94 + \left(\frac{35\,373}{75\,333} 3.35\right) = 6.69$$

$$p^*_{CO_2} = 800 \text{ mmHg}$$

Similarly,

$$\ln p^*_{H_2S} = \ln 47 + \left(\frac{35\,373}{75\,333} \ln \frac{1700}{47}\right) = 5.73$$

$$p^*_{H_2S} = 310 \text{ mmHg}$$

Actual partial pressures of H_2S and CO_2 can be computed from the inlet gas conditions:

$$p_{CO_2} = 0.10(1.38 \text{ MPa})(7502 \text{ mmHg/MPa}) = 1035 \text{ mmHg}$$

$$p_{H_2S} = 0.06(1.38 \text{ MPa})(7502) = 620 \text{ mmHg}$$

Thus for a 60°C MEA effluent the equilibrium pressures of both H_2S and CO_2 are lower than those in the inlet gas. The limiting MEA flow must be lower, corresponding to a higher effluent temperature. The CO_2 pressure apparently causes the limiting condition. Assuming $T_{MEA,\,out} = 61$°C, we find

$$r = 0.341 \text{ mol MEA/mol gas in} \qquad CO_2 \text{ loading} = 0.443 \text{ mol/mol MEA}$$

$$H_2S \text{ loading} = 0.206 \text{ mol/mol MEA} \qquad p^*_{CO_2} = 1200 \text{ mmHg}$$

The assumed temperature of 61°C is too high. Interpolating between 60 and 61°C, we have

$$T_{MEA,\,out,\,lim} = 60 + \frac{\ln(1035/800)}{\ln(1200/800)} 1\text{°C} = 60.6\text{°C} \qquad \text{and} \qquad r_{lim} = 0.346 \text{ mol MEA/mol gas in}$$

Despite the major extrapolations necessary with the equilibrium data, this value is relatively well known because of the high sensitivity of the equilibrium pressures to temperature and solute loading.

Taking the operating MEA circulation rate as 1.20 times the minimum, we have

$$r_{op} = 1.20(0.346) = 0.415 \text{ mol/mol gas in}$$

Determination of the equilibrium-stage requirements At this point it is very important to investigate the problem description. The compositions, temperatures, and relative flows of the two inlet streams are now fixed, as is the column pressure. There is one other variable which can be set by construction or manipulation, namely the number of equilibrium stages. We shall replace this variable by a separation variable, the concentration of *one* of the solutes in the effluent gas. The effluent-gas concentration of the other solute must be estimated in order to start the calculation. In this problem the stage-to-stage approach is facilitated by the fact that the recovery fractions of both solutes in the effluent MEA are quite high. Hence the size of the estimated solute concentration in the effluent gas will have little percentwise effect on the concentration of that solute in the effluent MEA. As a result the stage-to-stage calculation can readily be started at the rich end of the tower, and the estimated solute concentration will introduce a significant error only on the top stages, where the effect on solute concentrations in the MEA will be small.

It is also helpful to investigate the degree of approach to equilibrium at the tower top. A rough calculation shows that the maximum allowable effluent-gas partial pressures of CO_2 and H_2S are substantially above the equilibrium pressures over the inlet MEA. Therefore it is not immediately clear whether the H_2S concentration of the gas will fall below the maximum allowable effluent-gas contamination before the CO_2 level does, or vice versa. A logical procedure is to set both solute mole fractions or partial pressures in the effluent gas at the maximum allowable values, that is, $p_{CO_2, \text{out}} = 5.2$ mmHg and $p_{H_2S, \text{out}} = 2.1$ mmHg, and calculate stages until it is clear which solute will reach the maximum allowable mole fraction last. The estimated overhead mole fraction of the nonlimiting solute can then be adjusted.

We can now establish the solute loadings and temperature of the effluent MEA solution from the problem specification:

$$(\Delta T)_{\text{MEA}} = \frac{22.6}{1.2} = 18.8°C \qquad \text{and} \qquad T_{\text{MEA, out}} = 56.8°C$$

CO_2 loading: $0.150 + \dfrac{0.1000 - 0.0005}{0.415} = 0.390$ mol/mol MEA

H_2S loading: $0.030 + \dfrac{0.0600 - 0.0002}{0.415} = 0.174$ mol/mol MEA

The first step of the stage-to-stage solution working up the column is to compute the equilibrium partial pressures over the effluent MEA. These will be $p_{j,1}$ in the gas leaving the bottom equilibrium stage.

The partial pressures in equilibrium with 0.390 mol CO_2/mol MEA, 0.174 mol H_2S/mol MEA are taken from Figs. 10-2 to 10-5:

	Partial pressure, mmHg	
	At 25°C	At 100°C
p_{CO_2}	35	2000
p_{H_2S}	68	600

Interpolating, as before, to 56.8°C gives

$$p_{CO_2, 1} = 240 \text{ mmHg} \qquad p_{H_2S, 1} = 195 \text{ mmHg}$$

The liquid leaving stage 2 above the bottom passes the vapor rising from stage 1; hence the composition of the liquid from stage 2 comes from a mass balance. Because of the relatively high total pressure, the partial pressure of water vapor is not important.

$$CO_2 \text{ in gas from stage } 1 = \frac{240 \text{ mol}}{10{,}360 - (240 + 195) \text{ mol inert}} \left(0.84 \frac{\text{mol inert}}{\text{mol gas in}}\right)$$

$$= 0.0204 \text{ mol/mol gas in}$$

$$CO_2 \text{ assumed in effluent gas} = (0.0005)(0.84) = 0.00043 \text{ mol/mol gas in}$$

Therefore

$$CO_2 \text{ in MEA from stage } 2 = CO_2 \text{ in entering MEA} + CO_2 \text{ absorbed from gas above stage } 1$$

$$= 0.150 + \frac{0.0204 - 0.00043}{0.415} = 0.198 \text{ mol/mol MEA}$$

Similarly, H_2S in gas from stage 1 = 0.0166 mol/mol gas in

$$H_2S \text{ in MEA from stage } 2 = 0.070 \text{ mol/mol MEA}$$

The temperature of stage 2 now comes from an enthalpy balance around stage 2 and all stages above. The heat liberated by absorption of gases on stage 2 and above is given by

$$(\Delta H)_{\text{abs}} = [(0.0204 - 0.0004) \text{ mol } CO_2 \text{ abs/mol gas in}][(1.92)(44) \text{ kJ/mol}]$$

$$+ [(0.0166 - 0.0002) \text{ mol } H_2S \text{ abs/mol gas in}][(1.91)(34) \text{ kJ/mol}]$$

$$= 2.75 \text{ kJ/mol gas in}$$

The rise in temperature of the liquid between the MEA entry and the MEA leaving stage 2 is therefore

$$\Delta T = \frac{2.75 \text{ kJ/mol gas in}}{(1.59 \text{ kJ/mol MEA} \cdot {}^\circ C)(0.415 \text{ mol MEA/mol gas in})} = 4.2^\circ C$$

$$T_2 = 42.2^\circ C$$

We now repeat the calculation process for stage 2:

	Partial pressure, mmHg	
	At 25°C (extrap)	At 100°C
p_{CO_2}	0.1	32
p_{H_2S}	0.6	36

$$p_{CO_2, 2} = 0.48 \text{ mmHg}$$

CO_2 in gas from stage 2 = 0.00004 mol/mol gas in.

$$p_{H_2S, 2} = 1.82 \text{ mmHg}$$

H_2S in gas from stage 2 = 0.00015 mol/mol gas in. The maximum allowable moles per mole gas in are 0.00042 for CO_2 and 0.00017 for H_2S. Hence the H_2S is slightly under the maximum and the CO_2 is well under the maximum. Slightly under two equilibrium stages are required. The calculated results for part (a) are summarized in Table 10-3.

Table 10-3 Summary of calculated results for Example 10-2

	Temp. °C	p_{H_2S}, mmHg	p_{CO_2}, mmHg	H_2S in MEA, mol/mol MEA	CO_2 in MEA, mol/mol MEA
A. Equilibrium stages					
Gas, entering	25.0	620	1036		
Leaving stage 1	56.8	195	240		
Leaving stage 2	42.2	1.82	0.48		
Liquid, leaving stage 1	56.8	....	....	0.174	0.390
Leaving stage 2	42.2	....	....	0.070	0.198
Entering	38.0	....	....	0.030	0.150
B. $E_{MV} = 40\%$ for H_2S, 15% for CO_2					
Gas, entering	25.0	620	1036		
Leaving stage 1	56.8	450	916		
Leaving stage 2	53.0	292	784		
Leaving stage 3	49.6	180	557		
....					
Leaving stage 33	~38	≪2.1	<5.2		
Liquid, leaving stage 1	56.8	....	....	0.174	0.390
Leaving stage 2	53.0	....	....	0.130	0.354
Leaving stage 3	49.6	....	....	0.093	0.320
....					
Entering	38.0	....	....	0.030	0.150

From this computation one might conclude (1) that the separation is relatively simple and does not require large towers, and (2) that it is more difficult to remove H_2S to a given level than it is to remove CO_2. Both these deductions are erroneous, as shown in the solution to part (*b*).

(*b*) *Determination of actual plate requirement* The Murphree vapor efficiencies are given as 15 percent for CO_2 and 40 percent for H_2S. These efficiencies are determined to a major extent by the rate of reaction of the solutes with the MEA in the liquid phase. Since the reaction rate is much faster in the case of H_2S, the Murphree vapor efficiency for H_2S is considerably higher. The rate phenomenon affecting the efficiencies is distinct from the equilibrium phenomena governing the solubilities.

The computational approach used in part (*a*) can be followed for a solution based upon the Murphree efficiencies provided the step in which the equilibrium vapor composition is computed is modified properly. The absorbent flow and the bottom-plate temperature are also unchanged from those found in part (*a*) (why?). Thus the effluent MEA contains 0.390 mol CO_2/mol MEA and 0.174 mol H_2S/mol MEA and has a temperature of 56.8°C. Following the definition of Murphree efficiency,

Stage 1:
$$E_{MV, CO_2} = 0.15 = \frac{y_{CO_2, out} - y_{CO_2, in}}{y^*_{CO_2} - y_{CO_2, in}} = \frac{p_{CO_2, 1} - p_{CO_2, in}}{p^*_{CO_2, 1} - p_{CO_2, in}}$$

where $y^*_{CO_2}$ and $p^*_{CO_2}$ are values in equilibrium with the liquid leaving the stage in question. From part (*a*), $(p^*_{CO_2})_1 = 240$ mmHg. Substituting, we have

$$0.15 = \frac{p_{CO_2, 1} - 1036}{240 - 1036}$$

$$p_{CO_2, 1} = 1036 - (0.15)(796) = 916 \text{ mmHg}$$

Similarly, for H_2S

$$0.40 = \frac{p_{H_2S,1} - 620}{195 - 620}$$

$$p_{H_2S,1} = 620 - (0.40)(425) = 450 \text{ mmHg}$$

Proceeding as in part (a), we get

$$CO_2 \text{ in } V_1 = \frac{916}{8994} \, 0.84 = 0.0852 \text{ mol/mol gas in} \qquad H_2S \text{ in } V_1 = 0.0419 \text{ mol/mol gas in}$$

Stage 2: $\qquad CO_2 \text{ in } L_2 = 0.150 + \frac{0.0848}{0.415} = 0.354 \text{ mol/mol MEA}$

$$H_2S \text{ in } L_2 = 0.030 + \frac{0.0417}{0.415} = 0.130 \text{ mol/mol MEA}$$

$$(\Delta H)_{abs} \text{ above plate } 1 = (0.0848)(1.92)(44) + (0.0417)(1.91)(34) = 9.87 \text{ kJ/mol gas in}$$

$$T_2 = 38 + \frac{9.87}{(1.59)(0.415)} = 53.0°C$$

Results for calculations proceeding upward are shown in Table 10-3, part B. Above stage 3 the equilibrium solute pressures are not significant in the computation of p_j. Further, since the specified maximum allowable solute partial pressures in the outlet gas are far above the values in equilibrium with the regenerated absorbent, it is apparent that the H_2S will die out considerably faster than the CO_2 because of its higher E_{MV}. Thus the stage requirement will be governed by CO_2 absorption.

Writing Eq. (8-16) for $m = 0$, we have

$$N = -\frac{\ln(y_{A,in}/y_{A,out})}{\ln(1 - E_{MV})}$$

where N is the number of intervening stages with the specified E_{MV}. For our case this equation written for the stages *above* stage 3 becomes

$$N - 3 = -\frac{\ln(p_{CO_2,3}/p_{CO_2,N})}{\ln(1 - E_{MV,CO_2})} = -\frac{\ln(667/5.2)}{\ln 0.85} = 30$$

Therefore $N = 33$, and 33 plates are required in the absorption tower. With a 60-cm tray spacing, the tower would be on the order of 20 m high. ☐

It was assumed in the solution to part (b) of Example 10-2 that the gas and liquid streams leaving a plate have the same temperature (thermal efficiency = 100 percent). This is not necessarily true. The gas and liquid equilibrate thermally through a heat-transfer process; if the heat transfer is not rapid enough, the exit gas and exit liquid will not have achieved identical temperatures. From basic mass- and heat-transfer theory it can be deduced that thermal stage efficiencies generally will be equal to or greater than mass-equilibrium efficiencies. In Example 10-2 the low Murphree efficiencies for H_2S and CO_2 are caused by the fact that the full chemical solubility of each species is not available as an interfacial mass-transfer driving force. This limitation does not occur for heat transfer; hence it is probable that the thermal-equilibration efficiencies are relatively high. In any event, incomplete thermal equilibration on the plates would not change the plate requirement substantially, since the equilibrium partial pressures of CO_2 and H_2S are important on only the bottom three plates.

In Example 10-2 it was assumed that all the heat of absorption is carried down with the liquid phase and that the sensible heat of the vapor is negligible. Because of the high liquid-to-gas mass ratio this assumption is permissible for the overall enthalpy balance through which the effluent liquid temperature is found; however, the temperature profile for intermediate plates in the column can be influenced by the vapor heat capacity, and a more precise computation should take this into account. As already noted in Chap. 7, a maximum temperature can develop partway along an absorption column if the counterflowing gas and liquid have comparable products of flow rate and heat capacity and/or if the solvent has appreciable volatility. This high-temperature region can provide the controlling pinch (closest approach of operating and equilibrium curves) for the absorption; thus it is important to model this effect correctly. In a stage-to-stage calculation this can call for an overall iteration loop on the temperature of the exit liquid.

Rowland and Grens (1971) have investigated stage-to-stage calculations for acid-gas absorbers in some detail. They find that the method works well as long as the product of flow rate and heat capacity of the liquid exceeds that of the gas by a factor of 2 or more. If these products are of approximately the same magnitude, iteration on the exit-liquid temperature is necessary and may require damping of temperature changes between iterations in order to gain stability in the computation. If the product of vapor-flow rate and heat capacity substantially exceeds that for the liquid (an unusual case) errors in stage temperatures can build up prohibitively in a bottom-up calculation. In such cases a successive-approximation solution, of the type discussed later in this chapter, becomes preferable.

Part (b) of Example 10-2 was worked assuming constant values of E_{MV}. As is amplified in Chap. 12, the extreme curvature of the equilibrium data for systems like this can cause E_{MV} (or E_{ML}) to vary greatly across a column. Rowland and Grens (1971) present a calculation where E_{MV} varies from 67 percent on the upper stages to 4 percent on the lower stages. Problem 12L considers the calculation of an H_2S–CO_2 absorber where the Murphree efficiency varies substantially from stage to stage.

Kent and Eisenberg (1976) have correlated available equilibrium data for H_2S and CO_2 in solutions of monoethanolamine and diethanolamine in equation forms that are convenient for use with computer calculations.

TRIDIAGONAL MATRICES

If we search for ways to combine Eqs. (10-1) to (10-5) algebraically to simplify the system of equations, the most obvious step is to combine Eqs. (10-1) with the other equations to eliminate either all the $v_{j,p}$ or all the $l_{j,p}$. This will remove $N \times R$

equations and $N \times R$ unknowns. Arbitrarily, we shall eliminate the $v_{j,p}$ and retain the $l_{j,p}$. The sum of Eqs. (10-2) over all components and over the stages from p to either end of the column provides the mass balance

$$V_p - L_{p+1} - \Sigma F_{\geq p} = 0 \qquad (10\text{-}11)$$

where $\Sigma F_{\geq p}$ is the sum of all feeds entering the column on stage p or below, less all products leaving on stage p or below. Equation (10-11) can then be used to eliminate V_p or L_p from the remaining equations; again arbitrarily, we shall eliminate L_p and retain V_p.

Equations (10-2) then become $N \times R$ component-mass-balance equations of the form

$$C_{j,p}(l_{j,p+1}, l_{j,p}, l_{j,p-1}, V_p, V_{p-1}, T_p, T_{p-1}) = 0 \qquad (10\text{-}12)$$

given by

$$-\frac{K_{j,p-1}V_{p-1}}{L_{p-1}} l_{j,p-1} + \left[1 + \frac{S_{Lp} + K_{j,p}(V_p + S_{Vp})}{L_p}\right]l_{j,p} - l_{j,p+1} - z_{j,p}F_p = 0 \qquad (10\text{-}13)$$

where L_p and V_p are related through Eq. (10-11). S_{Lp} and S_{Vp} represent molar flows of sidestreams of liquid and vapor, respectively, leaving stage p. The term $z_{j,p}F_p$ represents the moles of j in feeds entering stage p.

Each set of N equations for each component can be solved for the N values of l_j if all values of S_L, S_V, z_jF, V, and K_j are known. In general, the values of K_j are dependent upon compositions as well as temperature and pressure, and this serves to make Eqs. (10-13) nonlinear. However, for cases where K_j does not depend upon composition, e.g., distillation of ideal mixtures, a knowledge of all T_p and V_p and all feed and product flows will serve to fix the K_j and make Eqs. (10-13) a set of linear equations, which are easily solved. Usually either or both the T_p and/or the V_p are unknown, and we shall have to iterate upon values of those variables. This will result in Eqs. (10-13) being a set of linear equations to be solved within each iteration. However, their linearity is still a major advantage.

If the K_j values depend upon composition but only weakly, it is possible to remove the nonlinearity by evaluating the K_j for the compositions obtained in the last previous iteration on T_p and/or V_p.

Once the $K_{j,p}$ and V_p have been assumed or set, Eqs. (10-13) for any component become a family of linear equations of the form

$$\begin{aligned}
B_1 l_{j1} &+ C_1 l_{j2} &&= D_1 \\
A_2 l_{j1} &+ B_2 l_{j2} + C_2 l_{j3} &&= D_2 \\
&\cdots\cdots\cdots\cdots\cdots\cdots\cdots\cdots\cdots\cdots\cdots\cdots && \qquad (10\text{-}14)\\
A_p l_{j,p-1} &+ B_p l_{j,p} + C_p l_{j,p+1} &&= D_p \\
&\cdots\cdots\cdots\cdots\cdots\cdots\cdots\cdots\cdots\cdots\cdots\cdots \\
A_N l_{j,N-1} &+ B_N l_{j,N} &&= D_N
\end{aligned}$$

where stages are numbered upward and

$$A_p = -\frac{K_{j,\,p-1}V_{p-1}}{L_{p-1}} \qquad 2 \leq p \leq N \tag{10-15}$$

$$B_p = \begin{cases} 1 + \dfrac{S_{Lp} + K_{j,\,p}(V_p + S_{Vp})}{L_p} & 2 \leq p \leq N - 1 & (10\text{-}16) \\[2ex] 1 + \dfrac{K_{j1}(V_1 + S_{V1})}{L_1} & p = 1 & (10\text{-}17) \\[2ex] 1 + \dfrac{S_{LN} + K_N V_N}{L_N} & p = N & (10\text{-}18) \end{cases}$$

$$C_p = -1 \qquad 1 \leq p \leq N - 1 \tag{10-19}$$

$$D_p = z_{j,\,p} F_p \qquad 1 \leq p \leq N \tag{10-20}$$

with L_p obtained from V_p via Eq. (10-11). For a distillation column, L_1 in Eqs. (10-15) and (10-17) is b if there is an equilibrium reboiler and V_N in Eq. (10-18) is D if there is an equilibrium partial condenser. For a total condenser as "stage" N, all equations remain the same except for Eq. (10-18), which becomes

$$B_N = 1 + \frac{d}{L_N} \tag{10-21}$$

Written in matrix form, Eqs. (10-14) become

$$\begin{bmatrix} B_1 & C_1 & & & & \\ A_2 & B_2 & C_2 & & & \\ \hline & & A_p & B_p & C_p & \\ \hline & & & A_{N-1} & B_{N-1} & C_{N-1} \\ & & & & A_N & B_N \end{bmatrix} \begin{bmatrix} l_{j,\,1} \\ l_{j,\,2} \\ \cdot \\ l_{j,\,p} \\ \cdot \\ l_{j,\,N-1} \\ l_{j,\,N} \end{bmatrix} = \begin{bmatrix} D_1 \\ D_2 \\ \cdot \\ D_p \\ \cdot \\ D_{N-1} \\ D_N \end{bmatrix} \tag{10-22}$$

A matrix like the ABC matrix in Eq. (10-22) which has entries only on the main diagonal B and the two adjacent diagonals A and C is called a *tridiagonal* matrix. Highly efficient methods exist for solving sets of linear equations represented by a tridiagonal matrix. Perhaps the most suitable of these is the *Thomas method*, presented by Bruce et al. (1953) and others (Lapidus, 1962; Varga, 1962; Wang and Henke, 1966).

The Thomas method involves the calculation of three different quantities (w_p, g_p, and u_p) for each row, advancing forward through the matrix:

$$w_1 = B_1 \tag{10-23}$$

$$u_1 = \frac{C_1}{w_1} \tag{10-24}$$

$$w_p = B_p - A_p u_{p-1} \qquad 2 \leq p \leq N \tag{10-25}$$

$$u_p = \frac{C_p}{w_p} \qquad 2 \le p \le N - 1 \qquad (10\text{-}26)$$

$$g_1 = \frac{D_1}{w_1} \qquad (10\text{-}27)$$

$$g_p = \frac{D_p - A_p g_{p-1}}{w_p} \qquad 2 \le p \le N \qquad (10\text{-}28)$$

Values of $l_{j,p}$ can then be obtained by working back up the rows of the matrix:

$$l_{j,N} = g_N \qquad (10\text{-}29)$$

$$l_{j,p} = g_p - u_p l_{j,p+1} \qquad (10\text{-}30)$$

One such matrix solution is required for each component.

The Thomas method is rapid and easily programmed and does not require much computer memory. Wang and Henke (1966) and Billingsley (1966) show that, except under very unusual circumstances, this method of solving Eqs. (10-13) does not lead to any significant buildup of computer truncation errors. The only possible exception occurs in the subtraction step of Eq. (10-25), and then only in the case of many stages coupled with a component that has $K_j > 1$ in one section of a cascade and $K_j < 1$ in another section. Boston and Sullivan (1972) present a modification of the Thomas method which can be used in such circumstances but requires more computing time. Birmingham and Otto (1967) have demonstrated, in the context of absorber computations, that the Thomas method is much faster than earlier methods of solving Eqs. (10-13) which used a stage-by-stage calculation.

Example 10-3 Natural fats occur as esters of fatty acids with glycerol, known as triglycerides. In the manufacture of fatty acids, fatty alcohols, and soaps the triglycerides are split chemically, and the fatty acids are separated, typically by vacuum distillation.

Another approach for separation would be fractional extraction of the triglycerides themselves. Chueh and Briggs (1964) measured equilibrium distribution coefficients for triolein and trilinolein between heptane and furfural. Triolein and trilinolein are triglycerides of oleic acid, $CH_3(CH_2)_7CH{=}CH(CH_2)_7COOH$, and linoleic acid,

$$CH_3(CH_2)_4CH{=}CHCH_2CH{=}CH(CH_2)_7COOH,$$

respectively. Smoothed results at high dilution of the acids in the furfural phase and as a function of temperature are:

Temp., °C	$K_D = \dfrac{\text{wt fraction in heptane}}{\text{wt fraction, in furfural}}$ at high dilution	
	Triolein	Trilinolein
60	21.8	9.4
62.5	18.6	8.1
65	16.1	6.9
67.5	13.8	5.9
70	12.2	5.1

Source: Data smoothed from Chueh and Briggs (1964).

Suppose that a mixer-settler extraction with five equilibrium stages is used to separate two feed mixtures of triolein and trilinolein:

Feed stage	Feed flow rate, kg/unit time	
	Triolein	Trilinolein
2	0.5	1.0
4	2.0	1.0

Pure heptane enters stage 1, and pure furfural enters stage 5, both at flow rates that are more than an order of magnitude higher than the feed flows. The mass flow ratio of furfural to heptane is 10.0.

A temperature gradient is imposed on the extraction cascade, with stage 1 at 60°C, stage 5 at 70°C, and a linear variation of temperature in between. The purpose of this is to help remove trilinolein from the triolein product through high temperature and lower K_D and to help remove triolein from the trilinolein product through low temperature and higher K_D.

Find the recovery fractions of the two triglycerides in the two product streams leaving the terminal stages.

SOLUTION If the temperature were constant, giving constant values of K_D, the problem could be solved using the multiple-section version of the Kremser-Souders-Brown equation. However, the changing temperature makes the extraction factors for each component (the equivalent of $K_j V/L$) different on each stage.

The temperatures are specified independently; the K_j's are independent of composition because of the high dilution; and the total flow rates of the phases are known because the high dilution and the immiscibility of furfural and heptane keep the phase flows effectively constant from stage to stage. Hence all the coefficients in Eqs. (10-13) are established, and we can use the Thomas method to solve the resulting set of linear tridiagonal equations.

Values of $A_{j,p}$, $B_{j,p}$, $C_{j,p}$, and $D_{j,p}$ are obtained from Eqs. (10-15) to (10-20). When we let V correspond to the mass flow rate of heptane and L to the mass flow rate of furfural, these equations become

$$A_{j,p} = -0.1K_{j,p-1} \qquad 2 \le p \le 5$$
$$B_{j,p} = 1 + 0.1K_{j,p} \qquad 1 \le p \le 5$$
$$C_{j,p} = -1 \qquad 1 \le p \le 4$$
$$D_{j,p} = f_j \qquad 1 \le p \le 5$$

Since the successive stage temperatures are 60, 62.5, 65, 67.5, and 70°C, we can substitute values of $K_{j,p}$ from the table in the problem statement, as well as values of f_j:

Stage	Triolein				Trilinolein			
	A	B	C	D	A	B	C	D
1	....	3.18	−1	0	....	1.94	−1	0
2	−2.18	2.86	−1	0.5	−0.94	1.81	−1	1
3	−1.86	2.61	−1	0	−0.81	1.69	−1	0
4	−1.61	2.38	−1	2.0	−0.69	1.59	−1	1
5	−1.38	2.22	....	0	−0.59	1.51	−1	0

The Thomas-method parameters, in order of calculation, are then:

	Triolein	Trilinolein		Triolein	Trilinolein		Triolein	Trilinolein
w_1	3.18	1.94	w_4	1.462	0.9505	g_4	1.636	1.463
u_1	-0.3145	-0.5155	u_4	-0.6838	-1.052	$g_5 = l_5$	1.770	0.9710
w_2	2.174	1.325	w_5	1.276	0.8892	l_4	2.846	2.484
u_2	-0.4599	-0.7545	g_1	0	0	l_3	1.866	2.869
w_3	1.755	1.079	g_2	0.2230	0.7547	l_2	1.081	2.920
u_3	-0.5699	-0.9269	g_3	0.2438	0.5666	l_1	0.3400	1.505

Values of v_j can then be obtained as $(K_j V/L)_p l_{j,\,p}$:

Stage	Triolein l	v	Trilinolein l	v
1	0.3400	0.7412	1.505	1.415
2	1.081	2.011	2.920	2.365
3	1.866	3.004	2.869	1.980
4	2.846	3.927	2.484	1.466
5	1.770	2.159	0.9710	0.4952

The recovery fraction of triolein in the heptane product is then $v_5/(f_2 + f_4) = 2.159/2.5 = 0.86$. The recovery fraction of trilinolein in the furfural product is $l_1/(f_2 + f_4) = 1.505/2 = 0.75$. ☐

The calculation in Example 10-3 carried four significant figures, which is more than the precision of the data for K_j, in order to indicate the accuracy of the Thomas solution. One can add v_5 and l_1 for both solutes and compare with f_2 and f_4, giving a mass-balance nonclosure of only 0.04 percent for triolein and 0.01 percent for trilinolein.

Holland (1975) gives a sample solution by the Thomas method for a multicomponent distillation with assumed values of T_p and V_p.

Equations (10-13) are generated from Eqs. (10-1) and (10-2) and are therefore valid for equilibrium stages. If the calculation is based upon Murphree efficiencies, Eqs. (10-1) must be replaced by Eqs. (10-7) if E_{MV} is used or by its analog for E_{ML}. Substituting Eqs. (10-7) into Eqs. (10-2) should give a set of equations to replace Eqs. (10-13). If the v_j's are to be eliminated in this substitution, as in the foregoing analysis, a more complex procedure will be required since both $v_{j,\,p}$ and $v_{j,\,p-1}$ appear in each of Eqs. (10-7). On the other hand, one might note that the l_j from only one stage appear in each of Eqs. (10-7), and thus we could solve for $l_{j,\,p}$ directly in terms of v_j's and substitute into Eqs. (10-2), giving

$$-\frac{L_{p+1}}{K_{j,\,p+1}V_{p+1}E_{MVj,\,p+1}}\,v_{j,\,p+1} + \left[1 + \frac{L_p}{K_{j,\,p}V_p E_{MVj,\,p}} - \frac{L_{p+1}}{K_{j,\,p+1}V_p}\right.$$

$$\left.\times\left(1 - \frac{1}{E_{MVj,\,p+1}}\right)\right]v_{j,\,p} + \left[\frac{L_p}{K_{j,\,p}V_{p-1}}\left(1 - \frac{1}{E_{MVj,\,p}}\right) - 1\right]v_{j,\,p-1} - f_{j,\,p} = 0$$

$$(10\text{-}31)$$

If E_{MVj}, K_j, V, and L were known for each stage, Eqs. (10-31) would be a tridiagonal linear set, solvable by the Thomas method. However, it appears that instabilities in such a solution can occur from the way the E_{MV} terms enter the equations. This might be expected from the fact that solution of Eqs. (10-7) for $l_{j,\,p}$ does not directionally represent a physical cause-and-effect situation. In general, therefore, it appears best to handle Murphree efficiencies in a way that gives up the computatational efficiency of the tridiagonal matrix.

Huber (1977) discusses ways of using the properties of a supertriangular matrix (all nonzero elements located on the main diagonal or above) to handle a calculation with specified Murphree efficiencies and/or with recycle, bypass, or interconnections between different separators. His allowance for Murphree vapor efficiencies involves repeated substitution of Eqs. (10-7), giving each vapor flow as a linear function of the liquid flows on all stages below.

Also, as noted before, there are only $R - 1$ independent values of Murphree efficiency for each stage if the interstage streams are dew-point and bubble-point vapors and liquids. This problem is avoided if all E_{MVj} are taken to be the same on any stage, since the dependent E_{MV} will then be equal to the others. Since there are not yet any good bases for predicting individual-component E_{MVj} values in a multi-component system, the question of handling different E_{MVj} for different components appears not to have been addressed systematically, and therefore all E_{MVj} on a stage have usually been assumed to be equal, by default.

DISTILLATION WITH CONSTANT MOLAL OVERFLOW; OPERATING PROBLEM

If constant molal overflow is postulated, the flows on each stage of a distillation can be regarded as fixed if the product and reflux or boil-up flows are fixed. Hence solution of a problem reduces to solution of the E, M, and S equations [Eqs. (10-1), (10-2), (10-4) and (10-5)], the enthalpy-balance equations no longer being needed.

Constant molal overflow also generally implies that the K_j are at most only weak functions of phase compositions; otherwise heat-of-mixing effects should cause appreciable changes in interstage flows. In that case Eqs. (10-1) and (10-2) can be combined into Eqs. (10-13), which can be made linear and solved by the Thomas method provided the total number of stages is known, as in an operating problem. Fixing the values of K_j to make Eqs. (10-13) linear requires assuming the temperatures of all stages and, if necessary, basing activity coefficients on the phase compositions obtained in a previous iteration or on assumed values. The principal aspect of the problem is then converging the stage temperatures to the correct values. The solution is iterative, i.e., assuming a set of stage temperatures, followed by solving Eqs. (10-13) by the Thomas method to obtain values of $l_{j, p}$, followed by using some form of Eqs. (10-4) and/or (10-5) to obtain new values of stage temperatures, etc. Solutions of this type are sometimes called *Thiele-Geddes methods*, after the original paper based upon such an approach (Thiele and Geddes, 1933).

The most obvious procedure to use for correcting the stage temperatures in an overall convergence loop is a simple bubble-point computation, one for each stage, using the values of $l_{j, p}$ computed for that stage in the last previous iteration. The bubble-point temperature so calculated would then be the postulated temperature for the next iteration. This constitutes a direct-substitution convergence method (Appendix A).

There are three drawbacks to the use of a simple bubble-point convergence loop for each stage:

1. There is a tendency for persistence of a temperature profile which is initially uniformly too high or too low. For example, a predominantly too high temperature profile will increase the $K_{j, p}$ unrealistically and will tend to place too much of the heavy components into the overhead product. This will make all stages too rich in the heavy components and, in turn, cause the bubble points to be too high. As a result the too high temperature profile is carried into the next iteration.
2. The bubble-point calculation is itself iterative, adding another inner loop for each stage, which serves to lengthen the computational time considerably.
3. Direct substitution of the calculated bubble point for each stage does not account for the effect of temperature corrections to one stage on the component flows and hence the bubble points of adjacent stages, coming through Eqs. (10-13).

We shall examine approaches to overcoming all three of these problems.

Persistence of a Temperature Profile That Is Too High or Too Low

One approach to the problem of persistence of an erroneously high or low temperature profile is to adjust the individual component flows on each stage before the bubble points are computed. One of the signs of a uniformly too high or too low temperature profile is a computed bottoms flow rate [$\Sigma l_{j, 1}$ from the Thomas-method solution of Eqs. (10-13)] that is either too low or too high, respectively. Since the product flow rates are usually taken to be specified, a logical step is to adjust the individual-component flows in the bottoms and in the distillate to satisfy the total-flow specifications. Holland (1963) and Hanson et al. (1962) have achieved considerable success by correcting the ratio b_j/d_j for each component by a single factor θ. The necessary value of θ is obtained from an iterative solution of the equation

$$d_{\text{spec}} = \sum_j \frac{f_j}{1 + \theta(b_j/d_j)_{\text{calc}}} \tag{10-32}$$

The corrected ratio b_j/d_j is equal to θ times the value of b_j/d_j computed in the solution of the tridiagonal matrix in the last previous iteration [$(b_j/d_j)_{\text{corr}} =$

$\theta(b_j/d_j)_{\text{calc}}]$. θ found so as to satisfy Eq. (10-32) will then satisfy the specified d and b.

To correct the liquid compositions on each stage before the bubble-point calculation, Holland (1963) proposed correcting the component flows by the ratio $(d_j)_{\text{corr}}/(d_j)_{\text{calc}}$ and then normalizing; i.e.,

$$x_{j,\,p} = \frac{(l_{j,\,p})_{\text{calc}}(d_j)_{\text{corr}}/(d_j)_{\text{calc}}}{\sum_j (l_{j,\,p})_{\text{calc}}(d_j)_{\text{corr}}/(d_j)_{\text{calc}}} \tag{10-33}$$

This correction is exact for total reflux (Holland, 1975). On the other hand Hanson, et al. (1962) and Seppala and Luus (1972) have noted that at finite reflux ratios a feed stage exerts an appreciable dampening effect on the amount of correction required. Hanson et al. proposed using for each component a correction factor which decreases geometrically from either end of a column to the feed stage, reaching unity at the feed stage. Seppala and Luus have proposed similarly based correction factors, anchored to unity at the feed and changing either linearly or quadratically toward $(d_j)_{\text{corr}}/(d_j)_{\text{calc}}$ and $(b_j)_{\text{corr}}/(b_j)_{\text{calc}}$ at either end of the column. They also present results from calculations of a number of cases showing that such variable correction factors do accomplish a substantial acceleration of convergence.

Accelerating the Bubble-Point Step

Various methods have been suggested for reducing the time required for predicting new stage temperatures in the bubble-point-calculation step. These follow the general line of the discussion of bubble- and dew-point calculations in Chap. 2. One attractive possibility is to use either a direct, noniterative calculation or a single iteration followed by one correction step to generate the new temperature. This will be an approximation rather than a fully converged bubble point, but it can be quite satisfactory as long as the correction function is stated in a way that will ultimately converge to the correct set of temperatures. This single-step correction can be made and/or acceleration of bubble-point convergence can be helped in any of several ways, e.g., using the relative insensitivity of relative volatilities (as opposed to the K_j's themselves) to temperature and/or using a functional form which takes advantage of the near linearity of $\ln K_j$ or $\ln \alpha_j$ in $1/T$ or T [see Example 2-3 and Eqs. (2-3), (2-4), (2-7), and (2-8)]. One can also work in terms of dew points, just as well as bubble points, having put Eqs. (10-13) into the form involving the v_j's rather than the l_j's, or one can converge $\Sigma y_j - \Sigma x_j$ or $\ln(\Sigma y_j/x_j)$ to zero, having computed both the v_j's and l_j's by using the tridiagonal-matrix solution followed by Eqs. (10-1). This last procedure works well when the components cover a wide span of volatilities. Some specific examples of implementations of one or a combination of these approaches are given by Holland (1963, 1975), Billingsley (1970), Boston and Sullivan (1974), and Lo (1975).

Allowing for the Effects of Changes on Adjacent Stages

In order to allow for the effects of changes in the temperature of adjacent stages on the converged temperature for a stage, Newman (1963) has proposed a convergence scheme in which Eq. (10-4) is evaluated at each stage and a multivariate Newton procedure is used to find corrections to the temperatures of all stages simultaneously. Equations (A-11) (Appendix A) are employed, with x_j replaced by T_s and with $f_k(T_1, T_2, \ldots, T_s, \ldots)$ replaced by $\Sigma l_{j,p} - L_p$. The subscript s also represents stage number. Both p and s are used as subscripts for this purpose to emphasize that $\partial l_{j,p}/\partial T_s$ represents the partial derivative of the liquid flow of j on one stage with respect to the temperature of another stage. The partial derivatives in Eqs. (A-11) become

$$\frac{\partial f_k}{\partial x_j} = \frac{\partial \sum_j l_{j,p}}{\partial T_s} = \sum_j \frac{\partial l_{j,p}}{\partial T_s} \qquad (10\text{-}34)$$

If we ignore sidestreams and differentiate Eqs. (10-13) with respect to T_s, we find that

$$-\frac{\partial l_{j,p+1}}{\partial T_s} + \left(1 + \frac{K_{j,p} V_p}{L_p}\right)\frac{\partial l_{j,p}}{\partial T_s} - \frac{K_{j,p-1} V_{p-1}}{L_{p-1}}\frac{\partial l_{j,p-1}}{\partial T_s}$$

$$+ \frac{V_s}{L_s} l_{j,s} \frac{\partial K_{j,s}}{\partial T_s}(\delta_{s,p} - \delta_{s,p-1}) = 0 \qquad (10\text{-}35)$$

$\delta_{s,p}$ is the Kronecker delta, which is 1 if the two subscripts are equal and 0 otherwise. The quantities other than the derivatives in Eqs. (10-35) are evaluated with the temperatures and component flows corresponding to the last previous solution of the individual component mass balances. The last term involves $\partial K_{j,s}/\partial T_s$, which must be obtained somehow for each component on each stage. If equilibrium data are available in simple algebraic form, it may be possible to use analytical expressions for the $\partial K_{j,s}/\partial T_s$; otherwise they must be obtained from finite-difference calculations at two slightly different temperatures.

Equations (10-35) for each component can be placed in the tridiagonal matrix form of Eqs. (10-22), forming $N \times R$ such matrices corresponding to each combination of component j and stage s. The elements A_p, B_p, and C_p are the same as in Eqs. (10-15) to (10-19), the $l_{j,p}$ matrix is replaced by a corresponding $\partial l_{j,p}/\partial T_s$ matrix, and the D_p terms become

$$D_p = \frac{V_s}{L_s} l_{j,s} \frac{\partial K_{j,s}}{\partial T_s}(\delta_{s,p-1} - \delta_{s,p}) \qquad 1 \le p \le N \qquad (10\text{-}36)$$

Note that the same $\partial K_{j,s}/\partial T_s$ are included in the matrices for component j for all values of s. Hence the number of these derivatives to be evaluated is $N \times R$. The $N \times R$ tridiagonal matrices corresponding to Eqs. (10-22) with elements given by Eqs. (10-15) to (10-19) and (10-36) can all be solved by the Thomas method. The resulting partial derivatives are then summed by Eq. (10-34), and the new values of T_s for the $(i + 1)$th iteration are obtained by solving Eqs. (A-11). If the initial estimates are far from the ultimate solution, it may be desirable to place an upper limit on the allowable $|T_{s,i+1} - T_{s,i}|$.

Since this procedure allows for the effects of changes in the temperatures of adjacent stages and for the specified overall mass balance, it does not require that the θ corrections be applied first to the component flows calculated in the previous iteration. Since the Newton method of convergence is second order, it will approach the final solution faster than direct substitution or a first-order method in the later iterations. Because of this, the Newton method of Newman requires fewer iterations to converge, but it does require more computation per iteration. Seppala and Luus (1972) have compared the computing times for the Newman Newton method and the θ direct-substitution method for a number of sample problems and have found the times to be roughly comparable.

One approach for shortening the computing time for a Newton convergence is to hold the matrix of partial derivatives in Eq. (A-11) unchanged for several iterations before computing it again. It also will be possible in many cases to obtain all the $\partial K_{j, s}/\partial T_s$ for a given component by computing one such derivative at one temperature and assuming that

$$\frac{\partial \ln K_{j, s}}{\partial(1/T_s)} = -\frac{T_s^2}{K_{j, s}}\frac{\partial K_{j, s}}{\partial T_s}$$

is a constant with respect to temperature. In that case only R derivatives need be evaluated, preferably at a temperature near that of a middle stage. Yet another effective method for reducing computing time is to update the matrix of partial derivatives in Eq. (A-11) by using residuals calculated in the previous iteration (Broyden, 1965).

By analogy to a bubble-point calculation, the $\Sigma l_{j, p}$ in Eqs. (10-4) and (10-34) are closer to logarithmic than linear functions of T_s. Therefore $f_k(T_s)$ in the Newton method could be made $\ln (\Sigma l_{j, p}/L_p)$ instead of $\Sigma l_{j, p} - L_p$. This should result in fewer iterations but would have to balanced against the computing time required for determining logarithms.

The Newton convergence method can be divergent for a particularly poor guess of the initial temperature profile. However, it can be shown that small corrections made in the indicated directions should be convergent. Hence one effective approach is to search for the value of a fraction f $(0 \leq f \leq 1)$ of the indicated temperature correction for each stage which will minimize the sum of the squares of Eqs. (10-4) written for each stage (Broyden, 1965).

Wang and Henke (1966), Tierney and Bruno (1967), Billingsley (1970), and Billingsley and Boynton (1971) also discuss ways of implementing the Newton method for converging the temperature profile in distillation.

The assumption of constant molal overflow can be relaxed through use of the modified-latent-heat-of-vaporization (MLHV) method, as described in Chap. 6 and used in other calculation methods described previously.

Example 10-4 A distillation column contains three stages, each providing simple equilibrium, and is equipped with a total condenser returning saturated liquid reflux and an equilibrium reboiler. Constant molal overflow may be postulated. A feed of 100 mol/h is fed to the middle stage of the column as a saturated liquid. The distillate and reflux flows are each 50 mol/h. The feed composition and the relative volatilities of the individual components, assumed independent of temperature, are:

Component	Relative volatility	f_j, mol/h
A	1	33.3
B	2	33.3
C	3	33.4

K_A as a function of temperature is given by

$$\ln K_A = 5.6769 - \frac{2117.5}{T}$$

where T is expressed in kelvins.

For a first trial, it will be assumed that all stage temperatures are 373 K. This makes $K_A = 1$ and therefore makes all K_j equal to or greater than 1. Therefore this is obviously too high a temperature, but it will be used to indicate the capabilities of the methods. For this assumption, a solution of the tridiagonal matrix for individual component flows, taken from a similar problem presented by Holland (1975) gives:

Component	b_j	l_{j1}	l_{j2}	l_{j3}	$d_j = r_j$
A	16.650	49.950	49.950	16.650	16.650
B	4.261	21.306	32.669	14.519	29.039
C	1.421	9.949	21.319	10.660	31.979

Stages are numbered from the bottom, not counting the reboiler. (*a*) Use the θ method and direct substitution through bubble-point calculations to obtain stage temperatures for the second iteration. (*b*) Indicate how the Newman method based on the Newton convergence technique would be used to derive a new set of stage temperatures, as an alternative to the method of part (*a*).

SOLUTION From the problem statement, the specified total flows are

$$d = 50 \qquad b = 50 \qquad F = 100 \qquad r = 50$$

$$L_1 = 150 \qquad L_2 = 150 \qquad L_3 = 50$$

$$V_R = 100 \qquad V_1 = 100 \qquad V_2 = 100 \qquad V_3 = 100$$

From the solution to the tridiagonal matrix, the calculated total liquid flows are

$$b = \Sigma b_j = 22.332 \qquad L_1 = \Sigma l_{j1} = 81.205 \qquad L_2 = \Sigma l_{j2} = 103.938$$

$$L_3 = \Sigma l_{j3} = 41.829 \qquad r = d = \Sigma d_j = 77.668$$

All except d and r are too low, whereas d and r are too high. These results are all in line with the fact that the assumed stage temperatures were all too high.

(*a*) We first determine θ by solution of Eq. (10-32):

$$50 = \frac{33.3}{1 + \theta(16.650/16.650)} + \frac{33.3}{1 + \theta(4.261/29.039)} + \frac{33.4}{1 + \theta(1.421/31.979)}$$

By trial and error (which could use Newton convergence)

$$\theta = 5.6215$$

For using Eq. (10-33), we need values of $(d_j)_{corr}$:

$$\frac{(b_j)_{corr}}{(d_j)_{corr}} = \theta \frac{(b_j)_{calc}}{(d_j)_{calc}}$$

$$(d_j)_{corr} = \frac{f_j}{1 + (b_j/d_j)_{corr}}$$

Component	$\left(\dfrac{b_j}{d_j}\right)_{calc}$	$\left(\dfrac{b_j}{d_j}\right)_{corr}$	f_j	$(d_j)_{corr}$
A	1.00000	5.6215	33.3	5.029
B	0.14673	0.8248	33.3	18.248
C	0.04444	0.2498	33.4	26.724
				50.001

Each $l_{j,p}$ will now be increased by a factor of $(d_j)_{corr}/(d_j)_{calc}$ [numerator of Eq. (10-33)], and these values will be normalized [denominator of Eq. (10-33)] so that the calculated $x_{j,p}$ add to unity on each stage:

Component	$\dfrac{(d_j)_{corr}}{(d_j)_{calc}}$	b_j†	$x_{j,b}$	$l_{j,1}$‡	$x_{j,1}$§	$l_{j,2}$	$x_{j,2}$	$l_{j,3}$	$x_{j,3}$
A	0.3020	28.271	0.5654	15.085	0.4100	15.085	0.2823	5.028	0.2180
B	0.6284	15.052	0.3011	13.389	0.3640	20.529	0.3842	9.124	0.3957
C	0.8357	6.676	0.1335	8.314	0.2260	17.816	0.3335	8.909	0.3863
			1.0000	36.788	1.0000	53.430	1.0000	23.061	1.0000

† Obtained as $f_j - (d_j)_{corr}$.
‡ Obtained as $(l_{j,1})_{calc} \times (d_j)_{corr} / (d_j)_{calc}$.
§ Obtained as $l_{j,1} / \Sigma l_{j,1}$.

Notice that the result of the θ-method corrections has been to increase mole fractions of the light component C and decrease mole fractions of the heavy component A in comparison to the tridiagonal-matrix solution given in the problem statement. This keeps a too high temperature profile from persisting into successive iterations.

New temperatures should then be obtained from bubble-point calculations. Because of the constant values of x_j, we can use the method of Example 2-3 and thereby avoid iteration in obtaining the bubble points:

$$K_A^{-1} = \sum_j x_j x_j \quad \text{and} \quad T(K) = \frac{2117.5}{5.6769 - \ln K_A}$$

Stage	$\Sigma x_j x_j = K_A^{-1}$	K_A	T, K
R	1.5681	0.6377	345.6
1	1.8160	0.5507	337.5
2	2.0512	0.4875	331.1
3	2.1683	0.4612	328.3

These temperatures would be used to provide values of K_j for the second iteration. Different temperatures and presumably more rapid convergence would have been obtained by making the component-flow corrections through one of the methods of Hanson et al. (1962) or Seppala and Luus (1972).

(b) Use of the Newman method requires that we first obtain the coefficients of Eqs. (10-35), put in the form of tridiagonal matrices [Eqs. (10-22)]. For $j = A$ and $s = 1$, for example, these coefficients are

$$
\begin{bmatrix}
3 & -1 & 0 & 0 & 0 \\
-2 & 1.667 & -1 & 0 & 0 \\
0 & -0.667 & 1.667 & -1 & 0 \\
0 & 0 & -0.667 & 3 & -1 \\
0 & 0 & 0 & -2 & 2
\end{bmatrix}
\begin{bmatrix}
\partial l_{A,0}/\partial T_1 \\
\partial l_{A,1}/\partial T_1 \\
\partial l_{A,2}/\partial T_1 \\
\partial l_{A,3}/\partial T_1 \\
\partial r_A/\partial T_1
\end{bmatrix}
=
\begin{bmatrix}
0.5067 \\
-0.5067 \\
0 \\
0 \\
0
\end{bmatrix}
$$

The matrices contain a row for $p = 0$, corresponding to the reboiler.
The values of A_p, B_p, and C_p come from Eqs. (10-15) to (10-19). The values of D_0 and D_1 come from Eq. (10-36), where

$$
\frac{\partial K_{A,1}}{\partial T_1} = -\frac{K_{A,1}}{T_1^2}\frac{\partial \ln K_{A,1}}{\partial(1/T_1)} = -\frac{1.0(-2117.5)}{(373)^2} = 0.01522
$$

and

$$
\frac{V_1}{L_1}l_{A,1}\frac{\partial K_{A,1}}{\partial T_1} = \frac{100}{150}(49.950)(0.01522) = 0.5067
$$

This matrix would be solved for values of $\partial l_{A,p}/\partial T_1$, and the other $N \times (R-1)$ matrices would be solved for the other values of $\partial l_{j,p}/\partial T_s$. The resulting partial derivatives would all be substituted into Eq. (A-11), which would be inverted to solve for all the temperature corrections.

Because of the logarithmic dependence of K_j on $1/T$, a convergence scheme based upon $\ln (\Sigma l_{j,p}/L_p)$ as $f_k(T_s)$ should converge even faster than the form used here, which is based upon $\Sigma l_{j,p} - L_p$ as $f_k(T_s)$. $\qquad\square$

MORE GENERAL SUCCESSIVE-APPROXIMATION METHODS

In general, successive-approximation methods work by assuming values of total flows, temperatures, and possibly also individual stage compositions. Equations are then solved as needed to obtain values for additional unknown variables, and the remaining equations are then used as check functions in a convergence scheme to obtain new values of the assumed variables. The θ direct-substitution and multivariate Newton methods for converging the temperature profile in constant-molal-overflow distillation are special cases of the successive-approximation class of methods where only the temperature profile needs to be assumed and converged upon.

More complex successive-approximation methods can be categorized according to whether or not highly nonideal solutions are allowed for. If solutions are ideal or only mildly nonideal, it is convenient to retain the efficiency of solving for stage compositions by means of the Thomas method for tridiagonal matrices. If solutions are highly nonideal, K_j values depend strongly upon composition and it is more effective to include the stage compositions along with temperatures and total flows as variables to be assumed and converged by successive approximation. We shall consider this latter situation first.

Nonideal Solutions: Simultaneous-Convergence Method

For highly nonideal solutions Eqs. (10-13) become highly nonlinear because of the dependence of the $K_{j,p}$ upon the individual flows of all components, i.e., solution composition. Newman (1967, 1968) and Naphtali and Sandholm (1971), among others, have presented efficient ways of applying the multivariate Newton convergence scheme in such situations. This is sometimes known as the *simultaneous convergence* (SC) method.

As described by Naphtali and Sandholm (1971), the method involves using all $l_{j,p}$, all $v_{j,p}$, and all T_p [$N(2R + 1)$ variables] as convergence variables. The check functions are Eqs. (10-2), (10-3), and (10-7), [$N(2R + 1)$ equations]. Use of Eqs. (10-7) rather than Eqs. (10-1) allows for the use of specified or calculated Murphree efficiencies. The method then involves solving for partial derivatives of the check functions with respect to all the convergence variables through a large set of simultaneous linearized equations, as in the earlier Newman method for the convergence of the temperature profile alone.

A very important feature of the method, in terms of computational efficiency, is the fact that the simultaneous equations, when grouped by stage, form a block-tridiagonal matrix, which can be solved by a very efficient logical extension of the Thomas method (Newman, 1968). A program for solving such a matrix is given in Appendix E, where the properties of these systems of equations are explored further. A block-tridiagonal matrix is a tridiagonal matrix whose elements are themselves smaller submatrices of partial derivatives.

Once the block-tridiagonal matrix is solved for all the partial derivatives, the derivatives are assembled into a matrix in the form of Eq. (A-11), which is then inverted to give indicated corrections to the values of each of the convergence variables. These variables are used in the relinearization of the equations for the next iteration.

Again, this multivariate Newton method can be divergent if the initial assumed values for the convergence variables are poor and/or if the system shows strong nonidealities. As for the convergence of the temperature profile in constant-molal-overflow distillation, stability can be gained by the method suggested by Broyden (1965), where the indicated corrections to the convergence variables are multiplied by a positive fraction, less than 1, chosen by a search procedure so as to minimize the sum of the squares of the nonclosures of the check functions.

This approach involves many partial derivatives and therefore requires both a large amount of computer storage and a substantial amount of time for algebraic manipulation within each iteration. As is characteristic of multivariate Newton methods, convergence is very rapid as the final solution is approached. Computing time can be reduced by formulating the check equations in terms of different variables or functions which take advantage of the insensitivity of particular functions to particular variables. Some equations for the temperature functionality which take advantage of the reduced sensitivity of relative volatilities to temperature and/or the likelihood that $\ln K_j$ is more nearly linear in T or $1/T$ were mentioned in the earlier discussion of constant-molal-overflow distillation. Boston and Sullivan (1974) present a formulation in terms of different variables taking advantage of the insensitivity of relative volatilities to temperature for distillation and invoking a variant of the modified-latent-heat-of-vaporization (MLHV) method for total flow rates. Although their formulation of the variables and equations was made in the context of a BP arrangement of the convergence loops (see below), it should work as well for the more general multivariate Newton solution considered here. Hutchison and Shewchuk (1974) also present an alternate arrangement of the equations for distillation which utilizes the insensitivity of relative volatilities and develops the enthalpy equations in terms of the activity coefficients in a way that makes the computation efficient for regular solutions (zero excess entropy of mixing). Holland (1975) gives a method for using "virtual" values of the partial molar enthalpy for obtaining enthalpies of mixtures. Computing efficiency can also be obtained by holding the matrix of partial derivatives unchanged for several iterations at a time (Orbach et al, 1972) or updating it using residuals computed in the previous iteration (Broyden, 1965).

Grouping the equations by stage allows a more efficient solution to the matrix than grouping by component (Goldstein and Stanfield, 1970) as long as a separation has more stages than components.

The block-tridiagonal matrix form is obtained for an operating problem in which the number of stages and end flows and/or reboiler and condenser duties are specified (Naphtali and Sandholm, 1971), as well as for a number of other specifications (see Appendix E). Some other specifications and situations with interlinked separators can cause other elements to appear in the matrix used for solving for the partial derivatives. Methods of handling such problems where a few elements appear off the main three diagonals are discussed by Kubiček et al. (1976) and Hofeling and Seader (1978).

Fredenslund et al. (1977a, 1977b) discuss combination of the full multivariate Newton solution with the UNIFAC method for predicting activity coefficients and hence K_j. Fredenslund et al. (1977b) also give a full listing of the Naphtali-Sandholm program for such calculations. A simpler program using SC convergence with less complex equilibrium expressions has been presented by Newman (1967) and is reproduced in Appendix E.

Ideal or Mildly Nonideal Solutions; $2N$ Newton Method

When solution nonidealities are either nonexistent or relatively weak, the values of $K_{j, p}$ will not be strong functions of solution compositions and it is advantageous to use a calculation method which solves the individual-component mass balances [Eqs. (10-13)] directly by the Thomas method within each iteration. The convergence variables then become all the stage temperatures and all the total flows of one of the counterflowing streams, and the check functions become the energy balances and the flow-summation equations [Eqs. (10-3) and (10-4) or (10-5)] for each stage. Within each iteration values of $K_{j, p}$ are obtained as functions of temperature alone or are computed on the basis of the stage compositions computed in the last previous iteration. It is generally preferable to place the summation equations in a form based upon $\Sigma y_{j, p} - \Sigma x_{j, p} = 0$,

$$\sum_j (K_{j, p} - 1)l_{j, p} = 0 \tag{10-37}$$

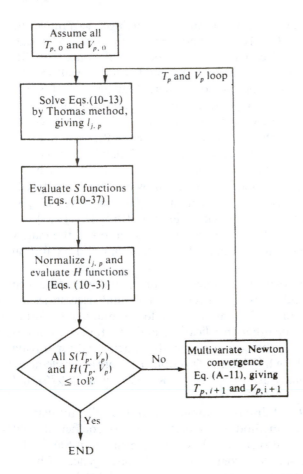

Figure 10-7 General simultaneous convergence method for solutions not highly nonideal.

since this gives improved convergence properties for reasons similar to those favoring use of Eq. (2-25) for single-stage equilibrium calculations.

Tomich (1970) describes a calculational method of this type, where a multivariate Newton method is used for the convergence of stage temperatures and total flows. The computation follows the scheme shown in Fig. 10-7. The number of convergence variables is much less ($2N$) than in the case of a full multivariate SC Newton solution, and consequently there is much less calculation per iteration. Therefore the total calculation time can be much less than for the full multivariate SC Newton method if solution ideality is sufficient for the nonlinearity of Eqs. (10-13) not to require many additional iterations or provide instability.

The full multivariate SC method is probably preferable whenever Murphree efficiencies are to be allowed for, since Eqs. (10-13) are then changed to a form which loses the efficiency of the tridiagonal matrix, as discussed earlier in the context of tridiagonal matrices. Murphree efficiencies can readily be handled in the Naphtali-Sandholm formulation, leading to the block-tridiagonal matrix.

Calculations using the multivariate Newton method to converge the $2N$ variables when the individual-component mass balances are solved directly can also be accelerated and/or made more stable by the methods discussed earlier for convergence of the temperature profile alone or for the full multivariate Newton SC approach. This includes selecting variables and formulating check functions to reduce the sensitivity to changes in the variables, searching for a fraction of the indicated changes in the convergence variables which minimizes the sum of the squares of the nonclosures of the check functions, and either holding the matrix of partial derivatives unchanged for a few iterations or correcting the partial derivatives through residuals calculated in the previous iteration (Broyden, 1965).

Pairing convergence variables and check functions In most cases particular convergence variables can be paired with particular convergence functions. This is also known as *partitioning* the matrix of partial derivatives, since it amounts to neglecting the influences of convergence variables on the check functions with which they are not paired. Solving for and handling fewer partial derivatives makes each iteration take less time but at the expense of an ability of the computational algorithm to handle a wide variety of problems.

Friday and Smith (1964) made a fundamental contribution by analyzing the circumstances under which different pairings are appropriate in problems where the stage temperatures and the total flows of one stream compose $2N$ convergence variables, the individual component flows being generated directly by solution of Eqs. (10-13) during each iteration. Their analysis is an extension of their approach for single-stage calculations, presented in Chap. 2. Two arrangements are possible, matching the temperatures and flows with enthalpy balances and summation equations in different combinations. These are known as the *BP* and *SR arrangements*.

BP arrangement The BP arrangement matches the enthalpy balances with the total flows and the summation equations with the stage temperatures, ignoring cross effects. It is shown in Fig. 10-8. The solid lines represent a sequential, or nested, scheme in which the temperatures are converged in an inner loop for each set of values of the flows. The dashed line, replacing the solid line labeled "sequential," represents a paired-simultaneous scheme in which both the temperatures and the flows are changed in each iteration. This latter scheme is comparable to that shown in Fig. 10-7 except that the effects of the T_p on the enthalpy balances and the effects of the V_p on the summation equations are not considered.

When the loops are nested, the temperature loop would logically be the inner one, since converged values of stage compositions are useful for reliable calculations of stream enthalpies. The nested-loop arrangement will take more computation per iteration of the outer loop but will require fewer iterations of the outer loop.

The BP pairing of variables should be applied to a situation where the stage temperatures are physically determined more by compositions ($\Sigma y_{j,p} = \Sigma x_{j,p} = 1$ restrictions) than by enthalpy balances and where total flows are determined more by enthalpy balances than by composition. This will be the case when latent-heat effects predominate over sensible-heat effects in the enthalpy-balance equations, since net transfer of material from one phase to another involves latent-heat effects. These criteria are satisfied by separations such as close-boiling distillations. For the distillation of close-boiling mixtures, latent-heat differences set the total vapor and liquid

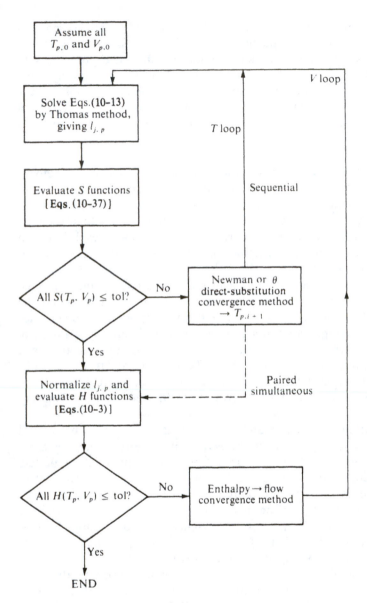

Figure 10-8 BP arrangement of convergence loops.

flow rates through the enthalpy balance, and the stage temperatures are more sensitive to composition.

Temperature loop Selecting a convergence method for the temperature loop is the same problem as selecting a convergence method for the stage temperatures in distillation with constant molal overflow, discussed previously. Either the θ direct-substitution method [Eqs. (10-32) and (10-33), followed by a temperature-prediction method] or the Newman multivariate N-variable Newton method can be used, with various methods already described for accelerating and stabilizing convergence.

Total-flow loop The enthalpy-balance nonclosures are used to obtain new values of the total flows in the BP arrangement. Most situations to which the BP convergence arrangement should be applied involve specifications which fix the net interstage flow of enthalpy at some point in the separation cascade. For example, in distillation the specifications of distillate and reflux flow fix the net upward flow of enthalpy in the rectifying section of the column, which we shall call ΔQ_R. If the feed variables are also fixed, the net upward flow of enthalpy ΔQ_S in the stripping section is also fixed. The total interstage flows which satisfy the enthalpy balances, Eqs. (10-3), for a given set of T_p and stream compositions can be obtained from a rearrangement of Eqs. (10-3):

$$L_{p,i+1} = \frac{\Delta Q - P_+ H_{p-1}}{H_{p-1} - h_p} \tag{10-38}$$

with $V_{p,i+1}$ then obtained from an overall mass balance; P_+ is the net molar upward flow of all species in the section under consideration. Using these total flows for the next iteration amounts to a direct-substitution procedure for correcting the interstage flows and has been found to work well for problems where the ΔQ are set by specification (Friday and Smith, 1964; Hanson et al., 1962; Holland, 1963; Wang and Henke, 1966; etc.). One difficulty that can occur in unusual circumstances is for the denominator of Eq. (10-38) to become small and produce instabilities. Holland (1963, 1975) suggests a rearrangement of Eq. (10-38) for such situations (see also Friday and Smith, 1964). This replaces h_p in the denominator of Eq. (10-38) by $\Sigma y_{j,p-1} h_{j,p}$, where $h_{j,p}$ is the partial molal enthalpy, or "virtual" partial molal enthalpy (Holland, 1975) of component j on stage p. The numerator is altered accordingly, to satisfy Eq. (10-38). This is known as the *constant-composition form*.

SR arrangement The SR arrangement matches the enthalpy balances with the stage temperatures and the summation equations with the total flows, ignoring cross effects. It is shown in Fig. 10-9. Once again, the solid lines correspond to the sequential, or nested, scheme. The dashed line, replacing the solid line labeled "sequential," represents the paired-simultaneous scheme in which both the temperatures and flows are changed in each iteration. When the loops are nested, the flow loop is logically the inner one, since converged values of the stage compositions are again useful for reliable calculation of stream enthalpies.

The SR pairing of variables should be applied to situations where the total flows are determined more by composition than by enthalpy balances, where temperatures are determined more by enthalpy balances than by composition, and where sensible-heat effects dominate the enthalpy balances. Gas absorbers and strippers are characterized by these criteria, and it has been found that the SR convergence-loop arrangement will work well for absorbers and strippers, whereas the BP scheme will not. The liquid flow in a gas absorber is determined by the amount of solute that has been absorbed, while the temperature is fixed by an enthalpy balance relating the heat of absorption to the increase in sensible heat of the counterflowing streams

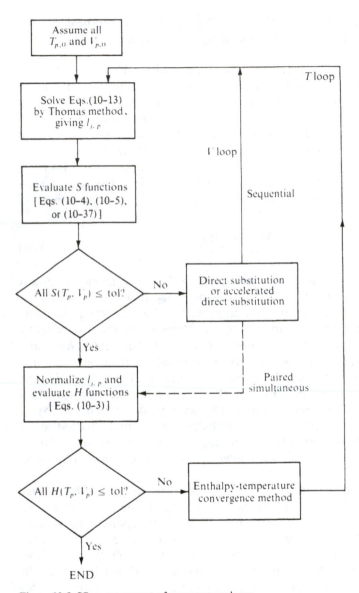

Figure 10-9 SR arrangement of convergence loops.

Distillation of a wide-boiling mixture also tends to favor the SR method. There is necessarily a wide spread in column temperature, which in turn means that sensible-heat effects will be substantial. Friday and Smith (1964) have suggested that a parameter Δ_{DB}, representing the difference between the dew point and bubble point of the total feed to a vapor-liquid separation column, be used as a criterion of whether to use the BP or SR arrangement of convergence loops. If $\Delta_{DB} < 55°C$, the BP arrangement should work well, whereas for relatively high Δ_{DB} the SR arrangement will work well. There is potentially a difficult intermediate region of Δ_{DB} where forcing and damping procedures may be required for either arrangement or where it may be necessary to resort to the $2N$ multivariate Newton simultaneous approach.

For chemical absorbers, like those considered in Example 10-2, the phase equilibrium is sufficiently complex for a full multivariate SC Newton convergence, including the individual component flows, to be desirable unless the specifications are such that stage-to-stage methods can be used (design problem, high recovery fractions of all absorbed solutes).

Total-flow loop In the SR arrangement new values of V_p are to be obtained using a summation equation as a check function. A direct-substitution approach is a simple matter here. Once the $l_{j,p}$ are known, the $L_{p,i+1}$ can be computed directly from the $\left(\sum_j l_{j,p}\right)_i$ by means of Eq. (10-4). This procedure has been used by Friday and Smith (1964) for extraction, absorption, and wide-boiling distillation problems and has been found to be effective.

Holland (1975) describes the use of forcing factors to adjust the individual-component flow ratios $l_{j,p}/v_{j,p}$ on each stage before obtaining new values of L_p and V_p by summation of the individual-component flows and direct substitution. Two different approaches are presented for this. In one, called the *single-θ method*, a single factor is found which serves to minimize the indicated nonclosures of the check functions employed, and corrections to component ratios on different stages are determined in a somewhat complex way from this factor. In the other approach, called the *multi-θ method* (see also Holland et al., 1975) correction factors θ_p are applied to the component flow ratios on each stage such that $(l_{j,p}/v_{j,p})_{corr} = \theta_p(l_{j,p}/v_{j,p})_{calc}$. The summation equations, in a form similar to Eqs. (10-37), are differentiated with respect to θ_p for each stage. The resultant tridiagonal matrix is solved for values of the partial derivatives generated thereby. The matrix of partial derivatives [in a form similar to Eq. (A-11)] is then inverted to give indicated corrections to each of the θ_p. These values of θ_p are then used to correct all the $l_{j,p}/v_{j,p}$ and thereby generate new values of $l_{j,p}$ and $v_{j,p}$, which are added to give the indicated new values of L_p and V_p. Values of θ used in these approaches should not be confused with the parameters θ used in Eqs. (10-32) and (10-33) for correcting the temperature profile in the BP pairing. However, both these methods do involve forcing functions to aid convergence.

Like the temperature profile in the BP pairing, the total interstage flows in the SR pairing can also be corrected by a multivariate N-variable Newton method, taking interactions between stages into account and using the Thomas solution of the tridiagonal matrix to generate the partial derivatives of the summation equations with respect to the interstage flows. The partial derivatives are then used through Eq. (A-11) to generate corrections to the total flows. Such a method is described by Tierney and Bruno (1967).

Hanson et al. (1962) suggest a method for total phase flow correction in an SR problem, deriving their procedure in the context of a multistage multicomponent liquid-liquid extraction problem. A mass-balance equation for each component on each stage is written in terms of the component flows in the ith iteration and the total phase flows in the $(i + 1)$th iteration:

$$\left(\frac{l_{j,p}}{\Sigma l_{j,p}}\right)_i (L_{p,i+1} + K_{j,p} V_{p,i+1}) - \frac{l_{j,p+1}}{\Sigma l_{j,p+1}} L_{p+1,i+1}$$

$$- K_{j,p-1} \frac{l_{j,p-1}}{\Sigma l_{j,p-1}} V_{p-1,i+1} - \Sigma f_{j,p} = 0 \quad (10\text{-}39)$$

Elimination of L_p and L_{p+1} by means of Eq. (10-11) gives a linear equation relating $V_{p-1,i+1}$, $V_{p,i+1}$, and $V_{p+1,i+1}$ with coefficients involving quantities evaluated in the ith iteration. There are N of these equations, forming a tridiagonal matrix which can be solved by the Thomas method or any other suitable technique to give the total interstage flows for the next iteration. This method will be rapidly convergent to the extent that the mole fractions represented by $l_{j,p}/\Sigma l_{j,p}$ do not change greatly from iteration to iteration.

For the dual-solvent extraction problem involving water, ethanol, acetone, and chloroform discussed in Chap. 7, Tierney and Bruno (1967) show that their method requires 6 iterations to achieve the same degree of convergence obtained by Hanson et al. (1962) in 19 iterations. On the other hand, the Hanson et al. convergence method should take considerably less time per iteration than the Tierney-Bruno approach. The simple summation of $(l_{j,p})_i$ to obtain $L_{p,i+1}$ by direct substitution should take even less time per iteration.

Temperature loop Surjata (1961) and Friday and Smith (1964) propose a convergence method for obtaining new T_p from enthalpy balances in the SR approach. The method is a multivariate Newton approach with N variables in which Eqs. (10-3) for constant composition and total flows are approximated by

$$[\bar{H}_p(T_{p+1}, T_p, T_{p-1})]_{i+1} - [\bar{H}_p(T_{p+1}, T_p, T_{p-1})]_i$$

$$= \frac{\partial \bar{H}_p}{\partial T_{p+1}} \Delta T_{p+1} + \frac{\partial \bar{H}_p}{\partial T_p} \Delta T_p + \frac{\partial \bar{H}_p}{\partial T_{p-1}} \Delta T_{p-1} \quad (10\text{-}40)$$

Here $\bar{H}_p$ represents the nonclosure of the enthalpy balance [left-hand side of Eq. (10-3)] for stage p.

The partial derivatives are given by

$$\frac{\partial \bar{H}_p}{\partial T_{p+1}} = -L_{p+1} c_{p+1} \quad (10\text{-}41)$$

$$\frac{\partial \bar{H}_p}{\partial T_p} = L_p c_p + V_p C_p \quad (10\text{-}42)$$

$$\frac{\partial \bar{H}_p}{\partial T_{p-1}} = -V_{p-1} C_{p-1} \quad (10\text{-}43)$$

where c_p and C_p are the heat capacities of the liquid and the vapor, respectively, leaving stage p. These heat capacities can be evaluated analytically or from computations of enthalpies at incrementally different temperatures.

When $[\bar{H}_p(T_{p+1}, T_p, T_{p-1})]_{i+1}$ is set equal to zero, Eqs. (10-40) represent a set of N linear equations, once again forming a tridiagonal matrix, solvable by the Thomas method. The results of the solution are the values of ΔT_p, which represent corrections to be added onto the old T_p such that

$$T_{p, i+1} = T_{p, i} + \Delta T_p \qquad (10\text{-}44)$$

Friday and Smith (1964) report that this method has worked well in a number of applications of the SR arrangement.

RELAXATION METHODS

Another approach to the solution of the various equations involved in multicomponent multistage separation processes is *relaxation*. In principle, relaxation proceeds by following the transient behavior of the separation process as it approaches steady-state operation. A set of interstage flows and stage compositions and temperatures is first assumed. The variables corresponding to each stage are then altered so as to relieve imbalances in enthalpy and component flows entering and leaving each stage. The parameters for the $(i + 1)$th iteration are obtained from the imbalance of flows in the ith iteration.

As an example, upon allowing for transient operation Eqs. (10-2) become

$$U_p \frac{dx_{j, p}}{dt} = L_{p+1}x_{j, p+1} + V_{p-1}y_{j, p-1} - L_p x_{j, p} - V_p y_{j, p} + f_{j, p} \qquad (10\text{-}45)$$

if we write the stage compositions as mole fractions. U_p is the moles of liquid present on stage p, which is regarded as being well mixed. Because of the lower density, vapor holdup is neglected in Eqs. (10-45). If Eq. (10-45) is put in finite-difference form $(dx_{j, p}/dt$ replaced by $\Delta x_{j, p}/\Delta t)$, we can solve for $\Delta x_{j, p}$, which will be used to update the assumed stage compositions from one iteration to the next through the relationship

$$x_{j, p, i+1} = x_{j, p, i} + \Delta x_{j, p} \qquad (10\text{-}46)$$

In order to do this we substitute all the values from the ith iteration into the right-hand side of Eqs. (10-45) and solve for $\Delta x_{j, p}$.

Since the interstage flows and the compositions and temperatures of the adjacent stages will have changed from the ith iteration to the $(i + 1)$th iteration, there will still be an imbalance, necessitating that parameters for the $(i + 2)$th iteration be computed from the parameters from the $(i + 1)$th iteration, etc.

The time increment used in the solution of the equations can be set more or less arbitrarily, within limits, and the quantity $\omega_p = \Delta t/U_p$ thereby becomes an important parameter, known as the *relaxation factor*, which governs the convergence properties of the solution. For low values of the relaxation factor the steady-state solution is reached very slowly. However, for too high values of ω_p the solution can become oscillatory from iteration to iteration.

Relaxation methods are highly stable because of the analog to a physically realizable transient start-up process. However, for the same reason, they converge relatively slowly. Their high stability can be helpful when one is confronted with a problem where the $K_{j,\,p}$ are strongly dependent upon composition; however, it is also effective to attack such problems by using the successive-approximation methods presented earlier in this chapter. Because of their slow but steady convergence, relaxation methods should probably be reserved for use with particularly difficult problems which cannot be handled effectively by other means.

Jelínek et al. (1973a) discuss the application of relaxation methods to multistage multicomponent separations, including such questions as forward- vs. backward-difference forms, optimal values of the relaxation factor, forcing and extrapolation procedures for accelerating convergence, and use of second-order difference equations. Relaxation methods can be used for converging all the different types of check functions in a problem, or they can be used for one or more of the classes of check functions, e.g., the component mass balances, while some other approach is used for converging the other functions.

The usual approach has been to apply relaxation separately and successively to the equations of different types. This implies a BP pairing of variables for distillation problems (Ball, 1961; Jelínek et al., 1973a) and an SR pairing for absorber problems (Bourne et al., 1974). However, the added stability of the relaxation method appears to make these pairings stable and convergent for wider ranges of problems than has been encountered for the paired multivariate Newton convergence schemes. Even so, for complex problems it would probably be desirable to take cross effects into account. Hanson et al. (1962) present a method of solving the enthalpy balances by relaxation that allows for the effects upon both stage temperatures and interstage flows. Seader (1978) has suggested writing the relaxation equations for $\Delta l_{j,\,p}/\Delta t$ as the component mass balances [Eqs. (10-2)], for $\Delta v_{j,\,p}/\Delta t$ as the summation equations in a form involving $K_{j,\,p}$ [i.e., Eqs. (10-37)], and for $\Delta T_p/\Delta t$ as the enthalpy balances [Eqs. (10-3)]. Total flows would be replaced by the sums of individual-component flows. The resultant equations should then be amenable to simultaneous solution, using the block-tridiagonal-matrix approach.

COMPARISON OF CONVERGENCE CHARACTERISTICS; COMBINATIONS OF METHODS

The various multivariate Newton convergence schemes are second order (see. Appendix A) and thereby converge at an accelerated rate as the solution is approached. The other methods described so far do not converge as rapidly in the vicinity of the solution but compensate for this by requiring less calculation per iteration. Relaxation methods can move toward the ultimate solution at a rate comparable to or better than other methods during the first few iterations but converge much more slowly than other methods as the ultimate solution is approached. Pairing of convergence variables and check functions in any method serves to reduce the calculation per iteration but requires more iterations to the extent that the neglected cross-term interactions are significant. For the paired arrangements (BP and SR) nesting the loops in the sequential scheme tends to increase the number of iterations of the inner loop required but to decrease the number of iterations of the outer loop in comparison to the paired-simultaneous scheme. No general statement about relative speeds of convergence can be made.

Stability is the ability of a convergence method to approach the ultimate solution in a monotonic fashion, without either oscillations or divergence. Relaxation is the most stable of the methods which have been described. The Newton methods are highly stable as the solution is approached and are usually satisfactorily stable from the start, but they can be divergent for a poor estimate of initial conditions. The Broyden (1965) procedure of searching for a fraction of the indicated changes which serves to minimize the sum of the squares of the discrepancy functions makes for a much more stable solution with the Newton methods. Pairing of convergence variables and check functions must be done so as to relate the variables and functions which have the strongest cause-and-effect relationships; otherwise stability is severely impaired. There appear to be no general statements regarding the relative stabilities of the sequential and simultaneous schemes with pairing of variables and check functions.

For highly nonlinear and complex problems an effective combination is to use relaxation for the first several iterations and to use a multivariate SC Newton successive-approximation method thereafter. This combination gains the greater stability of the relaxation method for the earlier iterations, where the Newton methods can be unstable, and the greater convergence rate of the Newton methods for the later iterations, where the relaxation methods converge very slowly.

DESIGN PROBLEMS

The discussion so far of successive-approximation and relaxation methods has assumed that the number of stages and the feed location are known. This corresponds to the specifications in an operating problem but not to those in a design problem. In a design problem the specifications of total stages and some other variable, e.g., reboiler boil-up rate, are replaced by specifications of two separation variables for distillation. In addition, the overhead reflux rate is often set at some multiple of the minimum reflux for the specified separation, and the feed location is usually set by some optimization criterion. Solution of such design problems by successive-approximation and relaxation methods is not straightforward since the number of stages and the feed location are not known a priori.

One approach for design problems is to solve a number of operating problems with different specifications and interpolate between the results. This can be a lengthy procedure, however.

Ricker and Grens (1974) describe a *design successive-approximation* (DSA) *procedure*, in which the column configuration for a multicomponent distillation design problem is changed continually to meet specifications during a successive-approximation solution. The procedure, shown schematically in Fig. 10-10, combines modification of the column configuration with the Naphtali-Sandholm full multivariate SC Newton successive-approximation convergence procedure, described earlier. The design problem is defined through all feed variables, column pressure, recovery fractions of two key components, the reflux ratio being some set

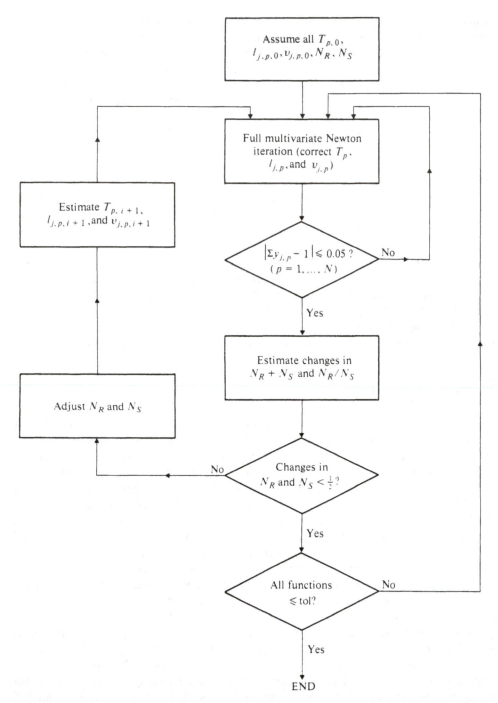

Figure 10-10 DSA approach for solving a design distillation problem using the full multivariate Newton successive-approximation method. (*Adapted from Ricker and Grens, 1974, p. 242; used by permission.*)

multiple of the minimum, and a feed-stage selection criterion based upon the ratio of the key components in the feed-stage liquid being the same as in the liquid portion of the feed (see below).

As an inner loop, the specifications of key-component ratio on the feed stage and of reflux ratio are exchanged for estimated values of the numbers of stages in the rectifying and stripping sections. This allows iterations by the SC Newton multivariate successive-approximation method to be made. The specifications of the recovery fractions of the two key components are retained and still produce the efficient block-tridiagonal matrix form for the successive-approximation iterations. A key point is that the successive-approximation iterations in this inner loop are converged to only a very loose tolerance, corresponding to a nonclosure of 5 percent or less in the summation equations; this typically takes only one or a very few iterations.

Before additional successive-approximation iterations the column configuration is changed to satisfy the design specifications better. The discrepancies in the key-component ratio in the feed-stage liquid and in the reflux ratio (as a percent of minimum) are used to determine new values of N_R and N_S, the numbers of stages in the rectifying and stripping sections. Here it is possible to decouple the effects of the variables by estimating $N_R + N_S$ and N_R/N_S separately, as follows:

1. The total stage requirement $N_S + N_R$ reflects primarily the reflux ratio for the fixed key-component recovery fractions. The total stage requirement is relatively independent of the feed location as long as the feed is not badly misplaced. The indicated change in $N_S + N_R$ is obtained from the difference between the calculated and specified reflux rates through a linearization of the Erbar-Maddox correlation (Fig. 9-2).
2. The ratio of stages in the two sections N_R/N_S governs primarily the ratio of the key components in the liquid on the feed stage. That ratio is insensitive to the reflux rate. The new ratio N_R/N_S is calculated from a secant-method convergence based upon the relationship between the number of stages in either section and the change in the key-component ratio over that section, as given by the Fenske equation (9-24).

The next step is to estimate new temperature and component-flow profiles for the altered column configuration, so as to preserve as much as possible the amount of convergence obtained in previous iterations of the inner successive-approximation loop. This involves scaling the individual-component flows in proportion to the changed reflux and allowing for the change in the number of stages in a section, as well as scaling the temperatures in the middle portion of the column and holding temperatures unchanged near the ends of either section.

These three basic steps are repeated until the indicated changes in the numbers of stages are less than 1, at which point the successive-approximation solution is repeated until a tighter convergence is obtained, the numbers of stages being checked during this procedure to determine that they do not change.

For solutions of six varied distillation problems, Ricker and Grens report that this procedure took an amount of computer time ranging from 1.2 to 2.6 times that required for a single solution of the equivalent operating problem. This amount of time is considerably less than would be required for a procedure of converging the answers to a number of different operating problems and interpolating between the results to derive the solution to a design problem.

Optimal Feed-Stage Location

For a design problem, the optimal feed-stage location would usually be that which requires the least reflux for a given number of stages to create a given separation of the keys or that which requires the least stages for a given reflux flow to accomplish a given separation. Such a criterion leads to a search which would typically require a substantial number of problem solutions for different specifications in order to surround the best configuration. It is obviously desirable to establish more efficient ways of determining the optimum feed location.

The simplest and most commonly used rule of thumb for feed location is that the ratio of key-component mole fractions in the liquid on the feed stage should be as close as possible to the ratio of key-component mole fractions in the liquid portion of the feed, flashed if necessary to tower pressure in a distillation. This is one way of extending the known result for binary systems and is the criterion used in the DSA procedure for distillation design problems, discussed above. Another way of extending the binary result is to say that the ratio of the key-component mole fractions in the feed-stage liquid should be the same as that corresponding to the intersection of the operating lines based on total flows, given by Eq. (8-109). This was the policy followed for the approximate solution of the Underwood equation in Chap. 8.

Hanson and Newman (1977) have used the Underwood equations for calculation of optimal feed locations in numerous distillations assuming constant relative volatilities and constant molal overflow. They present several general conclusions regarding the optimal feed location:

1. Although the two rules of thumb work very well for many cases, there are frequently cases for which they do not work well. There are even instances where a separation is feasible with a suitable feed location but becomes infeasible at feed locations given by the rules of thumb. The most substantial deviations from the rules of thumb tend to occur at very low multiples of the minimum reflux ratio, which are becoming more characteristic in column design.
2. Light nonkeys tend to raise the optimum ratio of light key to heavy key on the feed stage, while heavy nonkeys tend to lower it. As a corollary, greater deviations from the rules of thumb occur when there is a large preponderance of light nonkeys over heavy nonkeys, or vice versa, and when the amount of nonkeys rivals or exceeds the amount of the keys in the feed.
3. Nonkeys very close to, or very removed from, the keys in volatility have less effect in shifting the optimum ratio of the keys away from the rules of thumb than nonkeys moderately removed from the keys.
4. The ratio of the keys in the liquid portion of the feed tends to work better for systems with a preponderance of light nonkeys, while the ratio given by the operating-line intersection tends to work better for systems with a preponderance of heavy nonkeys.

Some of the ways suggested for improving the two simple rules of thumb are the following:

1. In stage-to-stage calculations, one can test for the most desirable feed location at various stages during the calculation, using maximum enrichment of the keys as the criterion for feed introduction. This was the policy followed in Example 10-1. However, stage-to-stage

calculations are effectively limited to systems containing either no light nonkeys or no heavy nonkeys.

2. Robinson and Gilliland (1950) developed equations to allow for the effects of light and heavy nonkeys on the optimum ratio of key components in an approximate fashion. They are complicated to use and can still lead to nonoptimal feed locations.

3. Hanson and Newman (1977) suggest carrying out an Underwood solution which precisely determines the optimal feed location as a first step. Since the Underwood solution is based upon approximations of constant relative volatility and constant molal overflow and is not able to incorporate Murphree efficiencies, it does not give the true stage requirement; however, it should be effective for approaching the optimal feed location in terms of the ratio of the keys, N_R / N_S, or some other parameter.

4. Tsubaki and Hiraiwa (1971) recommend equating the ratio of the key components in the feed-stage liquid to that at minimum reflux. They provide a method for obtaining the ratio of the keys on the feed stage at minimum reflux through an extension of the Underwood

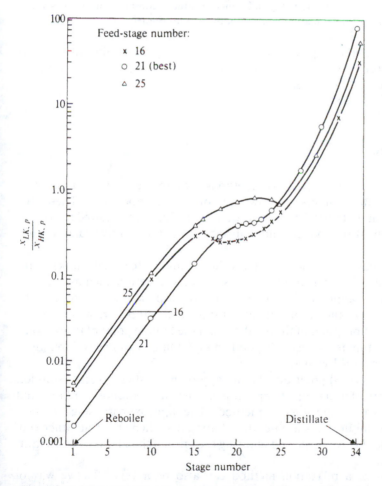

Figure 10-11 Effect of feed location on stage-to-stage enrichment of the keys in a hydrocarbon distillation. (*Adapted from Maas, 1973, p. 97; used by permission.*)

equations. This approach thereby assumes constant relative volatility and constant molal overflow between the zones of constant composition, as in the Underwood equations for determining minimum reflux. It can be expected to work well for systems operating at low multiples of minimum reflux.

5. Maas (1973) observed that attainment of the optimal feed location can be related empirically to the shape of a plot of $x_{LK,\,p}/x_{HK,\,p}$ vs. stage number in the vicinity of the feed. An example presented by Maas of a multicomponent hydrocarbon fractionator, where n-butane and isopentane are the keys, is shown in Fig. 10-11. As discussed in Chap. 7, the behavior of nonkeys near the feed can make the keys undergo reverse fractionation, the enrichment of the keys actually becoming less from stage to stage upward. This behavior is particularly pronounced close to minimum reflux. Misplacement of the feed increases the amount of this reverse fractionation. As shown in Fig. 10-11, too high a feed causes much reverse fractionation below the feed, and too low a feed produces much reverse fractionation above the feed. To allow for these effects, the empirical criterion put forward by Maas is that the feed should be moved in the direction in which $[d \ln(x_{LK,\,p}/x_{HK,\,p})/dN]$ is least (most negative) until a feed location is found which produces composition profiles where $[d \ln(x_{LK,\,p}/x_{HK,\,p})/dN]$ is most nearly equal on both sides of the feed stage.

The empirical criterion proposed by Maas is physically reasonable and appears to account for the observed effects of nonkeys. It should also be very easy to implement in the DSA method for design problems.

INITIAL VALUES

Successive-approximation and relaxation methods require that initial estimates be made of stage temperatures, interstage flows, and/or stage compositions. These estimates are important in that they determine the amount of change required to achieve convergence and can also produce initial instability in the multivariate Newton convergence methods.

The simplest method of generating initial values would be to set all temperatures at some intermediate value and to set all flows by some criterion such as constant molal overflow. A better approach is to take the temperature and flow profiles to be linear between conditions known or estimated for the end stages, e.g., a linear trend between the dew or bubble point of the distillate and the bubble point of the bottoms in distillation. Such estimates are usually good enough but can still be well removed from the ultimate converged values.

Ricker and Grens (1974) point out that an approximate stage-to-stage solution can be an effective way of initializing temperatures, flows, and compositions for a full multivariate SC Newton convergence method. The approximate stage-to-stage method assumes that all nonkeys are nondistributing and then starts at either end, calculating onward to the feed stage and ignoring the mismatch of nonkey concentrations at the feed stage.

A few iterations of a relaxation method can also be a very effective way of initializing a successive-approximation calculation as well as accelerating and stabilizing the overall convergence.

APPLICATIONS TO SPECIFIC SEPARATION PROCESSES

Distillation

For distillation involving strongly nonideal mixtures the full multivariate Newton SC successive-approximation approach, as developed by Naphtali and Sandholm and Ricker and Grens, among others, appears to combine stability and computational speed in the best way. This includes cases of azeotropic and extractive distillation, except for design problems where light nonkeys or heavy nonkeys are entirely absent, so that a stage-to-stage method can be used reliably. For very strong nonidealities and/or particularly poor initial estimates the full multivariate Newton method can present problems of initial stability; in that case an effective combination is to use a relaxation method for the early iterations, followed by the successive-approximation method.

For distillation with ideal or only mildly nonideal solutions, the multivariate Newton convergence of stage temperatures and total interstage flows ($2N$ variables), as implemented by Tomich (1970), is effective for handling feed mixtures with any sort of volatility characteristics. This method should be faster than the full multivariate SC Newton method because of the computational efficiency of the Thomas solution of the tridiagonal matrix. However, when Murphree efficiencies are to be incorporated, it is probably best to return to the full multivariate Newton method since the benefit of the tridiagonal matrix is lost.

For distillations that do not involve feeds with very wide boiling ranges or dumbbell characteristics, pairing the convergence variables and check functions in the BP arrangement can further accelerate the computation.

Seppala and Luus (1972) report computation times for 16 different combinations of convergence procedures for 9 different distillation operating problems involving relatively ideal solutions. Their results show that pairing in the BP arrangement does serve to accelerate convergence (consuming an average of about 73 percent as much time for the examples tested). They also show that the direct-substitution method with the θ forcing factor and the Newman method based upon multivariate Newton convergence are about equally effective for the temperature loop in the BP pairing and that the technique of using lesser corrections to the individual-component flows for the stages nearer to the feed is effective in accelerating the θ method of convergence for the temperature loop. For their examples they find little difference between using the approach of Eq. (10-38) and the constant-composition approach for converging the flow-enthalpy loop in the BP arrangement; they also find little difference between using and not using the θ-based corrections to the individual-component flows in the direct-substitution approach for the temperature loop. These two results suggest that it is still advantageous to use the θ corrections and the constant-composition method for enthalpy convergence to handle cases for which they are most needed.

Seppala and Luus found that nesting the loops led to less computation time than the paired-simultaneous approach for the BP pairing in their examples. However Ajlan (1975) compared computation times for the six distillation problems defined by Ricker and Grens (1974), with operating specifications, and found the paired-simultaneous approach to be faster than nested loops; no general statement regarding the advantage or disadvantage of nesting in the paired arrangement appears to be possible. Ajlan also found that the full multivariate Newton approach was at least as fast as paired approaches for problems with few stages and components but that the paired arrangements became more rapid as the numbers of stages and components grew, leading to very large matrices of partial derivatives. Boston and Sullivan (1974) studied 23 distillation problems without strong nonidealities and found that pairing and redefinition of the convergence variables and check functions to minimize sensitivities resulted in much faster solutions than were achievable with the Tomich $2N$ multivariate Newton method.

Block and Hegner (1976) considered distillations that are relatively close-boiling but contain sufficient nonideality to result in two liquid phases on some stages. They found that a BP pairing arrangement with a block-tridiagonal-matrix solution of the component mass balances and equilibrium equations was effective.

Relaxation methods are hardly ever needed for distillation calculations, but Jelínek and Hlaváček (1976) show that they are effective for calculating distillations involving kinetically limited chemical reactions on the plates.

Holland and Kuk (1975), Hess and Holland (1976), and Kubíček et al. (1974) discuss efficient ways of obtaining solutions for distillations with the same column configuration and the same feeds at a number of different operating conditions.

Absorption and Stripping

Computations of simple absorbers and strippers without strong nonidealities are handled well and are converged rapidly by means of the SR pairing of the $2N$ variables for stage temperatures and total flows. Holland (1975) and Holland et al. (1975) demonstrate this for a number of different absorption calculations involving hydrocarbons. The multivariate Newton method described by Sujata, and the single- and multi-θ methods described by Holland are all rapid for the flows loop in the SR pairing for these examples. In a number of cases the single-θ method is significantly faster.

Reboiled absorbers and some absorbers and strippers handling close-boiling mixtures combine the characteristics of simple absorbers with those of distillation. In such cases it is not advisable to pair the convergence variables and check functions. Instead, if the solutions are not strongly nonideal and Murphree efficiencies are not to be accounted for, the Tomich approach using multivariate Newton convergence of the $2N$ temperature and total-flow variables is more reliable and should not take substantially longer.

Absorbers and strippers involving strong solution nonidealities (such as chemical absorbers) and/or taking Murphree efficiencies into account should best be calculated by the full multivariate SC Newton successive-approximation method unless the problem has design specifications which permit stage-to-stage methods to be used reliably.

As noted in Chap. 7, an internal temperature maximum is a common characteristic of absorbers and can be very important in calculating their performances. Bourne et al. (1974) have demonstrated that relaxation methods of computation predict and handle the temperature profile effectively. However, the other methods described appear to handle this phenomenon well, too, and it does not seem useful, except in unusual cases, to resort to the much slower relaxation methods.

Extraction

Multistage multicomponent solvent-extraction problems have several distinguishing characteristics:

1. Since there is usually no important latent heat of phase change between liquid phases, temperature and enthalpy-balance effects tend not to be important. Thus, in effect, the SR pairing of total flows as convergence variables with summation equations as check functions is already made by the physical situation, no temperature-enthalpy balance convergence generally being needed. Since extractors are often staged as discrete units, e.g., mixer-settlers, stage temperatures are sometimes controlled at different values as independent variables, as in Example 10-3.
2. Extraction processes of necessity involve highly nonideal solutions, since it is the nonideality that generates the separation factor between components [Eq. (1-16)]. Computational methods must allow for these strong nonidealities.
3. Accurate values of activity coefficients are needed for generating separation factors and phase-miscibility relationships. Approaches such as the Margules, NRTL, UNIQUAC, and UNIFAC equations (Reid et al., 1977) can in principle be used to generate and correlate activity coefficients, but the lack of underlying data and approximations in these methods can cause significant errors. Fredunslund et al. (1977a) note that the UNIFAC group-contribution method is usually not suitable for extraction calculations for this reason, although it can predict phase splitting well enough to handle most problems of heterogeneous azeotropic distillation.

Four options exist for handling the strong nonideality in extraction systems:

1. Activity coefficients can be obtained from the compositions generated in the previous iteration. This is sometimes called the *composition-lag approach*. It slows convergence and is probably suitable only where activity coefficients for one component show only a mild dependence upon concentrations of that component and others in the system, e.g., for relatively dilute systems.
2. Activity coefficients can be converged as functions of phase compositions in a separate, nested loop. This consumes additional computing time.
3. Compositions and activity coefficients can be converged simultaneously with total flows in a full multivariate SC Newton method.
4. Relaxation methods can be used.

The SR pairing with solution of Eqs. (10-13) as a tridiagonal matrix and with direct substitution for convergence of the total flows was used by Friday and Smith (1964). Holland (1975) outlines the use of the single-θ and multi-θ forcing-function methods instead of direct substitution for converging the total flows. In order to allow solution of Eqs. (10-13) as a tridiagonal matrix, he generated composition-lag approaches for both methods, as well as a nested-loop approach for converging activity coefficients with the multi-θ method. Tierney and Bruno (1967) presented an

N-variable Newton method for converging the flows in such an arrangement, with activity coefficients determined by composition lag. Bouvard (1974) found that the composition-lag approach could be accelerated by performing a single-stage equili-bration calculation for each stage to obtain equilibrium products from the indicated entering feeds obtained in the ith iteration and then basing the activity coefficients for the $(i + 1)$th iteration on these equilibrium-product compositions. This procedure ensures that the activity coefficients will be generated from thermodynamically saturated stream compositions.

A full multivariate SC Newton approach for extraction was first presented by Roche (1969). Bouvard (1974) extended the method to design problems in a way similar to the approach of Ricker and Grens (1974) for distillation.

A relaxation procedure was developed for extraction processes by Hanson et al. (1962, method II). In it, the relaxation calculations are carried out as successive single-stage equilibration calculations, using the old component flows for the extract phase and the new component flows for the raffinate phase proceeding in the direc-tion of raffinate flow and then using the old component flows for the raffinate phase and the new ones for the extract phase proceeding in the direction of the extract flow, etc. Jelínek and Hlaváček (1976) considered improvements in the relaxation approach and investigated the effects of changes in the relaxation factor. Bouvard (1974) investigated combining relaxation for initial iterations with the full multivar-iate SC Newton method for later iterations and found the combination to be highly stable and rapidly convergent for a variety of problems. Bouvard also found that special formulations of the end stage equations are necessary to handle extract reflux in both relaxation and SC approaches.

Until recently, nearly all tests of computational methods for extraction were made with the problem presented by Hanson et al. (1962) and described in Chap. 7. This problem is inherently quite stable compared with other extraction problems because the dual-solvent type of process tends to cause solutes to prefer one phase strongly over the other. More severe problems are presented by refluxed extractions and/or those where certain components have $K_j V/L$ (or the equivalent) near unity or even above and below unity in different portions of a cascade. There is a more complex effect of solution nonidealities on phase equilibria and total flows in such cases.

Holland (1975) compared direct substitution, the single-θ method, the multi-θ method, and the N-variable Newton method—all with the SR pairing and solution of Eqs. (10-13) as a tridiagonal matrix—for solving the dual-solvent problem of Hanson et al. (1962). The multi-θ method required the fewest iterations, but it takes longer per iteration and no comparison of actual times was presented. Jelínek and Hlaváček (1976) found both relaxation and the SC Newton method to be effective for four extraction problems involving dual solvents, extract reflux, and multiple feeds in various combinations.

Bouvard (1974) compared relaxation, SC multivariate Newton, and both the direct-substitution and multi-θ SR methods for six extraction problems, covering dual-solvent, single-section, and refluxed extractions. He also investigated the use of several iterations of relaxation as an initial method for the other three approaches. The multi-θ method was found to require more computing time than the direct-substitution SR method, but both experienced divergency with refluxed extraction problems. Presumably this was the result of components with $K_j V/L$ near unity. Relaxation initiation was not particularly effective in hastening convergence of the two SR approaches since their limitations are not caused by initiation problems. The full multivariate SC Newton approach was found to be effective and generally convergent. Use of four to seven iterations of relaxation calculations was found to cause the SC Newton method to converge in only two or three more iterations; this combination resulted, on the average, in total computation times which were only 63 percent as long as for the uncombined SC method and which were only an average of 31 percent longer than the times taken by the direct-substitution SR method for the same problems.

It can be concluded that the full multivariate SC Newton method, preferably initialized by several iterations of relaxation, is the surest and most efficient general method for extraction problems. Some gain in computing time can be made by using the direct-substitution SR method for problems where it is known to work well, e.g., most dual-solvent extractors.

PROCESS DYNAMICS; BATCH DISTILLATION

The computations considered so far in this chapter are for steady-state operation. Dynamic or unsteady-state behavior of separation processes poses an additional degree of complexity. Approaches to calculation of dynamic behavior of multistage separations have been considered and reviewed by Amundsen (1966) and Holland (1966). Two basic approaches are those of Mah et al. (1962), where the matrix describing steady-state operation is considered to hold unchanged over a time increment, and of Sargent (1963), where the matrix elements are considered to vary linearly over a time increment. Control loops can provide additional off-diagonal matrix elements. Relaxation methods, as used for steady-state calculations, can provide dynamic information through the results of successive iterations.

Batch distillation is an example of a process run under unsteady-state conditions. One approach to calculating batch-distillation problems is to generate a succession of steady-state solutions corresponding to various points in time; this neglects the effects of liquid holdup. More correct approaches for computing batch distillation have been considered by Distefano (1968), among others. Stewart et al. (1973) discuss the effects of different design parameters upon batch distillation and also review earlier work.

REVIEW OF GENERAL STRATEGY

The closest thing to a general-purpose, efficient calculation method that will handle all fluid-phase multicomponent multistage separation processes is the full multivariate SC Newton convergence method initialized by several iterations of a relaxation method. The use of a good but shorter initialization method and the Broyden search

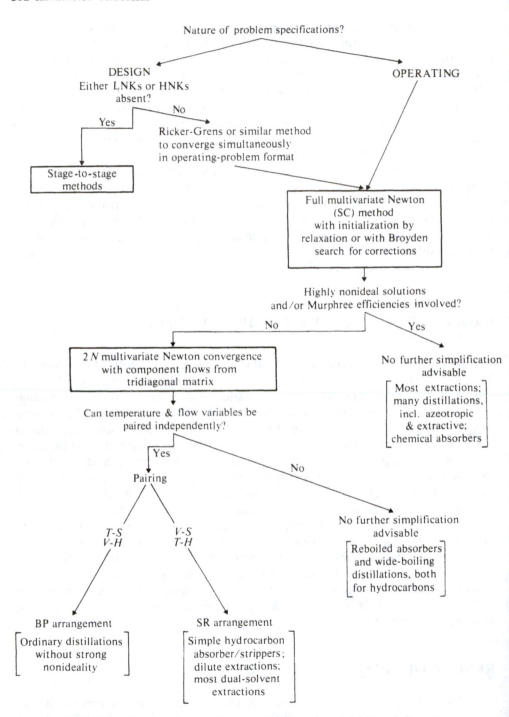

Figure 10-12 Computation methods for multicomponent multistage separation processes.

for the optimal fraction of indicated corrections with the SC Newton method should do nearly as well. These methods are directly applicable to operating problems, where the number of stages and feed locations are known. For design problems, iteration on the number of stages and feed locations can be efficiently combined with the solution for compositions, flow, and temperature profiles by methods similar to that developed by Ricker and Grens for multicomponent distillation.

Although the general method described above is not slow, faster methods can be used for specific problems where certain criteria are met. An outline of such simplifications is shown in Fig. 10-12, where slanting lines denote necessary choices and vertical lines denote optional choices leading to more rapid methods. This diagram summarizes many of the points developed in this chapter.

Exact solutions for multistage separations will not always be desirable. For example, poor knowledge of phase-equilibrium data, stage efficiencies, and/or feed compositions may not warrant such precision. Although the computing time for even the most general of the methods is small for one or a few solutions, computing time may become excessive when a very large number of solutions is to be made, as in optimization of the design of large chemical processes. In these cases more approximate methods can be appropriate, e.g., those based on the Gilliland or Erbar-Maddox correlations. Group methods are useful for dilute systems and/or those with relatively constant separation factors and molal overflows. Like the correlations, they can also be used for initial orientation in process decisions.

AVAILABLE COMPUTER PROGRAMS

A tabulation of specific computer programs available, as of 1978, for distillation, absorption, extraction, evaporation, and crystallization processes is given by Peterson et al. (1978).

REFERENCES

Ajlan, M. H. (1975): M. S. thesis in chemical engineering, University of California, Berkeley.
Amundsen, N. R. (1966): " Mathematical Methods in Chemical Engineering: Matrices and Their Application," Prentice-Hall, Englewood Cliffs, N.J.
Ball, W. E. (1961): *Am. Inst. Chem. Eng. Natl. Meet., New Orleans.*
Billingsley, D. S. (1966): *AIChE J.,* **12**:1134.
——(1970): *AIChE J.,* **16**:441.
—— and G. W. Boynton (1971): *AIChE J.,* **17**:65.
Birmingham, D. W., and F. D. Otto (1967): *Hydrocarbon Process.,* **46**(10):163.
Block, U., and B. Hegner (1976): *AIChE J.,* **22**:582.
Boston, J. F., and S. L. Sullivan (1972): *Can. J. Chem. Eng.,* **50**:663.
—— and —— (1974): *Can. J. Chem. Eng.,* **52**:52.
Bourne, J. R., U. von Stockar, and G. C. Coggan (1974): *Ind. Eng. Chem. Process Des. Dev.,* **13**:115.
Bouvard, M. (1974): M. S. thesis in chemical engineering, University of California, Berkeley.
Broyden, C. G. (1965): *Math. Comput.,* **19**:577.
Bruce, G. H., D. W. Peaceman, H. H. Rachford, and J. D. Rice (1953): *Trans. AIME,* **198**:79.
Chueh, P. L., and S. W. Briggs (1964): *J. Chem. Eng. Data,* **9**:207.
Distefano, G. P. (1968): *AIChE J.,* **14**:190.

Fredenslund, A., J. Gmehling, M. L. Michelsen, P. Rasmussen, and J. M. Prausnitz (1977a): *Ind. Eng. Chem. Process Des. Dev.*, **16**:450.

——, ——, and P. Rasmussen (1977b): "Vapor-liquid Equilibria Using UNIFAC," Elsevier, Amsterdam.

Friday, J. R., and B. D. Smith (1964): *AIChE J.*, **10**:698.

Gerster, J. A. (1963): Distillation, in R. H. Perry, C. H. Chilton, and S. D. Kirkpatrick (eds.), "Chemical Engineers' Handbook," 4th ed., sec. 13, McGraw-Hill, New York.

Goldstein, R. P., and R. B. Stanfield (1970): *Ind. Eng. Chem. Process Des. Dev.*, **9**:78.

Green, S. J., and R. E. Vener, (1955): *Ind. Eng. Chem.*, **47**:103.

Hader, R. N., R. D. Wallace, and R. W. McKinney (1952): *Ind. Eng. Chem.*, **44**:1508.

Hanson, D. N., J. H. Duffin, and G. F. Somerville (1962): "Computation of Multistage Separation Processes," Reinhold, New York.

——, and J. S. Newman (1977): *Ind. Eng. Chem. Process Des. Dev.*, **16**:223.

Hess, F. E., and C. D. Holland (1976): *Hydrocarbon Process.*, **55**(6):125.

Hofeling, B. S., and J. D. Seader (1978): *AIChE J.*, **24**:1131.

Hoffman, E. J. (1964): "Azeotropic and Extractive Distillation," Interscience, New York.

Holland, C. D. (1963): "Multicomponent Distillation," Prentice-Hall, Englewood Cliffs, N.J.

—— (1966): "Unsteady-State Processes with Applications to Multicomponent Distillation," Prentice-Hall, Englewood Cliffs, N.J.

—— (1975): "Fundamentals and Modeling of Separation Processes," Prentice-Hall, Englewood Cliffs, N.J.

—— and M. S. Kuk (1975): *Hydrocarbon Process.*, **54**(7):121.

——, G. P. Pendon, and S. E. Gallun (1975): *Hydrocarbon Process.*, **54**(1):101.

Huber, W. F. (1977): *Hydrocarbon Process.*, **56**(8):121.

Hutchison, H. P., and C. F. Shewchuk (1974): *Trans. Inst. Chem. Engr.*, **52**:325.

Jelinek, J. and V. Hlaváček: (1976): *Chem. Eng. Comm.*, **2**:79.

——, ——, and M. Kubíčik (1973a): *Chem. Eng. Sci.*, **28**:1825.

——, ——, and Z. Křivský (1973b): *Chem. Eng. Sci.*, **28**:1833.

Kent, R. L., and B. Eisenberg (1976): *Hydrocarbon Process.*: **55**(2):87.

Kohl, A. L., and F. C. Riesenfeld (1979): "Gas Purification," 3d ed., Gulf Publishing, Houston.

Kubíček, M., Hlaváček, and J. Jelínek (1974): *Chem. Eng. Sci.*, **29**:435.

——, ——, and F. Procháska (1976): *Chem. Eng. Sci.*, **31**:277.

Lapidus, L. (1962): "Digital Computation for Chemical Engineers," McGraw-Hill, New York.

Lewis, W. K., and G. L. Matheson (1932): *Ind. Eng. Chem.*, **24**:444.

Lo, C. T. (1975): *AIChE J.*, **21**:1223.

Maas, J. H. (1973): *Chem. Eng.*, Apr. 16, p. 96.

Mah, R. S. H., S. Michaelson, and R. W. H. Sargent (1962): *Chem. Eng. Sci.*, **17**:619.

Maxwell, J. B. (1950): "Data Book on Hydrocarbons," Van Nostrand, Princeton, N.J.

Muhlbauer, H. G., and P. R. Monaghan (1957): *Oil Gas J.*, **55**(17):139.

Naphtali, L. M., and D. P. Sandholm (1971): *AIChE J.*, **17**:148.

Newman, J. S. (1963): *Hydrocarbon Process. Petrol. Refin.*, **42**(4):141.

—— (1967): *USAEC Lawrence Rad. Lab. Rep. UCRL-17739*, August.

—— (1968): *Ind. Eng. Chem. Fundam.*, **7**:514.

Orbach, O., C. M. Crowe, and A. I. Johnson (1972): *Chem. Eng. J.*, **3**:176.

Peterson, J. N., C.-C. Chen, and L. B. Evans (1978): *Chem. Eng.*, July 3, p. 69.

Reid, R. C., J. M. Prausnitz, and T. K. Sherwood (1977): "The Properties of Gases and Liquids," 3d ed., McGraw-Hill, New York, 1977.

Ricker, N. L., and E. A. Grens (1974): *AIChE J.*, **20**:238.

Robinson, C. S., and E. R. Gilliland (1950): "Elements of Fractional Distillation," 4th ed., chaps. 9 and 10, and pp. 245–249, McGraw-Hill, New York.

Roche, E. C. (1969): *Br. Chem. Eng.*, **14**:1393.

Rowland, C. H., and E. A. Grens (1971): *Hydrocarbon Process.*, **50**(9):201.

Sargent, R. W. H. (1963): *Trans. Inst. Chem. Eng.*, **41**:52.

Seader, J. D., University of Utah, Salt Lake City (1978): personal communication.

Seppala, R. E., and R. Luus (1972): *J. Franklin Inst.*, **293**:325.

Smith, B. D. (1963): "Design of Equilibrium Stage Processes," McGraw-Hill, New York.
Stewart, R. R., E. Weisman, B. M. Goodwin, and C. E. Speight (1973): *Ind. Eng. Chem. Process Des. Dev.*, **12**:130.
Surjata, A. D. (1961): *Petrol. Ref.*, **40**(12):137.
Thiele, E. W., and R. L. Geddes (1933): *Ind. Eng. Chem.*, **25**:289.
Tierney, J. W., and J. A. Bruno (1967): *AIChE J.*, **13**:556.
Tomich, J. F. (1970): *AIChE J.*, **16**:229.
Tsubaki, M., and H. Hiraiwa (1971): *J. Chem. Eng. Jap.*, **4**:340.
Vaněk, T., V. Hlaváček, and M. Kubíček (1977): *Chem. Eng. Sci.*, **32**:839.
Varga, R. S. (1962): "Matrix Iterative Analysis," p. 194, Prentice-Hall, Englewood Cliffs, N.J.
Wang, J. C., and G. E. Henke (1966): *Hydrocarbon Process.*, **45**(8):155.

PROBLEMS

10-A$_1$ Verify that the values of $l_{j,p}$, b_j, and d_j used in Example 10-4 are a consistent solution of the individual-component mass-balance equations for the temperatures initially assumed.

10-B$_1$ Repeat both parts of Example 10-2 for an absorber with cooling coils on each plate and capable of operating isothermally at 25°C.

10-C$_1$ Carry out the Thomas-method solution to obtain the values of $l_{B,p}$, b_B, and d_B in Example 10-4.

10-D$_2$ Consider an extractive distillation of methylcyclohexane and toluene with phenol as solvent, as discussed in Prob. 7-E. Equilibrium data are given in Fig. 7-31. If a very high fraction of the toluene in a binary feed containing 55 mol % methylcyclohexane and 45% toluene is to be recovered in a product containing no more than 2% of the methylcyclohexane fed, if phenol is added well above the feed in an amount of 6.0 mol/mol of hydrocarbon feed, and if the boil-up ratio V'/b in the reboiler is fixed at 2.5, find the number of equilibrium stages required in the stripping section. Assume constant molal overflow, saturated liquid feed, optimal feed location and $(/P)_b = 1$.

10-E$_3$ Formaldehyde is manufactured commercially from methanol by the reactions

$$CH_3OH + \tfrac{1}{2}O_2 \rightarrow HCOH + H_2O \qquad \text{and} \qquad CH_3OH \rightarrow HCOH + H_2$$

The reaction employs a supported silver catalyst and takes place at about 600°C. Formaldehyde and unreacted methanol are absorbed from the reactor effluent gases into a circulating liquid stream of methanol, formaldehyde, and water. A portion of this liquid is continuously withdrawn and fed to a distillation column which removes methanol and provides a product solution of 37 to 45 wt % formaldehyde in water. The product formaldehyde can be no more concentrated than this because it would polymerize rapidly. It is also necessary to retain between 1 and 7 wt % methanol as a polymerization inhibitor. Further process information is given by Hader et al. (1952).

Vapor-liquid equilibrium data for the ternary methanol-water-formaldehyde system at 1 atm total pressure are shown in Fig. 10-13. The data are shown on a triangular diagram, the coordinates of which are the weight percent of each of the components in the liquid. The curves shown are for different constant weight percent of formaldehyde (dashed curves) and of water (solid curves) in the vapor. Thus, for example, a liquid containing 20% formaldehyde, 40% methanol, and 40% water is in equilibrium with a vapor containing 9% formaldehyde, 26% water, and (by difference) 65% methanol.

Suppose that an atmospheric-pressure distillation column is to be designed to produce a typical product solution containing 37 wt % formaldehyde, 0.8% methanol with the remainder water. The formaldehyde recovery fraction in the product will be 0.990. The weight ratio of methanol to formaldehyde in the tower feed (saturated liquid) is 0.70, and water is present in the proper proportion in the feed to give the desired formaldehyde product dilution. The feed is saturated liquid, and constant *mass* overflow may be assumed in the tower. If the overhead reflux flow is to be 1.50 times the minimum, find the number of equilibrium stages required and indicate a desirable feed location.

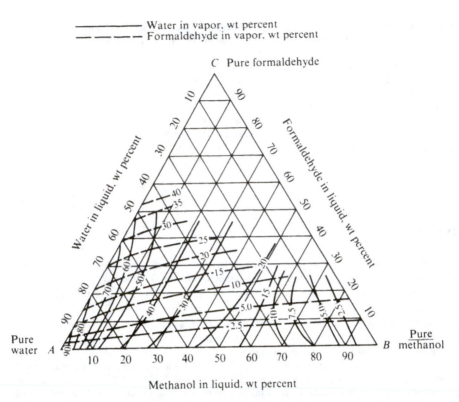

──────── Water in vapor, wt percent
── ── ── Formaldehyde in vapor, wt percent

Figure 10-13 Vapor-liquid equilibria for the methanol-water-formaldehyde system. (*Adapted from Green and Vener, 1955, p. 107; used by permission.*)

10-F₂ Consider a distillation tower providing three equilibrium stages plus a total condenser and an equilibrium kettle-type reboiler. The feed is a saturated vapor containing:

	Mole fraction	K_j at 127°C
Benzene	0.20	3.12
Toluene	0.40	1.34
Xylenes	0.40	0.60

and 20 mole percent of the feed is taken as distillate. The feed is injected to the reboiler. The reflux ratio L/d is held at 5.0, and the tower pressure is 17 lb/in² abs. Assume the temperatures on all stages and in the reboiler are equal to the feed dew-point temperature, which is 127°C. Values of K_j are given in the table. Use the θ-convergence method with the BP arrangement to accomplish *one* iteration toward a solution for stage compositions and predict a new temperature profile which could be used for a second iteration. Constant molal overflow may be assumed. Vapor-pressure data can be taken from Perry's "Handbook."

10-G₂ Repeat the solution to Example 10-3 for triolein if the temperature profile is exactly reversed. Account physically for the changes in the calculated recovery fraction.

10-H₁ Suggest the most efficient calculation method for part (*d*) of Prob. 6-E if allowance is made fully for the fact that the product of flow rate and heat capacity in the liquid phase is of the same order of magnitude of that in the vapor phase. Assume that the equilibrium and enthalpy data are implemented for computer by means of algebraic expressions.

10-I₃ Make a logic diagram for a computer calculation which will use a pseudo-steady-state assumption to calculate the maximum percent of component A which can be recovered at a given purity y_A in a batch distillation of a three-component mixture (A, B, and C) in which A is the most volatile component. Assume that relative volatilities are constant, that a Murphree vapor efficiency E_{MV} is specified and is the same for each component, that the number of stages is specified, and that the overhead reflux ratio r/d is specified as a function of x_A in the still pot. Neglect holdup on the plates of the column but allow for holdup in the liquid in the still pot, assumed to be well mixed. The column is equipped with a total condenser.

10-J₂ Suppose that a simulation of the carbonating tower and ammonia recovery tower of a Solvay plant is desired. (Problem 7-G describes the Solvay process and these two towers.) The number of stages, feed locations, and Murphree efficiencies will be provided, along with the duties of various associated heat exchangers, the tower pressures, and the conditions of all feeds to both towers. In the carbonation tower the temperature profile is specified. Reaction equilibrium data and vapor-liquid equilibrium data are available as subroutines. The simulation is to provide the product flows and product compositions, and also the temperature profile in the ammonia recovery tower. Indicate which of the different approaches described in this chapter would be most appropriate for each of these towers. Explain your answer briefly.

ELEVEN

MASS-TRANSFER RATES

Our attention so far has been focused for the most part on separation processes involving discrete stages, which either operate with equilibrium between the product streams from a stage or can be analyzed through a stage efficiency. Mass-transfer rates determine the degree of equilibration occurring on a stage, and hence the stage efficiency (Chap. 12). They govern the separation obtained in continuous contacting equipment, such as packed towers, and in fixed-bed operations. They also completely define the separation obtained in rate-governed processes.

The field of mass transfer is broad and complex and has been the subject of much research over the past 50 years and more. A much fuller treatment of the subject is given by Sherwood et al. (1975). The reader will find most of the topics in this chapter developed in much more detail in that reference.

MECHANISMS OF MASS TRANSPORT

Matter can move spontaneously from one place to another through a number of mechanisms, including:

1. *Molecular diffusion*, which results from the thermal motion of molecules, limited by collisions between molecules
2. *Convection*, or bulk flow, which occurs under a pressure gradient or other imposed external force
3. *Turbulent mixing*, where macroscopic packets of fluid, or eddies, move under inertial forces

As an example, in a typical stirred vessel of liquid large-scale convective currents are set up by the agitator. If the agitation is intense enough, turbulent eddies will be shed off from the flow and will result in macroscopic mixing between material in different streamlines of the convective flow. Molecular diffusion will smooth out concentration differences over short distances on the microscale.

MOLECULAR DIFFUSION

The most common equations for analysis of transport by molecular diffusion in a binary mixture, in the notation of Bird et al. (1960, chap. 16), are either

$$J_A^* = -cD_{AB} \nabla x_A \tag{11-1}$$

or

$$J_A^V = -D_{AB} \nabla c_A \tag{11-2}$$

where J_A^*, J_A^V = fluxes, defined below
$\quad D_{AB}$ = molecular diffusivity, m^2/s
$\quad c$ = molar density of medium, mol/m^3
$\quad \nabla x_A$ = gradient of mole fraction x of component A, m^{-1}
$\quad \nabla c_A$ = gradient of concentration c of component A, $(mol/m^3)/m = mol/m^4$

For one-dimensional transport ∇x_A and ∇c_A become $\partial x_A/\partial z$ and $\partial c_A/\partial z$, where z is the distance variable.

J_A^* in Eq. (11-1) is the molar flux of component A across a plane which is normal to ∇x_A and moving at the mole-average velocity of the medium. In order to clarify this concept, let us define as N_A the flux of component A across a stationary plane normal to the gradient in x_A. The units of N_A could be moles per second and per square meter since it is the molar flow across this plane per unit cross-sectional area and per unit time. In general, both components A and B will have nonzero fluxes across this stationary plane. The mole-average velocity is defined as

$$v^* = \frac{1}{c}(N_A + N_B) \tag{11-3}$$

Since the flux of A caused by the flow at the mole-average velocity is $c_A v^*$, and since $c_A/c = x_A$, J_A^* is related to N_A and N_B through

$$J_A^* = N_A - x_A(N_A + N_B) \tag{11-4}$$

J_A^* has the same units as N_A and N_B, for example, moles per second and per square meter. Combining Eqs. (11-1) and (11-4) leads to an expression for N_A in terms of the diffusivity

$$N_A = -cD_{AB} \nabla x_A + x_A(N_A + N_B) \tag{11-5}$$

In Eq. (11-2), J_A^V is the molar flux of component A across a plane which is normal to ∇c_A and is moving at the volume-average velocity of the medium. Here the volume-average velocity is defined as

$$v^V = \bar{V}_A N_A + \bar{V}_B N_B \tag{11-6}$$

where $\bar{V}_i$ is the molar volume of component i (the partial molal volume if $\bar{V}_i$ is a function of composition). Since the flux of A due to the volume-average velocity is $c_A v^V$, J_A^V is related to N_A and N_B through

$$J_A^V = N_A - c_A(\bar{V}_A N_A + \bar{V}_B N_B) \tag{11-7}$$

Combining Eqs. (11-2) and (11-7) leads to another expression for N_A in terms of the diffusivity:

$$N_A = -D_{AB} \nabla c_A + c_A(\bar{V}_A N_A + \bar{V}_B N_B) \tag{11-8}$$

It can be shown (Lightfoot and Cussler, 1965) that Eqs. (11-1) and (11-2) [and Eqs. (11-5) and (11-8)] are identical for systems at constant temperature and pressure; also D_{AB} used in Eqs. (11-1) and (11-5) is equal to D_{AB} used in Eqs. (11-2) and (11-8). For an ideal gas it can readily be seen that the equations are identical under those conditions, since $c \nabla x_A = \nabla c_A$, $v^* = v^V$, and $\bar{V}_A = \bar{V}_B = 1/c$.

Equation (11-1) is generally used along with the assumption that c is independent of composition; this is a good assumption for *gas mixtures* at low and moderate pressures. Equation (11-2) is generally used along with the assumption that $\bar{V}_A$ and $\bar{V}_B$ are independent of composition but without requiring that c be constant. This is a better idealization for most liquid mixtures.

Equations (11-1), (11-2), (11-5), and (11-8) are various forms of *Fick's law* for diffusion. There is a direct parallel in form between Eqs. (11-1) and (11-2) and Newton's law for viscous flow and Fourier's law for heat conduction if J_A^* or J_A^V is made analogous to the shear stress τ and the heat flux q, if D_{AB} is made analogous to the kinematic viscosity μ/ρ and the thermal diffusivity $k/\rho C_p$, and if c_A is made analogous to mass velocity ρu and the thermal energy $\rho C_p T$ (Bird et al., 1960). Here μ is viscosity, ρ is density, k is thermal conductivity, C_p is heat capacity, u is flow velocity, and T is temperature.

In the Fick's law expressions for fluxes with reference to stationary coordinates [Eqs. (11-5) and (11-8)] the right-hand side is the sum of two terms. The first, involving D_{AB}, is sometimes called the *diffusive flux*, and the second, involving the sum of the fluxes, is sometimes called the *convective flux*. When the convective flux is negligible, there is a direct parallel between Fick's law and Newton's and Fourier's laws written for stationary coordinates, but when the convective flux is important, the analogy is less direct.

An example of the distinction between the two terms and the importance of each is shown in Fig. 11-1, which represents the transport processes occurring near the membrane in a reverse-osmosis process for desalination of salt water. Here pressure is applied to force water through the membrane. Since the membrane is highly selective for water over salt, the salt does not pass through. Salt must thereby build up in concentration adjacent to the membrane surface c_{S1} to a value larger than the salt concentration in bulk solution c_{SL}. If A is salt and the membrane is completely selective, $N_A = 0$. If the positive direction of distance z is taken to be to the left, as shown in Fig. 11-1, then N_B, the flux of water across the membrane, will be negative, making the second (convective) term on the right-hand side of Eq. (11-8) negative. On the other hand, the first (diffusive) term will be positive, since ∇c_A ($= dc_A/dz$) is negative. Also this first term will be equal to the second term in absolute magnitude if N_A is to be zero and there are no additional transport effects. Salt is brought to the

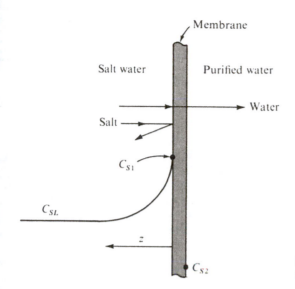

Salt water

Purified water

$\longrightarrow$ Water

Salt $\longrightarrow$

C_{S1}

C_{SL}

z

C_{S2}

Figure 11-1 Mass-transfer processes occurring during reverse osmosis of salt water.

membrane by convection with the permeating water but returns to the bulk solution by diffusion along the resulting concentration gradient. The diffusive and convective fluxes are exactly equal and opposite in sign at all values of z.

For systems containing more than two components, the diffusion equations become more complex. Multicomponent diffusion is reviewed by Cussler (1976), among others. An important special case is that where all components except one are dilute; it is then possible to use the binary equations to obtain the flux of any one of the minor components, the major component being taken as component B.

Prediction of Diffusivities

Gases The kinetic theory of gases, coupled with the Lennard-Jones intermolecular potential, leads to the following equation for D_{AB} in binary gas mixtures at low and moderate pressures (Hirschfelder et al., 1964):

$$D_{AB} = \frac{1.882 \times 10^{-22} T^{3/2} (M_A^{-1} + M_B^{-1})^{1/2}}{P \sigma_{AB}^2 \, \Omega_D} \qquad \text{m}^2/\text{s} \qquad (11\text{-}9)$$

where T = temperature, K
$\quad M_i$ = molecular weight of component i, g/mol
$\quad P$ = total pressure, Pa
$\quad \sigma_{AB}$ = collision diameter, m
$\quad \Omega_D$ = diffusion collision integral, dimensionless

Ω_D is a function of kT/ϵ_{AB}, where ϵ_{AB} is a measure of the relative strength of intermolecular attraction. This function is shown graphically in Fig. 11-2 and is tabulated

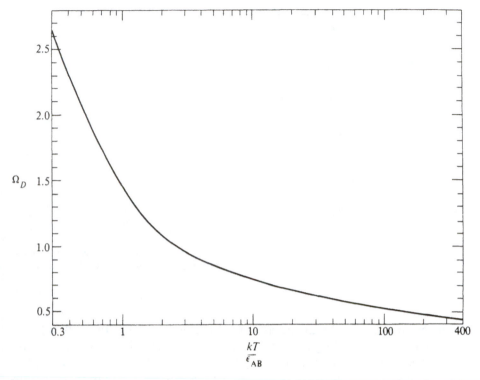

Figure 11-2 Diffusion collision integral as a function of kT/ϵ_{AB}.

by Bird et al. (1960, app. B) and Sherwood et al. (1975), among others. σ_{AB} and ϵ_{AB}/k are usually obtained from pure-component parameters by the mixing rules

$$\sigma_{AB} = \tfrac{1}{2}(\sigma_A + \sigma_B) \tag{11-10}$$

$$\frac{\epsilon_{AB}}{k} = \left(\frac{\epsilon_A}{k}\frac{\epsilon_B}{k}\right)^{1\,2} \tag{11-11}$$

Values of σ_i and ϵ_i/k for various substances are tabulated by Bird et al. (1960, app. B) and Sherwood et al. (1975), among others. σ_i is usually given in angstrom units $(1\ \text{Å} = 1 \times 10^{-10}\ \text{m})$.

Experimental data, such as the extensive tabulation by Marrero and Mason (1972), agree closely with Eq. (11-9).

From Eq. (11-9) it can be seen that D_{AB} in gases is inversely proportional to total pressure, this being the result of more frequent molecular collisions at higher pressure. There is some extra temperature dependence from Ω_D, with the result that D_{AB} tends to increase with about the 1.75 power of temperature; this is primarily the result of the increase in molecular velocity with increasing temperature. D_{AB} tends to be lower for larger molecules (lower molecular velocity), the lower of the two molecular weights exerting the dominant influence. D_{AB} at low and moderate pressures is independent of composition, as given by Eq. (11-9).

Deviations from Eq. (11-9) occur at high pressures, D_{AB} generally becoming lower than predicted by Eq. (11-9) (Bird et al., 1960, chap. 16).

In gas-filled porous solids, D_{AB} for gas transport becomes less because of reduced cross-sectional area and a more tortuous path for transport. Also, at low pressures and/or for media with very fine pores, collisions of the molecules with pore walls become significant, reducing D_{AB}; this is known as the *Knudsen regime*. Offsetting these two factors can be added transport within an adsorbed layer on pore walls (Satterfield, 1970).

Example 11-1 Calculate the diffusivity of ammonia in nitrogen at 358 K and 200 kPa and compare with the experimental value of 1.66×10^{-5} m^2/s (Sherwood et al., 1975). Values of the Lennard-Jones parameters are:

	σ_i, Å	ϵ_i/k, K
Ammonia	2.900	558.3
Nitrogen	3.798	71.4

SOURCE: Data from Sherwood et al. (1975).

SOLUTION From Eqs. (11-10) and (11-11)

$$\sigma_{AB} = \tfrac{1}{2}(2.900 + 3.798) \times 10^{-10} \text{ m} = 3.349 \times 10^{-10} \text{ m}$$

$$\frac{\epsilon_{AB}}{k} = [(558.3)(71.4)]^{1/2} \text{ K} = 199.7 \text{ K}$$

Hence

$$\frac{kT}{\epsilon_{AB}} = \frac{358}{199.7} = 1.79$$

and from Fig. 11-1

$$\Omega_D = 1.118$$

Substituting into Eq. (11-9) gives

$$D_{AB} = \frac{(1.882 \times 10^{-22})(358)^{3/2}[(1/17.03) + (1/28.02)]^{1/2}}{(200 \times 10^3)(3.349 \times 10^{-10})^2(1.118)} \text{ m}^2/\text{s}$$

$$= 1.56 \times 10^{-5} \text{ m}^2/\text{s}$$

This is 6 percent below the experimental value. □

Liquids Diffusivities in the liquid phase are much lower than those in gases because of the much smaller intermolecular distances. For liquid mixtures of simple mixtures where one solute A is dilute in a solvent B it has been found theoretically and experimentally that the dimensional group $D_{AB}\mu_B/T$, where μ_B is solvent viscosity, tends to be relatively independent of temperature and correlatable with solute size for a given solute. The most common correlation is that of Wilke and Chang (1955),

$$D_{AB} = 7.4 \times 10^{-12} \frac{(\phi M_B)^{1/2} T}{\mu_B V_A^{0.6}} \qquad \text{m}^2/\text{s} \qquad (11\text{-}12)$$

where M_B = solvent molecular weight
$\quad\quad T$ = temperature, K
$\quad\quad \mu_B$ = solvent viscosity, mPa·s (=cP)
$\quad\quad V_A$ = molal volume of pure solute at normal boiling point, cm³/mol

Values of V_A can be computed from the LeBas group contributions (Reid et al., 1977; Sherwood et al., 1975) or from experimental data. ϕ is an *association parameter* for the solvent, set at 2.6 for water, 1.9 for methanol, 1.5 for ethanol, and 1.0 for benzene, diethyl ether, hydrocarbons, and nonassociated solvents in general. For nonaqueous solvents the equation of King et al. (1965) seems to work somewhat better (Reid et al., 1977):

$$D_{AB} = 4.4 \times 10^{-12} \frac{T}{\mu_B}\left(\frac{V_B}{V_A}\right)^{1/6}\left(\frac{\Delta H_B}{\Delta H_A}\right)^{1/2} \quad \text{m}^2/\text{s} \quad\quad (11\text{-}13)$$

where $\quad T$ = temperature, K
$\quad\quad \mu_B$ = solvent viscosity, mPa·s = cP
$\quad\quad V_A, V_B$ = solute and solvent molal volumes
$\Delta H_A, \Delta H_B$ = solute and solvent molar latent heats of vaporization at normal boiling point

Experimental data for liquid-phase diffusivities have been collected by Reid et al. (1977) and by Ertl et al. (1973), among others.

From Eqs. (11-12) and (11-13) it can be seen that D_{AB} is independent of pressure in liquids except at very high pressure. D_{AB} increases much more sharply with increasing temperature in liquids than in gases; for aqueous systems near ambient temperature the increase is about 2.6 percent per kelvin. The principal temperature-sensitive term on the right-hand side of Eqs. (11-12) and (11-13) is μ_B.

D_{AB} from Eq. (11-12) or (11-13) should be interpreted as the diffusivity for A at high dilution in B. It is *not* the same as D_{BA}, the diffusivity of B at high dilution in A. The effect of concentration level on liquid-phase diffusivities is complex but seems to reflect solution nonidealities as the dominant factor (Reid et al., 1977; Sherwood et al., 1975).

Prediction and analysis of diffusivities in electrolyte solutions involves separate allowance for the ionic mobilities of independent ions through the Nernst-Haskell equation, as well as consideration of the effect of concentration (Sherwood et al., 1975; Reid et al., 1977; Newman, 1967a).

Solids Diffusivities in solids cover a wide range of values, becoming quite low for dense and/or crystalline materials. Analysis of diffusion in solids is also made more complicated if the solid is heterogeneous and/or nonisotropic, such as wood, most foods, composite materials, and the like.

Diffusion in polymer materials is reviewed by Crank and Park (1968) and summarized by Sherwood et al. (1975). Diffusion in metals is treated by Bugakov (1971), among others. Diffusivities in various solid materials have been compiled and discussed by Barrer (1951), Jost (1960), and Nowick and Burton (1975).

SOLUTIONS OF THE DIFFUSION EQUATION

Solutions of the diffusion equation for various geometries are given by Crank (1975) and Barrer (1951). Solutions to the heat-conduction equation in stationary media are given by Carslaw and Jaeger (1959). These can be applied to diffusion by direct analogy as long as the convective term in Eq. (11-5) or (11-8) is insignificant. The convective term will be negligible in either of two special cases:

1. $N_A = -N_B$ [in Eq. (11-5)], or $\bar{V}_A N_A = -\bar{V}_B N_B$ [in Eq. (11-8)]. These cases are known as *equimolar* and *equivolume counterdiffusion*, respectively, and would occur, for example, for diffusion in a nonuniform gas mixture in a closed container.
2. A becomes very dilute in B, and N_B is either zero or very small. In this case the convective term in either equation will be the product of two quantities (concentration of A and flux), each of which approaches zero.

When the convective term is insignificant, Eq. (11-8) becomes

$$N_A = -D_{AB} \nabla c_A \tag{11-14}$$

For transient diffusion in a stagnant medium, a mass balance on a differential element gives

$$\frac{\partial c_A}{\partial t} = -\nabla N_A \tag{11-15}$$

Combining Eqs. (11-14) and (11-15) gives, for constant D_{AB}

$$\frac{\partial c_A}{\partial t} = D_{AB} \nabla^2 c_A \tag{11-16}$$

For one-dimensional transport, Eq. (11-16) becomes

$$\frac{\partial c_A}{\partial t} = D_{AB} \frac{\partial^2 c_A}{\partial z^2} \tag{11-17}$$

Solutions of Eqs. (11-16) and (11-17) give the fraction of the ultimate concentration change which has occurred as a unique function of the dimensionless group $D_{AB} t/L^2$, where t is elapsed time and L is an appropriate length variable. In turn, if the concentrations at all points are integrated to give an average concentration, this average concentration can be related to the same group, where L is now an appropriate dimension of the entire medium.

Figure 11-3 shows the solutions to the transient-diffusion equation for a one-dimensional slab, an infinite circular cylinder, and a sphere (Sherwood et al., 1975; Carslaw and Jaeger, 1959), expressed as $(c_{Af} - c_{A,av})/(c_{Af} - c_{A0})$ vs. $D_{AB} t/L^2$. Here L is the half-thickness of the slab and the radius of the cylinder and sphere. $c_{A,av}$ is the average concentration; c_{A0} is the initial concentration of the medium, assumed to be uniform; and c_{Af} is the concentration reached after an infinite time, assumed to be the value at which the surface of the medium is held throughout the diffusion process. Over much of the range the solutions form straight lines on the semilogarithmic plot. Detailed numerical values are given by Sherwood et al. (1975) and are needed for

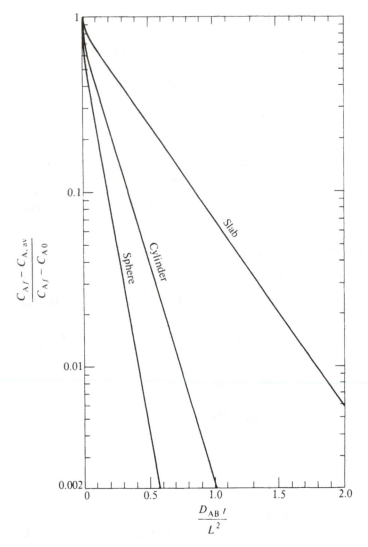

Figure 11-3 Solutions of the transient-diffusion equation for simple shapes (constant surface concentration $= C_{Af}$). *(Data from Sherwood et al., 1975.)*

precision at very short times. Other solutions apply for other boundary conditions (Carslaw and Jaeger, 1959), and solutions for more complex shapes can be obtained by superposition of the solutions for simple shapes (Sherwood et al., 1975).

Example 11-2 One of the original processes for decaffeination of coffee involved solvent extraction, or leaching, of caffeine from whole coffee beans (Moores and Stefanucci, 1964). Beans were first steamed to free caffeine and provide an aqueous transport medium for it inside the bean. The beans were then contacted with a suitable organic solvent, into which the caffeine was leached.

   Assume that the quantity and agitation of the solvent are sufficient to reduce the concentration of caffeine at the bean surface to zero at all times during the leaching. Assume also that the combination of percent moisture, voidage and tortuosity of the diffusion path inside a bean serve to reduce the diffusivity of caffeine to 10 percent of the value in pure water. Take the temperature to be 350 K and the beans to be equivalent to spheres with diameters of 0.60 cm. Caffeine has the structure

(a) Calculate the contact time with solvent required to reduce the caffeine content of a group of beans to 3.0 percent of the initial value. (b) Would halving the average bean dimension by cutting the beans serve to reduce the required contact time? If so, by how much? (c) Would a change of extraction temperature alter the required contact time? If so, in what direction should the temperature be changed to reduce the required contact time? What would be the effect of a change of 10 K? (d) What would be the directional effect on the required contact time if there were a slow rate of solubilization of the caffeine in the beans? If cell-wall membranes within the beans presented a significant additional resistance to mass transport?

SOLUTION The diffusivity of caffeine in water is estimated from Eq. (11-12). The molar volume of caffeine is obtained by the group-contribution method of LeBas, using the tabulation of Sherwood et al. (1975, p. 31):

8 carbons	$8 \times 14.8 =$	118.4
10 hydrogens	$10 \times 3.7 =$	37.0
2 oxygens	$2 \times 7.4 =$	14.8
4 nitrogens	$4 \times 15.6 =$	62.4
		$\overline{232.6}$ cm^3/mol

The association parameter for water is 2.6. The viscosity of water at 350 K (77°C) is 0.38 mPa·s (Perry and Chilton, 1973).

$$D_{AB} = 7.4 \times 10^{-12} \frac{(2.6 \times 18)^{1\,2}(350)}{(0.38)(232.6)^{0.6}} = 1.77 \times 10^{-9} \text{ m}^2/\text{s}$$

The value assumed to apply inside the coffee beans is then $(0.1)(1.77 \times 10^{-9}) = 1.77 \times 10^{-10}$ m^2/s.

(a) From Fig. 11-3, for $(c_{A0} - c_{A,av})/(c_{A0} - c_{Af}) = 0.030$ (97 percent removal of caffeine), $D_{AB} t/L^2$ for a sphere equals 0.305. Since the sphere radius L is $\frac{1}{2}(0.6$ cm$) = 0.3$ cm, we have

$$t = \frac{0.305L^2}{D_{AB}} = \frac{(0.305)(0.003)^2}{1.77 \times 10^{-10}} = 1.55 \times 10^4 \text{ s} = \frac{1.55 \times 10^4}{3600} = 4.3 \text{ h}$$

(b) Halving the bean dimension would reduce L by a factor of 2. Since $D_{AB} t/L^2$ should be the same for 97 percent removal, t will be one-fourth as much as calculated in part (a), or 1.1 h.

(c) Changing temperature changes the diffusivity of caffeine within the beans. Higher temperature gives higher D_{AB} and hence a lower t for the desired $D_{AB} t/L^2$, corresponding to the desired degree of removal. Increasing the temperature to 360 K would decrease μ_{H_2O} from 0.38 to 0.32 mPa·s. Hence D_{AB} increases by a factor of

$$\frac{(0.38)(360)}{(0.32)(350)} = 1.22$$

and the required contact time decreases by the same factor.

(d) A slow rate of solubilization would reduce the concentration of caffeine in solution inside the beans, reducing the driving force for diffusion, slowing the diffusion process, and taking a longer contact time. Additional resistance from the cell-wall membranes would reduce the transport rate for a given driving force and again would lengthen the required contact time. □

MASS-TRANSFER COEFFICIENTS

Solutions of the diffusion equation often become quite complex or impossible when mass transfer occurs in a flowing system, in a turbulent medium, and/or in a complicated geometry. For this reason it is common practice to define mass-transfer coefficients, which relate fluxes of matter to known differences in mole fraction or concentration. These are analogous to heat-transfer coefficients, which are used to relate heat fluxes to known temperature differences.

Dilute Solutions

As we have seen, for mass transfer in dilute solutions we can neglect the convective terms of Eqs. (11-5) and (11-8) and consider only the terms involving D_{AB}, as long as N_B is not large enough in absolute magnitude to invalidate this simplification. Furthermore, for a multicomponent system in which all components but one are dilute, we can analyze the fluxes of each of the minor components using Eqs. (11-5) or (11-8), D_{AB} being the diffusivity for the binary system of that minor component and the major component. Again, the convective terms can be neglected unless N_B is large enough to preclude this.

For situations where the convective terms can be neglected, it is universal practice to define the mass-transfer coefficient as the ratio of N_A to an appropriate measure of the difference in composition across the region where the mass-transfer process is occurring. For mass transfer between a bulk fluid of uniform composition and an interface where the same fluid phase has a different composition, several different definitions of the mass-transfer coefficient are possible:

$$N_A = k_y(y_{AG} - y_{Ai}) \tag{11-18}$$

$$N_A = k_x(x_{AL} - x_{Ai}) \tag{11-19}$$

$$N_A = k_G(p_{AG} - p_{Ai}) \tag{11-20}$$

$$N_A = k_c(c_{AL} - c_{Ai}) \tag{11-21}$$

Here the subscript A refers to component A, the subscript i refers to the interface, the subscript G refers to bulk gas, the subscript L refers to bulk liquid, p is partial pressure, y and x are gas and liquid mole fractions, and c is concentration. k_y, k_x, k_G, and k_c are alternative forms of the mass-transfer coefficient, with typical units of mol/s·m², mol/s·m², mol/s·m²·Pa, and m/s [(mol/m²·s)/(mol/m³)], respectively. k_y and k_G would be used for gas-phase processes, and at constant total pressure $k_y = k_G P$. k_x and k_c would be used for liquid-phase processes, and at constant molar density $k_x = k_c c$. The concentration-based mass-transfer coefficient k_c is sometimes also applied to a gas phase, in which case c_{AL} in Eq. (11-21) would be replaced by

c_{AG}. Common practice is to write the driving force so as to give the mass-transfer coefficient a positive sign; i.e., if the mass transfer were from the interface to the bulk fluid and N_A were considered positive in that direction, the Δy, Δx, Δp, and Δc terms in Eqs. (11-18) to (11-21) would be reversed in sign.

For a few simple or idealized cases mass-transfer coefficients can be obtained from simple theory; in most other cases they must be obtained experimentally and are then correlated through the assistance of dimensionless groups.

Film model The film model, which originated with Nernst (1904), is based upon the observation that the concentration of a transferring solute usually changes most rapidly in the immediate vicinity of the interface and is relatively uniform in bulk fluid away from the interface. For an agitated or flowing fluid near an interface the assumption is then made that all the concentration change occurs over a thin region immediately adjacent to the interface. This region is called the *film* and is considered to be stationary and so thin that steady-state diffusion is immediately established across it. In that case, Eq. (11-17) applied to the film becomes

$$\frac{\partial^2 c_A}{\partial z^2} = 0 \tag{11-22}$$

which with the boundary conditions $c_A = c_{Ai}$ at $z = 0$ and $c_A = c_{AL}$ (or c_{AG}) at $z = \delta$ becomes

$$c_A = c_{Ai} + \frac{z}{\delta}(c_{AL} - c_{Ai}) \tag{11-23}$$

Coupled with Eq. (11-14), this becomes

$$N_A = \frac{D_{AB}}{\delta}(c_{AL} - c_{Ai}) \tag{11-24}$$

if N_A is taken positive toward the interface. Comparison with Eq. (11-21) gives

$$k_c = \frac{D_{AB}}{\delta} \tag{11-25}$$

Equation (11-25) could be used for predicting mass-transfer coefficients if δ could somehow be predicted a priori or if δ were relatively independent of flow rates and other conditions. Neither is true. Also, Eq. (11-25) predicts that k_c varies with the first power of D_{AB}, which is almost never observed; this discrepancy results from the assumption of a discontinuity in transport conditions at the outer film boundary.

Even though the film model cannot be used effectively for prediction and correlation, it is useful (because of its mathematical simplicity) for predicting and analyzing the effects of such additional complicating factors as simultaneous chemical reaction near the interface and high solute concentrations and fluxes. It is a useful model for a membrane, which obeys the film assumptions, and is a reasonable first approximation for highly turbulent fluids near fixed interfaces, where a thin, relatively stagnant boundary layer can exist.

Penetration and surface-renewal models For many situations it is a reasonable assumption to postulate that a mass of fluid is exposed at an interface for an identifiable amount of time before being swept away and remixed with bulk fluid. If there is no gradient of velocity within this mass of fluid during the exposure, we can analyze the mass-transfer process by following the fluid mass and solving the diffusion equation for the transient-diffusion process that occurs. The resulting model then follows the *penetration* of the solute concentration profile into the fluid mass, away from the interface.

The appropriate form of the diffusion equation is Eq. (11-17), with the boundary conditions $c_A = c_{Ai}$ at $t > 0$ and $z = 0$; $c_A = c_{AL}$ at $t = 0$ (when the fluid mass is brought to the interface); and $c_A \rightarrow c_{AL}$ as $z \rightarrow \infty$, far removed from the interface. The solution for the concentration profile is

$$\frac{c_A - c_{Ai}}{c_{AL} - c_{Ai}} = 1 - \text{erf}\left(\frac{z}{2\sqrt{D_{AB}t}}\right) \tag{11-26}$$

where $\text{erf}(x)$ is the error function [see discussion following Eq. (8-51)]. The mass-transfer coefficient is obtained by applying Eqs. (11-14) and (11-21) at the interface ($z = 0$), and is

$$k_c = \left(\frac{D_{AB}}{\pi t}\right)^{1/2} \tag{11-27}$$

at time t. Notice that at $t = 0$, k_c becomes infinite, reflecting the step change in concentration from c_{Ai} to c_{AL} at the interface as the exposure of the fluid mass begins. k_c given by Eq. (11-27) is known as the *instantaneous mass-transfer coefficient*. An *average mass-transfer coefficient* over an entire exposure interval, from $t = 0$ to $t = \theta$, can also be obtained

$$k_{c,\text{av}} = \frac{1}{\theta} \int_0^\theta k_c \, dt = 2\left(\frac{D_{AB}}{\pi\theta}\right)^{1/2} \tag{11-28}$$

The average coefficient for the exposure is twice the instantaneous coefficient at the end of the exposure.

The concentration-difference driving force in Eq. (11-21) for k_c defined by Eqs. (11-27) and (11-28) is the difference between the *initial* bulk-liquid concentration and the interface concentration, since the bulk concentration in a semi-infinite medium does not change as diffusion occurs.

The penetration model is applicable to a situation where (1) a fluid mass is suddenly brought to the interface and is just as suddenly mixed back into the bulk after time θ, (2) there is no velocity gradient within the fluid mass, and (3) the depth of the fluid mass is sufficient to ensure that the solute concentration profile does not penetrate far enough to reach a bounding surface or a region with turbulent transport or a different velocity. Assumption 2 is usually well met by liquid in the vicinity of a gas-liquid interface, since the viscosity of a liquid is usually much greater than that of a gas, meaning that drag from the gas phase will create little gradient of velocity in the liquid near the interface. In liquids diffusivities are also low enough for the depth of penetration to be small. For example, for $D_{AB} = 1 \times 10^{-9}$ m²/s and $t = 1$ s, the solute concentration will have changed 98 percent of the way from the

Liquid flow

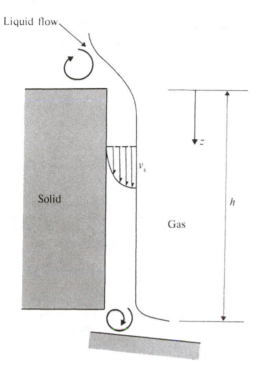

Solid

v_s

z

Gas

h

Figure 11-4 Flow of liquid over a short solid surface.

interfacial concentration to the bulk concentration only 100 μm away from the interface.

One situation to which the penetration model applies well is desorption or absorption of a volatile solute from or into a liquid flowing over a short solid surface as a film, with the outside of the film exposed to gas. The flow pattern tends to mix the liquid at the top and bottom of the solid wall, giving a distinct beginning and end to the exposure of surface liquid to the gas, as shown in Fig. 11-4. The liquid velocity is relatively uniform near the gas interface (the result of a semiparabolic profile) because of the low drag of the gas on the liquid. Hence mass-transfer coefficients for transfer between the gas-liquid interface and the bulk liquid can be estimated well from Eqs. (11-27) and (11-28), t being z/v_s and θ being h/v_s, where h is the fall height and v_s the surface velocity of the liquid. There is an obvious similarity between the situation shown in Fig. 11-4 and the flow situation in an irrigated packed tower (Fig. 4-13), where liquid flows over packing of the sort shown in Fig. 4-14 and contacts a gas which flows in the interstices. Indeed, it has been found that k_c for the liquid phase in packed gas absorbers does vary with the liquid diffusivity to the $\frac{1}{2}$ power, as predicted by Eqs. (11-27) and (11-28), and that values of k_c for the liquid phase are of the magnitude predicted by Eq. (11-28) with $\theta = h/v_s$, where h is the height of an individual piece of packing.

In many other situations where a liquid is exposed to a gas it is not so apparent how to predict the value of θ for the penetration model. An example would be a stirred-vessel absorber where an impeller causes eddies of liquid to come up to the surface, stay for a time, and then be mixed back into the bulk liquid. As one approach to such situations, Danckwerts (1951) has proffered a model based upon random surface renewal, leading to

$$k_c = (D_{AB} s)^{1/2} \tag{11-29}$$

where s is the fraction of the surface renewed by fluid masses from the bulk per unit time. There does not appear to be a good way of correlating s with observable flow properties, however.

Example 11-3 Show that the penetration model also predicts the behavior of the curves in Fig. 11-3 for low values of $D_{AB} t / L^2$.

SOLUTION At short times the concentration profiles for transient diffusion into a slab, cylinder, or sphere will have penetrated only a short distance and hence should be describable by the equations for transient diffusion into a semi-infinite medium, on which Eqs. (11-27) and (11-28) are based. The average concentration would come from a mass balance relating the flux across the surface to the capacity of the object:

$$V \frac{dc_{A, av}}{dt} = N_A A = k_c A (c_{Ai} - c_{A0}) \tag{11-30}$$

where V is the volume and A the surface area of the object. The driving force for the mass-transfer coefficient is taken as $c_{Ai} - c_{A0}$, since the penetration model postulates a semi-infinite medium where c_A is c_{A0} far from the surface.

Integrating Eq. (11-30) with the boundary condition $c_A = c_{A0}$ at $t = 0$ leads to

$$\frac{c_{A, av} - c_{A0}}{c_{Ai} - c_{A0}} = \frac{A}{V} \int_0^t k_c \, dt$$

Substituting Eq. (11-27) and integrating and then subtracting each side from 1 yields

$$\frac{c_{Ai} - c_{A, av}}{c_{Ai} - c_{A0}} = 1 - 2 \frac{A}{V} \left(\frac{D_{AB} t}{\pi} \right)^{1/2} \tag{11-31}$$

For the slab $A/V = 1/L$; for the infinite cylinder $A/V = 2/L$; and for the sphere $A/V = 3/L$, recalling that L is the half thickness for the slab and the radius for the cylinder and sphere. The left-hand side

Table 11-1 Values of $(c_{A, av} - c_{A0})/(c_{Ai} - c_{A0})$ from Fig. 11-3 compared with those computed from Eq. (11-31)

	Fig. 11-3			Eq. (11-31)		
$D_{AB} t / L^2$	Slab	Cyl.	Sphere	Slab	Cyl.	Sphere
0.005	0.922	0.843	0.774	0.920	0.840	0.761
0.01	0.890	0.784	0.690	0.887	0.774	0.661
0.02	0.839	0.698	0.579	0.840	0.681	0.521
0.04	0.773	0.558	......	0.774	0.549	......
0.06	0.725	0.512	......	0.724	0.447	......
0.10	0.643	......	......	0.643	......	......
0.20	0.497	......	......	0.495	......	......
0.30	0.388	......	......	0.382	......	......
0.50	0.236	......	......	0.202	......	......

of Eq. (11-31) is the same as the vertical coordinate of Fig. 11-3, and the second term of the right-hand side is a multiple of the $\frac{1}{2}$ power of the horizontal coordinate, being $2n/\sqrt{\pi}$ times $(D_{AB} t/L^2)^{1/2}$, where $n = 1, 2,$ and 3 for the slab, cylinder, and sphere, respectively.

From Table 11-1 it can be seen that the agreement with the penetration model is excellent at the shortest times for all three geometries. For the slab model the penetration approximation remains very good for values of the concentration factor down to 0.40 (representing 60 percent equilibration with the surface). For the cylinder the penetration approximation deviates significantly below a higher value of the concentration factor, and for the sphere the critical value of the concentration factor becomes higher yet. This trend from one geometry to another results because the penetration model considers diffusion into a unidirectional medium. This assumption is obeyed for the slab, but for the cylinder and sphere the cross section for diffusion becomes less as one proceeds inward from the surface, causing less mass transfer to occur than is predicted by the unidirectional penetration model. □

Diffusion into a stagnant medium from the surface of a sphere When molecular diffusion occurs from the surface of a sphere of constant size into a surrounding stagnant medium of infinite extent, a steady-state situation is eventually set up because of the increasing cross section proceeding away from the sphere. To solve for the rate of diffusion it is necessary to put Eq. (11-16) into spherical coordinates and set $\partial c_A/\partial t = 0$. The solution (Sherwood et al., 1975, p. 215) is

$$N_{A, r=r_0} = \frac{D_{AB}}{r_0}(c_{Ai} - c_{A0}) \tag{11-32}$$

where r_0 is the radius of the sphere. Combining Eqs. (11-21) and (11-32) gives

$$k_c = \frac{D_{AB}}{r_0} = \frac{2D_{AB}}{d_0} \tag{11-33}$$

if d_0 is the sphere diameter. For short times, before this steady state is established, it can be shown that the flux is the sum of the steady-state value and a transient term (Sherwood et al., 1975, p. 70):

$$N_A = \left[\frac{D_{AB}}{r_0} + \left(\frac{D_{AB}}{\pi t}\right)^{1/2} \right](c_{Ai} - c_{A0}) \tag{11-34}$$

leading to

$$k_c = \frac{D_{AB}}{r_0} + \left(\frac{D_{AB}}{\pi t}\right)^{1/2} \tag{11-35}$$

The dimensionless group $k_c d_0/D_{AB}$ is known as the *Sherwood number* (Sh); hence the solution corresponds to

$$\text{Sh} = 2 + (\pi \, \text{Fo})^{-1/2} \tag{11-36}$$

where the Fourier number Fo is $D_{AB} t/d_0^2$.

If flow, turbulence, or other factors are also present, k_c can be expected to be higher than given by Eq. (11-35). However, Eq. (11-35) leads to very high mass-transfer coefficients for very small spheres because of the presence of the sphere radius in the denominator. Because of this, the molecular-diffusion terms dominate for smaller spheres, and below some critical radius Eq. (11-35) will describe the mass transfer, even in the presence of flow and/or turbulence.

The analysis leading to Eq. (11-35) postulates an unchanging sphere diameter. If the diameter changes slowly, it is permissible to apply Eq. (11-35) for k_c as a quasi-steady-state approximation. However, for rapid changes in the sphere diameter, e.g., many cases of bubble growth, it is necessary to take account of the effect of the surface motion.

Dimensionless groups We have already seen that the mass-transfer coefficient k_c can be combined with D_{AB} and a length variable into a dimensionless group known as the Sherwood number. The Sherwood group can also be expressed in terms of the other mass-transfer coefficients defined in Eqs. (11-18) to (11-20):

$$\text{Sh} = \frac{k_c L}{D_{AB}} = \frac{k_x L}{cD_{AB}} = \frac{k_y L}{cD_{AB}} = \frac{k_G RTL}{D_{AB}} \tag{11-37}$$

The ideal-gas law has been invoked in writing the form involving k_G. The length variable is designated as L in each case.

For systems of mass transfer to and from interfaces in flowing systems we can expect the mass-transfer coefficient to be influenced by the fluid mainstream velocity u, the fluid density ρ, the viscosity μ, the diffusivity D_{AB}, and one or more characteristic lengths L. If the coefficient varies locally, it will also be influenced by a position variable x. Dimensional analysis applied to this collection of variables leads to the Sherwood group and three other independent dimensionless groups:

$$\frac{Lu\rho}{\mu}, \qquad \frac{\mu}{\rho D_{AB}}, \qquad \text{and} \qquad \frac{x}{L}$$

The first of these is the *Reynolds number* Re (ratio of inertial fluid forces to viscous forces), and the second is the *Schmidt number* Sc (ratio of viscous transport of momentum to diffusional transport of matter). When transient phenomena are involved, time enters, leading to the Fourier number $D_{AB}t/L^2$. Any additional factors lead to additional groups; e.g., if gravitational forces are important, as in motion by natural convection (convection driven by naturally occurring density differences), a new group involving g enters, typically the Rayleigh number or the Grashof number (Bird et al., 1960, p. 645).

For gaseous mixtures, the Schmidt number is of the order of unity, but for liquid mixtures the Schmidt number is much higher, typically of the order of 10^3 to 10^4.

For complex situations, the mass-transfer coefficient can be correlated in terms of these variables through the functionality

$$\text{Sh} = f\left(\text{Re}, \text{Sc}, \frac{x}{L}, \dots\right) \tag{11-38}$$

The use of dimensionless groups reduces the number of experiments required and makes it easier to work with the resulting functionalities.

The use of dimensionless groups also facilitates extension of heat-transfer relationships to the corresponding mass-transfer situation for dilute solutions. The *Nusselt number* hL/k, where h is the heat-transfer coefficient and k is the thermal conductivity, converts into the Sherwood number; the Reynolds number remains the same; and the *Prandtl number* $C_p \mu/k$, where C_p is heat capacity and k is thermal conductivity, converts into the Schmidt number. The Fourier number for heat transfer is $kt/\rho C_p L^2$.

Laminar flow near fixed surfaces For mass transfer between a fluid in laminar flow within a circular tube and the tube wall, the classical Graetz solution leads to the following expression for the mass-transfer coefficient averaged over the tube-wall surface (Eckert and Drake, 1959):

$$\frac{k_{c,\,\mathrm{av}} d_p}{D_{\mathrm{AB}}} = 3.65 + \frac{0.0668(d_p/x)(d_p u \rho/\mu)(\mu/\rho D_{\mathrm{AB}})}{1 + 0.04[(d_p/x)(d_p u \rho/\mu)(\mu/\rho D_{\mathrm{AB}})]^{2/3}} \tag{11-39}$$

provided the solute concentration at the tube wall is kept constant. d_p is the diameter of the tube, and x is the length of the tube surface over which the mass-transfer process occurs. Since the bulk concentration of solute in the fluid will change along the tube length, it is important to identify the concentration-difference driving force to be used with k_c from this expression. In this case it is the logarithmic mean of the inlet and outlet driving forces; i.e., if $A = (c_{AL} - c_{Ai})_{\mathrm{inlet}}$ and $B = (c_{AL} - c_{Ai})_{\mathrm{outlet}}$, the appropriate value of $c_{AL} - c_{Ai}$ to use in Eq. (11-21) with the value of k_c from Eq. (11-39) is $(A - B)/[\ln (A/B)]$.

Near the entry of the tube, Eq. (11-39) becomes simpler:

$$\frac{k_{c,\,\mathrm{av}} d_p}{D_{\mathrm{AB}}} = 1.67 \left(\frac{d_p}{x} \frac{d_p u \rho}{\mu} \frac{\mu}{\rho D_{\mathrm{AB}}} \right)^{1/3} \tag{11-40}$$

where x is the distance from the tube inlet and $k_{c,\,\mathrm{av}}$ is the average k_c over the distance from $x = 0$ to $x = x$. Because of the $-\frac{1}{3}$ power on distance, the local, or instantaneous, k_c at any x is 2/3 times the average coefficient from $x = 0$ up to that position, as can be shown by an integration similar to Eq. (11-28). This leads to a form of Eq. (11-40) where k_c replaces $k_{c,\,\mathrm{av}}$ and the constant is changed from 1.67 to 1.11.

Equation (11-40) is one form of the Lévêque solution (Knudsen and Katz, 1958) for mass transfer between a fixed surface and a semi-infinite fluid which flows over the surface with a well-established linear velocity gradient near the surface and zero velocity at the surface:

$$\frac{k_c x}{D_{\mathrm{AB}}} = 0.538 \left(\frac{a x^2}{D_{\mathrm{AB}}} \right)^{1/3} \tag{11-41}$$

Here a is the slope of the linear velocity profile ($a = du/dy$, where y is distance from the fixed surface). Equation (11-41) gives the local coefficient. The average coefficient $k_{c,\,\mathrm{av}}$ between the start of the mass-transfer process and x is $\frac{3}{2}$ times the value given by Eq. (11-41), changing the constant to 0.807 if $k_{c,\,\mathrm{av}}$ replaces k_c.

For large values of L/d_p (far downstream) Eq. (11-39) shows that the Sherwood group asymptotically reaches a lower limit of 3.65, provided the logarithmic-mean driving force is used. This particular asymptotic value of Sh is specific to the circular-tube geometry and the boundary condition of constant wall concentration along the tube. Limiting values of Sh for triangular passages and for flow between parallel plates, as well as for other boundary conditions, e.g., constant wall flux rather than constant wall concentration and/or some of the surfaces insulated or inactive for transfer, are given by Rohsenow and Choi (1961), Gröber et al. (1961), and Knudsen and Katz (1958). The limiting value of the Sherwood number is the same as the limiting value of the Nusselt number in these references, which consider heat transfer.

For laminar flow near the leading edge of a sharp flat plate, with flow parallel to the plane of the plate, the use of laminar-boundary-layer theory (Schlichting, 1960; Sherwood et al., 1975) leads to the following expression for the local k_c at any distance from the leading edge of the plate, assuming that the plate surface has constant solute concentration:

$$\frac{k_c x}{D_{AB}} = 0.332 \left(\frac{u\rho x}{\mu}\right)^{1/2} \left(\frac{\mu}{\rho D_{AB}}\right)^{1/3} \tag{11-42}$$

Because of the $-\frac{1}{2}$-power dependence of k_c upon distance, the average value of k_c is twice the local, changing the constant to 0.664. The physical situation pictured here is different from that in the Lévêque model, in that Eq. (11-42) is based on a uniform flow upstream from the plate which causes a developing (and changing) velocity profile along the plate. Equation (11-41) is based upon a linear velocity gradient which is already fully developed when mass transfer begins. For both Eqs. (11-41) and (11-42) it is appropriate to use a driving force based upon the difference between the surface concentration and the initial fluid concentration.

Equation (11-42) will also apply to a situation where the bulk flow is turbulent as long as the boundary layer remains laminar, which will occur for $u\rho x/\mu$ up to about 300,000.

Beek and Bakker (1961) and Byers and King (1967) have considered mass transfer in situations where there is a finite surface velocity and either an established velocity gradient or a developing boundary layer, as can occur in laminar contacting of immiscible fluids.

Turbulent mass transfer to surfaces If pressure drop is due to skin friction rather than form drag (Bird et al., 1960, p. 59), measurements of pressure drop can, in principle, be converted into heat-transfer and mass-transfer coefficients through appropriate analogies based upon Newton's, Fourier's, and Fick's laws. For turbulent systems, especially for flow in circular tubes and over a flat surface, various efforts have been made to produce such analogies by assuming that an eddy diffusivity for turbulent transport is additive with the molecular diffusivity (Sherwood et al., 1975, pp. 156–171). The eddy diffusivity is assumed to vary in some specified way with distance from a fixed or free surface across which mass transfer occurs.

Probably the most successful of the analogies is one that is entirely empirical, based upon qualitative knowledge of the various phenomena involved. This is the *Chilton-Colburn analogy* (Chilton and Colburn, 1934), which states that

$$j_D = j_H = \frac{f}{2} \tag{11-43}$$

where

$$j_D = \frac{k_{c, \text{av}}}{u}\left(\frac{\mu}{\rho D_{AB}}\right)^{2/3}$$ (11-44)

$$j_H = \frac{h}{C_p \rho u}\left(\frac{C_p \mu}{k}\right)^{2/3}$$ (11-45)

and f is the Fanning friction factor (see, for example, Perry and Chilton, 1973). The probable genesis of this analogy is discussed by Sherwood et al. (1975, p. 167).

As one use of the Chilton-Colburn analogy, k_c for mass transfer between the wall and a fluid in turbulent flow through a smooth tube can be calculated from the friction factor for flow in smooth tubes. Alternatively, one can use the $j_H = j_D$ portion of the analogy to convert the Colburn (1933) equation for heat transfer into the mass-transfer form:

$$\frac{k_c d_p}{D_{AB}} = 0.023\left(\frac{d_p u \rho}{\mu}\right)^{0.8}\left(\frac{\mu}{\rho D_{AB}}\right)^{1/3}$$ (11-46)

Except very near the entrance, k_c for turbulent flow is independent of downstream distance. Similar approaches are possible for turbulent flow over a flat surface (Sherwood et al., 1975, pp. 201–203).

The $\frac{2}{3}$ exponent on the Schmidt and Prandtl numbers in Eqs. (11-44) and (11-45) matches the observed effects of k and D_{AB} in a large number of cases.

A more general version of the same approach to correlating mass transfer between turbulent fluids and interfaces has been taken by Calderbank and Moo-Young (1961), who correlate k_c in terms of the energy dissipation per unit volume and per unit time E_v:

$$k_c\left(\frac{\mu}{\rho D_{AB}}\right)^{2/3} = 0.13\left(\frac{E_v \mu}{\rho^2}\right)^{1/4}$$ (11-47)

Here E_v would have units such as joules per cubic meter and per second. Equation (11-47) fits data for mass transfer to interfaces in such diverse situations as turbulent flow in tubes, agitated vessels with suspended solids, and flow through packed beds (Calderbank and Moo-Young, 1961). However, it should not be regarded as more than a first approximation for k_c.

Packed beds of solids Sherwood et al. (1975, pp. 242–245) have reviewed data for mass-transfer and heat-transfer coefficients between a flowing fluid and beds of particles (mostly spherical or cylindrical). They recommend the following equation for values of the Reynolds number between 10 and 2500:

$$j_D = 1.17\left(\frac{d_s u_s \rho}{\mu}\right)^{-0.415}$$ (11-48)

where d_s is the diameter of a sphere having the same surface area as the particle and u_s is the superficial velocity of the fluid, defined as the velocity for the same flow rate in an empty bed. The data leading to Eq. (11-48) show appreciable scatter and are mostly for beds in which the void fraction ϵ is about 0.42. Based upon other results for beds with different void fractions, one approach for allowing for changes in void fraction is to include a factor of $0.58/(1 - \epsilon)$ in the Reynolds number and to include a factor of $\epsilon/0.42$ in the definition of j_D in Eq. (11-48).

Simultaneous chemical reaction The effects of a simultaneous chemical reaction on a mass-transfer process are discussed by Sherwood et al. (1975, chap. 8), Astarita (1966), and Danckwerts (1970), among others. The chemical reaction can alter both the concentration-difference driving force and the mass-transfer coefficient. For an absorption process, a mass-transfer coefficient based upon the physical (unreacted) solubility of a solute will be *increased* by a chemical reaction occurring near the interface unless the reaction rate is very small. A mass-transfer coefficient based upon the full solubility (long-time equilibrium measurement) will be reduced by the chemical reaction unless the reaction is very fast.

Interfacial Area

To obtain a rate of mass transfer per unit time it is necessary to multiply the flux N_A by the interfacial area over which the mass transfer occurs. For mass transfer between a solid surface and a fluid the interfacial area is usually easily determined from the geometry of the system, but for most instances of mass transfer between a liquid and a gas or between immiscible liquids the interface is highly mobile, often broken apart in a dispersion, and may be constantly disappearing and reforming. In such cases it is necessary to develop some sort of correlation for the interfacial area itself. Often the problems of correlating the interfacial area and correlating the mass-transfer coefficient are combined, the product $k_c a$ being correlated, where a is the interfacial area per unit equipment volume in units such as m^{-1}.

Examples of common contacting situations in separation processes where the interfacial area is difficult to determine are packed columns for absorption, stripping, and distillation; agitated or sparged vessels contacting a liquid with either a gas or another liquid; and various extraction devices. Methods of correlating k_c and a or $k_c a$ for these devices are covered in various sections of Perry and Chilton (1973), as well as in certain specialized references, e.g., Valentin (1968). Analysis of mass-transfer rates in the gas-liquid dispersion on plates in plate columns is considered in Chap. 12.

Effects of High Flux and High Solute Concentration

We have so far considered mass-transfer coefficients for dilute solutes and with low N_B, where the convective flux in Eqs. (11-5) or (11-8) is not important. Two effects frequently enter to cause the convective flux to be important, namely, *high solute concentration* and *high flux*. High flux can also be viewed as a large difference in solute concentration across the mass-transfer zone. The two effects often occur together, but we shall first identify situations where each is present separately.

High solute concentration without high flux occurs when the solute concentration is large but the difference in solute concentrations is small. If we continue to assume that N_B is not significant, Eqs. (11-5) and (11-8) for this case become

$$N_A(1 - x_A) = -cD_{AB} \nabla x_A \qquad (11\text{-}49)$$

and

$$N_A(1 - c_A \bar{V}_A) = -D_{AB} \nabla c_A \qquad (11\text{-}50)$$

Since the concentration is so nearly uniform, x_A and c_A on the left-hand sides of these equations may be taken to be constant. Therefore the flux N_A predicted by any of the analyses or correlations for mass transfer in dilute systems can be corrected for the high concentration level by dividing it by $1 - x_A$ [Eq. (11-5)] or $1 - c_A \bar{V}_A$ [Eq. (11-8)]. $c_A \bar{V}_A$ is the *volume fraction* of component A. Hence the two correction factors equal the mole fraction and the volume fraction of component B.

This analysis indicates that as long as $N_B \to 0$ high solute concentration with a low concentration difference serves to increase the flux per unit concentration-difference driving force in inverse proportion to the mole or volume fraction of the nondiffusing substance. In the limit of a pure substance (x_B or $c_B \bar{V}_B \to 0$), no concentration difference is required if A moves.

The other extreme is the sort of situation presented in connection with Fig. 11-1, where N_B was large and N_A was small in a reverse-osmosis process. If c_A is small, this is a case of low solute concentration but high flux (from N_B), which converts Eqs. (11-5) and (11-8) into

$$N_A - x_A N_B = -cD_{AB} \nabla x_A \qquad (11\text{-}51)$$

and

$$N_A - c_A \bar{V}_B N_B = -D_{AB} \nabla c_A \qquad (11\text{-}52)$$

Here it is apparent that the high flux will serve to distort the solute concentration profile because the gradient term is now equal to a quantity on the left-hand side that is very different from N_A in absolute magnitude. Distortion of the concentration profile will alter the concentration gradient at the interface and thereby alter the mass-transfer coefficient. In particular, it turns out that a high flux of mass *into* a phase serves to reduce the solute concentration gradient and thereby reduce the diffusive flux, whereas a high mass flux *out of* a phase has the opposite effect.

Before considering the general case, it is useful to point out again that there is another special case where the k_c values from dilute-solution relationships can be used directly. That is the case of equal-molar or equal-volume counterdiffusion ($N_A = -N_B$ or $\bar{V}_A N_A = -\bar{V}_B N_B$) where the convective terms in Eqs. (11-5) and (11-8), respectively, become zero. This case is often a good approximation; recall, for example, that in distillation it is often assumed that $N_A = -N_B$ in connection with the assumption of equimolal overflow.

There are two definitions of the mass-transfer coefficient in use for the general case where there can be high solute concentration and/or high flux. One of these (see, for example, Sherwood, et al., 1975) retains the definitions of Eqs. (11-18) to (11-21) making the coefficient the ratio of N_A to the concentration difference. We shall call such coefficients k_{yc}, k_{yx}, etc. The other definition, used by Bird et al. (1960, chap. 21) defines the coefficient as the ratio of J_A^* or J_A^y to the concentration difference, substituting J_A^* for N_A in Eqs. (11-18) and (11-19), following Eq. (11-1), and substituting J_A^V for N_A in Eqs. (11-20) and (11-21), following Eq. (11-2). We shall call such coefficients k_{Jc}, k_{Jx}, etc. These coefficients reflect the diffusive flux only and are therefore simpler to relate to high-concentration and high-flux effects in most cases. The definitions of the k_J coefficients thereby become

$$N_A = k_{Jx}(x_{Ai} - x_{AL}) + x_{Ai}\left(\sum_j N_j\right) \tag{11-53}$$

$$N_A = k_{Jc}(c_{Ai} - c_{AL}) + c_{Ai}\left(\sum_j \bar{V}_j N_j\right) \tag{11-54}$$

and so forth. The summations of fluxes in the second terms on the right-hand sides extend the definition to multicomponent as well as binary systems. With the definitions given by Eqs. (11-53) and (11-54) and similar forms, it now becomes very important to keep track of the signs of the concentration differences and fluxes. The convention adopted here is to make N_A positive if the net flux of component A is in the direction from the interface to the bulk.

Bird et al. (1960, chap. 21) have summarized correction factors to be used in converting values of k_x, k_c, etc., for dilute systems into values of k_{Jx}, k_{Jc}, etc., for systems with high flux. They are derived from the film and penetration models and the laminar-boundary-layer theory for flow over the leading edge of a flat plate. The solutions are expressed in terms of three dimensionless groups for the analyses based upon Eqs. (11-5) and (11-53):

$$\theta_A = \frac{k_{Jx}}{k_x} \quad \text{for component A} \tag{11-55}$$

$$\phi_A = \frac{\sum_j N_j}{k_x} \tag{11-56}$$

$$R_A = \frac{x_{Ai} - x_{AL}}{\left(N_A \Big/ \sum_j N_j\right) - x_{Ai}} \tag{11-57}$$

Only two of these groups are independent, since by Eq. (11-53) $\theta_A = \phi_A/R_A$.

The solutions for the various models are plotted in Figs. 11-5 and 11-6. (For the boundary-layer model, ϕ_A in Fig. 11-5 can be obtained as $\phi_A = \theta_A R_A$ from Fig. 11-6.) Also included is the result obtained by Clark and King (1970) for the Levêque model with $R_A > 1$, which is very close to the penetration solution. The results for the laminar-boundary-layer model also depend upon the Schmidt number ($Sc = \mu/\rho D_{AB}$), but the results for the film, penetration, and Levêque models are independent of Schmidt number.

By direct analogy, these dimensionless groups and the solutions shown in Figs. 11-5 and 11-6 can be extended to analyses based upon Eqs. (11-8) and (11-54)

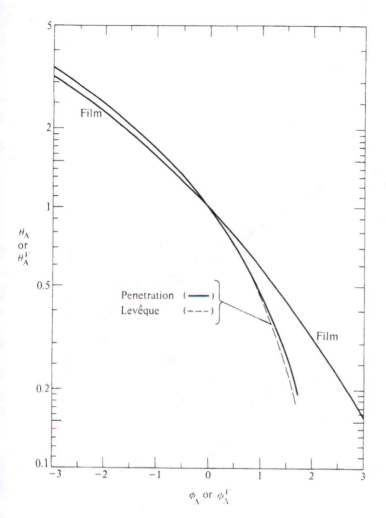

Figure 11-5 High-flux correction factors for mass-transfer coefficients. (*Adapted from Bird et al., 1960, p. 675; used by permission.*)

by redefining the dimensionless groups as

$$\theta_A^V = \frac{k_{Jc}}{k_c} \tag{11-58}$$

$$\phi_A^V = \frac{\Sigma \bar{V}_j N_j}{k_c} \tag{11-59}$$

$$R_A^V = \frac{\bar{V}_A c_{Ai} - \bar{V}_A c_{AL}}{(\bar{V}_A N_A / \Sigma \bar{V}_j N_j) - \bar{V}_A c_{Ai}} \tag{11-60}$$

Notice that θ_A is greater than unity for a net flux out of the stream; this reflects the effect of the flux in increasing the concentration gradient at the interface. Con-

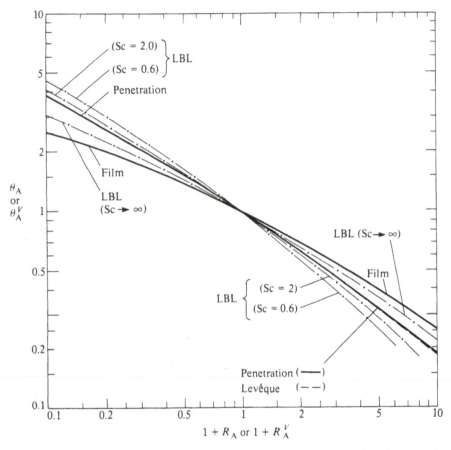

Figure 11-6 High-flux correction factors for mass-transfer coefficients. (*Adapted from Bird et al., 1960, p. 675; used by permission.*)

versely, θ_A is less than unity for a net flux into the stream, reflecting the effect of the flux in decreasing the interfacial concentration gradient in such a case.

No solution is presented for cases of mass transfer between an interface and a turbulent liquid. For such a case the film-model result is the best prediction because of the approach of a turbulent-flow situation to the conditions postulated by the film model.

The solution for the film model is also given by simple analytical relationships

$$\theta_A = \frac{\phi_A}{e^{\phi_A} - 1} \tag{11-61}$$

and

$$\theta_A = \frac{\ln (R + 1)}{R} \tag{11-62}$$

For the film model the expression for k_{Nx}, k_{Nc}, etc., is also obtainable analytically (Wilke, 1950) and is

$$k_{Nx} = \frac{k_x}{x_{Af}} \tag{11-63}$$

$$k_{Nc} = \frac{k_c}{(\bar{V}_A c_A)_f} \tag{11-64}$$

etc., where x_{Af}, $(\bar{V}_A c_A)_f$, etc., are called *film factors* and defined by

$$x_{Af} = \frac{(1 - t_A x_{AL}) - (1 - t_A x_{Ai})}{\ln\left[(1 - t_A x_{AL})/(1 - t_A x_{Ai})\right]} \tag{11-65}$$

$$(\bar{V}_A c_A)_f = \frac{(1 - t_A^V \bar{V}_A c_{AL}) - (1 - t_A^V \bar{V}_A c_{Ai})}{\ln\left[(1 - t_A^V \bar{V}_A c_{AL})/(1 - t_A^V \bar{V}_A c_{Ai})\right]} \tag{11-66}$$

where $t_A = \sum_j N_j/N_A$ and $t_A^V = \sum_j \bar{V}_j N_j/\bar{V}_A N_A$. For the special case of $N_{j,\,j \neq A} = 0$, $t_A = 1$ and x_{Af} is $(1 - x_A)_{LM}$, where the subscript LM represents the logarithmic mean between interface and bulk conditions. Similarly, for the special case of $\bar{V}_j N_{j,\,j \neq A} = 0$, $t_A^V = 1$ and $(\bar{V}_A c_A)_f$ is $(1 - \bar{V}_A c_A)_{LM}$. These conditions would be expected to hold for many absorption and stripping operations, and one therefore often sees the term y_{BM}^{-1} or P/p_{BM} in correlations for gas-phase mass-transfer coefficients (k_{Ny} and k_{NG}) for absorbers. These refer to the reciprocal logarithmic-mean mole fraction or partial pressure of the nontransferring component of the gas B. The expression is unique to the film model and the case where only component A transfers across the interface. The solutions for k_N from other models, e.g., the penetration model (Bird et al., 1960, pp. 594–598), are much more complex.

The solutions presented in Figs. 11-5 and 11-6 are exact for binary systems and represent good approximations in most cases in multicomponent systems. Major exceptions occur in multicomponent systems, however. For example, for dilute components the appropriate diffusivity to use in a correlation for k_x, k_c, etc., can vary substantially and can be difficult to determine; it can even become negative in some cases (Cussler, 1976).

Reverse osmosis As indicated in Fig. 1-29 and the surrounding discussion, desalination of water by reverse osmosis requires that the feedwater be put under a pressure which exceeds the osmotic pressure of the feed, thereby creating a difference in chemical potential of water and causing it to flow through the membrane. As already discussed in Chap. 1, the fluxes of water and salt through the membrane are usually described by

$$N_W = k_w(\Delta P - \Delta \pi) \tag{1-22}$$

$$N_S = k_S(C_{S1} - C_{S2}) \tag{1-23}$$

where ΔP is the difference in total pressure across the membrane and $\Delta \pi$ is the difference in osmotic pressure (both computed as feed side minus product side) and C_{S1} and C_{S2} are the salt concentrations on the high- and low-pressure sides of the

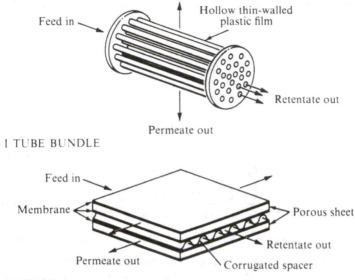

Feed in

Hollow thin-walled
plastic film

Retentate out

Permeate out

I TUBE BUNDLE

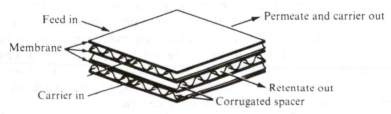

Feed in

Membrane

Porous sheet

Retentate out

Permeate out

Corrugated spacer

II STACK

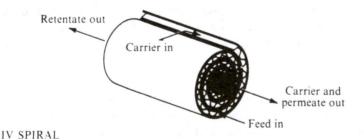

Feed in

Permeate and carrier out

Membrane

Carrier in

Retentate out

Corrugated spacer

III BI-FLOW STACK

Retentate out

Carrier in

Carrier and
permeate out

Feed in

IV SPIRAL

Figure 11-7 Membrane-permeation designs (retentate = high-pressure product; permeate = low-pressure product). *(Amicon Co., Lexington, Massachusetts.)*

membrane. respectively. k_w and k_S are empirically determined proportionality constants, usually taken to be independent of concentration and flux levels.

Figure 11-1 depicts the transport processes occurring in reverse osmosis used for desalination of water. The buildup of salt concentration near the membrane because of preferential passage of water through the membrane, known as *concentration polarization*, has two deleterious effects: (1) the increase in C_{S1} serves to increase the driving force for salt transport through the membrane in Eq. (1-23) and thereby engender more salt leakage into the product water, and (2) the increase in C_{S1} increases $\Delta\pi$ in Eq. (1-22) and thereby necessitates a greater applied total pressure to produce a given water flux across the membrane. π increases in direct proportion to C_S if salt activity coefficients do not change.

There have been two primary goals in the design of reverse-osmosis equipment: (1) incorporation of a large amount of membrane area per unit equipment volume, to increase the amount of water-product flow per unit volume, and (2) provision of thin channels and high-velocity flow to increase k_L for salt transport back into the bulk liquid from the membrane surface and reduce the increase of C_{S1} above C_{SL} (Fig. 11-1). Figure 11-7 shows some of the flow configurations used for reverse osmosis, ultrafiltration, and dialysis.

Example 11-4 For seawater, the osmotic pressure is 2.5 MPa. The principal solute is NaCl, for which the diffusivity in water at 18.5°C is about 1.2×10^{-9} m²/s (Reid et al., 1977). Assume that a reverse-osmosis desalting process is carried out using turbulent flow through a tubular 1.0-cm-diameter membrane with a system temperature of 18.5°C. Only a small fraction of the feedwater is taken as freshwater product, so that the bulk concentration does not change significantly through the process. Seawater contains 3.5 weight percent dissolved salts, which for the purposes of this problem can be considered to be entirely NaCl. The feedwater flows inside the membrane tube at a velocity of 1 m/s. The volumetric flux of water through the membrane is 3×10^{-5} m/s (m³ s·m²), and the applied feed pressure is 8.0 MPa greater than the product-water pressure. The membrane is highly selective for water over salt. (a) Calculate the percent increase in salt concentration in the product water, referred to the hypothetical case where there is no concentration polarization and the water flux is the same. (b) Calculate the percentage reduction in feed pressure that would be possible in the absence of concentration polarization if the water flux were the same. (c) Which of the following factors would be effective in reducing the degree of concentration polarization if the water flux is held constant: (1) reduced temperature; (2) reduced tube diameter with the same mass flow rate of seawater; and (3) recirculation of the seawater with the same tube size and length?

SOLUTION Since this is a liquid solution where salt and water will have unequal partial molal volumes, it is preferable to use equations based upon Eq. (11-2) rather than Eq. (11-1). Because k_c for salt will be influenced by the high flux of water relative to salt, it is appropriate to use Eq. (11-54) to describe the flux of salt (A = salt, B = water). Taking N_A to be very small in comparison with each of the two right-hand terms and taking $|N_B| \gg |N_A|$ gives

$$k_{Jc}(c_{Ai} - c_{AL}) = -c_{Ai}\bar{V}_B N_B \qquad (11-67)$$

N_B is negative since the direction of the water flux is from the liquid bulk toward the interface. $\bar{V}_B N_B$ is a volumetric flux, which from the problem statement is equal to -3×10^{-5} m³/m²·s. c_{AL} is obtained by converting the weight percent salt to molar-concentration units, using a solution density of 1020 kg m³ (Perry and Chilton, 1973)

$$c_{AL} = \frac{(1020 \text{ kg soln m}^3)(3.5 \text{ kg NaCl } 100 \text{ kg soln})}{0.05848 \text{ kg NaCl mol NaCl}}$$

$$= 6.1 \times 10^2 \text{ mol/m}^3$$

For the limit of very low water flux k_c can be obtained from Eq. (11-46). The viscosity of seawater will be taken, as an approximation, to be equal to that of pure water, which is 1.00 mPa·s at 18.5°C (Perry and Chilton, 1973). Hence

$$\text{Re} = \frac{d_p u \rho}{\mu} = \frac{(0.010 \text{ m})(1.0 \text{ m s})(1020 \text{ kg m}^3)}{10^{-3} \text{ Pa·s}} = 10,200$$

(This is within the turbulent region.)

$$Sc = \frac{\mu}{\rho D_{AB}} = \frac{10^{-3}\ Pa\cdot s}{(1020\ kg/m^3)(1.2 \times 10^{-9}\ m^2/s)} = 817$$

$$k_c = \frac{(1.2 \times 10^{-9}\ m^2/s)}{(0.01\ m)}(0.023)(10{,}200)^{0.8}(817)^{1\ 3} = 4.15 \times 10^{-5}\ m/s$$

Since we know $\bar{V}_B N_B$ and k_c, k_{Jc} is obtained from Fig. 11-5 or Eq. (11-61). (The film model is appropriate for the correction factor. since this is a turbulent-flow situation.)

$$\phi_A^V = \frac{\bar{V}_B N_B}{k_c} = \frac{-3.0 \times 10^{-5}\ m/s}{4.15 \times 10^{-5}\ m/s} = -0.723$$

$$\theta_A^V = \frac{-0.723}{e^{-0.723} - 1} = 1.40$$

$$k_{Jc} = \theta_A^V k_c = (1.40)(4.15 \times 10^{-5}\ m/s) = 5.81 \times 10^{-5}\ m/s$$

We can now substitute into Eq. (11-67) to determine c_{Ai}:

$$(5.81 \times 10^{-5})(c_{Ai} - 6.1 \times 10^2) = -c_{Ai}(-3.0 \times 10^{-5})$$

$$c_{Ai} = 1.26 \times 10^3\ mol/m^3$$

The salt concentration adjacent to the membrane is about twice that in the bulk solution.

(a) If we assume that the salt concentration in the product is very low compared with that in seawater. the ratio of concentration-difference driving forces in Eq. (1-23) is simply the ratio of values of C_{S1} ($= c_{Ai}$). Hence the salt concentration in the product will increase by a factor of $(1.26 \times 10^3)/(6.1 \times 10^2) = 2.07$.

(b) Taking osmotic pressure directly proportional to salt concentration, we find that the concentration-polarization effect must raise the osmotic pressure from 2.5 to $(2.07)(2.5) = 5.18$ MPa. Hence the driving force for water transport in Eq. (1-22) is $8.0 - 5.18 = 2.72$ MPa. In the absence of concentration polarization this same driving force would be required to produce the same flux, by Eq. (1-22). Hence the difference in total pressure between feed and product should be $2.72 + 2.5 = 5.22$ MPa. This is 35 percent less than the 8.0 MPa required in the presence of concentration polarization.

(c) Factors that raise k_c and hence k_{Jc} will serve to reduce the degree of concentration polarization. Lower temperature increases the viscosity and lowers the diffusivity. k_c varies as $\mu^{-0.47}$ and $D_{AB}^{2.3}$; hence lower temperature produces a lower k_c and makes concentration polarization more severe. *Higher* temperature would alleviate concentration polarization.

Reducing tube diameter with the same mass flow rate of water will raise u by a factor equal to the square of that by which d_p was reduced. with the result that the Reynolds number increases. Also d_p in the Sherwood number will make k_c increase as d_p is reduced. Both these effects increase k_c, and so concentration polarization will be reduced.

Recirculation of the seawater will increase u while leaving other factors unchanged. This increases Re and hence k_c; it therefore alleviates concentration polarization but does so at the expense of much more pumping power (more flow and more pressure drop). $\square$

INTERPHASE MASS TRANSFER

Consider, for example, a process in which a substance A is being absorbed from a gas stream into a liquid solvent.† In order for component A to travel across the interface from the gas phase to the liquid phase by a diffusional mass-transfer mechanism

---

† For convenience the following discussion will be conducted for gas-liquid contacting; however, there is a logical extension to liquid-liquid and fluid-solid systems.

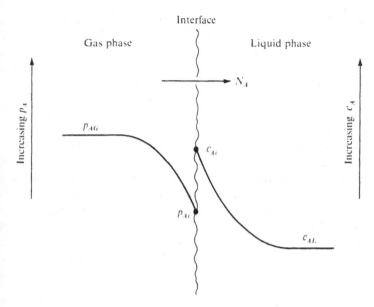

Figure 11-8 Concentration and partial-pressure gradients in interface gas-liquid mass transfer.

there must be a gradient in concentration of component A in *both* phases adjacent to the interface. This situation is shown schematically in Fig. 11-8. The partial pressure of component A in the gas and the concentration of component A in the liquid at the interface (p_{Ai} and c_{Ai}) will be in equilibrium with each other unless there are extraordinarily high rates of mass transfer. Since component A is traveling from gas to liquid, p_{AG}, the bulk-gas partial pressure of A will be higher than p_{Ai} and c_{Ai} will be higher than c_{AL}, the bulk-liquid concentration of A. Diffusional transfer of a component in a binary mixture within a phase must occur in the direction of decreasing partial pressure or concentration of that component.

The hydrodynamic conditions within each phase and the solute diffusivities combine to give certain rate coefficients for mass transfer within the two phases. These are the individual-phase mass-transfer coefficients k_x, k_y, k_G, and/or k_c defined by Eqs. (11-18) to (11-21). For situations of high concentration and/or high flux, they may be either the k_J or the k_N coefficients. In any event, there are individual-phase coefficients for both phases, which relate the flux of component A to the difference between interfacial and bulk compositions in either phase.

Since generally the interfacial partial pressure and concentration of A are not readily measurable in a separation device, it is usually more convenient to define and work in terms of *overall* mass-transfer coefficients, defined for dilute systems as

$$N_A = K_G(p_{AG} - p_{AE}) \tag{11-68}$$

$$N_A = K_L(c_{AE} - c_{AL}) \tag{11-69}$$

Here p_{AE} represents the gas-phase partial pressure of A which would be in equilibrium with the prevailing concentration of A in the bulk liquid c_{AL}, and c_{AE} represents the liquid-phase concentration of A which would be in equilibrium with the prevailing partial pressure of A in the bulk gas p_{AG}. For our example of A transferring from gas to liquid, c_{AL} is necessarily less than c_{Ai}; hence it follows that p_{AE} is necessarily less than p_{Ai}, assuming that increasing concentration of A gives increasing equilibrium partial pressure of A. As a result, the partial-pressure-difference driving force in Eq. (11-68) is greater than that in Eq. (11-20), and K_G is necessarily less than k_G. Similarly K_L in Eq. (11-69) is necessarily less than k_c in Eq. (11-21).

The overall coefficients comprise contributions from both of the individual phase coefficients (k_G and k_L†), and the individual phase coefficients are related to the hydrodynamic conditions and solute diffusivities in their respective phases. It is the overall coefficients which are most readily used in the design and analysis of separation devices, but it is the individual phase coefficients for which correlations against hydrodynamic conditions and diffusivities are best made. As a result, it is necessary to use equations or graphical relationships for predicting k_G and k_L, and then to obtain K_G or K_L from these individual phase coefficients. The equations relating K_G, K_L, k_G, and k_L can be obtained by linearizing the equilibrium relationship to the form

$$p_A = HC_A + b \qquad (11\text{-}70)$$

and then by combining Eqs. (11-20), (11-21), and (11-68) to (11-70) to give

$$\frac{1}{K_G} = \frac{1}{k_G} + \frac{H}{k_L} \qquad (11\text{-}71)$$

and

$$\frac{1}{K_L} = \frac{1}{HK_G} = \frac{1}{Hk_G} + \frac{1}{k_L} \qquad (11\text{-}72)$$

From Eqs. (11-71) and (11-72) it can be seen that when H is very large, that is, when A is a relatively insoluble component in the liquid, the term H/k_L will outweigh $1/k_G$ and as a result K_G will very nearly equal k_L/H and K_L will very nearly equal k_L. In this case we say the mass-transfer process is *liquid-phase-controlled*, since the individual liquid-phase mass-transfer coefficient affects the mass-transfer rate directly, whereas the mass-transfer rate is essentially independent of the value of k_G. In the converse situation of a very low H (high solubility of A in the liquid), we find that K_L very nearly equals Hk_G, and K_G very nearly equals k_G. Here we have a *gas-phase-controlled* mass-transfer process, wherein the mass-transfer rate is directly proportional to k_G but is essentially independent of k_L. For a fully liquid-phase-controlled process we need only ascertain k_L and equate it directly to K_L, and for a fully gas-phase-controlled process we need only ascertain k_G and equate it directly to K_G.

Equations (11-71) and (11-72) are frequently called the *addition-of-resistances equations* because of the similarity to the equation for compounding resistances in series in an electric circuit. A similar relationship holds for relating overall heat-transfer coefficients to individual-phase heat-transfer coefficients. However, an important distinction in the mass-transfer case is the presence of the equilibrium

---

† We shall use k_L to represent k_c in a liquid system.

solubility, which can change greatly from one solute to another. It is more often the solute itself than the flow conditions which determine whether a mass-transfer system is gas-phase- or liquid-phase-controlled.

If mole-fraction driving forces are used for the two phases, the coefficients k_y and k_x are used for systems with dilute solutes and Eqs. (11-71) and (11-72) become

$$\frac{1}{K_y} = \frac{1}{k_y} + \frac{K'}{k_x} \tag{11-73}$$

and

$$\frac{1}{K_x} = \frac{1}{K'K_y} = \frac{1}{K'k_y} + \frac{1}{k_x} \tag{11-74}$$

where K' is dy_A/dx_A at equilibrium. For a system with constant K_A $(= y_A/x_A$ at equilibrium), $K' = K$, but for a system with variable K_A the two are not, in general, equal. K_y and K_x are overall coefficients defined by

$$N_A = K_y(y_{AG} - y_{AE}) \tag{11-75}$$

and

$$N_A = K_x(x_{AE} - x_{AL}) \tag{11-76}$$

where the subscript E has the same meaning as before.

When there are high-flux and/or high-solute-concentration effects, the procedure to be used for compounding individual-phase resistances depends upon whether the k_J or k_N coefficients are used. For k_N coefficients the definitions lead directly to the same equations as used for dilute systems at low flux [Eqs. (11-71) to (11-74)]; however, the individual-phase coefficients themselves must be corrected for high flux and/or high concentration. As noted earlier, this is straightforward only for the case of high concentration without high flux or when corrections are made by the film model.

If k_J coefficients are used, the individual-phase coefficients are corrected using the relationship between θ_A and ϕ_A or R_A (or θ_A^V and ϕ_A^V or R_A^V) given by Figs. 11-5 and 11-6 [or Eqs. (11-61) and (11-62) for the film model], with due attention to maintaining the proper signs in Eqs. (11-53) and (11-54). The convention employed by Bird et al. (1960, chap. 21) is to define the overall coefficients as

$$N_A = K_{Jx}(x_{AE} - x_{AL}) + x_{AE}\left(\sum_j N_j\right) \tag{11-77}$$

$$N_A = K_{Jc}(c_{AE} - c_{AL}) + c_{AE}\left(\sum_j \bar{V}_j N_j\right) \tag{11-78}$$

etc. Algebraic combination of these equations with Eqs. (11-53), (11-54), etc., for the case of all N_j except N_A being zero leads to

$$\frac{1 - x_{AE}}{K_{Jx}} = \frac{1 - x_{Ai}}{k_{Jx}} + \frac{1 - y_{Ai}}{K'k_{Jy}} \tag{11-79}$$

$$\frac{1 - \bar{V}_A c_{AE}}{K_{Jc}} = \frac{1 - \bar{V}_A c_{Ai}}{k_{Jc}} + \frac{1 - P_{Ai}/P}{H k_{JG}} \tag{11-80}$$

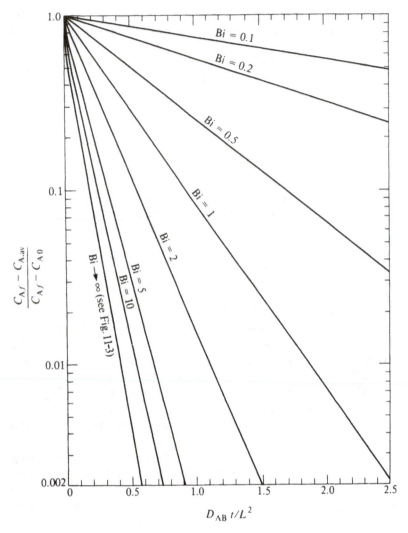

Figure 11-9 Transient diffusion in a sphere with a mass-transfer resistance in the surrounding phase.

etc. When N_j for other components is nonzero, the numerators in the various terms in Eqs. (11-79) and (11-80) become $1 - (\Sigma N_j/N_A)x_{AE}$, etc. Since the addition equation itself involves both N_j's and interfacial compositions, it is more difficult to obtain overall coefficients in high-flux and high-concentration situations.

Transient Diffusion

The solutions of the diffusion equation for transient diffusion in a stagnant medium, given in Fig. 11-3, were based upon the assumption of constant solute concentration at the surface. For an interphase mass-transfer situation, such as leaching from a spherical solid particle into a surrounding fluid medium, this implies that the mass-transfer process is completely controlled by resistance to diffusion within the solid without being influenced significantly by mass-transfer resistance in the surrounding medium. Solutions to the diffusion equation can also be obtained when the boundary condition $c_{Ai} = c_{Af} = $ cons at $t > 0$ is replaced by the boundary condition

$$K'' k_c (c_A - c_{Af}) = -D_{AB} \frac{\partial c_A}{\partial z} \qquad \text{at } z = \text{interface} \qquad (11\text{-}81)$$

Here z is taken positive outward in the solid medium. K'' is the equilibrium partition coefficient [$= (c_A$ in surrounding fluid)/(c_A in solid at equilibrium)]. c_{Af} is the value of c_A in the solid that would be in equilibrium with the prevailing value of c_A in the surrounding fluid, which is presumed not to change.

Figure 11-9 shows such solutions for a solid sphere with a mass-transfer resistance in the surrounding phase. The results are plotted for the average solute concentration within the sphere as a function of the Fourier group (Fo $= D_{AB} t/L^2$) and the Biot group (Bi $= K'' k_c L/D_{AB}$), where D_{AB} is the diffusivity within the sphere, k_c is the mass-transfer coefficient in the surrounding medium, L is the sphere radius, and K'' is the equilibrium partition coefficient, defined above. For short times the solution is taken by analogy from the corresponding heat-transfer solutions [Gröber et al. (1961, fig. 3.12) or a corresponding plot against Fo and Bi from Hsu (1963)] and for longer times from the first term of eq. 3.57a from Gröber et al. (1961), rearranged to give fraction remaining. Solutions for the slab and infinite-cylinder geometries are also given in those references.

It is apparent from Fig. 11-9 that the presence of the external resistance slows the diffusion process to an extent that is greater the larger the ratio of the external resistance to the internal diffusivity (Bi^{-1}). This is entirely analogous to the effect of a resistance in the second phase in the addition-of-resistances equations.

Example 11-5 Returning to Example 11-2, suppose that extractive decaffeination of coffee beans is carried out in a packed bed with a solvent which provides an equilibrium partition coefficient of 0.10 [$= $ (caffeine concentration in solvent)/(concentration of caffeine in beans at equilibrium)]. The solvent has a density of 800 kg/m^3 and a viscosity of 2.00 mPa · s and flows at a superficial velocity of 0.010 m/s. The bed height is low enough to prevent appreciable buildup of caffeine in the solvent phase. The diffusivity of caffeine in the solvent is 0.25×10^{-9} m^2/s. (a) By what factor does the presence of mass-transfer resistance in the solvent phase increase the time required to reduce the caffeine content of a bed of beans to 3.0 percent of the initial level? (b) What will be the influence on the ratio of internal resistance to external resistance (Bi) of (1) decreasing particle size by cutting the beans, (2) increasing the solvent flow rate, and (3) finding a solvent which gives a higher partition coefficient into the solvent phase without changing any other solvent properties? Consider each effect separately.

SOLUTION (a) The Biot number is needed for Fig. 11-9. From Example 11-2, D_{AB} inside the beans $= 1.77 \times 10^{-10}$ m^2/s and $L = 0.0030$ m. k_c comes from Eqs. (11-44) and (11-48), using the properties of the solvent phase. The beans are assumed to be spherical and to provide a bed voidage of about 0.42.

$$j_D = 1.17 \left[\frac{(0.0030)(0.010)(800)}{0.0020} \right]^{-0.415}$$

$$= 1.17(12)^{-0.415} = 0.417$$

$$k_c = j_D u_s \, \text{Sc}^{-2/3} = (0.417)(0.010) \left[\frac{0.0020}{(800)(0.25 \times 10^{-9})} \right]^{-2/3}$$

$$= (0.417)(0.010)(10,000)^{-2/3} = 8.98 \times 10^{-6} \text{ m s}$$

$$\text{Bi} = \frac{K'' k_c L}{(D_{AB})_{\text{bean}}} = \frac{(0.10)(8.98 \times 10^{-6})(0.0030)}{1.77 \times 10^{-10}} = 15.2$$

For high values of Bi in Fig. 11-9, we can interpolate by taking linear increments in Bi^{-1} between the curves shown. Using the values of $D_{AB} t / L^2$ for $\text{Bi}^{-1} = 0$ and $\text{Bi}^{-1} = 0.10$ at the desired percentage removal for the interpolation, we have

$$\frac{D_{AB} t}{L^2} = 0.305 + \frac{(15.2)^{-1}}{0.10} (0.402 - 0.305) = 0.369$$

which represents an increase of $(0.369 - 0.305)/0.305 = 21$ percent in the time required for decaffeination to a residual caffeine content equal to 3 percent of the original.

It should be noted that extreme values of several parameters had to be used in order to generate even this much contribution from the external resistance. Any effective solvent should give a higher value of K'', and a substantially higher solvent velocity would be likely. For the situation described, the bed height would have to be very low to avoid a significant buildup of caffeine concentration in the solvent, which would affect the driving force for mass transfer.

(b) Decreasing particle size will decrease Bi in proportion to L^1 through the effect of L directly in Bi and will increase Bi in proportion to $L^{-0.415}$ through the effect of j_D (and hence k_c) on Bi. The net effect is to decrease Bi, thereby making external resistance more important.

Increasing the solvent flow rate will decrease j_D in proportion to $u_s^{-0.415}$ but will increase k_c in proportion to $u_s^{1-0.415} = u_s^{0.585}$. This will therefore increase Bi and lessen the importance of external resistance.

Increasing K'' will increase Bi in direct proportion, thereby lessening the importance of external resistance. □

Combining the Mass-Transfer Coefficient with the Interfacial Area

In most contacting devices the interfacial area between phases is not easily ascertained, and it is the product $K_G a$ or $K_L a$ (where a is the interfacial area per unit volume of equipment) which is required for design. Equations (11-71) and (11-72) are often converted into the forms

$$\frac{1}{K_G a} = \frac{1}{(k_G a)^*} + \frac{H}{(k_L a)^*} \tag{11-82}$$

and

$$\frac{1}{K_L a} = \frac{1}{H(k_G a)^*} + \frac{1}{(k_L a)^*} \tag{11-83}$$

where $(k_G a)^*$ is the product of K_G and a obtained from a mass-transfer experiment in which liquid-phase resistance is either suppressed or absent and $(k_L a)^*$ is the product of K_L and a measured in a mass-transfer experiment in which gas-phase resistance is either suppressed or absent. Often $(k_G a)^*$ is obtained from measurements of evaporation rates of pure liquids, where only gas-phase mass-transfer resistance can be present, or from measurements of the rate of absorption of ammonia (a very soluble gas) into water. Frequently $(k_L a)^*$ is obtained from measurements of the rate of absorption of sparingly soluble gases, e.g., carbon dioxide and oxygen, into water from air.

In order for Eqs. (11-82) and (11-83) to be valid, a number of criteria must be met (King, 1964):

1. H must be a constant, or if it is not, the value of the equilibrium curve slope at the properly defined value of C_A must be used.
2. There must be no significant resistance present other than those represented by $(k_G a)^*$ and $(k_L a)^*$; for example, p_{Ai} must be in equilibrium with C_{Ai}.
3. The hydrodynamic conditions (interfacial area, etc.) for the case in which the resistances are to be combined must be the same as for the measurements of the individual phase resistances. Similarly, the solute diffusivities must be the same. These factors are usually taken into account through correlations for $(k_G a)^*$ and $(k_L a)^*$, usually expressed in terms of a dimensionless Sherwood group [for example, $(k_L a)^* d^2 / D$, where d is an important length variable and D is diffusivity] as a function of the Schmidt and Reynolds groups. In a number of instances the interfacial mass-transfer rate can be accelerated by interfacial mixing cells, which alter the hydrodynamic conditions and are dependent upon the size of the interfacial flux, the direction of mass transfer, and surface tensions.
4. The mass-transfer resistances of the two phases must not interact; i.e., the magnitude of k_L must not depend upon the magnitude of k_G or vice versa.
5. The ratio Hk_G / k_L must be constant at all points of interface.

These five criteria are violated to one extent or another in nearly all mass-transfer systems. In some cases, e.g., gas-liquid mass transfer in a stirred vessel (Goodgame and Sherwood, 1954), the effects seem to be unimportant or to cancel each other out, and Eqs. (11-82) and (11-83) hold well. In other cases, including packed and plate columns (King, 1964), these effects are more important, and Eqs. (11-82) and (11-83) do not work well when data for the vaporization of water and for the absorption or desorption of a relatively insoluble gas are used to predict absorption rates for a gas of intermediate solubility; they overpredict $K_G a$ and $K_L a$ because criterion 5 is severely violated.

Despite these shortcomings of the addition-of-resistances equations incorporating the interfacial area, our knowledge of the complicating factors is so slim that these equations are really the only tools available for design. They have been found to work best when data for absorption of a highly soluble gas, e.g., ammonia into water, are used rather than solvent vaporization data to predict $K_G a$ or $K_L a$ for a solute where the resistances of both phases are significant in Eqs. (11-82) and (11-83).

Example 11-6 To illustrate the effects of different factors on the degree of gas-phase or liquid-phase control in a gas-liquid contacting system, consider the following cases involving mass transfer between a gas and a liquid in an irrigated packed column. For desorption of oxygen from water into air at 25°C the column provides $K_L a = 0.0100$ s^{-1} at the flow conditions used. Measurements made for absorption of ammonia from air into water, corrected for liquid-phase resistance, give $(k_G a)^* = 0.00100$ mol s·m^3·Pa for the same flow conditions at 25°C and atmospheric pressure. For each case below, determine the fraction gas-phase control, expressed as

$$f_G = \frac{K_G a}{(k_G a)^*}$$

This is a measure of the relative importance of the terms in Eq. (11-82). The fraction liquid-phase control is

$$f_L = \frac{K_L a}{(k_L a)^*}$$

and can easily be shown to equal $1 - f_G$. The diffusivity of oxygen in water at 25°C is 2.4×10^{-9} m²/s, and that of ammonia in air at 25°C is 2.3×10^{-5} m²/s. All systems considered operate at 25°C and atmospheric pressure. (a) Absorption of ammonia from air into water at high dilution, for which the ammonia solubility is 0.77 mole fraction/atm (1 atm = 101.3 kPa) and the diffusivity of ammonia in water is 2.3×10^{-9} m²/s. (b) Absorption of carbon dioxide from air into water. The solubility of carbon dioxide can be obtained from Fig. 6-6. Diffusivities are 1.59×10^{-5} m²/s for CO_2 in air and 2.0×10^{-9} m²/s for CO_2 in water. (c) Absorption of CO_2 from air into a chemically reacting base which increases $k_L a$ for carbon dioxide by a factor of 100, referred to the unreacted solubility as a driving force. (d) Absorption of ethanol from a dilute mixture in air into a water solvent. Diffusivities are 1.18×10^{-5} m²/s for ethanol in air and 1.24×10^{-9} m²/s for ethanol in water. The solubility of ethanol vapor in water at high dilution and 25°C is 2.1 mole fraction/atm ("International Critical Tables").

SOLUTION Values of $(k_G a)^*$ and $(k_L a)^*$ must be corrected for changes in solute diffusivity. To do this, we shall invoke the penetration model [Eq. (11-28)] for the liquid phase and the j_D correlation [Eq. (11-44)] for the gas phase, giving $k_L \approx (D_{AB})_L^{0.5}$ and $k_G \approx (D_{AB})_G^{2/3}$.

For desorption of oxygen we take $(k_L a)^* = K_L a$, since the solubility of oxygen in water is very low.

(a) $$(k_L a)^* = (K_L a)_{O_2} \left(\frac{D_{NH_3}}{D_{O_2}} \right)^{0.5} = (0.0100) \left(\frac{2.3}{2.4} \right)^{0.5} = 0.0098 \text{ s}^{-1}$$

$$(k_G a)^* = (k_G a)_{NH_3}^* = 0.00100 \text{ mol/s} \cdot \text{m}^3 \cdot \text{Pa}$$

$$H = \frac{p_{NH_3}}{c_{NH_3}} = \frac{(1 \text{ atm})(1.013 \times 10^5 \text{ Pa/atm})(18 \times 10^{-6} \text{ m}^3/\text{mol } H_2O)}{0.77 \text{ mol } NH_3/\text{mol } H_2O}$$

$$= 2.37 \text{ m}^3 \cdot \text{Pa/mol}$$

$$\frac{1}{(k_G a)^*} = 1000 \text{ m}^3 \cdot \text{Pa} \cdot \text{s/mol}$$

$$\frac{H}{(k_L a)^*} = \frac{2.37}{0.0098} = 241 \text{ m}^3 \cdot \text{Pa} \cdot \text{s/mol}$$

[By Eq. (11-82)]

$$\frac{1}{K_G a} = \frac{1}{(k_G a)^*} + \frac{H}{(k_L a)^*} = 1241 \text{ m}^3 \cdot \text{Pa} \cdot \text{s/mol}$$

$$f_G = \frac{K_G a}{(k_G a)^*} = \frac{1000}{1241} = 0.81$$

The system is primarily gas-phase-controlled, but the influence of liquid-phase resistance is significant.

(b) $$(k_L a)^* = (0.0100) \left(\frac{2.0}{2.4} \right)^{0.5} = 0.0091 \text{ s}^{-1}$$

$$(k_G a)^* = (0.00100) \left(\frac{1.59}{2.3} \right)^{2/3} = 0.00078 \text{ mol/m}^3 \cdot \text{Pa} \cdot \text{s}$$

Solubility of CO_2 in water at 25°C is 0.00060 mole fraction/atm.

From Fig. 6-6

$$H = \frac{(1.013 \times 10^5)(18 \times 10^{-6})}{0.00060} = 3040 \text{ m}^3 \cdot \text{Pa/mol}$$

$$\frac{H}{(k_L a)^*} = \frac{3040}{0.0091} = 3.34 \times 10^5 \text{ m}^3 \cdot \text{Pa} \cdot \text{s/mol}$$

$$\frac{1}{(k_G a)^*} = \frac{1}{0.00078} = 1280 \text{ m}^3 \cdot \text{Pa} \cdot \text{s/mol}$$

$$f_G = \frac{1280}{1280 + 3.34 \times 10^5} = 0.004$$

The system is highly liquid-phase-controlled.

(c) $(k_L a)^*$ increases by a factor of 100, giving $H/(k_L a)^* = 3.34 \times 10^3$ m$^3 \cdot$ Pa·s/mol. Everything else is unchanged. Hence

$$f_G = \frac{1280}{1280 + 3340} = 0.28$$

The large increase in k_L has turned the system from highly liquid-phase-controlled to a situation where resistances in both phases are important.

(d)

$$(k_L a)^* = (0.0100)\left(\frac{1.24}{2.4}\right)^{0.5} = 0.00719 \text{ s}^{-1}$$

$$(k_G a)^* = (0.00100)\left(\frac{1.18}{2.3}\right)^{2.3} = 0.000641 \text{ mol/m}^3 \cdot \text{Pa} \cdot \text{s}$$

$$H = \frac{(1.013 \times 10^5)(18 \times 10^{-6})}{2.1} = 0.868 \text{ m}^3 \cdot \text{Pa/mol}$$

$$\frac{H}{(k_L a)^*} = \frac{0.868}{0.00719} = 120.7 \text{ m}^3 \cdot \text{Pa} \cdot \text{s/mol}$$

$$\frac{1}{(k_G a)^*} = \frac{1}{0.000641} = 1560 \text{ m}^3 \cdot \text{Pa} \cdot \text{s/mol}$$

$$f_G = \frac{1560}{1560 + 120.7} = 0.93$$

The system is highly gas-phase-controlled, with a slight contribution from liquid-phase resistance.

□

A principal point made in Example 11-6 is that effects of changes in the partition coefficient (solubility) and effects of chemical reactions on the partition coefficient or k_L are the most important factors in determining which phase controls the mass-transfer process. Changes in k_L and k_G due to changes in diffusivity have relatively little effect. Changes in the ratio of $k_G a$ to $k_L a$ due to changes in flow conditions are also usually much less important than changes in the partition coefficient or effects from a simultaneous chemical reaction.

SIMULTANEOUS HEAT AND MASS TRANSFER

Although rates of mass transfer necessarily govern rates of equilibration and stage efficiencies in separation processes, rates of heat transfer are sometimes important or even dominant, as well. This is particularly true for processes which involve an appreciable latent heat of phase change, since that heat must be supplied and/or removed for sustained mass transfer between phases to take place. Common situations involving interactions between heat transfer and mass transfer are evaporation and drying processes. Other situations are distillation (discussed in Chap. 12) and absorption or stripping (see, for example, Fig. 7-6 and surrounding discussion).

Evaporation of an Isolated Mass of Liquid

Figure 11-10 depicts the heat- and mass-transfer processes taking place in the vicinity of an isolated mass of a pure liquid undergoing evaporation. Mass transfer of evaporated liquid will occur outward from the liquid surface; consequently p_{Ai} must be greater than p_{AG}. The latent heat of vaporization must be transferred from the bulk gas phase to the evaporating surface. Hence T_G must exceed T_i. If a steady state is reached, the rate of heat input must equal the rate of heat consumption by evaporation:

$$hA(T_G - T_i) = \Delta H_v k_G A(p_{Ai} - p_{AG}) \tag{11-84}$$

where ΔH_v = latent heat of vaporization
A = interfacial area
h = heat-transfer coefficient

Furthermore, if the liquid mass is isolated from other heat sources or sinks, the entire liquid mass will reach the temperature T_i.

If the evaporation flux is low enough, a heat-transfer coefficient from some appropriate standard correlation can be used as h and a mass-transfer coefficient for a system at low flux can be used for k_G. For higher evaporation fluxes the effect of high flux on both the heat- and mass-transfer coefficients should be taken into account. Methods for doing so are discussed by Sherwood et al. (1975, chap. 7) and by Bird et al. (1960, chap. 21).

For ordinary rates of evaporation p_{Ai} will be the equilibrium vapor pressure of the liquid at T_i. Since T_i is lower than T_G, p_{Ai} will be lower than it would be if the surface temperature were equal to T_G. This reduces the rate of evaporation. Con-

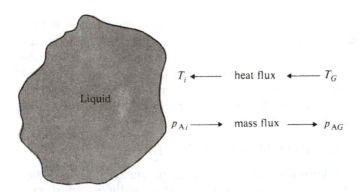

Figure 11-10 Transport processes occurring during evaporation of an isolated mass of liquid.

sideration of both heat transfer and mass transfer is needed in order to predict the rate of evaporation.

The *wet-bulb thermometer* is a device making use of the depression of T_i below T_G to measure p_{AG} or, equivalently, to measure the *relative humidity* of the surrounding gas r, where

$$r = \frac{p_{AG}}{P_w^0} \qquad (11\text{-}85)$$

and P_w^0 is the vapor pressure of pure water at T_G.

Example 11-7 (a) Determine the temperature of a small drop of water held stationary in stagnant air at 300 K when the relative humidity of the air is 25 percent. (b) By what factor is the rate of evaporation of the drop reduced compared with the rate for a drop temperature equal to T_G?

SOLUTION The air is stagnant, and the mole fraction of water in the air (presumed to be at atmospheric pressure) is low; hence we need not allow for effects of high flux and high concentration.

(a) The mass-transfer coefficient comes from Eq. (11-33) as Sh = 2, or

$$k_c = \frac{2D_{AB}}{d_0}$$

The analogous heat-transfer expression is Nu = 2, or

$$h = \frac{2k}{d_0}$$

where k is thermal conductivity. Combining these two equations, along with the fact that $k_G = k_c\,RT$ for an ideal gas. gives

$$\frac{k_G}{h} = \frac{D_{AB}}{kRT}$$

Substituting into Eq. (11-84). we have

$$\frac{T_G - T_i}{p_{Ai} - p_{AG}} = \frac{D_{AB}\,\Delta H_t}{kRT}$$

Since temperature varies between the bulk gas and the interface. it is necessary to assume an average temperature in order to evaluate the physical properties. We shall assume 285 K. in which case

$$D_{AB} = 2.37 \times 10^{-5}\ \text{m}^2\,\text{s} \qquad \text{water vapor in air}$$

$$k = 0.0250\ \text{J s·m·K} \qquad \text{pure air}$$

$$\Delta H_t = 44.5\ \text{kJ mol} \qquad \text{water}$$

These values are taken from Perry and Chilton (1973. pp. 3-223. 3-216. and 3-206). R is 8.314 J mol·K. Hence

$$\frac{T_G - T_i}{p_{Ai} - p_{AG}} = \frac{(2.37 \times 10^{-5})(44.5 \times 10^3)}{(0.0250)(8.314)(285)} = 0.0178\ \text{K Pa} = 17.8\ \text{K/kPa}$$

The vapor pressure of water at 300 K is 26.6 mmHg (Perry and Chilton. 1973), or 3.55 kPa. Hence p_{AG} is $(0.25)(3.55) = 0.886$ kPa.

$$\frac{300 - T_i}{p_{Ai} - 0.886} = 17.8 \qquad \begin{array}{l} T_i \text{ in K} \\ p_{Ai} \text{ in kPa} \end{array}$$

This expression should be solved jointly with the vapor-pressure expression for water, which relates T_i and p_{Ai}. This can be done graphically or by trial and error:

T_i, K	p_{Ai}, kPa	$\dfrac{300 - T_i}{p_{Ai} - 0.886}$
285	1.393	29.6
290	1.924	9.6
287	1.587	18.5
287.2	1.608	17.7

Thus T_i is 287.2 K. Recomputation with the properties determined at $(287.2 + 300)/2 = 293.6$ K would produce little change.

(b) The effect of the change in the partial-pressure driving force will outweigh the effect of changes in physical properties on k_G. Hence the factor by which the evaporation rate is depressed can be calculated from the change in $p_{Ai} - p_{AG}$:

$$\frac{(p_{Ai} - p_{AG})_{\text{with heat transfer}}}{(p_{Ai} - p_{AG})_{\text{without heat transfer}}} = \frac{1.608 - 0.886}{3.55 - 0.886} = 0.271$$

The evaporation rate of the drop is only 27 percent as great as it would be if the drop assumed the bulk-air temperature. □

Example 11-8 A wet-bulb thermometer is made by wrapping a wet wick around the bulb of an ordinary thermometer. Air is blown at high velocity over the wick. The bulk-air temperature and relative humidity are 300 K and 25 percent, respectively. Heat conduction along the stem of the thermometer can be neglected. Find the indicated wet-bulb temperature of the air once the thermometer reaches steady state.

SOLUTION This problem is similar to Example 11-7 except that h and k_G should be related through the Chilton-Colburn analogy $(j_H = j_D)$ rather than through the equations for steady-state transport into a stagnant medium from a sphere (Nu = Sh). From Eqs. (11-44) and (11-45), we have for $j_H = j_D$ and with $k_G = k_c/RT$

$$\frac{k_G}{h} = \frac{1}{RTC_p\rho}\left(\frac{D_{AB}C_p\rho}{k}\right)^{2/3}$$

Substituting into Eq. (11-84) yields

$$\frac{T_G - T_i}{p_{Ai} - p_{AG}} = \frac{\Delta H_v}{RTC_p\rho}\left(\frac{D_{AB}C_p\rho}{k}\right)^{2/3}$$

Since for an ideal gas the molar density is P/RT, we have

$$\frac{T_G - T_i}{p_{Ai} - p_{AG}} = \frac{\Delta H_v}{PC_p}\left(\frac{D_{AB}C_p\rho}{k}\right)^{2/3} \tag{11-86}$$

Again we shall assume $T_{av} = 285$ K, giving $D_{AB} = 2.37 \times 10^{-5}$ m^2/s, $k = 0.0250$ J/s·m·K, and $\Delta H_v = 44.5$ kJ/mol. Also,

$$C_p = 993 \text{ J/kg·K} = 28.8 \text{ J/mol·K} \qquad \text{and} \qquad \rho = 1.238 \text{ kg/m}^3$$

Both these properties are for pure air and are taken from Perry and Chilton (1973, pp. 3-134 and 3-72). Hence

$$\frac{D_{AB}C_p\rho}{k} = \frac{(2.37 \times 10^{-5})(993)(1.238)}{0.0250} = 1.17$$

Equation (11-86) requires the molar C_p, since the density was taken to be molar in replacing $RT\rho$ with P:

$$\frac{300 - T_i}{p_{Ai} - 0.886} = \frac{(44.5 \times 10^3 \text{ J/mol})(1.17)^{2 \; 3}}{(1.013 \times 10^5 \text{ Pa})(28.8 \text{ J/mol} \cdot \text{K})}$$

$$= 0.0169 \text{ K Pa} = 16.9 \text{ K/kPa}$$

Coincidentally, this is close to the value obtained for the sphere in a stagnant medium in Example 11-7.

Again, solving jointly with the vapor-pressure relationship for water gives:

T_i, K	p_{Ai}, kPa	$\dfrac{300 - T_i}{p_{Ai} - 0.886}$
287.2	1.608	17.7
287.4	1.629	17.0

so that the wet-bulb temperature is 287.4 K, a depression of 12.6 K below the temperature of the bulk gas. □

The dimensionless group $D_{AB} C_p \rho / k$ in Eq. (11-86) is the ratio of the Prandtl number to the Schmidt number, known as the *Lewis number*. It is also the ratio of the mass diffusivity to the thermal diffusivity. The Lewis number should be of order unity for a gas mixture, on the basis of the kinetic theory of gases. It is fortunate that the Lewis number is so close to unity for the water-vapor–air system. With the assumption that $\text{Le}^{2/3} = 1$ in Eq. (11-86), the equation becomes identical to the equation for determining the *adiabatic-saturation temperature* T_{as} of an air-water mixture. T_{as} and $p_{A,as}$ replace T_i and p_{Ai} in Eq. (11-86) if the Lewis-number term is omitted. The adiabatic-saturation temperature is the temperature that an air mass would assume if water were to be evaporated into it adiabatically until the gaseous mixture of water vapor and air became thermodynamically saturated.

If the adiabatic-saturation temperature and the wet-bulb temperature are taken to be equal, the common psychrometric chart (or *humidity chart*) [see, for example, Perry and Chilton (1973, p. 20-6)] can be used to perform the simultaneous solution of Eq. (11-86) and the vapor-pressure relationship for water. Sometimes psychrometric charts have separate curves for determining the adiabatic-saturation temperature and the wet-bulb temperature.

The wet-bulb thermometer and its analysis have an interesting history, related by Sherwood (1950).

The fact that water (or any volatile liquid) will cool when caused to evaporate is the basis for the many large *cooling towers* built by the electric-power and other industries. In a cooling tower, process water is cooled by being contacted (usually countercurrently) with ambient air that has a relative humidity less than 100 percent. Water then evaporates into the air, bringing the air toward thermodynamic saturation, and this serves to cool the large bulk of the water, which does not evaporate. In a countercurrent operation the effluent water can reach a temperature no lower than the wet-bulb temperature provided by the inlet air. Hence cooling towers are most effective in a dry climate.

Drying

Rate-limiting factors Drying moist solids is a common situation involving simultaneous *interphase* heat and mass transfer. In a typical dryer moist solids, divided into particles, are placed inside and heated by circulating air, heated walls, radiation, or the like. Heat must pass from the heat source to the particle surface and through the particle to wherever evaporation of water occurs. The water vapor generated must then travel to the piece surface and from the piece surface to a moisture sink, which may be a condenser, a desiccant, an exhaust of humid air, etc. The different transfer processes which can occur in various situations for each of these steps are shown in Fig. 11-11. The possible mechanisms of moisture transport within the solid have been reviewed by McCormick (1973), among others. For most substances that are commonly dried the most important internal mass-transfer mechanism from among those possible has not been identified.

The representation of a drying process in Fig. 11-11 can be compared to a network of electrical resistances, each possible transport mechanism being an individual resistor. Thus for mechanisms in parallel, we add the resistances reciprocally or add the conductances directly, that is, $k_{c1} + k_{c2}$, etc., since mass-transfer coefficients are analogous to conductances. For mechanisms which must operate in series we add the resistances directly or the conductances reciprocally, that is, $(1/k_{c1}) + (1/k_{c2})$, etc. Thus the overall coefficient for heat transfer U can be computed by

$$\frac{1}{U} = \frac{1}{(h_{\text{cond}} + h_{\text{conv}} + h_{\text{rad}})_{\text{ext}}} + \frac{1}{(h_{\text{cond}} + h_{\text{rad}})_{\text{int}}} \qquad (11\text{-}87)$$

with a similar equation for an overall mass-transfer coefficient.

This concept can be extended to identify the *rate-limiting factor* in various cases.

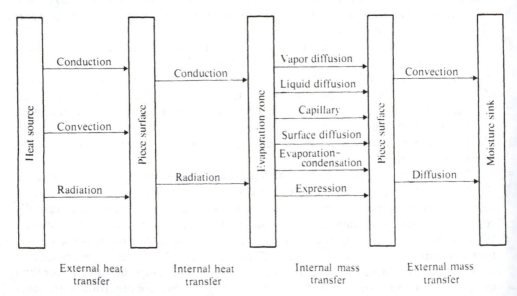

Figure 11-11 Rate factors in drying moist solids.

The rate-limiting factor in the electrical analogy is the largest resistance (or smallest conductance) among the set composed of the lowest resistances (or highest conductances) for each of the steps occurring in series. Similarly, for the mass- and heat-transfer processes in drying, the rate-limiting factor poses the smallest heat- or mass-transfer coefficient among the collection of largest coefficients for each of the steps which must occur in series. The rate-limiting factor for heat transfer will require the greatest temperature drop (akin to electric potential) for the various steps in series, and the rate-limiting factor for mass transfer will require the greatest Δc_A, Δx_A, etc., of the various mass-transfer steps in series. Such an analysis parallels the development of the addition-of-resistances equation for a gas-liquid system [Eq. (11-71), etc.]. In a gas-phase-controlled system, k_G is the rate-limiting factor; in a liquid-phase-controlled system, k_L is the rate-limiting factor.

It is important to identify the rate-limiting factor because accelerating the overall rate process most effectively requires that the rate-limiting factor be accelerated. Increasing the rate of some step that is not rate-limiting will have little or no effect. As an example, if the following individual heat-transfer coefficients occur in Eq. (11-87)

Internal: $\qquad\qquad\qquad\qquad h_{cond} = 50 \qquad h_{rad} = 2$

External: $\qquad\quad h_{cond} = 1 \qquad h_{conv} = 10 \qquad h_{rad} = 1$

all in consistent units, the overall heat-transfer coefficient U is found to be 9.75. The largest internal h is h_{cond} (50), and the largest external h is h_{conv} (10). The smallest of those two is h_{conv}, and it is therefore the rate-limiting factor. Notice that the overall coefficient (9.75) is very nearly equal to h_{conv} (10).

The controlling influence of the rate-limiting factor on the overall rate can be ascertained by calculating the individual effects of doubling each of the individual coefficients. Doubling h_{conv} increases U to 15.5, by a factor of 1.6. The increases in U from doubling the other coefficients are much less, 7 percent for $h_{rad, ext}$ and $h_{cond, ext}$, 10 percent for $h_{cond, int}$, and 0.7 percent for $h_{rad, int}$.

Experimentally, the rate-limiting step for heat transfer or mass transfer can be determined by seeing whether the greatest temperature difference (or concentration difference) occurs between the heat source and the piece surface or within the piece. The location with the larger driving force is more rate-limiting, since the same flux must equal the product of the coefficient and the driving force in both the internal and external steps. The step with the lower coefficient (the rate-limiting step) will therefore have the larger driving force.

It is also possible for drying processes to be rate-limited by a heat-transfer step or by a mass-transfer step. Because of the different phenomena and units involved, heat- and mass-transfer coefficients are not directly additive, but a comparison can be made if heat- and mass-transfer rates and driving forces are linked through the latent-heat and the vapor-pressure relationships. Considering only the fastest of each of the parallel mechanisms for each step, using Eq. (11-84), and if the vapor-pressure relationship is linearized through the Clausius-Clapeyron equation

$$\frac{dP_w^0}{dT} = \frac{H_v P_{w, av}^0}{RT_{av}^2} \qquad\qquad (11\text{-}88)$$

where P_w^0 is the vapor pressure of water, we can relate Δp_A for any mass-transfer step to a hypothetical equivalent temperature difference through

$$\Delta p_A = \frac{N_A}{k_G} = \frac{q}{\Delta H_v\, k_G} = \frac{\Delta H_v\, P_{w,\,av}^0}{RT_{av}^2}\, \Delta T \tag{11-89}$$

for a steady-state process in which free water is present within the drying solid. q is the heat flux, which for any heat-transfer step is $h\, \Delta T$. If we now define a fictitious overall temperature driving force for the drying process as the temperature of the heat source minus the temperature that would give an equilibrium partial pressure of water equal to the actual partial pressure of water at or in the moisture sink, we can write this overall ΔT driving force as the sum of the temperature drops for each of the individual heat- and mass-transfer steps, expressed by Eq. (11-89) or the heat-transfer relationship:

$$\frac{(\Delta T)_{\text{overall}}}{q} = \frac{1}{h_{\text{ext}}} + \frac{1}{h_{\text{int}}} + \left[\frac{RT_{av}^2}{(\Delta H_v)^2 P_{w,\,av}^0\, k_G}\right]_{\text{int}} + \left[\frac{RT_{av}^2}{(\Delta H_v)^2 P_{w,\,av}^0\, k_G}\right]_{\text{ext}} \tag{11-90}$$

The rate-limiting factor is now the largest of the terms in Eq. (11-90), since that term consumes the largest fraction of the overall temperature difference. Which factor that is will determine whether the rate can be accelerated most readily by augmenting internal or external heat transfer or mass transfer.

It should be pointed out that the derivative in Eq. (11-88) will vary substantially with temperature, since P_w^0 is a nonlinear function of T.

A similar analysis for the rate-limiting factor can be applied to any interphase simultaneous heat- and mass-transfer process.

Drying rates Typical trends for drying rates during batch drying of moist solids in commercial dryers are shown qualitatively in Fig. 11-12. After an initial transient A to B, the rate of moisture removal is typically constant for a time B to C and then falls off to lower values as time goes on C to D. Period B to C is called the *constant-rate period* and C to D the *falling-rate period*.

For a sufficiently moist solid, early in a batch drying process the water is able to move fast enough over the short distances below the surface to keep the surface of the solid entirely wet. Under such circumstances the internal resistances to heat and mass transfer are not important, and the drying process is rate-limited by either the external heat- or mass-transfer resistance or both. The situation is then analogous to evaporation of an isolated mass of liquid if the only heat input is by convection or conduction from the gas phase. In that case the surface will assume the wet-bulb temperature of the surrounding gas. This will fix and keep constant the driving forces for external heat and mass transfer, giving a constant rate of drying with respect to time or residual-moisture content. Determination of this rate is analogous to the calculation in Example 11-7, using appropriate heat- and mass-transfer-coefficient expressions. If there is heat input from other sources, the surface temperature will be higher than the wet-bulb temperature.

When enough water has been removed for internal transport to be unable to keep the surface wet, the locus of vaporization retreats into the solid and/or there will

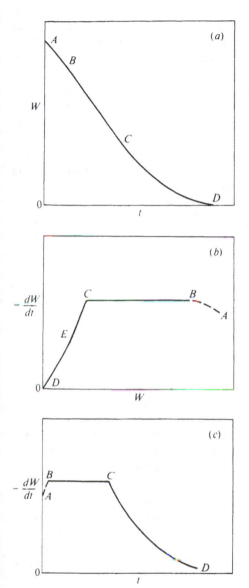

Figure 11-12 Typical drying rates for moist solids; W = average moisture content, kilograms H_2O per kilogram of dry matter, and t = time since start of drying: (a) change in moisture content vs. time; (b) drying rate vs. water content; and (c) drying rate vs. time. (*From McCormick, 1973, p. 20-10; used by permission.*)

be a significant resistance to transport of liquid water to the surface. In either case, the internal mass- and/or heat-transfer terms in Eq. (11-90) become important. This produces another significant resistance in series and thereby lowers the drying rate. This is the beginning of the falling-rate period.

As time goes on, the internal resistances increase, but the external resistances are unchanged. Hence the drying process swings over to where the rate-limiting factor is internal mass transfer and/or internal heat transfer. For consolidated media, such as wood, foods in an unfrozen state, polymer beads, etc., the relatively high solid density will make the ratio of thermal conductivity to moisture-transport coefficient sufficiently high for internal mass transfer to be more rate-limiting than internal heat transfer. For some special cases, e.g., freeze drying foods (King, 1971), the solid becomes so porous that internal heat transfer is a more significant rate limit than internal mass transfer.

If the diffusion equation is used to analyze the internal mass-transfer process, reference to Fig. 11-3 shows that $(d \ln W)/dt$ should become constant. Taking the derivative of the logarithm, this implies that $|dW/dt|$ should decrease linearly in W as W drops. Such a behavior is shown for the period C to E in Fig. 11-12b and is observed experimentally in many instances. However, it is rare for the water-transport mechanism within the solid to be simple homogeneous diffusion. Other mechanisms usually enter, and many of them are capable of making the rate vary as $-dW/dt = A - BW$.

The drying rate often varies in a simple fashion with changing size for particles of the same substance. For the constant-rate period, a differential mass balance indicates that

$$-\rho_s V \frac{dW}{dt} = N_A A = M_A k_G A(p_{Ai} - p_{AG}) \tag{11-91}$$

where ρ_s = dry particle density
M_A = molecular weight of transferring solute
V = particle volume
A = particle surface area

Since the mass flux remains constant throughout the constant-rate period, dW/dt is proportional to A/V The ratio A/V is $6/d_s$ if d_s is the equivalent-sphere diameter. Hence the drying rate, expressed as fraction water loss vs. time, varies inversely as the *first* power of the particle size during the constant-rate period. On the other hand, if we apply the transient-diffusion model (Fig. 11-3) to the portion of the falling-rate period where internal resistances are dominant, the time required to reach a given W varies as d_p^2, through the Fourier group. Hence, when internal resistance controls, the drying rate, expressed as fraction water loss vs. time, varies inversely as the *second* power of the particle size. A similar conclusion comes from most other potential mechanisms of internal moisture transport. Therefore the dependence of drying rate upon particle size gives another way of distinguishing between internal and external rate limits.

Drying rates and dryer designs are covered in much more detail by Keey (1978).

Example 11-9 Moist extruded catalyst particles are placed as a packed bed in a through-circulation dryer, in which air at 300 K and 25 percent relative humidity is passed at a superficial velocity of 0.5 m/s through the particle bed. The equivalent-sphere diameter of the particles is 1.30 cm, and the dry particle density is 1500 kg/m^3. The particles initially contain 1.80 kg H$_2$O per kilogram of solid.

The drying rate experienced for these particles is depressingly low, even in the early period of drying, only about 20 percent of the moisture removed per *hour*. (a) What is the rate-limiting factor? (b) Is the observed rate during the initial drying period reasonable in view of the operating conditions? (c) Evaluate the relative desirability of each of the following suggestions for increasing the drying rate: (i) halve the particle size. (ii) double the air velocity. (iii) desiccate the inlet air, and (iv) heat the inlet air to increase its temperature by 50 K.

SOLUTION (a) Since the rate is low and apparently relatively constant at the beginning of drying, the probable rate-limiting factor then is a combination of external heat- and mass-transfer resistances. External coefficients will be used as the basis for the calculation in part (b). If they do not substantially overpredict the drying rate, external resistances control during the initial period.

(b) The air temperature and relative humidity are the same as in Example 11-8, and heat is received by convection only. Hence T_i and p_{Ai} will be the values calculated in Example 11-8 during the period when external resistances to heat and mass transfer control. For the packed bed, j_D can be calculated from Eq. (11-48):

$$j_D = 1.17 \left(\frac{d_s u_s \rho}{\mu} \right)^{-0.415}$$

If we assume once again that $T_{av} = 285$ K, μ is found to be 1.78×10^{-5} Pa·s (Perry and Chilton, 1973, p. 3-210). Other physical properties come from Examples 11-7 and 11-8:

$$j_D = 1.17 \left[\frac{(0.0130 \text{ m})(0.5 \text{ m/s})(1.238 \text{ kg/m}^3)}{1.78 \times 10^{-5} \text{ Pa·s}} \right]^{-0.415}$$

$$= 1.17(452)^{-0.415} = 0.0925$$

From Eq. (11-44),

$$k_c = \frac{j_D u_s}{Sc^{2/3}}$$

$$Sc = \frac{1.78 \times 10^{-5}}{(1.238)(2.37 \times 10^{-5})} = 0.607$$

$$k_c = \frac{(0.0925)(0.5 \text{ m/s})}{(0.607)^{2/3}} = 0.0645 \text{ m/s}$$

$$k_G = \frac{k_c}{RT} = \frac{0.0645}{(8.314)(285)} = 2.72 \times 10^{-5} \text{ mol/m}^2 \cdot \text{Pa·s}$$

From Eq. (11-91), substituting $A/V = 6/d_s$, we have

$$-\frac{d(W/W_0)}{dt} = \frac{6 M_A k_G (p_{Ai} - p_{AG})}{W_0 \rho_s d_s}$$

where W_0 is the initial water content in kilograms per kilogram of solids.

$$-\frac{d(W/W_0)}{dt} = \frac{(6)(0.018 \text{ kg/mol})(2.72 \times 10^{-5})(1629 - 886 \text{ Pa})}{(1.80)(1500)(0.013)}$$

$$= 6.22 \times 10^{-5} \text{ s}^{-1}$$

This is the fraction of the initial water removed per second. The fraction removed per hour is

$$(6.22 \times 10^{-5})(3600) = 0.224$$

This agrees reasonably well with the observation of about 20 percent of the water removed per hour and explains the low rate. In addition, the calculation confirms that external resistances are indeed rate-limiting at this point.

(c) (i) Halving the particle size halves the Reynolds number, increases j_D by $(0.5)^{-0.415} = 1.33$, and increases k_G by the same factor. The combined effects of k_G and d_s in Eq. (11-91) serve to increase the drying rate by a factor of $(2)(1.33) = 2.67$.

(ii) Doubling the air velocity doubles the Reynolds number, decreases j_D by a factor of 1.33, and therefore increases k_G by a factor of $2/1.33 = 1.50$. Through Eq. (11-91), this increases the drying rate by a factor of 1.50.

(iii) Drying the inlet air completely would reduce the wet-bulb temperature. Repeating the calculation of Example 11-8 gives a wet-bulb temperature of 281 K, with $p_{Ai} = 1.065$ kPa. The mass-transfer driving force is thereby increased from $1629 - 886 = 743$ Pa to 1065 Pa. If the effect of the small temperature change on physical properties is neglected, the drying rate increases by a factor of $1065/743 = 1.43$.

(iv) Raising the air temperature to 350 K will increase T_i and hence p_{Ai} and the driving force for mass transfer. If we neglect the change in physical properties over this larger range of temperature, as an approximation, we find, by the method of Example 11-8, a wet-bulb temperature of 300.5 K. (Note that p_{AG} would remain unchanged, since no water vapor is added or subtracted upon heating the inlet air.) The corresponding p_{Ai} is 3650 Pa. The increase in driving force, and increase in drying rate, is a factor of $(3650 - 886)/743 = 3.7$.

Comparing these alternatives, we see that heating the air is by far the most effective avenue unless the catalyst material is heat-sensitive to such an extent that the higher air temperature cannot be used. If the catalyst is not heat-sensitive, a much higher air temperature than 350 K would be even more attractive. Comparing the other alternatives, drying the inlet air to get a maximum of 43 percent rate increase seems unattractive, since the drying step would be expensive. Increasing the air velocity and halving the particle size both increase the pressure drop and power required to circulate the air. It may not be possible to reduce the particle size because of specifications from the process(es) where the catalyst will be used.   □

DESIGN OF CONTINUOUS COUNTERCURRENT CONTACTORS

As observed in Chap. 4, continuous countercurrent contactors are often used as an alternative to discretely staged countercurrent contactors. An example is the irrigated packed column for gas-liquid contacting, which was compared with a plate column and with a countercurrent heat exchanger in Figs. 4-13 and 4-16.

As long as each of the counterflowing streams passes through the contactor in *plug flow*, the mass-balance equations for a continuous contactor are the same as for a multistage contactor and the operating line or curve on a yx diagram is the same. The equilibrium curve is, of course, unchanged as well, and the only difference is in how the operating diagram is used to estimate the contactor height required.

Plug flow implies that all fluid elements in a stream move at the same forward velocity and that there is no mixing in forward or backward directions due to turbulence, local flow patterns, etc. For relatively tall packed columns and many other contactors which prevent gross fluid circulation it is a good assumption. However, for a number of situations it is necessary to allow for departures from plug flow. This is usually done through the concept of *axial dispersion* or *axial mixing*, which serves to change the operating line or curve. In extreme cases, e.g., the continuous phase of spray extractors and absorbers and bubble-column absorbers, axial-mixing characteristics can dominate the separation obtained.

We shall consider design methods for plug flow of both phases first and then consider allowance for axial dispersion.

Plug Flow of Both Streams

The equilibrium and operating curves are the same for staged and plug-flow continuous processes, but the analyses from that point on should be different. In the continuous-contact process, equilibrium is not attained, and *rate* effects are control-

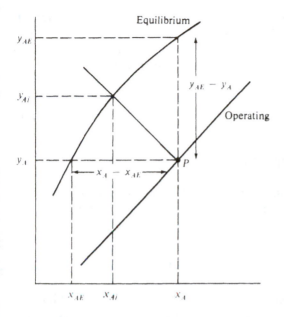

Figure 11-13 Driving forces for continuous countercurrent stripping.

ling, whereas equilibrium conditions alone determine the separation in a discretely staged equilibrium-stage device. The height of the separation device for a continuous contactor must be determined from a consideration of the rate of mass transfer, just as the area of a heat exchanger is determined from a consideration of the rate of heat transfer.

Allowance for the rate of mass transfer leads to the use of mass-transfer coefficients, usually overall coefficients obtained by applying the additivity-of-resistances concept to individual-phase coefficients obtained from correlations and calibrated where necessary by experiment.

If we consider first a stripping process involving a solute that is dilute in both gas and liquid, a portion of the operating diagram is shown in Fig. 11-13. The driving forces in the overall-coefficient mass-transfer-rate expressions, e.g., Eqs. (11-75) and (11-76), are related to the operating and equilibrium curves. Let us presume that point P on an operating line in Fig. 11-13 represents the bulk vapor and liquid compositions passing each other at some given level in our packed tower. Compositions corresponding to y_A, y_{AE}, x_A, and x_{AE} are marked in Fig. 11-13. The driving force for Eq. (11-75), $y_A - y_{AE}$, is the vertical distance between the equilibrium curve and the operating line at P, while the driving force for Eq. (11-76), $x_{AE} - x_A$, is the horizontal distance between the equilibrium curve and the operating line, also at P. The driving forces are negative since the equations are written for A going from gas to liquid, whereas stripping corresponds to A going from liquid to gas.

The interfacial compositions are also shown in Fig. 11-13. Combination of Eqs. (11-18) and (11-19) shows that the slope of the line from point P to the interfacial point is $-k_x/k_y$. If $k_x/K'k_y$ ($K' = dy_A/dx_A$ at equilibrium) is much less than unity, the system is liquid-phase-controlled, $x_A - x_{Ai}$ is nearly equal to $x_A - x_{AE}$, and $y_{Ai} - y_A$ is much less than $y_{AE} - y_A$. If, on the other hand, $k_x/K'k_y$ is much greater than unity, the system is gas-phase-controlled, $y_{Ai} - y_A$ is nearly equal to $y_{AE} - y_A$, and $x_A - x_{Ai}$ is much less than $x_A - x_{AE}$.

The rate of mass transfer can also be related to the changes in bulk composition of the two counterflowing streams from level to level in the tower. For the stripping process we have

$$\frac{V}{A}\frac{dy_A}{dh} = \frac{L}{A}\frac{dx_A}{dh} = \text{rate of mass transfer of A into vapor, mol/s} \cdot (\text{m}^3 \text{ tower volume})$$

(11-92)

where h = tower height (measured upward)

A = tower cross-sectional area

V = vapor flow, assumed constant (no $y_A \, dV/dh$ term, etc.), moles

L = liquid flow, assumed constant (no $x_A \, dL/dh$ term, etc.), moles

Combining Eqs. (11-68) and (11-92), along with $p_A = y_A P$, we obtain

$$\frac{V}{A}\frac{dy_A}{dh} = K_G Pa(y_{AE} - y_A)$$

(11-93)

where a is the interfacial area between phases, expressed as square meters of interface per cubic meter of total tower volume. The negative of Eq. (11-68) is used since Eq. (11-92) represents transfer of A into the vapor. The height of packing required then comes from an integration of Eq. (11-93) over the range of y_A to be experienced in that packed height:

$$h = \int_0^h dh = \frac{V}{A} \int_{y_{A,\,in}}^{y_{A,\,out}} \frac{dy_A}{K_G Pa(y_{AE} - y_A)}$$

(11-94)

Equation (11-94) is most simply integrated if $K_G P$ is constant throughout the tower. Also K' or $H \, (= K'P/\rho_M$, where ρ_M is the liquid molar density) may vary from point to point. If we can expect the individual phase coefficients k_G and k_L to be relatively constant, K_G will tend to be constant for a case of gas-phase-controlled mass transfer where $K_G \approx k_G$. K_L would be less constant in that case since $K_L \approx Hk_G$ and H varies. Conversely, for a case of liquid-phase-controlled mass transfer, K_L is usually more constant than K_G, since $K_L \approx k_L$ and $K_G \approx k_L/H$.

For the gas-phase-controlled case, assuming that K_G, P, a, and V are constant, Eq. (11-94) becomes

$$\frac{hAK_G Pa}{V} = \int_{y_{A,\,in}}^{y_{A,\,out}} \frac{dy_A}{y_{AE} - y_A}$$

(11-95)

Transfer units The quantity on the right-hand side of Eq. (11-95) is commonly called *number of transfer units* (NTU), an expression originally coined by Chilton and Colburn (1935). Because the equation is based on the driving force between bulk-vapor composition and that vapor composition in equilibrium with the bulk liquid in *gas-phase* units, we have in this case $(\text{NTU})_{OG}$, or *overall gas-phase* transfer units. The integral

$$(\text{NTU})_{OG} = \int_{y_{A,\,in}}^{y_{A,\,out}} \frac{dy_A}{y_{AE} - y_A}$$

(11-96)

a measure of the amount of separation obtained, is the ratio of the change in bulk-gas composition, $y_{A,\,out} - y_{A,\,in}$, to the average effective driving force, $y_{AE} - y_A$. $(NTU)_{OG}$ is the number of properly averaged overall gas-phase driving forces by which the bulk-gas composition changes.

If the degree of separation is represented by the number of transfer units, we can obtain the tower height as

$$h = \frac{(NTU)_{OG} V}{K_G PaA} \qquad (11\text{-}97)$$

If K_G, P, a, and V are constant from one level to another in the tower, the height of packing required must be directly proportional to the number of transfer units $(NTU)_{OG}$ involved in the separation. The number of transfer units is thus also a measure of the height requirement in continuous-contacting equipment, just as the number of equilibrium stages is a measure of number of plates, and hence tower height, in a plate tower.

The *height of a transfer unit* $(HTU)_{OG}$ is defined as the combination of flow and mass-transfer coefficient which give one transfer unit of separation:

$$(HTU)_{OG} = \frac{V}{K_G PaA} \qquad (11\text{-}98)$$

The subscripts O and G once again refer to the fact that this transfer-unit expression is based upon the overall gas-phase driving force. A greater $K_G Pa$ or a lesser V will reduce the height requirement per transfer unit of separation.

The definition of $(HTU)_{OG}$ converts Eq. (11-97) into

$$h = (HTU)_{OG}(NTU)_{OG} \qquad (11\text{-}99)$$

If the desired separation $(NTU)_{OG}$ is known and $(HTU)_{OG}$ is obtained from correlations for $k_L a$ and $k_G a$ combined to give $K_G a$, the tower height required can be obtained from Eq. (11-99). Alternatively, if a tower height gives a degree of separation converted into $(NTU)_{OG}$, the corresponding value of $(HTU)_{OG}$ can be obtained from Eq. (11-99).

In the event that the mass-transfer process is liquid-phase-controlled rather than gas-phase-controlled, one can anticipate that K_L will be more nearly constant than K_G, since $K_L \approx k_L$ but $K_G \approx k_L/H$. A train of thought parallel to the development of Eq. (11-95) yields

$$\frac{h K_L \rho_M a A}{L} = \int_{x_{A,\,in}}^{x_{A,\,out}} \frac{dx_A}{x_{AE} - x_A} \qquad (11\text{-}100)$$

where ρ_M is the molar density of the liquid.

Again the right-hand side can be used to define a number of transfer units, this time $(NTU)_{OL}$, based upon the overall liquid-phase driving force,

$$(NTU)_{OL} = \int_{x_{A,\,in}}^{x_{A,\,out}} \frac{dx_A}{x_{AE} - x_A} \qquad (11\text{-}101)$$

as a measure of the separation obtained. Likewise, we can define another height of a transfer unit $(HTU)_{OL}$ as

$$(HTU)_{OL} = \frac{L}{K_L \rho_M a A} \tag{11-102}$$

so that

$$h = (NTU)_{OL}(HTU)_{OL} \tag{11-103}$$

$(HTU)_{OG}$ and $(HTU)_{OL}$ are in general different numerically [as are $(NTU)_{OG}$ and $(NTU)_{OL}$], since the driving forces used in the defining expressions are different.

Because predicting values of $(HTU)_{OG}$ and $(HTU)_{OL}$ by Eqs. (11-98) and (11-102) requires combining $k_L a$ and $k_G a$ by the additivity-of-resistances relations, individual-phase heights of a transfer unit are sometimes defined by

$$(HTU)_G = \frac{V}{k_G a P A} \tag{11-104}$$

and

$$(HTU)_L = \frac{L}{k_L a \rho_M A} \tag{11-105}$$

Substituting these into the additivity-of-resistances equations (11-82) and (11-83) and into Eqs. (11-98) and (11-102) yields

$$(HTU)_{OG} = (HTU)_G + \frac{HV \rho_M}{LP} (HTU)_L \tag{11-106}$$

and

$$(HTU)_{OL} = (HTU)_L + \frac{LP}{HV \rho_M} (HTU)_G \tag{11-107}$$

Sometimes correlations report $(HTU)_G$ and $(HTU)_L$ instead of $k_G a$ and $k_L a$, in which case the resulting $(HTU)_G$ and $(HTU)_L$ can be compounded through Eqs. (11-106) and (11-107). When $HV \rho_M / LP$ varies, $(HTU)_{OG}$ and $(HTU)_{OL}$ can become variable themselves.

Sometimes, also, $(HTU)_G$ and $(HTU)_L$ are used together with the contactor height to generate numbers of individual-phase transfer units provided by the contactor, i.e.,

$$(NTU)_G = \frac{h}{(HTU)_G} \quad \text{and} \quad (NTU)_L = \frac{h}{(HTU)_L} \tag{11-108}$$

In that case the number of overall gas- or liquid-phase transfer units provided by the contactor is given by

$$\frac{1}{(NTU)_{OG}} = \frac{(HTU)_{OG}}{h} = \frac{1}{(NTU)_G} + \frac{HV \rho_M / LP}{(NTU)_L} \tag{11-109}$$

or

$$\frac{1}{(NTU)_{OL}} = \frac{(HTU)_{OL}}{h} = \frac{1}{(NTU)_L} + \frac{LP / HV \rho_M}{(NTU)_G} \tag{11-110}$$

In general, the transfer-unit integrals, Eqs. (11-96) and (11-101), must be evaluated graphically. For Eq. (11-96) this is done by relating an x_A to every y_A through the operating-line expression and then obtaining y_{AF} in equilibrium with that x_A

through the equilibrium expression. Similarly, for Eq. (11-101) a y_A is related to every x_A through the operating-line expression, and the corresponding x_{AE} is then obtained from the equilibrium relationship. When K_G, K_L, V, and/or L change from one position to another, they should be retained under the integral sign. This precludes the separation of variables implied by Eqs. (11-99) and (11-103).

When high-concentration and/or high-flux effects are important, they must be included in the analysis. If k_J coefficients are used, the additivity-of-resistances equations should be used in the form of Eqs. (11-79) and (11-80) or generalizations of them when some N_i besides N_A are nonzero. The high-flux corrections should be incorporated into the k_J coefficients. For k_N coefficients Eqs. (11-63) to (11-66) can be used if the film model is invoked for the effects of high concentration and/or high flux. These introduce the film factors $[x_{Af}, (\bar{V}_A c_A)_f$, etc.] into the expressions for the mass-transfer coefficients, and this will generally serve to make the mass-transfer coefficients variable. If these are the only factors making the mass-transfer coefficients variable, the separation of variables implicit in the transfer-unit analysis can be retained by including the film factor in the numerator of the transfer-unit integral and separating the rest of the mass-transfer coefficient $[k_L = k_{NL}(\bar{V}_A c_A)_f$, etc.] out of the integral into the HTU expression. Similarly, if the total molar flow changes, the variable portion can be retained in the integral, and a constant multiplier, e.g., the flow of nontransferring inerts, can be taken into the HTU expression (Sherwood et al., 1975; Wilke, 1977).

The following example illustrates the use of the transfer-unit integral for a continuous countercurrent contactor.

Example 11-10 (*a*) Find the number of overall gas-phase transfer units required for the distillation operation solved in Example 5-1 if it is carried out in a packed tower to give the same separation. (*b*) Find the packed height required if $(\text{HTU})_{OG} = 0.50$ m.

SOLUTION (*a*) In Example 5-1 the separation was specified as

$$x_{A,F} = 0.5 \qquad x_{A,d} = 0.90 \qquad r F = 0.5$$

$$h_F = \text{saturated liquid} \qquad N = 5 \text{ equilibrium stages (all above feed) besides reboiler}$$

$$P \text{ to give } \alpha_{AB} = 2 \qquad T_c = \text{saturated liquid reflux}$$

Solving, we found

$$\frac{d}{F} = 0.187 \qquad \frac{b}{F} = 0.813 \qquad x_{A,b} = 0.408$$

For our purposes in this problem we replace N as a specification by one of the three separation variables for which we originally solved. The other separation variables then remain the same through mass balances. We now solve for $(\text{NTU})_{OG}$ instead of N since the operation is carried out in a packed tower.

Distillation operations with a narrow volatility gap tend to be limited by the mass-transfer resistance in the gas phase (see Chap. 12). Therefore K_G tends to be more nearly constant than K_L, and it is most convenient to analyze the distillation through $(\text{NTU})_{OG}$, using the integral expressed by Eq. (11-96).

In Fig. 11-14 the driving forces $y_{AE} - y_A$ for each value of y_A are indicated by the series of arrows. Figure 11-15 shows a graphical integration carried out on a plot where the horizontal axis is y_A at any point on the operating line and the vertical axis is the reciprocal of the driving force at that

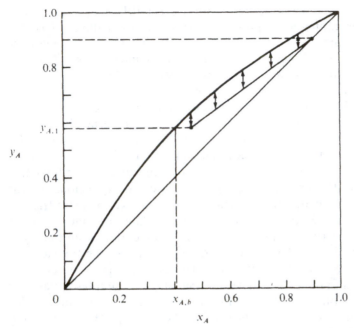

Figure 11-14 Driving force for packed-tower distillation, Example 11-10.

point. We still retain the reboiler as an equilibrium stage; hence the lower limit on y_A is 0.580, corresponding to equilibrium with $x_{A,b}$. The area under the curve is 6.2 units, hence

$$(\text{NTU})_{OG} = 6.2$$

Note that this value is different from the number of equilibrium stages (five) above the reboiler.

(b) Using Eq. (11-99), we get

$$h = (\text{NTU})_{OG}(\text{HTU})_{OG} = (6.2)(0.50 \text{ m}) = 3.1 \text{ m of packing height required} \qquad \square$$

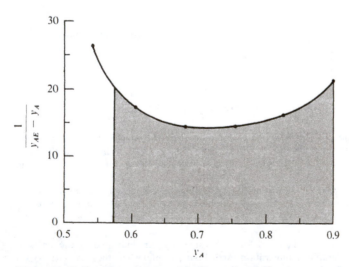

Figure 11-15 Transfer-unit integral for Example 11-10.

Analytical expressions If y_{AE} is either constant or linear in y_A, as would occur for straight equilibrium and operating lines, we can integrate Eq. (11-96) to give

$$(\text{NTU})_{OG} = \frac{y_{A,\,out} - y_{A,\,in}}{(y_{AE} - y_A)_{LM}} \tag{11-111}$$

where the subscript LM refers to the logarithmic mean:

$$(y_{AE} - y_A)_{LM} = \frac{(y_{AE} - y_A)_{in} - (y_{AE} - y_A)_{out}}{\ln\left[(y_{AE} - y_A)_{in}/(y_{AE} - y_A)_{out}\right]} \tag{11-112}$$

The direct analogy between Eq. (11-111) and the use of the logarithmic-mean-temperature driving force in the analysis of a simple heat exchanger should be apparent.

Also, for x_{AE} constant or linear in x_A,

$$(\text{NTU})_{OL} = \frac{x_{A,\,in} - x_{A,\,out}}{(x - x_{AE})_{LM}} \tag{11-113}$$

When either the terminal compositions or V/L are unknown, it is convenient to use another form of Eq. (11-111). Equation (11-96) can be put in the form

$$(\text{NTU})_{OG} = \int_{y_{A,\,in}}^{y_{A,\,out}} \frac{dy_A}{mx_A + b - y_A} \tag{11-114}$$

using Eq. (8-1) to linearize the equilibrium expression. When applied to continuous countercurrent equipment, the mass balance expressed by Eq. (8-2) becomes

$$y_A = y_{A,\,out} + \frac{L}{V}(x_A - x_{A,\,in}) \tag{11-115}$$

Combining Eqs. (11-114) and (11-115) gives

$$(\text{NTU})_{OG} = \int_{y_{A,\,in}}^{y_{A,\,out}} \frac{dy_A}{mx_{A,\,in} + b + (mV/L)(y_A - y_{A,\,out}) - y_A} \tag{11-116}$$

Integrating, we have

$$\begin{aligned}
(\text{NTU})_{OG} &= \frac{1}{1 - (mV/L)} \ln \frac{mx_{A,\,in} + b + (mV/L)(y_{A,\,in} - y_{A,\,out}) - y_{A,\,in}}{mx_{A,\,in} + b - y_{A,\,out}} \\
&= \frac{\ln\left\{[1 - (mV/L)][(y_{A,\,in} - mx_{A,\,in} - b)/(y_{A,\,out} - mx_{A,\,in} - b)] + (mV/L)\right\}}{1 - (mV/L)}
\end{aligned} \tag{11-117}$$

or $\quad (\text{NTU})_{OG} = \dfrac{\ln\left\{[1 - (mV/L)][(y_{A,\,in} - y^*_{A,\,out})/(y_{A,\,out} - y^*_{A,\,out})] + (mV/L)\right\}}{1 - (mV/L)} \tag{11-118}$

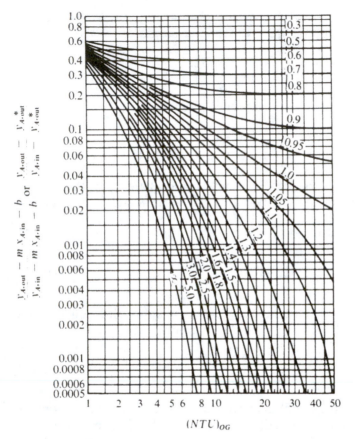

Figure 11-16 Plot of Eqs. (11-118) and (11-119) for continuous countercurrent contactor. Parameter is L/mV.

This equation was originally developed by Colburn (1939). Similar equations can be obtained containing any three terminal concentrations. Equation (11-118) can also be rearranged to a form explicit in $y_{A, out}$ provided $y_{A, in}$ and $y^*_{A, out}$ are known:

$$\frac{y_{A, in} - y^*_{A, out}}{y_{A, out} - y^*_{A, out}} = \frac{e^{(NTU)_{OG}(1 - mV/L)} - (mV/L)}{1 - (mV/L)} \quad (11\text{-}119)$$

Equations (11-117) to (11-119) are plotted in Fig. 11-16.

The reader should note the similarity of Eq. (11-118) to Eq. (8-15), which was developed for discrete equilibrium stages. The only difference occurs in the denominator of the right-hand term. Similarly, Fig. 11-16 has a form similar to that of Fig. 8-3, the plot of the Kremser-Souders-Brown equation. The essential difference, of course, is that one is specific to discretely staged contactors while the other is specific to continuous countercurrent contactors.

Like the KSB equations, Eqs. (11-117) to (11-119) and Fig. 11-16 can be put into forms involving x_A by substituting x_A for y_A, $(NTU)_{OL}$ for $(NTU)_{OG}$, $1/m$ for m,

L for V, V for L, etc. Thus Fig. 11-16 can also be used as a plot of $(x_{A, out} - x^*_{A, out})/(x_{A, in} - x^*_{A, out})$ vs. $(NTU)_{OL}$ with mV/L as the parameter.

Furthermore, for straight equilibrium and operating lines $(NTU)_{OL}$ can be used instead of $(NTU)_{OG}$ in the y_A form of these equations and Fig. 11-16 through the substitution

$$(NTU)_{OG} = \frac{L}{mV} (NTU)_{OL} \qquad (11\text{-}120)$$

which follows from Eqs. (11-109) and (11-110). Similarly, for mV/L constant, Eqs. (11-106) and (11-107) allow interchangeability between $(HTU)_{OG}$ and $(HTU)_{OL}$ through

$$(HTU)_{OG} = \frac{mV}{L} (HTU)_{OL} \qquad (11\text{-}121)$$

$HV\rho_M/LP$ in Eqs. (11-106) to (11-110) is equivalent to mV/L.

The use of Fig. 11-16 and of Eqs. (11-117) to (11-119) is very similar to that of the KSB equations, as well. Maximum precision in solutions is obtained if the equations are used so as to place the solution in the lower region of Fig. 11-16, that is, the y_A form for L/mV greater than 1 and the x_A form for mV/L greater than 1. The same reasoning holds with regard to the desirability of making $L/mV > 1$ for an absorber and $mV/L > 1$ for a stripper, to effect high solute removal. Multiple-section forms of the Colburn equation can also be derived for continuous countercurrent contactors, similar to those for staged contactors.

Sherwood et al. (1975, pp. 447–466) present ways of extending the Colburn equation in approximate fashion to allow for variations in $K_G a$ and V because of a concentrated gas and for slight to moderate degrees of curvature in the operating and equilibrium lines.

Example 11-11 A stream of air containing 0.2 mol $\%$ ammonia and saturated with water is contacted countercurrently with water in a packed tower. Operation is isothermal at 25°C and is at atmospheric pressure. The tower diameter is 0.80 m, and the packing is 1.0-in (2.54-cm) Raschig rings. The water flow rate is 1.36 kg/s, and the air flow rate is 0.41 kg/s. These are the flow conditions (2.7 kg/s per square meter of tower cross section for water, and 0.82 kg/s per square meter of tower cross section for air) for which the mass-transfer coefficients were determined in Example 11-6.

Find the height of packing required to remove 99.0 percent of the ammonia in the inlet air if the inlet water contains no dissolved ammonia. Assume plug flow of both streams.

SOLUTION First we find the transfer-unit requirement. From Example 11-6, the solubility of ammonia at high dilution such as this is 0.77 mole fraction/atm. At atmospheric pressure, this converts into $K_{NH_3} = K'_{NH_3} = 1 \ 0.77 = 1.30 = m$. The molar flow ratio is obtained as

$$\frac{L}{V} = \frac{1.36}{0.41} \frac{29}{18} = 5.34$$

Hence $L \ mV = 5.34 \ 1.30 = 4.11$.

Since there is no NH_3 in the inlet water, $y^*_{A, out} = 0$. $y_{A, out}/y_{A, in}$ is specified to be 0.01. By Eq. (11-118) or Fig. 11-16, $(NTU)_{OG} = 5.7$.

From part (b) of Example 11-6, $1 \ K_G a = 1241 \ m^3 \cdot Pa \cdot s/mol$; hence $K_G a = 1/1241 = 8.06 \times 10^{-4} \ mol \ m^3 \cdot Pa \cdot s$. Substituting into Eq. (11-98), we have

$$(HTU)_{OG} = \frac{V}{K_G a P A} = \frac{(0.82 \ kg \ m^2 \cdot s)(1 \ mol/0.029 \ kg)}{(8.06 \times 10^{-4} \ mol \ m^3 \cdot Pa \cdot s)(1.013 \times 10^5 \ Pa)} = 0.346 \ m$$

By Eq. (11-99),

$$h = \text{packed height} = (NTU)_{OG}(HTU)_{OG} = (5.7)(0.346) = 1.97 \text{ m} \qquad \square$$

Minimum contactor height In continuous countercurrent separation processes there will be a minimum number of transfer units required under conditions of infinite interstage flow. The derivation of the appropriate equations is a relatively simple matter. For a process in which one flow can become infinite in concept while the other flow remains finite, the number of transfer units must be based on the flow which remains finite. For a packed absorber receiving a solute-free absorbent, the condition of an infinite solvent-to-feed-gas ratio gives

$$(NTU)_{OG,\,\text{min}} = \int_{y_{A,\,\text{in}}}^{y_{A,\,\text{out}}} \frac{dy_A}{-y_A} = \ln \frac{y_{A,\,\text{in}}}{y_{A,\,\text{out}}} \qquad (11\text{-}122)$$

upon substitution into Eq. (11-96). For a packed binary distillation column, on the other hand, both flows become infinite together, and total reflux corresponds to $L/V = 1$ and $y = x$ for the passing streams. Substituting into Eq. (11-96) gives

$$(NTU)_{OG,\,\text{min}} = \int_{x_{A,\,b}}^{x_{A,\,d}} \frac{dx_A}{y_{AE} - x_A} \qquad (11\text{-}123)$$

The right-hand side of this equation is identical to the Rayleigh equation (3-16) for a single-stage semibatch separation. Thus Eqs. (3-17) and (3-18) apply for constant K_A and constant binary α_{AB}, respectively, if $\ln (L'/L_0')$ is replaced by $(NTU)_{OG,\,\text{min}}$:

$$(NTU)_{OG,\,\text{min}} = \frac{1}{K_A - 1} \ln \frac{x_{A,\,d}}{x_{A,\,b}} \qquad (11\text{-}124)$$

$$(NTU)_{OG,\,\text{min}} = \frac{1}{\alpha_{AB} - 1} \ln \frac{x_{A,\,d}(1 - x_A)_b}{x_{A,\,b}(1 - x_A)_d} + \ln \frac{1 - x_{A,\,b}}{1 - x_{A,\,d}} \qquad (11\text{-}125)$$

More complex cases Additional complications which can enter the analysis of continuous countercurrent processes include (1) complex phase equilibria, possibly involving chemical reactions; (2) multiple transferring solutes, which may interact with each other in phase equilibria and mass-transfer coefficients; (3) simultaneous heat effects; (4) partial phase miscibility in extraction processes; and (5) effects of high flux and/or high solute concentration.

Multivariate Newton convergence If the mass-transfer coefficients can be predicted for prevailing compositions and fluxes, these more complex cases can be handled effectively and efficiently through a numerical computer approach leading to tridiagonal or block-tridiagonal matrices of the sort considered for multistage separations in Chap. 10 and Appendix E. The method is outlined by Newman (1967b, 1968, 1973) and is suitable for any system of coupled first- or second-order ordinary differential equations involved in boundary-value problems, where the boundary conditions may themselves involve first derivatives. The method is closely related to the full multivariate Newton SC method for discretely staged processes.

Similar to Eqs. (10-1) to (10-5), we can tabulate the equations for a multi-component continuous countercurrent process; z is column height, measured upward.

1. Component mass balances **M** (R equations):

$$\frac{d(l_j)}{dz} - \frac{d(v_j)}{dz} = 0 \qquad (11\text{-}126)$$

2. Enthalpy balances **H**:

$$\frac{d(hL)}{dz} - \frac{d(HV)}{dz} - q = 0 \qquad (11\text{-}127)$$

3. Summation equations **S** (two equations):

$$\Sigma l_j = L \qquad (11\text{-}128)$$

$$\Sigma v_j = V \qquad (11\text{-}129)$$

4. Mass-transfer rate expressions **R** (R^* equations, where R^* is the number of transferring components):

$$\frac{d(v_j)}{dz} + \frac{K_{Gj}aP}{V}(v_j - v_{jE}) = 0 \qquad (11\text{-}130)$$

5. Equilibrium equations **E** (R^* equations):

$$v_{jE} = \frac{K_j V}{L} l_j \qquad (11\text{-}131)$$

The **M**, **H**, and **R** equations are now first-order differential equations and are in general nonlinear. The rate expressions enter because of the rate effects dominating mass transfer in continuous countercurrent equipment; however, equilibrium expressions are still needed to provide the driving forces for the mass-transfer expressions. Equations (11-130) could equivalently be replaced by equations involving $K_L a$, $K_y a$, or $K_x a$. These equations are coupled with boundary conditions corresponding to the specification of the problem, e.g., specifications regarding feed and product locations, compositions, and/or flows at various values of z.

The derivatives in the **M**, **H**, and **R** equations can be converted into finite-difference form if the column height is broken up into sections of length Δz, such that $z = n\,\Delta z$ with $n = 0, 1, \ldots, N$, where $N\,\Delta z$ is the total column height:

$$\left[\frac{df(z)}{dz}\right]_n = \frac{f(z)_{n+1} - f(z)_{n-1}}{2\,\Delta z} \qquad (11\text{-}132)$$

Substitution of Eq. (11-132) for the various first derivatives leads to a set of N simultaneous equations replacing each single equation in Eqs. (11-126) to (11-131). These equations relate conditions at only three adjacent positions, $n + 1$, n, and $n - 1$; hence the equations form a tridiagonal or block-tridiagonal matrix once they are linearized by assuming values for all dependent variables. They therefore are solvable by the full multivariate Newton SC method described for staged separators in Chap. 10 and Appendix E. Furthermore, in various special subcases the equations can be handled by the hierarchy of partitioning and simplification methods outlined for staged contactors in Fig. 10-12 and all the associated discussion. For design problems, as opposed to operating problems, simultaneous convergence of the column height is appropriate, similar to the method of Ricker and Grens for multi-stage distillation, described in Chap. 10.

The equations for continuous countercurrent contactors are similar in form and method of solution to those for staged contactors, but it is also important to stress the two essential differences between them: (1) The subscript p, representing stage number, is replaced by the subscript n, representing column height in arbitrary divisions. The changes from stage p to stage $p + 1$ in a staged contactor are in general not equivalent to the changes from level n to level $n + 1$ in a continuous contactor. (2) The R equations appear in the set for continuous countercurrent contactors, whereas rate effects do not enter in the analysis of an equilibrium-stage contactor.

An example of the use of the full multivariate Newton SC method for analyzing vacuum steam stripping of gases from water in a packed column is given by Rasquin (1977) and Rasquin et al. (1977).

Relaxation Once Eqs. (11-126) to (11-131) are put in finite-difference form, they are also subject to solution by relaxation methods, provided terms are included to account for transient changes associated with liquid holdup. Thus, terms for $(U_n/L) \times (dl_j/dt)$ are required in Eqs. (11-126), where U_n is the amount of liquid holdup in one of the incremental column sections. The resulting equation is analogous to Eq. (10-45) for staged contactors. A similar term is needed in Eq. (11-127), involving transient changes in liquid enthalpy. The methods for using relaxation techniques to solve the resulting equations are then analogous to those discussed for staged contactors in Chap. 10.

Stockar and Wilke (1977a) describe a relaxation method for analyzing continuous countercurrent gas absorbers with heat effects.

As for staged contactors, it should be effective to combine a relaxation solution for the first several iterations with a multivariate Newton SC method for subsequent iterations in analyzing complex continuous countercurrent contactors.

Limitations Overall mass-transfer coefficients are required for any of the approaches for calculating the performance of continuous countercurrent contactors. Prediction of these must allow for hydrodynamic effects (usually through correlations) and effects of high flux and/or high concentration level, if important. Interfacial areas are also required and often must be obtained by correlation, sometimes together with the mass-transfer coefficients. Departures from simple additivity of resistances because of varying ratios of k_G to k_L over the contacting interface can also complicate analysis.

Multicomponent diffusion is complex (Cussler, 1976; etc.), and mass-transfer coefficients for solutes in systems where several transferring components are present in substantial concentrations are often not simple extensions of mass-transfer coefficients measured in binary or dilute systems. This is the result of interaction of component fluxes in the basic diffusion equations.

Short-cut methods Stockar and Wilke (1977*b*) have developed an approximate method for relating the separation to the column height in packed gas absorbers where there is a significant heat effect leading to an internal temperature maximum. The approach is to predict the magnitude of the maximum increase in temperature through a semiempirical correlation, to use this value to predict the entire temperature profile, and then to use the resultant temperature profile through either a transfer-unit integral or a modification of Eq. (11-118) and Fig. 11-16, allowing for the curved equilibrium line. When the product of flow rate and heat capacity in one phase considerably exceeds the product in the other phase, an even simpler approach can be used, awarding the entire heat of absorption to the phase with the higher product of flow rate and heat capacity and thereby calculating the temperature increase of that phase as it passes through the column (see also Wilke, 1977).

Eduljee (1975) proposes a correlation for transfer units in continuous contactors for distillation, similar to the Gilliland correlation (Fig. 9-1) for equilibrium-stage contactors.

Height equivalent to a theoretical plate (HETP) Since methods for analyzing distillation and other countercurrent separations in terms of equilibrium stages are so well developed, another approximate approach toward analysis of continuous countercurrent contactors has used the concept of the height of a theoretical plate ($=$ equilibrium stage) HETP. The column height for a given separation is then obtained as $h =$ HETP $\times N$, where N is the number of equilibrium stages required for the separation. Various correlations have been put forward for predicting HETP in distillation (see, for example, Perry and Chilton, 1973, p. 18-49). In general, however, it can be expected that HETP would change considerably with respect to operating conditions, liquid properties, etc., since it would be determined by a complex combination of many different factors.

If the HETP concept is to be used, a more appropriate technique is that described by Sherwood et al. (1975, pp. 518–524), where HETP is related to $(HTU)_{OG}$ through a linearization of the operating- and equilibrium-curve expressions, giving

$$\text{HETP} = (\text{HTU})_{OG} \frac{\ln (mV/L)}{(mV/L) - 1} \tag{11-133}$$

Values of $(HTU)_{OG}$ are predicted from values of $K_G a$ through Eq. (11-98) in the usual way and are then converted into HETP through Eq. (11-133). Since mV/L will change considerably throughout a typical distillation, HETP will change with respect to composition, even though $(HTU)_{OG}$ may not. In such a case, it is advisable to calculate a new value of HETP for each equilibrium stage. For concentrated absorbers and strippers it is also necessary to allow for x_{Af} [Eq. (11-65)] or its equivalent in the prediction of $(HTU)_{OG}$.

Since the contactor height must be the same, Eq. (11-133) can also be converted into a form relating the equilibrium-stage requirement N and the overall gas-phase transfer-unit requirement $(NTU)_{OG}$ for a given separation:

$$(\text{NTU})_{OG} = N \frac{\ln (mV/L)}{(mV/L) - 1} \tag{11-134}$$

One could as well use Eq. (11-134) to obtain an equivalent number of transfer units for each stage during a calculation of a continuous countercurrent contactor by equilibrium-stage equations.

From Eqs. (11-133) and (11-134) it can be seen that $(NTU)_{OG}$ will be greater than N for a given separation and $(HTU)_{OG}$ will be less than HETP if $mV/L < 1$. The reverse is true if $mV/L > 1$.

Allowance for Axial Dispersion

The methods presented so far for analysis of continuous countercurrent contactors have been based upon the assumption of plug flow of the counterflowing streams. This leads to operating lines or curves identical to those for the same flow rates in staged equipment. Plug flow corresponds to forward movement of all elements of a stream at the same linear velocity, with no mixing in a forward or backward direction.

Departures from plug flow can occur for any or all of several reasons:

1. Longitudinal mixing can occur because of turbulence or because of the presence of well-mixed pockets along the flow path, e.g., large void spaces in a packed column.
2. Drag from the motion of one of the counterflowing streams can cause local reverse flow of the other stream. An example is countercurrent contacting of a liquid at a high flow rate with a gas at a low flow rate, where there is resultant local reverse flow of the gas. Another example is a spray contactor, where the motion of the dispersed droplets causes large-scale mixing motions in the continuous phase.
3. Fluid elements can move forward at locally different velocities because of velocity gradients or because of inhomogeneities in a packing, e.g., near a wall. Even in laminar flow in a tube the fluid at the center moves at a much greater axial velocity than the fluid near the walls. Extreme forms of this phenomenon are known as *channeling.*

Mixing in the radial direction, perpendicular to the overall direction of flow, serves to reduce the amount of apparent mixing or dispersion in the direction of flow. Differences in composition which develop over a cross section because of channeling, longitudinal mixing, etc., are ironed out by mixing or diffusion across the cross section. This leads to the interesting situation, known as *Taylor dispersion,* where the apparent diffusion coefficient for axial or longitudinal dispersion in laminar flow varies *inversely* with the molecular diffusion coefficient (Sherwood et al., 1975, pp. 81–82). This follows since the velocity profile causes the axial spread of solute whereas molecular diffusion in the radial direction serves to remix the fluid and reduce axial dispersion.

Departures from plug flow due to axial-dispersion effects are most severe (1) when a design calls for a change in solute concentration by a very large factor in a separator, e.g., 99.9 percent solute removal, (2) when a relatively low $(HTU)_{OG}$ or $(HTU)_{OL}$ means that a relatively short contactor accomplishes a substantial number of transfer units, (3) when large eddies or circulation patterns can develop in a continuous phase because of a lack of flow constrictions, (4) when there is a wide distribution of drop sizes in the dispersed phase of a gravity-driven contactor, and/or (5) when there is a very large or very small flow ratio. Allowance for axial dispersion

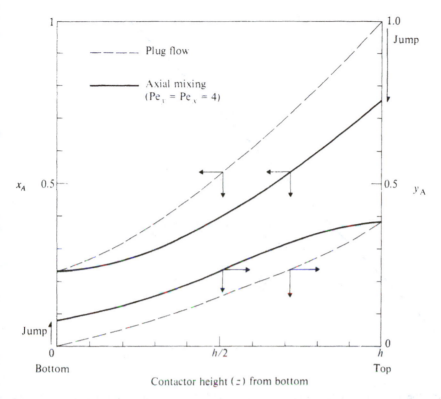

Figure 11-17 Concentration profiles for a continuous countercurrent stripping operation, with and without axial mixing. *(Adapted from Pratt, 1975, p. 75; used by permission.)*

is particularly important in the analysis of most column-form liquid-liquid extractors, gas-liquid spray columns, fixed-bed separation processes such as chromatography, and the cross-flow contacting on the individual plates of a plate column, in addition to other situations.

The effect of axial dispersion upon the performance of a continuous countercurrent contactor is shown in Figs. 11-17 and 11-18, which show a solution for axial dispersion described by effective axial diffusion coefficients in both streams. Figure 11-17 shows concentration profiles of the two counterflowing streams vs. contactor length, and Fig. 11-18 is the resulting yx operating diagram. Curves are shown both for the absence and presence of axial mixing. From Fig. 11-17 it can be seen that axial mixing produces two effects: (1) a general reduction in the concentration gradients along the column length, resulting from concentrations being evened out by the axial mixing process, and (2) a jump in concentration at the inlet of each stream. The concentration at the feed level within the column is different from the concentration of the feed itself because of the dilution of the feed by material brought from farther within the column by the axial-mixing effect. The concentration jump at the feed inlet is specific to mechanisms 1 and 2, mentioned at the beginning of this section, but does not occur for the third mechanism of differences in forward velocity.

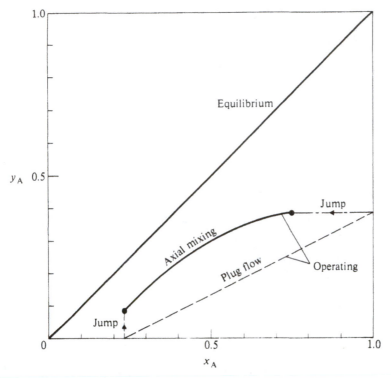

Figure 11-18 Operating diagram for stripping operation of Fig. 11-17, with and without axial mixing. *(Adapted from Pratt, 1975, p. 75: used by permission.)*

The axial-mixing effects necessarily draw the curves for x_A vs. z and for y_A vs. z closer together in Fig. 11-17, reducing the concentration-difference driving force for mass transfer between the counterflowing streams. This effect can also be seen on the equivalent operating diagram (Fig. 11-18). The inlet-concentration jumps displace the ends of the operating curve inward toward the equilibrium line from the plug-flow operating line, and the entire operating curve with axial mixing is located closer to the equilibrium line than in plug flow. This reduction in concentration-difference driving force decreases the denominator of Eqs. (11-95) and (11-96) [or Eq. (11-101)], making more transfer units and more contactor height necessary to accomplish a given separation. Alternatively, less separation is accomplished with a given contactor height. The greater the amount of axial mixing the greater the effect.

Models of axial mixing For the most part, two basic models have been used to analyze the effect of axial mixing on the performance of countercurrent contactors. These are the *differential* model, treating axial mixing as a diffusion process, and the *stagewise backmixing* model, treating axial mixing as a succession of mixed stages or mixing cells with both forward flow and backflow between stages.

Differential model When axial mixing is described as a diffusion process with an equivalent axial-diffusion coefficient in either phase, Eqs. (11-92) and (11-93) for y_A

and for x_A are modified by the addition of axial-diffusion terms, denoting the difference between the diffusive fluxes of component A out of and into a differential slice of column height as $dN_A/dz = -Ec\,d^2x_A/dz^2$, where E is an effective axial diffusion coefficient. E is at least as large as the molecular diffusion coefficient but usually is orders of magnitude greater. It must be determined and correlated experimentally for different types of contacting equipment.

The resulting equations for the y and x phases, assuming constant total flows, are

$$\frac{V}{A}\frac{dy_A}{dz} - E_y c_y \frac{d^2 y_A}{dz^2} = \frac{K_L a c_x}{m}(y_{AE} - y_A) \tag{11-135}$$

$$-\frac{L}{A}\frac{dx_A}{dz} - E_x c_x \frac{d^2 x_A}{dz^2} = -K_L a c_x (x_A - x_{AE}) \tag{11-136}$$

These are simultaneous second-order ordinary differential equations, coupled through $y_{AE} = mx_A + b$ and $y_A = mx_{AE} + b$. c_y and c_x are the molar densities in the y and x phases. The boundary conditions most often used (see, for example, Miyauchi and Vermeulen, 1963) are

$$\frac{V}{A}(y_A - y_{AF}) = E_y c_y \frac{dy_A}{dz} \qquad \text{at } z = 0 \tag{11-137}$$

and

$$\frac{L}{A}(x_{AF} - x_A) = E_x c_x \frac{dx_A}{dz} \qquad \text{at } z = h \tag{11-138}$$

for the stream inlets (subscript F = feed compositions), and

$$E_y \frac{dy_A}{dz} = 0 \qquad \text{at } z = h \tag{11-139}$$

and

$$E_x \frac{dx_A}{dz} = 0 \qquad \text{at } z = 0 \tag{11-140}$$

at the stream outlets. Equations (11-137) and (11-138) give the inlet concentration jumps directly. Figure 11-17 shows that dy_A/dz and $dx_A/dz \to 0$ at the stream outlets, corresponding to Eqs. (11-139) and (11-140).

Solutions to Eqs. (11-135) and (11-136) with boundary conditions given by Eqs. (11-137) to (11-140) necessarily involve seven dimensionless groups: (1) a dimensionless y-phase concentration, such as $(y_A - y_{AF})/(y_{AE,\,x_A=x_{AF}} - y_{AF})$, (2) a dimensionless x-phase concentration, (3) a y-phase column Péclet number $Pe_y = Vh/AE_y c_y$, (4) an x-phase column Péclet number $Pe_x = Lh/AE_x c_x$, (5) the stripping or extraction factor mV/L, (6) the number of transfer units provided in the absence of axial mixing, $h/(HTU)_{OL} = hK_L ac_x A/L$, or, instead, the related $h/(HTU)_{OG}$ expression, and (7) fractional column height z/h.

Stagewise backmixing model Figure 11-19 shows the assumptions of the stagewise-backmixing model, as applied to a three-stage contactor. f_L is the fraction of the net forward-flowing liquid stream that backmixes to the previous stage, and f_V is the fraction of the net forward-flowing vapor stream that backmixes. This leads to two

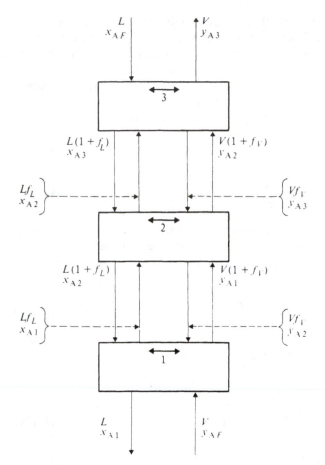

Figure 11-19 Stagewise-backmixing model for three-stage contactor.

sets of difference equations, where the subscripts $p - 1$, p, $p + 1$, etc., refer to stage numbers:

$$(1 + f_L)x_{A,\,p+1} - (1 + 2f_L)x_{A,\,p} + f_L x_{A,\,p-1} = \frac{h K_L a c_x A}{NL} (x_{A,\,p} - x_{AE,\,p}) \quad (11\text{-}141)$$

and

$$f_V y_{A,\,p+1} - (1 + 2f_V)y_{A,\,p} + (1 + f_V)y_{A,\,p-1} = \frac{h K_L a c_x A}{mNV} (y_{A,\,p} - y_{AE,\,p}) \quad (11\text{-}142)$$

N is the total number of stages. Boundary conditions have usually been obtained by adding fictitious end stages in which settling (but no mass transfer) occurs. This gives

$$x_{AF} + f_L x_{A,\,N} = (1 + f_L)x_{A,\,N+1} \quad (11\text{-}143)$$

and

$$y_{AF} + f_V y_{A,\,1} = (1 + f_V)y_{A,\,0} \quad (11\text{-}144)$$

at the phase inlets, and

$$x_{A,0} = x_{A,1} \qquad \qquad (11\text{-}145)$$

and
$$y_{A,N} = y_{A,N+1} \qquad \qquad (11\text{-}146)$$

at the phase outlets.

Solutions to the stagewise-backmixing model involve eight dimensionless groups, which are the same as those for the differential model, except that the two column Péclet numbers are replaced by the two fraction backmixing parameters, f_L and f_V. The added group is N.

The differential model should be more appropriate for devices such as packed columns which are the same throughout the contacting height, while the stagewise-backmixing model resembles more closely the physical characteristics of compartmentalized column extractors, e.g., the rotating-disk contactor (RDC) shown in Fig. 4-22. Notice that the stagewise-backmixing model, as described here, allows for rate limitations on mass transfer within a stage [Eqs. (11-141) and (11-142)]. It is also possible to use a backmixing model with equilibrium stages or with specified Murphree efficiencies.

Both the differential and stagewise-backmixing models postulate that an element of fluid is as likely to go forward as backward relative to the average forward flow of a stream. It is therefore not too surprising that solutions to the two models become the same in form for a large number of stages N, with the following interchange of variables:

$$\frac{N}{f_L + \frac{1}{2}} \rightarrow \mathrm{Pe}_x \qquad \qquad (11\text{-}147)$$

$$\frac{N}{f_V + \frac{1}{2}} \rightarrow \mathrm{Pe}_y \qquad \qquad (11\text{-}148)$$

(Mecklenburgh and Hartland, 1975).

Other models Mecklenburgh and Hartland (1975) describe additional modeling approaches taking into account differences in forward velocities and cross mixing between such streams. Kerkhof and Thijssen (1974) present a modeling approach based upon a series of mixing cells that is a different number for each phase with no backmixing between cells.

Analytical solutions Analytical solutions to both the differential and stagewise-backmixing models are generally quite complex, even for a linear equilibrium relationship and constant total flows. As is shown clearly, by Pratt (1975) for example, when Eqs. (11-135) and (11-136) are combined for the differential model, a fourth-order ordinary differential equation results. The solution to this equation, for x_A or y_A as a function of z, is a summation of exponential terms, the coefficients in the exponents themselves being implicit roots of a characteristic equation. Furthermore the coefficients of the terms themselves are determined from simultaneous solution of four equations involving the boundary conditions. Similarly (see, for example, Pratt,

1976b), the stagewise-backmixing model reduces to a fourth-order difference equation in y_A or x_A, the solution being another sum of exponential terms with the coefficients in the exponents determined from another characteristic equation and the coefficients of the terms coming from simultaneous equations. If the problem is a design problem rather than an operating problem, the situation is complicated even further by the fact that the column height appears in the Péclet numbers, which enter strongly into the characteristic equations, necessitating a complicated iterative solution.

Mecklenburgh and Hartland (1975) have compiled and analyzed solutions to both the basic models for countercurrent contacting, considering many simpler subcases of the general problem. They present convenient algorithms which can be used for attacking design and operating problems under various circumstances. Miyauchi and Vermeulen (1963) have also summarized solutions to the differential model for both the general case (with linear equilibrium) and various subcases.

Pratt (1975) has presented an approximate method which is satisfactory for design problems where mV/L lies between 0.5 and 2 and where the contactor length exceeds 1.3 m and $(NTU)_{ox}$ and $(NTU)_{oy}$ exceed 2. The method involves solving the cubic characteristic equation restated in terms of local Péclet numbers, $Pe'_y = Vd_p/AE_y c_y$ and $Pe'_x = Ld_p/AE_x c_x$, and then using the roots of that equation directly in approximate algebraic expressions. Pe'_y and Pe'_x involve a local characteristic dimension, e.g., the packing size d_p, instead of the unknown column height h; these local Péclet numbers are functions of packing geometry and flow conditions alone, determined experimentally. $Pe_y = Pe'_y h/d_p$, and $Pe_x = Pe'_x h/d_p$. Pratt (1976a) suggests handling cases of curved equilibrium by dividing the column into two or three subsections and applying the linear-equilibrium analysis to each. A similar approach can be applied for the stagewise-backmixing model (Pratt, 1976b).

Rod (1964) describes a graphical method involving a modified operating diagram suitable for cases of curved equilibrium and axial mixing in only one phase. It is difficult to extend this method to cases with axial mixing in both phases, however (Mecklenburgh and Hartland, 1967).

One situation which arises with some frequency and for which there is a relatively simple analytical solution is the case where L/mV is effectively zero and there is axial mixing in the x phase. This could correspond to a situation where V/L is very large or where x_{AE} is effectively zero or constant throughout the contactor, perhaps as a result of an irreversible reaction of A in the y phase. Since $L/mV \to 0$ and x_{AE} does not change along the contactor, axial mixing in the y phase is unimportant. The equation for the outlet concentration of the x phase (Miyauchi and Vermeulen, 1963) is

$$\frac{x_{A,\,out} - x_{AE,\,out}}{x_{A,\,in} - x_{AE,\,out}} = \frac{4 v e^{Pe_x/2}}{(1+v)^2 e^{(v\,Pe_x)/2} - (1-v)^2 e^{-(v\,Pe_x)/2}} \tag{11-149}$$

where
$$v = \left[1 + \frac{4h}{(HTU)_{Ox}\,Pe_x} \right]^{1/2} \tag{11-150}$$

Another extreme occasionally encountered is that where there is essentially complete axial mixing of one phase and negligible axial mixing in the other phase. The rise of uniform bubbles through a short height of liquid or the fall of drops through a short height of gas can approach this situation. If it is the liquid that is well mixed and the vapor that is unmixed at all points $x_A = x_{A, \text{out}}$ and $y_{AE} = mx_{A, \text{out}} + b$. Substituting into Eq. (11-95) and integrating gives

$$\frac{hAK_G aP}{V} = \frac{h}{(\text{HTU})_{OG}} = \ln \frac{(y_{AE} - y_{A, \text{in}}) - (y_{A, \text{out}} - y_{A, \text{in}})}{y_{AE} - y_{A, \text{in}}} \qquad (11\text{-}151)$$

or

$$\frac{y_{A, \text{out}} - y_{A, \text{in}}}{y_{AE} - y_{A, \text{in}}} = 1 - e^{-h/(\text{HTU})_{OG}} \qquad (11\text{-}152)$$

Modified Colburn plots For linear equilibrium and any combination of Pe_x and Pe_y it is possible to depict the solution of the differential model for effects of axial dispersion graphically, in the same form used in Fig. 11-16. This can also be done for the stagewise-backmixing model for any combination of f_L, f_V, and N, with linear equilibrium.

Figure 11-20 shows such a plot for the case of a contactor where $\text{Pe}_x = 10$ and $\text{Pe}_y = 20$, such as might typify the operation of an RDC extractor. The plug-flow solution is presented for comparison. From the figure it is apparent (1) that the axial dispersion serves to reduce the separation obtained with a given contactor height and (2) that the effect of axial mixing in reducing the separation is particularly severe for mV/L of the order of unity and slightly above. There is only a small effect for the asymptotic curves occurring for $mV/L < 1$.

Numerical solutions The equations for the stagewise-backmixing model [Eqs. (11-141) and (11-142)] are both tridiagonal. If f_L, f_V, m, and $K_L ac_x/L$ are not functions of composition, and if multiple transferring solutes do not interact through phase equilibrium or mass-transfer expressions, the equations can be solved by the Thomas method for each solute.

If the coefficients are dependent upon composition and/or if the solutes do interact, the equations can still be handled as a set of simultaneous nonlinear equations which will take the block-tridiagonal form (Chap. 10 and Appendix E) upon successive linearization in a successive approximation solution. McSwain and Durbin (1966) describe an approach of this type, using a pentadiagonal matrix to solve a problem with one transferring component. Ricker et al. (1979) extend the method to multiple transferring solutes, allowance for mass-transfer resistances in both phases, more complex phase equilibria, and systems described by the diffusion model.

Equations (11-137) and (11-138) for the differential model can be converted into difference equations by dividing the column height into a succession of slices and replacing the derivatives by Eq. (11-132) for the first derivative and

$$\left[\frac{d^2 f(z)}{dz^2} \right]_n = \frac{f(z)_{n+1} - 2f(z)_n + f(z)_{n-1}}{(\Delta z)^2} \qquad (11\text{-}153)$$

for the second derivative. This converts Eqs. (11-137) and (11-138) into simultaneous sets, each composed of tridiagonal equations. In fact, the resulting equations are very

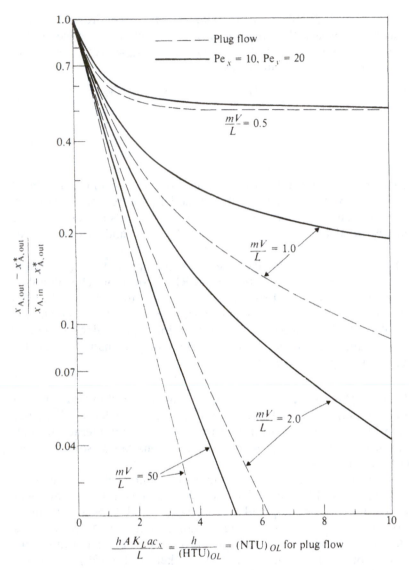

$$\frac{h A K_L a c_x}{L} = \frac{h}{(\text{HTU})_{OL}} = (\text{NTU})_{OL} \text{ for plug flow}$$

Figure 11-20 Modified Colburn plot, showing effects of axial dispersion for $\text{Pe}_x = 10$ and $\text{Pe}_y = 20$. *(Adapted from Earhart, 1975.)*

nearly the same as those for the stagewise-backmixing model, except for how the first derivatives are approximated. The equations resulting from putting the differential model into difference form can then also be solved by the block-tridiagonal-matrix method. Newman (1967b, 1968) shows how the first derivatives in the boundary conditions can be handled through image points. The equations are linear if E_x, E_y, m, and $K_L a c_x / L$ are not functions of composition and solutes do not interact. If any or all of those parameters do vary with composition, a successive-approximation solution can be made by the full multivariate Newton SC method.

These solutions, as described, are suitable for operating problems, where h is known. For design problems with unknown h, the solution can be obtained as an interpolation between successive operating problems. If successive approximation is required to solve the operating subproblems because of nonlinearity, one can then devise a method similar to that of Ricker and Grens for staged distillation (Chap. 10) to converge the column height simultaneously with the compositions.

Example 11-12[†] Sherwood and Holloway (1940) report data for the desorption of oxygen from water into air flowing at atmospheric pressure in a 0.51-m-diameter column packed with 5.1-cm Raschig rings to a height of 15.5 cm. For water flows and air flows of 5.4 and 0.31 kg/s·m², respectively, the value of $(HTU)_{OL}$ reported at 25°C was 0.29 m, calculated assuming plug flow of both phases. For the same packing and flow conditions Dunn et al. (1977) report $Pe'_x = 0.21$. (a) What was the true $(HTU)_{OL}$ in the Sherwood and Holloway experiment, calculated allowing for axial mixing? (b) Calculate the removal of oxygen for a column with the same packing and flow conditions but a packed height of 2.5 m. Express the removal as the percent of the total removal achievable if equilibrium were obtained with air. By how much does axial mixing increase the height requirement for this removal?

SOLUTION Because of the very low solubility of oxygen in water (Fig. 6-6) the system is completely liquid-phase-controlled for mass transfer ($K_L a \approx k_L a$), and the amount of oxygen buildup in the gas phase from the desorption process is negligible. Hence $L\,mV \to 0$, and Eq. (11-149) can be used to analyze the effect of axial dispersion.

(a) For the 15.5-cm packed height, $Pe_x = Pe'_x\,h\,d_p = (0.21)(15.5/5.1) = 0.64$. Substituting into Eq. (11-103) for plug flow, we have

$$[(NTU)_{OL}]_{\text{plug flow}} = \frac{h}{[(HTU)_{OL}]_{\text{plug flow}}} = \frac{0.155}{0.29} = 0.53$$

From Eq. (11-101),

$$[(NTU)_{OL}]_{\text{plug flow}} = \ln \frac{x_{AE} - x_{A,\text{in}}}{x_{AE} - x_{A,\text{out}}}$$

and so

$$\frac{x_{A,\text{out}} - x^*_{A,\text{out}}}{x_{A,\text{in}} - x^*_{A,\text{out}}} = e^{-0.53} = 0.586$$

Substituting into Eq. (11-149) gives

$$0.586 = \frac{4ve^{0.32}}{(1+v)^2 e^{0.32v} - (1-v)^2 e^{-0.32v}}$$

By trial and error,

$$v = 2.27$$

From Eq. (11-150),

$$\frac{h}{(HTU)_{OL}} = \frac{Pe_x}{4}(v^2 - 1) = \frac{(0.64)[(2.27)^2 - 1]}{4} = 0.66$$

$$(HTU)_{OL} = \frac{0.155}{0.66} = 0.235 \text{ m}$$

Allowance for axial mixing served to reduce $(HTU)_{OL}$ to $0.235/0.29 = 81$ percent of the apparent plug-flow value.

[†] Adapted from Sherwood et al., 1975, pp. 615–616; used by permission.

(b) For the 2.5-m packed height, $Pe_x = (0.21)(250/5.1) = 10.3$. Substituting into Eq. (11-150) gives

$$v = \left[1 + \frac{(4)(2.5)}{(0.235)(10.3)}\right]^{1\,2} = 2.27$$

The value is the same as in part (a) since h/Pe_x is constant.
    Substituting into Eq. (11-149) yields

$$1 - R = \frac{(4)(2.27)e^{10.3\,2}}{(3.27)^2 e^{(2.27)(10.3)\,2} - (-1.27)^2 e^{-(2.27)(10.3)/2}}$$

where R is the fraction of the equilibrium removal achieved. Solving, we have

$$1 - R = 0.0012$$

so that the removal is 99.88 percent of the equilibrium removal.
    The transfer-unit requirement for the same removal in the plug-flow case can be obtained from Fig. 11-16 or Eq. (11-118), put in the form involving x_A and $(NTU)_{OL}$, in which case $mV/L \to \infty$, giving $(NTU)_{OL} = 6.73$. The height if plug flow prevailed would then be

$$h_{\text{plug flow}} = (6.73)(0.235) = 1.58 \text{ m}$$

Axial dispersion has increased the required packed height by a factor of $2.5/1.58 = 1.58$, or by 58 percent. $\qquad\square$

Example 11-12 illustrates the upper range of effects that can be expected from axial mixing in packed gas-liquid contactors, since $(HTU)_{OL}$ is relatively low. In part (a) the small packed height made Pe_x relatively small (0.64), so that there was an amount of mixing large enough to affect the separation even though the liquid solute concentration did not change by much of a factor through the column. In part (b) the value of Pe_x was much higher, signifying a much smaller amount of axial mixing. However, the effect of this smaller amount of axial mixing on the height requirement was even greater than in part (a) because the liquid concentration changed by a very large factor through the column.

DESIGN OF CONTINUOUS COCURRENT CONTACTORS

Continuous-contactor separation processes requiring the action of less than one equilibrium stage to accomplish the desired separation can be operated in cocurrent, as well as countercurrent-flow configurations. Figure 11-21a shows a packed gas-liquid contactor operated with countercurrent flow, while Fig. 11-21b shows a packed gas-liquid contactor with cocurrent flow. As is further discussed in Chap. 12, cocurrent flow can give higher throughput and more rapid interphase mass transfer but does not give the benefits of multiple staging.
    The analysis of a continuous cocurrent contactor is quite similar to that of a countercurrent contactor. For plug flow the rate expressions, Eqs. (11-75) and (11-76), are the same, and the mass balance is changed by a minus sign. For gas-liquid cocurrent flow, the equivalent of Eq. (11-92) is

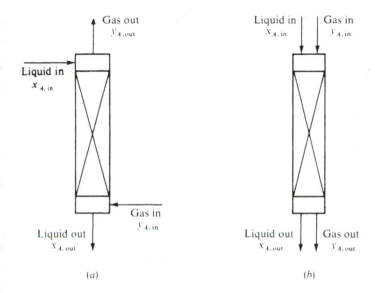

Figure 11-21 (*a*) Countercurrent and (*b*) cocurrent packed gas-liquid contactors.

$$-\frac{V}{A}\frac{dy_A}{dh} = \frac{L}{A}\frac{dx_A}{dh} = \text{rate of mass transfer of A into vapor, mol/s} \cdot (\text{m}^3 \text{ tower volume})$$

(11-154)

Equations (11-92) and (11-154) differ only by a minus sign on the first term.

Carrying through for cocurrent flow the same derivation that led to Eq. (11-94), we find

$$h = -\frac{V}{A}\int_{y_{A,\,\text{out}}}^{y_{A,\,\text{in}}} \frac{dy_A}{K_G\,Pa(y_{AE} - y_A)}$$

(11-155)

which is identical to Eq. (11-94). Similarly, Eq. (11-100) involving K_L is unchanged.

The two types of contactor differ, however, in the functionality between y_{AE} and y_A, as is shown in Fig. 11-22 for a stripping process in which a solute is removed from liquid into a gas. The operating line for the countercurrent case is given by

$$Vy_A - Lx_A = Vy_{A,\,\text{out}} - Lx_{A,\,\text{in}}$$

(11-156)

while that for cocurrent flow is

$$Vy_A + Lx_A = Vy_{A,\,\text{in}} + Lx_{A,\,\text{in}}$$

(11-157)

The operating lines in Fig. 11-22 have been set up so that the terminal gas and liquid compositions are the same in the cocurrent and countercurrent cases. For any value of y_A, the mass-transfer driving force $y_A - y_{AE}$ is given by the vertical arrows shown in Fig. 11-22. Clearly $y_A - y_{AE}$ at a given y_A is different for cocurrent flow than for countercurrent flow, because of the different placement of the operating line. Consequently, the transfer-unit integrals will have different values, and the packed

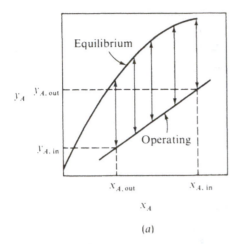
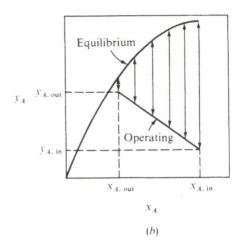

Figure 11-22 Driving forces for (a) countercurrent and (b) cocurrent strippers.

heights required for a given separation will be different for the two cases, even though $K_G Pa$ may be the same.

The integral in Eq. (11-155) can be used again to define a number of transfer units through Eq. (11-96). With straight operating and equilibrium lines an analytical solution can be obtained but differs from that for countercurrent flow because of the different operating-line expression. For cocurrent flow, the equivalent of Eq. (11-118) is

$$(\text{NTU})_{OG} = \frac{\ln \left\{ [1 + (mV/L)][(y_{A, in} - y^*_{A, out})/(y_{A, out} - y^*_{A, out})] - (mV/L) \right\}}{1 + (mV/L)} \quad (11\text{-}158)$$

For more complex situations involving curved equilibria, variable $K_G a$, multicomponent systems, varying total flows, interacting solutes, etc., the block-tridiagonal matrix solution can be used in the same way as for complex cases with countercurrent plug flow. However, cocurrent-flow computations are usually more readily accomplished as initial-value problems, analogous to stage-by-stage methods for multistage separation processes. The computation should start at the feed end of the column and proceed forward, increment by increment. This initial-value approach is not suitable for most countercurrent contactors for reasons entirely analogous to those for the unsuitability of stage-to-stage methods for most multicomponent multistage separations, i.e., errors in assumed terminal concentrations tend to build up during the calculation. However, initial-value formulations are also suitable for countercurrent design problems where either heavy nonkeys or light nonkeys are entirely absent.

Effects of axial mixing can also be handled for cocurrent contactors in ways analogous to those used for countercurrent contactors. Mecklenburgh and Hartland (1975) present analytical solutions for a variety of cocurrent flow cases, using both the differential and stagewise-backmixing models. For more complex situations, involving curved equilibria, multicomponent systems, variable parameters, interacting solutes, etc., block-tridiagonal-matrix approaches analogous to those for countercurrent contactors can be used; again, however, cocurrent systems are usually more efficiently handled as initial-value problems, starting calculations from the feed end (Mecklenburgh and Hartland, 1975).

DESIGN OF CONTINUOUS CROSSCURRENT CONTACTORS

Methods for the design and analysis of continuous crosscurrent contactors are a logical extension of the methods for countercurrent and cocurrent contactors and have been reviewed by Thibodeaux (1969).

FIXED-BED PROCESSES

Fixed-bed processes, such as adsorption, ion exchange, and column chromatography, can also be analyzed for concentration profiles and the separation obtained using concepts of mass-transfer coefficients, transfer units, and axial mixing. Reviews of approaches for the design and analysis of mass transfer in fixed-bed contactors are given by Vermeulen et al. (1973), Vermeulen (1977), Giddings (1965), and Sherwood et al. (1975, chap. 10).

SOURCES OF DATA

Data for k_G, k_L, and/or $k_G a$ and $k_L a$ in various gas-liquid and liquid-liquid contactors are reported by Perry and Chilton (1973), along with various correlations. Additional predictive methods are given by Sherwood et al. (1975, chap. 11), for gas-liquid contactors, and by Hanson (1971) for extractors. Approaches for plate contactors are covered in Chap. 12 of this book. Bolles and Fair (1979) evaluated existing predictive methods for gas-liquid contacting in packed columns in the light of a data bank of 545 experimental measurements; they also present an improved correlation.

Vermeulen et al. (1966) and Hanson (1971) summarize data for axial mixing in extraction devices. Additional data are given by Haug (1971) and Boyadzhiev and Boyadjev (1973). Data for axial mixing in packed columns contacting gas and liquid are given by Dunn et al. (1977), Woodburn (1974), and Stiegel and Shah (1977). Mecklenburgh and Hartland (1975, chap. 2) show how to determine Péclet numbers from experimental concentration profiles in countercurrent or cocurrent contactors. Axial-mixing data are usually reported as Pe'_x and Pe'_y, involving d_p as the length dimension. These can be converted into Pe_x and Pe_y for a given contactor height by multiplying by h/d_p.

REFERENCES

Astarita, G. (1966): "Mass Transfer with Chemical Reaction," Elsevier, Amsterdam.
Barrer, R. M. (1951): "Diffusion in and through Solids," Macmillan, New York.
Beek, W. J., and C. A. P. Bakker (1961): *Appl. Sci. Res.*, **A10**:241.
Bird, R. B., W. E. Stewart, and E. N. Lightfoot (1960): "Transport Phenomena," Wiley, New York.

Bolles, W. L., and J. R. Fair (1979): in Distillation—1979, *Inst. Chem. Eng. Symp. Ser.*, **56**.

Boyadzhiev, L., and C. Boyadjev (1973): *Chem. Eng. J.*, **6**:107.

Bugakov, V. Z. (1973): "Diffusion in Metals and Alloys," trans. C. Nisenbaum, Israel Program for Scientific Translations, Jerusalem, 1971.

Byers, C. H., and C. J. King (1967): *AIChE J.*, **13**:628.

Calderbank, P. H., and M. Moo-Young (1961): *Chem. Eng. Sci.*, **16**:39.

Carslaw, H. S., and J. C. Jaeger (1959): "Conduction of Heat in Solids," 2d ed., Oxford University Press, Oxford.

Chilton, T. H., and A. P. Colburn (1934): *Ind. Eng. Chem.*, **26**:1183.

———— and ———— (1935): *Ind. Eng. Chem.*, **27**:255.

Clark, M. W., and C. J. King (1970): *AIChE J.*, **16**:64.

Colburn, A. P. (1933): *Trans. AIChE*, **29**:174.

———— (1939): *Trans. AIChE*, **35**:211.

Crank, J. (1975): "The Mathematics of Diffusion," 2d ed., Clarendon Press, Oxford.

———— and G. S. Park (eds.) (1968): "Diffusion in Polymers," Academic, New York.

Cussler, E. L., Jr. (1976): "Multicomponent Diffusion," Elsevier, Amsterdam.

Danckwerts, P. V. (1951): *Ind. Eng. Chem.*, **43**:1460.

———— (1970): "Gas-Liquid Reactions," McGraw-Hill, New York.

Dunn, W. E., T. Vermeulen, C. R. Wilke, and T. T. Word (1977): *Ind. Eng. Chem. Fundam.*, **16**:116.

Earhart, J. P. (1975): Ph.D. dissertation in chemical engineering, University of California, Berkeley.

Eckert, E. R. G., and R. M. Drake (1959): "Heat and Mass Transfer," McGraw-Hill, New York.

Eduljee, H. E. (1975): *Hydrocarbon Process.*, **54**(9):120.

Ertl, H., R. K. Ghal, and F. A. L. Dullien (1973): *AIChE J.*, **19**:881.

Giddings, J. C. (1965): "Dynamics of Chromatography," pt. 1, Dekker, New York.

Goodgame, T. H., and T. K. Sherwood (1954): *Chem. Eng. Sci.*, **3**:37.

Gröber, H., S. Erk, and U. Grigull (1961): "Fundamentals of Heat Transfer," trans. by J. R. Moszynski, McGraw-Hill, New York.

Hanson, C. (ed.) (1971): "Recent Advances in Liquid-Liquid Extraction," Pergamon, New York.

Haug, H. F. (1971): *AIChE J.*, **17**:585.

Henrion, P. N. (1964): *Trans. Faraday Soc.*, **60**:72.

Hirschfelder, J. O., C. F. Curtiss and R. B. Bird (1964): "Molecular Theory of Gases and Liquids," Wiley, New York.

Hsu, S. T. (1963): "Engineering Heat Transfer," pp. 83–85, Van Nostrand Co., Princeton, N.J.

Jost, W. (1960): "Diffusion in Solids, Liquids, Gases," Academic, New York.

Keey, R. B. (1978): "Introduction to Industrial Drying Operations," Pergamon, Oxford.

Kerkhof, P. J. A. M., and H. A. C. Thijssen (1974): *Chem. Eng. Sci.*, **29**:1427.

King, C. J. (1964): *AIChE J.*, **10**:671.

———— (1971): "Freeze-Drying of Foods," CRC Press, Cleveland, Ohio.

————, L. Hsueh, and K. W. Mao (1965): *J. Chem. Eng. Data*, **10**:348.

Knudsen, J. G., and D. L. Katz (1958): "Fluid Dynamics and Heat Transfer," McGraw-Hill, New York.

Lightfoot, E. N., and E. L. Cussler, Jr. (1965): *Chem. Eng. Prog. Symp. Ser.*, **61**(58):66.

Lu, C. H., and B. M. Fabuss (1968): *Ind. Eng. Chem. Process Des. Dev.*, **7**:206.

Marrero, T. R., and E. A. Mason (1972): *J. Phys. Chem. Ref. Data*, **1**:3.

McCormick, P. Y. (1973): Solids-Drying Fundamentals and Solids-Drying Equipment, pp. 20-1 to 20-64, in R. H. Perry and C. H. Chilton (eds.), "Chemical Engineers' Handbook," 5th ed., McGraw-Hill, New York.

McSwain, C. V., and L. D. Durbin (1966): *Separ. Sci.*, **1**:677.

Mecklenburgh, J. C., and S. Hartland (1967): *Inst. Chem. Eng. Symp. Ser.*, **26**:115.

———— and ———— (1975): "The Theory of Backmixing," Wiley-Interscience, New York.

Miyauchi, T., and T. Vermeulen (1963): *Ind. Eng. Chem. Fundam.*, **2**:113.

Moores, R. G., and A. Stefanucci (1964): Coffee, in R. E. Kirk and D. F. Othmer (eds.), "Encyclopedia of Chemical Technology," 2d ed., vol. 5, Interscience, New York.

Nernst, W. (1904): *Z. Phys. Chem.*, **47**:52.

Newman, J. S. (1967a): in C. W. Tobias (ed.), "Advances in Electrochemistry and Electrochemical Engineering," vol. 5, Wiley-Interscience, New York.

——— (1967*b*): *USAEC Rep.* UCRL-17739.

——— (1968): *Ind. Eng. Chem. Fundam.*, **7**:514.

——— (1973): "Electrochemical Systems," Prentice-Hall, Englewood Cliffs, N.J.

Nowick, A. S., and J. J. Burton (eds.) (1975): "Diffusion in Solids," Academic, New York.

Perry, R. H., and C. H. Chilton (eds.) (1973): "Chemical Engineers' Handbook," 5th ed., McGraw-Hill, New York.

Porter, M. C. (1972): *Ind. Eng. Chem. Prod. Res. Dev.*, **11**:235.

Pratt, H. R. C. (1975): *Ind. Eng. Chem. Process Des. Dev.*, **14**:74.

——— (1976*a*): *Ind. Eng. Chem. Process Des. Dev.*, **15**:34.

——— (1976*b*): *Ind. Eng. Chem. Process Des. Dev.*, **15**:544.

Rasquin, E. A. (1977): M. S. thesis in chemical engineering, University of California, Berkeley.

———, S. Lynn, and D. N. Hanson (1977): *Ind. Eng. Chem. Fundam.*, **16**:103.

Reid, R. C., J. M. Prausnitz, and T. K. Sherwood (1977): "The Properties of Gases and Liquids," 3d ed., chap. 11, McGraw-Hill, New York.

Ricker, N. L., F. Nakashio, and C. J. King (1979): *Am. Inst. Chem. Eng.*, Boston mtg., August.

Rod, V. (1964): *Br. Chem. Eng.*, **9**:300.

Rohsenow, W. M., and H. Y. Choi (1961): "Heat, Mass and Momentum Transfer," Prentice-Hall, Englewood Cliffs, N.J.

Satterfield, C. N. (1970): "Mass Transfer in Heterogeneous Catalysis," M.I.T. Press, Cambridge, Mass.

Schlichting, H. (1960): "Boundary Layer Theory," McGraw-Hill, New York.

Sherwood, T. K. (1950): *Chem. Can.*, June, pp. 19–21.

———, P. L. T. Brian, and R. E. Fisher (1967): *Ind. Eng. Chem. Fundam.*, **6**:2.

——— and F. A. L. Holloway (1940): *Trans. AIChE*, **36**:39.

———, R. L. Pigford, and C. R. Wilke (1975): "Mass Transfer," McGraw-Hill, New York.

Stockar, U. von, and C. R. Wilke (1977*a*): *Ind. Eng. Chem. Fundam.*, **16**:88.

——— and ——— (1977*b*): *Ind. Eng. Chem. Fundam.*, **16**:94.

Stiegel, G. J., and Y. T. Shah (1977): *Ind. Eng. Chem. Process Des. Dev.*, **16**:37.

Thibodeaux, L. J. (1969): *Chem. Eng.*, June 2, p. 165.

Valentin, F. H. H. (1968): "Absorption in Gas-Liquid Dispersions," Spon, London.

Vermeulen, T. (1977): Adsorption (Design) in J. J. McKetta and W. A. Cunningham (eds.), "Encyclopedia of Chemical Processing and Design," vol. 2, Dekker, New York.

———, G. Klein, and N. K. Hiester: Adsorption and Ion Exchange, sec. 16 in R. H. Perry and C. H. Chilton (eds.), "Chemical Engineers' Handbook," 5th ed., McGraw-Hill, New York.

———, J. S. Moon, A. Hennico, and T. Miyauchi (1966): *Chem. Eng. Prog.*, **62**(9):95.

Wendel, M. M., and R. L. Pigford (1958): *AIChE J.*, **4**:249.

Wilke, C. R. (1950): *Chem. Eng. Prog.*, **46**:95.

——— (1977): Absorption, in R. E. Kirk and D. F. Othmer (eds.), "Encyclopedia of Chemical Technology," 3d ed., vol. 1, Wiley-Interscience, New York.

——— and P. Chang (1955): *AIChE J.*, **1**:264.

Woodburn, E. T. (1974): *AIChE J.*, **20**:1003.

Yoshida, F., and Y. Miura (1963): *AIChE J.*, **9**:331.

PROBLEMS

11-A₁ Estimate the diffusion coefficients of (*a*) chlorine in nitrogen at 45°C and 150 kPa abs and (*b*) *n*-butane at high dilution in liquid water at 45°C and 500 kPa abs.

11-B₁ One process that has been suggested for food dehydration involves soaking pieces of the food in a solvent, such as ethanol, and then boiling off the mixture of solvent and residual water under vacuum at ambient temperatures (U.S. patent 3,298,199). One drawback of such a process is the relatively long time required for the solvent to soak into the food and displace water. Suppose that ethanol is to be used as the solvent for dehydration of pieces of steak, in the form of cubes 1.5 cm on a side. Steak contains about 65 vol % water, which for purposes of this problem may be considered to be accessible by a nontortuous path, so that the diffusivity is reduced to simply 65 percent of the free-liquid value.

(a) Estimate the time required for displacement of 98 percent of the initial water. Assume an effective diffusivity equal to the arithmetic average of the two infinite-dilution values. Temperature = 25°C.

(b) Other than the slow rate, what drawbacks would you foresee for this process?

11-C$_2$ Proceed to the nearest wash basin and turn on the water gently enough to give a sustained, laminar flow. Assuming that the entering water contains no dissolved air, calculate a good estimate of the percentage aeration of the water impinging upon the basin at the bottom of the falling jet of water. Percentage aeration is defined as 100 × [(average dissolved-air content)/(equilibrium dissolved-air content)]. Use as a basis for the calculation whatever *simple* measurements and observations of the falling stream of water are pertinent.

11-D$_2$ The corrosion of copper in contact with aerated dilute sulfuric acid is believed to occur as follows:

$$2Cu + H^+ + O_2 \rightarrow 2Cu^+ + HO_2^-$$
$$HO_2^- + 2Cu^+ + 3H^+ \rightarrow 2Cu^{2+} + 2H_2O$$

Various studies have shown that the corrosion rate is rate-limited by mass transfer of dissolved oxygen to the copper surface.

Consider the flow of an aerated, 10 wt % solution of sulfuric acid in water at 25°C through a long copper pipe 5.00 cm in diameter. The inlet acid is equilibrated with air at atmospheric pressure (101.3 kPa), and there is no nucleation of air bubbles within the pipe. The flow rate of acid is 1400 kg/h, and operation is continuous.

Data Assume that the diffusivity of O_2 in 10% H_2SO_4 is the same as that in water. The viscosity of 10% H_2SO_4 at 25°C is 1.10 mPa·s. The density of 10% H_2SO_4 at 25°C is 1064 kg/m^3; that of copper is 8920 kg/m^3. The Bunsen coefficient for pure oxygen dissolved in 10% H_2SO_4 is 0.0230 at 25°C. [The Bunsen coefficient is the volume of gas (measured at 273 K and 101.3 kPa) which dissolves in one volume of liquid at the temperature in question.]

Calculate the average corrosion rate of the copper pipe, expressed as millimeters per year.

11-E$_1$ Calcium sulfate is the least soluble compound present in seawater which has been pretreated by acidification to prevent deposition of $CaCO_3$ and/or $Mg(OH)_2$. At ambient temperature, the solubility limit of $CaSO_4$ is reached when seawater becomes concentrated by a factor of 3.0 over the natural concentration; see, for example, Lu and Fabuss (1968). Consider a reverse-osmosis process for desalination, in which seawater is recycled so that the feed contains seawater already concentrated by a factor of 1.5. Tubular membranes (2-mm diameter) are used, with the water in laminar flow through the tubes at a Reynolds number of 200. The length-to-diameter ratio of the tubes is 50, for which it has been found that the Levêque solution still describes the mass-transfer coefficient. The density of the seawater is 1060 kg/m^3, and the viscosity may be taken to be 1.1 mPa·s at the temperature of operation. The diffusivity of $CaSO_4$, calculated from the Nernst-Haskell equation neglecting the other salts, is 0.91×10^{-9} m^2/s.

(a) What location in the tubes will be most susceptible to deposition of solid $CaSO_4$ on the membrane surface?

(b) For this critical location, what is the maximum water flux through the membrane that can occur without deposition of $CaSO_4$?

11-F$_2$ Sherwood et al. (1967) studied liquid-phase mass-transfer limitations on the desalination of water by reverse osmosis. The membrane was mounted on a porous rotating cylinder, with salt water outside the cylinder and purified water withdrawn from the cylinder inside the membrane. The mass-transfer coefficient in the salt solution adjacent to the membrane surface was varied by changing the rotation speed of the cylinder. Figure 11-23 shows the water fluxes and product water compositions observed at various cylinder rotation speeds for a feedwater containing 165 mol m^3 NaCl, which gives an osmotic pressure of 0.773 MPa vs. pure water. The applied total-pressure difference across the membrane was 4.17 MPa.

(a) Why does the product-water salt content decrease with increasing stirrer speed?

(b) What is the cause of the apparent asymptotes for water flux and product-water salt content at high stirrer speeds?

(c) Calculate the apparent mass-transfer coefficient k_{Jc} for salt between the membrane surface and the bulk feed solution at 100 r/min stirrer speed.

(d) What is the apparent value of the low-flux mass-transfer coefficient k_c for salt at 100 r/min stirrer speed?

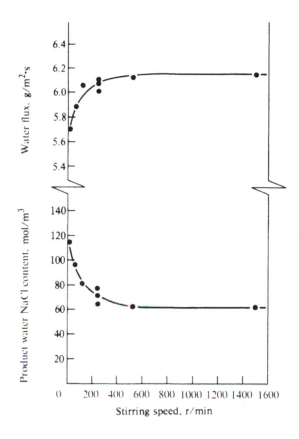

Figure 11-23 Water flux and product-water salt content vs. rotation speed for cylinder-mounted membrane. *(Adapted from Sherwood et al., 1967, p. 10; used by permission.)*

11-G₂ *Ultrafiltration* is a membrane separation process in which solvent is removed from solutions containing high-molecular-weight solutes such as proteins. The principle is similar to that of reverse osmosis, in that pressure is applied to the solution on the feed side of a supported membrane and solvent passes through the membrane. The high-molecular-weight solutes cannot pass through the membrane. One difference from ordinary reverse osmosis is that the osmotic pressure caused by the solutes, even at high concentrations, is usually felt to be negligible because of the high solute molecular weight. Another difference is that the solutes may have only a limited solubility, so that a layer of precipitated solutes, or gel, can readily form adjacent to the membrane surface on the feed side.

The performance of ultrafiltration devices has been successfully interpreted in terms of rate limitations on the solvent flux due to the resistances to solvent flow from both the membrane itself and varying thicknesses of gel or precipitated solutes on the surface of the membrane. The thickness of this gel layer and its consequent resistance to solvent permeation represent a steady-state balance between the rate at which solute is brought to the membrane surface by convection with the permeating solvent, on the one hand, and the rate of mass transfer of solute back into the bulk feed solution, on the other; see, for example, Porter, (1972).

Porter (1972) presents water fluxes (in cubic centimeters of water per minute and per square centimeter of membrane area) observed in a stirred ultrafiltration device of the sort shown in Fig. 11-24 when the feeds were aqueous solutions containing varying concentrations of bovine serum albumin; these data are shown in Fig. 11-25. In the case of 0.9% saline solution (equivalent to the albumin solutions with zero concentration of albumin) the membrane was sufficiently "open" for nothing to be retained; the salt passes freely through the membrane with the water.

(*a*) In terms of the steady-state thickness of accumulated albumin gel, explain why the flux curves in

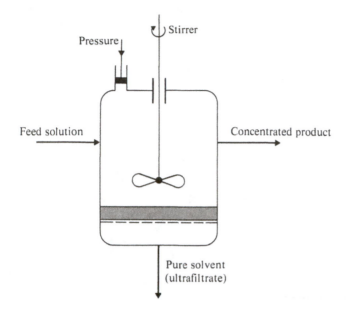

Figure 11-24 Ultrafiltration device.

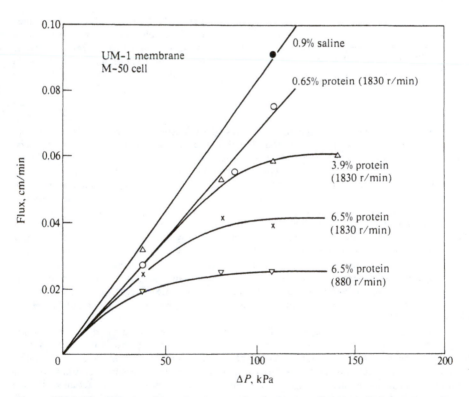

Figure 11-25 Ultrafiltration fluxes for aqueous bovine serum albumin solutions. *(From Porter, 1972, p. 235; used by permission.)*

Fig. 11-25 reach a horizontal asymptote at high transmembrane pressure drops; i.e., once this asymptote is reached, why can't more pressure drop give more water flux? The water flux is $N_w \bar{V}_w$, the ultimate solubility of the protein is c_{A_g}, the feed concentration of protein is c_{AF}, and the mass-transfer coefficient of albumin at the membrane surface is k_{J_c}.

(b) For the 6.5$^\circ_0$ albumin feed, why is the water flux at 1830 r/min higher than that at 880 r/min?

(c) Using the film model to allow for high-flux effects, derive an *analytical* expression relating the water-permeation flux in the presence of a gel layer to the following variables and no others: the low-flux mass-transfer coefficient for albumin k_c, the solubility of albumin c_{A_g}, and the concentration of albumin in the bulk feed solution c_{AF}.

(d) Using the result of part (c) and the observed water fluxes at 1830 r/min for both 3.9 and 6.5$^\circ_0$ albumin in the feed, estimate c_{A_g}. Assume that k_c is unaffected by any changes in viscosity or diffusivity which come from changing albumin concentration and assume that $\bar{V}_w$ is independent of solute concentration.

11-H$_2$ A short packed column is used to remove dissolved gases from a downflowing water stream by desorption into upflowing air. Consider the airflow rate, the water flow rate, the temperature, and the pressure to be constant. In one case the inlet water contains a small amount of dissolved ammonia, and in another case the inlet water contains a small amount of dissolved carbon dioxide. Both situations correspond to less than 0.1 percent dissolved gas in the water. The airflow is sufficient for the desorbed gases not to build up to a level that is significant compared with the equilibrium partial pressure over the aqueous solution. Do you expect that the *percentage removal* of dissolved ammonia will be greater or less than the percentage removal for the carbon dioxide or that they will be about the same? Explain your answer. Note that this is a qualitative problem, not seeking a quantitative answer.

11-I$_3$ Drops of sucrose solution are being dried in a spray at atmospheric (101.3 kPa) pressure. Assume that the drops are spherical and noncirculating from the start of the drying process and that they do not move relative to the air phase. The drop temperature is 25°C, and the effective drop diameter is 60 μm. Assume that sucrose solutions obey Raoult's law and that the partial molal volumes of sucrose and water are constant and equal to the pure-component volumes. The molecular weight of sucrose is 342, and its density is 1588 kg/m^3. The vapor pressure of water at 25°C is 3170 Pa. Henrion (1964) reports diffusivities of sucrose in water at 25°C to be 0.54×10^{-9} m^2/s at high dilution of sucrose in water, and 0.21×10^{-9} m^2/s at 45 wt $^\circ_0$ sucrose in water.

For evaporation of water from (a) a 0.1 wt $^\circ_0$ solution of sucrose in water and (b) a 45 wt $^\circ_0$ solution of sucrose in water indicate which phase is rate-limiting for mass transfer and find the time required for the removal of the first 2 percent of the water present, assuming that the diffusivity is uniform throughout the drop.

11-J$_2$ Freeze-drying of foods removes water by sublimation, i.e., direct transition of water from ice to water vapor. The process is usually carried out by loading frozen food particles batchwise into large shallow trays stacked in a vacuum chamber. Heat for the sublimation is supplied by conduction and radiation from heating platens, on which the trays rest. The water vapor evolved is taken up as solid ice on chilled condenser tubes or plates, on the side of or outside the drying chamber. As drying occurs, a frozen core retreats inward within each particle. This core is surrounded by a nearly dry layer, through which incoming heat must be conducted and through which outgoing water vapor must pass.

Drying rates are slow, with particles 1.0 cm in size typically taking 4 h or more to dry. The drying rate is limited by one of two constraints: (1) the frozen core must not exceed its apparent melting point $T_{f.\,max}$, and (2) the outer, dry particle surface cannot exceed whatever temperature will cause thermal damage to it $T_{s.\,max}$. The typical absolute pressure range in the dryer is 10 to 100 Pa.

There may or may not be a short constant-rate period at the start of a drying cycle. Subsequently the rate continually decreases throughout the cycle as ice is sublimed. Measured values of T_f and T_s tend to be relatively constant during this period, however. For relatively low chamber pressures, the rate of heat input is typically limited by the constraint involving $T_{s.\,max}$, and T_f is close to the condenser temperature. If the total pressure is raised by allowing inerts to accumulate in the chamber, T_f rises above the condenser temperature and the drying rate becomes limited by the $T_{f.\,max}$ constraint above some critical pressure, often about 1.3 kPa. As the total pressure is further increased toward atmospheric, the rate typically drops, in approximate inverse proportion to pressure. The drying rate in kilograms per hour at these higher pressures is usually independent of the particle loading density on the trays and the particle size.

(a) Explain what causes the rate to decrease throughout the drying cycle even though T_f and T_s are relatively constant.

(b) What is the rate-limiting factor for drying at low total pressures?

(c) What appears to be the rate-limiting factor at higher pressures? Explain how this factor is consistent with the observed effects of pressure, particle size, and loading density.

(d) Suggest a design change to accelerate drying at higher total pressures.

(e) Microwave heating has been suggested and confirmed as a way of accelerating rates of freeze drying at lower chamber pressures. Why could it provide a higher rate? (If necessary, consult a reference to determine the physical basis of microwave heating.)

11-K₁ Rework Prob. 6-C if a packed column is used, with a height of 2.1 m and $(HTU)_{OL} = 0.30$ m. Axial dispersion is negligible.

11-L₂ Repeat Prob. 5-B, if the distillation is to be carried out in a packed column providing $(HTU)_{OG} = 0.46$ m.

11-M₃ An absorption tower packed with 2.5-cm Raschig rings is to be designed to recover NO_2 from a gas stream which is essentially air at atmospheric pressure (101.3 kPa), containing fixed nitrogen as both NO_2 and N_2O_4. Dilute NaOH solution will be used as the absorbing liquid. The mechanism by which nitrogen dioxide, NO_2, is absorbed by water and dilute caustic solution can be described by the reactions

$$2NO_2 \rightleftharpoons N_2O_4 \qquad \text{gas phase}$$

$$N_2O_4(g) \rightleftharpoons N_2O_4(l) \qquad \text{Henry's law equilibrium}$$

$$N_2O_4 + H_2O \rightarrow HNO_2 + HNO_3 \qquad \text{liquid phase}$$

In the case of dilute caustic, the acids are rapidly neutralized as they are formed.

Many investigators have established that the rate of absorption is directly proportional to the partial pressure of N_2O_4 at the gas-liquid interface. In water and in dilute caustic solution the hydrolysis occurs at a finite rate and is pseudo first order and reversible. The gas-phase reaction is so rapid that NO_2 and N_2O_4 are always in equilibrium. At 25°C the equilibrium constant is 6.5×10^{-5} Pa^{-1} $(= p_{N_2O_4}/p^2_{NO_2})$.

Wendel and Pigford (1958) used a short wetted-wall column 8.72 cm long and 2.54 cm ID to investigate the absorption of NO_2 by water. At 25° they found an absorption rate of 1.2×10^{-2} g atom of fixed nitrogen per second and per square meter for an interfacial partial pressure of N_2O_4 of 1010 Pa, in equilibrium with 3950 Pa of NO_2. This rate of absorption was found to be independent of both gas and liquid flow rates.

Yoshida and Miura (1963) have shown for the dilute caustic–air system at a liquid flow of 2.71 kg/m²·s and a gas flow of 0.39 kg/m²·s that the total gas-liquid interfacial area for 2.5-cm Raschig rings is 73 m² per cubic meter of packing.

(a) In the short wetted-wall column, why is the rate of absorption independent of both gas and liquid flow rates?

(b) Estimate the height of packing required to reduce the concentration of NO_2 in an airstream at 25°C from 1 to 0.2 mole percent if the liquid and gas flows are 2.71 and 0.39 kg/m²·s, respectively (per tower cross-sectional area). It is important to recognize that the transferring species here is different from the principal form of nitrogen oxide in the gas phase.

(c) The height calculated in part (b) is large for the relatively modest NO_2 removal achieved. As a good engineer, what suggestions do you have for improving the process?

11-N₂ Suppose that a packed column is used to humidify air by contact with water. The water rapidly reaches the wet-bulb temperature and remains isothermal throughout the column. For the flow conditions and packing size (2.5 cm) used, Dunn et al. (1977) report Pe'_x and Pe'_y equal to 0.14 and 1.0, respectively. Suppose that independent experiments have shown that the value of $(HTU)_{OG}$ for these conditions is 0.30 m, axial dispersion being allowed for properly. Calculate the packing height required to bring the airstream from an initial water-vapor content of zero up to 99.8 percent of the partial pressure corresponding to equilibrium with the water at the wet-bulb temperature.

TWELVE

CAPACITY OF CONTACTING DEVICES; STAGE EFFICIENCY

Most of the discussion so far has been concerned with means of determining the product compositions from a separation device employing one or more contacting stages or with means of determining the stage requirement for a given degree of separation. For separations based upon contacting immiscible phases it is often assumed that each stage provides equilibrium between the product streams or that a *stage efficiency* is used to account for the lack of equilibrium. In addition to stage efficiency, which we have not yet considered, another important design parameter is the throughput *capacity* of a stage or contacting device of a given size, which is the amount of feed that can be processed per unit time. Alternatively, we may want to ascertain the size of a given type of contacting device, diameter of a column, etc., necessary to process a given amount of feed per unit time.

Stage efficiency and throughput capacity are related variables since they both reflect the internal configuration of the contacting device. In a distillation tower they are both influenced by the nature of the trays used, the weir height, the tray spacing etc.; in a mixer-settler contactor they are both influenced by the stirrer speed and the settler geometry. Hence it is appropriate to consider factors influencing efficiency and capacity together, and that is the purpose of this chapter.

FACTORS LIMITING CAPACITY

Most contacting devices fall into some one of the following categories of flow configuration: (1) countercurrent flow, (2) crosscurrent flow, (3) cocurrent flow, and (4) well-mixed vessel. Although the same basic factors influence capacity for these

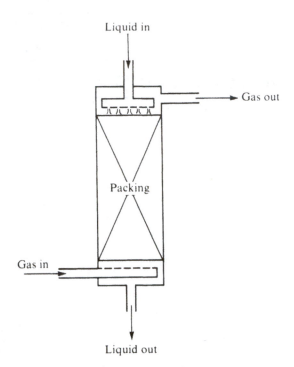

Liquid in

Gas out

Packing

Gas in

Liquid out

Figure 12-1 Countercurrent packed column for gas-liquid contact.

different flow configurations, we shall see that their relative importance can vary widely from situation to situation. Attention will be focused, however, on countercurrent plate and packed columns.

Flooding

Any countercurrent-flow separation device is subject to a capacity limitation due to *flooding*. The phenomenon is related to the ability of the two phases to flow in sufficient quantity in opposite directions past each other within the confines of the contacting device. If we consider the countercurrent packed gas-liquid contacting column shown in Fig. 12-1, we find that the gas phase will pass upward through the column under the impetus of a pressure drop necessitated by friction and form drag against both the packing and the falling liquid. The liquid must fall downward against this pressure drop under the impetus of gravitational force. Generally a packed tower is designed or operated to provide a certain ratio of phase flows, i.e., a fixed L/V, corresponding to a set reflux ratio in distillation or a set solvent-to-gas ratio in absorption. For a tower of given diameter, as the flow rates are increased, the gas pressure drop will increase because of a greater drag force against the packing and the falling liquid. At some point the pressure drop will become so great that it balances the gravity head for liquid flow. At this point the liquid cannot fall down through the packing at a rate equal to the desired feed rate. As a result, a layer of liquid builds up above the packing and the gas flow is seriously reduced and may surge with respect to time. The tower has become unstable and cannot handle the feed rates. A larger-diameter tower is necessary.

To allow for unavoidable variations in flow rate and to provide some extra capacity, countercurrent towers are usually designed to operate at 50 to 85 percent of their flooding limit. Too low a velocity will require a more expensive tower and can result in channeling, which gives ineffective contacting between the phases. Hence the design throughputs are usually not removed by more than a factor of 2 from a controlling flooding limit.

Packed columns A flooding correlation (Sherwood et al., 1938; Leva, 1954; Peters and Timmerhaus, 1968) for gas-liquid contacting in packed towers is shown in Fig. 12-2. Note that the capacity is higher for a tower containing regularly stacked packing of the same type and voidage than for one containing randomly dumped packing. This follows from the greater continuity of the flow channels for regularly stacked packing. Note also that the flooding gas mass flow rate per unit area G increases with decreasing L/G ratio, with decreasing liquid viscosity (film thickness),

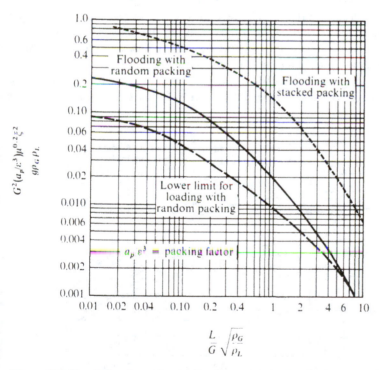

Figure 12-2 Correlation for estimating flooding rate with gas-liquid systems in packed towers, where L = liquid flow rate, lb·h·(ft^2 of empty tower cross-section area); G = vapor flow rate, lb/h·(ft^2 of empty tower cross-sectional area); a_p = surface area of packing per unit tower volume, ft^{-1}; ε = fractional void volume of dry packing; μ = liquid viscosity, cP; g = local acceleration due to gravity = 4.17×10^8 ft/h^2; ξ = (density of water) (density of liquid); ρ_G = density of gas, lb/ft^3; and ρ_L = density of liquid, lb/ft^3. *(From Peters and Timmerhaus, 1968, p. 648; used by permission.)*

with increasing voidage. and with decreasing packing surface area. These trends are all in accord with the picture of flooding being caused by the drag of the gas upon the packing and the falling liquid. The *loading* curve in Fig. 12-2 represents the point at which the pressure drop starts to increase more rapidly with increasing G than it does at lower gas flows.

Plate columns Flooding also will occur at too high a vapor velocity for gas-liquid contacting in a plate column. In this case flooding occurs because the tray-to-tray pressure drop and the liquid flow rate are so large that the downcomers cannot pass the liquid from tray to tray without causing the liquid level in the downcomers to exceed the tray spacing. Flooding capacities of gas-liquid contacting plate columns are usually analyzed through use of the Souders-Brown equation

$$U_{\text{flood}} = K_v \sqrt{\frac{\rho_L - \rho_G}{\rho_G}} \tag{12-1}$$

where U_{flood} is the flooding gas velocity in cubic feet of gas per second and per square foot of active tray area (tower cross-sectional area minus downcomer cross-sectional areas, inlet and outlet) and K_v is a "constant" related to a large number of variables. Figure 12-3 shows a correlation (Fair and Matthews, 1958; Van Winkle, 1967; Fair, 1973) of K_v vs. tray spacing L/G and density ratio for sieve plates and bubble-cap-plates. Note that the *molar* gas flow at flooding is proportional to $U_{\text{flood}}\,\rho_G$, and hence increases with increasing ρ_G except at high values of the abscissa in Fig. 12-3.

The correlation of Fig. 12-3 should be used only for a first approximation of the flooding limit. A more comprehensive design will allow for all the factors influencing

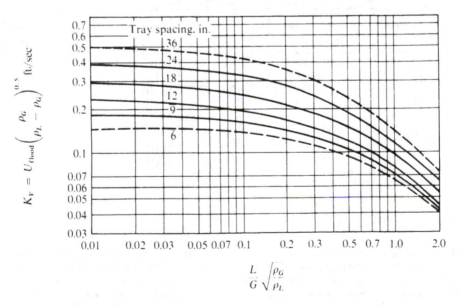

Figure 12-3 Flooding limits for bubble-cap and perforated plates. Notation is given in Fig. 12-2 and in text. (*Adapted from Fair and Matthews, 1958, p. 153; used by permission.*)

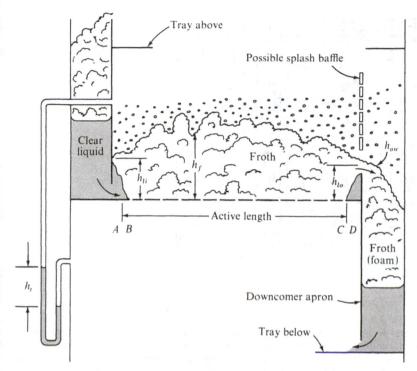

Figure 12-4 Tray-dynamics schematic diagram for froth regime. (*Adapted from Bolles and Fair, 1963, p. 542; used by permission.*)

tray-to-tray pressure drop (see below) and all the factors causing liquid to back up in the downcomer. Such an analysis is described for the froth regime by Fair (1973), Peters and Timmerhaus (1968), and Van Winkle (1967), among others. The factors to be considered are summarized in Fig. 12-4; the liquid backup in the downcomer, expressed as a height H of clear liquid is given by

$$H = h_t + \beta' h_w + h_{ow} + \Delta + h_{da} \tag{12-2}$$

where h_t = tray-to-tray pressure drop, expressed as height of clear liquid [see Eq. (12-4)]

h_w = weir height

β' = aeration factor for dispersion adjacent to the weir (= vol. fraction liquid; $\beta' \leq 1$); Fair (1973) tacitly assumes $\beta' = 1$.

h_{ow} = clear liquid crest over weir (h_{lo} in Fig. 12-4 = $\beta' h_w + h_{ow}$)

Δ = hydraulic gradient across tray, $h_{li} - h_{lo}$, expressed as clear-liquid height difference, Δh_l

h_{da} = friction head loss for flow through downcomer and under downcomer apron

By "clear" liquid height we mean the height to which the aerated froth would settle if the gas in the froth were somehow removed. h_{lo} and h_{li} are values of h_l (shown in Fig. 12-4) at either end of the liquid flow path.

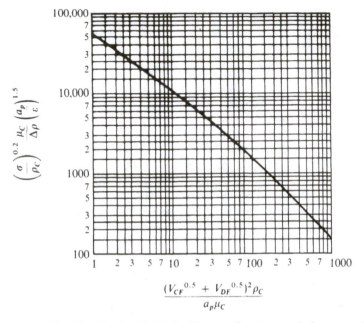

$$\frac{(V_{CF}^{0.5} + V_{DF}^{0.5})^2 \rho_c}{a_p \mu_c}$$

Figure 12-5 Flooding for liquid-liquid contacting in a packed tower, where σ = interfacial tension, lb/h^2; ρ_c = density of continuous phase, lb/ft^3; μ_c = viscosity of continuous phase, lb/ft·h; $\Delta\rho$ = density difference between phases, lb/ft^3; a_p = (surface area of dry packing)/(unit tower volume), ft^{-1}; ε = voidage of dry packing; V_{CF} = flow rate of continuous phase, ft^3/h·(ft^2 empty tower cross-sectional area); and V_{DF} = flow rate of dispersed phase, ft^3/h·(ft^2 empty tower cross-sectional area). (*Adapted from Crawford and Wilke, 1951, p. 428; used by permission.*)

Usual design practice calls for the downcomer liquid height H (based upon clear-liquid density) during operation to be 50 percent or less of the tray spacing. This is necessary since the liquid will be aerated with a significant fraction of vapor in the upper portion of the downcomer.

Liquid-liquid contacting For liquid-liquid contacting in counterflow systems, different flooding correlations and analyses are required because of the greater similarity of densities of the two phases. Figure 12-5 shows a correlation for flooding during liquid-liquid contacting in counterflow packed columns (see also Treybal, 1963). Notice that the phase velocities possible without flooding increase with decreasing packing surface area, increasing voidage, increasing density difference between phases, and decreasing interfacial tension, as would be expected from a consideration of flooding mechanism.

Flooding analyses for liquid-liquid contacting in other types of countercurrent apparatus often require a more fundamental consideration of drop dynamics within the system (Treybal, 1963, 1973).

Entrainment

Entrainment is the incomplete physical separation of product phases from each other. In a plate tower for gas-liquid contacting the gas stream rising to the tray above may sweep liquid droplets along with it, thus entraining liquid to the stage above. In a mixer-settler contactor, if the settler is undersized, there will be drops or bubbles of either phase entrained in the other. Entrainment often represents a capacity limit in separation devices, both because of its detrimental effect upon stage efficiency and because it increases the interstage flows above those which would occur with no entrainment. Hence entrainment in a distillation tower can increase the downward liquid flow so much that it causes flooding of the downcomers.

Plate columns Figure 12-6 shows a correlation of the available data for entrainment in bubble-cap and sieve-plate gas-liquid contacting columns (Fair and Matthews, 1958; Fair, 1973; Bolles and Fair, 1963). The entrainment is expressed as ψ, moles of entrained liquid per mole of gross downflowing liquid (net flow plus return of entrainment). The parameter (percent of flood) is the actual vapor velocity divided by the flooding vapor velocity at the same L/G. Entrainment increases with decreasing tray spacing; this effect is accounted for in Fig. 12-6 by making the "percent of flood" a function of tray spacing (Fig. 12-3).

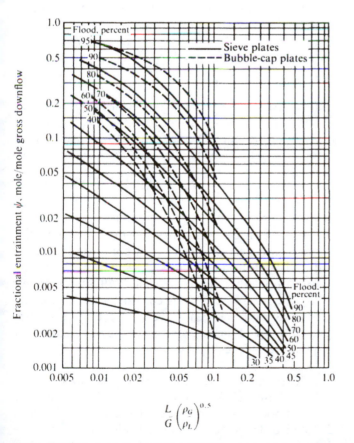

$$\frac{L}{G}\left(\frac{\rho_G}{\rho_L}\right)^{0.5}$$

Figure 12-6 Entrainment correlation. (*Bubble-cap data adapted from Fair and Matthews, 1958, p. 153; sieve-plate data adapted from Bolles and Fair, 1963, p. 547; both used by permission.*)

The rate of entrainment increases sharply with increasing tower loading; hence it is dangerous to operate under conditions where the amount of entrainment is substantial, lest an aberration from normal operating conditions cause the amount of entrainment to be so great that loss of product purity and/or flooding of the tower result. An upper limit of $\psi = 0.15$ is probably advisable for this purpose, and ψ should usually be substantially less.

Another interesting point can be made from Fig. 12-6: entrainment is a more important limiting factor for low values of the group $(L/G)(\rho_G/\rho_L)^{1/2}$. At the higher values of this group, flooding is approached at vapor velocities well below those where entrainment becomes important, but for lower values of the group entrainment will become serious before flooding is approached. Flooding is the more important limit for high L/G and for high-pressure columns (high ρ_G). Entrainment is most important in vacuum columns. The greater tendency toward entrainment at low L/G or low pressure probably relates to the dispersion becoming gas-continuous, rather than liquid-continuous (spray regime vs. froth regime—see below).

Another phenomenon related to entrainment in plate columns is *priming*, wherein the dispersion height on a plate becomes so high that it fills the space between trays and causes the liquid from the tray below to come through the perforations or caps and mix with the liquid on the tray above. The greatest tendency toward priming occurs for naturally foamy liquids and or small tray spacings. Van Winkle (1967) discusses criteria for avoiding priming, the simplest being

$$u_G(\rho_G)^{1/2} \leq 2.3 \tag{12-3}$$

where u_G is gas velocity (ft^3/s · ft^2 active area) and ρ_G is gas density (lb/ft^3).

The effect of entrainment on the separation obtained is usually taken into account by including it in the stage efficiency. Alternatively, entrainment can be included in the mass-balance equations (10-2), which then retain their tridiagonal form (Loud and Waggoner, 1978).

Pressure Drop

Another factor closely related to capacity is the pressure drop within the contacting device. This pressure drop generally will necessitate pump or compressor work at some point outside the separation vessel. In a vacuum system there will be some upper limit to the possible pressure drop within the device, which will often represent the controlling capacity limit; e.g., the pressure drop in a column cannot exceed the total pressure at the bottom. Also, as we have seen in Eq. (12-2), the tray-to-tray pressure drop is an important contributor to the liquid height in the downcomer of a plate tower, and hence a large pressure drop can cause flooding.

Packed columns A pressure-drop correlation for countercurrent gas-liquid contacting in packed columns (Leva, 1954; Fair, 1973) is shown in Fig. 12-7. The coordinates are the same as those in the flooding correlation for plate towers shown in Fig. 12-2. The curves marked A and B delineate the zone of *loading*, defined above. Notice that the pressure drop begins to increase more rapidly with increasing G in the loading region and increases still more rapidly as flooding is approached.

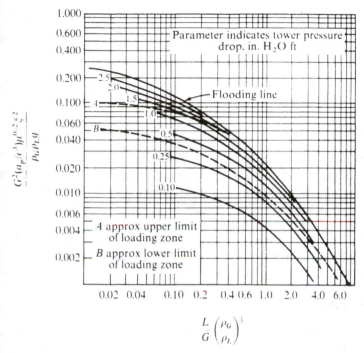

Figure 12-7 Generalized pressure-drop correlation for randomly packed, irrigated columns. Notation identical to that for Fig. 12-2. *(Adapted from Leva, 1954, p. 57; used by permission.)*

Plate columns The pressure drop from tray to tray in a plate column is made up of a number of contributing factors. Referring to Fig. 12-4 and the notation under Eq. (12-2), we find that the tray-to-tray pressure drop h_t expressed as height of clear liquid is given by a sum of terms reflecting static head and additional factors

$$h_t = \beta' h_w + h_{ow} + \tfrac{1}{2}\Delta + h_F \qquad (12\text{-}4)$$

where h_F is the pressure drop due to gas flow through the gas-dispersing unit (the holes in the case of a sieve tray) and the dispersion. h_F may be less than the dry-tray pressure drop since the expansion portion of the latter quantity may become buoyancy supporting the liquid on a wet tray. For a bubble-cap tray the term $\beta' h_w$ should be replaced by the depth of clear-liquid seal over the slots of the caps (a height less than h_w). The factor $\tfrac{1}{2}\Delta$ results from the extra liquid depth due to the hydraulic gradient and equals half the total gradient across the tray. Methods for estimating the various terms in Eq. (12-4) are given in a number of sources (Peters and Timmerhaus, 1968; Van Winkle, 1967; Fair, 1973; Bolles and Fair, 1963; etc.). Most of these authors replace h_F in Eq. (12-4) by a term for dry-tray pressure drop, and some multiply the static-head terms by another aeration factor, β.

Combining Eq. (12-4) with Eq. (12-2) for the liquid backup in the downcomer, we have

$$H = 2(\beta' h_w + h_{ow}) + \tfrac{3}{2}\Delta + h_F + h_{da} \qquad (12\text{-}5)$$

Notice that the terms representing h_{lo} in Fig. 12-4 contribute doubly to the liquid backup.

Pressure drop is discussed in more detail by Davy and Haselden (1975) for sieve trays and by Thorngren (1972) and Bolles (1976a) for valve trays.

Residence Time for Good Efficiency

Yet another factor which can govern the size of a contacting device or limit the throughput of a device of given size is the fluid residence time required for an adequate stage efficiency. In the following discussion of efficiencies it will be apparent that higher efficiencies are gained, in general, by allowing the contacting phases to stay in the contacting device longer. As flow rates through a stage increase, the stage efficiency usually decreases and a point is eventually reached where the stage efficiency becomes so low that the stage or series of stages cannot provide the degree of separation required. This shortcoming is evidenced by unsatisfactory product purities. Poor product purities can also be caused by entrainment, priming, and flooding, in addition to inadequate residence times.

Flow Regimes; Sieve Trays

The flow situation on a plate for vapor-liquid contacting is one of intense agitation and phase dispersion. A typical view is shown in Fig. 12-8, from which it is apparent that it would be very difficult to describe the hydrodynamics by any simple model.

Figure 12-8 A view of typical vapor-liquid contacting on a sieve tray. (*Fractionation Research, Inc., South Pasadena, California.*)

Most analyses of tray hydraulics and efficiencies have been based on the concept of an aerated liquid froth flowing across the tray, as shown schematically in Fig. 12-4. More recently, it has been established that many commercial sieve trays operate instead in a spray regime. The dispersion in the spray regime is mostly vapor-continuous, while in the froth regime it is liquid-continuous (Fane and Sawistowski, 1969; Porter and Wong, 1969; Pinczewski and Fell, 1974). Transition between the regimes appears to be associated with the change from chain bubble formation to more steady jetting of vapor at the holes in a tray (Pinczewski et al., 1973). The transition from the froth regime to the spray regime is favored by high gas velocities and gas densities, larger holes, and greater fraction hole area in the tray (Pinczewski and Fell, 1972; Loon et al., 1973). These are also the current directional trends in tray design.

Trends in tray operating characteristics undergo changes with the transition from the froth to the spray regime. Although tray pressure drop continues to increase with increasing vapor velocity, the difference between the wet and dry tray pressure drops tends to decrease in the spray regime while increasing or staying relatively constant with increasing vapor velocity in the froth regime. Entrainment is more severe and varies more sharply with vapor velocity in the spray regime, reflecting a change from a mechanism of vapor drag on droplets in the froth regime to a mechanism of sustained droplet inertia in the spray regime. As a result, the entrainment correlation given in Fig. 12-6 is less reliable in the spray than in the froth regime (Pinczewski et al., 1975). Regular oscillations of the vapor-liquid dispersion back and forth across small-diameter sieve trays have been observed under some conditions (Biddulph and Stephens, 1974; Biddulph, 1975a); one such oscillation pattern has been associated with the transition from the froth regime to the spray regime (Pinczewski and Fell, 1975).

Range of Satisfactory Operation

Plate columns The capacity limits mentioned so far place an upper limit upon the flow rates allowable within a separation device. There usually will be some factors which place lower limits on the flows too. Figure 12-9 shows schematically the zone of satisfactory operation of a sieve tray for gas-liquid contacting, along with the range of flows in which various different factors can cause unsatisfactory performance (Bolles and Fair, 1963). The coordinates of Fig. 12-9 are similar to those of Figs. 12-3 and 12-7.

As we have already seen, for most values of $(L/G)(\rho_G/\rho_L)^{1/2}$ the capacity limit coming from too high a vapor rate will be *flooding*. For low $(L/G)(\rho_G/\rho_L)^{1/2}$, such as for vacuum towers, the capacity limit corresponding to too high a vapor rate comes from *entrainment*. At very high vapor velocities and relatively low L/G, the efficiency may drop markedly because of *blowing*, wherein the tray is blown clear of liquid in the immediate vicinity of the vapor distributors. When L/G is high, the quantity of liquid flow across the plate may require a very high *liquid gradient* in order to drive the flow. In such a case $\Delta = h_{li} - h_{lo}$ in Fig. 12-4 will be quite large, with possible tendencies toward flooding [Eq. (12-5)] or too high a pressure drop [Eq. (12-4)]. Another result of too high a liquid gradient can be *phase maldistribution*, wherein the

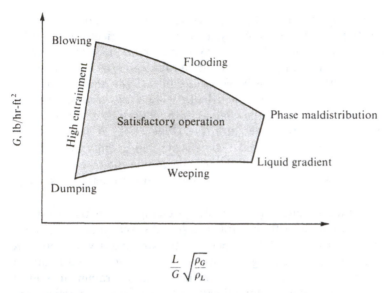

$$\frac{L}{G}\sqrt{\frac{\rho_G}{\rho_L}}$$

Figure 12-9 Effects of vapor and liquid loadings on sieve-tray performance. (*From Bolles and Fair, 1963, p. 556; used by permission.*)

vapor flows preferentially through the perforations near the liquid outlet and the liquid flows in part downward through the perforations near the liquid inlet where the liquid depth is greatest. This flow of liquid downward through the perforations rather than through the downcomer is known, somewhat colorfully, as *weeping* and is favored by relatively low gas-phase flow rates where the gas velocity in the perforations is not large enough to hold the liquid out of the perforations. Massive weeping, known as *dumping*, results in particularly severe phase maldistribution. Within the shaded range of satisfactory operation, the upper portion corresponds to the spray regime and the lower to the froth regime.

The problem of a high liquid gradient is particularly severe for plate columns of large diameter, where there is a long liquid flow path across a plate. One way to prevent a large liquid gradient is to use a *split-flow* tray. As shown in Fig. 12-10a and b, split flow involves dividing the liquid flow in half on each tray, with a central

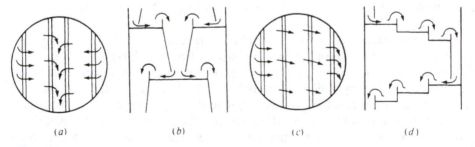

Figure 12-10 Liquid flow patterns for reducing detrimental effects of hydraulic gradient: split flow in (a) top and (b) side view, and cascade cross flow in (c) top and (d) side view. (*From Peters and Timmerhaus, 1968, p. 612; used by permission.*)

Figure 12-11 A 2.9-m-diameter split-flow tray containing type V-1 ballast caps, also known as valve caps. (*Fritz W. Glitsch & Sons, Inc., Dallas, Texas.*)

downcomer and two side downcomers on alternate trays (Peters and Timmerhaus, 1968). Figure 12-11 shows a split-flow valve-cap tray. Multipass trays extend the split-flow concept by dividing the liquid into more than two portions, using multiple downcomers. Bolles (1976*b*) discusses good design practices for split-flow and multipass trays. Another approach for minimizing detrimental gradient effects is the use of *cascade* trays, as shown in Fig. 12-10*c* and *d*. In this case the liquid flows from one level to another along a tray, and the lengths of continuous liquid flow paths are shortened. Figure 12-12 shows a 12-m-diameter cascade tray during assembly. Still other techniques for overcoming flow maldistributions associated with hydraulic gradients on large-diameter trays are the use of bubbling promoters at the liquid inlet and slotted trays which direct the vapor flow horizontally in the direction of liquid flow. These modifications are described by Weiler et al. (1973) and Smith and Delnicki (1975).

The allowable range of vapor velocities in a tower is indicated by the *turndown* ratio, which is the ratio of the maximum allowable vapor velocity to the minimum allowable vapor velocity. For sieve trays this ratio is approximately 3 (Bolles and Fair, 1963; Gerster, 1963; Zuiderweg et al., 1960; Hengstebeck, 1961).

Some of the additional practical factors which enter into tray selection and column design (accessibility, supports, etc.) are discussed by Interess (1971). Frank (1977) discusses a number of different aspects of tray design.

Figure 12-12 Assembly of a 12-m-diameter cascade tray containing ballast, or valve caps. The bottom portion of the plate, grid and matting, is a mist eliminator to remove entrainment from the vapors rising from below. *(Fritz W. Glitsch & Sons, Inc., Dallas, Texas.)*

Comparison of Performance

There have been relatively few comprehensive comparisons of the capacity and efficiency of various types of plates and packing for gas-liquid contacting. One exception is the study reported by Zuiderweg et al. (1960), in which efficiency and capacity measurements were made for four different types of plate (bubble-cap, valve, sieve, and Kittel) and two types of packing (Spraypak and Pall rings) in a 46-cm-diameter column with a 41-cm tray spacing and 7.6-cm weir height, carrying out a benzene-toluene distillation at total reflux.

Table 12-1 shows a qualitative comparison of the suitability of various types of trays and packings by different criteria. The counterflow trays included in the table are downcomerless trays, such as Turbogrid, ripple, and Kittel trays. High-void packings include Pall rings and grid packing, while "normal" packings include Raschig rings, Berl and Intalox saddles, etc. (see Figs. 4-14 and 4-15). Table 12-2 shows a comparison of trays and packings with regard to more specific service needs and includes tower internals of the alternating disk and doughnut type.

Zuiderweg et al. (1960) found the stage efficiencies of bubble-cap, sieve, and valve-cap trays to be very nearly the same. Others (Hengstebeck, 1961; Lockhart and Leggett, 1958; Procter, 1963) have reported that sieve trays and valve trays provide a stage efficiency 10 to 20 percent above that of bubble-cap trays at optimal column loadings. The performances of sieve trays and valve trays have been compared by Bolles (1976a) and Anderson et al. (1976). Another important factor is cost. Valve trays and sieve trays cost about 50 to 70 percent as much as bubble-cap trays, installed (Gerster, 1963; Hengstebeck, 1961). Valve trays and sieve trays are the most common trays used currently for new column construction.

Several other important differences between plate and packed towers should be

Table 12-1 Relative performance ratings† of contacting devices for distillation

	Trays				Packings	
	Bubble-cap	Sieve	Valve	Counterflow	High-void	Normal
Vapor capacity	3	4	4	4	5	2
Liquid capacity	4	4	4	5	5	3
Efficiency (separation per unit column height)	3	4	4	4	5	2
Flexibility (turndown ratio)	5	3	5	1	2	2
Pressure drop	3	4	4	4	5	2
Cost	3	5	4	5	1	3
Design reliability, based on published literature	4	4	3‡	2	2	3

† 5 = excellent: 4 = very good: 3 = good: 2 = fair: 1 = poor.
‡ Probably better now (1978).
Source: From Fair and Bolles, 1968: used by permission.

Table 12-2 Selection guide† for distillation-column internals

	Trays			Packed columns		
	Sieve or valve	Bubble-cap	Counterflow	Random	Stacked	Disk and doughnut
Pressure, low (<13 kPa)	2	1	0	2	3	1
moderate	3	2	1	2	1	1
high (>50% of critical)	3	2	2	2	0	0
High turndown ratio	2	3	0	1	2	1
Low liquid flow rates	1	3	0	1	2	0
Foaming systems	2	1	2	3	0	1
Internal tower cooling	2	3	1	1	0	0
Suspended solids	2	1	3	1	0	1
Dirty or polymerizing solution	2	1	3	1	0	2
Multiple feeds or sidestreams	3	3	2	1	0	1
High liquid flow rates	2	1	3	3	0	2
Small-diameter columns	1	1	1	3	2	1
Column diameter 1 to 3 m	3	2	2	2	2	1
Larger-diameter columns	3	1	2	2	1	1
Corrosive fluids	2	1	2	3	1	2
Viscous fluids (at column T)	2	1	1	3	0	0
Low pressure drop (efficiency unimportant)	1	0	0	2	2	3
Expanded column capacity	2	0	2	2	3	0
Low cost (performance unimportant)	2	1	3	2	1	3
Reliability of design	3	2	1	2	1	1

† 0 = do not use. 1 = evaluate carefully. 2 = usually applicable. 3 = best selection.
Source: Adapted from Frank. 1977. p. 117: used by permission.

brought out, in addition to those indicated in Tables 12-1 and 12-2. Plate columns tend to have a greater liquid holdup per unit tower volume than packed columns. This can be of value when a slow liquid-phase chemical reaction is involved. The total weight of a dry plate tower is usually less than that of a dry packed column, but if the liquid holdup during operation is taken into account, the weights are usually about the same. The construction of a plate column is such that stage efficiencies most often fall in the range of 50 to 90 percent (see below), with the result that the proportion of tower height to equivalent equilibrium stages does not vary widely for many common distillations. The height of a packed column equivalent to an equilibrium stage varies more widely and often becomes greater for larger tower diameters because of liquid-distribution problems. When large temperature changes are involved, as in many distillations, there is the threat of thermal expansion or contraction crushing the packing in packed towers. Finally, packed columns often provide less pressure drop than plate columns for a given separation (Tables 12-1 and 12-2). This advantage, plus the fact that the packing serves to lessen the possibility of tower-wall collapse, makes packed towers particularly useful for vacuum operations (top row of Table 12-2).

Large-scale comparison studies of different trays and packings are made by Fractionation Research, Inc. (FRI), So. Pasadena, California, but the results are confidential to companies which subscribe to FRI. Another large-scale comparative testing facility has been built at the University of Manchester in Great Britain (Standart, 1972).

Example 12-1 Consider the acetone-water distillation specified in Examples 6-4 and 6-5:

$$z_{A, F} = 0.50 \qquad x_{A, d} = 0.91 \qquad x_{A, b} = 0.022$$

$$\frac{d}{F} = 0.538 \qquad \frac{r}{d} = 0.298 \qquad \Delta V \text{ across feed tray} = 0.55F$$

Pressure $= 1$ atm abs   Overhead temperature $= 135°F$ Bottoms temperature $= 186°F$

Assume that the feed flow rate is to be 500 lb mol h. Estimate the tower diameter required if the distillation is carried out in (a) a sieve-plate column with a tray spacing of 24 in and (b) a packed column containing 1-in ceramic Raschig rings randomly dumped.

SOLUTION As a preliminary step, it is important to determine the point in the column at which a capacity limit is most likely to occur. From Example 6-5 we know that the liquid and vapor flows decrease downward in the column. The vapor load in the rectifying section is greater than that in the stripping section, but the liquid load is less. The vapor density is least at the bottom of the column where the water-vapor mole fraction is highest and the temperature is greatest. Because of these competing factors, it is a good idea to calculate $(L/V)(\rho_G/\rho_L)^{1/2}$ and $V(\rho_G)^{-1/2}$ at the tower top, just above the feed, just below the feed, and at the tower bottom. The first of these factors is the abscissa of Figs. 12-2 and 12-3, while the second is proportional to the ordinate of these figures and contains those variables which change most. At the tower top

$$L = (0.298)(0.538)(500) = 80 \text{ lb mol h} \qquad V = (1.298)(0.538)(500) = 349 \text{ lb mol h}$$

Since the molecular weight of acetone is 58,

$$\rho_G = [(0.91)(58) + (0.09)(18)] \frac{1}{359} \frac{492}{595} = 0.126 \text{ lb ft}^3$$

Since the specific gravity of acetone is 0.791 and the density of water is 61 lb/ft^3 at 160°F (Weast, 1968)

$$\rho_L = \frac{61[(0.91)(58) + (0.09)(18)]}{[(0.91)(58)/0.791] + [(0.09)(18)/1.00]} = 48.5 \text{ lb/ft}^3$$

Hence
$$\frac{L}{V}\left(\frac{\rho_G}{\rho_L}\right)^{1\,2} = \frac{80}{349}\left(\frac{0.126}{48.5}\right)^{1\,2} = 0.0117$$

and
$$V(\rho_G)^{-1\,2} = 349\left(\frac{1}{0.126}\right)^{1\,2} = 981 \text{ lb mol/h}\cdot(\text{ft}^3/\text{lb})^{1/2}$$

The flows just above and below the feed and at the tower bottom can be obtained from Fig. 6-16. The term H_{diff} for the stripping section has been set at -4700 Btu/lb mol. At the feed point the passing vapor and liquid streams have compositions of $y_A = 0.775$ and $x_A = 0.170$. The enthalpies of these streams are $+13,900$ and -1000 Btu/lb mol, respectively. Since $b = (0.462)(500) = 231$ lb mol/h, we have, just below the feed,

$$L' - V' = 231 \qquad \text{and} \qquad (-1000)L' - (13,900)V' = (-4700)(231)$$

Solving, we find

$$L' = 288 \text{ lb mol/h} \qquad \text{and} \qquad V' = 57 \text{ lb mol/h}$$

Similarly, just above the feed tray,

$$L = 63 \text{ lb mol/h} \qquad \text{and} \qquad V = 332 \text{ lb mol/h}$$

The flows just above the reboiler have compositions $y_A = 0.45$ and $x_A = 0.10$. By the above type of analysis,

$$L' = 284 \text{ lb mol/h} \qquad \text{and} \qquad V' = 53 \text{ lb mol/h}$$

Computing densities, we can make the following table (M_v = vapor molecular weight):

	L	V	ρ_L	ρ_G	M_v	$\dfrac{L}{V}\left(\dfrac{\rho_G}{\rho_L}\right)^{1\,2}$	$V(\rho_G)^{-1\,2}$
Tower top	80	349	48.5	0.126	55	0.0117	981
Just above feed	63	332	53	0.111	49	0.0088	996
Just below feed	288	57	53	0.111	49	0.230	171
Tower bottom	284	53	61	0.076	36	0.189	193

Referring to Figs. 12-2 and 12-3, we find that as V is increased proportionately throughout the tower, the capacity limit will come for the conditions just above the feed. We are at low values of $(L/G) \times (\rho_G \, \rho_L)^{1\,2}$, where entrainment may well be an important factor in plate columns (Fig. 12-6). This again points to the tray just above the feed as the capacity limit. The very high vapor rate in the top section is the dominant factor in this case.

   (a) For the sieve-plate column, if we were to operate at 80 percent of flooding for $(L/G) \times (\rho_G \, \rho_L)^{1\,2} = 0.0088$, we would find from Fig. 12-6 that the entrainment ψ would be 0.28 mol per mole of gross downflow. This is above the suggested maximum ψ of 0.15. If we choose to limit ψ to 0.09, we find from Fig. 12-6 that we must design our column for 58 percent of flooding. The exact ψ

chosen does not affect this figure greatly, as long as ψ is on the order of 0.10 or less. Returning to Fig. 12-3, we find that for a 24-in tray spacing at $(L/G)(\rho_G/\rho_L)^{1/2} = 0.0088$

$$K_v = U_{\text{flood}}\left(\frac{\rho_G}{\rho_L - \rho_G}\right)^{1/2} = 0.39$$

hence $\quad G_{\text{flood}} = 0.39\rho_G\left(\frac{\rho_L - \rho_G}{\rho_G}\right)^{1/2} \quad$ based on active area

$$= (0.39)(0.111)\left(\frac{52.9}{0.111}\right)^{1/2} = 0.945 \text{ lb s}\cdot\text{ft}^2 = 0.945\frac{3600}{49} = 69.4 \text{ lb mol/h}\cdot\text{ft}^2$$

Since we have chosen to operate at 58 percent of flooding, we have, if we estimate that 70 percent of the tower cross-sectional area will be active tray area,

$$\text{Tower cross-sectional area} = \frac{332 \text{ lb mol h}}{(69.4 \text{ lb mol h}\cdot\text{ft}^2)(0.58)(0.70)} = 11.8 \text{ ft}^2 = \frac{\pi d^2}{4}$$

$$\text{Tower diameter } d = \left[\frac{4(11.8)}{3.14}\right]^{1/2} = 3.88 \text{ ft}$$

Rounding to the next highest half foot, we would estimate a 4-ft required diameter for a sieve-plate column with a 24-in tray spacing.

(b) From Fig. 12-2 at $(L/G)(\rho_G/\rho_L)^{1/2} = 0.0088$

$$\frac{G_{\text{flood}}^2(a_p/\epsilon^3)\mu^{0.2}\zeta^2}{g\rho_G\rho_L} = 0.25$$

For liquid water at 145°F, $\mu = 0.48$ cP, while for liquid acetone at 145°F, $\mu = 0.23$ cP (Perry et al., 1963). Since μ is raised to a low fractional power, the exact value of μ is not critical; we therefore estimate $\mu = 0.44$ cP and get $\mu^{0.2} = 0.85$. The density ratio ζ is equal to approximately 1.18.

From Perry et al. (1963, p. 18-28), Van Winkle (1967), or Peters and Timmerhaus (1968), we obtain a_p/ϵ^3 for 1-in dumped ceramic Raschig rings as

$$\frac{a_p}{\epsilon^3} = \frac{58}{(0.73)^3} = 150 \text{ ft}^{-1}$$

Hence $\quad G_{\text{flood}} = \left[\frac{(0.25)(4.17 \times 10^8)(0.111)(53)}{(150)(0.85)(1.18)^2}\right]^{1/2} = (3.53 \times 10^6)^{1/2} = 1858 \text{ lb/h}\cdot\text{ft}^2$

$$= \frac{1858}{49} = 37.9 \text{ lb mol/h}\cdot\text{ft}^2$$

If we operate at 80 percent of the flooding G,

$$\text{Tower cross-sectional area} = \frac{332}{37.9(0.80)} = 10.9 \text{ ft}^2 = \frac{\pi d^2}{4}$$

$$\text{Tower diameter } d = \left[\frac{4(10.9)}{3.14}\right]^{1/2} = 3.73 \text{ ft}$$

Rounding to the next highest half foot, we would call for a 4-ft-diameter tower. ☐

FACTORS INFLUENCING EFFICIENCY

Although two-phase separation processes are often analyzed on the basis of hypothetical equilibrium stages, it is important to realize that in all probability any real single-stage contacting device will not give product streams which are in equilibrium with each other. This lack of equilibrium is usually taken into account through *stage efficiencies*. Several definitions of stage efficiency are possible; the most useful definition for the analysis of multistage separation processes with cross flow on each stage is probably the *Murphree efficiency*, discussed briefly in Chap. 3:

$$E_{M1i} = \frac{x_{1i,\,out} - x_{1i,\,in}}{x_{1i}^* - x_{1i,\,in}} \tag{3-23}$$

where E_{M1i} is the Murphree efficiency for component i based upon mole fractions in phase 1 and x_{1i}^* is the mole fraction of i in phase 1 which would be in equilibrium with the actual outlet composition of phase 2. Equation (3-23) written for the gas phase relates the actual gas composition exiting a stage and the gas composition which would be in equilibrium with the existing liquid. Equation (3-23) can therefore be incorporated into various computation approaches for multistage processes in a relatively straightforward manner. Still simpler is the *overall efficiency*, which relates the actual number of stages to the number of equilibrium stages required for an equivalent separation. It is difficult to develop sound predictive methods for the overall efficiency, however.

The factors causing a departure from equilibrium between product streams from a stage were discussed in Chap. 3 and include (1) mass- and heat-transfer limitations, (2) incomplete separation of the product phases, and (3) flow configuration and mixing effects. In this chapter we explore these various factors in more detail and consider the quantitative expressions which have been obtained experimentally for vapor-liquid contacting on bubble-cap, sieve and valve trays.

Empirical Correlations

Two empirical correlations have seen considerable use. The correlation of Drickamer and Bradford (1943) was based upon experimental data for 84 distillations separating hydrocarbon mixtures in petroleum refineries. It relates the overall stage efficiency to the mole-average viscosity of the feed at feed conditions. The overall efficiency decreases with increasing feed viscosity, presumably reflecting a poorer dispersion for higher-viscosity feeds.

The O'Connell (1946) correlation (Fig. 12-13) modified the Drickamer-Bradford correlation by changing the correlating parameter to the product of the relative volatility of the key components and the viscosity of the feed mixture, both evaluated at the arithmetic mean of the top and bottom column temperatures. Data were included for distillation of alcohol-water mixtures and chlorinated-hydrocarbon mixtures as well as refinery hydrocarbon mixtures. The reduction in stage efficiency at higher relative volatility may correspond to the increasing importance of liquid-phase resistance to mass transfer in such cases, as rationalized by the AIChE approach, discussed below. O'Connell (1946) generated a second correlation for absorbers, using a solubility function instead of the relative volatility.

Mechanistic Models

The most extensive coordinated study of efficiencies of bubble-cap and sieve trays available in the open literature is that carried out under the sponsorship of the

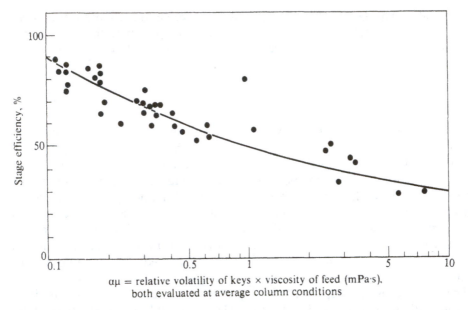

$\alpha\mu$ = relative volatility of keys × viscosity of feed (mPa·s), both evaluated at average column conditions

Figure 12-13 Correlation for bubble-cap distillation columns. *(From O'Connell, 1946, p. 751; used by permission.)*

Research Committee of the American Institute of Chemical Engineers in the 1950s (AIChE, 1958). The approach developed for analysis and prediction of efficiencies involves first accounting for the gas- and liquid-phase mass-transfer rates, so as to generate a number of overall gas-phase transfer units $(NTU)_{OG}$ provided in the vertical direction at any location on a contacting tray. This number of transfer units is then converted into a *point efficiency* E_{OG} by use of the simple Murphree model for bubbles rising through a well-mixed liquid (Murphree, 1925). Changes in the liquid composition across the tray are then accounted for through a tray-mixing model, to convert the point efficiency into a Murphree vapor efficiency for the entire stage E_{MV}. Finally effects of entrainment are estimated and used to convert the Murphree vapor efficiency into an apparent efficiency E_a, which is used if no other corrections are made in the distillation calculation for the effects of entrainment.

   In the 20 years since the AIChE study, considerable additional data have been obtained for stage efficiencies of commercial-sized columns with various sorts of trays. Most of these have been obtained by Fractionation Research, Inc., and are therefore not available in the open literature; but with certain exceptions the basic calculation approach and equation forms of the AIChE method have not been changed; instead the parameters resulting from the AIChE analysis have been updated. The exceptions have to do with allowance for liquid-mixing effects in the conversion from E_{OG} to E_{MV} and for the inherent differences between the froth and spray regimes on sieve and valve trays. The AIChE method was predicated on a froth-regime model.

   Our approach will be to develop the AIChE model with some of the more important updatings that appear in the open literature. This provides a basis for mechanistic understanding of the factors influencing stage efficiencies for vapor-liquid contacting on various sorts of plates and at the same time provides a framework of analysis which can be updated on the basis of more current information.

Mass-Transfer Rates

Following the addition-of-resistances concept [Eq. (11-82)], the AIChE approach first computes the number of overall gas-phase transfer units $(NTU)_{OG}$ provided vertically as the gas flows through the liquid at a point on the tray. This is done by computing numbers of individual gas- and liquid-phase transfer units [$(NTU)_G$ and $(NTU)_L$, Eqs. (11-108a) and (11-108b)] and then adding these reciprocally to obtain $(NTU)_{OG}$ by Eq. (11-109):

$$\frac{1}{(NTU)_{OG}} = \frac{1}{(NTU)_G} + \frac{\lambda}{(NTU)_L} \tag{12-6}$$

The parameter λ is $HV\rho_M/LP$ [Eq. (11-109)], or it is $K_i V/L$ if K_i is y_i/x_i for component i at equilibrium and is constant. Otherwise K_i should be interpreted as dy_i/dx_i at equilibrium.

For most common distillation systems the gas-phase term in Eq. (12-6) is dominant, and the process is thereby largely gas-phase-controlled. Liquid-phase resistance to mass transfer becomes important for large values of λ, for many absorption systems, and for situations of a slow chemical reaction in the liquid phase, among other cases.

In the AIChE study (AIChE, 1958) experimental measurements of gas-phase-controlled systems provided values of $(NTU)_G$ which were correlated empirically as a sum of linear terms involving different operating variables:

$$(NTU)_G = (0.776 + 0.116W - 0.290F + 0.0217L)/(Sc)^{1/2} \tag{12-7}$$

where W = outlet weir height, in

$F = u_G\sqrt{\rho_G}$ = product of gas flow rate, $ft^3/s \cdot ft^2$ active bubbling area, and the square root of the gas density, lb/ft^3 (square root of gas kinetic energy per unit volume)

L = liquid flow, gal min·ft of average liquid-flow-path width

Sc = gas-phase Schmidt number $\mu_G/\rho_G D_G$ (μ_G = gas viscosity, ρ_G = gas density, D_G = gas-phase diffusivity)

Gas-phase Schmidt numbers are on the order of unity.

In the same study the individual liquid-phase resistance was correlated as

$$(NTU)_L = (1.065 \times 10^4 \times D_L)^{1/2}(0.26F + 0.15)t_L \tag{12-8}$$

where D_L = solute diffusivity in liquid, ft^2/h

t_L = residence time of liquid on active zone of tray, s

The term t_L was defined as

$$t_L = \frac{37.4(Z_c)(Z_l)}{L} \tag{12-9}$$

where Z_c = holdup on tray, $in^3/(in^2$ tray bubbling area)

L = liquid rate, gal/min · (ft average liquid-flow-path width)

Z_l = length of liquid travel across active zone of tray, ft (distance between inlet and outlet weirs)

37.4 = conversion factor, gal · s/min · in · ft^2

Z_c was determined experimentally as

$$Z_c = 1.65 + 0.19W + 0.020L - 0.65F \qquad (12\text{-}10)$$

where the terms have been defined previously. Alternatively, Z_c can be estimated by a method outlined by Fair (1973, pp. 18-9 and 18-15), which involves the aeration factor and the dynamic seal.

With the individual phase resistances given by Eqs. (12-7) and (12-8), the overall resistance expressed as $(NTU)_{OG}$ is then obtained from Eq. (12-6).

Gerster (1963) summarizes the results of mass-transfer measurements for sieve trays, which appear to give $(NTU)_{OG}$ approximately 15 to 25 percent higher than that given by the preceding equations for bubble-cap trays. However, Fair (1973), on the basis of accumulated experience, states that Eq. (12-7) " appears to be equally applicable to bubble-cap, sieve and valve plates," and Bolles (1976a) indicates that the AIChE equations have been found to give satisfactory results when used directly for valve trays.

Point Efficiency E_{OG}

The original analysis leading to the concept of the Murphree vapor efficiency (Murphree, 1925) was based upon a picture of individual, discrete bubbles rising through a pool of liquid on a plate, as shown in Fig. 12-14. In the AIChE approach,

Gas out

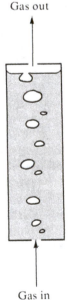

Gas in

Figure 12-14 Gas bubbling through well-mixed liquid.

this concept is retained to generate the point efficiency E_{OG} at any local position on a tray from $(NTU)_{OG}$. It is assumed that the gas phase passes through in plug flow with no backmixing and that the liquid is totally mixed vertically because of its short height and its intense agitation as the continuous phase. The analysis is identical to that leading to Eq. (11-152), and thereby gives

$$\frac{y_{A,\,out} - y_{A,\,in}}{y_{A,\,out,\,E} - y_{A,\,in}} = E_{OG} = 1 - e^{-(NTU)_{OG}} \tag{12-11}$$

It should be apparent from views like that in Fig. 12-8 that the contacting situation is not nearly as simple as that depicted in Fig. 12-14. The Murphree model does seem a reasonable first approximation for the froth regime, however, since in that regime the liquid phase tends to be continuous. However, Raper, et al. (1977) have shown that the gas flow in the froth regime for trays of industrial size can be uneven, with a substantial fraction of the gas passing through the dispersion as large slugs or jets. This effect is not taken into account by the simple Murphree model.

There has been much less work directed toward analysis and prediction of the mass-transfer situation for the spray regime; also, reliable experimental data are relatively limited. Hai et al. (1977; see also Fell and Pinczewski, 1977) have found that the Murphree vapor efficiency for absorption of ammonia from air into water in the spray regime increases with increasing F factor [compare the decrease with increasing F in the froth regime, corresponding to the minus sign on F in Eq. (12-7)], increases with increasing hole diameter, and decreases with increasing free area (hole area per plate active area). Fane and Sawistowski (1969) have outlined a mass-transfer model for the spray regime in which measured or correlated drop-size distributions are used as the basis for analyzing mass transfer to and from individual drops independently. Hai et al. (1977) and Raper et al. (1979) add to this the concept that droplets form several different times from a mass of liquid as it travels along a plate and find that the predictions of such a model agree at least qualitatively with experiment. However, more extensive data and analysis and improvement of models are required before a reliable predictive method will be available for the spray regime.

Flow Configuration and Mixing Effects

The basic flow pattern on a cross-flow plate is shown in Fig. 12-15. Although the liquid concentration changes from inlet to outlet, as equilibration with the gas phase occurs, how the liquid composition changes with respect to location is complex to analyze because of different forward velocities of the liquid at different points, mixing caused by the agitation in directions both parallel and transverse to the overall direction of flow, and even local backward flow under some conditions.

Bell (1972) used a fiber-optic technique to identify the residence-time distributions and flow patterns of liquid across commercial-scale sieve trays. The results showed a wide distribution of residence times, coupled with a pattern of more rapid flow of liquid along the center of the tray than near the walls. Furthermore, there is a tendency for retrograde, or backward, flow of liquid near the walls, which can result in closed-circulation cells, shown schematically in Fig. 12-16. Related studies of flow

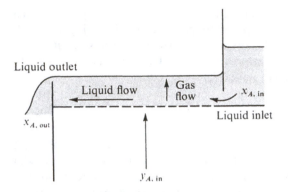

Figure 12-15 Flow pattern on a plate.

nonuniformity of liquid across large distillation trays carried out by Alexandrov and Vybornov (1971), Porter et al. (1972), and Weiler et al. (1973) all point to the same features of the flow, with a tendency for backflow near the walls and circulation cells to become more pronounced as the width of the flow path, i.e., column diameter, increases.

A number of mathematical models have been proposed to analyze the effects of

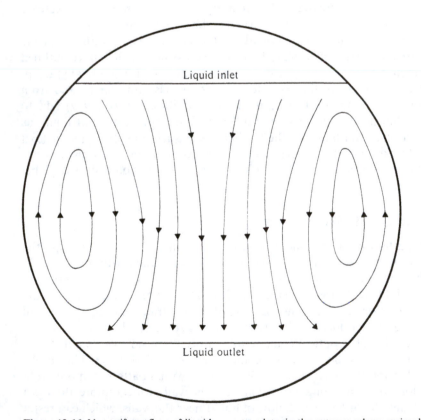

Figure 12-16 Nonuniform flow of liquid across a plate, in the extreme where recirculation cells form.

the liquid-flow pattern across a plate. The results are best understood if the effects are superimposed on each other, starting from the simplest cases.

Complete mixing of the liquid If the liquid is totally mixed in the direction of flow, as well as in the vertical direction, the liquid composition at all points must be uniform and equal to $x_{A,out}$. If $y_{A,in}$ is uniform across the plate, and if $(NTU)_{OG}$ and hence E_{OG} are uniform across the plate, Eq. (12-11) indicates that $y_{A,out}$ will be constant across the plate. If the vapor composition in equilibrium with the liquid exiting the stage is denoted by $y_{AE,x_{out}}$, the Murphree vapor efficiency for the entire stage should be defined as

$$E_{MV} = \frac{y_{A,out,av} - y_{A,in}}{y_{AE,x_{out}} - y_{A,in}} \quad (12\text{-}12)$$

For complete liquid mixing in the direction of flow $y_{A,out,E}$ at all points will equal $y_{AE,x_{out}}$, and we have

$$E_{MV} = E_{OG} = 1 - e^{-(NTU)_{OG}} \quad (12\text{-}13)$$

The Murphree vapor efficiency for the entire plate is equal to the point efficiency.

No liquid mixing: uniform residence time In the other extreme, plug flow of the liquid along the plate, the liquid composition will vary continuously from $x_{A,in}$ at the liquid inlet to $x_{A,out}$ at the liquid outlet. The relationship between E_{MV} and E_{OG} [or $(NTU)_{OG}$] for this case was first obtained by Lewis (1936). Considering a differential fraction of the total gas flowing upward through the liquid at some point along the liquid-flow path (see Fig. 12-17), we can write

$$(y_{A,out} - y_{A,in})\, dG = L\, dx_A \quad (12\text{-}14)$$

assuming that enough of component B travels in the other direction across the interface to hold L constant.

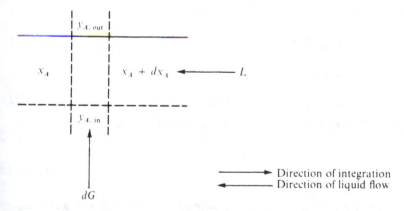

Figure 12-17 Mass transfer in a differential slice of liquid on a plate.

Substituting a linearized equilibrium expression $y_{AE} = H\rho_M x_A/P$ into Eq. (12-14) gives

$$(y_{A,\,out} - y_{A,\,in})\,dG = \frac{LP}{H\rho_M}\,dy_{AE} \tag{12-15}$$

or introducing $\lambda = HG\rho_M/LP$ leads to

$$\frac{\lambda}{G_M}(y_{A,\,out} - y_{A,\,in})\,dG = dy_{AE} \tag{12-16}$$

where G_M is the total molar gas-phase flow rate per unit area and is not a variable. If $y_{A,\,in}$ is uniform across the plate, Eq. (12-11) can be differentiated to yield

$$E_{OG}\,dy_{AE} = dy_{A,\,out} \tag{12-17}$$

Combining Eqs. (12-16) and (12-17) gives

$$\frac{\lambda E_{OG}}{G_M}\,dG = \frac{dy_{A,\,out}}{y_{A,\,out} - y_{A,\,in}} \tag{12-18}$$

Integrating Eq. (12-18) from the liquid *outlet* back to any point along the flow path, we have

$$\lambda E_{OG}\int_0^f df = \int_{y_{A,\,out,\,x_{out}}}^{y_{A,\,out}} \frac{dy_{A,\,out}}{y_{A,\,out} - y_{A,\,in}} \tag{12-19}$$

where f is the fraction of the total gas flow which passes to the left, i.e., toward the liquid outlet, of the point under consideration $(df = dG/G_M)$. Equation (12-19) becomes

$$\lambda E_{OG}\,f = \ln\frac{y_{A,\,out} - y_{A,\,in}}{y_{A,\,out,\,x_{out}} - y_{A,\,in}} \tag{12-20}$$

Solving for $y_{A,\,out}$ as a function of f, we have

$$y_{A,\,out} = y_{A,\,in} + e^{\lambda E_{OG}f}(y_{A,\,out,\,x_{out}} - y_{A,\,in}) \tag{12-21}$$

The average outlet-gas composition leaving the plate is

$$y_{A,\,out,\,av} = \int_0^1 y_{A,\,out}\,df = y_{A,\,in} + (y_{A,\,out,\,x_{out}} - y_{A,\,in})\frac{e^{\lambda E_{OG}} - 1}{\lambda E_{OG}} \tag{12-22}$$

Applying Eq. (12-11) to the liquid outlet point gives

$$y_{A,\,out,\,x_{out}} - y_{A,\,in} = E_{OG}(y_{AE,\,x_{out}} - y_{A,\,in}) \tag{12-23}$$

Substituting Eq. (12-23) into Eq. (12-22) and substituting the resultant equation into Eq. (12-12) gives

$$E_{MV} = \frac{e^{\lambda E_{OG}} - 1}{\lambda} \tag{12-24}$$

Figure 12-18 is a plot of E_{MV}/E_{OG} vs. λE_{OG} following Eq. (12-24), showing that E_{MV} is always greater than E_{OG} for this case of no mixing in the direction of liquid flow

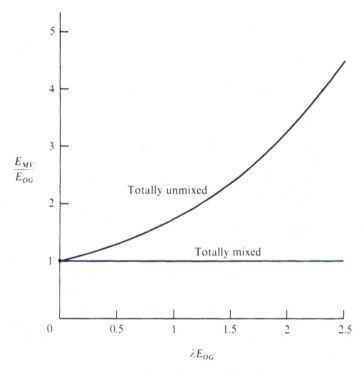

Figure 12-18 Relation between E_{MV} and E_{OG} for liquid totally mixed and for uniform residence time with liquid totally unmixed in the direction of flow.

and uniform liquid residence time. Comparing with Eq. (12-13), we see that the lack of liquid mixing has increased E_{MV} for a given $(NTU)_{OG}$.

Equation (12-24) and the curve in Fig. 12-18 correspond to the vapor entering a tray having uniform composition, as would occur for full mixing of vapor between trays. Lewis (1936) also examined two other cases corresponding to the extreme of no lateral mixing of the vapor between plates. The ratio E_{MV}/E_{OG} is improved some-what over Eq. (12-24) if the liquid flows in the same direction across successive plates, and it is lessened somewhat if the liquid-flow direction alternates from plate to plate, which is the usual case. Smith and Delnicki (1975) describe a design which achieves parallel liquid-flow directions on successive trays.

Full mixing of the liquid and uniform residence time with no mixing represent the extremes between which the results for real flow and mixing conditions should lie.

No liquid mixing: distribution of residence times Bell and Solari (1974) have analyzed theoretically the separate effects of a nonuniform liquid velocity field and retrograde flow on the ratio E_{MV}/E_{OG} in the absence of liquid-mixing effects. Both factors serve to reduce the ratio E_{MV}/E_{OG} below the predictions of Eq. (12-24) and the curve in Fig. 12-18. The effect of retrograde flow is particularly severe.

Several approaches have been pursued in efforts to narrow the distribution of liquid residence times and thereby increase E_{MV}/E_{OG} with large trays. Weiler et al.

(1973) report that slotting sieve trays to introduce vapor with a horizontal velocity component in the direction of liquid flow serves to reduce the hydraulic gradient and narrow the liquid-residence-time distribution, while at the same time discouraging the development of retrograde flow. Smith and Delnicki (1975) present results of several instances where sieve-plate vacuum columns with diameters in the range of 6 to 9 m were retrayed with trays which had a variable slot density and variable slot directions, chosen to make the liquid residence time more uniform. The stage efficiency was found to increase by 8 to 60 percent upon retraying. Yanagi and Scott (1973) report the use of unusual designs for the inlet downcomer baffle and the outlet weir on sieve trays 1.2 and 2.4 m in diameter, developed to narrow the residence-time distribution for the liquid considerably. It is interesting that no increase in stage efficiency was observed for two different distillation systems. This may be because the tray diameter was much smaller than those cited by Smith and Delnicki, or it may be the result of operation in the spray regime, where retrograde flow is less likely.

Partial liquid mixing Liquid mixing can occur in the directions parallel and perpendicular to the overall direction of liquid flow. The former is called *longitudinal*, or *axial, mixing*, and the latter is called *transverse*, or *radial, mixing*. The two forms of mixing tend to affect the ratio E_{MV}/E_{OG} in different ways. Longitudinal mixing serves to reduce the change in liquid composition along the length of the flow path and make the liquid composition at all locations closer to the outlet-liquid composition. This moves the system directionally from the "totally unmixed" curve in Fig. 12-18 toward the $E_{MV} = E_{OG}$ line for total mixing and serves to reduce the ratio E_{MV}/E_{OG}. On the other hand, transverse mixing serves to reduce the differences in liquid composition created by nonuniform residence times and retrograde flow. In the case of no longitudinal mixing, this should increase E_{MV}/E_{OG} above the predictions of Bell and Solari (1974) for nonuniform residence time, toward the case of uniform residence time. For the same reason, transverse mixing should also increase E_{MV}/E_{OG} in the presence of partial longitudinal mixing.

The AIChE model considers only the effect of longitudinal mixing, coupled with a uniform liquid residence time. A certain effective diffusivity D_E is assumed to be the cause of mixing in the direction of liquid flow. Allowing for mixing as a diffusion mechanism, one must modify the mass balance given by Fig. 12-17 and Eq. (12-14) to include a term accounting for the gain or loss of component A by diffusion

$$(y_{A,\,out} - y_{A,\,in})\frac{dG}{dz} = L\,\frac{dx_A}{dz} - D_E A \rho_M \frac{d^2 x_A}{dz^2} \qquad (12\text{-}25)$$

where z is the distance in the direction of liquid flow and A the cross-sectional area of liquid in the direction of flow. In the solution of this equation (Gerster, 1958) a *Péclet number* Pe arises, given by

$$\text{Pe} = \frac{LZ_l}{D_E A \rho_M} = \frac{Z_l^2}{D_E t_L} \qquad (12\text{-}26)$$

where Z_l is the length of the liquid-flow path and t_L is given by Eq. (12-9). If t_L is expressed in seconds and Z_l in feet (or meters), D_E must be given in square feet (or square meters) per second.

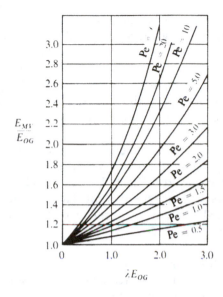

Figure 12-19 E_{MV}/E_{OG} as a function of λE_{OG} and Pe for diffusion liquid-mixing model. (*From AIChE, 1958, p. 48; used by permission.*)

The solution to Eq. (12-25) with the appropriate boundary conditions is

$$\frac{E_{MV}}{E_{OG}} = \frac{1 - e^{-(\eta + \text{Pe})}}{(\eta + \text{Pe})\{1 + [(\eta + \text{Pe})/\eta]\}} + \frac{e^{\eta} - 1}{\eta\{1 + [\eta/(\eta + \text{Pe})]\}} \qquad (12\text{-}27)$$

where

$$\eta = \frac{\text{Pe}}{2}\left[\left(1 + \frac{4\lambda E_{OG}}{\text{Pe}}\right)^{1/2} - 1\right] \qquad (12\text{-}28)$$

Again, it is assumed that $(NTU)_{OG}$ and hence E_{OG} is constant across the plate. Figure 12-19 shows Eq. (12-27) as E_{MV}/E_{OG} vs. λE_{OG} with Pe as a parameter. Notice that the extremes of Pe $= 0$ and Pe $\to \infty$ correspond to the two cases shown in Fig. 12-18.

For 3-in bubble caps on a 4.5-in triangular pitch and for sieve trays, the AIChE tray-efficiency study (AIChE, 1958) found that D_E could be correlated as

$$(D_E)^{0.5} = 0.0124 + 0.0171u_G + 0.00250L + 0.0150W \qquad (12\text{-}29)$$

for D_E in square feet per second; u_G is the superficial gas velocity, expressed as cubic feet per second of vapor flow divided by the active bubbling area in square feet, and W and L are the same as used previously. For sieve trays Gerster (1960) recommends using values of D_E from Eq. (12-29) multiplied by 1.25.

The effect of transverse mixing with no longitudinal mixing has been studied by Solari and Bell (1978) by means of a theoretical model. The results give a basis for analyzing the quantitative effect of transverse mixing in increasing E_{MV}/E_{OG} for nonuniform residence times and/or retrograde flow. Their results show that circulation patterns from retrograde flow should still have a strong reducing effect on E_{MV}/E_{OG}. The results also indicate that transverse mixing will often be a more important effect than longitudinal mixing if the effective diffusion coefficients D_E for both processes are about the same.

Porter et al. (1972) have given a model considering a combination of nonuniform flow and partial mixing in various directions. The tray area is divided into a straight-through rectangular flow zone with uniform liquid velocity, adjoined by stagnant circulating pools on either side of the main flow path. This is an idealization of the flow pattern shown in Fig. 12-16. They use diffusional-mixing models within the circulating pools and for longitudinal mixing in the main flow area as well as for interchange between the two types of zones. Insufficient experimental data are available to allow one to develop correlations for the parameters of this model or for the Solari and Bell model.

Porter et al. (1972) point out that another deleterious effect can stem from a lack of full mixing of vapor between trays, since for common tray designs the stagnant-pool zones on different trays will be located in a vertical line. This can lead to channeling of the vapor through several trays without effective distillation. The model of Porter et al. (1972) has been extended to cover consecutive plates in a column (Lockett et al., 1973) and split-flow plates (Lim et al., 1974).

Discussion The analyses of Porter et al. and of Solari and Bell predict that the ratio E_{MV}/E_{OG} should go through a maximum as tray diameter is increased for single-pass trays or as the length of a flow path is increased for multipass trays. For very small tray diameters the liquid will be fully mixed. For larger tray diameters incomplete longitudinal mixing will increase E_{MV}/E_{OG}, and transverse mixing will keep nonuniform liquid velocity from exerting a strong negative effect. However, above some critical large tray diameter, transverse mixing should no longer be able to counteract the effects of nonuniform liquid velocity and backflow, and E_{MV}/E_{OG} should begin to decrease again. Thus E_{MV}/E_{OG} can go through a maximum as a function of tray diameter, rather than continually increasing with increasing tray diameter as would be concluded from the model of longitudinal mixing with uniform residence time (Fig. 12-19). Even below this maximum, nonuniform liquid velocities can make E_{MV}/E_{OG} substantially less than predicted by Eq. (12-27) and Fig. 12-19. Porter et al. (1972) predict that the maximum in E_{MV}/E_{OG} should occur for single-pass tray diameters in the range of 1.5 to 6 m, depending upon values of various parameters. The fragmentary results of Smith and Delnicki (1975) and Yanagi and Scott (1973), mentioned above, agree with this conclusion qualitatively.

Improvements in methods for analyzing and predicting flow configuration and mixing effects on plates should continue. For trays of small to moderate diameter (less than 2 m) it is probably still appropriate to use the longitudinal-mixing correction from the AIChE model, being wary of any predicted values of E_{MV}/E_{OG} greater than about 1.20. It will be appropriate to use mixing and velocity-distribution models allowing for more different effects as experimental data are obtained in sufficient amounts to give satisfactory ways of predicting and correlating the parameters in these models.

Entrainment

As pointed out in Chap. 3, entrainment necessarily reduces the quality of separation obtained in a stage. The effect of entrainment upon stage efficiencies in a countercurrent cascade of discrete stages has been analyzed by Colburn (1936). For $\lambda = 1$ (that is, for parallel operating and equilibrium lines) the apparent Murphree vapor efficiency in the presence of entrainment E_a is related to the Murphree efficiency in the absence of entrainment E_{MV} by

$$E_a = \frac{E_{MV}}{1 + (eE_{MV}/L)} \tag{12-30}$$

where e is the entrainment of liquid upward with the rising vapor reaching the next stage above and L is the net liquid downflow, both in moles per unit time.

The entrainment for bubble and sieve trays at various vapor loadings can be taken from Fig. 12-6. The quantity on the ordinate of that figure is ψ, the moles of entrainment per mole of gross liquid downflow (net downflow plus entrainment return). Substituting ψ into Eq. (12-30) gives

$$E_a = \frac{E_{MV}}{1 + [\psi E_{MV}/(1 - \psi)]} \tag{12-31}$$

Equations (12-30) and (12-31) account for the effect of entrainment satisfactorily for cases where λ is not too far removed from 1.0 and the variation of liquid composition across the plate is not unusually large, i.e., for E_{MV}/E_{OG} near unity. The Colburn derivation assumes that the entrainment from a stage has the composition of the exit liquid from that stage.

Danly (1962) has examined the effect of entrainment when λ is substantially different from 1.0. Kageyama (1969) has explored effects of entrainment and weeping upon efficiency when there is partial liquid mixing in the direction of liquid flow across the plates. Entrainment can also be taken into account through modification of the mass-balance equations instead of the efficiency (Loud and Waggoner, 1978).

Summary of AIChE Tray-Efficiency Prediction Method

Table 12-3 summarizes the AIChE prediction method for tray efficiencies.

Table 12-3 Summary of AIChE procedure for prediction of tray efficiency

1. Predict a value for $(NTU)_G$, the number of gas-phase transfer units, from

$$(NTU)_G = \frac{0.776 + 0.116W - 0.290F + 0.0217L}{(Sc)^{1/2}} \tag{12-7}$$

where Sc = dimensionless gas-phase Schmidt number
   W = height of outlet weir, in
   F = F factor, defined as product of gas rate, ft^3/s·(ft^2 of tray bubbling area), and square root of gas density, lb/ft^3
   L = liquid rate, gal min·(ft of average column width)

2. Compute liquid holdup on tray Z_c expressed as inches of clear liquid:

$$Z_c = 1.65 + 0.19W + 0.020L - 0.65F \tag{12-10}$$

(continued)

Table 12-3 (*continued*)

3. Compute average liquid contact time t_L on the tray in seconds:

$$t_L = \frac{37.4 Z_c Z_l}{L} \qquad (12\text{-}9)$$

where Z_l is distance in feet traveled on the tray by liquid and may be taken as the distance between inlet and outlet weirs

4. Predict a value for $(NTU)_L$, the number of liquid-phase transfer units, by

$$(NTU)_L = (1.065 \times 10^4 D_L)^{1/2}(0.26F + 0.15)t_L \qquad (12\text{-}8)$$

where D_L = liquid-phase diffusivity, ft²/h

5. Combine $(NTU)_G$ and $(NTU)_L$ to predict point efficiency E_{OG}:

$$\frac{1}{-\ln(1 - E_{OG})} = \frac{1}{(NTU)_{OG}} = \frac{1}{(NTU)_G} + \frac{\lambda}{(NTU)_L} \qquad (12\text{-}6), (12\text{-}13)$$

where λ = ratio of slopes of equilibrium curve and operating line KG_M/L_M or $H\rho_M G_M/PL_M$. L_M is in the same (molar) units as G_M.

6. Compute a value for effective diffusivity in direction of liquid flow:

$$(D_E)^{1/2} = 0.0124 + 0.017u_G + 0.00250L + 0.0150W \qquad (12\text{-}29)$$

where u_G = gas rate, ft³/s·(ft² tray bubbling area)
 D_E = effective diffusivity, ft²/s

This equation is valid for round-cap bubble trays having cap diameters of 3 in or less; for 6.5-in round bubble caps increase value of D_E by 33% and for sieve trays multiply D_E from Eq. (12-29) by 1.25

7. Compute Péclet number Pe

$$\text{Pe} = \frac{Z_l^2}{D_E t_L} \qquad (12\text{-}26)$$

8. Obtain ratio E_{MV}/E_{OG} from Fig. 12-19 or Eq. (12-27); use of figure requires knowledge of E_{OG}, λ, and Pe; beware of any value of E_{MV}/E_{OG} greater than 1.2, and evaluate the behavior of trays with diameters above 2 m in the light of recent studies (see text)

9. Obtain quantity of entrainment ψ from Fig. 12-6

10. Correct resulting tray efficiency for effect of entrainment by Colburn's equation

$$E_a = \frac{E_{MV}}{1 + [\psi E_{MV}/(1 - \psi)]} \qquad (12\text{-}31)$$

which relates E_{MV}, the efficiency obtainable in the absence of entrainment, to E_a, the efficiency obtained in presence of ψ mol of entrainment/mol of gross liquid downflow

---

Example 12-2 One of the original two processes for the production of heavy water D_2O was the distillation of natural water, which contains 0.0143 atom % deuterium. A flowsheet of the Morgantown, West Virginia, heavy-water distillation plant constructed in 1943 is given in Fig. 13-22. The process is described further in Chap. 13, and the operating conditions are summarized in Table 13-2.

Because of the very large equipment costs for this plant, the stage efficiency for the distillation was of paramount importance. Bubble-cap trays were used, and design efficiencies (Murphree vapor) were set at 80 percent, according to the best estimates of that day, but in operation the efficiencies turned out to vary between 50 and 75 percent. It is interesting to compare these results with the predictions of the more recently developed AIChE method.

The following tray characteristics applied to two of the towers in the distillation train:

	Tower 2A	Tower 3
Pressure, mmHg abs	126	126
Tower diameter, in	126	40
Plate spacing, in	12	12
Vapor flow, lb/h	21,200	3050
Type of tray	Bubble cap	Bubble cap
Cap OD, in	3	3
Slot width, in	$\frac{3}{32}$	$\frac{3}{32}$
Slot height, in	$\frac{13}{16}$	$\frac{13}{16}$
Submergence, top of slot to top of weir, in	$\frac{3}{8}$	$\frac{3}{8}$
Weir height, in†	2	2
Active bubbling area per tray (% of tower cross-sectional area)†	65	65
Length of liquid flow path (% of tower diameter)†	75	75

† The weir height and detailed tray layouts are not given, but it may be assumed that these values are close to correct.

Source: Data from Murphy et al. (1955).

Because of the low relative volatility (about 1.05), the towers operated very close to total reflux. The tower pressures correspond to a temperature of 133°F. The properties of D_2O may be considered essentially equal to those of H_2O. Reid et al. (1977) give the diffusivity of D_2O as 4.75×10^{-5} cm^2/s at 45°C. Assume that the Schmidt number for the vapor mixture is about 0.50.

Using the AIChE method, estimate Murphree vapor efficiencies for the top few trays of towers 2A and 3. Compare the observed values with your prediction.

SOLUTION As the first step, compute F for tower 2A.

$V = 21,200$ lb/h (given)   Vapor density $= \dfrac{18}{359}\dfrac{126}{760}\dfrac{492}{593} = 0.00689$ lb/ft^3

Bubbling area $= \dfrac{\pi}{4}(10.5)^2(0.65) = 56.4$ ft^2 Vapor velocity $= \dfrac{21,200}{0.00689(3600)(56.4)} = 15.1$ ft/s

$F = (15.1)(0.00689)^{1/2} = 1.25$

A similar calculation for tower 3 gives $u_G = 21.7$ ft/s and $F = 1.80$.

As the next step we shall compute L for tower 2A. The average width of liquid flow path is obtained from the analysis shown in Fig. 12-20.

Max width of flow path $=$ diam $= 126$ in

$\cos \alpha = 0.75$ (Fig. 12-20) $\alpha = 41.5°$

Min width of flow path $= (\sin \alpha)(\text{diam}) = 0.662 \times$ diam

Av width of liquid-flow path $\approx 0.85 \times$ diam $= 8.93$ ft

Take $L = V = 21,200$ lb/h, with liquid density equal to 8.2 lb/gal.

$$L = \frac{21,200}{(8.2)(60)(8.93)} = 4.82 \text{ gal/min} \cdot \text{ft}$$

A similar analysis for tower 3 gives $L = 2.18$ gal/min · ft.

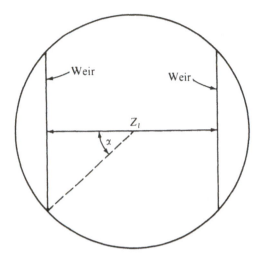

Figure 12-20 Tray geometry for Example 12-2.

Next we compute $(NTU)_G$. For tower 2A,

$$(NTU)_G(0.500)^{1\,2} = 0.776 + (0.116)(2) - (0.290)(1.25) + (0.0217)(4.82)$$

$$= 0.776 + 0.232 - 0.362 + 0.105 = 0.751$$

$$(NTU)_G = 1.06$$

For tower 3 the same calculation gives $(NTU)_G = 0.753$.

As a next step we compute Z_c for tower 2A.

$$Z_c = (0.19)(2) - (0.65)(1.25) + (0.020)(4.82) + 1.65 = 0.38 - 0.812 + 0.096 + 1.65 = 1.31 \text{ in}$$

The same calculation for tower 3 gives $Z_c = 0.90$ in.

As a next step we compute t_L for tower 2A:

$$t_L = \frac{37.4 Z_c Z_l}{L} = \frac{37.4(1.31)(0.75)(10.5)}{4.82} = 80 \text{ s}$$

Performing the same calculation for tower 3, we find that $t_L = 38.4$ s.

Next we can compute $(NTU)_L$ for tower 2A, but in order to do so it is necessary to convert the reported diffusivity to the correct temperature and the correct units. The operating temperature of 133°F corresponds to 56°C, and aqueous diffusivities increase approximately 2.5 percent per Celsius degree (Reid et al., 1977). Hence

$$D_L = (4.75 \times 10^{-5})[1 + (56 - 45)(0.025)] = 6.2 \times 10^{-5} \text{ cm}^2\text{ s} = 6.2 \times 10^{-5} \frac{3600}{(30.5)^2}$$

$$= 2.4 \times 10^{-4} \text{ ft}^2/\text{h}$$

$$(D_L)^{1\,2} = 1.55 \times 10^{-2}$$

$$0.26F + 0.15 = (0.26)(1.25) + 0.15 = 0.475$$

$$(NTU)_L = (1.03 \times 10^2)(1.55 \times 10^{-2})(0.475)(80) = 61$$

The same calculation for tower 3 gives $(NTU)_L = 38$.

For combining the individual phase transfer units we need a value of λ. Because of the very high reflux ratio necessitated by the relative volatility being close to 1.0, λ can be set as equal to 1.00.

within about 2 percent. Hence from Eq. (12-6) we find that $(NTU)_{OG}$ is essentially equal to $(NTU)_G$ for both towers. In other words, the system is highly gas-phase-controlled.

$$(NTU)_{OG} = \begin{cases} 1.04 & \text{tower 2A} \\ 0.74 & \text{tower 3} \end{cases}$$

Next we can compute E_{OG} from Eq. (12-13):

$$E_{OG} = \begin{cases} 1 - e^{-1.04} = 0.646 & \text{tower 2A} \\ 1 - e^{-0.74} = 0.522 & \text{tower 3} \end{cases}$$

It is next necessary to allow for partial liquid mixing and thereby relate E_{MV} to E_{OG}. For tower 2A,

$$(D_E)^{1\,2} = 0.0124 + (0.017)(15.1) + (0.00250)(4.82) + (0.0150)(2)$$

$$= 0.0124 + 0.256 + 0.012 + 0.030 = 0.310$$

$$D_E = 0.0962 \text{ ft}^2/\text{s}$$

Similarly, $D_E = 0.173 \text{ ft}^2/\text{s}$ for tower 3. For tower 2A,

$$Pe = \frac{(0.75 \times 10.5)^2}{0.0962(80)} = 8.05$$

For tower 3 we get $Pe = 0.94$, the much lower value coming from the smaller tower diameter. Following Fig. 12-19, we get:

Tower	λE_{OG}	Pe	E_{MV}/E_{OG}	E_{OG}	E_{MV}
2A	0.646	8.05	1.29	0.646	0.833
3	0.522	0.94	1.05	0.522	0.548

Next we need to assess the effect of entrainment in lowering E_{MV} during operation. Computing the abscissa of Fig. 12-6 for both towers, we get

$$\frac{L}{G}\left(\frac{\rho_G}{\rho_L}\right)^{1\,2} = 1\left(\frac{0.00689}{61}\right)^{1\,2} = 0.0106$$

From Fig. 12-3 for a 12-in tray spacing

$$K_V = U_{\text{flood}}\left(\frac{\rho_G}{\rho_L}\right)^{1\,2} = 0.23 \text{ ft s} \quad \text{and} \quad U_{\text{flood}} = \frac{0.23}{0.0106} = 22 \text{ ft/s}$$

For tower 2A,

$$\text{Percent of flooding} = 100\,\frac{15.1}{22} = 69\%, \quad \psi = 0.25$$

From Eq. (12-49)

$$E_a = \frac{0.833}{1 + [(0.25)(0.833)/0.75]} = \frac{0.833}{1.278} = 0.65$$

For tower 3, the indicated percent of flooding is very nearly 100 percent. We shall presume that the detailed tray layout and/or the operating conditions were such as to reduce this to, perhaps, 90 percent of flooding, in which case from Fig. 12-6 we find that $\psi = 0.50$ and

$$E_a = \frac{0.548}{1 + [(0.50)(0.548)/0.50]} = 0.35$$

If the operation of tower 3 were at 80 percent of flooding, we would get $\psi = 0.33$ and $E_a = 0.43$. We can summarize as:

Tower	E_{OG}	E_{MV}	E_a	Experimental range
2A	0.65	0.83	0.65	0.50–0.70
3	0.52	0.55	0.35–0.43	

Therefore our conclusion is that the low observed plate efficiencies would have been predicted with this design if the AIChE results and prediction method had been available when this plant was built.

□

Although the AIChE method for the prediction of gas-liquid plate efficiencies is elaborate and accounts for a number of effects which are to be expected theoretically, it does not always give good agreement with observed efficiencies. To some extent this is a result of the simplicity of the model for liquid mixing and the fact that the experimental data incorporated in the correlations were limited to certain ranges of operation; however, there are also a number of other effects which are known to have an influence upon plate efficiencies and which are not accounted for in the AIChE method. Several of these are considered in the following sections.

One particular observation that is not rationalized by the AIChE model is that distillation, absorption, and stripping columns designed to reach unusually high product purities (very low concentration of a solute or a key component) have often been found to yield an unexpectedly low stage efficiency. In some cases (but not all) this can reflect a large influence of $\lambda/(NTU)_{OL}$ in Eq. (12-6). Another possible explanation is given under Surface-Tension Gradients, below.

Chemical Reaction

Murphree efficiencies are based on a comparison of the actual exit composition of one phase leaving a stage to the composition of that phase which would be in equilibrium with the exiting composition of the other phase. If a chemical reaction is involved in the equilibration procedure within the stage, as in the absorption of carbon dioxide by basic solutions, it is necessary to account for the rate of this reaction in predicting and analyzing stage efficiencies. Phase-equilibrium data are based upon relatively long-time measurements wherein full chemical equilibrium is attained. The shorter times of contact in a continuous separation device can often result in the reaction proceeding to a lesser extent than represented by the equilibrium data. Thus the effect of a chemical reaction of finite rate is necessarily to *reduce* the stage efficiency or else to leave it unchanged. If the solute reacts completely and immediately achieves equilibrium upon entering the phase wherein it reacts, the process will essentially be the same as a purely physical mass-transfer process, in which the full concentration-difference driving force for diffusion is operative throughout the reacting phase. The efficiency then will be similar to the efficiencies for situations which are not complicated by chemical reactions. If, on the other hand, the solute must cross the interface under the impetus of only its physical solubility or a solubility corresponding to only partial reaction and then reacts in the bulk phase, the driving force in the denominator of the transfer-unit expressions will be reduced compared with the desired change in bulk composition, and a lesser amount of equilibration will occur with a given interfacial area and given mass-transfer coefficient. The driving force for mass transfer is small, but the amount of composition change to be accomplished is large.

This phenomenon is the reason for the low stage efficiencies noted in Example 10-2 for the carbon dioxide–ethanolamine absorption system. In order for equilibrium to be reached in the liquid phase, it is necessary to overcome the physical resistance to diffusion *and* the additional resistance afforded by the finite rate of chemical reaction. The rate of reaction between H_2S and ethanolamines is much more rapid than the reaction rate between CO_2 and ethanolamines; hence the observed stage efficiencies for H_2S absorption into ethanolamines are substantially greater than those for CO_2 absorption into ethanolamines.

Ways of allowing for the effect of simultaneous chemical reaction upon stage efficiencies and upon mass-transfer processes in general are discussed by Danckwerts and Sharma (1966), Astarita (1966), and Danckwerts (1970).

Surface-Tension Gradients: Interfacial Area

It has been found that the flow pattern on a plate during distillation can have very different froth and spray characteristics depending upon the relative surface tensions of the species being separated. This phenomenon was first explored in detail and analyzed by Zuiderweg and Harmens (1958), who also examined the same phenomenon for gas-liquid contacting in packed, wetted-wall, and spray columns.

When the more volatile component has the lower surface tension in the distillation of a binary mixture in the froth regime, the froth is more substantial and more stable than when the more volatile component has the higher surface tension. The explanation for this phenomenon lies in a consideration of the role of surface-tension gradients in governing the stability of froths and foams. The liquid in a froth will become percentwise more depleted in the more volatile component during distillation in local regions where the liquid film is thin. In a distillation where the more volatile component has the lesser surface tension, called a *positive system*, this greater depletion will mean that the liquid surface tension is higher in the thin-film regions than at surrounding points. As shown in Fig. 12-21a, the resultant surface-tension gradient along the surface sets up a surface-energy driving force, causing liquid flow from the low-surface-tension region to the high-surface-tension region. This flow is favored energetically because of the reduction it will cause in the total surface energy of the system. As a result of this flow, thin regions which would otherwise break are made thicker and reinforced. Thus froth stability is promoted.

In a system where the more volatile component has the higher surface tension (a *negative* system), thin regions of the froth will have a lower surface tension and, as shown in Fig. 12-21b, there will be a flow away from the thin regions, reducing the total surface energy. Thus thin regions will tend to break even more readily

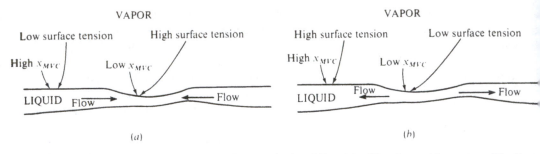

Figure 12-21 Effect of surface-tension gradients on froth stability: (a) self-heating positive system; (b) self-destructive negative system.

than they would in the absence of any surface-tension gradient, and the froth is unstable.

Zuiderweg and Harmens (1958) cite data for the effect of this phenomenon on contacting efficiency for the distillation of a number of different mixtures in different devices. For example, in a 1-in-diameter Oldershaw sieve-plate column with vapor velocities in the range of 0.2 to 2 ft/s the system n-heptane–toluene (a positive system) gave plate efficiencies of 80 to 90 percent, whereas the system benzene–n-heptane (a negative system) gave efficiencies of 50 to 55 percent. This increase in efficiency is obtained at the expense of some loss in capacity, however. For the heptane-toluene system, froth heights of 4 to 6 cm were found, as opposed to 1 to 2 cm for the benzene-heptane system. Thus a positive system would be expected to show greater tendencies toward entrainment and flooding.

Hart and Haselden (1969) found similar influences of surface-tension gradients upon froth heights and stage efficiencies and offer additional interpretations. They used a quite small column, as did Zuiderweg and Harmens. The effect has also been observed in a number of other studies of distillation in the froth regime.

The effects of positive and negative systems are reversed in the spray regime. Bainbridge and Sawistowski (1964) found higher stage efficiencies for negative systems than for positive systems for a sieve-tray column operating in the spray regime (see also Fane and Sawistowski, 1968). They attributed this to the fact that spray droplets are formed by a liquid-necking mechanism, shown schematically in Fig. 12-22. As a mass of liquid is thrust outward from the liquid bulk, the narrow neck connecting this incipient droplet will become depleted in the more volatile component, because of the high surface-to-volume ratio of the neck. In a positive system this causes the neck liquid to have a higher surface tension, and there will consequently be a healing flow from the surrounding liquid, reducing this surface tension. The drop therefore tends not to break away. On the other hand, for a negative system the liquid in the neck will have a lower surface tension than the bulk, and a flow will be set up whereby this low-surface-tension liquid is taken into the bulk liquid, lessening its surface tension. This promotes breakage of the neck and formation of the drop. Photographs supporting this mechanism are shown by Boyes and Ponter (1970). Higher efficiencies in the spray regime for negative systems, as opposed to positive or neutral systems, have also been found by several other investigators.

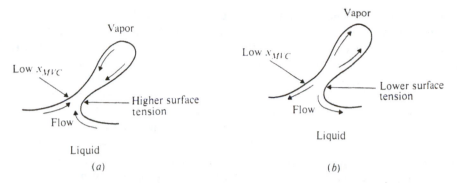

Figure 12-22 Effect of surface-tension gradients on drop formation: (a) positive and (b) negative system. *(Adapted from Bainbridge and Sawistowski, 1964, p. 993; used by permission.)*

These differences between the froth and spray regimes lead to a suggested design strategy (Fell and Pinczewski, 1977) whereby for surface-tension-positive systems one would design sieve trays to operate at a relatively low vapor velocity, consistent with acceptable turndown ratios, so as to operate in the froth regime. The tray spacing could be kept low (0.30 to 0.45 m) because of the consequent low tendency toward entrainment. Small holes and low hole areas would be used, since they favor high efficiency in the froth regime. For surface-tension-negative systems, one would design sieve trays to operate at a relatively high vapor velocity, with large hole diameter and greater hole area, since they all favor the transition to the spray regime and increase efficiency in that regime. Tray spacing would be greater (0.45 to 0.60 m) to accommodate the larger tendency toward entrainment. In some cases one might choose the spray regime for a positive system in order to reduce the column diameter. For severely foaming systems one might choose the froth regime for the lower vapor velocity and/or greater ease of providing large downcomer volumes for phase disengagement.

The froth and spray stabilizing and collapsing effects of positive and negative systems should be enhanced by factors which increase the local gradients in surface tension, i.e., larger differences in surface tension between the pure components, high relative volatility, and any other factors which increase the composition change from stage to stage. Multicomponent systems are as susceptible as binary systems to these effects and can more readily lead to different behavior in different sections of a column.

As mentioned previously, unexpectedly low stage efficiencies are often found in distillation, stripping, and absorption systems where very high purities are sought and the component being separated is present at very low concentrations. In such a case surface-tension gradients become insignificantly small; positive systems will no longer give the froth-stabilizing effects in the froth regime, and negative systems will no longer give the neck-rupture effect in the spray regime. This may account for at least some of the reports of stage efficiencies which become much lower at extremes of the composition range.

Zuiderweg and Harmens (1958) show that surface-tension-gradient effects are important in packed towers and wetted-wall columns, as well. The liquid spreads more readily over the solid surface and provides more interfacial area (and hence a greater efficiency) for a positive system than for a negative one. The reasoning is essentially the same as that shown in Fig. 12-21. In positive systems thin regions of liquid become more depleted in the more volatile component and are healed by surface tension-driven flow in from thicker regions. In negative systems the same phenomena cause liquid to flow out of thin regions. Norman (1961) gives a vivid evidence of this phenomenon from measurements of the minimum flow necessary to wet the wall of a wetted-wall column totally during distillation of n-propanol–water mixtures. As shown in Fig. 12-23, the n-propanol–water system forms an azeotrope, n-propanol being more volatile at low mole fractions of n-propanol, and water being more volatile at high mole fractions of n-propanol. Since water has a greater surface tension than n-propanol, the system is positive for mole fractions of n-propanol below the azeotrope and is negative for mole fractions of n-propanol above the azeotrope. The walls are much more readily wet during distillation at positive-system compositions than in the range of negative-system compositions.

The same phenomenon is observed in glass wetted-wall columns used for HCl absorption into water from air. In the absence of HCl gas one can set the water rate to achieve full wetting of the walls, but when HCl is introduced to the system, the liquid film breaks and falls into rivulets. In this case thin regions of liquid film are richer in HCl and are hotter because of the large heat of absorption. The presence of dissolved HCl reduces the surface tension of water, as does increasing temperature; thus the absorption of HCl into water is a negative system.

Surface-tension-gradient effects in separation processes have been reviewed in more detail by Berg (1972).

Density and Surface-Tension Gradients: Mass-Transfer Coefficients

A number of investigators have observed the occurrence of interfacial mixing cells in a two-phase fluid system undergoing an interphase mass-transfer process. This phen-

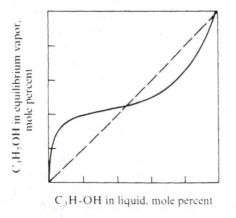
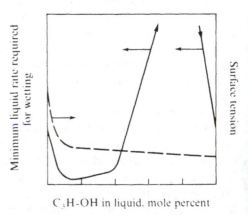

Figure 12-23 Minimum wetting rates for n-propanol–water distillation in a wetted-wall column. (*Adapted from Norman and Binns, 1960, p. 296; used by permission.*)

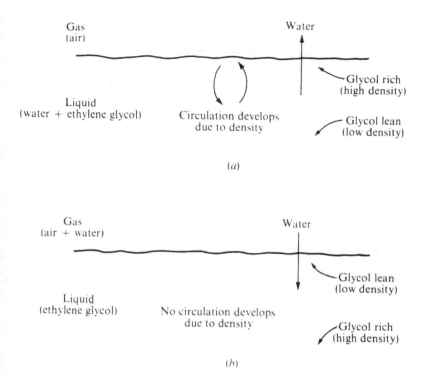

Figure 12-24 Density-driven interfacial mixing: (a) desorption, unstable system; (b) absorption, stable system.

omenon can result from gradients in either density or surface tension. The density-driven phenomenon is illustrated in Fig. 12-24a, where it is presumed that a lighter substance, e.g., water, is being desorbed from a heavier, less volatile solvent, e.g., ethylene glycol, into a gas phase which lies above the liquid phase. As the light substance evaporates, a region of greater density develops near the interface and as a result a region of high-density liquid occurs above a region of low-density liquid. This is an unstable situation which will tend to be relieved through cellular motion in which heavy interface liquid flows downward and lighter bulk liquid flows upward. This circulation increases the liquid-phase mass-transfer coefficient and is analogous to the action of natural convection in heat transfer.

Figure 12-24b depicts a stable situation, wherein water vapor is absorbed from humid air into ethylene glycol. Here a region of lower density develops above a region of higher density; this is a stable configuration and no density-driven circulation develops.

Surface-tension-driven interfacial mixing is illustrated in Fig. 12-25a and b. The surface tension of pure water (72 dyn/cm at 25°C) is greater than that of pure ethylene glycol (48 dyn/cm at 25°C). When there is absorption of water vapor from humid air into glycol, a water-rich region develops near the interface, compared with the bulk liquid. Consequently this means that the liquid near the interface has a higher inherent surface tension than the bulk liquid. As a result, circulation cells

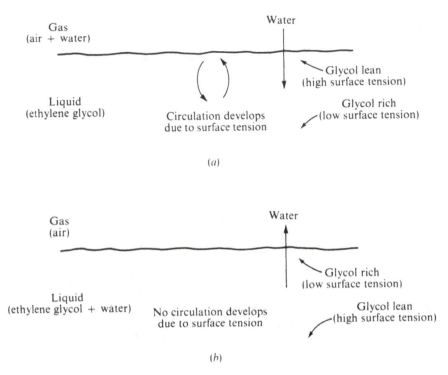

Figure 12-25 Surface-tension-driven interfacial mixing: (*a*) absorption, unstable system; (*b*) desorption, stable system.

which remove liquid from the region near the surface and replace it with bulk liquid are energetically favored, since they will lower the surface energy of the system. Again, as a result of these circulation patterns, the liquid-phase mass-transfer coefficient will be increased. In the reverse situation where the liquid near the interface has a lower inherent surface tension than the bulk liquid the situation is stable, and no surface-tension-driven circulation develops.

It should be stressed that interfacial mixing can occur for mass transfer in different directions in different systems, depending upon the relative densities and surface tensions of the species present. Density-driven interfacial circulation can occur in a gas phase as well as a liquid phase, but surface-tension-driven circulation is unlikely in a gas phase, except for what may be caused by drag from the liquid phase, because surface tensions are quite insensitive to the nature or composition of the gas phase. The density-driven phenomenon is dependent upon density gradients in the direction of gravity, while the surface-tension-driven phenomenon is dependent upon surface-tension gradients in the direction normal to the interface.

The quantitative effects of density-driven circulation in increasing rates of mass transfer have been summarized by Lightfoot et al. (1965). Berg (1972) reviews experimental measurements of enhancement of mass transfer by interfacial mixing.

The accelerating effects of surface-tension-driven cellular convection upon mass-transfer rates have all been measured for laminar or stagnant systems, however, and it is still an open question whether such cellular motion will affect mass-transfer rates significantly under the highly turbulent conditions of most commercial separation devices.

Surface-Active Agents

Surfactants are substances that markedly lower the surface tension of a liquid when added in small quantities. Aqueous surfactants are typically amphiphilic molecules, having one portion which is polar and water-loving and another portion which is nonpolar, e.g., hydrocarbon, and is less compatible with water. An example of an aqueous surfactant is hexadecanol, $CH_3(CH_2)_{14}CH_2OH$, in which the OH group is polar but the rest of the molecule is hydrocarbon.

When surfactants are added to fluid-phase separation devices, they can have a marked influence upon mass-transfer rates. Because of the decrease in surface tension, the addition of surfactants generally makes the liquid tend to spread on a wetted wall or a packing more readily. Also, surfactants impart an *elasticity* to a liquid film wherein any disruptions to the film will cause a locally lower surfactant concentration and hence a higher surface tension. This in turn will give a tendency for flow *into* the disrupted region, which will tend to keep the film from breaking.

Thus Francis and Berg (1967) found that $K_G a$ for the distillation of formic acid and water in a packed column was increased by as much as a factor of 1.5 by the addition of a surfactant, 1-decanol. This is a gas-phase-controlled system for mass transfer, and it appears that the increased efficiency comes from an increased interfacial area caused by better spreading of the liquid on the packing. Bond and Donald (1957) found a similar beneficial effect from the addition of a surfactant to water absorbing ammonia from a gas phase in a wetted-wall column. In the presence of the surfactants the walls became fully wet much more readily. Ponter et al. (1976) have interpreted data for packed-column distillation of butylamine and water in terms of the system wetting properties.

Because of their film-stabilizing properties, many surfactants serve to generate and promote foams. Foaming in plate columns for gas-liquid contacting can increase stage efficiencies (Bozhov and Elenkov, 1967), but more often than not foaming causes a serious problem of entrainment, priming, and/or early flooding. Therefore it is usually avoided. Ross (1967) analyzes causes of foaming in distillation columns and means of controlling it, e.g., antifoam agents.

Surfactants can influence the amount of surface area in a froth or spray, even if a foam, as such, is not formed. Brumbaugh and Berg (1973) found that 1-decanol increases froth height and stage efficiency for distillation of the azeotropic system formic acid–water in the composition range where it is a negative system. In the positive range the froth height increased, but no change in efficiency was detectable.

Injecting a surfactant into a gas-liquid or liquid-liquid contacting system often results in a reduction of liquid-phase mass-transfer coefficients, in addition to whatever effect it may have on interfacial area. Usually this lowering of the mass-transfer coefficient is the result of hydrodynamic factors, wherein the surfactant suppresses large-scale fluid motions in the vicinity of the interface (Davies, 1963; Davies et al., 1964) or causes surface stagnation (Merson and Quinn, 1965) because the replacement of the surface liquid layer with bulk liquid would result in an elevation of the surface energy of the system. Here again, however, it has not been confirmed that these effects would be important in the intensely agitated situation on a distillation plate. The possibility of an interfacial resistance to mass transfer caused by a reduced solute solubility or diffusivity in a surfactant layer at the interface has been the subject of controversy for a number of years. Careful measurements (Sada and Himmelblau, 1967; Plevan and Quinn, 1966) indicate that such a resistance probably will be significant only for surfactant molecules which form a rigid semi-solid film at the interface. Thus hexadecanol can provide a significant interfacial resistance to mass transfer in aqueous systems, but naturally occurring surfactants in water usually do not.

Berg (1972) has reviewed the effects of added surfactants.

Heat Transfer

The AIChE method ignores effects of heat transfer, even though the vapor and liquid entering a plate have different temperatures and must also equilibrate thermally. Kirschbaum (1940) suggested that plate efficiencies in distillation should be analyzed as a heat-transfer process or in terms of driving forces for both heat and mass transfer (1950). Danckwerts et al. (1960) and Liang and Smith (1962) have discussed how simultaneous heat transfer can affect the rate of equilibration. Two effects are possible: one involves the tendencies of the bulk phases to become supersaturated during the equilibration, and the other involves the need for net evaporation or net condensation at the interface.

Figure 12-26 shows a temperature-composition diagram for a binary system. The

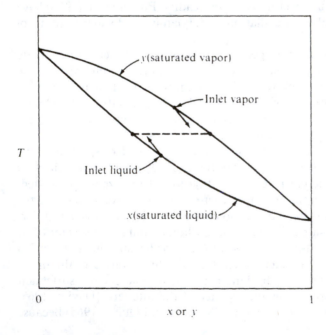

Figure 12-26 Effect of simultaneous heat and mass transfer on bulk phase compositions in distillation.

saturated-vapor (dew-point) and saturated-liquid (bubble-point) curves are shown, along with postulated temperatures and compositions of the vapor and liquid entering a plate. If equilibrium between these streams were reached, the two exit phases would have the same temperature and the vapor and liquid compositions would be those corresponding to the ends of the dashed line. When one considers the comparative rates of heat and mass transfer which occur between the phases, it turns out that the ratio of heat transfer to mass transfer should in most cases be large enough for the two phases to become supersaturated. This tendency is shown by the arrows leading into the two-phase region. As the phases become supersaturated, fog or mist can form in the vapor phase and either bubbles can form in the liquid or bulk liquid can flash-vaporize when it comes to the phase interface through large-scale mixing actions. Material which has been formed in equilibrium with the bulk vapor can join the bulk liquid, and material which has formed in equilibrium with the bulk liquid can join the bulk vapor. As a result plate efficiencies should be increased.

Fog formation in distillation towers has been observed by Haselden and Sutherland (1960) and others. Boiling or flashing in a plate distillation column is difficult to detect visually, but bubble formation has been noted on the surface of the packing in packed distillation columns (Norman, 1960; etc.). Further confirmation that evaporation and condensation occur and relieve phase supersaturation during distillation comes from the measurements made by Liang and Smith (1962) and Haselden and Sutherland (1960), who found that the liquid, and probably also the vapor, leaving a plate or flowing in a packed column is at the temperature corresponding to thermodynamic saturation for the particular composition of the vapor or liquid stream.

Although the additional equilibration in distillation caused by the evaporation and condensation resulting from simultaneous heat transfer can no doubt be significant, it should not be an overwhelming effect because of the usually large values of the latent heat of vaporization.

Heat transfer can also occur across the metallic surfaces of the downcomers and plates in a distillation column. The effect of this type of heat transfer upon apparent plate efficiencies has been measured and analyzed by Warden (1932) and Ellis and Shelton (1960), who found it to be most significant at low vapor flow rates. For the large-diameter columns usually employed in practice the effect should be relatively small, however.

The second way simultaneous heat transfer can affect the equilibration rate on a distillation plate is through preferential evaporation or condensation at the interface. Figure 12-27 shows the temperature and composition profiles in the vapor and liquid phases on either side of the interface. In the absence of heat transfer the interfacial composition tends to achieve a value such that the mass flux N_A will be the same in each phase, avoiding accumulation at the interface. Similarly, in the absence of mass transfer the interfacial temperature will achieve such a value as to make the heat-transfer rates to and from the interface equal. The interface temperature will thus be the average of the bulk-phase temperatures weighted by the heat-transfer coefficient of either phase. The liquid-phase heat-transfer coefficient is usually substantially greater than the gas-phase heat-transfer coefficient because of the higher thermal conductivity, and as a result the interface temperature will be close to the liquid-phase temperature.

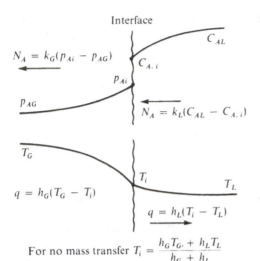

Interface

$$N_A = k_G(p_{Ai} - p_{AG})$$

C_{AL}

$C_{A,i}$

p_{Ai}

p_{AG}

$$N_A = k_L(C_{AL} - C_{A,i})$$

T_G

T_i

T_L

$$q = h_G(T_G - T_i)$$

$$q = h_L(T_i - T_L)$$

For no mass transfer $T_i = \dfrac{h_G T_G + h_L T_L}{h_G + h_L}$

Figure 12-27 Factors controlling net evaporation or condensation at the interface.

When heat and mass transfer occur simultaneously, the interfacial composition and temperature must be in equilibrium with each other following a phase diagram like Fig. 12-26. In order to maintain this condition there must be a net evaporation or condensation of material at the interface so as to make the heat flux different in the two phases. Because the liquid-phase heat-transfer coefficient is usually much greater than that in the gas, the necessary ΔT's usually will tend to require that the heat flux away from the interface into the liquid be greater than that to the interface from the gas. Therefore there should usually be a net condensation of material at the interface to release heat, which will then be removed through the liquid. This net condensation will affect the gas- and liquid-phase mass- and heat-transfer coefficients somewhat and will tend to produce a supersaturation of the liquid, which would then be relieved by subsequent flashing of material brought to the interface from the bulk through large-scale mixing action.

A number of experimental results show plate efficiencies and packed-column efficiencies increasing in a range of composition where temperature driving forces are large and have been interpreted in terms of the added efficiency due to simultaneous heat transfer (Liang and Smith, 1962; Sawistowski and Smith, 1959). The systems for which the largest effects of this sort have been found (methanol-water, cyclohexane-toluene, acetone-benzene, heptane-toluene, acetone-chlorobenzene) are also positive systems which have a surface-tension-vs.-composition relationship which favors spreading of liquid films and froth stability. When thermal driving forces are large, the surface-tension gradients are also large. Consequently it is difficult to separate the surface-tension effect from the heat-transfer effect. One can *see* the interfacial area effects, and they have been shown to exist and to be of considerable importance. The same cannot be said for the heat-transfer effects.

Multicomponent Systems

The stage equilibration process in a multicomponent system must be characterized by $R - 1$ Murphree efficiencies if there are R components. There is no need for these individual efficiencies to equal each other. The importance of the factor λ in the various equations underlying the AIChE method for binaries strongly suggests that components with different K_j values (and hence different λ_j) will have different values of E_{OGj} and/or E_{MVj}. Through calculations using tray mixing models, Biddulph (1975b, 1977) has demonstrated that even equal values of E_{OGj} in a ternary system should lead to variable and different values of E_{MVj} for different components at different column locations. Theories of multicomponent diffusion also indicate that E_{OGj} should be different for different components in a mixture of dissimilar substances.

Measurements of E_{OGj} and/or E_{MVj} for multicomponent distillations have confirmed that the efficiencies for different components tend to be different, except for E_{OGj} of quite similar components under conditions of gas-phase control and E_{MV}/E_{OG} near unity (Nord, 1946; Qureshi and Smith, 1958; Free and Hutchison, 1960; Haselden and Thorogood, 1964; Diener and Gerster, 1968; Miskin et al., 1972; Young and Weber, 1972; etc.). However, insufficient data are available to allow the development of a reliable predictive method.

Two factors complicate allowance for different efficiencies for different components in multicomponent-distillation calculations, the lack of data and the difficulty of making the computation. The computational difficulty arises from the fact that the Murphree efficiency of one component is a dependent variable. For most multicomponent separation processes it is the compositions of the key components which are of the most interest, the nonkey components rather rapidly approaching their limiting concentrations. Thus one Murphree efficiency corresponding to the key-component separation can be used satisfactorily for all components in most cases where the actual distribution of nonkeys is not of interest. This efficiency can often be predicted by binary methods, since the key components frequently constitute a large fraction of the interstage flows and contribute most of the interfacial mass flux.

ALTERNATIVE DEFINITIONS OF STAGE EFFICIENCY

Criteria

A number of different expressions for plate and stage efficiency have been proposed over the years. To some extent they are interchangeable and can be related through equations involving λ ($= KV/L$) and other parameters. There are, however, two criteria which should be met by a definition of plate or stage efficiency in order for it to be most useful: (1) The defined efficiency should be usable in a computation sequence for the separation device under consideration with a minimum of complexity and iteration in the calculation; (2) the magnitude of the efficiency should reflect primarily the size of heat- and mass-transfer coefficients and should be relatively independent of the value of λ, the solute concentration level, and the size of the driving force for equilibration. Under these conditions the efficiency should not vary greatly from stage to stage, and it may be possible to use a single value of the efficiency throughout a separation cascade.

The Murphree vapor efficiency meets these criteria well for situations where

1. The liquid phase can be considered well mixed.
2. The vapor flows through the liquid in plug flow.
3. The mass-transfer process is gas-phase-controlled.
4. The stages are part of a countercurrent cascade for which calculations are being made along the cascade in the direction of vapor flow from stage to stage.

These four conditions apply reasonably well to common distillation processes. With the liquid well mixed in the direction of vapor flow and with the vapor in plug flow, E_{OG} is given by Eq. (12-13) and is a function of $(\text{NTU})_{OG}$ alone; $(\text{NTU})_{OG}$ is determined by $(\text{NTU})_G$, which reflects mass-transfer parameters and is independent of λ if the system is gas-phase-controlled, as shown by Eq. (12-6). If the liquid is well mixed in the direction of liquid flow, E_{MV} will equal E_{OG} and will depend solely upon $(\text{NTU})_{OG}$ if E_{OG} does. If the liquid is not totally mixed in the direction of flow, some dependence of E_{MV} upon λ is introduced through the functionality shown in Fig. 12-19. If the stages in a countercurrent cascade are calculated sequentially in the direction of vapor flow, it is possible to obtain the composition of the vapor leaving a stage directly from the composition of the liquid leaving that stage without trial and error if the Murphree vapor efficiencies are known.

Murphree Liquid Efficiency

In order for the Murphree liquid efficiency, defined as

$$E_{ML} = \frac{x_{\text{A, out, av}} - x_{\text{A, in}}}{x_{\text{AE, y}_{\text{out}}} - x_{\text{A, in}}} \tag{12-32}$$

to be as useful, the system would have to be liquid-phase-controlled for mass transfer and should have plug flow of liquid through a well-mixed vapor. These conditions are not well met in plate distillation columns but may be reasonable for gas-liquid processes carried out in relatively short spray chambers or similar devices. The Murphree liquid efficiency can be related to basic mass-transfer parameters by interchanging vapor and liquid terms in the equations already presented for the Murphree vapor efficiency.

From Eqs. (12-12) and (12-32) solved for $(1/E_{MV}) - 1$ and $(1/E_{ML}) - 1$, it can be shown that the relationship between E_{MV} and E_{ML} for a linear equilibrium and constant vapor and liquid flows is given by

$$\lambda \left(\frac{1}{E_{ML}} - 1 \right) = \frac{1}{E_{MV}} - 1 \tag{12-33}$$

where λ is equal to $K_i V/L$ and K_i is the equilibrium constant for the component under consideration. Inspection of Eq. (12-33) shows that E_{MV} will be substantially less than E_{ML} when λ is large, i.e., when the system tends to be liquid-phase-controlled, provided the efficiencies are less than 1.00. Similarly, E_{ML} will be substantially less than E_{MV} when λ is much less than unity, which corresponds to the system tending toward gas-phase control.

Overall Efficiency

The efficiency most commonly used for quick and rough calculations is the *overall efficiency* E_o, defined simply as the ratio of the number of equilibrium stages required for a specified quality of separation at a specified reflux ratio to the actual number of stages required for the separation

$$E_o = \frac{N_{eq}}{N_{actual}} \qquad (12\text{-}34)$$

This is the form of efficiency easiest to use for calculations since it necessitates only a solution to the equilibrium-stage problem without the worry of applying an efficiency in the computation of each individual stage. On the other hand, the overall efficiency has the drawback of trying to represent the complex equilibration processes on each stage by means of a single parameter which bears no direct relationship to fundamental heat- and mass-transfer parameters. Also, use of the overall efficiency with an equilibrium-stage analysis cannot yield reliable nonkey splits. Prediction and correlation of overall efficiencies for plate distillation towers is safest for cases where all the towers considered treat similar substances at similar temperatures and similar reflux ratios and with similar tray diameters and designs.

The relationship between the overall efficiency and the Murphree vapor efficiency for constant total phase flows and a linear equilibrium relationship in a binary system (Lewis, 1936) is

$$E_o = \frac{\ln\left[1 + E_{MV}(\lambda - 1)\right]}{\ln \lambda} \qquad (12\text{-}35)$$

Note that the parameter λ affects this relationship strongly.

Vaporization Efficiency

Holland (1963) and coworkers have employed yet another definition of plate efficiency, called *vaporization efficiency*, which can be used in a simple fashion for computations. The efficiency E_{ip} for component i on plate p is defined as

$$E_{ip} = \frac{(y_{i,\,out,\,p})_{av}}{K_{ip}(x_{i,\,out,\,p})_{av}} \qquad (12\text{-}36)$$

Thus the "effective" K_{ip} to be used in a computation allowing for lack of complete equilibration is equal to $E_{ip} K_{ip}$. Holland further suggests that

$$E_{ip} = \bar{E}_i \beta_p \qquad (12\text{-}37)$$

where $\bar{E}_i$ is characteristic of component i and has the same value on all plates and β_p is characteristic of plate p and is the same for all components. Although this approach is simple to use, it does not correspond in a direct fashion to fundamental mass- and heat-transfer phenomena. As a result it can be expected that values of E_{ip} will be difficult to predict independently or to correlate and that the indicated values of $\bar{E}_i$ and β_p may vary substantially. Consideration of the use of efficiencies defined by Eq. (12-36) for a binary distillation shows that E_{ip} for the more volatile component must generally increase as its concentration increases and must be very nearly equal to unity near the top of the column. Thus the value of E_{ip} will change throughout the column even though the heat- and mass-transfer coefficients do not change appreciably.

Hausen Efficiency

Hausen (1953) and others have defined an efficiency based upon the approach to the products from a stage which would have been obtained if equilibrium had been achieved with the given feeds:

$$E_i = \frac{y_{A,\,out,\,av} - y_{A,\,in}}{(y_{A,\,out})^* - y_{A,\,in}} = \frac{x_{A,\,out,\,av} - x_{A,\,in}}{(x_{A,\,out})^* - x_{A,\,in}} \tag{12-38}$$

Here $(y_{A,\,out})^*$ and $(x_{A,\,out})^*$ are the compositions which would have been obtained if the given feed(s) to the stage had achieved complete equilibrium. This definition is different from the definition of the Murphree vapor or liquid efficiency. The denominator of the E_i expression is based upon the vapor composition which would have been in equilibrium with the liquid composition occurring in an *equilibrium* flash of the feeds, whereas the denominator of the Murphree vapor efficiency is based upon the vapor composition which would be in equilibrium with the *actual* exiting liquid.

Standart (1965) has examined this definition of efficiency at length and has modified the expression for E_i to take into account any changes in total phase flow rates which may occur across the stage:

$$E_i = \frac{V_p y_{A,\,out,\,av} - V_{p+1} y_{A,\,in}}{V_p^*(y_{A,\,out})^* - V_{p+1} y_{A,\,in}} = \frac{L_p x_{A,\,out,\,av} - L_{p-1} x_{A,\,in}}{L_p^*(x_{A,\,out})^* - L_{p-1} x_{A,\,in}} \tag{12-39}$$

Here V_p^* and L_p^* are the total vapor and liquid flows which would leave stage p if full equilibrium were obtained with the given feeds.

One advantage of the definition of efficiency given by Eq. (12-38) for constant molal flows and by Eq. (12-39) for varying molal flows is that the expressions are the same whether vapor or liquid compositions are used for the definition.

The term E_i is somewhat more difficult to use than E_{MV} when a countercurrent cascade is being analyzed, since a determination of the denominators of Eqs. (12-38) and (12-39) involves the feeds entering the stage from both directions. On the other hand, when a single-stage separation is being analyzed, E_i can be used directly once the equilibrium solution has been obtained, whereas the use of E_{MV} or E_{ML} requires iteration.

In any real contacting situation, E_i will most likely be substantially influenced by λ; E_i is based in concept upon the maximum change in composition which can be achieved either in cocurrent plug flow or in a vessel where both phases are well mixed. For both these situations, however, E_i depends upon λ. For example, for cocurrent plug flow with a linear equilibrium expression, a binary mixture, and constant phase flows it can be shown that

$$E_i = 1 - e^{-(\lambda + 1)(NTU)_{OG}} \tag{12-40}$$

When both phases are well mixed,

$$E_i = \frac{(1 + \lambda)(\text{NTU})_{OG}}{1 + (1 + \lambda)(\text{NTU})_{OG}} E_{MV} \tag{12-41}$$

The relationship between E_{MV} and E_i is given by

$$E_i = \frac{(1 + \lambda)E_{MV}}{1 + \lambda E_{MV}} \tag{12-42}$$

COMPROMISE BETWEEN EFFICIENCY AND CAPACITY

In the design and operation of any separation device it is necessary to strike a compromise between factors promoting efficiency or degree of separation, on the one hand, and factors promoting a high flow capacity, on the other. A high stage efficiency is obtained through high mass-transfer coefficients, and high mass-transfer coefficients, in turn, are obtained through intensive agitation and mixing, which bring with them a high pressure drop per unit length of flow path. High stage efficiencies can also be obtained by providing a long contact time between phases in the separation device, but a long contact time corresponds to larger equipment volumes and to longer flow paths. Longer flow paths also give a greater pressure drop.

For gas-liquid contacting in a plate tower, a higher plate efficiency can be obtained with a greater weir height, but this increases the pressure drop per stage and gives a greater tendency toward flooding because of the greater backup of liquid in the downcomer. The intensity of contact and hence the stage efficiency generally increase with increasing vapor velocity for the spray regime until the entrainment and/or flooding limits are approached. A factor favoring high plate efficiency in the froth regime is a greater froth height, like that obtained with a surface-tension-gradient positive system. This greater froth height will give increased tendencies toward flooding and entrainment, however.

The various expressions relating stage efficiency to the number of transfer units show a decreasing value of additional transfer units as the number of transfer units becomes greater. Typically the stage efficiency varies as the group $1 - e^{-(\text{NTU})_{OG}}$. Since providing additional transfer units generally means creating a greater pressure drop and a more severe capacity limit, it is advisable to provide a number of transfer units in a stage which brings the system to the point of diminishing returns and no farther. For example, it generally works well to give $1 - e^{-(\text{NTU})_{OG}}$ a value in the range of 0.6 to 0.85. Making this term larger will require a substantially greater number of additional transfer units per absolute gain in stage efficiency. Providing so few transfer units that this term comes out much less than 0.6 to 0.85 probably will make it necessary to use a significantly greater number of stages, and it is usually cheapest to provide a given number of transfer units in as few stages as possible.

Choosing an overdesign factor to allow for uncertainties in stage efficiency and column capacity is discussed in Appendix D.

Cyclically Operated Separation Processes

Cyclic operation involves continual change of operating parameters, e.g., flow rates, so that a process never achieves a steady state. For extraction, distillation, and some similar processes cycling can lead to higher capacity and/or greater stage efficiency.

In cyclic distillation there is a period during which vapor flows upward but liquid does not flow, followed by a period when liquid flows downward but the vapor does not flow. A similar procedure is used for cyclic extraction. Experiments (Szabo et al., 1964; Schrodt, 1967; Schrodt et al., 1967; Gerster and Schull, 1970; Breuer et al., 1977) show that a marked increase in the capacity of a given size column is possible and that an increase in the degree of separation provided by a given column height can be obtained for extraction and, in some cases at least, for distillation. The capacity increase is associated with the lack of a need for continuous counterflow of the contacting phases, with a reduction in flooding tendencies which is more than enough to offset the fact that each phase flows for only a fraction of the time. Theoretical analyses of the apparent stage efficiency during cyclical operation (Robinson and Engel, 1967; Sommerfeld et al., 1966; May and Horn, 1968; Horn and May, 1968; Rivas, 1977) show that an enhanced separation provided by a given number of stages can result from gradients in composition in the liquid flowing across a plate, much as incomplete mixing of the liquid in the direction of flow can cause the Murphree vapor efficiency to be greater than the point efficiency. Belter and Speaker (1967) and Lövland (1968) have shown that the analysis of a cyclically operated multistage extraction process is similar to that for the fully countercurrent version of the Craig distribution apparatus, counter-double-current distribution (CDCD; see Fig. 4-43).

Cyclic operation of a large-scale distillation or extraction column presents a number of major control difficulties; in fact, it is the control problem which has primarily held back the use of cyclically operated separation processes. Wade et al. (1969) discuss some approaches to control of these operations.

Pulsing is a form of rapid cycling which has proved effective for decreasing axial mixing and increasing contacting efficiency in extractors where equipment volume is of prime concern (Treybal, 1973).

Countercurrent vs. Cocurrent Operation

A cocurrent packed column can give at best the degree of separation corresponding to one equilibrium stage, whereas a countercurrent packed column can give a degree of separation corresponding to a large number of equilibrium stages. Countercurrent devices, however, are subject to the capacity limit of flooding, whereas this phenomenon does not occur in cocurrent systems. Therefore cocurrent contacting can be more desirable when only a single stage, or less, of contacting is needed.

Cocurrent contacting may also be desirable when the action of more than one equilibrium stage is required but the number of equilibrium stages is not great. Absorption with simultaneous chemical reaction in the liquid phase is a case in point. As noted in Example 10-2, Murphree vapor efficiencies are very low for the absorption of carbon dioxide into ethanolamines in plate towers. Thus about 30 plates may be required in practice for a carbon dioxide–ethanolamine absorber. even though the separation required corresponds to only two or three equilibrium stages. as was the case in Example 10-2. This very low Murphree vapor efficiency results from the large amount of mass transfer required. in comparison to the small driving force provided by the physical solubility of carbon dioxide.

The efficiency of contacting cannot be increased greatly in a countercurrent packed or plate column because of the capacity limit caused by flooding. One alternative absorber configuration would be a countercurrent arrangement of perhaps three smaller packed columns. each operated with the gas and liquid in cocurrent flow within the tower. Thus we have a countercurrent cascade of cocurrent stages. With cocurrent flow the superficial gas and liquid velocities can be a factor of 10 or more greater than is possible with countercurrent flow. Reiss (1967) and others have shown that much higher mass-transfer coefficients are obtained under these conditions because of the intense agitation due to the greater flow velocities. It is possible that in a number of cases the smaller volume of equipment required would more than offset the complexity of arranging a few cocurrent packed towers in such a way as to give countercurrent flow between the towers. Zhavoronkov et al. (1969) and others have proposed distillation devices wherein cocurrent contacting of vapor and liquid is achieved on each stage of a countercurrently staged single column.

A Case History

The separation of ethylbenzene from styrene (the monomer for the manufacture of polystyrene plastics) by distillation represents an interesting case where a crucial compromise must be made between factors governing efficiency and capacity of a distillation column. As shown in Figs. 12-28 and 12-29. styrene is manufactured from ethylbenzene by catalytic dehydrogenation (Stobaugh. 1965). Fresh and recycle ethylbenzene are mixed with superheated steam and fed to a catalyst-containing reactor at 650 to 750°C and a pressure near atmospheric. In the reactor ethylbenzene is converted into hydrogen and styrene at a conversion of 35 to 40 percent per pass:

Ethylbenzene   Styrene

Cooling steps following the reactor separate condensed steam from hydrocarbon product. and then separate condensed aromatics from the hydrogen product and other light hydrocarbon gases. The reaction selectivity is over 90 percent to styrene; however, some benzene and toluene are formed as cracking by-products and must be removed as a first distillation step. The following towers separate styrene from unconverted ethylbenzene and from heavier tars (polymerization by-products).

The separation of ethylbenzene from styrene presents unique difficulties. Styrene polymerizes readily and can therefore foul the reboiler. bottom trays. etc. Even in the presence of polymerization inhibitors, styrene polymerizes at temperatures greater than about 100°C. As a result it is necessary to run the ethylbenzene-styrene column under vacuum to hold temperatures down. On the other hand, the relative volatility

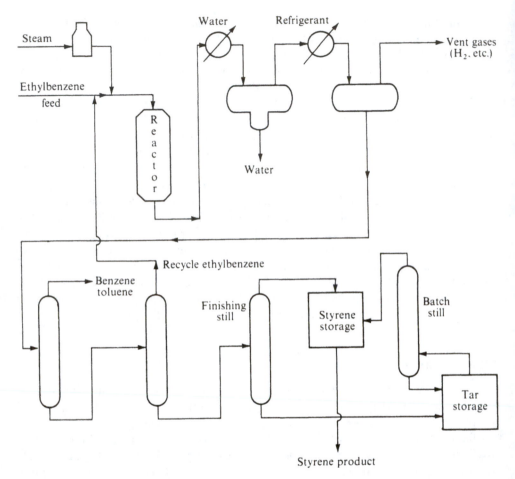

Figure 12-28 Typical process for manufacture of styrene from ethylbenzene. (*Adapted from Stobaugh,* *1965, p. 140; used by permission.*)

of ethylbenzene to styrene is not great, and so a large number of plates is required for the distillation. Consequently there is a large pressure drop through the tower, and this factor places a lower limit on the absolute pressure in the reboiler and hence on the reboiler temperature. If steps are taken to reduce the pressure drop per plate, the plate efficiency may also drop, with the result that more plates are required and the pressure drop goes back up.

A history of efforts to cope with the efficiency and capacity problems associated with ethylbenzene-styrene distillation has been given by Frank (Frank, 1968; Frank et al., 1969) and is reproduced here.†

---

† Joseph C. Frank, Early Developments in Styrene Process Distillation Column Design, in "Professors' Workshop on Industrial Monomer and Polymer Engineering," The Dow Chemical Company. Midland, Michigan, 1968. Reprinted with permission of The Dow Chemical Company.

Figure 12-29 A section of Dow Chemical's styrene complex. (*Dow Chemical USA, Midland, Michigan.*)

In the development of the Dow styrene process using the catalytic dehydrogenation of ethylbenzene, one of the most important process unit operations is distillation. In the alkylation section distillation columns separate the benzene and polyethylbenzenes for recycle and produce an ethylbenzene of over 99 weight percent purity.

With the ethylbenzene dehydrogenation step giving a crude product of about 40 weight percent styrene. distillation columns are used to separate and purify the benzene and toluene, separate for recycle the ethylbenzene and purify a styrene monomer to ever increasing purity specifications.

The most difficult fractionation problem encountered is the separation of styrene from the unreacted ethylbenzene. With an atmospheric boiling point for ethylbenzene of 136.2°C and for styrene of 145.2°C. the temperature difference for this distillation is 9°C, and the relative volatility is less than 1.3. Vacuum operation improves this relative volatility to the range of $\alpha = 1.34$ to 1.40. Today this would be considered an easy distillation column design problem, but in the early 1930's it was a very difficult design problem. In addition, styrene monomer as the bottom product of this distillation step polymerizes rapidly at the temperatures encountered in the distillation column even at the best vacuum conditions commercially available.

The first step in solving this problem was the development of an efficient inhibitor to this polymerization and adding this inhibitor to the distillation column with the feed and reflux streams. Sulfur was the inhibitor used and adding it high in the distillation column was one of the basic Dow patents on this process.

The first commercial styrene plant had a single shower deck low-pressure-drop column to make the ethylbenzene-styrene separation. Because of poor efficiency. this column proved inadequate and a second section was added. Later, a third section was added with 3-inch bubble cap tray design. Even operating with these three sections in series. the separation was inadequate with 2 to 5% ethylbenzene in the bottom product. This ethylbenzene had to be removed in the batch finishing stills.

A careful study of the problem at this point showed that, to make the required separation between ethylbenzene and styrene. at least 70 of the most efficient design bubble cap trays available were necessary. Even with the use of small 3-in. diameter caps and low slot immersion, the pressure drop with this number of trays was too high. With the minimum overhead vacuum of 35 mm Hg which would allow for condensing of the ethylbenzene in a water-cooled condenser, the column pressure drop was too high to give a satisfactory reboiler temperature.

From laboratory checks of the rate of polymerization of styrene monomer and of the reaction rate for the sulfur-styrene reaction under conditions encountered in the reboiler, it was decided that the bottoms temperature in this column must be held below 90°C. Later experience and data have shown that the operation is satisfactory at a much higher temperature if the residence time is kept low, but, at that time, 90°C was set as the maximum design temperature.

Efficient bubble cap trays could be designed for 3 mm Hg per tray pressure drop; therefore for 70 actual trays, this would give a column pressure drop of 210 mm Hg. If the minimum top pressure is 35 mm Hg then the reboiler pressure would be 245 mm Hg. The resultant temperature was 108 to 110°C and was much too high.

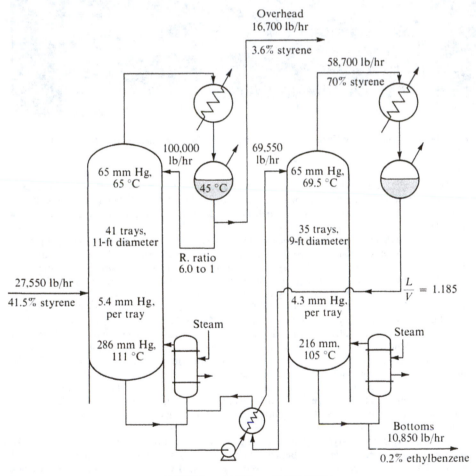

$$((12\text{-}30)) \qquad \text{Steam requirement} \frac{(100,000 + 16,700 + 58,700)(163)}{(940)(10,850)} = 2.80 \text{ lb/lb styrene}$$

Figure 12-30 Primary-secondary column system for styrene. (*Adapted from Frank et al., 1969, p. 80; used by permission.*)

Figure 12-31 Three sets of primary-secondary column distillation units for styrene at the Dow Midland Plant. *(Dow Chemical USA, Midland, Michigan.)*

After the study of several schemes, it was decided to split the required trays into two columns operating in series with complete condensing of the overhead vapors of each column so that vacuum of 35 mm Hg could be maintained at the top of each column. This split was made with 41 trays in the primary column and 35 trays in the secondary column since the most critical bottom temperature was that in the secondary column.

[Figure 12-30] shows this primary and secondary column set up with a typical set of operating data. A photograph of three such two-tower units is shown in [Fig. 12-31]. The first unit was put into operation at the Midland styrene plant in 1941. The operation of this system was an immediate success. With 76 bubble cap trays and 6 to 1 reflux ratio (and lower) they gave good separation with the ethylbenzene being removed from the bottoms so that the first overhead product from the batch stills was specification styrene. and within a few years we were finishing most of the styrene monomer in a continuous finishing still feeding secondary bottoms.

Additional plants using the primary and secondary column system came rapidly as the World War II Rubber Program styrene plants were built and started up in 1943 with eight sets of these stills in the Texas plant. four sets in the Los Angeles plant. and two sets in the Sarnia plant of Polymer Corporation. Also. all of our styrene know-how was furnished through government agencies (Rubber Reserve Co.) to our competitors at that time.

At the same time we were installing a second unit in the Midland plant and after the end of World War II. a third larger unit was installed at Midland using 76 bubble cap trays with 41 trays in the primary and 35 trays in the secondary. The operating data are shown in [Fig. 12-30]. Also in 1961 a new distillation unit was installed in the Texas plant with a primary and secondary column, with 50 dual flow trays in each column.

The primary-secondary column system with condensing and reboiling of the vapors between the columns takes more steam and cooling water (or air) utilities than a single column. If the reflux ratio or L/V below the feed tray is maintained the same in the secondary as in the primary column, then the steam required will be twice that required in a single column system.

When designing the primary and secondary column system. it was noted from study of the

McCabe-Thiele diagram that the reflux in the secondary column could be much lower with the requirement of only two or three more trays in this section. Almost all secondary columns have been designed in this manner with the steam load on the secondary at about 65% of that required for the primary column. You will also note [Fig. 12-30] that the secondary column is smaller in diameter, i.e., 9-foot diameter as compared with an 11-foot diameter for the primary column.

The use of a single column for the ethylbenzene-styrene separation had often been discussed after our styrene know-how became more extensive, and it was found that these columns could be operated at higher pressure and higher bottoms temperatures. The sieve tray or valve tray could be designed for lower pressure drop (in the range of 2 mm Hg per tray), but we were never convinced that tray efficiencies could be obtained in a high enough range to give the required separation. The sieve tray design was very difficult because most design data were extrapolated from atmospheric pressure correlations. Also with the low design pressure the sieve tray is very close to the weeping range and, furthermore, requires a minimum foam (or liquid) depth on the tray. Either of these conditions can give poor tray efficiency.

We had several reports in the 1950's of our competitors using a single column for this separation with valve tray design, but results were not available to us, and reports on operation did not appear to be very good.

Figure 12-32 Two duplicate single-column distillation units for the separation of ethylbenzene and styrene each using 70 Linde sieve trays. (*Dow Chemical USA, Midland, Michigan.*)

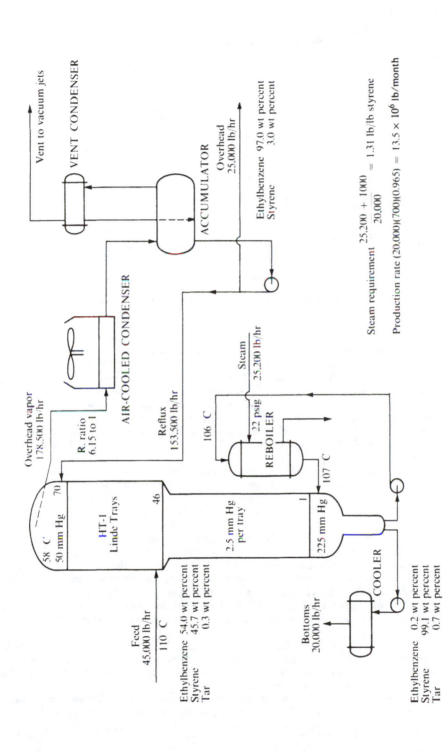

Figure 12-33 Single-column system for styrene with performance at maximum rate. (*Adapted from Frank et al., 1969, p. 82: used by permission.*)

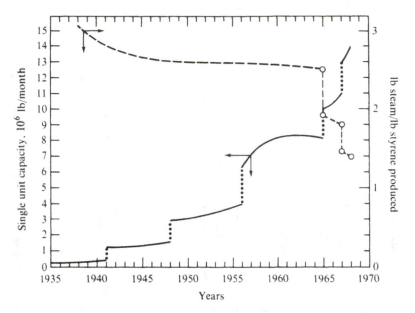

Figure 12-34 Learning curves for Midland plant styrene-ethylbenzene distillation units. *(Dow Chemical USA, Midland, Michigan.)*

In 1963 the Linde Division of Union Carbide announced that they were offering for sale their know-how on sieve tray design which had been developed over the years in their design of oxygen and liquid air plants. The Dow Sarnia plant was at this time actively working on a styrene plant expansion and sent out an inquiry to Linde among others. Mr. Garrett and Mr. Bruckert of Linde came to Sarnia in January 1964 and outlined their tray design know-how and made preliminary proposals for design of a single column unit for the ethylbenzene-styrene separation. Linde required a secrecy agreement before making a formal proposal. The agreement was made and, after a formal proposal was made, the first order with Linde for a single column using Linde Trays was placed for the Sarnia plant.

In March 1964, Linde was invited to come to Midland to present their story and make proposals for two columns for Midland's planned plant modernization. The Linde proprietary additions to the standard sieve tray along with their design experience and engineering know-how in tray design appeared to be the break-through required for the successful design of a single column for the ethylbenzene-styrene separation. Linde had already designed a single column unit for the Union Carbide styrene monomer plant at Seadrift, Texas, which would be in operation before our design was finalized. There was extensive discussion and study of the Linde Tray design by Dow Engineers which was climaxed by a demonstration by Linde at their Tonawanda Laboratory comparing the weeping tendency and stability of the Linde Tray as compared to a more standard sieve tray. This demonstration was convincing enough so that we gave Linde the go-ahead approval on the Midland columns in the summer of 1964, and the formal order was placed in January 1965. Also added to the same agreement was an order for one column in Texas and two columns for the Terneuzen styrene plant.

The two units for single column ethylbenzene-styrene distillation were started up in Midland in late 1965 and have met all production plant requirements from that date up to the present time. [Figure 12-32] is a photograph of these columns, and [Fig. 12-33] shows typical operating data for the columns at maximum production rates.

In summary, we like to show our improvements in chemical process know-how in what we call a " Learning Curve." [Figure 12-34] shows our learning curve improvements in the styrene distillation process.

Acknowledgment. The accomplishments discussed here have come from the cooperative efforts of many engineers and scientists, and the author gratefully acknowledges their contributions and wishes to thank The Dow Chemical Company for permission to publish this discussion.

The improvements which make the Linde sieve trays particularly suitable for this service are the bubbling promoters, slotted trays, and design for parallel flow on consecutive trays described by Weiler et al. (1973) and Smith and Delnicki (1975). Winter and Uitti (1976) describe another instance of poor tray performance in ethylbenzene-styrene distillation where a problem of weeping and liquid-flow maldistribution was solved by using a froth initiator (similar to the bubbling promoter) and a larger inlet weir. Stage (1970) explores tray-design alternatives and resulting performance for ethylbenzene-styrene distillation in considerable detail.

REFERENCES

AIChE (1958): " Bubble-Tray Design Manual," American Institute of Chemical Engineers, New York.

Alexandrov, I. A., and V. G. Vybornov (1971): *Teor. Osn. Khim. Tekhnol.*, **5**:339.

Anderson, R. H., G. Garrett, and M. Van Winkle (1976): *Ind. Eng. Chem. Process Des. Dev.*, **15**:96.

Astarita, G. (1966): " Mass Transfer with Chemical Reaction," Elsevier, Amsterdam.

Bainbridge, G. S., and H. Sawistowski (1964): *Chem. Eng. Sci.*, **19**:992.

Bell, R. L. (1972): *AIChE J.*, **18**:491, 498.

———— and R. B. Solari (1974): *AIChE J.*, **20**:688.

Belter, P. A., and S. M. Speaker (1967): *Ind. Eng. Chem. Process Des. Dev.*, **6**:36.

Berg, J. C. (1972): Interfacial Phenomena in Fluid Phase Separation Processes, in N. N. Li (ed.), "Recent Developments in Separation Science," vol. 2, CRC Press, Cleveland.

Biddulph, M. W. (1975a): *AIChE J.*, **21**:41.

———— (1975b): *AIChE J.*, **21**:327.

———— (1977): *Hydrocarbon Process.*, **56**(10):145.

———— and D. J. Stephens (1974): *AIChE J.*, **20**:60.

Bolles, W. L. (1976a): *Chem. Eng. Prog.*, **72**(9):43.

———— (1976b): *AIChE J.*, **22**:153.

———— and J. R. Fair (1963): Tray Hydraulics, in B. D. Smith, " Design of Equilibrium Stage Processes," chaps. 14 and 15, McGraw-Hill, New York.

Bond, J., and M. B. Donald (1957): *Chem. Eng. Sci.*, **6**:237.

Boyes, A. P., and A. B. Ponter (1970): *Chem. Eng. Sci.*, **25**:1952.

Bozhov, I., and D. Elenkov (1967): *Int. Chem. Eng.*, **7**:316.

Breuer, M. E., C. Y. Yoon, D. P. Jones, and M. J. Murry (1977): *Chem. Eng. Prog.*, **73**(6):95.

Brumbaugh, K. H., and J. C. Berg (1973): *AIChE J.*, **19**:1078.

Colburn, A. P. (1936): *Ind. Eng. Chem.*, **28**:536.

Crawford, J. W., and C. R. Wilke (1951): *Chem. Eng. Prog.*, **47**:423.

Danckwerts, P. V. (1970): " Gas-Liquid Reactions," McGraw-Hill, New York.

————, H. Sawistowski, and W. Smith (1960): *Inst. Chem. Eng. Int. Symp. Distillation, London*, p. 7.

———— and M. M. Sharma (1966): *The Chem. Engr.*, no. 202, p. CE 244, October.

Danly, D. E. (1962): *Ind. Eng. Chem. Fundam.*, **1**:218.

Davies, J. T. (1963): Mass Transfer and Interfacial Phenomena, in T. B. Drew, J. W. Hoopes and T. Vermeulen (eds.), "Advances in Chemical Engineering," vol. 4, Academic, New York.

————, A. A. Kilner, and G. A. Ratcliff (1964): *Chem. Eng. Sci.*, **19**:583.

Davy, C. A. E., and G. G. Haselden (1975): *AIChE J.*, **21**:1218.

Diener, D. A., and J. A. Gerster (1968): *Ind. Eng. Chem. Process Des. Dev.*, **7**:339.

Drickamer, H. G., and J. R. Bradford (1943): *Trans. AIChE*, **39**:319.

Ellis, S. R. M., and J. T. Shelton (1960): *Inst. Chem. Eng. Int. Symp. Distillation, London*, p. 171.

Fair, J. R. (1973): Gas-Liquid Contacting, in R. H. Perry and C. H. Chilton (eds.), "Chemical Engineer's Handbook," 5th ed., pp. 18-3 to 18-57. McGraw-Hill, New York.
—— and W. L. Bolles (1968): *Chem. Eng.*, Apr. 22, p. 156.
—— and R. L. Matthews (1958): *Petrol. Refin.*, **37**(4):153.
Fane, A. G., and H. Sawistowski (1968): *Chem. Eng. Sci.*, **23**:943.
—— and —— (1969): in J. M. Pirie (ed.), "Distillation 1969," Institution of Chemical Engineers, London.
Fell, C. J. D., and W. V. Pinczewski (1977): *The Chem. Engr.*, January, p. 45.
Finch, R. N., and M. Van Winkle (1964): *Ind. Eng. Chem. Process Des. Dev.*, **3**:106.
Francis, R. C., and J. C. Berg (1967): *Chem. Eng. Sci.*, **22**:685.
Frank, J. C. (1968): Early Developments in Styrene Process Distillation Column Design, in "Professors' Workshop on Industrial Monomer and Polymer Engineering," Dow Chemical Company, Midland, Mich.
——, G. R. Geyer, and H. Kehde (1969): *Chem. Eng. Prog.*, **65**(2):79.
Frank, O. (1977): *Chem. Eng.*, Mar. 14, p. 111.
Free, K. W., and H. P. Hutchison (1960): *Inst. Chem. Eng. Int. Symp. Distillation, London*, p. 231.
Gerster, J. A. (1960): *Ind. Eng. Chem.*, **52**:645.
—— (1963): *Chem. Eng. Prog.*, **59**(3):35.
—— et al.: Research Committee, American Institute of Chemical Engineers, "Final Report, University of Delaware," 1958.
—— and H. M. Schull (1970): *AIChE J.*, **16**:108.
Hai, N. T., J. M. Burgess, W. V. Pinczewski, and C. J. D. Fell (1977): *Trans. 2d Austr. Conf. Heat Mass Transfer, Sydney*, p. 377.
Hart, D. J., and G. G. Haselden (1969): in J. M. Pirie (ed.), "Distillation 1969," Institution of Chemical Engineers, London.
Haselden, G. G., and J. P. Sutherland (1960): *Inst. Chem. Eng. Int. Symp. Distillation, London*, p. 27.
—— and R. M. Thorogood (1964): *Trans. Inst. Chem. Eng.*, **42**:T81.
Hausen, H. (1953): *Chem. Ing. Tech.*, **25**:595.
Hay, J. M., and A. J. Johnson (1960): *AIChE J.*, **6**:373.
Hengstebeck, R. J. (1961): "Distillation: Principles and Design Procedures," Reinhold, New York.
Holland, C. D. (1963): "Multicomponent Distillation," Prentice-Hall, Englewood Cliffs, N.J.
Horn, F. J. M., and R. A. May (1968): *Ind. Eng. Chem. Fundam.*, **7**:349.
Interess, E. (1971): *Chem. Eng.*, Nov. 15, p. 167.
Kageyama, O. (1969): in J. M. Pirie (ed.), "Distillation 1969," Institution of Chemical Engineers, London.
Kirschbaum, E. (1940): "Destillier- und Rektifiziertechnik," Springer-Verlag, Berlin.
—— (1950): "Destillier- und Rektifiziertechnik," 2d ed., Springer-Verlag, Berlin.
Leva, M. (1954): *Chem. Eng. Prog. Symp. Ser.*, **50**(10):51.
Lewis, W. K., Jr. (1936): *Ind. Eng. Chem.*, **28**:399.
Liang, S. Y., and W. Smith (1962): *Chem. Eng. Sci.*, **17**:11.
Lightfoot, E. N., C. Massot, and F. Irani (1965): *Chem. Eng. Prog. Symp. Ser.*, **61**(58):28.
Lim, C. T., K. E. Porter, and M. J. Lockett (1974): *Trans. Inst. Chem. Eng.*, **52**:193.
Lockett, M. J., C. T. Lim, and K. E. Porter (1973): *Trans. Inst. Chem. Eng.*, **51**:281.
Lockhart, F. J., and C. W. Leggett (1958): in A. Kobe and J. J. McKetta (eds.), "Advances in Petroleum Chemistry and Refining," vol. 1, Interscience, New York.
Loon, R. E., W. V. Pinczewski, and C. J. D. Fell (1973): *Trans. Inst. Chem. Eng.*, **51**:374.
Loud, G. D., and R. C. Waggoner (1978): *Ind. Eng. Chem. Process Des. Dev.*, **17**:149.
Lövland, J. (1968): *Ind. Eng. Chem. Process Des. Dev.*, **7**:65.
May, R. A., and F. J. M. Horn (1968): *Ind. Eng. Chem. Process Des. Dev.*, **7**:61.
Merson, R. L., and J. A. Quinn (1965): *AIChE J.*, **11**:391.
Miskin, L. G., U. Ozalp, and S. R. M. Ellis (1972): *Br. Chem. Eng. Process Technol.*, **17**:153.
Murphree, E. V. (1925): *Ind. Eng. Chem.*, **17**:747.
Murphy, G. M., H. C. Urey, and I. Kirshenbaum (eds.) (1955): "Production of Heavy Water," McGraw-Hill, New York.
Nord, M. (1946): *Trans. AIChE*, **42**:863.
Norman, W. S. (1961): "Absorption, Distillation and Cooling Towers," Longmans, London.

———— and D. T. Binns (1960): *Trans. Inst. Chem. Eng.*, **38**:294.

O'Connell, H. E. (1946): *Trans. AIChE*, **42**:741.

Perry, R. H., C. H. Chilton, and S. D. Kirkpatrick (eds.) (1963): "Chemical Engineers' Handbook," 4th ed., McGraw-Hill, New York.

Peters, M. S., and K. D. Timmerhaus (1968): "Plant Design and Economics for Chemical Engineers," 2d ed., chap. 15, McGraw-Hill, New York.

Pinczewski, W. V., N. D. Benke, and C. J. D. Fell (1975): *AIChE J.*, **21**:1210.

———— and C. J. D. Fell (1972):*Trans. Inst. Chem. Eng.*, **50**:102.

———— and ———— (1974): *Trans. Inst. Chem. Eng.*, **52**:294.

———— and ———— (1975): *AIChE J.*, **21**:1019.

————, H. K. Yeo, and C. J. D. Fell (1973): *Chem. Eng. Sci.*, **28**:2261.

Plevan, R. E., and J. A. Quinn (1966): *AIChE J.*, **12**:894.

Ponter, A. B., P. Trauffler, and S. Vijayan (1976): *Ind. Eng. Chem. Process Des. Dev.*, **15**:196.

Porter, K. E., M. J. Lockett, and C. T. Lim (1972): *Trans. Inst. Chem. Eng.*, **50**:91.

———— and P. F. Y. Wong (1969): in J. M. Pirie (ed.), "Distillation 1969," Institution of Chemical Engineers, London.

Procter, J. F. (1963): *Chem. Eng. Prog.*, **59**(3):47.

Qureshi, A. K., and W. Smith (1958): *J. Inst. Petrol.*, **44**:137.

Raper, J. A., J. M. Burgess, and C. J. D. Fell (1977): *4th Ann. Res. Meet., Inst. Chem. Eng., Swansea*, April.

————, N. T. Hai, W. V. Pinczewski, and C. J. D. Fell (1979): in "Distillation 1979," *Inst. Chem. Eng. Symp. Ser.*, 56, London.

Reid, R. C., J. M. Prausnitz, and T. K. Sherwood (1977): "Properties of Gases and Liquids," McGraw-Hill, New York.

Reiss, L. P. (1967): *Ind. Eng. Chem. Process Des. Dev.*, **6**:486.

Rivas, O. R. (1977): *Ind. Eng. Chem. Process Des. Dev.*, **16**:400.

Robinson, R. G., and A. J. Engel (1967): *Ind. Eng. Chem.*, **59**(3):23.

Ross, S. (1967): *Chem. Eng. Prog.*, **63**(9):41.

Sada, E., and D. M. Himmelblau (1967): *AIChE J.*, **13**:860.

Sawistowski, H., and W. Smith (1959): *Ind. Eng. Chem.*, **51**: 915.

Schrodt, V. N. (1967): *Ind. Eng. Chem.*, **59**(6):58.

————, J. T. Sommerfeld, O. R. Martin, P. E. Parisot, and H. H. Chien (1967): *Chem. Eng. Sci.*, **22**:759.

Sherwood, T. K., and R. L. Pigford (1950): "Absorption and Extraction," McGraw-Hill, New York.

————, ————, and C. R. Wilke (1975): "Mass Transfer," McGraw-Hill, New York.

————, G. H. Shipley, and F. A. L. Holloway (1938): *Ind. Eng. Chem.*, **30**:765.

Smith, V. C., and W. V. Delnicki (1975): *Chem. Eng. Prog.*, **71**(8):68.

Solari, R. B., and R. L. Bell (1978): *AIChE Meet., Atlanta*, February.

Sommerfeld, J. T., V. N. Schrodt, P. E. Parisot, and H. H. Chien (1966): *Separ. Sci.*, **1**:245.

Stage, H. (1970): *Chem. Z.*, **94**:271.

Standart, G. (1965): *Chem. Eng. Sci.*, **20**:611.

———— (1972): *Chem. Eng. Prog.*, **68**(10):66.

Stobaugh, R. B. (1965): *Hydrocarbon Process.*, **44**(12):137.

Szabo, T. T., W. A. Lloyd, M. R. Cannon, and S. M. Speaker (1964): *Chem. Eng. Prog.*, **60**(1):66.

Thorngren, J. T. (1972): *Ind. Eng. Chem. Process Des. Dev.*, **11**:428.

Treybal, R. E. (1963): "Liquid Extraction," 2d ed., McGraw-Hill, New York.

———— (1973): Liquid-Liquid Systems, in R. H. Perry and C. H. Chilton (eds.), "Chemical Engineers' Handbook," 5th ed., pp. 21-3 to 21-39, McGraw-Hill, New York.

Van Winkle, M. (1967): "Distillation," McGraw-Hill, New York.

Wade, H. L., C. H. Jones, T. B. Rooney, and L. B. Evans (1969): *Chem. Eng. Prog.*, **65**(3):40.

Walter, J. F., and T. K. Sherwood (1941): *Ind. Eng. Chem.*, **33**:493.

Warden, C. P. (1932): *J. Soc. Chem. Ind.*, **11**:405.

Weast, R. C. (ed.) (1968): "Handbook of Chemistry and Physics," 49th ed., CRC Press, Cleveland.

Weiler, D. W., W. V. Delnicki, and B. L. England (1973): *Chem. Eng. Prog.*, **69**(10):67.

Winter, G. R., and K. D. Uitti (1976): *Chem. Eng. Prog.*, **72**(9):50.

Yanagi, T., and B. D. Scott (1973): *Chem. Eng. Prog.*, **69**(10):75.

Young, G. C., and J. H. Weber (1972): *Ind. Eng. Chem. Process Des. Dev.*, **11**:440.

Zhavoronkov. N. M.. V. A. Malyusov. and N. A. Malafeev (1969): *Inst. Chem. Eng. Int. Symp. Distillation,* *Brighton.*

Zuiderweg. F. J.. and A. Harmens (1958): *Chem. Eng. Sci.,* **9**:89.

——, H. Verberg, and F. A. H. Gilissen (1960): *Inst. Chem. Eng. Int. Symp. Distillation, London,* p. 151.

PROBLEMS

12-A₁ Results have been reported for the performance of a new type of distillation contacting tray. Air was passed upward through a single tray and a large excess of pure ethylene glycol was passed in cross flow over the tray. The temperature of the feeds and of the entire tray was uniform at 53°C. at which the vapor pressure of ethylene glycol is 133 Pa. Measurements showed that the exit air contained a mole fraction of glycol equal to 0.00100. the pressure of operation being 101.3 kPa.

(*a*) What is the Murphree vapor efficiency of the tray?

(*b*) If a tower were built using these same trays and the same glycol and airflow rates, how many trays would be required to make the exit air 99.0 percent saturated with ethylene glycol? Neglect pressure drop in the tower and assume operation at 53°C and atmospheric pressure (101.3 kPa).

(*c*) What would be the *overall* tray efficiency of the tower of part (*b*)?

(*d*) What would be the effect on the number of trays required in part (*b*) if the degree of backmixing of glycol on each tray were markedly increased?

12-B₁ Figure 12-33 shows a single-column operation for vacuum distillation of ethylbenzene and styrene.

(*a*) What feature of the process causes the purity of the styrene product to be so much greater than the purity of the ethylbenzene product?

(*b*) The vapor flow above the feed does not differ much percentwise from the vapor flow below the feed. yet the design has made the tower diameter above the feed substantially greater than that below the feed. Why was this design chosen?

12-C₂ Account physically for the sign (plus or minus) of each of the terms in (*a*) Eq. (12-7). (*b*) Eq. (12-8). (*c*) Eq. (12-10). and (*d*) Eq. (12-29).

12-D₂ Derive (*a*) Eq. (12-42). (*b*) Eq. (12-40). and (*c*) Eq. (12-41).

12-E₂ What change or changes in tray design would be most effective for increasing the plate efficiency of one or both of the towers in Example 12-2 without excessive extra expense?

12-F₂ A processing modification being installed in your plant requires the quantitative removal of isobutane from a stream of hydrogen at 200 lb in² abs. Two packed columns currently idle in the plant are being considered for use in a scheme whereby the isobutane would be absorbed into a heavy hydrocarbon oil. The hydrocarbon oil would be regenerated by stripping with nitrogen and would be recirculated. One of the two towers is 18-in ID and can contain a packed-bed height of up to 20 ft. The other tower is 3 ft ID and can contain a packed height of up to 12 ft. Both can operate continuously at pressures from 20 to 200 lb/in² abs. No heat exchangers are available: hence it is proposed that both towers will operate at 80°F. For operation of hydrocarbon systems at these conditions it has been found that $(HTU)_{OG}$ is approximately 2 ft.

(*a*) What would be the capacity of this two-tower system. expressed as standard cubic feet (60°F) of purified hydrogen per hour?

(*b*) How sensitive is the capacity to the estimate of $(HTU)_{OG}$: that is. what would be the capacity if $(HTU)_{OG}$ were 3 ft?

Data and notes (1) The feed hydrogen contains 1.0 mole percent isobutane; the purified hydrogen must contain no more than 0.05 mole percent isobutane. (2) K ($= y/x$) for isobutane in hydrocarbon oils at 80°F is given by Sherwood and Pigford (1950. p. 191) as

Pressure, atm	0.5	1	2	5	10	25
K	7.2	3.6	1.85	0.81	0.46	0.27

(3) The gas rates in the towers should be no more than 75 percent of the flooding gas velocity at the prevailing L/G. The packing will be dumped 1-in. Pall rings for both towers; a/c^3 for this packing is

45 ft^2/ft^3. (4) For the hydrocarbon oil at 80°F, the specific gravity is 0.78 and the viscosity is 2.5 cP. The average molecular weight is 200.

12-G$_2$ Hay and Johnson (1960) studied the performance of sieve trays in the rectification of methanol-water mixtures in an 8-in-diameter five-tray column. From measurements made at *total reflux* they inferred values of both Murphree vapor and point efficiencies E_{MV} and E_{OG} as a function of average *vapor* composition. Results were as follows:

Av. mol % MeOH in vapor	10	20	30	40	60
E_{OG}	0.66	0.69	0.72	0.73	0.74
E_{MV}	1.04	0.95	0.87	0.83	0.82

    Source: Data from Hay and Johnson, 1960.

Explain, as best you can on the basis of limited data, (a) why E_{MV} is greater than E_{OG}, (b) why E_{OG} increases with increasing mole fraction methanol, and (c) why E_{MV} decreases with increasing mole fraction methanol.

12-H$_2$ Frank states in reviewing the primary-plus-secondary column system for styrene-ethylbenzene distillation[†]

> It was noted from study of the McCabe-Thiele diagram that the reflux in the secondary column could be much lower with the requirement of only two or three more trays in this section. Almost all secondary columns have been designed in this manner with the steam load on the secondary at about 65 percent of that required for the primary column.

Demonstrate the basis for this statement.

12-I$_2$ Finch and Van Winkle (1964) measured tray efficiencies for the evaporation of methanol from water into humidified air. They employed simple sieve plates made by boring a succession of holes (on the order of 5 mm diameter) into plates of 20-gauge stainless steel. Operation was isothermal at 33°C and atmospheric pressure, with the mole fraction of methanol in the effluent liquid held constant at about 0.04. They determined the effects of five variables, changing each independently:

Hole diameter d	1.6–8.0 mm
Vapor flow G	1.1–3.0 kg/s · (m^2 active bubbling area)
Liquid flow L	1.1–3.7 kg/s · (m liquid flow width)
Weir height W	2.5–12.5 cm
Tray length between inlet and outlet weirs Z_l	28–58 cm

They measured both Murphree vapor efficiencies (E_{MV}, based on gross inlet and outlet compositions and ranging from 75 to 97 percent) and point efficiencies (E_{OG}, based on compositions at a particular location on a tray and ranging from 69 to 93 percent). Over their range of investigation they found that

1. Both E_{MV} and E_{OG} decrease with increasing G.
2. Both E_{MV} and E_{OG} increase with increasing L.
3. Both E_{MV} and E_{OG} increase with increasing W.
4. E_{OG} increases slightly with increasing Z_l, whereas E_{MV} increases substantially more with increasing Z_l.

    (a) Does the plate efficiency in the range of conditions covered in this study appear to be predominantly gas-phase- or liquid-phase-controlled? Explain.
    (b) Why does E_{OG} increase with increasing Z_l?

---

† From Frank, 1968; used by permission.

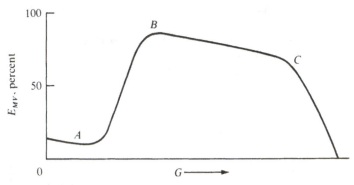

Figure 12-35 Variation of Murphree vapor efficiency with gas flow rate. (*Data from Finch and Van Winkle, 1964.*)

(c) Why does E_{MV} increase more rapidly with increasing Z_l then E_{OG} does?

(d) In relating their measurements to past studies of similar systems on sieve plates, Finch and Van Winkle indicate that as the gas rate is increased from zero (at fixed L, W, and Z_l), the efficiency is initially quite low (see Fig. 12-35). After a certain point A the efficiency rises sharply from almost zero to a maximum B. After passing the maximum the efficiency falls off slowly with increasing gas rate until a sudden rapid fall is reached at C as entrainment or flooding begins to reduce efficiency. Suggest causes for the indicated behavior below A, between A and B, and between B and C.

12-J$_2$ There have been virtually no tests reported for the applicability of the AIChE efficiency prediction method to high-pressure light-hydrocarbon systems. In addition, the extent to which the AIChE correlating equations for transfer units are applicable to trays other than bubble-cap trays has not been reported in any detail. A recent field test of a propylene-propane splitter (the one considered in Prob. 8-J) afforded the following results:

Average operating pressure = 1.86 MPa   Overhead temperature = 44°C,

Bottom temperature = 55°C   Reflux ratio $L/d = 21.5$

Propylene purity = 96.2 mol %   Propane purity = 91.1 mol %

Propylene in feed = 50.45 mol %   Feed rate = 530 bbl/day (satd liquid)

The tower diameter is 48 in with 90 sieve trays, the feed being introduced to the forty-fifth. Tray spacing is 18 in. Details of construction are the following, as shown in Fig. 12-36,

Weir length = 36.7 in   Downcomer width at bottom = 6.5 in

Weir height = 2.0 in $\frac{3}{16}$-in holes on $\frac{7}{16}$-in triangular pitch   4970 holes/tray

Analysis of the equilibrium-stage requirement (Prob. 8-J) reveals that 85 equilibrium stages are required to give the observed split with the given feed tray.

Compare the observed stage efficiency with that predicted by the AIChE bubble-tray design method.

Data and notes Barrels of feed (1 bbl = 42 gal) are measured at 15.5°C, where the specific gravities of propylene and propane are 0.522 and 0.508, respectively.

	Propylene	Propane
Critical temperature, °C	91.4	96.9
Critical pressure, MPa	4.60	4.25
Specific gravity satd liquid at 49°C	0.458	0.453
Viscosity of satd liquid, cP	0.086†	0.080‡
Vapor viscosity at 49°C and 1.86 MPa, cP	0.0108	0.0108
Vapor diffusivity at 49°C and 1.86 MPa, m²/s	3.9×10^{-7}	3.9×10^{-7}
Liquid-phase diffusivity on the order of 1×10^{-8} m²/s		

† At 45°C.   ‡ At 55°C.

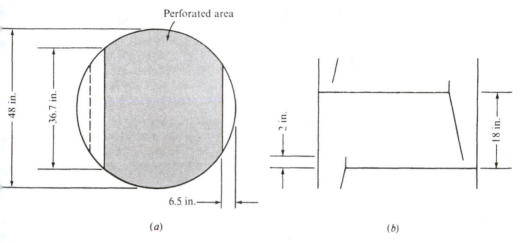

Perforated area

48 in.

36.7 in.

6.5 in.

2 in.

18 in.

(a)

(b)

Figure 12-36 Tray layout for Probs. 12-J and 12-K: (a) top and (b) side view.

12-K₂ At what percentage of the ultimate feed capacity at the given reflux ratio was the tower of Prob. 12-J being run during the field test of Prob. 12-J? Use the predictions of the correlations given in this chapter.

12-L₂ There are several different reboiler designs employed in current practice, the choice of a particular design being governed by the particular processing requirements. Among the many criteria influencing reboiler selection is the fact that certain reboiler designs are more effective than others in providing an additional full equilibrium stage of vapor-liquid contacting. (This point also has a direct bearing on the temperature to which liquid must be raised within the reboiler.) In Fig. 12-37 four common reboiler designs are shown which you should analyze and contrast by the above criterion of providing an additional equilibrium stage. In each of the systems L represents the overflow from the first tray, B is the bottoms product, F is the reboiler feed, V' is the reboiler vapor output, and u is the unvaporized liquid returning from the reboiler; LC refers to a level controller which regulates the withdrawal of net bottoms. Types 1, 2, and 3 are *thermosiphon* reboilers, which return liquid along with vapor and for good heat transfer can vaporize no more than 30 percent of the reboiler feed. The total volume liquid holdup is the same in all cases. Steam passes through the reboiler shell in cases 1 to 3 and through the tubes in case 4. Rank these schemes in the order of increasing additional separation obtained in the reboiler.

12-M₃ (a) Suppose that for the H_2S–CO_2 ethanolamine absorber of Example 10-2 the values of $(NTU)_G$ and $(NTU)_L$ were constant at 1.50 and 0.45, respectively, from plate to plate for CO_2. Determine what E_{MV} would be on the bottom plate and on the top plate, respectively.

(b) Develop a block diagram for a computer program which would perform the stage-to-stage calculation of part (b) of Example 10-2, using constant values of $(NTU)_G$ and $(NTU)_L$ for both solutes rather than average values of E_{MV}.

(c) Explain physically why $(NTU)_L$ might be so low.

12-N₃ It has frequently been observed that stage efficiencies for plate-type gas absorption columns tend to be significantly less than stage efficiencies for distillation in similar columns. Walter and Sherwood (1941) measured Murphree vapor efficiencies for a number of systems in a small 5-cm bench-scale bubble-cap column. As shown in Table 12-4, E_{MV} for the absorption of propylene and isobutylene into *gas oil*† ranged from 11 to 17 percent whereas E_{MV} for ethanol-water distillation ranged from 88 to 91 percent. All their runs were conducted well below the flooding point, and entrainment was not a significant factor. The gas diffusivities do not differ greatly between the systems. The liquid-phase diffusivity is approximately a factor of six higher in the distillation system than in the absorption system.

---

† Gas oil is a hydrocarbon mixture, bp = 230 to 350°C, average MW = 210, viscosity = 6.2 mPa·s, and sp gr = 0.86 at 25°C.

Table 12-4 Efficiency data reported by Walter and Sherwood (1941)

Distillation of ethanol-water:

Total reflux   Pressure = 101.3 kPa (atmospheric)
$x_{C_2H_5OH} = 0.05$ leaving plate

Vapor flow, mol s	0.0236	0.0236	0.0246	0.0337	0.0454
E_{MV}	0.88	0.91	0.89	0.88	0.88

Absorption of propylene and isobutylene into gas oil at 25°C:

Gas flow = 0.098 mol s = 3.05 g s (MW = 31)
Liquid flow = 0.047 mol s = 9.8 g s (MW = 210)
Pressure = 456 kPa (4.5 atm)

Solute	$(K = y/x)_{eq}$	E_{MV}
Propylene	2.4	0.110
Isobutylene	0.66	0.174

(a) Interpreting on the basis of the concepts involved in the AIChE method for analyzing stage efficiencies, indicate the *principal physical factor(s)* of difference between the absorption and distillation systems which probably cause(s) the values of E_{MV} for the absorption process to be so much less than E_{MV} for the distillation process.

(b) Why is E_{MV} for isobutylene absorption greater than E_{MV} for propylene absorption?

12-O₁ A countercurrent sieve-plate stripping column with reboiler is to be used to remove low concentrations (less than 0.5 mole percent) of *n*-butyl acetate from water. The operating pressure will be either atmospheric (101.3 kPa), or a moderate level of vacuum, say, 25 kPa. In either case, the temperature will be the thermodynamic saturation temperature of water. The relative volatility of *n*-butyl acetate to water in both cases is in the range 500 to 1000. The vapor rate in the stripper will be equal to 10 percent of the purified-water product flow rate in both cases. The Murphree vapor efficiency is substantially less than 100 percent.

(a) Is the Murphree vapor efficiency for this process likely to be gas-phase-controlled or liquid-phase controlled? Explain briefly.

(b) Which of the operating pressures under consideration should require the larger column diameter? Why?

12-P₂ An ethylene-ethane distillation column operates at an average pressure of 2.5 MPa and has a temperature range of 238 to 279 K. Sieve plates are used, with a hole diameter of 0.95 cm, an interplate spacing of 0.61 m, an outlet weir height of 6.3 cm, a tower diameter of 1.30 m, and an operating capacity 60 percent of flooding. Analysis of the plate operation using the AIChE plate-efficiency model gives the following results:

$$(NTU)_G = 1.90 \qquad (NTU)_L = 9.35 \qquad Pe = 0.42$$

$$\text{Entrainment} = 0 \qquad \lambda\,(= mG/L) \text{ ranges from 0.8 to 1.3}$$

(a) What is the range of Murphree vapor efficiencies in the column?

(b) On the basis of the information given and the AIChE model, indicate which of the following changes should serve to increase the Murphree vapor efficiency. Explain each answer briefly.

1. Increase the outlet weir height to 7.5 cm

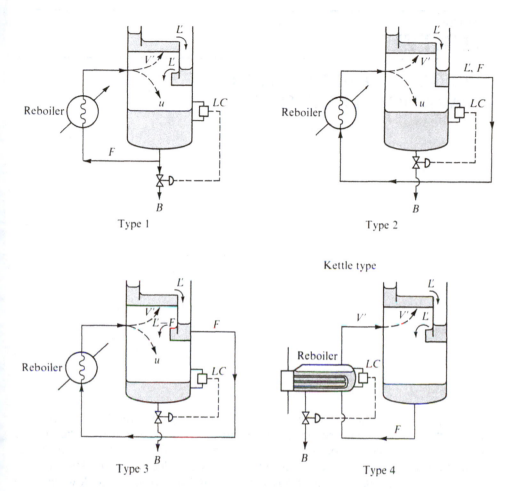

Figure 12-37 Reboiler flow configurations. Type 3 is for use when V'/L is quite low, as in strippers.

2. Increase u_G (the volumetric gas flow per unit active tray bubbling area) while holding L (liquid flow per unit flow width) constant, e.g., by decreasing the fraction of the tower cross section that is active bubbling area
3. Decrease L while holding u_G constant, e.g., by decreasing the fraction of the tower cross section that is active bubbling area and decreasing the reflux and boil-up ratios simultaneously.

THIRTEEN

ENERGY REQUIREMENTS OF SEPARATION PROCESSES

The energy consumption is often a critical process parameter for a large-scale separation.† The cost of energy supply is usually a major contributor to the process cost. Different classes of separation processes can have inherently different energy consumptions, and this can be a critical factor in their selection. Understanding the factors underlying energy consumption can often lead to ideas for lowering the energy consumption, and the cost, of a process.

In this chapter we first develop the thermodynamic minimum energy consumption for a specified separation and then explore the characteristics of different types of single-stage and multistage separation processes as related to energy consumption. This discussion is followed by consideration of ways of reducing energy consumption. Some of these approaches are extremely simple, and others require relatively complex designs.

---

† The discussion in this chapter postulates some familiarity with classical thermodynamics on the part of the reader. particularly with regard to the second law and outgrowths of it. The concepts of reversiblity, free energy, available energy. and entropy are developed at greater length by B. F. Dodge. "Chemical Engineering Thermodynamics." McGraw-Hill, New York. 1944; O. A. Hougen et al., " Chemical Process Principles." vol. II. " Thermodynamics." 2d ed.. Wiley. New York. 1959; J. M. Smith and H. C. Van Ness, " Introduction To Chemical Engineering Thermodynamics." 3d ed.. McGraw-Hill. New York. 1975; M. W. Zemansky. " Heat and Thermodynamics." 5th ed.. McGraw-Hill. New York. 1968; and H. C. Weber and H. P. Meissner. " Thermodynamics for Chemical Engineers." Wiley. New York. 1959; among others.

MINIMUM WORK OF SEPARATION

Mixing substances together is inherently an irreversible process. Substances can be mixed spontaneously, but separation of homogeneous mixtures into two or more products of different composition at the same temperature and pressure necessarily requires some sort of device which consumes work and/or heat energy.

The minimum possible work consumption for a separation, no matter what process is employed to accomplish it, is found by postulating a hypothetical reversible process. One of the consequences of the second law of thermodynamics is that any reversible process for accomplishing a given transformation has the same work requirement and that the work requirement of any real process for carrying out the separation is greater. The minimum reversible work requirement is dependent solely upon the composition, temperature, and pressure of the mixture to be separated and upon the desired composition, temperature, and pressure of the products; it is a state property.

Isothermal Separations

As a generalization of the analyses presented by Dodge (1944), Robinson and Gilliland (1950), and Hougen et al. (1959) the minimum (reversible) mechanical work required for separation of a homogeneous mixture into pure products at constant pressure and constant temperature T is

$$W_{\text{min. }T} = -RT \sum_j x_{jF} \ln \left(\gamma_{jF} x_{jF} \right) \tag{13-1}$$

where $W_{\text{min. }T}$ = minimum work consumption per mole of feed
R = gas constant
x_{jF} = mole fraction of component j in feed
γ_{jF} = activity coefficient of component j in feed mixture

The summation is over all components in the feed. If there are R components, there are R pure products and R terms in the series. The convention for definition of the activity coefficient is that $\gamma_j = 1$ in the pure state. Equation (13-1) applies to gas, liquid and solid mixtures. For gases γ_{jF} denotes the degree of departure from the ideal-gas law and the Lewis and Randall ideal-mixing rule.

For an ideal gas mixture or an ideal liquid solution Eq. (13-1) becomes

$$W_{\text{min. }T} = -RT \sum_j x_{jF} \ln x_{jF} \tag{13-2}$$

For a binary mixture Eq. (13-2) becomes

$$W_{\text{min. }T} = -RT[x_{AF} \ln x_{AF} + (1 - x_{AF}) \ln (1 - x_{AF})] \tag{13-3}$$

Comparing Eqs. (13-1) and (13-2), we find that if there are positive deviations from ideality, γ_A and γ_B will be greater than unity and the minimum isothermal work requirement for separation will be less than that for an ideal mixture. Similarly, a system with negative deviations from ideality requires a greater W_{min} than an ideal system. Negative systems involve preferential interactions between dissimilar molecules and are therefore more difficult to separate. In Eq. (13-1) if $\gamma_{jF} = 1/x_{jF}$ the system is totally immiscible and the work of separation is zero; otherwise the isothermal work of separation must be positive.

Although the minimum work for separation depends upon the degree of solution nonideality, it is important to note that it does not depend upon the separation factor of the actual process postulated. For example, if a liquid mixture is to be separated by a distillation process designed to be reversible, the work requirement of that process does not depend upon the relative volatility.

The minimum work of separation of a feed mixture into impure products at constant temperature and pressure can be computed by subtracting from Eq. (13-1) the minimum works for separation of those impure products into pure products, giving

$$W_{\text{min. }T} = -RT\left(\sum_j x_{jF}\ln\left(\gamma_{jF}x_{jF}\right) - \sum_i \phi_i \sum_j x_{ji}\ln\left(\gamma_{ji}x_{ji}\right)\right) \qquad (13\text{-}4)$$

where ϕ_i = molar fraction of feed entering product i
x_{ji} = mole fraction of component j in product i
γ_{ji} = activity coefficient of component j in product i

For a given feed mixture, the work requirement given by Eq. (13-4) for separation into impure products is necessarily less than that given by Eq. (13-1) for separation into pure products.

For a binary mixture, if activity coefficients are taken equal to unity for simplicity, and if the lever rule is used to generate values of ϕ_i, algebraic rearrangement of Eq. (13-4) yields

$$W_{\text{min. }T} = -\frac{RT}{x_{A1} - x_{A2}}\left|(x_{AF} - x_{A2})\left[x_{A1}\ln\frac{x_{AF}}{x_{A1}} + (1 - x_{A1})\ln\frac{1 - x_{AF}}{1 - x_{A1}}\right]\right.$$
$$\left. + (x_{A1} - x_{AF})\left[x_{A2}\ln\frac{x_{AF}}{x_{A2}} + (1 - x_{A2})\ln\frac{1 - x_{AF}}{1 - x_{A2}}\right]\right| \qquad (13\text{-}5)$$

The solid curve in Fig. 13-1 shows $W_{\text{min. }T}/RT$ for separation of an ideal binary mixture into pure products as a function of x_{AF}. Notice that an equimolal feed mixture requires more work per mole of feed for isothermal isobaric separation into pure components than a mixture of any other composition. The dashed curve in Fig. 13-1 gives $W_{\text{min. }T}$ as a function of x_{AF} for a binary feed where the product compositions are $x_{A1} = 0.95$ and $x_{A2} = 0.20$. Notice that the minimum work for the separation into impure products is substantially less than that for separation into pure products.

Equations (13-1) to (13-5) assume that the products have the same temperature and pressure as the feed. If pressure changes, one or more terms representing $\int V\,dP$ must be added to these expressions for $W_{\text{min. }T}$. For liquid mixtures at low pressures the contribution of such terms (the *Poynting effect*) is usually small. For an ideal gas with feed at pressure P_1 and products at pressure P_2, the expression for minimum work becomes

$$W_{\text{min. }T} = W_{\text{min. }T,\text{ isobaric}} + RT\ln\frac{P_2}{P_1} \qquad (13\text{-}6)$$

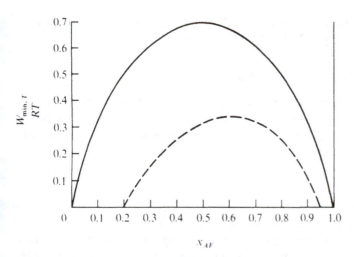

Figure 13-1 Minimum work of separation of a binary ideal liquid or gas mixture. (Solid curve = pure products; dashed curve = $x_{A1} = 0.95$, $x_{A2} = 0.20$.)

which shows that the minimum work requirement for an isothermal separation may be zero or negative if P_2 is less than P_1. In this case the work required for separation is derived from the work available from expansion. Since $W_{min, T, isobaric}$ is necessarily positive,

$$W_{min, T} > RT \ln \frac{P_2}{P_1} \tag{13-7}$$

indicating that the work available from a reversible expansion is reduced if a separation also occurs. Similarly, if there is a net compression, the minimum work requirement will be necessarily greater when a separation occurs than when no separation occurs. One can postulate a process in which helium is removed from natural gas available at 4 MPa from a gas well by selective diffusion across a glass or polymeric membrane or through a sintered-metal diaphragm. In this case a separation occurs without any heat input or compression work, but the separation is possible only because the natural gas is available at a pressure P_1, which is substantially greater than the pressure P_2 on the other side of the membrane or diaphragm, which could be as low as atmospheric.

The minimum isothermal work of separation is also necessarily equal to the increase in Gibbs free energy of the products over the feed. The Gibbs free energy G is defined as

$$G = H - TS \tag{13-8}$$

Therefore

$$\Delta G_{sep} = W_{min, T} = \Delta H - T \Delta S \tag{13-9}$$

where T = absolute temperature
$\quad \Delta H$ = enthalpy of products minus enthalpy of feed
$\quad \Delta S$ = entropy of the products minus entropy of feed

For the isothermal separation of a mixture of ideal gases the enthalpy change is zero, and the right-hand side of Eq. (13-2) represents the $-T \Delta S$ term of Eq. (13-9).

Nonisothermal Separations; Available Energy

When the products of a separation process are removed at temperatures different from the feed temperature, the minimum work required for separation can be obtained from the increase of *available energy* of the products with respect to the feed. The available energy B, sometimes called the *exergy* in the European literature, is defined as

$$B = H - T_0 S \tag{13-10}$$

where T_0 is the absolute temperature of the surroundings, from or to which we presume that heat can be transferred on essentially a *free* basis. Thus T_0 is the temperature of sea water or river water or is the prevailing atmospheric temperature.

The increase in available energy of products over the feed is a measure of the minimum work required for separation when heat sources and sinks are available only at temperature T_0

$$\Delta B_{\text{sep}} = W_{\text{min. } T_0} = \Delta H - T_0 \Delta S \tag{13-11}$$

This expression is different from Eq. (13-9) in that it allows for feeds and products being at different temperatures. Equation (13-11) also does not reduce to Eq. (13-9) for a separation giving products at the same temperature as the feed unless that temperature is T_0. This is a result of no longer considering that an infinite heat sink is available at T. The case of a heat sink available at T_0 is the more realistic consideration for engineering purposes.

For separation of a mixture of ideal gases into pure components ΔH and ΔS for use in Eq. (13-11) are given by

$$\Delta H = \sum_j x_{jF} \int_{T_F}^{T_j} C_{Pj} \, dT \tag{13-12}$$

$$\Delta S = \sum_j x_{jF} \left(\int_{T_F}^{T_j} \frac{C_{Pj}}{T} \, dT - R \ln \frac{P_j}{x_{jF} P_F} \right) \tag{13-13}$$

where C_{Pj} = heat capacities of various components
T_F, P_F = feed temperature and pressure
T_j, P_j = temperatures and pressures of various pure component products

For an isothermal separation with $\Delta H = 0$, combination of Eqs. (13-9) and (13-11) shows that $W_{\text{min. } T_0}$ is given by the various equations for $W_{\text{min. } T}$ with RT replaced by RT_0.

Significance of W_{min}

The minimum work of separation represents a lower bound on the energy that must be consumed by a separation process. In most cases the energy requirement for a real process will be many times greater than this minimum. Nonetheless, the relative sizes of the minimum work requirements for different separations are a first indication of the relative difficulties of the separations. In some cases, e.g., the desalinization of sea water on a large scale, the separation must be carried out with an energy consumption rather close to the minimum work of separation in order to be economical. In these situations the minimum work requirement is a highly significant quantity which must be kept in mind during the synthesis and evaluation of different designs.

The concept of *separative work* is commonly used for the analysis of isotope-enrichment processes (Benedict and Pigford, 1957) and is directly related to the reversible-work requirement for separation (Opfell, 1978). In fact, the value of enriched uranium in the nuclear-fuel market is commonly stated in terms of the number of separative work units (SWU) contained; they are quaintly known as *swoos*.

NET WORK CONSUMPTION

Often the energy to drive a separation process is supplied in the form of heat rather than mechanical work. In such cases it is convenient to speak of the *net work consumption* of the process, defined as the difference between the work that could have been obtained with a reversible heat engine from the heat entering the system, on the one hand, and the work that could be obtained in a reversible heat engine from the heat leaving the system, on the other. The other heat source or sink of the heat engines would be at the ambient T_0.

In the separation process shown schematically in Fig. 13-2 the process is driven by heat Q_H entering the system at a temperature T_H. An amount of heat Q_L leaves the system at a temperature T_L. If Q_H were supplied to a reversible heat engine rejecting heat at T_0, an amount of work equal to

$$Q_H \frac{T_H - T_0}{T_H}$$

could be obtained. Similarly, an amount of work equal to

$$Q_L \frac{T_L - T_0}{T_L}$$

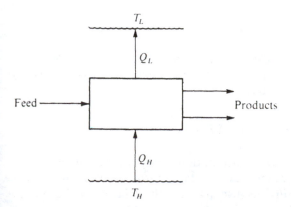

Figure 13-2 A separation process driven by heat input.

could be obtained from Q_L. The net work consumption of the process W_n is

$$W_n = Q_H \frac{T_H - T_0}{T_H} - Q_L \frac{T_L - T_0}{T_L} \qquad (13\text{-}14)$$

It can be shown that W_n for any real separation is necessarily greater than ΔB_{sep} and will be equal to it only in the limit of a reversible separation process. If any mechanical work is consumed by the process, it must be added into Eq. (13-14) directly.

If no mechanical work is involved in a separation process and the enthalpy difference between products and feed is negligible compared with the heat input, $Q_H = Q_L = Q$ and

$$W_n = Q T_0 \left(\frac{1}{T_L} - \frac{1}{T_H} \right) \qquad (13\text{-}15)$$

which is necessarily a positive quantity since T_H must be greater than T_L.

An ordinary distillation column is a good example of a separation process driven by heat input. An amount of heat equal to Q_R enters at the reboiler at temperature T_R. Heat in the amount Q_c is removed in the condenser at a temperature T_c. If the enthalpy of the products is not substantially different from that of the feed, Eq. (13-15) can be used to find W_n, with $T_H = T_R$ and $T_L = T_c$. When cooling water is used to remove heat in the condenser, $T_L = T_0$ and Eq. (13-15) becomes

$$W_n = Q \left(1 - \frac{T_0}{T_R} \right) \qquad (13\text{-}16)$$

T_R necessarily will be above ambient in such a case, and W_n is therefore positive; W_n also will be positive for a low-temperature refrigerated distillation column since T_H is still greater than T_L in Eq. (13-15).

THERMODYNAMIC EFFICIENCY

A thermodynamic efficiency η can be defined as the ratio of the minimum (reversible) work consumption to the actual work consumption of a separation process. For a separation driven by heat input at a high temperature and heat rejection at a lower temperature

$$\eta = \frac{W_{min, T_0}}{W_n} \qquad (13\text{-}17)$$

W_{min, T_0} is obtained from Eq. (13-11) [or from Eq. (13-4) if the process is isothermal and isobaric]. W_n is obtained from Eq. (13-14). Any mechanical work consumed by the process should be added to W_n directly.

SINGLE-STAGE SEPARATION PROCESSES

Separation processes in which the energy consumption is critical are usually carried out in multistage equipment, to reduce the amount of separating agent required. The separating-agent requirement, in turn, is usually directly related to the energy consumption. When the separation factor is quite large and when the equilibrium behavior is unusual, the separation can be accomplished commercially in a single stage. The following examples explore the energy requirements of three single-stage processes for hydrogen purification. One of these processes is an equilibration process with energy separating agent, another is an equilibration process with a mass separating agent, and the third is a rate-governed separation process.

Example 13-1 A stream of 50 mol $^{\circ}_{0}$ hydrogen and 50 mol $^{\circ}_{0}$ methane at 3.45 MPa pressure and 294 K is to be separated continuously into two gaseous products at the same temperature or less and at the same pressure as the feed. The hydrogen product should have a purity of at least 90 mol $^{\circ}_{0}$ and should contain at least 90°_{0} of the hydrogen present in the feed. (a) Find the net work consumption of a thermodynamically reversible process if heat sources and sinks are available at 294 K. (b) Find the net work consumption if the separation is to be accomplished by a continuous, single-stage partial condensation using a single refrigerant. Use heat exchange where warranted. Also obtain the thermodynamic efficiency.

SOLUTION (a) Assume the gas-phase activity coefficients to be unity. Since ΔH will be zero in the absence of any heat of mixing, we can use Eq. (13-5) for $W_{min. T_0}$, substituting RT_0 for RT [see note under Eq. (13-13)]:

$$W_{min. T_0} = - \frac{RT_0}{x_{A1} - x_{A2}} \left\{ (x_{AF} - x_{A2}) \left[x_{A1} \ln \frac{x_{AF}}{x_{A1}} + (1 - x_{A1}) \ln \frac{1 - x_{AF}}{1 - x_{A1}} \right] \right.$$

$$\left. + (x_{A1} - x_{AF}) \left[x_{A2} \ln \frac{x_{AF}}{x_{A2}} + (1 - x_{A2}) \ln \frac{1 - x_{AF}}{1 - x_{A2}} \right] \right\}$$

Since the minimum recovery fraction of hydrogen is 0.90, the maximum mole percent hydrogen in the methane product for a 90 mol $^{\circ}_{0}$ hydrogen product purity is 10 percent. Hence

$$W_{min. T_0} = - \frac{(8.31)(294) \text{ J/mol}}{0.90 - 0.10} \left[(0.50 - 0.10) \left(0.90 \ln \frac{0.50}{0.90} + 0.10 \ln \frac{0.50}{0.10} \right) \right.$$

$$\left. + (0.90 - 0.50) \left(0.10 \ln \frac{0.50}{0.10} + 0.90 \ln \frac{0.50}{0.90} \right) \right]$$

$$= - \frac{(8.31)(294)}{0.80} [(0.40)(2)(-0.530 + 0.161)] = + 902 \text{ J/mol feed}$$

(b) Binary equilibrium data are given in Fig. 2-23 (Prob. 2-D). Mollier diagrams for hydrogen and methane are given in Figs. 13-3 and 13-4. The English engineering units used in these figures will be converted into SI units as needed.

A schematic of the single-stage partial condensation process is shown in Fig. 13-5. Refrigeration is used to cool the feed mixture to the point where the desired degree of separation of hydrogen is obtained between the gas and the liquid condensed out. The cold products are used to provide as much of the feed cooling as possible.

At first glance it would seem that the products should be capable of providing *all* the cooling if the outlet product streams can be brought up to feed temperature in the heat exchanger. This cannot occur, of course, because the refrigerant duty represents the net work consumption of the process and must be a positive quantity. The products cannot provide all the cooling because much of the heat effect in the heat exchanger is latent heat rather than sensible heat. The latent heat of vaporization of the methane is released at the boiling point of methane at 3.45 MPa, neglecting the effect of the hydrogen remaining in the methane product. From Fig. 13-4, the boiling point of methane at 3.45 MPa is 181 K. Because of the presence of 50°_{0} hydrogen in the feed, the dew point of the feed will be substantially lower than 181 K. Thus the latent heat of vaporizing the methane product cannot be used to consume the latent heat of condensing that product from the feed, and an appreciable amount of refrigeration will be required.

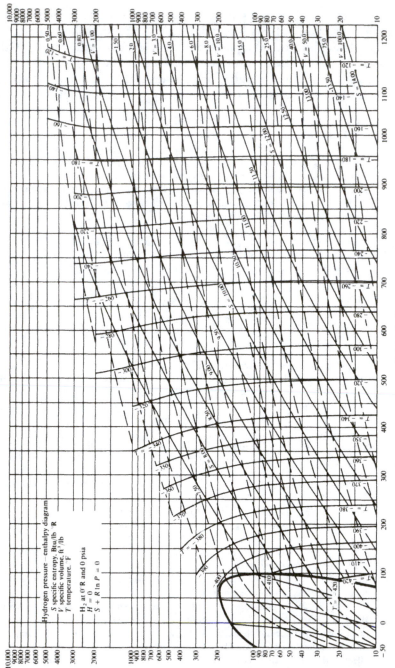

Hydrogen pressure – enthalpy diagram
S specific entropy, Btu/lb °R
V specific volume, ft³/lb
T temperature, °F

H₂ at 0°R and 0 psia
$H = 0$
$S + R \ln P = 0$

Enthalpy, Btu/lb

Pressure, psia

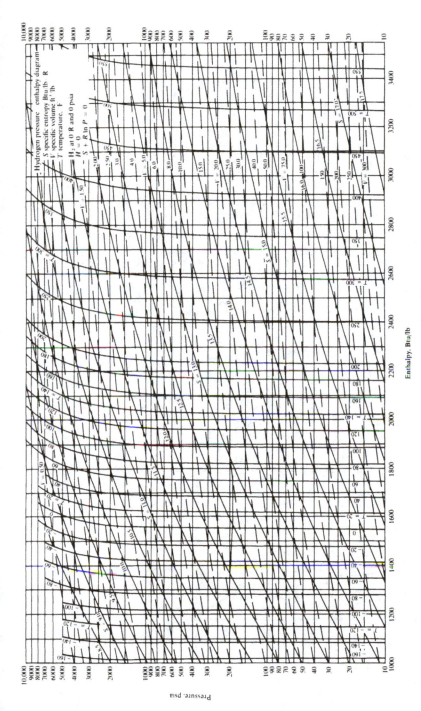

Figure 13-3 Mollier diagram for hydrogen. (From *Thermodynamic Properties and Reduced Correlations for Gases* by Lawrence N. Canjar and Francis S. Manning. Copyright © 1967 by Gulf Publishing Company, Houston, Texas. Used with permission. All rights reserved.)

669

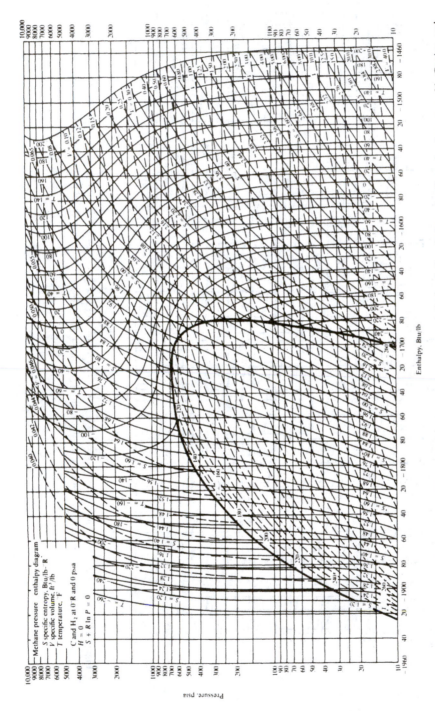

Figure 13-4 Mollier diagram for methane. (From *Thermodynamic Properties and Reduced Correlations for Gases* by Lawrence N. Canjar and Francis S. Manning. Copyright © 1967 by Gulf Publishing Company, Houston, Texas. Used with permission. All rights reserved.)

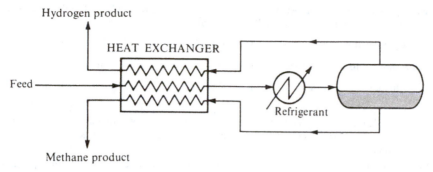

Figure 13-5 Separation of hydrogen and methane by partial condensation.

The equilibrium data must be used to determine the temperature to which the feed must be cooled by the refrigerant. For any assumed temperature we can obtain values of K_j from Fig. 2-23 and use Eqs. (2-12) and (2-14) to obtain x_j and V/F for this binary system; $y_j = K_j x_j$.

T							
°F	K	K_{H_2}	K_{CH_4}	x_{H_2}	y_{H_2}	V/F	$(/_{H_2})_l$
−150	172.2	9.0	0.92	†	†	†	1.000
−200	144.4	19	0.37	0.0338	0.642	0.769	0.984
−250	116.7	31	0.075	0.0300	0.93	0.522	0.971

† Above dew point.

The limitation comes from the purity of the hydrogen product, rather than from the necessary recovery fraction of hydrogen. (From the description rule it should be noted that the conditions of 90 percent hydrogen purity and 90 percent recovery both cannot be imposed. The more stringent of the two conditions sets the limit.) Interpolating, it will be necessary to cool to 119 K in order to obtain a 90 percent hydrogen purity.

To find the refrigeration duty, we shall compare the enthalpy increase of the products in going from 119 to 181 K, the methane remaining liquid, and the enthalpy decrease of the feed in going from vapor at 181 K to a two-phase mixture at 119 K. We do this because the closest temperature approach in the heat exchanger will come at 181 K, when the methane product has been raised to its boiling point but has not yet vaporized. This assumes that at some internal location in the heat exchanger the two streams have the same temperature, 181 K. It therefore postulates a heat exchanger of infinite area. For convenience we shall treat the products as pure streams in order to determine enthalpies.

Enthalpies of methane at 3.45 MPa (Fig. 13-4)

	Temp, K	MJ/kg
Vapor	181	−3.920
Liquid	181	−4.167
Liquid	119	−4.424

If the enthalpy increase of hydrogen in going from 119 to 181 K is ΔH_{H_2} kJ/mol, the cooling available from raising the products to 181 K with the methane still liquid is

$$\Delta H_{\text{prod}} = (0.5)(0.016)(-4.167 + 4.424) \times 10^3 + 0.5 \, \Delta H_{H_2} = 2.06 + 0.5 \, \Delta H_{H_2} \qquad \text{kJ/mol feed}$$

The cooling required to take the feed from vapor at 181 K to a two-phase mixture at 119 K is

$$\Delta H_{\text{feed}} = (0.5)(0.016)(-4.424 + 3.920) \times 10^3 - 0.5 \, \Delta H_{H_2} = -4.03 - 0.5 \, \Delta H_{H_2} \quad \text{kJ/mol feed}$$

The sum of these two quantitites $\Delta H_{\text{prod}} + \Delta H_{\text{feed}} = -1.97$ kJ per mole of feed is equal to the latent heat of condensation of the methane and represents the amount of refrigeration required in the refrigeration exchanger. In this process the refrigeration must be delivered at 119 K or less, if we use a single refrigerant.

If the refrigeration circuit is a reversible heat pump, the net work consumption corresponding to the refrigeration duty is

$$W_n = Q_{\text{ref}} \frac{T_0 - T_{\text{ref}}}{T_{\text{ref}}} \tag{13-18}$$

Thus the work consumption for 1.97 kJ per mole of feed is given by

$$W_n = 1.97 \frac{294 - 119}{119} = 2.90 \text{ kJ/mol feed}$$

From Eq. (13-17) the *thermodynamic efficiency* is

$$\eta = \frac{W_{\text{min}, \, T_0}}{W_n} = \frac{902}{2900} = 0.31$$

In any real situation the refrigeration cycle will be irreversible, the refrigerant must be at some temperature less than 119 K, and the products cannot be raised all the way to 181 K at the point in the exchanger where the feed has been cooled to 181 K. If the thermodynamic efficiency of the refrigeration cycle is 0.35, the overall efficiency of the process would be reduced to 0.35(0.31) = 0.11.

□

Often for a hydrogen purification process like that of Example 13-1 it is not necessary for the methane (or other contaminant removed) to be kept at the same high pressure as the feed. If the methane product can be reduced in pressure, cooling can be obtained by passing the liquid methane through an adiabatic expansion (Joule-Thomson) valve. This will take the methane to a lower temperature and will reduce the temperature at which the methane vaporizes. As a result, the feed can be cooled to a much lower temperature in the feed-products heat exchanger, and less auxiliary refrigeration is required. If enough methane is in the feed, the need for auxiliary refrigeration during steady-state operation may be eliminated.

The partial-condensation process described in Example 13-1 for separating hydrogen and methane requires refrigeration at a very low level. Other approaches to separation involve higher temperatures. Two such processes are considered in Examples 13-2 and 13-3.

Example 13-2 Equilibrium data for methane dissolving in a paraffinic oil of molecular weight 220 at $31 \pm 2°C$ are shown in Fig. 13-6. Suppose that an oil of these characteristics is used to separate methane and hydrogen by single-stage absorption at 31°C, with absorbent regeneration carried out by reducing the pressure. The feed conditions and product specifications are the same as in Example 13-1. (a) Devise a flowsheet for such a process. (b) Find the net work consumption and thermodynamic efficiency of the process. (c) By what amount could the net work consumption be reduced if the absorption were carried out with multiple stages?

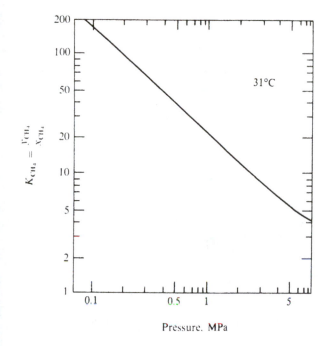

Figure 13-6 Equilibrium ratios for methane in paraffinic oil of molecular weight 220. (*Data from Kirkbride and Bertetti, 1943.*)

Pressure, MPa

SOLUTION (*a*) A flowsheet for the process is shown in Fig. 13-7. The feed is contacted with absorbent oil in the absorber vessel. The gas leaving this vessel is the hydrogen product. The liquid leaving the absorber is reduced in pressure through an expansion valve, and the gas which is formed is taken as the methane product, which must be recompressed to feed pressure. The regenerated absorbent oil is recirculated to the absorber. The process is presumed to be nearly isothermal at 31°C as a result of the large absorbent circulation rate required.

(*b*) The absorber will operate at 3.45 MPa, the feed pressure, and must give a gaseous product containing no more than 10 mol % methane. From Fig. 13-6 the value of K_{CH_4} at 3.45 MPa is 7.0; hence we can compute x_{CH_4} in the oil leaving the absorber as

$$x_{CH_4} = \frac{y_{CH_4}}{K_{CH_4}} = \frac{0.10}{7.0} = 0.0143$$

Because of the very low solubility of hydrogen in oils under these conditions, we can presume that x_{H_2} in the liquid leaving the absorber will be one or more orders of magnitude less and that the 90 percent recovery fraction of hydrogen specification is not a limit.

The regenerator must reduce the methane mole fraction in the absorbent to a value substantially less than the mole fraction in the liquid leaving the absorber. The mole fraction of methane in the methane product will be close to 1.00; hence, if x_{CH_4} is to be reduced by half to 0.0071, we must have

$$K_{CH_4} = \frac{y_{CH_4}}{x_{CH_4}} = \frac{1.00}{0.0071} = 140$$

From Fig. 13-6, this value of K occurs at 124 kPa, which we shall adopt as the regenerator pressure.

Note that the mole fraction of methane cannot be reduced by much more than a factor of 2 in the regenerator without necessitating a vacuum system. It would be possible to regenerate the absorbent at a higher pressure if some heat input (more net work) were introduced into the regenerator.

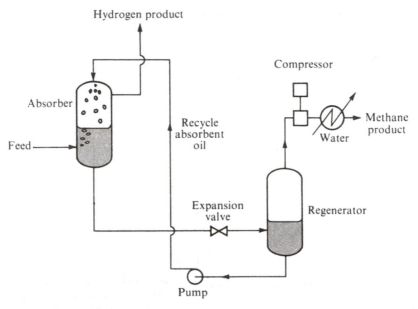

Figure 13-7 Single-stage absorption process for separation of hydrogen and methane at ambient temperature.

The absorbent recirculation rate A can be obtained by a mass balance on the absorber, noting that 0.45 mol of methane is to be removed per mole of feed:

$$(0.0143 - 0.0071)A = 0.45 \quad \text{and} \quad A = 62.5 \text{ mol/mol feed}$$

The energy consumption of this process comes primarily from the work of recompressing the methane product to 3.45 MPa and from the work of pumping the oil back up to 3.45 MPa. The work of recompressing the methane can be obtained from Fig. 13-4. If a single isentropic compressor is used, the enthalpy of the methane must increase from that at 124 kPa and 31°C to that at the same entropy and 3.45 MPa. From the Mollier diagram we see that this compression would result in a large enthalpy increase and would lead to a very high gas temperature, which would be off the chart. It is common practice to carry out such compressions in stages, with intercooling between the stages, to hold the gas temperature and work requirements down. We shall presume that the compression is carried out in four stages, with intercooling to 37.8°C (100°F) between stages, and we shall neglect mechanical inefficiencies. The overall compression ratio is 3.45/0.124 = 27.8; therefore we shall take the compression ratio per stage to be $(27.8)^{1/4} = 2.30$, giving interstage pressures of 0.284, 0.65, and 1.50 MPa (41, 95, and 218 lb/in² abs). The enthalpy increase in each stage is found from Fig. 13-4:

Stage		ΔH, Btu/lb CH$_4$
1	$-1454 + 1520 =$	66
2	$-1452 + 1515 =$	63
3	$-1454 + 1516 =$	62
4	$-1462 + 1518 =$	56
		$\overline{247}$

The net work consumption for methane compression is

$$W_{n.\,comp} = \left(247\,\frac{Btu}{lb}\right)\frac{1\;lb}{454\;g}\left(16\,\frac{g}{mol\;CH_4}\right)\left(0.5\,\frac{mol\;CH_4}{mol\;feed}\right)\left(1.055\,\frac{kJ}{Btu}\right) = 4.59\;kJ/mol\;feed$$

The density of a paraffinic oil with a molecular weight of 220 is about 750 kg/m³ (Perry and Chilton, 1973). The minimum work requirement for the pump is $V\,\Delta P$, where V is the volumetric flow rate of absorbent oil. Hence

$$W_{n.\,pump} = \left(62.5\,\frac{mol\;oil}{mol\;feed}\right)\left(0.220\,\frac{kg\;oil}{mol\;oil}\right)\frac{1\;m^3}{750\;kg}\,(3450 - 124\;kPa) = 61.1\;kJ/mol\;feed$$

Therefore
$$W_n = 4.59 + 61.1 = 65.7\;kJ/mol\;feed$$

and the thermodynamic efficiency is

$$\eta = \frac{W_{min.\,T_0}}{W_n} = \frac{0.902}{65.7} = 0.014$$

(c) If the absorber is staged, it will no longer be necessary for the exit liquid from the absorber to be in equilibrium with the exit gas. As long as the regenerated oil has been depleted in methane sufficiently for there to be a positive driving force at the top of the absorber, the minimum absorbent flow will correspond to equilibrium with the *feed* gas. If we again neglect temperature changes due to the heat of absorption, this gives

$$x_{CH_4} = \frac{y_{CH_4.\,F}}{K_{CH_4}} = \frac{0.50}{7.0} = 0.0714$$

in the rich absorbent.

The equilibrium x_{CH_4} at the top of the absorber will still be 0.0143; hence we shall still ask that the regenerator operate at 124 kPa and reduce the actual x_{CH_4} to 0.0071. Thus $W_{n.\,comp}$ will remain the same as in part (b).

The advantage of staging lies in the reduction of the separating-agent (oil) requirement in the absorber. By mass balance we have

$$A = \frac{0.45}{0.0714 - 0.0071} = 7.0\;mol/mol\;feed$$

Staging allows the oil to pick up as much as 10 times as much methane as in a single-stage absorber. The pump work now becomes reasonable:

$$W_{n.\,pump} = \frac{(7.0)(0.220)(3.45 - 0.12)(1000)}{750} = 6.84\;kJ/mol\;feed$$

Hence
$$W_n = 4.59 + 6.84 = 11.43\;kJ/mol\;feed$$

$$\eta = \frac{0.902}{11.43} = 0.079$$

There is considerable advantage to staging the process. □

The relatively high absorbent circulation work found in Example 13-2 is the result of the small solubility of methane in *any* solvent at ambient temperatures. An approach often used for hydrogen-methane separation is absorption into a hydrocarbon solvent at subambient temperatures. Because of the presence of the absorbent the temperatures required are not as low as those found for partial condensation in Example 13-1, and because of the lower temperature the absorbent circulation requirement is not as great as that found in Example 13-2. In addition, a lower-molecular-weight absorbent such as butane or hexane can be used; this will also reduce the absorbent pumping power because a given number of moles will correspond to less absorbent volume.

Example 13-3 Palladium metal has the unique property of allowing hydrogen to diffuse through it at significant rates under conditions where other gases are not transmitted to any appreciable amount. McBride and McKinley (1965) describe the operation of processes which use diffusion of hydrogen through thin palladium barriers in order to produce relatively pure hydrogen from streams containing mixtures of hydrogen and light hydrocarbons or hydrogen and carbon monoxide. A flow diagram of such a process for separating hydrogen and methane is shown in Fig. 13-8.

In order to prevent loss of hydrogen transport rate caused by adsorption of methane on the palladium surface, the diffuser must be operated at about 617 K. The feed is heated by the effluent methane and hydrogen and by a furnace. The diffuser must present a large amount of palladium barrier area in a compact volume; one design for accomplishing this would employ a number of supported palladium tubes in parallel inside a shell. The product hydrogen must be recompressed.

McBride and McKinley (1965) report the following operating conditions for one plant:

Hydrogen content of feed = 53 mol %   Feed pressure = 3.45 MPa
Product volume = 3.9×10^5 std m³/day   Hydrogen product purity = 99.2 mol %

Find the net work consumption and thermodynamic efficiency for the hydrogen-methane separation specified in Example 13-1 if the separation is carried out by the palladium diffusion process of Fig. 13-8.

SOLUTION Apparently there will be no difficulty meeting the separation specifications with this single-stage process. Although the recovery fraction of hydrogen for the preceding process is not reported, it should certainly be possible to obtain a 90 percent recovery without reducing the hydrogen purity below 90 percent.

It is necessary, however, that the product hydrogen pressure leaving the diffuser be less than the hydrogen partial pressure in the methane product from the diffuser to assure a positive driving force

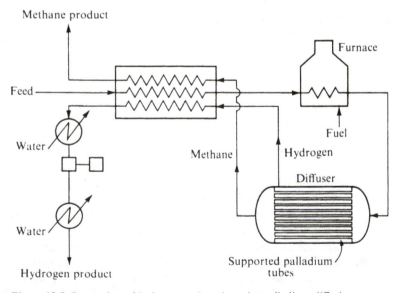

Figure 13-8 Separation of hydrogen and methane by palladium diffusion.

for mass transfer across the barrier. Since the hydrogen partial pressure in the methane product can at most be 345 kPa, we shall take the pressure of the hydrogen leaving the diffuser to be 40 percent of that value, or 138 kPa.

The work consumption of the hydrogen compressor can be determined by using the hydrogen Mollier diagram in Fig. 13-3. The overall compression ratio is $3450/138 = 25$, which we shall accomplish in four stages with intercooling to 38°C and individual compression rates of $(25)^{1\cdot4} = 2.24$, giving interstage pressures of 309, 690, and 1540 kPa (45, 100, and 224 lb/in² abs):

Stage		ΔH, Btu/lb H_2
1	$2400 - 1885 =$	515
2	$2400 - 1885 =$	515
3	$2400 - 1885 =$	515
4	$2410 - 1885 =$	525
		$\overline{2070}$

Hence
$$W_{n,\,comp} = \frac{2070\text{ Btu}}{\text{lb }H_2}\left(\frac{2.02\text{ g }H_2}{\text{mol }H_2}\right)\frac{0.5\text{ mol }H_2}{\text{mol feed}}\left(\frac{1.055\text{ kJ/Btu}}{454\text{ g/lb}}\right) = 4.86\text{ kJ/mol feed}$$

The furnace also involves net energy consumption. If the thermal driving force in the heat exchanger is 40 K, the furnace will have to raise the feed from 577 to 617 K. With heat capacities from Perry and Chilton (1973) we have

$$Q_F = (0.5C_{p_{H_2}} + 0.5C_{p_{CH_4}})(40\text{ K}) = [0.5(20.8) + 0.5(51.1)](40) = 1440\text{ J/mol feed}$$

The heat is to be supplied at an average temperature of 597 K. The equivalent work is

$$W_{n,\,turn} = Q_F\frac{T_F - T_0}{T_F} = 1440\frac{597 - 311}{597} = 690\text{ J/mol feed}$$

Hence
$$W_n = 4.86 + 0.69 = 5.55\text{ kJ/mol feed} \quad\text{and}\quad \eta = \frac{0.902}{5.55} = 0.16 \qquad \square$$

The net work consumption of these processes should not be the sole basis for comparison between them. For example, the palladium diffusion process requires substantial amounts of palladium metal, which is very expensive; hence the palladium process probably will require a greater initial capital investment than the other processes. Feed impurities are important. Carbon dioxide, water, and hydrogen sulfide all solidify in the cryogenic partial-condensation process and must be removed from the feed before it is chilled. Sulfur poisons palladium, and hence sulfur compounds must be removed from the feed to that process.

The choice between these processes is also influenced by the required product pressures and product purities. If the methane product must be at feed pressure but the hydrogen product can be at a lower pressure, the palladium process enjoys a relative advantage since the hydrogen compressor work for that process can be reduced or eliminated altogether. If the hydrogen is required at feed pressure but the methane can be taken to a lower pressure, the more common situation, the condensation and absorption processes are favored relative to the palladium process. The condensation process has a particular advantage in this case since much of the needed refrigeration can be obtained by expanding the liquid methane to a lower pressure. The condensation process then works best when removing a relatively large methane impurity since more methane refrigeration is available.

The palladium process gives very high product hydrogen purities and relatively high hydrogen recoveries and has an advantage when ultrahigh purity is desired. The cryogenic process, on the other hand, cannot easily provide hydrogen purities above 95 to 98 percent. Palazzo et al. (1957) describe a system for using absorption to improve the hydrogen purity obtained from a cryogenic partial-condensation process.

All three of the foregoing types of process are used commercially for separating hydrogen and methane in various situations. Another process sometimes used is *heatless adsorption* (Alexis, 1967; Stewart and Heck, 1969), in which methane is removed from hydrogen through adsorption, with regeneration accomplished by frequent lowerings of the pressure on the adsorbent beds. This process is most useful when hydrogen recoveries of 80 to 85 percent, or less, are acceptable and when the methane level in the feed is low.

MULTISTAGE SEPARATION PROCESSES

Benedict (1947) classified multistage separation processes into three categories, as follows, on the basis of the relative energy consumption for a specified separation at a given separation factor:

1. *Potentially reversible processes.* The net work consumption can, in principle, be reduced to W_{min, T_0}. This category generally includes those separation processes based upon equilibration of immiscible phases, which employ only *energy* as a separation agent. Examples are distillation, crystallization, and partial condensation.
2. *Partially reversible processes.* Most steps are potentially reversible except for one or two, e.g., the addition of solvent, which are inherently irreversible. These processes generally include those equilibration separation processes which employ a stream of *mass* as a separating agent. Examples are absorption, extractive distillation, and chromatography.
3. *Irreversible processes.* All steps require irreversible energy input for operation. These processes are generally *rate-governed* separation processes. Examples include membrane separation processes, gaseous diffusion, and electrophoresis.

The energy consumption of processes in each of these categories is explored in the ensuing discussion. It will be shown that for cases where α is near unity:

1. The potentially reversible or energy-separating-agent processes have a net work consumption which is, to the first approximation, independent of separation factor α, and an energy throughput inversely proportional to $\alpha - 1$.
2. The partially reversible or mass-separating-agent processes have a net work consumption varying, to the first approximation, inversely as $\alpha - 1$.

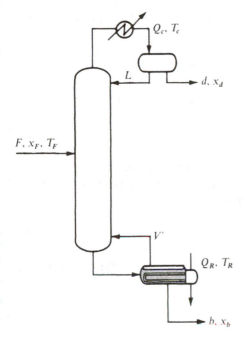

Q_c, T_c

L

d, x_d

F, x_F, T_F

V'

Q_R, T_R

b, x_b **Figure 13-9** Distillation of a close-boiling mixture.

3. The irreversible or rate-governed processes have a net work consumption varying, to a first approximation, inversely as $(x - 1)^2$.

We shall also find that the energy consumption for a given separation with a separation factor that is the same for all processes tends to increase in the ascending order of potentially reversible process < partially reversible process < rate-governed process, as long as the separation factor is in the range 0.1 to 10 and the separation requires staging.

Potentially Reversible Processes: Close-boiling Distillation

As an example of a potentially reversible process we consider the distillation of a close-boiling binary mixture, e.g., propylene-propane. The process and notation are shown in Fig. 13-9. The feed rate is F; the feed contains a mole fraction x_F of the more volatile component; the distillate rate and mole fraction are d and x_d; the bottoms rate and mole fraction are b and x_b. For convenience, we shall postulate that the liquid flow in the rectifying section is constant from plate to plate and equal to L; this is usually a good assumption for a close-boiling mixture. The latent heat of both species is λ, and latent-heat effects outweigh sensible-heat effects. The feed enthalpy and temperature T_F are chosen so that $Q_R = Q_c$. This will correspond closely to saturated liquid feed. The condenser and reboiler areas are assumed to be so large that Q_R is put in at T_R, the bubble-point temperature of the bottoms, and Q_c is removed at T_c, the bubble point of the distillate. Pressure drop through the tower is ignored.

Under these conditions the net work consumption of the distillation is given by Eq. (13-15) as[†]

$$W'_n = Q_c T_0 \left(\frac{1}{T_c} - \frac{1}{T_R} \right) \tag{13-19}$$

To find Q_c, we consider the case of minimum reflux. From Eq. (9-7) we have

$$L_{min} = \frac{[(x_d/x_F) - \alpha(1 - x_d)/(1 - x_F)]d}{\alpha - 1} \tag{13-20}$$

As noted in Chap. 9, the minimum reflux does not continue increasing as the products become highly pure but instead reaches an asymptotic value, given by Eq. (9-9), for all cases of relatively pure distillate:

$$L_{min} = \frac{1}{\alpha - 1} F \tag{9-9}$$

In Eq. (13-19) Q_c is then given by $Q_c = \lambda(L_{min} + d)$. For close-boiling mixtures L_{min} will be much larger than d, and Eq. (9-9) leads to

$$Q_c \approx \frac{\lambda}{\alpha - 1} F \tag{13-21}$$

The difference of reciprocal temperatures in Eq. (13-19) can also be estimated in terms of α and λ. The Clausius-Clapeyron equation gives

$$\frac{d \ln P^0}{d(1/T)} = -\frac{\lambda}{R} \tag{13-22}$$

where P^0 is vapor pressure. The overhead temperature T_c for a relatively complete separation corresponds closely to the boiling point of the more volatile component at the column pressure. Similarly, T_R corresponds closely to the boiling point of the less volatile component at the column pressure. The vapor pressure of the more volatile component at the bottoms temperature will then be α times the column pressure. Hence we can integrate Eq. (13-22) for the more volatile component between the overhead and bottoms temperature to obtain

$$\ln \alpha = \frac{\lambda}{R} \left(\frac{1}{T_c} - \frac{1}{T_R} \right) \tag{13-23}$$

Substituting Eqs. (13-21) and (13-23) into Eq. (13-19) and making use of the fact that $\ln \alpha \approx \alpha - 1$ for a close-boiling mixture by a Taylor-series expansion, we have

$$W'_n = RFT_0 \tag{13-24}$$

This result shows that W'_n tends to be independent of α for the distillation of close-boiling mixtures. The lack of dependence of W'_n on α is typical of the category of

---

[†] Equation (13-15) was developed for the net work consumption per mole of feed W_n. For a continuous-flow process with a given feed rate F we are interested in the net work consumption per unit time W'_n: W'_n is related to W_n through $W'_n = W_n F$.

potentially reversible processes, as mentioned previously. In a distillation this behavior is a consequence of two factors which offset each other: (1) as α becomes closer to unity, a higher reflux is required and hence the energy flow through the column, as represented by Q_c and Q_R, increases; (2) as α becomes closer to unity, the difference between T_R and T_c becomes less and the energy passing through the column is degraded to a lesser extent. Consequently the net work consumption, to a first approximation, is independent of α.

This conclusion was derived for minimum reflux but is also true for any actual operating reflux ratio as long as the products are relatively pure. For a reflux ratio above the minimum, readers should convince themselves that Eq. (13-24) would become

$$W'_n = \frac{d + L_{act}}{d + L_{min}} RFT_0 \qquad (13\text{-}25)$$

For the separation of a nearly ideal close-boiling mixture ΔH for the process streams entering and leaving is very nearly equal to zero. Hence, by Eqs. (13-3) and (13-1)

$$\Delta B_{sep} \approx -RT_0[x_F \ln x_F + (1 - x_F) \ln (1 - x_F)] \qquad (13\text{-}26)$$

From Eqs. (13-11) and (13-17) the *thermodynamic efficiency* η of the separation process is

$$\eta = \frac{\Delta B_{sep}}{W_n} = \frac{\Delta B_{sep} \, F}{W'_n} \qquad (13\text{-}27)$$

For close-boiling distillation ΔB_{sep} is given by Eq. (13-26) and W'_n by Eq. (13-24) or (13-25).

Figure 13-10 shows the thermodynamic efficiency of a close-boiling distillation of an ideal mixture giving relatively pure products when carried out at minimum reflux. Also included for comparison are results given by Robinson and Gilliland (1950) for a benzene-toluene distillation and an ethanol-water distillation, both carried out at atmospheric pressure. A complete separation was postulated in the benzene-toluene case. The ethanol-water distillation was taken to give 87 mol % and 0 % alcohol in the distillate and bottoms, respectively. Actual equilibrium and enthalpy data were used for these two cases. Note that nonideality does not necessarily imply a lower thermodynamic efficiency.

One factor neglected in this analysis is the pressure drop through the tower. For extremely close-boiling distillations, such as propylene-propane, the pressure drop may have as much or more effect on the difference between T_R and T_c as the composition change; however, in principle the pressure drop can be reduced through altered tray design, more open packings, and/or increased tower diameter. For an economic optimum design, though, the pressure drop can still be important. In such cases, a term in $(\alpha - 1)^{-2}$ is added to Eqs. (13-24) and (13-25), since the additional difference in $1/T$ is proportional to the pressure drop, the pressure drop is proportional to the number of stages N, N is approximately proportional to N_{min}, and N_{min}, by the Fenske equation (9-24), is proportional to $(\ln \alpha)^{-1}$ [or to $(\alpha - 1)^{-1}$ for a close-boiling distillation].

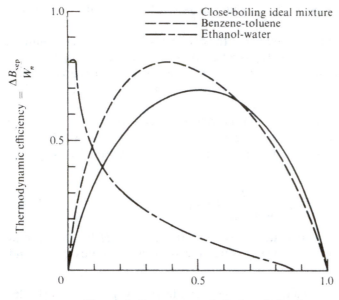

Figure 13-10 Thermodynamic efficiency of distillation of various mixtures at minimum reflux. (*Data from Robinson and Gilliland, 1950.*)

Also neglected in this analysis were the additional temperature drops for heat transfer across the reboiler and the condenser. Most distillation designs use a relatively large temperature drop across the reboiler and/or the condenser, and the resultant increase in $\Delta(1/T)$ can be substantial and sometimes dominant. For distillations using a steam source at fixed pressure to heat the reboiler and cooling water for the condenser, W_n' becomes directly proportional to Q_c and hence to $(\alpha - 1)^{-1}$, since the term in parentheses in Eq. (13-19) is then independent of α.

The additional components in a multicomponent distillation serve to increase W_n' in two ways. The temperature span across the column is greater than for the equivalent binary distillation of the keys alone; thus $\Delta(1/T)$ is greater. Also, the nonkeys increase the minimum reflux ratio and hence Q_c.

Fonyó (1974*b*) has analyzed the relative contributions of irreversibilities within the column, temperature drops across reboiler and condenser, and pressure drop to the energy requirements of a distillation separating ethylene from ethane, propane, and butane.

Partially Reversible Processes: Fractional Absorption

The contrast between the energy requirements of a potentially reversible (energy-separating agent) process and those of a partially reversible (mass-separating-agent) process can perhaps best be appreciated by replacing the overhead condenser and reflux system of a distillation tower with an entering stream of heavy absorbent liquid. In this case we have the absorber-stripper process shown in Fig. 13-11. The

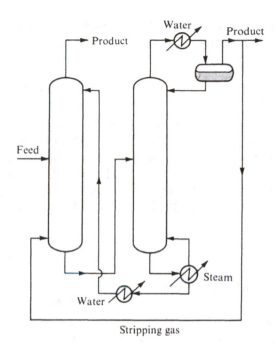

Water

Product

Product

Feed

Steam

Water

Stripping gas

Figure 13-11 Absorber-stripper column (*left*), with regeneration by distillation (*right*).

gaseous feed mixture to be separated enters midway along the column on the left. The upper portion of the column acts as an ordinary absorber, with absorbing liquid flowing downward countercurrent to the rising gas. The liquid leaving the bottom of the column is regenerated in the distillation column on the right. The regenerated heavy liquid is then recirculated to the top of the column. Some of the separated gas is used as a stripping medium providing vapor counterflow in the column, and the rest of the separated gas is taken as product. Both absorption and stripping sections are used to fractionate effectively between two components with significant solubilities (see discussion surrounding Fig. 4-24).

The analysis of the main absorber-stripper tower is similar to the analysis of a distillation tower. We shall presume that the feed contains two soluble components, A and B. The lighter component A appears primarily in the overhead product from the absorber. Component B appears primarily in the other product. The separation can be analyzed as an equivalent binary distillation of A and B provided we define the equivalent liquid flow L in this distillation as the total moles of dissolved (solvent-free) gases.

The feed gas enters the upflowing gas stream at the feed stage but has little effect on the moles of dissolved gas to be found in the downflowing liquid. Hence we have the equivalent of a saturated vapor feed to a binary distillation column. For relatively pure products, the overhead product gas flow rate will be $y_{A,F}F$, and a rearrangement of Eq. (9-10) gives

$$L_{min} = V_{min} - y_{A,F}F = \frac{\alpha_{AB}(1 - y_{A,F}) + y_{A,F}}{\alpha_{AB} - 1}F \qquad (13\text{-}28)$$

Again, L is the total moles of dissolved gases, not the total liquid flow. If the number of moles of heavy liquid required to dissolve 1 mol of gas at the feed stage is χ_A, the minimum absorbent liquid flow is

$$A_{min} = \frac{\chi_A}{\alpha_{AB} - 1} [\alpha_{AB}(1 - y_{A,F}) + y_{A,F}]F \qquad (13\text{-}29)$$

The work requirement of this absorption process comes in the separation of the dissolved gas from the liquid leaving the bottom of the column. The number of moles of liquid plus dissolved gas leaving the bottom of the column under conditions of minimum absorbent flow and relatively pure products is $A_{min} + (1 - y_{A,F})F$. For a difficult separation where α_{AB} is close to unity, the term in the brackets in Eq. (13-29) approaches 1, and A_{min} is much greater than $(1 - y_{A,F})F$. The molar feed rate to the regeneration distillation column F_D then becomes

$$F_D = \frac{F\chi_A}{\alpha_{AB} - 1} \qquad (13\text{-}30)$$

F_D is substantially greater than F, which would have been the feed rate to the distillation if it had been used directly to separate the mixture of A and B. There are two reasons for this: (1) χ_A is typically greater than 1, so as to provide sufficient absorption medium for the solute gases; (2) as $\alpha_{AB} \to 1$, the quantity $(\alpha_{AB} - 1)^{-1}$ becomes much greater than 1. Even in the case where the mole fraction of dissolved gases is substantial in the rich liquid from the absorber, a term involving $(\alpha_{AB} - 1)^{-1}$ must be a major contributor to the feed to the regenerator.

Because of the factor $(\alpha_{AB} - 1)^{-1}$ in Eq. (13-30), the energy throughput and net work consumption for this absorption process must vary with a more negative power of $\alpha_{AB} - 1$ than for a simple distillation process separating the same mixture. If the distillative regeneration for the absorption process operates at minimum reflux and follows Eq. (13-24), the net work consumption for the absorption process becomes

$$W'_n = \frac{RT_0 F\chi_A}{\alpha_{AB} - 1} \qquad (13\text{-}31)$$

The behavior shown for this absorption process is characteristic of that for mass-separating-agent processes where the absorbent or solvent is regenerated by distillation or by any other, similar process. The additional factor of $(\alpha_{AB} - 1)^{-1}$ stems from the fact that the necessary flow of mass separating agent varies with $(\alpha_{AB} - 1)^{-1}$ [Eq. (13-29)], and the mass separating agent must then be regenerated, e.g., by distillation. In the mass-separating-agent process there is no simple way in which the greater throughput of separating agent as $\alpha_{AB} \to 1$ can be compensated by less degradation in energy level, as is true for straight distillation.

Irreversible Processes: Membrane Separations

Rate-governed processes are characterized by the necessity of adding separating agent irreversibly to each stage. An example is the multistage membrane process shown in Fig. 13-12. In this process the pressure difference required to drive the

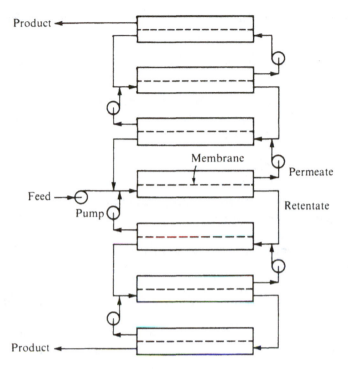

Product

Feed

Pump

Product

Membrane

Permeate

Retentate

Figure 13-12 Multistage membrane separation process.

permeate through the membrane must be resupplied to each permeate stream before it enters the next stage. The pumping work required to increase the pressures of these streams is the primary energy input to the process.

If the flow of permeate (moles per hour) through the membrane within a given stage is denoted as V_p (by analogy to the vapor streams in distillation), the required pressure drop across the membrane will be given by Eq. (1-22) as

$$\Delta P = \frac{V_p}{k_W A_p} + \Delta \pi \tag{13-32}$$

where k_W is the membrane permeability, in moles per unit time and per unit membrane area and per unit pressure drop, and A_p is the membrane area in stage p. The expression V_p/A_p is the same as N_W in Eq. (1-22). The term $\Delta \pi$ represents the difference in osmotic pressure across the membrane, occasioned by the difference in composition between the upstream and downstream sides, and corresponds to the minimum thermodynamic work requirement for the separation of that stage if there is infinite membrane area. For most real devices the first term on the right-hand side of Eq. (13-32) substantially outweighs the second term, in order to give sufficient permeation rate.

Assuming that the membrane pressure drop and the molar density of the permeate ρ_M are the same for each stage and that the $\Delta\pi$ term is negligible in Eq. (13-32), we have

$$W'_n = \frac{\Delta P}{\rho_M} \sum_{p=1}^{N} V_p \qquad (13\text{-}33)$$

The work requirement is thus proportional to both the permeate flow per stage and the number of stages. For a relatively complete separation the minimum permeate flow at the feed stage is given by Eq. (9-10) as

$$V_{F,\,\mathrm{min}} = \frac{\alpha_{AB}}{\alpha_{AB} - 1} F \qquad (13\text{-}34)$$

The permeate flow required in each stage will be proportional to this $V_{F,\,\mathrm{min}}$, but because of the need for separating-agent introduction to each stage V_p can readily change from stage to stage.

The minimum stage requirement for a given degree of separation is given by the Fenske-Underwood equation as

$$N_{\mathrm{min}} = \frac{\ln\left[(/_A)_d / (/_B)_d\right]}{\ln \alpha_{AB}} \qquad (13\text{-}35)$$

For α_{AB} close to 1, $\ln \alpha_{AB}$ can be replaced by $\alpha_{AB} - 1$, and N_{min} is proportional to $(\alpha_{AB} - 1)^{-1}$. Therefore, for α_{AB} close to 1, a combination of Eqs. (13-33) to (13-35) gives

$$W'_n = \frac{F}{(\alpha_{AB} - 1)^2} \frac{\Delta P}{\rho_M} \frac{N_{\mathrm{act}}}{N_{\mathrm{min}}} \ln \frac{(/_A)_d}{(/_B)_d} \qquad (13\text{-}36)$$

where we presume that $N_{\mathrm{act}}/N_{\mathrm{min}}$ is a factor which is independent of α and that the permeate flow in each stage will be equal to $V_{F,\,\mathrm{min}}$.

Several unique features of this category of rate-governed separation processes should be pointed out. First it should be noted that the net work consumption for potentially and partially reversible processes involved only thermodynamic variables such as R, T_0, and α. Equation (13-36) and any similar expression for any other rate-governed separation process involves a *rate-constant* characteristic of the device performing the separation. For the membrane process this rate constant is k_W, which enters as $V_p/A_p\Delta P$. Second, the rate-governed processes are more flexible than other types of processes with respect to the size of the interstage flows. Since separating agent must be introduced to each stage, it is readily possible to adjust the interstage flows at different interstage locations independently of each other. This is not such a simple matter with the potentially reversible and partially reversible processes. We shall see later that this flexibility usually leads to the use of smaller interstage flows at the product ends of the cascade for rate-governed processes compared with the feed stage. Even with this type of design, however, the interstage flows at all points will be proportional to that required at the feed stage, and W'_n will still vary as $(\alpha - 1)^{-2}$ for α near 1.

Since multistage membrane processes require a net energy consumption which increases as $(\alpha - 1)^{-2}$ as $\alpha \to 1$, they are usually not preferred for such separations.

Membrane separation processes find their greatest application when they provide a large separation factor and, as a result, one stage or very few stages at most will accomplish the separation.

In some cases, notably isotope separation processes, the rate-governed separation processes are the only processes which provide a separation factor of any appreciable size. An example is the separation of uranium isotopes, where gaseous diffusion provides a separation factor of 1.0043. Although this separation factor gives a seemingly low value of $\alpha - 1$, it is still orders of magnitude greater than the value of $\alpha - 1$ attainable with potentially reversible and partially reversible processes. Hence gaseous diffusion was selected in the World War II Manhattan Project as the large-scale process for separating uranium isotopes, even though 345 stages were required, as a minimum, with compressor power input necessary for each stage (Benedict, 1947; Benedict and Pigford, 1957).

REDUCTION OF ENERGY CONSUMPTION

Energy Cost vs. Equipment Cost

With high costs of energy it is advantageous to look for ways of reducing the energy consumption of a process. Reducing energy consumption often involves using additional equipment; a classical example is reducing energy consumption by adding more effects in multieffect evaporation (Appendix B). An economic optimization balancing energy and equipment costs to achieve the lowest cost is then appropriate (Appendix D). During the 1970s the cost of energy rose dramatically, and although the cost of equipment increased also, it increased less. Under these circumstances it is appropriate to seek methods of reducing energy consumption as a likely avenue back to the economic optimum. The relative incentives along these lines in future years will depend upon the relative scarcity and consequent cost increases of energy and of material resources.

General Rules of Thumb

Table 13-1 presents a number of rules of thumb to help reduce the energy requirements of separation processes. Many of them are rooted in the preceding discussion of factors controlling energy consumption in separations.

Mechanical separations (filtration, centrifugation, etc.) generally require much less energy than separation processes for homogeneous mixtures. Hence it is often advantageous to perform a mechanical separation first (rule 1) if part of the separation can be accomplished in that way.

Heat losses can be controlled through insulation (rule 2). The value of insulation increases with increasing departures from ambient temperature and with increasing surface-to-volume ratio of equipment. The percentage heat loss from a large distillation column is usually quite small, but insulation may still be used in order to reduce process upsets in response to changes in ambient conditions, e.g., rainstorms. Many drying processes involve release of hot effluent air to the atmosphere; there is incen-

Table 13-1 Approaches to decreased energy consumption in separations

No.	Rule
1	Perform mechanical separations first if more than one phase is present in feed mixture.
2	Avoid losses of heat, cold, or mechanical work; insulate where appropriate; avoid large hot or cold discharges of products, mass separating agent, etc.
3	Avoid overdesign and/or operating practices which unnecessarily lead to overseparation; for variable plant capacity, seek designs which allow efficient turndown.
4	Seek efficient control schemes which minimize excess energy consumption during transients and which reduce process disturbances due to interactions resulting from energy integration.
5	Look for those constituents of a process which have the largest changes in available energy (or largest costs) as prime candidates for reducing energy consumption through process modification.
6	Favor separation processes transferring the minor, rather than the major, component(s) between phases.
7	Use heat exchange where appropriate; where heat exchange is expensive, seek higher heat-transfer coefficients.
8	Endeavor to reduce flows of mass separating agents; favor agents giving high K_j, as long as selectivity can be achieved.
9	Favor high separation factors as long as they are useful.
10	Avoid designs which mix streams of dissimilar composition and/or temperature.
11	Recognize value differences of energy in different forms and of heat and cold at different temperature levels; add and withdraw heat at a temperature level close to that at which it is required or available; endeavor to use the full temperature difference between heat source and heat sink efficiently, e.g., multieffect evaporation.
12	For separations driven by heat throughput over relatively small temperature differences, investigate possible use of mechanical work in a heat pump.
13	Use staging or countercurrent flow where appropriate to reduce separating-agent consumption.
14	For cases of similar separation factors, favor energy-separating-agent processes over mass-separating-agent processes, and, if staging is necessary, favor equilibration processes over rate-governed processes.
15	Among energy-separating-agent processes, favor those with lower latent heat of phase change.
16	When pressure drop is an important contributor to energy consumption, seek efficient equipment internals which reduce pressure drop.

tive in these cases for using recycle and indirect heating of the air or other drying medium or using higher-temperature inlet air.

Often separation processes for which the feed or other process conditions have changed separate the products to a greater extent than necessary (rule 3). In a distillation column gains can frequently be made simply by reducing the boil-up and reflux ratio. Limited turndown capabilities of equipment lead to excessive energy consumption at reduced capacity. For example, excess vapor boil-up may be needed in a distillation to keep the trays within the region of efficient operation (see Fig. 12-9).

Process-control schemes play a central role in process energy consumption (rule 4). Often transient operation, including start-up and shutdown, leads to a much higher energy consumption per unit of throughput which could be avoided with an improved control scheme. Many methods for reducing energy consumption (such as heat exchange) lead to increased dynamic interactions within a process. Use of these methods is often held back by worries about reliable process control.

Available energy [B, Eq. (13-10)] is a state function, and, taken together for all components of a process, the change in available energy of process streams determines the net work consumption of the process. The greatest gains in reducing energy requirements can potentially be made by modification of those steps which involve the greatest loss of available energy (rule 5). A similar approach for overall cost reduction involves looking for modifications of the most expensive component of a process. King et al. (1972) explored systematic logic by which these concepts could be applied to evolutionary improvement of the designs of a demethanizer distillation column and a methane-liquefaction process.

The energy consumption of a separation process is often directly related to the amount of material which must change phase. Particularly when a dilute solution is to be separated, it is often effective to choose a process which will transfer the low-concentration component(s) between phases rather than the high-concentration component (rule 6). For example, ion exchange enjoys an energy advantage over evaporation for desalting slightly brackish water.

Heat exchange is a very direct form of energy conservation and should be considered where possible (rule 7), e.g., between products and the feed to a separation process at nonambient temperature, or to make the condenser of one distillation column serve as the reboiler for another. Also, whenever heat exchange is quite expensive but would be effective for conserving energy, there is considerable incentive for developing exchangers with high heat-transfer coefficients. Many of the advances in evaporative desalting of seawater have come about in that way.

Mass separating agents usually require regeneration, and the cost and energy consumption of that regeneration are directly proportional to the amount of agent used. Agents providing higher equilibrium distribution coefficients require lower circulation rates (rule 8).

High separation factors are useful in reducing the need for staging and in reducing the separating-agent requirement in a staged process (rule 9). Once a separation is achievable in a single stage, increased separation factor is of no further value unless it allows the flow of separating agent to be reduced.

Mixing dissimilar streams is a source of irreversibility and thereby tends to increase energy consumption (rule 10). Recycle streams should be introduced at the point where they are most similar to the prevailing process stream. Large driving forces for direct-contact heat-transfer, mass-transfer, and chemical-reaction steps should be avoided.

Because of Carnot inefficiencies electric energy and mechanical work have higher values per unit of energy than heat and refrigeration energy. The value of a particular form of energy also relates to the opportunity for using it in that form. The value of heat or refrigeration energy is greater the farther removed in temperature it is from ambient (the availability concept). Hence it is desirable to add heat from a source at a temperature not far above that at which the heat is needed and to use withdrawn heat at a temperature not much below the temperature at which it is withdrawn (rule 11). When a particular heat source and heat sink are most convenient, it is desirable to use the temperature difference between them as fully as possible. The multieffect principle accomplishes this for evaporation and is applicable to any vapor-liquid separation process. The forward-feed multieffect evaporation process for seawater discussed in Appendix B (Fig. B-2) is a good example of a design using heat effectively at its own level.

Energy-separating-agent processes which do not involve too large a temperature span between heat source and heat sink can also be operated through a *heat-pump principle*, in which mechanical work is used, as in a refrigeration cycle, to withdraw heat at a low temperature and supply it at a higher temperature (rule 12). There are several approaches for doing this, developed later for distillation. Vapor-recompression evaporators (see, for example, Casten, 1978; Bennett, 1978) are effective for using mechanical work when the boiling-point elevation in the evaporation is not too large.

Staging is effective for reducing the consumption of separating agent (rule 13), as well as increasing product purities.

Energy-separating-agent processes have two energy-related advantages over mass-separating-agent processes (rule 14): mass-separating-agent processes are only partially reversible, in the categories of Benedict (1947), and hence require more energy in a close separation than a potentially reversible process. Also, an energy separating agent can readily be removed and exchanged with another stream, but such an operation with a mass separating agent requires an additional separation. Rate-governed separations are irreversible by Benedict's classification and hence require even greater energy consumption for a close separation with a given separation factor.

Among energy-separating-agent processes, the energy throughput required for a given separation factor is directly proportional to the latent heat of phase change; this favors processes with low latent heats (rule 15). This incentive is reduced somewhat by the ease of building energy-separating-agent processes in multieffect configurations.

Finally, when the pressure drop within the separator is an important contributor to the overall energy requirement, there is an incentive to utilize internals which inherently give low pressure drop (rule 16).

Examples Two examples will illustrate how these guidelines can be used. The first involves a process reported by Bryan (1977) for dehydration of waste citrus peels to make them suitable for cattle feed (see Fig. 13-13). The energy consumption would be quite large (the latent heat of vaporization of all the water) if the dewatering were accomplished entirely in a dryer. Several energy economies are represented in the process shown in Fig. 13-13:

1. A mechanical separation (pressing) is used to remove much of the peel liquor before the peel is put into the dryer (rule 1).

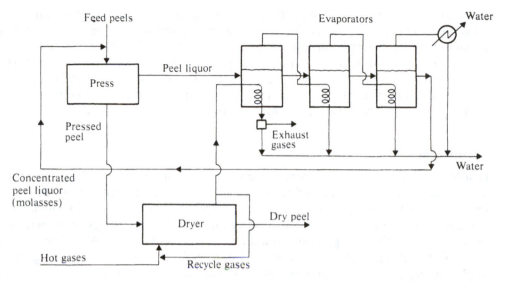

Figure 13-13 Process for conversion of citrus peel into cattle feed, using dehydration.

2. The peel liquor is concentrated separately in a multieffect evaporator (rule 11). No similar energy savings could be gained with common dryer designs. The concentrated peel liquor is returned to the feed before the press, so that the liquor remaining in the pressed peels will be as concentrated as possible. This lessens the load on the dryer, shifting it to the more energy-efficient evaporators.
3. The dryer is run with recirculation of the hot-air heating medium. This makes it possible to develop a high enough water-vapor content in the exhaust air from the dryer for this air stream to be used to drive the multieffect evaporator instead of being discharged (rule 2). The presence of inert gases in the moist air stream probably reduces the heat-transfer coefficient in the first effect of the evaporator, however.

   The second example is removal and recovery of citrus oil from the water effluent from a citrus-processing plant. The principal constituents of this water are terpenes, and their concentration (about 0.1 percent) considerably exceeds their solubility (about 15 ppm); hence they are predominantly emulsified. Candidate separation processes for recovering the oil are stripping, extraction, adsorption, freeze-concentration, and reverse osmosis. Relative to the other processes, freeze-concentration and reverse osmosis have the disadvantage that the major component (water) must change phases (rule 6). This disadvantage is less important for reverse osmosis since it can operate in one stage and involves no latent heat of phase change (rule 15). Stripping, extraction, and adsorption are all mass-separating-agent processes which require that the separating agent be regenerated. Hence there is a considerable incentive to find separating agents which provide a high equilibrium distribution coefficient for the oil (rule 8). There is an interesting contrast between extraction and adsorption if the waste water contains only dissolved oil. If the partition coefficient is independent of oil concentration, the solvent-to-water ratio required in the extraction will be independent of concentration and the energy required for solvent regeneration will not change significantly. On the other hand, in a fixed-bed adsorption process the frequency of regeneration and resultant energy consumption are directly proportional to the oil concentration. Therefore, from an energy viewpoint, extraction is favored for higher oil concentrations and adsorption for lower concentrations. Finally, the energy consumption for stripping and adsorption could be considerably reduced if a preliminary mechanical separation were made to remove suspended oil, e.g., by centrifugation or flotation (rule 1).

Distillation

Distillation is by far the most common separation technique used in the petroleum, natural-gas, and chemical industries. In an audit of distillation energy consumptions for the production of various large-volume chemicals, Mix et al. (1978) concluded that distillation consumes about 3 percent of the United States energy. A 10 percent savings in distillation energy would amount to a savings of about $500 million in the national energy cost. There is clearly a substantial incentive for developing and implementing ways of lessening the energy consumption of distillation.

Some of the more obvious approaches are direct extensions of several of the principles listed in Table 13-1, e.g., reducing reflux to the smallest necessary level, insulation, feed-product heat exchange, and energy-efficient control. Often a change in feed location will be effective in reducing reflux requirements for an older column. Mix et al. (1978) discuss in some detail the potential advantages of tray retrofit, i.e., substituting trays that are more efficient; they also identify industrial distillations for which tray retrofit should be most attractive. There are also rather direct ways to make use of reject heat at its own level, e.g., by making steam with pumparound loops in crude-oil distillation and using two-stage condensation for a wide-boiling column overhead (Bannon and Marple, 1978). Two-stage condensation can serve to preheat a stream or make steam in the first, hotter stage and then achieve the desired final level of cooling in the second stage.

Other approaches that can be effective for distillation involve improving the efficiency of using available heat sources and sinks (rule 11) and the reduction of irreversibility in the design of the distillation itself. We shall consider both these areas in more detail.

Heat Economy *Cascaded columns* For the atmospheric-pressure benzene-toluene distillation analyzed by Robinson and Gilliland (1950) and considered in Fig. 13-10, the condensation temperature of the benzene overhead is 80°C and the boiling temperature of the toluene bottoms is 111°C. In any practical situation cooling water would most likely be used to condense the overhead, and steam at some pressure above atmospheric would be used to reboil the bottoms. If the cooling water were available at 27°C and the steam were at 121°C, the temperature difference between heating medium and coolant would be 94°C, whereas a temperature drop of only 31°C is required by the distillation itself. The heat passing through the column would be degraded through a greater temperature range, and the term $1/T_c - 1/T_R$ in Eq. (13-23) would increase from $1/353 - 1/384 = 2.29 \times 10^{-4} \text{ K}^{-1}$ to $1/300 - 1/394 = 7.95 \times 10^{-4} \text{ K}^{-1}$, or by a factor of 3.5. Hence the net work consumption is raised by a factor of 3.5 over that given by Eq. (13-24), and the thermodynamic efficiencies for benzene-toluene in Fig. 13-10 would be decreased by a factor of 3.5.

Despite the greater degradation of energy one would still probably use steam and cooling water because they are the cheapest utilities available for the purpose in the plant. However, a multieffect or cascaded-column design can be considered for greater energy economy at the expense of added investment. Use of the multieffect principle is possible whenever the temperature difference between the heat source and heat sink required to drive a separation process is substantially less than the actual temperature difference between the available heat source and the available heat sink. In the case of evaporation the *necessary* difference in temperature between the heat source and heat sink equals the boiling-point elevation due to nonvolatile solute in solution, but the *available* temperature difference is typically that between steam and cooling water, which is much greater.

Figure 13-14*a* to *c* illustrates three ways in which the multieffect principle can be used to make the energy supply and removal to and from a distillation process more efficient (Robinson and Gilliland, 1950). In all three cases there are two distillation columns, the reboiler of one and the condenser of the other being combined into a single heat exchanger. The lower column in each case is operated at a higher pressure than the upper column. The pressures are chosen so that the condensation temperature of the overhead stream from the lower column is greater than the boiling temperature of the bottoms stream of the upper column. In this way the vapor generated in the reboiler of the lower column is used throughout both columns.

In Fig. 13-14*a* the upper and lower columns perform identical functions, both separating the same feed into relatively pure products. The only difference lies in the pressures of the columns. Thus in the situation of Fig. 13-14*a* we are able to process twice as much feed with a given amount of heat input to the process, but that heat energy is degraded over twice as great a temperature range as is needed for a single column. The net work consumption *within* the distillation column per mole of feed is the same (half as much energy, twice as much degradation), but the process of Fig. 13-14*a* is able to utilize a large temperature differential between heat source and heat sink more efficiently.

In the process of Fig. 13-14*b*, the lower column receives the entire feed and separates it into a relatively pure bottoms product and an overhead product somewhat enriched in the more volatile component. The upper column then takes this enriched feed and separates it into two relatively pure products. In this situation the heat energy need not be degraded to the same extent required in Fig. 13-14*a* since the temperature drop across the lower column is not as great. On the other hand, it is no longer possible to process twice as much feed per unit heat input. The heat energy is used twice throughout the full stripping sections of the columns, but the final purification of the distillate is accomplished using the vapor only once. Such an arrangement has some good potential features with regard to irreversibilities inside the columns, however, as will be shown subsequently.

One can also picture an operation which is the inverse of that shown in Fig. 13-14*b*. The lower column could manufacture a relatively pure distillate and a bottoms somewhat depleted in the more volatile component. This bottoms would then be fed to the upper column where it would be separated into relatively pure

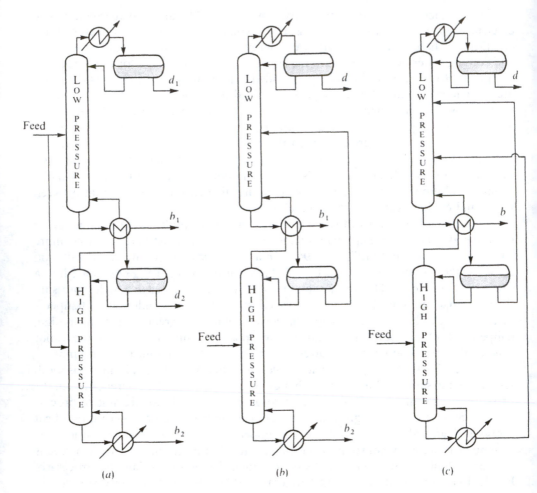

Feed

Figure 13-14 Multieffect distillation columns: (a) individual feeds; (b) forward feed of one product; (c) forward feed of two products.

products. In such a case, the portion of the distillation closest to the bottoms composition would be accomplished with smaller total interstage flows than the rest of the distillation.

Figure 13-14c shows a situation where the lower column makes two products which are only somewhat enriched in the components and where both products from the lower column are fed to appropriate points in the upper column, which manufactures relatively pure products. In this scheme the temperature range of the lower column is even less and the required degree of heat energy degradation is even less than for the other schemes. The generated vapor is used twice for those portions of the distillation with compositions just above and below that of the feed and is used only once for compositions nearer to the product composition.

One problem with cascading columns by linking the reboiler of one with the condenser of another is that dynamic process upsets propagate back and forth between the columns and the control task becomes more difficult. Tyreus and Luyben (1976) discuss the advantages and disadvantages of different control schemes for such systems.

Tyreus and Luyben (1975) present a case study of cascaded-column designs for propylene-propane distillation and for methanol-water distillation, both using steam and cooling water.

It is also possible to combine the reboiler of one column with the condenser of another for towers distilling entirely different mixtures, but it must be recognized that in such cases transient disturbances may recycle over much larger portions of the entire process.

In some cases it is possible to derive some or all of the reboiler duty of a column from the heat content of the feed mixture if the feed is available as a vapor at a pressure considerably above that of the column. Figure 13-15 shows a design where a feed gas at high pressure loses sensible heat and condenses partially to supply reboiler heat before being reduced to column pressure.

Heat pumps For close-boiling distillations, rule 12 leads to consideration of heat-pump designs (Null, 1976), three of which are shown in Fig. 13-16. In all three cases

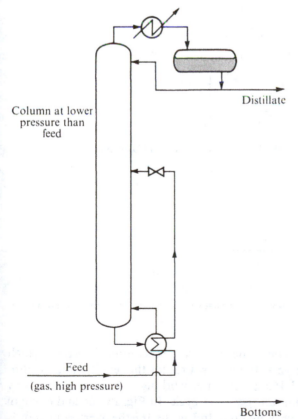

Column at lower pressure than feed

Distillate

Feed

(gas, high pressure)

Bottoms

Figure 13-15 Use of high-pressure feed as a reboiling medium.

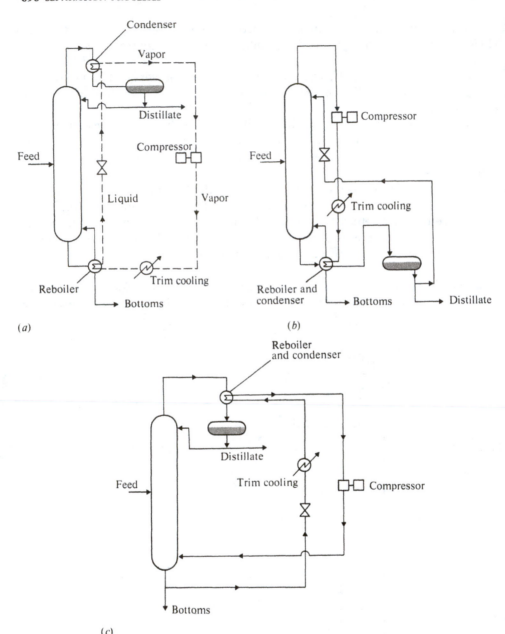

Figure 13-16 Heat-pump schemes for distillation: (a) external working fluid; (b) overhead vapor recompression; (c) reboiler liquid flashing.

compression work is used to overcome the adverse temperature difference which precludes having the condenser serve as the heat source for the reboiler in an ordinary distillation column. In Fig. 13-16a an external working fluid is used in a way entirely analogous to a compression refrigeration cycle. In Fig. 13-16b and c one of the process streams is used as the working fluid. In Fig. 13-16b the overhead vapor is compressed to a pressure high enough for its condensation temperature to exceed the bubble point of the bottoms; the heat of condensation of the overhead can then be used to reboil the bottoms. In Fig. 13-16c the stream to be vaporized at the tower bottom is expanded through a valve to a lower pressure at which its dew point is less than the bubble point of the overhead vapor; the heat of vaporization can then be used to condense the overhead. The resultant vapor is then compressed and used as boil-up at the tower bottom. Reboiler liquid flashing does not lead to as high pressures as overhead vapor recompression; this can be advantageous.

The external-working-fluid scheme requires some extra compression, since the temperature difference between vaporization and condensation of the fluid must be enough to overcome the temperature difference of the distillation and provide temperature-difference driving forces for two heat exchangers. In the other cases only the temperature-difference driving force for one exchanger need be provided, in addition to overcoming the temperature difference of the distillation. On the other hand, the external working fluid may be more suitable in terms of compression characteristics and other properties than the overhead and bottom streams from the distillation.

Some additional heat exchange is required in all three systems because of the general failure of the required condenser duty to match the required reboiler duty, because of the heat input from the compressor, for control purposes, and/or to make up for heat leaks. In Fig. 13-16 it is assumed that additional cooling will be required as a result of these factors, and a trim cooler using external refrigerant or cooling water is included in each case. In some instances net heat input would be required instead.

Use of heat pumps and mechanical energy derives the benefits of the temperature-difference term in Eq. (13-19) when the temperature difference is small, whereas use of fixed-temperature heat sources and sinks without column cascading does not. Heat-pump designs are most suitable for close-boiling distillations because the compression requirements become excessive if the temperature difference to be overcome is too great. For this reason, they are at a disadvantage for multicomponent distillations. Some of the incentive for this approach is also dampened by the fact that compressors themselves are not 100 percent efficient.

Use of a heat-pump design for a close-boiling distillation places a premium on low pressure drop for the vapor flowing up the column. Low-pressure-drop trays and open packings can therefore be of more interest than usual in such cases.

Specific cases of heat-pumped distillation have been analyzed by Null (1976), Kaiser et al. (1977), Petterson and Wells (1977), and Shaner (1978), among others.

Examples Figure 13-17 shows one of the industrial applications made of the principles developed in Figs. 13-14 to 13-16. The Linde double column, shown in Fig. 13-17, is commonly used for the fractionation of air into oxygen and nitrogen. This is a two-column arrangement with a low-pressure column situated physically above a higher-pressure column. Following the multieffect principle, the condenser of the high-pressure column is the reboiler of the low-pressure column. The high-pressure column follows the variant of Fig. 13-14b in which the distillate is relatively pure but the bottoms is only somewhat enriched in the less volatile component

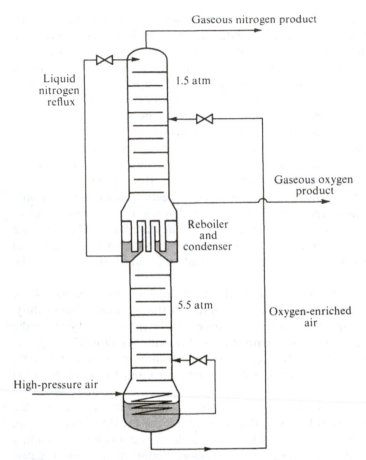

Gaseous nitrogen product

Liquid
nitrogen
reflux

1.5 atm

Gaseous oxygen
product

Reboiler
and
condenser

5.5 atm

Oxygen-enriched
air

High-pressure air

Figure 13-17 Linde double column for air separation.

(oxygen). The feed to the high-pressure column enters the process as air at a still higher pressure; hence this feed can be used as the reboiler heating medium following the scheme set forth in Fig. 13-15. Another interesting feature is that the liquid-nitrogen distillate from the high-pressure column is not taken as product but is used as reflux in the low-pressure column. Because there is more nitrogen than oxygen in air and because the nitrogen product is in many cases a waste stream which may be gaseous, no other source of overhead cooling is required. This is a major advantage.

More elaborate variations of the Linde double column have been developed, and descriptions and analyses of air-fractionation processes in general have been given by a number of authors (Bliss and Dodge, 1949; Ruhemann, 1949; Scott, 1959; Latimer, 1967).

Another example of industrial use of techniques for increasing the efficiency of heat supply and removal is the fractionation section of plants for the manufacture of ethylene and propylene. Two general approaches to this separation are followed in practice. In a *high-pressure process* the feed to the demethanizer column is raised to

3.5 MPa. A high-pressure process usually employs propane and ethylene refrigeration circuits, which are capable of providing cooling at temperatures down to $-100°C$. A *low-pressure process* employs a methane refrigeration circuit in addition to the ethylene and propane circuits. As a result the low-pressure process can provide cooling at much lower temperatures and consequently lower tower pressures are employed. The added expense of a methane circuit has made the high-pressure process more common, however.

A flow diagram of the demethanizer and C_2 splitter facilities of a typical low-pressure process as reported by Ruhemann and Charlesworth (1966) is shown in Fig. 13-18. The feed entering from the deethanizer tower consists of hydrogen, methane, ethylene, and ethane at a pressure of 1.2 MPa. The goal is to separate ethylene and ethane products from the hydrogen and methane. Following the scheme shown in Fig. 13-15, this feed passes through the reboilers of the demethanizer and one of the two C_2 splitter towers. The feed is the sole source of heat for the demethanizer and supplies a portion of the reboiler heat to the second C_2 splitter. Added refrigeration is available from the hydrogen plus methane *tail gas* leaving the process, which can be reduced to near-atmospheric pressure. The tail gas chills the feed further in a feed separator drum. Both gas and liquid phases exist in the feed under these conditions. The gas contains almost all the hydrogen, some methane, and almost no ethylene; hence it need not be fractionated further. The liquid from the separator contains part of the methane and almost all the ethylene and ethane; it is reduced in pressure and fed to the demethanizer, which operates at 520 kPa.

The overhead vapor from the demethanizer is essentially pure methane, which enters the methane refrigeration circuit directly. In the methane refrigeration circuit this vapor is compressed and is liquefied in a condenser cooled by the ethylene refrigeration circuit. Some of the liquid methane formed is returned as reflux to the demethanizer, and the remainder is used for feed prechilling. The use of a direct feed of vapor to the methane refrigeration compressor with return of condensed liquid from that compressor as reflux represents a variant of the vapor recompression scheme shown in Fig. 13-16b.

There are two C_2 splitter towers, cascaded in a variant of the Fig. 13-14b scheme. The high-pressure tower provides a relatively pure distillate (ethylene) and a bottoms somewhat enriched in ethane. This bottoms is fed to the low-pressure splitter where it is separated into relatively pure products. Condensing the overhead of the high-pressure column is a source of part of the reboil heat for the low-pressure column (multieffect principle).

The ethylene overhead from the low-pressure column is compressed and used to provide reboil heat to the high-pressure column. This is a direct application of the vapor-recompression principle shown in Figure 13-16b. The overhead vapor from the low-pressure column passes through two heat exchangers on the way to the compressor. These exchangers serve to help cool the compressed vapor and chill the reflux stream to the low-pressure splitter before that reflux stream is flashed down to column pressure.

Irreversibilities within the column; binary distillation In addition to pressure drop, irreversibilities within a binary distillation column result from the lack of equilibrium

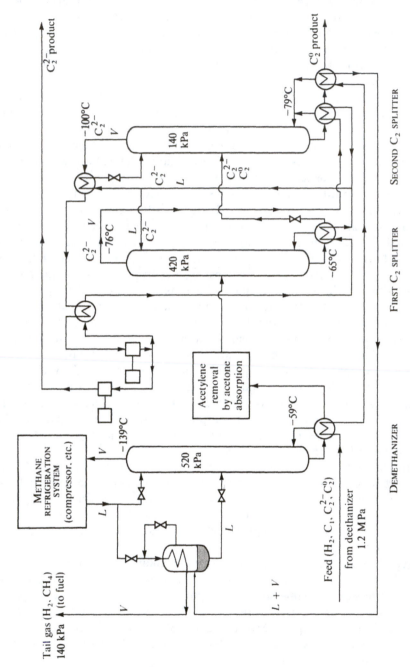

Figure 13-18 Light-hydrocarbon separation in a low-pressure ethylene plant.

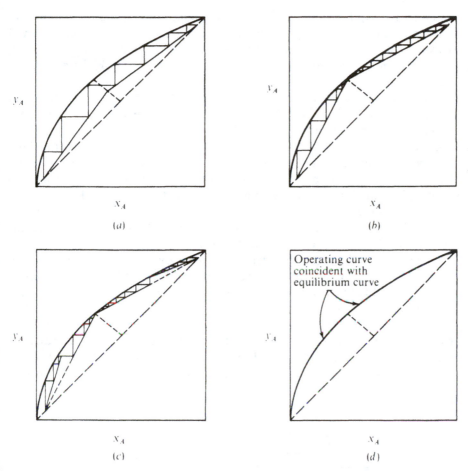

Figure 13-19 Increasing the reversibility within a binary distillation process: (a) ordinary distillation; finite stages; (b) ordinary distillation; minimum reflux; (c) intermediate reflux and intermediate boil-up; (d) totally reversible distillation.

between the vapor and liquid streams entering a stage (rule 10). The vapor enters from the stage below and hence is at a higher temperature than the liquid, which enters from the stage above. Also, the entering vapor will contain less of the more volatile component than corresponds to equilibrium with the entering liquid. Within a stage there is sensible heat transfer from vapor to liquid and mass transfer between the phases, both of which serve to dissipate available energy.

In order to reduce the net work consumption of a binary distillation it is necessary to lessen the driving forces for heat and mass transfer within the individual stages. This reduces to a problem of making the operating and equilibrium curves more nearly coincident. The point is illustrated in Fig. 13-19. Figure 13-19a represents an ordinary distillation run at a reflux substantially greater than the minimum. The driving forces for heat and mass transfer between the streams entering a stage $(T_{p-1} - T_{p+1}$ and $K_{A, p+1} x_{A, p+1} - y_{A, p-1})$ can be reduced by moving the operating lines closer to the equilibrium curve. The minimum-reflux condition shown in Fig. 13-19b corresponds to the upper and lower operating lines having been moved as close as possible to the equilibrium curve, and we have already seen [Eq. (13-25)] that W_n' for minimum reflux in ordinary distillation is lower than W_n' for any higher reflux ratio.

Even at minimum reflux there are still substantial driving forces for heat and mass transfer at compositions in the tower removed from the feed stage in a binary distillation. These irreversibilities can be reduced by using a different operating line in portions of the column where the irreversibilities with the original operating lines were more severe. Such a situation is shown in Fig. 13-19c, where we postulate that there are two operating lines applying to different parts of the stripping section and two operating lines applying to different parts of the rectifying section. The operating lines used closer to the feed have slopes nearer to unity; hence the liquid and vapor flows nearer the feed are larger than those at the ends of the column. Thus the situation shown in Fig. 13-19c corresponds to the use of a second reboiler midway up the stripping section and a second condenser midway down the rectifying section, as shown in Fig. 13-20. The conditions shown in Fig. 13-19c are still those corresponding to minimum reflux at the feed point; hence the interstage flows at the feed stage are the same in Fig. 13-19c as in Fig. 13-19b, and the overhead condenser duty corresponding to Fig. 13-19b must be the same as the sum of the duties of the two condensers above the feed corresponding to Fig. 13-19c. The gain in reversibility is not manifested as a reduced total heat duty but as a lesser degradation of the heat energy passing through the column. The heat energy supplied at the intermediate

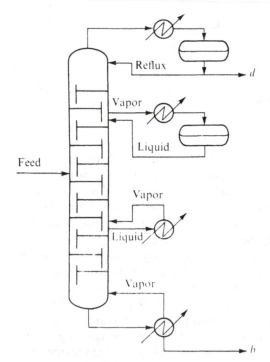

Figure 13-20 Distillation column with one intermediate condenser and one intermediate reboiler.

reboiler is supplied at a lower temperature than that to the reboiler at the bottom of the column, and the heat removed from the intermediate condenser is removed at a temperature higher than that of the column overhead. In order for this to be attractive, some way must be found to derive benefits from the differences in temperature between the two reboilers and/or between the two condensers.

The extreme of reducing thermodynamic irreversibilities within a distillation column would be to arrange the introduction of reflux to stages above the feed and of reboiled vapor to stages below the feed in such a way that the operating line at each stage is coincident with the equilibrium curve, as shown in Fig. 13-19d. A schematic of a device for carrying out such a process is shown in Fig. 13-21. The reflux must grow larger on each stage proceeding downward from the top of the column. As a result, there must be a condenser removing heat from each stage above the feed in

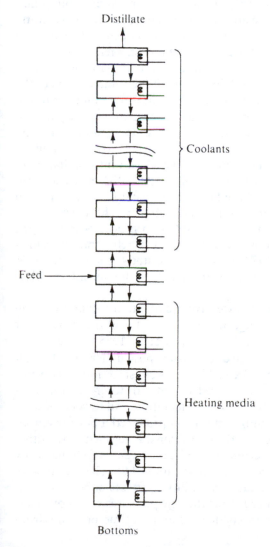

Figure 13-21 An approach to reversible distillation.

just the right amount to make the operating line coincident with the equilibrium curve at the composition corresponding to that stage. Similarly, each stage below the feed must be equipped with a reboiler to increase the vapor flow upward in the required pattern. Each reboiler and condenser must employ a heating or cooling medium with a temperature equal to that of the particular stage.

This hypothetical situation of " reversible " distillation, where the operating and equilibrium curves are the same, would require an infinite number of stages for any finite amount of separation. As the equilibrium and operating curves come closer together, there is less progress per stage along the yx diagram. Thus there is considerable expense involved in altering an ordinary distillation to increase its reversibility. The number of stages required for a given separation becomes greater, and the required heat duty must be split up between the terminal reboiler and condenser and those reboilers and condensers necessary to generate the intermediate boilup and reflux. Offsetting this need for considerable additional capital outlay are two factors: (1) The heat energy used in the distillation is degraded to a lesser extent. Much of the reboil heat can be added at temperatures lower than the bottoms temperature, and much of the heat removal can be effected at temperatures warmer than the overhead temperature. (2) The reduced vapor and liquid flows toward the product ends of the cascade may make it possible to reduce the tower diameter at those points or to use towers of different diameters when so many stages are required for the separation that more than one tower must be employed. In practice the opportunity for using lower-pressure steam or any other lower-temperature heating medium in intermediate reboilers does not seem to carry enough incentive to warrant installation of intermediate reboilers in any but unusual cases. One such is the ethylene-plant deethanizer described by Zdonik (1977), where hot quench water from elsewhere in the plant is used to provide heat for an intermediate reboiler. The incentive for generating intermediate reflux in low-temperature distillation processes is stronger, since the intermediate reflux can utilize a less severe level of refrigeration than is required for the overhead. King et al. (1972) present a case study of a demethanizer from a high-pressure ethylene plant, evaluating the incentive for an intermediate condenser.

One multiple-tower system which made extensive use of intermediate boil-up in order to gain the tower diameter advantage (item 2, above) was the process for manufacture of heavy water D_2O by distillation (Murphy et al., 1955). Figure 13-22 shows a flow diagram for the plant constructed for this purpose under the Manhattan Project at Morgantown, West Virginia. Table 13-2 gives construction and operating details of the plant. The plant received a feed of natural water (also used as reboiler steam to tower 1B) containing 0.0143 atom % deuterium. All towers together served as one very large distillation stripping section and produced a bottoms product containing 89 atom % deuterium. Most of the feedwater was rejected in the other product, and the recovery fraction of deuterium was quite low even though the purity was high. Under these conditions the effective $\alpha_{H_2O-D_2O}$ is about 1.05. The towers were run under moderate vacuum since the relative volatility increases substantially as pressure and hence temperature are reduced. Note that the pressures were not so low as to preclude the use of cooling water in condensers and that the pressure drops through the towers were sizable.

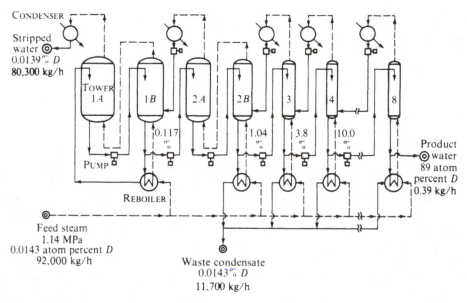

Figure 13-22 Morgantown water distillation plant. (*Adapted from Benedict and Pigford, 1957, p. 418; used by permission.*)

Intermediate boil-up was used at the bottom of each tower. This did not reduce the net work consumption of the process but did allow the vapor rate to vary by a factor of 885 through the cascade, from 91 kg/h at the deuterium oxide product end to 80,300 kg/h at the feed end (all to make 0.39 kg/h of 89% D_2O!). This in turn allowed the tower diameter to be reduced from five 4.6-m-diameter towers in parallel

Table 13-2 Construction and operating conditions of the Morgantown heavy-water distillation plant (data from Benedict and Pigford, 1957)

Tower	No. in parallel	Diameter, m	No. of plates	Tower vol., m³	Vapor flow, kg/h	Pressure, kPa Top	Pressure, kPa Bottom	D in bottom, atom %
1A	5	4.6	80	2010	(80,300)	8.9	31.8	
1B	5	3.7	90	1450	80,300	31.8	71.4	0.117
2A	1	3.2	72	176	(9,600)	17.2	45.3	
2B	1	2.4	83	118	9,600	45.3	86.0	1.40
3	1	1.0	72	17	1,380	16.5	45.7	3.8
4	1	0.46	72†	3.5	329	16.9	58.6	10.0
5	1	0.25	72†	4.5	86	16.9	45.3	11.5
6	1	0.25	72†	4.5	84	16.5	43.7	21.2
7	1	0.25	72†	4.5	91	16.5	44.4	56.4
8	1	0.25	72†	4.5	91	16.9	40.4	89
Total	18		757	3790	92,000			

† Number of equivalent equilibrium stages in packed column.

at the feed end to one 25-cm-diameter tower at the product end. The result was a major saving in capital cost.

Another advantage lay in the smaller water holdup in the towers at the product end. There was so much water in this plant that it took 90 days to level out at new steady-state conditions once the operating parameters were changed. Without the reduction in tower size at the product end this time would have been greater yet.

It is interesting to explore the reduction in net work consumption of a binary distillation process which can be accomplished by including a single intermediate condenser in a distillation column. Benedict (1947) considered the distillation of a binary close-boiling mixture containing 10 mole percent of the more volatile component in the feed ($x_{A,F} = 0.10$). The products were assumed to be relatively pure. If the distillation is run at minimum reflux with no intermediate condenser, $W'_n = R T_0 F$ [Eq. (13-24)]. If the distillation is run at 1.25 times the minimum reflux, $W'_n = 1.25 R T_0 F$. If the distillation is made totally reversible (Fig. 13-21), $W'_n = 0.325 R T_0 F$. If one intermediate condenser is used at $x_A = 0.30$, and if the reflux is 1.25 times the minimum at the feed and 1.10 times the minimum at the intermediate reflux point, $W'_n = 0.655 R T_0 F$. Hence, in this case, one intermediate condenser reduces W'_n by 64 percent of the amount which could be conserved by going to a totally reversible distillation. To realize this benefit it would still be necessary to find a use for the higher-level heat energy removed at the intermediate condenser.

It is also interesting to explore the behavior of the thermodynamic-efficiency curve for an ordinary ethanol-water distillation given in Fig. 13-10 in the light of this discussion. The thermodynamic efficiency is high for low ethanol mole fractions in the feed, and the efficiency is low when there are high ethanol mole fractions in the feed. Figure 13-23 depicts an analysis given by Robinson and Gilliland (1950). As is evident from Fig. 13-23, the ethanol-water system with $x_d = 0.87$ is one wherein the minimum reflux is determined by a tangent pinch in the upper portion of the tower. Minimum reflux operating lines for saturated-liquid-feed ethanol mole fractions of 0.56, 0.31, 0.15, and 0.04 are denoted as 1, 2, 3, and 4, respectively. It is apparent that greater gaps between the equilibrium curve and the lower operating line exist for higher feed mole fractions of ethanol; thus the thermodynamic efficiency is very low at high feed mole fractions. For a high feed mole fraction a large portion of the heat could be introduced in an intermediate reboiler at a temperature only slightly above that of the condenser.

Other approaches besides intermediate reboilers and condensers can be used to derive the same benefits. Pumparounds (liquid withdrawals, cooled and returned to the column) are used extensively in wide-boiling hydrocarbon distillations (see, for example, Bannon and Marples, 1978) and serve the same purpose as an intermediate condenser. A feed preheater provides some but not all of the benefit of an intermediate reboiler; a comparison of those two alternatives for a specific case is presented by Petterson and Wells (1977). Similarly, for low-temperature distillations, prechilling of the feed, which can result in multiple feeds, derives some but not all of the benefits of an intermediate condenser. These alternatives are explored for a demethanizer column by King et al. (1972).

A prefractionator column (see Fig. 5-34 and Prob. 5-K) can be used to generate partly enriched feeds for a subsequent main distillation column. The reboiler and

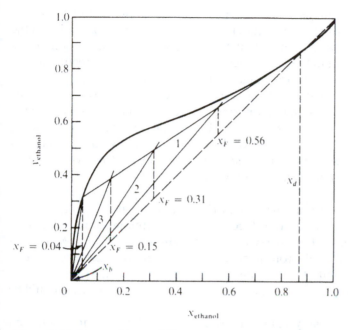

Figure 13-23 Degree of irreversibility in an ethanol-water distillation operated at minimum reflux. (*Data from Robinson and Gilliland, 1950.*)

condenser of the prefractionator serve the roles of an intermediate reboiler and an intermediate condenser and provide additional vapor and liquid flows in the vicinity of the feed composition. Use of a prefractionator therefore results in an energy savings if one can take advantage of the less extreme temperature levels of the reboiler and condenser of the prefractionator. Cascaded designs in which the first column produces only partially enriched products (Fig. 13-14b and c) are in fact prefractionator designs. The first column supplies extra flows in the vicinity of the feed composition, and the smaller temperature span of the first column results in less overall degradation of energy level than the design of Fig. 13-14a.

Freshwater (1961) has pointed out that a heat pump can be combined with an ordinary distillation design to provide extra flows in the vicinity of the feed composition. The design of Fig. 13-16b can be modified to withdraw the vapor feed to the compressor from a plate midway in the rectifying section, the condensed liquid from the reboiler-condenser being returned to the stage of vapor withdrawal. An ordinary condenser is then added for the column overhead. This modification provides higher flows below the vapor-withdrawal stage than above it and reduces the compression ratio required for the compressor. Similarly, a compressor receiving overhead vapor or vapor from an intermediate stage in the rectifying section could discharge to an intermediate reboiler, providing similar advantages but requiring the addition of a reboiler at the tower bottom. Similar modifications can be made to the heat-pump schemes in Fig. 13-16a and c.

Gunther (1974) and Mah et al. (1977) point out that operating the rectifying section of a distillation process at a pressure sufficiently higher than that of the stripping section would make transfer of heat possible between individual plates in the rectifying section and individual plates in the stripping section. Thus plates high in the rectifying section could exchange heat with plates high in the stripping section, plates low in the stripping section could exchange heat with plates low in the rectifying section, and intermediate plates could exchange heat with intermediate plates. The net result would be to give additional boil-up on some, most, or all of the plates below the feed and to give additional condensation on some, most, or all of the plates above the feed—in the direction of the process shown in Fig. 13-21. This form of cascading would reduce the internal irreversibilities of the distillation and would require less temperature span than the configurations shown in Fig. 13-14. Control of such a distillation would become more complex, however.

Isothermal distillation Distillation is usually carried out at a relatively uniform pressure with the temperature varying from stage to stage to maintain saturation conditions. In principle, it is possible to carry out a distillation with both pressure and temperature varying substantially from stage to stage or, in another extreme, to carry out a distillation with essentially the same temperature on each stage but with pressure varying from stage to stage to maintain saturation. Such a process could be called *isothermal distillation.*

Distillation at an essentially constant pressure is by far the most common approach because it lends itself to the common distillation-column configuration where the vapor phase travels upward under the sole impetus of the pressure drop from plate to plate. If the temperature were to be maintained constant from stage to stage, it would bé necessary for the pressure to *increase* from stage to stage in the direction of vapor flow. As a result, there would have to be compressors between all stages to move the vapor to a higher pressure. The expense associated with building the individual stages would probably also be greater, since it is necessary to isolate the stages more from each other. The expense associated with the compression is usually prohibitive because compressors are relatively costly to build and operate.

One situation where isothermal distillation has been used to advantage is in the process for manufacturing ethylene and propylene from refinery gases and naphtha, outlined in Fig. 13-24. In the high-pressure version of these plants the gas stream typically must be compressed from approximately atmospheric pressure up to about 3.5 MPa before entering the demethanizer column, which removes hydrogen and methane. The compression is generally carried out in four stages to prevent high temperatures which might cause polymerization within the compressors and to reduce the work input required.

Figure 13-25 shows a scheme proposed by Schutt and Zdonik (1956) for accomplishing some product separation during the course of this multistage compression. The operation amounts to an isothermal distillation. The effluent from each stage of compression is cooled to perhaps 43°C in a water-cooled heat exchanger. This cooling causes some hydrocarbon material to liquefy after each stage since the stream has been raised to a higher pressure within each stage. This liquid is removed in a separator drum and is made to flow countercurrent to the gas stream by flashing it into the separator drum at the next lower pressure. The result is a four-stage isothermal distillation, equivalent to four stages in the rectifying section of a distillation

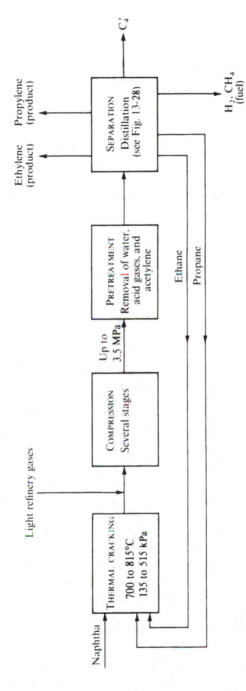

Figure 13-24 Overall processing scheme for manufacture of ethylene and propylene.

FIRST STAGE   SECOND STAGE   THIRD STAGE   FOURTH STAGE

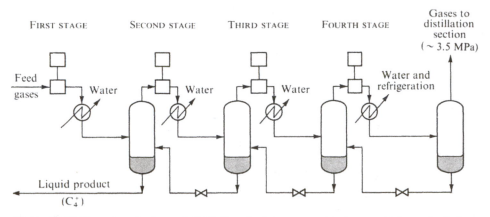

Figure 13-25 Four-stage isothermal distillation during compression in manufacture of ethylene and propylene.

column. The hydrocarbon liquid leaving the lowest-pressure separator is in equilibrium with the gas phase at the lowest interstage pressure, and the four stages of distillation have served to remove heavier hydrocarbons efficiently. The result is a lesser amount of C_4^+ material entering the demethanizer, deethanizer, and depropanizer towers of Fig. 13-28, with a commensurate reduction of the heat input required in those columns.

The inclusion of this isothermal distillation into the compressor sequence increases the vapor flow through the compressors somewhat, but the refluxing liquid stream is relatively small and the increased compressor-capacity requirement does not usually offset the gain made by the four-stage distillation. This is a rather unusual situation where the energy being put into the vapor stream for another purpose may be partially used to accomplish some separation at the same time.

Multicomponent distillation The conditions under which reversibility can be approached in multicomponent separations have been explored by Grunberg (1956), Petyluk et al. (1965), and Fonyó (1974a), among others. The need for reversible addition and removal of heat over the boiling range of the mixture in reboilers and condensers is apparent, and the reduction of energy consumption through the use of side reboilers and condensers at the appropriate temperature is a direct extension from binary distillation. The most interesting result, however, is that a reversible separation of a multicomponent mixture into its constituents requires that each column section remove only one component from the product of that section. For example, for a four-component mixture ABCD, where A has the greatest volatility and D the least, the rectifying section of the first distillation column would remove D from ABC, and the stripping section would remove A from BCD. Thus the two products would be a mixture of all A with some B and C (distillate) and a mixture of all D with some B and C (bottoms). This is different from conventional distillation practice, where one makes sharp separations between components of *adjacent* volatility, i.e., separating AB in the first column from CD, or A from BCD, or ABC from D.

Approaching reversibility requires separating components of *extreme* volatility instead. This introduces another dimension of possibilities for designing sequences of columns for multicomponent mixtures; earlier columns in the sequence can, as their main function, separate any pair of components, not necessarily those adjacent in volatility.

Alternatives for ternary mixtures Figure 13-26 shows eight different alternative distillation configurations for separating a mixture of three components into relatively pure single-component products. Configurations 1 and 2 separate one component from the other two in a first column and then separate the remaining binary mixture in a subsequent column. Configuration 3 follows the concept of separating the extreme components first, with B appearing to a substantial extent in both products. The two resulting binaries could then be separated in two different subsequent columns (not shown); however, these separations can be made as well in a single column, where B can be obtained as an intermediate sidestream in any purity required. Configuration 3 extends the prefractionator concept to the ternary separation. Configuration 4 differs from configuration 3 in that the first column does not have a reboiler and a condenser; instead it communicates with the second column

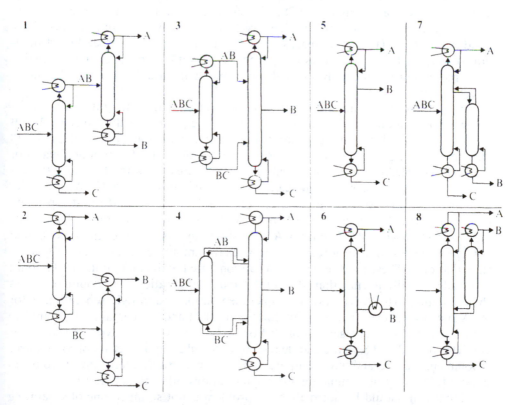

Figure 13-26 Alternative configurations for separating a ternary mixture by distillation.

through both vapor and liquid streams at each end. This has been called a *thermally coupled configuration* (Stupin and Lockhart, 1972). In configurations 5 and 6 the intermediate product is taken from a single column as a sidestream above (liquid) and below (vapor) the feed, respectively. As shown in Chap. 7, the sidestream product must contain a significant amount of A in configuration 5 and C in configuration 6. This then leads to the use of a sidestream stripper to purify the sidestream withdrawn above the main feed (configuration 7) or a sidestream rectifier to purify the sidestream withdrawn below the main feed (configuration 8). The configurations in Fig. 13-26 are shown with total condensers and liquid products. Partial condensers and/or some vapor products could be used as well. The number of possibilities would become even larger if cascading or side reboilers and condensers were considered.

Rod and Marek (1959) and Heaven (1969) have explored the relative advantages of configurations 1 and 2 and find that the scheme which removes the component (A or C) present in greater amount first is preferred, with some advantage for configuration 2 when A and C are roughly equal in amount. The result is also affected by the component volatilities and the desired product purities. Petyluk et al. (1965) explored various attributes of configurations 1 to 4 for a close-boiling ternary mixture. Stupin and Lockhart (1972) explored costs of configurations 1, 2, and 4 for a particular ternary distillation, finding the thermally coupled scheme most advantageous. Doukas and Luyben (1978) compared configurations 1, 2, 3, 5, and 6 in detail for distillation of a benzene-toluene-xylene mixture with varying compositions. Tedder and Rudd (1978a) compared all configurations except number 4 for distillation of various ternary mixtures of hydrocarbons, with set product characteristics. The product purities themselves can also be important variables affecting the choice of configuration.

From these studies and intuition it can be inferred that configuration 5 is attractive when the amount of A is small and/or when the purity specifications for A in B are not tight. Similarly, configuration 6 is attractive when the amount of C is small and/or when the purity specifications for C in B are not tight. The prefractionator or thermally coupled schemes (configurations 3 and 4) are often attractive when there is a large amount of B, with significant amounts of both A and C. With smaller amounts of B and significant amounts of A and C, the sidestream stripper and rectifier schemes (configurations 7 and 8) can be favored; configuration 7 would be more attractive when the amount of A is substantially less than the amount of C, and configuration 8 would be more attractive when the amount of A is substantially more than that of C. These schemes must also be compared with configuration 1 (amount of C substantially greater than that of A) and configuration 2 (amount of C less than or similar to that of A). A very complete separation and/or tight separation factor for A and B gives extra incentive for configurations 1 and 7, whereas a very complete separation and/or tight separation factor for B and C gives extra incentive for configurations 2 and 8. In the middle region where all components are of comparable amounts in the feed and have comparable recovery fractions, there appears to be no good a priori way of eliminating any configurations other than 5 and 6.

Finally, it should be noted that the problem is not so much one of separating components as it is one of separating products. Thus the same type of analysis of the column configurations in Fig. 13-26 can be applied to the separation of any mixture of many components into three different products.

Sequencing distillation columns When more than three products are to be separated, the number of possibilities becomes far greater than the alternatives shown in Fig. 13-26 for three products. Studies of techniques for generating multicolumn sequences for four or more products have for the most part been limited to sequences of simple columns separating adjacent components and having no sidestreams. This is not to say that such schemes are best, and for any complex distillation problem one should contemplate a number of different cases, including some with sidestreams, thermal coupling, prefractionators, sidestream strippers and/or rectifiers, cascaded columns, heat pumps, and/or side reboilers and condensers.

Even for sequences of simple distillation columns separating adjacent components, the number of possibilities becomes large. If a mixture is to be separated into R products by $R - 1$ simple distillation columns, we can develop a recurrence relationship for the number of possible column sequences S_R as a function of R. The first column which the feed enters will take j of the products overhead and hence will take $R - j$ products in the bottoms. There will be S_j sequences by which the j overhead products can be separated in subsequent distillations. Similarly there are S_{R-j} sequences by which the bottoms products can be separated subsequently. Hence the number of different column sequences which separate R products by taking j products overhead in the first column is $S_j S_{R-j}$. Allowing now for all possible separations that could have been performed by the first column in the sequence, we have

$$S_R = \sum_{j=1}^{R-1} S_j S_{R-j} \tag{13-37}$$

Starting with the known facts that $S_1 = 1$ (so as to count sequences in which one product is isolated in the first column) and $S_2 = 1$, we can generate the values of S_R shown in Table 13-3 from Eq. (13-37) (Heaven, 1969). The number of possible column sequences rises rapidly as the number of components and products rises. Table 13-3 can also be generated from a closed-form equation

$$S_R = \frac{[2(N-1)]!}{N!(N-1)!} \tag{13-38}$$

(Thompson and King, 1972).

Several studies have been made of ways of systematically identifying the best or one of the best sequences from among the many possibilities for a multiproduct system. Hendry and Hughes (1972) used a dynamic-programming technique to

Table 13-3 Number of column sequences S_R for separating a mixture into R products

Products R	2	3	4	5	6	7	8	9	10	11
Column sequences S_R	1	2	5	14	42	132	429	1430	4862	16,796

locate the optimum path through a tree of separation possibilities, assuming that the optimum design of each individual separator possibility is independent of its location in the sequence. Tedder and Rudd (1978*b*) explore suboptimizations of individual distillation columns and maintain that this is a relatively good assumption. Rathore et al. (1974) have explored the extension of this approach to the case where cascaded column design is included. For large problems the computing requirements for these dynamic-programming approaches become quite large. In an effort to eliminate at least some of the search space, Westerberg and Stephanopoulos (1975) suggested a branch-and-bound strategy for screening alternative configurations.

Because of the large combinatorial problem resulting when many products are to be made, simplification of the screening procedure by incorporating one or more *heuristics*, or rules of thumb, can be attractive. Thompson and King (1972) investigated the policy of identifying those candidate distillations which could lead to the desired final products and then selecting as the next step in a sequence that candidate distillation which had the lowest predicted costs. The sequencing procedure was repeated iteratively, the predicted costs for different separations being updated on the basis of more complete designs of separators used in previous iterations, proportioned according to the equilibrium-stage requirement. Rodrigo and Seader (1975) combined heuristic and branched-search methods by backtracking and branching the search for the best sequence, following the order dictated by the heuristic of including the cheapest candidate separator next. The number of sequences to be considered was reduced by means of an updated upper bound on the cost of the best sequence. Gomez and Seader (1976) found that a further improvement was to rely upon the heuristic that a separation is least expensive when conducted in the absence of nonkey components, so as to predict a lower-bound cost for sequences beginning with a particular next-included separator. Groups of possibilities whose lower bound exceeds the cost of a known sequence can then be eliminated.

The distillation-sequencing problem is a two-level problem, where the design of each column should be optimized, as well as the sequence being optimized. Solving both levels of problem simultaneously is quite complex, although methods exist to cope with this (see, for example, Westerberg and Stephanopoulos, 1975). It is probably best to optimize individual column designs after the few best candidate sequences have been identified. Optimization of reflux ratio, pressure, and recovery fractions is discussed in Appendix D, along with the optimal degree of overdesign. For sequenced columns, the recovery fractions in individual columns can also be optimized for components that are keys in more than one column.

For initial design and screening and for distillation situations with many products, it is usually inefficient to use a rigorous or highly systematic method to generate the most attractive candidate sequences. Instead, it is easier to use a few simple heuristics to generate some sequences which should be near optimal. Studies of the relative costs of different sequences of simple distillation columns for three-, four-, and five-product systems have been made by Lockhart (1947), Harbert (1957), Heaven (1969), Nishimura and Hiraizumi (1971), Freshwater and Henry (1975), and Freshwater and Ziogou (1976), in addition to the studies mentioned earlier for ternary systems. From these results, four simple heuristics can be inferred for sequencing simple distillation columns:

Heuristic 1. Separations where the relative volatility of the key components is close to unity should be performed in the absence of nonkey components. In a distillation, W_n' was shown to be proportional to the *product* of the interstage flow and the difference in reciprocal temperature between the reboiler and the condenser. Therefore, when one is selecting a sequence of distillation columns to accomplish the separation of a multicomponent mixture into relatively pure products, it is usually best to avoid column sequences wherein large interstage flows appear in a column which has a large temperature difference between reboiler and condenser, to avoid making W_n' the product of two large numbers. Since interstage flows are roughly proportional to $(\alpha_{LK-HK} - 1)^{-1} F$ [Eq. (13-21)], this indicates that it is desirable to select column sequences which do not cause nonkey components to be present in columns where the keys are close together in volatility. In this way the temperature drops and internal flows in these towers are kept as low as possible. In other words, the most difficult separations should be reserved until last in a sequence.

Heuristic 2. Sequences which remove the components one by one in column overheads should be favored. Returning to the Underwood equation for minimum reflux [Eq. (8-94)],

$$V_{\min} = \sum_{i=1}^{R} \frac{\alpha_i x_{i,d} d}{\alpha_i - \phi} \tag{8-94}$$

we see that adding nonkey components to the overhead of a column necessarily causes the minimum required interstage vapor flow to increase. The vapor flow is directly proportional to both the reboiler duty and the condenser duty. If any effect of partially vaporized feeds is neglected, it is then advantageous to have as few components as possible in the distillate from a tower, since this will enable the vapor flow to be as low as possible. This line of reasoning leads to the *direct* sequence of towers shown in Fig. 13-27 for separating multicomponent mixtures. The components are taken as overhead products one at a time, in the order of descending volatility.

When some of the components being separated have boiling points below ambient temperature, some of the columns must run under pressure and/or use refrigerant as a condenser cooling medium. The sequence of towers shown in Fig. 13-27 avoids the presence of a light diluent in any of the overheads and hence gives the least stringent conditions of pressure or refrigeration possible in the towers past the first. On the other hand, this ordering scheme causes the reboiler temperatures to be the highest possible, on the average, thus requiring higher-temperature heating media. This is not usually an important factor, however.

Heuristic 3. A product composing a large fraction of the feed should be removed first, or, more generally, sequences which give a more nearly equimolal division of the feed between the distillate and bottoms product should be favored. The overhead reflux flow and the vapor flow from the reboiler cannot both be adjusted independently in a distillation which gives fixed distillate and bottoms flows. Fixing the reflux flow fixes the reboiler boil-up rate. If the distillate molar flow rate is much less than the bottoms molar flow rate, the value of L/V in the rectifying section will be much closer to unity than the value of V'/L' in the stripping section. In such a case the rectifying

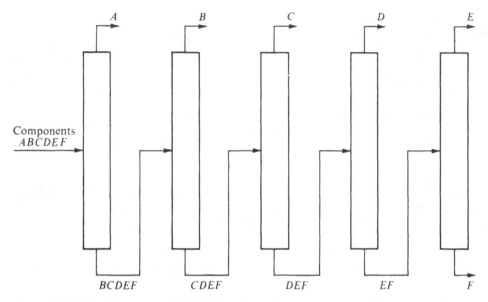

Figure 13-27 "Direct" sequence of distillation columns for separating a multicomponent mixture.

section will most likely be running at a much higher reflux ratio than is necessary for the separation, and because of the resulting high temperature and composition driving forces the operation of the rectifying section will be highly irreversible thermodynamically. If the bottoms product is substantially less than the distillate product, the reasoning is reversed and the stripping section will be highly irreversible thermodynamically. When the amounts of overhead and bottoms products are about the same, the reflux ratios in the sections above and below the feed will be better balanced and the operation will be more reversible. As a result the energy requirement (steam or refrigeration) for the separation should be less.

Heuristic 4. Separations involving very high specified recovery fractions should be reserved until late in a sequence. High product purities do not require higher reflux ratios but do require a greater number of stages, as we have seen in Chap. 9. Hence a particular separation of key components which requires very high recovery fractions of these components in their respective products will require a large number of stages without requiring any greater reflux requirements. If nonkey components are present when this separation is made, the necessary column diameter will be greater and the extra stages needed to provide the high product purities will all be larger in diameter. Hence there is an advantage in reduced equipment size to be obtained by reserving separations with high specified purities or recovery fractions until late in a multicomponent distillation column sequence. This heuristic can be combined with the first to become "perform the most difficult separations last."

These four heuristics for column sequencing often conflict with each other, one heuristic leading to one particular column sequence and another heuristic leading to another. In any real design situation it may well be necessary to examine several different sequences in order to see which of these heuristics is dominant. The real value of the heuristics comes in reducing the number of logical alternative sequences which should be examined, because a large number of the possible sequences will not be favored to any substantial degree by any of the heuristics. Seader and Westerberg (1977) suggest a systematic way of applying these heuristics in an evolutionary design.

The single heuristic of including the cheapest candidate separator next in a sequence has been found to be rough but effective in the work mentioned above on systematic screening of sequencing alternatives. This is a generalization of the first and fourth heuristics, and (to a much lesser extent) the other two as well.

These heuristics apply to sequences of simple distillation columns. Freshwater and Ziogou (1976) have shown that allowance for column cascading can alter the optimal sequence. Control factors also become important in determining the best sequence when column cascading is used.

Example: Manufacture of Ethylene and Propylene Figure 13-24 gives a schematic outline of the thermal cracking process which is used on a very large scale to manufacture ethylene and propylene from other hydrocarbons. Ethylene and propylene produced by such plants form the core of the petrochemical industry. The feed to such a plant may be a mixture of light refinery hydrocarbon gases and/or a naphtha stream consisting of hydrocarbons in the molecular-weight range of 80 to 150. Ethylene and propylene are formed by the thermal cracking of ethane, propane, and/or the naphtha hydrocarbons. The complex hydrocarbon mixture emanating from the cracking step must then be separated into

1. Relatively pure ethylene and propylene products
2. Ethane and propane for recycle as cracking feedstocks for the manufacture of additional ethylene and propylene
3. Methane and hydrogen for use as fuel
4. Products heavier than propane which can ultimately be used for gasoline or other purposes

The separation involves the distillation of low-boiling gas mixtures, and as a result a high pressure is required. Since cracking is favored by low pressure, the compression step occurs between the cracking and separation steps. Before distillation the gas mixture is pretreated to remove H_2O, CO_2, H_2S, and acetylene by a number of different separation processes.

When the feeds to the process are primarily refinery gases, a typical feed to the separation train is as shown in Table 13-4. The sequence of distillation towers most commonly used for isolating the products specified above when the feed is primarily refinery gases is shown in Fig. 13-28. This corresponds to a high-pressure plant, where there is no methane refrigeration.

Notice that the column arrangement in Fig. 13-28 represents two changes from the simple direct sequence shown in Fig. 13-27 (heuristic 2). Both these changes have been made to reserve a difficult separation until last so it can be performed as a binary distillation (heuristic 1). In this case the two difficult separations are ethylene from ethane and propylene from propane, both of which have relative volatilities quite close to 1. The ethylene-ethane and propylene-propane separations also

Table 13-4 Typical feed (data from Schutt and Zdonik, 1956)

Component	%	Component	%
Hydrogen, H_2	18	Propylene, C_3^{2-}	14
Methane, C_1	15	Propane, C_3^0	6
Ethylene, C_2^{2-}	24	Heavies, C_4^+	8
Ethane, C_2^0	15		

have very high purity requirements (heuristic 4) and require large towers in both diameter and height.

In addition to providing the benefits of the direct tower sequence (heuristic 1), placing the deethanizer before the depropanizer also provides more nearly equal distillate and bottoms flows from the tower following the demethanizer (heuristic 3). Cracking a naphtha feed provides more heavy products. In naphtha-cracking plants the deethanizer is sometimes placed before the demethanizer in the scheme of Fig. 13-28 (heuristic 3).

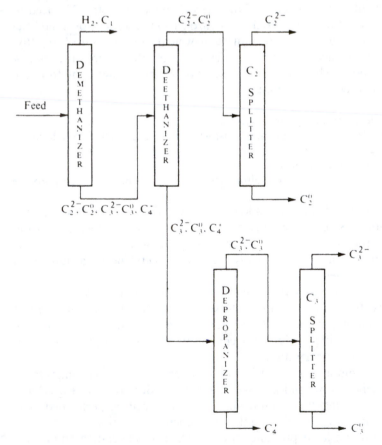

Figure 13-28 Typical distillation column sequence for separation of products in light olefin manufacture.

Figure 13-29 A panoramic view of the Sinclair-Koppers 230 million kilogram per year ethylene plant built by Pullman Kellogg at Houston, Texas. The pyrolysis furnaces are the rectangular units in the far left background. The tall tower at the right is the demethanizer. The tallest tower, painted in two colors, is the C_2 splitter. Just to the left of it are two large towers which compose the C_3 splitter, operating in series. (*Pullman Kellogg, Inc., Houston, Texas.*)

In a low-pressure plant (Fig. 13-18) the deethanizer column typically precedes the demethanizer, as opposed to the high-pressure sequence in Fig. 13-28. This is done because the demethanizer feed must be cooled to a temperature so much lower in the low-pressure process that it is worthwhile to cool only that portion of the total feed which necessarily requires the very low temperature for distillation.

The distillation section of a typical ethylene plant is shown in Fig. 13-29. More details on processes for the manufacture of ethylene and propylene are given by Schutt and Zdonik (1956) and Frank (1968).

Sequencing multicomponent separations in general We have so far discussed criteria for sequencing distillations which create several products out of a multicomponent mixture. If other separation processes can be used as well, the possibilities become still more complex: (1) the number of possible sequences (Table 13-3) grows in proportion to the number of different separation processes considered; (2) different separation processes generally produce different orderings of individual-component separation factors, making different groupings of components in products possible; and (3) when mass separating agents are added, additional components are introduced and must usually be separated subsequently for recycle.

Most of the systematic approaches suggested for sequencing distillations can be extended to the case where more than one separation method is considered to be available; in fact, many of the examples which have been considered allow extractive distillation, extraction, and/or other processes as alternatives to distillation. The dynamic-programming approach (Hendry and Hughes, 1972; etc.) becomes more complex if mass separating agents are allowed to be recovered more than one step after they are introduced, but the other methods mentioned previously (Thompson and King, 1972; Westerberg and Stephanopoulos, 1975; Rodrigo and Seader, 1975; Gomez and Seader, 1976) handle this possibility.

Of the sequencing heuristics listed for distillation, number 2 is specific for distillation and some other separation processes, but the other three extend readily to separations in general, as does the rough criterion of "cheapest first." In addition, three more sequencing heuristics can be generated for cases where several different candidate separation processes are considered.

Heuristic 5. Favor sequences which yield the minimum necessary number of products. Equivalently, avoid sequences which separate components which should ultimately be in the same product. This results in the minimum number of separators, which in most cases is best. Thompson and King (1972) explore this criterion in more detail and propose a product-separability matrix as a systematic method of keeping track of feasible product splits.

Heuristic 6. When alternative separation methods are available for the same product split, (1) *discourage consideration of any method giving a separation factor close to unity,* e.g., less than 1.05, and (2) *compare the separation factors attainable with the alternative methods in the light of previous experience with those separation methods* (Seader and Westerberg, 1977). For example, Souders (1964) compares the typical improvements in separation factor needed to make extraction and/or extractive distillation preferable to distillation, using economic factors for that time.

Heuristic 7. When a mass separating agent is used, favor recovering it in the next separation step unless it improves separation factors for candidate subsequent separations. This is an extension of heuristic 3, since a mass separating agent is usually present in large proportions.

Reducing Energy Consumption for Other Separation Processes

Much less attention has been paid to means of reducing the energy consumption of separations other than distillation, both because distillation is so common and because the energy separating agent in distillation is so easily exchanged between points in a process.

Mass-separating-agent processes In distillation, reversibility could be approached by adding or removing heat reversibly from additional stages so as to make the succession of operating lines become more nearly coincident with the equilibrium curve, or by cascading columns for more efficient heat utilization. If we consider an analogous approach to make a separation with mass separating agent more reversible, the matter becomes more difficult.

For example, in the fractionating absorber of Fig. 13-11, moving the operating line above the feed closer to the equilibrium curve would require that absorbent liquid be added to each stage above the feed. In addition, in order to conceive a reversible process we would have to postulate that the absorbent liquid stream added to each stage was saturated with respect to the light components being separated at the composition of that stage. This would require some scheme for presaturating the liquid entering each stage with these components, e.g., expanding portions of the top gas product and regenerated bottoms gas product reversibly to saturation conditions while recovering the work generated. This would be highly complex and is never done. It would also be necessary to *remove* absorbent liquid reversibly from stages below the feed point in the fractionating absorber. This would be an even more difficult procedure requiring, for example, reversible distillation processes treating each of these streams. It is much easier to exchange energy than it is to exchange matter in an analogous fashion.

In some cases of single-solute transfer and highly curved operating lines, another approach which can reduce the energy consumption of absorber-stripper operations is to remove a portion of the partly regenerated absorbent from a level midway down the regenerator and then add it to the absorber at an appropriate level midway down that column [see part (f) of Prob. 6-N]. This has the effect of making the curvature of the operating lines match that of the equilibrium curve to a greater extent.

Rate-governed processes; the ideal cascade Rate-governed separation processes differ from equilibration processes in that the separating agent cannot be reused from stage to stage in a multistage system. This means that energy must be introduced at every stage in proportion to the interstage flows in a rate-governed process. Consequently, the energy consumption becomes infinite in both extremes—minimum stages and infinite interstage flows, on the one hand, and minimum flows and infinite stages, on the other. The minimum total energy consumption occurs at an intermediate condition.

Energy consumption for the original gaseous-diffusion process for separating uranium isotopes was very large, and as a result the theoretical analysis yielding the design conditions for minimum energy consumption in multistage rate-governed processes was worked out at that time (Cohen, 1951; Benedict and Pigford, 1957). Pratt (1967) and Wolf et al. (1976) summarize more recent improvements in the analysis. The theory, presented also in the first edition of this book, leads to the concept of the so-called "ideal" cascade, which minimizes both total energy consumption and total equipment volume. In the ideal cascade the interstage flows at any point are exactly twice the minimum flows which would give a pinch at that point. Thus the flows are least at either end of the cascade and are largest in the vicinity of the feed stage for separation of a binary mixture. For separation factors close to unity, e.g., in isotope separations, the ideal cascade requires a number of stages equal to twice the minimum number that would correspond to infinite flows.

REFERENCES

Alexis, R. W. (1967): *Chem. Eng. Prog.*, **63**(5):69.
Bannon, R. P., and S. Marple, Jr. (1978): *Chem. Eng. Prog.*, **74**(7):41.

Benedict, M. (1947): *Chem. Eng. Prog.*, **43**(2):41.

———— and T. H. Pigford (1957): "Nuclear Chemical Engineering," McGraw-Hill, New York.

Bennett, R. C. (1978): *Chem. Eng. Prog.*, **74**(7):67.

Bliss, H., and B. F. Dodge (1949): *Chem. Eng. Prog.*, **45**:129.

Bromley, L. A., et al. (1974): *AIChE J.*, **20**:326.

Bryan, W. L. (1977): *AIChE Symp. Ser.*, **73**(163):25.

Casten, J. W. (1978): *Chem. Eng. Prog.*, **74**(7):61.

Cohen, K. (1951): "The Theory of Isotope Separation as Applied to the Large-Scale Production of U^{235}," McGraw-Hill, New York.

Davison, J. W., and G. E. Hays (1958): *Chem. Eng. Prog.*, **54**(12):52.

Dodge, B. F. (1944): "Chemical Engineering Thermodynamics," p. 596. McGraw-Hill, New York.

Doukas, N., and W. L. Luyben (1978): *Ind. Eng. Chem. Process. Des. Dev.*, **17**:272.

Fonyó, Z. (1974a): *Int. Chem. Eng.*, **14**:18.

———— (1974b): *Int. Chem. Eng.*, **14**:203.

Frank, S. M. (1968): chap. 3 in S. A. Miller (ed.), "Ethylene and Its Industrial Derivatives," Benn, London.

Freshwater, D. C. (1961): *Br. Chem. Eng.*, **6**:388.

———— and B. D. Henry (1975): *The Chem. Engr.*, no. 301, p. 533.

———— and E. Ziogou (1976): *Chem. Eng. J.*, **11**:215.

Gomez, A., and J. D. Seader (1976): *AIChE J.*, **22**:970.

Grunberg, J. F. (1956): *Proc. 4th Cryogen. Eng. Conf.*, Boulder, p. 27.

Gunther, A. (1974): *Chem. Eng.*, Sept. 16, p. 140.

Harbert, W. D. (1957): *Petrol Refiner*, **36**(3):169.

Heaven, D. L. (1969): M.S. thesis in chemical engineering, University of California, Berkeley.

Hendry, J. E., and R. R. Hughes (1972): *Chem. Eng. Prog.*, **68**(6):69.

Hougen, O. A., K. M. Watson, and R. A. Ragatz (1959): "Chemical Process Principles," 2d ed., vol. 2, pp. 968–969, Wiley, New York.

Kaiser, V., P. Daussy, and O. Salhi (1977): *Hydrocarbon Process.*, **56**(1):123.

King, C. J., D. W. Gantz, and F. J. Barnés (1972): *Ind. Eng. Chem. Process. Des. Dev.*, **11**:271.

Kirkbride, C. G., and J. W. Bertetti (1943): *Ind. Eng. Chem.*, **35**:1242.

Labine, R. A. (1959): *Chem. Eng.*, Oct. 5, pp. 118–121.

Latimer, R. E. (1967): *Chem. Eng. Prog.*, **63**(2):35.

Lockhart, F. J. (1947): *Petrol. Refin.*, **26**(8):104.

Mah, R. S. H., J. J. Nicholas, Jr., and R. B. Wodnik (1977): *AIChE J.*, **23**:651.

McBride, R. B., and D. L. McKinley (1965): *Chem. Eng. Prog.*, **61**(3):81.

Mix, T. W., J. S. Dweck, M. Weinberg, and R. C. Armstrong (1978): *Chem. Eng. Prog.*, **74**(4):49.

Murphy, G. M., H. C. Urey, and I. Kirschenbaum (eds.) (1955): "Production of Heavy Water," chap. 3, McGraw-Hill, New York.

Nishimura, H., and Y. Hiraizumi (1971): *Int. Chem. Eng.*, **11**:188.

Null, H. R. (1976): *Chem. Eng. Prog.*, **72**(7):58.

Opfell, J. B. (1978): *AIChE J.*, **24**:726.

Palazzo, D. F., W. C. Schreiner, and G. T. Skaperdas (1957): *Ind. Eng. Chem.*, **49**:685.

Perry, R. H., and C. H. Chilton (eds.) (1973): "Chemical Engineers' Handbook," 5th ed., McGraw-Hill, New York.

Petterson, W. C., and T. A. Wells (1977): *Chem. Eng.*, Sept. 26, p. 79.

Petyluk, F. B., V. M. Platanov, and D. M. Slavinskii (1965): *Int. Chem. Eng.*, **5**:555.

Pratt, H. R. C. (1967): "Countercurrent Separation Processes," Elsevier, Amsterdam.

Rathore, R. N. S., K. A. Van Wormer, and G. J. Powers (1974): *AIChE J.*, **20**:491, 940.

Robinson, C. S., and E. R. Gilliland (1950): "Elements of Fractional Distillation," 4th ed., pp. 162–174, McGraw-Hill, New York.

Rod, V., and J. Marek (1959): *Coll. Czech. Chem. Comm.*, **24**:3240.

Rodrigo, F. R., and J. D. Seader (1975): *AIChE J.*, **21**:885.

Ruhemann, M. (1949): "The Separation of Gases," 2d ed., Oxford University Press, London.

———— and P. L. Charlesworth (1966): *Br. Chem. Eng.*, **11**:839.

Schutt, H. C., and S. B. Zdonik (1956): *Oil Gas J.*, **54**:98 (Feb. 13); 99 (Apr. 2); 149 (May 14); 92 (June 25); 171 (July 30); 133 (Sept. 10).

Scott, R. B. (1959): "Cryogenic Engineering," chap. 3, Van Nostrand, Princeton, N.J.

Seader, J. D., and A. W. Westerberg (1977): *AIChE J.*, **23**:951.

Shaner, R. L. (1978): *Chem. Eng. Prog.*, **74**(5):47.

Souders, M. (1964): *Chem. Eng. Prog.*, **60**(2):75.

Stewart, H. A., and J. L. Heck (1969): *Chem. Eng. Prog.*, **65**(9):78.

Stoughton, R. W., and M. H. Lietzke (1965): *J. Chem. Eng. Data*, **10**:254.

Stupin, W. J., and F. J. Lockhart (1972): *Chem. Eng. Prog.*, **68**(10):71.

Tedder, D. W., and D. F. Rudd (1978*a*): *AIChE J.*, **24**:303.

―――― and ―――― (1978*b*): *AIChE J.*, **24**:316.

Thompson, R. W., and C. J. King (1972): *AIChE J.*, **18**:942.

Tyreus, B. D., and W. L. Luyben (1975): *Hydrocarbon Process.*, **54**(7):93.

―――― and ―――― (1976): *Chem. Eng. Prog.*, **72**(9):59.

Westerberg, A. W., and G. Stephanopoulous (1975): *Chem. Eng. Sci.*, **30**:963.

Wolf, D., J. L. Borowitz, A. Gabor, and Y. Shranga (1976): *Ind. Eng. Chem. Fundam.*, **15**:15.

Zdonik, S. B. (1977): *Chem. Eng.*, July 4, p. 99.

PROBLEMS

13-A₁ Draw qualitative operating diagrams for distillation of a relatively ideal mixture by each of the two-tower schemes shown in Fig. 13-14. Show the various operating lines for both towers on the same diagram in each case, i.e., *one* diagram for each scheme.

13-B₁ Perry and Chilton (1973, p. 13-41) report that at 101.3 kPa and 64.86°C the system ethanol-benzene-water forms an azeotrope containing 22.8 mol ° ethanol, 53.9 mol ° benzene, and 23.3 mol ° water. Taking advantage of the fact that the composition of the equilibrium vapor is the same as that of the liquid at the azeotrope, and using available vapor-pressure data, find the minimum possible work consumption of an isothermal process separating this azeotropic liquid mixture into three relatively pure liquid products at 64.86°C.

13-C₂ Stoughton and Lietzke (1965) report the following data† for the boiling-point elevation caused by the dissolved species in seawater at different degrees of concentration at a total pressure of 3.14 kPa. The boiling point of pure water at this pressure is 25.0°C.

Dissolved salts, wt °	2.0	3.45	4.0	6.0	8.0	12.0
bp elevation, °C	0.177	0.311	0.363	0.564	0.783	1.285

Natural seawater contains 3.45 weight percent dissolved salts.

(*a*) Find the minimum possible work requirement for recovering 1 m³ of fresh water from a very large volume of natural seawater with feed and products at 25°C. If energy is supplied as electric power at 0.8 cent per megajoule, find the minimum possible energy cost as cents per cubic meter of fresh water. (A target *total* cost for equipment, energy, labor, etc., for a successful seawater conversion process is 25 cents per cubic meter of fresh water.)

(*b*) Find the minimum possible energy cost if the products of the process are fresh water and doubly concentrated brine (6.9 weight percent dissolved salts).

13-D₂ Referring to the data for Loeb reverse osmosis membranes shown in Table 1-2, find the energy consumption per cubic meter to recover purified water from a very large volume of water containing 5000 ppm NaCl at 25°C, with ΔP across the membrane equal to 4.15 MPa. Assume (somewhat unrealistically) that the net work consumption is entirely associated with the pressure drop across the membrane, that work can be recovered from the exit brine completely, and that the energy input required to combat concentration polarization is negligible. Compute the thermodynamic efficiency of this process, assuming that the properties of NaCl solutions are the same as those reported for seawater solutions in Prob. 13-C.

† Additional thermodynamic data for seawater solutions are given by Bromley et al. (1974).

13-E₂ Figure 13-30 shows schematics for two processes for making fresh water from seawater by freezing. In process *a*, seawater is partially frozen using an external refrigerant at a single temperature level. In process *b*, seawater is sprayed into a vacuum at a pressure such that the boiling point of the brine formed is less than the freezing point of the brine. The cooling required to freeze a portion of the feed seawater comes from evaporation of another portion of the water. The evaporated water vapor is compressed and fed to a melter operating at a higher pressure such that the condensation temperature of the vapor is higher than the melting point of the ice; hence the heat of condensation of the vapor is removed through the heat of fusion of the ice. Compute the energy consumption of each process and the cost of that energy (cents/per cubic meter of fresh water) subject to the following assumptions:

1. The feed seawater is fully cooled to the freezer temperature by heat exchange against the products. Only latent heat effects need be considered in the freezers and the melter.
2. Energy requirements for pumping and agitation (except for vapor compression) may be ignored.
3. Additional refrigeration to remove heat from heat leaks into the system and to remove heat input from the compressor may be neglected.
4. The feed seawater contains 3.45 percent dissolved salts and the product brine contains 6.9 percent dissolved salts. The freezing-point depressions for these two concentrations are 1.95 and 4.1°C, respectively.
5. The freezers and the melter provide simple equilibrium between liquid and solid.
6. The filters provide a complete separation of phases.
7. The equilibrium condensation temperature of the vapor in the melter must be 1°C above the freezing point of the liquid (scheme *b*).
8. The equilibrium condensation temperature of the vapor in the freezer must be 1°C below the freezing point of the liquid (scheme *b*).

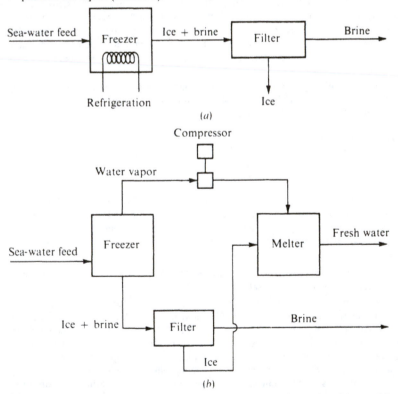

Figure 13-30 Freezing processes for seawater conversion: (*a*) simple refrigerated freezer; (*b*) evaporative freezing.

9. In scheme a, the refrigerant must be supplied at a temperature 3°C below the freezing point of the liquid. The thermodynamic efficiency of the refrigeration system is 40 percent.

10. The cost of energy is 0.8 cent per megajoule. In scheme b the compressor is adiabatic, with a work efficiency of 90 percent.

13-F$_2$ Staging proved to be of benefit in Example 13-2 for reducing the separating agent requirement. Would there be any benefit to staging in (a) Example 13-1 or (b) Example 13-3? Explain your answers.

13-G$_2$ (a) Repeat part (b) of Example 13-1 if the methane-rich product can now be expanded isenthalpically to 138 kPa.

(b) Find the product compositions from the process of Fig. 13-5 if *no* auxiliary refrigeration is used in steady-state operation and if the methane-rich product may be expanded isenthalpically to 138 kPa.

13-H$_2$ Find the thermodynamic efficiency of the heavy-water distillation process described in Fig. 13-22 and Table 13-2. Assume that the separation is between H_2O and D_2O; neglect the existence of HDO.

13-I$_3$ Fig. 13-31 is the flow diagram for a demethanizer column following a deethanizer and receiving a feed of hydrogen, methane, ethylene, and ethane in an ethylene manufacturing plant. The process scheme uses ethylene and methane refrigerants at the approximate temperatures indicated in Fig. 13-31. The tower itself operates at approximately 620 kPa.

(a) What is gained by creating two separate liquid feeds to the column and by having the feed pass through two separator drums?

(b) What is gained by the use of the expander and of the Joule-Thomson valve? (The expander produces shaft work from a gas expansion, but generally that work is not used gainfully elsewhere.)

(c) Explain the logic leading to the particular sequence of heat exchangers which is employed.

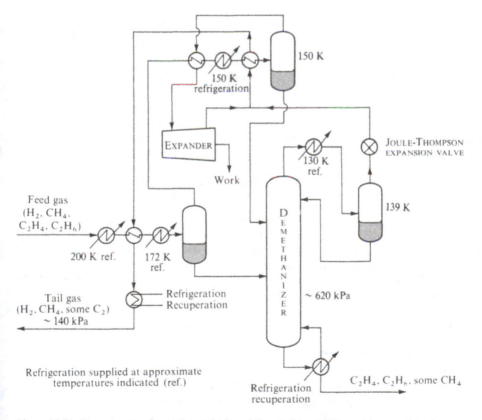

Figure 13-31 Demethanizer flow scheme. (*Adapted from Lahine, 1959; used by permission.*)

(d) What is gained by having the vapor from the 150-K separator drum bypass the demethanizer altogether?

(e) Suppose that a product more enriched in methane than the tail gas were desired. Where could it be obtained?

13-J₂ Figure B-2† shows a four-effect evaporation system for seawater conversion. Notice that the process includes heat exchangers of two kinds. One variety of exchanger contacts the feed to the system with the combined condensate from previous effects. The other contacts a portion of the vapor from each effect with the feed to the system. The vapor condenses in these exchangers.

(a) Why are these two types of heat exchangers included in the process?

(b) Why aren't the feed-vs-combined-condensate exchangers sufficient in themselves?

(c) The use of a portion of the vapor from each effect for feed preheat results in a loss of evaporation capacity in succeeding effects. This seems detrimental to the process. What incentive is there for the use of these vapor streams for feed preheat?

(d) The feed encounters exchangers of the two types alternately. Why is this arrangement employed rather than taking the vapor stream for feed preheat from just the first effect?

(e) Once this process has been built and is in operation, how many variables can be independently adjusted during operation? List a complete set of independent variables (in addition to those which have been set by construction).

(f) Suggest a control scheme for this multieffect evaporation process.

13-K₃ The following mixture of alcohols is available as a saturated liquid, and you have been asked to devise a sequencing scheme for distillation towers to separate the alcohols into relatively pure single-component streams.

	Feed, mole fraction	Relative volatility			Feed, mole fraction	Relative volatility
Methanol	0.15	3.58		n-Propanol	0.30	1.00
Ethanol	0.10	2.17		Isobutanol	0.10	0.669
Isopropanol	0.20	1.86		n-Butanol	0.15	0.412

Propose a distillation sequencing scheme to accomplish this separation. Present a flowsheet showing the appropriate layout and support your conclusion that this particular sequence is best. Your selection criterion should be *minimum total vapor generation* in the reboilers. Ground rules:

1. All towers will be run at 1.25 times minimum reflux. All feeds are saturated liquid.
2. No sidestreams may be employed, nor may any condensers be combined with reboilers into a single exchanger.
3. Since the boss wants the answer in half an hour, it is important that you develop a scheme of logic which will lead you to the optimal solution with a minimum of calculation.

13-L₂ A vapor-recompression distillation of the type shown in Fig. 13-16b is used to fractionate a mixture of ethylene and ethane. The feed is saturated liquid and contains 60 mol % ethylene; the recovery fractions of ethylene in the distillate and ethane in the bottoms are high (>0.99); the tower contains 100 plates; and the measured temperatures and pressures are as follows:

Location	Temp., K	Pressure, MPa
Vapor returning to tower from reboiler	263.3	1.862
Overhead vapor, before compressor	238.3	1.655
Condensing vapor, in tubes of reboiler	268.9	3.72

† Review of Appendix B in connection with this problem should be helpful.

The relative volatility of ethylene to ethane at 1.65 to 1.86 MPa as a function of liquid composition has been determined as follows:

$x_{C_2H_4}$	0.0	0.2	0.4	0.6	0.8	1.0
$\alpha_{C_2H_4\ C_2H_6}$	1.64	1.60	1.56	1.54	1.52	1.46

SOURCE: Data from Davison and Hays (1958).

Vapor pressures of the pure components are as tabulated below. The relative volatility is less than the ratio of the vapor pressures because of vapor-phase nonidealities.

	Vapor pressure, MPa	
Temperature, K	Ethylene	Ethane
235	1.49	0.82
250	2.32	1.30
265	3.39	1.94

(a) If compressor inefficiencies are neglected, the energy input to the process (compressor work) is needed (1) to overcome pressure drop associated with vapor flow from plate to plate, (2) to supply a driving force for heat transfer in the combined reboiler-condenser, and (3) to provide the thermodynamic minimum work of separation and supply driving forces for heat and mass transfer on the plates. Calculate the percentages of the total energy input required for each of these three purposes, using the data given.

(b) A suggested modification of the design to reduce energy consumption involves using two different vapor-recompression condenser-reboiler loops, each with its own compressor. One of these would withdraw overhead vapor and compress it sufficiently to allow this condensing vapor to supply the heat for the bottoms reboiler, as shown in Fig. 13-16b. The resulting liquid would be used as overhead reflux and distillate. The other loop would withdraw vapor partway up in the rectifying section and compress it sufficiently to allow it to supply the heat for a side reboiler, located partway down in the stripping section. The resultant liquid, formed from the condensed vapor, would then be fed back onto the plate from which the vapor was withdrawn. Does this modification have the potential for reducing the total compressor work required to achieve a given degree of separation in this distillation, for operation at a specified multiple of the minimum reflux ratio? Explain briefly.

(c) If the two compressor feeds were reversed in part (b), would the resulting scheme have a potential for lowering the energy consumption compared with the base case? Explain briefly.

(d) Vapor recompression is often used for ethylene-ethane distillation in ethylene plants but has not been used for demethanizer columns in such plants. Why is vapor recompression less attractive for demethanizers?

FOURTEEN

SELECTION OF SEPARATION PROCESSES

In this chapter the logic leading to the selection of particular processes as candidates for carrying out a particular separation is explored. Most of this book has been concerned with the analysis or design of a given separation process once the means of separation, the type of equipment, and the separating agent have been chosen. Very often, however, it is not immediately clear what separation process will work best for a mixture in a particular set of circumstances, and many important advances are made by generating improved approaches for the separation of a mixture of practical importance.

FACTORS INFLUENCING THE CHOICE OF A SEPARATION PROCESS

The pertinent factors to be considered in the selection of a processing approach for separating a particular mixture vary greatly from case to case, and it is difficult to tabulate any reliable pattern of thought that should be followed in solving such problems. There are a number of rules of thumb which can be followed, and with them it should be possible to identify a few separation processes which ought to be particularly strong candidates for use in any given problem situation. It will be convenient to refer to Table 1-1, which lists separation processes and their underlying physical or chemical principles and phenomena.

Feasibility

First and foremost, any separation process to be considered in a given situation must be feasible; i.e., it must have the potential of giving the desired result. Quite a large amount of screening of separation processes can be accomplished on the basis of feasibility alone. For example, if we are confronted with the need to separate a pair of nonionic organic compounds, such as acetone and diethyl ether, it is immediately apparent that the process cannot be carried out by ion exchange, magnetic separation, or electrophoresis, since the physical phenomena underlying these processes are not useful in connection with such a mixture. The molecules do not differ in those ways. It should also be apparent from the start that these molecules do not differ enough in surface activity for foam or bubble fractionation to be useful.

Often the question of process feasibility will have to do with the need for *extreme processing conditions*. Here the dividing lines between what is extreme and what is not extreme are not easy to draw, but the general idea is that a process which requires very high or very low pressures or temperatures, very high voltage gradients, or other such conditions will suffer in comparison with one which does not require extreme conditions. For example, separation of an acetone–diethyl ether mixture by any process which requires a solid feed, e.g., leaching, freeze-drying, or zone melting, would require that the feed be frozen, which in turn would require low-temperature refrigeration. If it is possible to avoid refrigeration, it will probably be desirable to do so. As another example, separation of a mixture of sodium chloride and potassium chloride by distillation or evaporation would require extremely high temperatures and extremely low pressures because of the very low volatility of these substances, and it follows that some other process will most likely work better.

Another way in which process feasibility enters into consideration occurs when a mixture of many components is to be separated into relatively few different products. In such a case it will be necessary for *each component to enter the proper product*. Since different separation processes accomplish the separation on the basis of different principles, it is quite possible for different processes to divide the various components in different orders between products. As a simple example, let us consider the separation of a mixture of propylene, $H_2C=CH-CH_3$; propane, $H_3C-CH_2-CH_3$; and propadiene, $H_2C=C=CH_2$. This can be an important problem in practice (Prob. 8-L), where it may be desired to obtain the propylene in a relatively pure product while leaving the propane and the propadiene in the other product. In a distillation, propylene is the most volatile component, and it is possible to take the propylene overhead in a large distillation column while removing most of the propane and propadiene as bottoms; thus the separation could be made as specified. In an extractive distillation process, on the other hand, a polar solvent would be added and would serve to increase the volatility of propane while increasing the volatility of the propylene less and increasing the volatility of the propadiene still less. Propane would have to be the distillate product, while propadiene would necessarily appear in the bottoms; hence the separation could not be performed to give the desired product splits. Similarly, an extraction process would most likely have to employ an immiscible solvent which was polar, thereby exerting an affinity for propadiene (two relatively polarizable double bonds), propylene (one double bond), and propane, in that order. Here again the propadiene cannot concentrate in the propane product.

Another interesting practical example of this sort has been cited by Oliver (1966). In the manufacture of various aromatic hydrocarbons it is often necessary to obtain

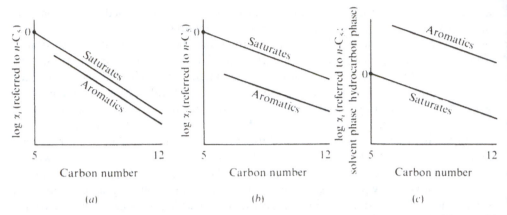

Figure 14-1 Separation factors for C_5 to C_{12} saturates and aromatics: (a) ordinary distillation; (b) extractive distillation; (c) extraction. (*Adapted from Oliver, 1966, p. 139; used by permission.*)

the aromatics (benzene, toluene, xylenes, cumenes, etc.) from a mixture of saturated, unsaturated, and aromatic hydrocarbons of a wide range of molecular weights leaving a gasoline reforming plant. Figure 14-1a qualitatively shows the relative volatility of various saturated and aromatic compounds relative to n-pentane as a function of the number of carbon atoms in the compound. From Fig. 14-1a it is apparent that the aromatics cannot be separated from the saturates in a mixture of C_5 to C_{12} saturates and aromatics in a single distillation column because the boiling points of the various aromatics overlap those of the various saturates. n-Pentane and perhaps n-hexane could be recovered as an overhead product, but the next most volatile component would be benzene, which is an aromatic.

Figure 14-1b shows the result of adding a solvent to turn a distillation into an extractive distillation process. The relative volatility between each saturate and its corresponding aromatic has become greater but not quite great enough to allow the separation of all the saturates from all the aromatics to be accomplished in a single tower. Nonetheless, it has now become possible to cut the mixture into perhaps three main portions through ordinary distillation and separate each of these portions by extractive distillations in different columns. Such a process would be expensive but workable.

A liquid-liquid extraction solvent is usually more selective than an extractive-distillation solvent because the extraction process operates under conditions of enough nonideality to give immiscible liquid phases, whereas the extractive-distillation process does not. Hence the separation factors between saturates and their corresponding aromatics shown in Fig. 14-1c are even greater than those for extractive distillation shown in Fig. 14-1b. The extraction solvent is most likely polar and interacts preferentially with the mobile π electrons in the benzene rings of the aromatics. Here the entire separation can very nearly be done in a single multistage extraction process.

For feeds of a very wide boiling range, a combination of extraction and extractive distillation could be more attractive than either process by itself. As shown in

Fig. 14-1, light saturate compounds (low molecular weight) are most easily removed from the rest of the mixture in an extractive-distillation process, whereas heavy saturates are most easily removed from the rest of the mixture in an extraction process. Oliver (1966) suggests such a hybrid process whereby extraction is first used to separate out the heavy saturates and then extractive distillation is used to separate the remaining saturates from the aromatics.

Product Value and Process Capacity

The economic value of the products being isolated influences the choice of a separation process. Fresh water obtained from seawater has a value of about 0.025 cent per kilogram, ethylene is worth about 25 cents per kilogram, silicone oils are worth about $3 per kilogram, and numerous fine chemicals (vitamins, pharmaceuticals, etc.) have values of many dollars per kilogram. Clearly many separation processes which would be suitable for substances with a high value cannot be considered for a substance with a low value. The lower the economic value of the product, the more important it will be to select a process with a relatively low energy consumption compared with other processes and to select a process where the unit cost of any added mass separating agent is relatively low. A small loss of a costly mass separating agent can be an important economic penalty to a process.

A process for manufacture of a substance with a low economic worth per unit quantity will most likely be a large-capacity process, since there will probably be a large market for the substance and processing economies can be realized through large-scale operation. The plant capacity can be an important factor in separation-process selection, since some processes, e.g., chromatography, mass spectrometry, and field-flow fractionation, are difficult to carry out on a very large scale.

Damage to Product

Often the question of avoiding damage to the product can be a major consideration in the selection of a separation process. Artificial kidneys and lungs are cases in point, since human blood is quite sensitive to the nature of surfaces with which it comes in contact and can be damaged by heat or by the addition of foreign substances. As another example, the addition of a mass separating agent in food processing is unusual, again because of the possibility of *contamination* from any residual separating agent in the product. Agents which are added generally have the approval of the Food and Drug Administration.

Often it is necessary to take special steps to avoid *thermal damage* to a product. Thermal damage may be manifested through denaturation, formation of an unwanted color, polymerization, etc. When thermal damage is a factor in separation by distillation, a common approach is to carry out the distillation under vacuum to keep the reboiler temperature as low as possible. Frequently evaporators and reboilers are given special design to minimize the holdup time at high temperature of material passing through them.

Oxygen present in a stripping gas may be detrimental to easily oxidizable substances. Freezing can also cause irreversible damage to biological materials, although the problem is usually not as severe as that of thermal damage and can be minimized by using well-chosen freezing conditions.

Classes of Processes

Some generalizations can be drawn between different classes of separation processes insofar as their advantages and disadvantages for various processing applications are concerned.

We have already met with the distinction between energy-separating-agent equilibration processes, mass-separating-agent equilibration processes, and rate-governed processes. For a multistage separation without a particularly large separation factor the energy consumption of the process increases as we go from energy separating agent to mass separating agent to rate-governed processes. Therefore a multistage rate-governed separation process should ordinarily give a better separation factor than an equilibration process if the rate-governed process is to be considered, and a mass-separating-agent process should ordinarily give a better separation factor than an energy-separating-agent process if the mass-separating-agent process is to be considered. The higher energy consumption of the mass-separating-agent process for a given separation factor is associated with the introduction of yet another component (a mixing process) and the need for removing this component from at least one of the products. Souders (1964) has given a generalized plot of separation factors required for extractive distillation and/or extraction to be attractive over distillation. The necessary separation factors increase in the order distillation < extractive distillation < extraction.

For single-stage separations the relative disadvantages of the rate-governed processes is less, since a separating agent need not be put into more than one stage. With the exception of some membrane processes, rate-governed separation processes will still tend to have a large energy requirement, or a low thermodynamic efficiency, because of their inherent dependence upon a transport phenomenon giving selective rate differences.

Separation processes which involve handling a solid phase have a disadvantage in continuous operation relative to processes in which all phases are fluid. This disadvantage stems from the difficulty of handling solids in continuous flow. To resolve this problem either more complex equipment is needed or fixed-bed configurations are used. Fixed-bed operation is inherently not totally continuous, and it is usually necessary to allow for intermittent bed regeneration by switching between beds, etc. Fixed-bed processes also lose some of the benefits of continuous countercurrent flow of contacting streams. Fixed-bed processes tend to be most attractive when a substance present at a low concentration in the fluid phase is to be taken up on or into the solid phase. The lower the concentration of the transferring solute in the fluid phase, the less often it will be necessary to regenerate the bed or the smaller the bed required. Rapid pressure-swing adsorption mitigates some of these problems, as does the rotating-bed approach (Figs. 4-32 and 4-33); hence these techniques have also been used with relatively concentrated mixtures on a large scale.

Another consideration which may become important is the ease of staging various types of separation processes. As we have seen, membrane separation processes and other rate-governed processes are difficult (but by no means impossible) to stage because of the need of adding separating agent to each stage and also because it is frequently necessary to house each stage in a separate vessel. A distillation column, on the other hand, can provide many stages within a single vessel. Some other processes are best suited for those separations which require multiple staging. An example is chromatography in any form. A chromatographic flow configuration is not worth constructing and operating for a single-stage separation, but the cost of providing many additional stages or transfer units in a chromatographic device is relatively small; one simply uses a longer column. Hence chromatography finds an application for separations where the separation factor is close enough to unity and the purity requirements are high enough for many stages to be required. Membrane processes, on the other hand, find greatest application for systems where they can provide a relatively large separation factor.

The comparisons between different classes of separation processes so far lead rather strongly toward distillation. Distillation is an energy-separating-agent equilibration process and hence desirable from an energy-consumption viewpoint when staging is required. Since distillation involves no solid phases, it enjoys an advantage relative to crystallization, which is another energy-separating-agent equilibration process. No contaminating mass separating agent is added in distillation, and it is easily staged within a single vessel. Because of this favorable combination of factors, it is no accident that distillation is the most frequently used separation process in practice, at least for large-scale petroleum-refining and heavy-chemical operations. In fact, a sound approach to the selection of appropriate separation processes is to begin by asking: Why not distillation? Unless there is some clear reason why distillation is not well suited, distillation will be a leading candidate. Factors most often operating against distillation are thermal damage to the product, a separation factor too close to unity, and the need for extreme conditions of temperature and/or pressure if distillation is to be used.

Keller (1977) has evaluated the processes likely to be most seriously considered as alternatives to distillation in the petrochemical and chemical industries, as energy costs continue to increase. He concludes that extractive and azeotropic distillation, extraction (including liquid-phase ion exchange), and pressure-swing adsorption are all very likely to see markedly increased use, whereas crystallization and solid-phase ion exchange may see some increased use and all other processes will not see much use in these industries in the near future.

Separation Factor and Molecular Properties

For most separation processes the separation factors reflect differences in measurable bulk, or macroscopic, properties of the species being separated. For distillation the pertinent bulk property is the vapor pressure, as modified by activity coefficients in solution. For extraction and absorption the pertinent bulk property is the solubility in an immiscible liquid. These differences in bulk, or macroscopic, properties must in turn result from differences in properties attributable to the molecules themselves, which we shall call *molecular properties*. Determining the relationship of bulk properties to molecular properties is one of the frequent goals of physical and chemical research. Means of predicting various bulk properties (vapor pressure, latent heat of vaporization, surface tension, viscosity, diffusivity, etc.) from known molecular properties have been well summarized by Reid et al. (1977).

In addition to molecular weight the following molecular properties are important in governing the size of separation factors attainable in various separation processes.

Molecular volume Molecular volumes are usually taken from the molar volume of the substance in the liquid state at the normal (1 atm) boiling point. This is a measured quantity or can be predicted from additive contributions of atomic volumes (Reid et al., 1977). Another measure of the molecular volume is the Lennard-Jones collision diameter, obtained from measurements of gas-phase transport properties or of virial coefficients from PVT data (Reid et al., 1977; Bird et al., 1960). Lennard-Jones collision diameters are available for fewer substances than molar volumes at the normal boiling point.

Molecular shape This is a qualitative property related to the question of whether the molecule is long and thin, nearly spherical, branched, etc. It is best judged from molecular models constructed using known bond angles.

Dipole moment and polarizability These properties characterize the strength of intermolecular forces between molecules (Moore, 1963). The dipole moment is a measure of the permanent separation of charge within a molecule, or of its *polarity*. Groups like O—H and C=O in molecules are polar, in that the electrons of the bond within the group tend to be more associated with the oxygen atom than with the other atom of the group. When such polar groups are present in an asymmetrical fashion within a molecule, the molecule exhibits a finite *dipole moment*. Polar molecules interact more strongly with other molecules than nonpolar molecules of the same size do. As a result a polar substance, e.g., water, has a lower vapor pressure than a nonpolar species of about the same molecular weight, e.g., methane, and polar substances are dissolved more readily into polar solvents. Dipole moments are tabulated by Weast (1968) or can be predicted from known dipole-moment contributions of different groups and a knowledge of the geometrical structure of a particular molecule (Moore, 1963). An extreme manifestation of dipolar interaction between molecules is *hydrogen bonding* between electropositive H and electronegative O, Cl, etc., atoms of adjacent molecules. Even a molecule with no net dipole moment can act as a polar molecule if offsetting polar groups are present locally within the molecule.

The *polarizability* of a substance reflects the tendency for a dipole to be *induced* in a molecule of that substance due to the presence of a nearby dipolar molecule. Polarizability depends upon the size of a molecule and the mobility of electrons in various bonds within the molecule. Electrons in aromatic rings and, to a lesser extent, olefinic bonds are more mobile than the electrons in single covalent bonds and therefore impart a greater polarizability to a molecule. A more polarizable molecule will tend to have a lower vapor pressure and will have a greater solubility in a polar solvent. Thus diethylene glycol, a polar solvent, dissolves aromatics more readily than paraffins and olefins and can be used to recover aromatics from a mixed-hydrocarbon stream through extraction. Polarizabilities are also tabulated by Weast (1968). The tabulated polarizability reflects what may be an average of different polarizabilities in different directions with respect to the molecular axis.

The dielectric constant is a measure of the combined effects of the dipole moment and the polarizability (Moore, 1963). The depth ϵ of the potential-energy well in the Lennard-Jones model of intermolecular forces is also a measure of the strength of intermolecular forces between like molecules and is tabulated in several references (Bird et al., 1960; Reid et al., 1977). Yet another measure of the degree of polarity and the strength of intermolecular forces is the solubility parameter δ, discussed later in this chapter.

Molecular charge Molecules can carry a net charge in liquid solution or in ionized gases. Protein molecules contain both acidic ($-COOH$) and basic ($-NH_2$) groups which ionize to different extents depending upon the pH of the solution in which they are present. At a given pH different proteins will have different net charges in solution. Simple ions also differ in charge, allowing separation by processes depending upon charge or charge-to-mass ratio.

Chemical reaction Many separations are based upon the difference between molecules in their ability to take part in a given chemical reaction.

Table 14-1 indicates the importance of these different molecular properties in determining the value of the separation factor for various separation processes. No categorization such as this can be relied upon to be exact because of the numerous exceptions which arise and because the boundaries between strong and weak influences of a given property are often quite nebulous. Nevertheless, basic differences in the importance of different molecular properties in determining the separation factors for different separation processes are apparent from Table 14-1. For example, the separation factor in distillation reflects vapor pressures, which in turn reflect primarily the strength of intermolecular forces. The separation factor in crystallization, on the other hand, reflects primarily the ability of molecules of different kinds to fit together, and simple geometric factors of size and shape become much more important.

Classification of separation processes in terms of the molecular properties primarily governing the separation factor can be quite useful for the selection of candidate processes for separating any given mixture. Processes which emphasize molecular properties in which the components differ to the greatest extent should be given special attention. For example, if the components of a mixture have a substantially different polarity from each other, the likely processes are distillation or, if the volatilities are not very different, extraction or extractive distillation with a polar solvent. If the more polar molecule is present in low concentration, fixed-bed adsorption with a polar adsorbent could be attractive.

Chemical Complexing

Mass-separating-agent processes offer the additional dimension of choosing the solvent or other similar agent. This applies, for example, to absorption, extraction,

Table 14-1 Dependence of separation factor upon difference in molecular properties†

Separation process	Properties of pure substances					Interaction with mass separating agent or barrier		
	Molecular weight	Molecular volume	Molecular shape	Dipole moment and polarizability	Molecular charge	Chemical-reaction equilibrium	Molecular size and shape	Dipole moment and polarizability
Distillation	2	3	4	2	0	0	0	0
Crystallization	4	2	2	3	2	0	0	0
Clathration (as interior molecule)	0	0	0	0	0	3	1	3
Solvent extraction and absorption	0	0	0	0	0	2	3	2
Ordinary adsorption	0	0	0	0	0	2	2	2
Adsorption with molecular sieves	0	2	0	0	0	0	1	3
Dialysis; gel permeation	0	0	3	0	0	0	1	3
Ultrafiltration	0	0	4	0	0	0	1	0
Gaseous diffusion	1	0	0	0	0	0	0	0
Sweep diffusion	2	2	0	0	0	0	0	0
Ultracentrifugation	1	0	0	0	0	0	0	0
Electrophoresis	2	3	3	0	1	0	0	0
Electrodialysis	0	0	0	0	1	0	2	0
Ion exchange	0	0	0	0	0	1	2	0

† 1 = primary effect; necessary for any separation, 2 = primary effect, 3 = secondary effect (perhaps through another property), 4 = small effect, 0 = no effect.

extractive and azeotropic distillation, ion exchange (both solid and liquid), adsorption, adductive crystallization, and foam and bubble processes, among others. It also applies to membrane processes, where the interaction of the membrane or a membrane component with the species to be separated determines the solubility and the driving force for diffusion of that species across the membrane.

Physically interacting solvents are often used for absorption, extraction, and extractive distillation; they interact with the feed mixture through van der Waals intermolecular forces. Physically interacting solvents are usually easily regenerated but do not exert a strong or specific selectivity between the substances to be separated. Chemically interacting (or "complexing") solvents offer much better selectivity in many cases and hence tend to give higher separation factors. Chemical-complexing agents tend to be more difficult to regenerate, however.

Effective chemical-complexing solvents and other agents tend to give reaction bond energies falling in a certain critical range. Figure 14-2 shows this range and gives a number of examples of classes of chemical interactions with bond energies within that range. Bond energies for chemical complexing will usually be somewhat greater than those typical of van der Waals forces but should be substantially less than those for covalent bonds because of the need for regeneration and avoiding decomposition of the complexing agent itself. Some examples of currently used sep-

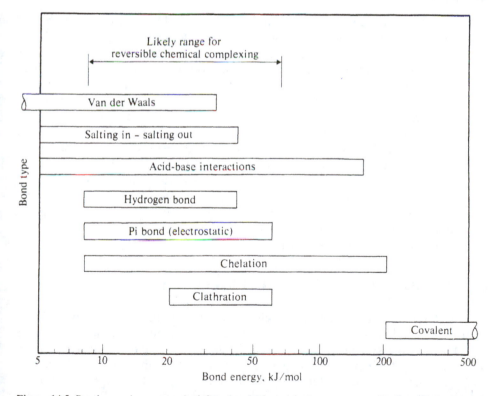

Figure 14-2 Bond energies most suited for chemical-complexing processes. (*Keller, 1977, courtesy of Dr. Keller.*)

arations involving chemical-complexing agents are mineral-oil dewaxing by crystallization involving urea adduct formation, absorption of CO_2 with ethanolamines, use of a salting-out process to overcome the HCl–water azeotrope in recovery of hydrogen chloride, and the use of cuprous ammonium acetate for extractive distillation of butadiene and butenes.

It has been noted by Keller (1977) and Mix et al. (1978) that high-cost large-scale separations in the chemical and petrochemical industries tend to fall in certain basic categories, i.e., more saturated and less saturated (paraffin and olefin, olefin and diene or acetylene); mixtures of isomers; mixtures of water and polar organics; and mixtures with overlapping boiling ranges, e.g., aromatics and nonaromatics. Complexing agents that have been found effective for one separation in a category would be logical candidates for another separation in the same category.

Some of the potential problems with chemical-complexing processes are gaining the right degree of reaction reversibility, avoiding side reactions and instability of the complexing agent, achieving sufficient reaction capacity for the material to be separated, obtaining a fast enough reaction rate for stage efficiency or tower height not to be affected too adversely, and maintaining a reasonable cost of the complexing agent (Keller, 1977).

Experience

Development of a new separation process necessarily requires research and laboratory-scale testing. Additional research and development will probably also be required when a known separation process is used for a new mixture. Installation of the first large-scale unit for separating a mixture on a commercial scale by a new process will involve a certain amount of uncertainty in design and reliability of plant operation. In view of these factors there is an understandable tendency for designers to stick with the better-known separation processes or with those which have been proved in the past for a particular application. As new processes become more developed and have been used successfully on more occasions, they become a more important component of the spectrum of separation processes to be considered for new plants. However, before that point, the projected value added by a relatively untried separation process must more than offset the costs of additional testing, development, and uncertainty. Industrial economics tend to work out so that initial installations of a new processing approach are more attractive on a smaller scale, and perhaps in some quite different service, such as waste treatment.

GENERATION OF PROCESS ALTERNATIVES

The introduction of novel separation techniques can be limited by conception of the initial idea as well as by the need for extensive development before large-scale use. Several qualitative, systematic techniques exist for structuring one's thinking to facilitate conception of new processing ideas: these include morphological analysis, functional analysis, and evolutionary techniques (King, 1974a). Morphological analysis involves systematic generation of alternative ways of fulfilling the various functions which must be included in any process for the goal desired and then looking at all combinations of those alternatives. An example is given in the discussion of fruit-juice concentration and dehydration, below. For separations, it is helpful to think of new property differences on which separations can be based, as well as new flow configurations and types of equipment internals. In many cases innovations which have been made in one type of process can be carried over to quite different classes of processes through a form of technology transfer.

Rochelle and King (1978; see also Rochelle, 1977) explore the use of morphological analysis, technology transfer, and evolutionary techniques to generate new possibilities for desulfurization of flue gases from fossil-fuel-fired power plants.

ILLUSTRATIVE EXAMPLES

In this section we consider two important separation problems, with the aim of identifying the important characteristics of the separation problem and determining the differences between the molecules to be separated which led to the choice of particular separation processes as most suitable for that application. The examples are chosen to be quite different from each other. They are typical of other separation problems except that distillation plays a less prominent role than it ordinarily would. Separations for which distillation has drawbacks have been chosen in order to bring a wide variety of processes into consideration.

Separation of Xylene Isomers

Xylenes are dimethylbenzenes and exist as three distinct isomers, depending upon the relative positions of the methyl groups on the aromatic ring:

Ortho	Meta	Para

As shown in Fig. 1-8, the xylenes are obtained commercially from the mixed hydrocarbon stream manufactured in naphtha reforming units in oil refineries. There is also some (but much less) production from coke-oven gas in steel mills. p-Xylene production in 1977 was estimated to be about 1.6×10^9 kg/year in the United States (Debreczeni, 1977), with an approximate value of 29 cents per kilogram (*Chem. Mark. Rep.*, 1977). p-Xylene is used as a raw material for the manufacture of terephthalic acid and dimethyl terephthalate, both used to manufacture polyester synthetic fibers:

Terephthalic acid

Dimethyl terephthalate

o-Xylene was produced to the extent of about 5.5×10^8 kg/year in the United States in 1977, with a value of about 25 cents per kilogram. It is used as a raw material for the manufacture of phthalic anhydride,

which is used in turn for the manufacture of dioctyl phthalate and other phthalates, which are used as plasticizers for polyvinyl chloride. Plasticizers are incorporated into polyvinyl chloride goods in order to impart flexibility and elasticity. Phthalic anhydride is also made from naphthalene. m-Xylene is almost entirely used for gasoline blending and conversion into the other isomers through isomerization, although some uses for petrochemical production are being explored.

p-Xylene has the greatest demand in relation to the availability of the various isomers from refinery streams. o-Xylene is a relatively pure side product from the purification of p-xylene. Since its production along with p-xylene exceeds the current demand for o-xylene, some o-xylene is returned to gasoline blending. There is a rapid growth in the demand for p- and o-xylenes, since they are so central to the plastics industry (Debreczeni, 1977).

Various properties of the three xylene isomers and of ethylbenzene,

which is also an isomer of the xylenes, are shown in Table 14-2. The free energies of formation of the various isomers are such that m-xylene is the most abundant isomer

Table 14-2 Properties of xylenes and ethylbenzene†

	o-Xylene	m-Xylene	p-Xylene	Ethylbenzene
Amount in equilibrium mixture at 1000 K, %	23	43	19	15
Boiling point, K	417.3	412.6	411.8	409.6
Freezing point, K	248.1	225.4	286.6	178.4
Change in boiling point with change in pressure, 10^{-4} K/Pa	3.73	3.68	3.69	3.68
Dipole moment, 10^{-28} C/molecule	2.1	1.2	0	
Polarizability, 10^{-31} m³	141	141.8	142	
Dielectric constant	2.26	2.24	2.23	2.24
Surface tension at 293 K, mJ/m²	30.03	28.63	28.31	29.04
Molecular weight	106.16	106.16	106.16	106.16
Density at 293 K, Mg/m³	0.8802	0.8642	0.8610	0.8670
Density at critical point, Mg/m³	0.28	0.27	0.29	0.29
Latent heat of vaporization at boiling point, kJ/kg	347	343	340	339

† Data from " Handbook of Chemistry and Physics," " Encyclopedia of Chemical Technology," " Chemical Engineers' Handbook," and Landolt-Börnstein.

in the equilibrium mixture. Since the species are isomers of each other, they have identical molecular weights, and hence any separation process dependent upon molecular-weight differences for the separation factor will fail. The three xylene isomers do differ somewhat in polarity because the bond between the methyl group and the aromatic ring is somewhat polar. In p-xylene these two dipolar bonds oppose each other and the net dipole moment is zero. In o-xylene the dipolar bonds are aligned in nearly the same direction and the net dipole moment is greatest. Even though there is a difference in dipole moments, it is not great (the dipole moment of phenol is 4.8×10^{-28} C, for example). The strength of intermolecular forces between the various xylene isomers does not vary greatly, as indicated by the very similar dielectric constants. Hence processes dependent upon differences in intermolecular forces can be expected to provide separation factors close to unity, although it may be possible to make some use of the difference in dipole moments.

The boiling points are quite close together. From the boiling points and the changes in boiling points with respect to pressure it can be computed that the relative volatility of m- or p-xylene to o-xylene is about 1.16, whereas the relative volatility of p-xylene to m-xylene is 1.02. The ortho isomer is different enough in volatility for a separation of o-xylene from m-xylene by distillation to be practicable, although a reflux ratio of 15 to 1 and 100 or more plates are required to accomplish the distillation. The difference in volatilities between the ortho isomer and the other isomers in this case is a reflection of the difference in dipole moments; the higher dipole moment of o-xylene causes some preferential alignment of the molecules in the liquid phase and reduces the volatility somewhat in comparison with the other two isomers. Figure 14-3 shows a xylene splitter distillation column used to separate o-xylene from the other isomers.

The relative volatility between p- and m-xylene is so slight that separation by distillation is out of the question. Considering the headings of the different molecular property columns in Table 14-1, it is apparent that the only property in which the two isomers differ is the molecular shape. p-Xylene is a narrow molecule, with the methyl groups at either end. m-Xylene is more nearly spherical because of the position of the methyl groups. The separation process most dependent upon molecular shape at a fixed molecular volume is *crystallization*. The difference in molecular shapes has two effects: (1) p-xylene molecules can stack together more readily into a crystal structure because of their symmetrical shape, and as a result p-xylene has a much higher freezing point (286.6 K) than any of the other isomers; (2) the difference in shape between p- and m-xylene means that m-xylene molecules cannot fit easily into the p-xylene crystal structure in the solid phase. As a result, the solid phase formed by partial freezing of a mixture of the two isomers contains essentially pure p-xylene, and the separation factor for a crystallization process is very high indeed.

Crystallization has classically been the most common method for separating p-xylene from m-xylene commercially, following removal of o-xylene by distillation. A number of different crystallization processes and plants for p-xylene manufacture have been described (Anon., 1955; Findlay and Weedman, 1958; Anon., 1963; McKay et al., 1966; Brennan, 1966; Anon., 1968).

The phase diagram for binary p-xylene–m-xylene mixtures is shown in Fig. 14-4. The eutectic compositions in a ternary mixture of all three xylene isomers are shown

Figure 14-3 A xylene splitter column at the Richmond, California, refinery of Standard Oil Company of California. This column takes *m*- and *p*-xylenes as distillate and *o*-xylene as bottoms. Notice the large size compared with other columns. (*Chevron Research Company, Richmond, California.*)

in Fig. 14-5. For the binary system the eutectic contains 13 percent *p*-xylene and freezes at 221 K. Figure 14-5 shows the phases which exist in equilibrium during partial freezing of mixtures of various compositions. The ternary eutectic (minimum-freezing) mixture occurs at 30.5% ortho, 61.4% meta, and 8.1% para and freezes at 208 K. Until a binary eutectic line is reached, the solid phase consists of a pure isomer. No solid solutions are formed.

The presence of the eutectic makes it difficult to recover a high fraction of the entering *p*-xylene in a crystallization process. If the *o*-xylene is removed from the high-temperature equilibrium mixture of isomers and the ethylbenzene is removed or was absent in the first place, the resulting binary mixture of *p*- and *m*-xylene contains 31% *p*-xylene. Since the binary eutectic contains 13% *p*-xylene, the maximum percentage of the *p*-xylene in the feed which can be frozen out before the eutectic starts to form can be calculated as

$$(f_{\text{para}}) = \frac{0.87 - 0.69}{0.87(0.31)} = 67\%$$

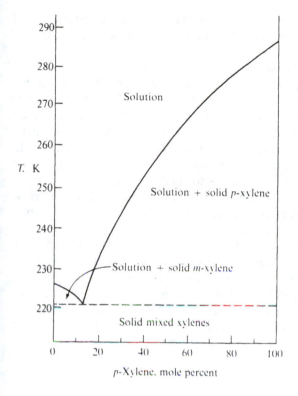

Figure 14-4 Phase diagram for *p*-xylene–
m-xylene system. (*Data from Egan and
Luthy, 1955.*)

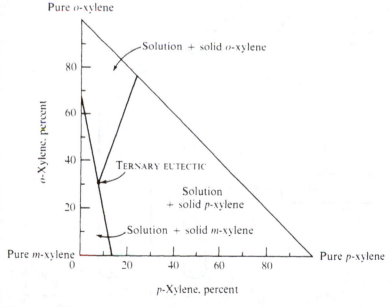

Figure 14-5 Ternary eutectic diagram for mixed xylenes. (*Data from Pitzer and Scott, 1943.*)

Because of the need for avoiding too close an approach to the eutectic and because of incomplete physical separation of the crystals from the supernatant liquid, the recovery of *p*-xylene will in reality be less.

Since the separation of xylene isomers by crystallization provides such a high separation factor, and since molecular shape is the only substantial difference between the isomers, most efforts for increasing *p*-xylene recovery until recently have involved improvements on the basic crystallization process. The most common modification has involved recycle of the supernatant residual xylenes to a catalytic isomerization reactor operating at 640 to 780 K (Brennan, 1966; Prescott, 1968). In the isomerization reactor there is a net conversion of *m*-xylene into *p*-xylene. If *o*-xylene is also recycled to the isomerization reactor and the crystallization step is repeated, an essentially complete conversion of the mixed-xylenes feed to a *p*-xylene product is possible. A flow diagram of a combined crystallization-isomerization process (Prescott, 1968) is shown in Fig. 14-6.

Referring again to Table 14-1, we see that another way to take advantage of differences in molecular shape is through partial solidification processes involving interaction with an appropriate mass separating agent. An example is the process of *clathration*, which has been investigated for the separation of xylene isomers (Schaeffer and Dorsey, 1962; Schaeffer et al., 1963). It has been found (Schaeffer and Dorsey, 1962) that nickel[(4-methylpyridine)$_4$(SCN)$_2$] selectively clathrates *p*-xylene, giving a recovery of 92 percent of the *p*-xylene in a 64 percent pure product in a single-stage

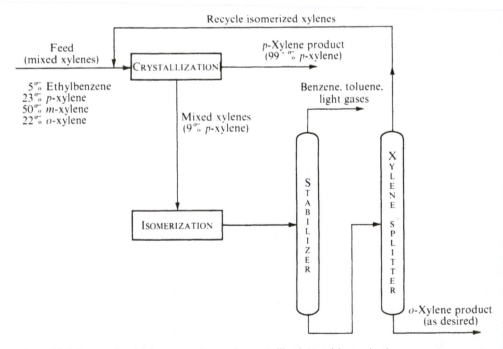

Figure 14-6 Process for *p*-xylene manufacture by crystallization and isomerization.

process. The expense of the clathrating agent has discouraged application of such a process. Another crystallization process involving addition of a mass separating agent is *adductive crystallization*, in which a substance is added which will form a solid compound preferentially with one of the species being separated (clathration can be considered a special case of adductive crystallization). Egan and Luthy (1955) have found that carbon tetrachloride will form a stoichiometric compound with p-xylene ($CCl_4 \cdot p$-xylene). If CCl_4 is added to the binary p-xylene–m-xylene eutectic in such a proportion as to give a solution containing 54% CCl_4, 6% p-xylene, and 40% m-xylene, the mixture will begin to freeze at 233 K and will deposit crystals of the $CCl_4 \cdot p$-xylene compound until reaching a eutectic freezing at 197 K and containing 54% CCl_4, 45% m-xylene, and 1% p-xylene. Thus a greater recovery of p-xylene in the crystallization step is possible at the expense of lower refrigeration temperatures and an additional step separating p-xylene from CCl_4. It has also been found that antimony trichloride will form a crystal preferentially with p-xylene (Meek, 1961).

Yet another approach for modifying crystallization through addition of a mass separating agent is *extractive crystallization*, in which a third component is added to alter the position of the binary eutectic without actually taking part in the solid phase. Findlay and Weedman (1958) describe how n-pentane can be added to a mixture of the p- and m-xylene isomers to shift the eutectic point. Subsequent removal of n-pentane restores the binary eutectic, and it is possible to achieve a complete separation by combining a crystallization with pentane and a crystallization without pentane into one process.

As noted in Table 14-1, it is also possible to generate a separation factor based on differences in molecular shape in membrane separation processes. Choo (1962) reports separation factors on the order of 2.0 for p-xylene over o-xylene and on the order of 1.3 for p-xylene over m-xylene for preferential passage through a low-density polyethylene membrane. Membrane separation processes have a disadvantage when applied to xylene mixtures, however, because of the difficulty of staging rate-governed processes and because of complications needed in order to provide a driving force which will cause p-xylene to cross the membrane (Michaels et al., 1967; see also Prob. 14-J).

Because of the lack of a marked difference in dipole moments and polarizabilities between xylene isomers, separations dependent upon the addition of a physically interacting solvent to modify vapor-liquid or liquid-liquid equilibria have not been particularly useful. For example, Wilkinson and Berg (1964) examined 40 different entrainers for azeotropic distillation and found that the best relative volatility between p- and m-xylene was 1.029, compared with 1.019 in the absence of any entrainer.

Since adsorption processes are more influenced by molecular-shape factors (Table 14-1), one would expect adsorption using a well-designed adsorbent to give a better separation factor for xylene isomers than can be obtained with physically interacting liquid solvents. Certain types of synthetic zeolites (molecular sieves) have been found particularly effective for this purpose (Anon., 1971; Broughton, 1977) because of the controlling effect of sizes and shapes of internal apertures on their adsorption properties. This discovery has been coupled with the development of improved means of approaching a continuous-flow adsorption process on a large scale (Broughton, 1977; Otani, 1973; see also Fig. 4-33). Over a period of less than 10 years the result has been at least 22 new industrial xylene-separation units (as of 1978) based upon molecular-sieve technology (Broughton, 1977). Thus adsorption is assuming much of the role formerly played by crystallization.

The only other successful approach to the separation of the xylene isomers has been to take advantage of differences in the ability of different isomers to take part in certain chemical reactions. These differences in reactivity are associated with steric effects resulting from the different relative positions of the methyl groups on the aromatic ring in the three isomers. A common organic-chemistry laboratory technique for separating the three isomers involves sulfonation with H_2SO_4 (Whitmore, 1951). In cold, concentrated sulfuric acid the ortho and meta isomers are sulfonated while the para isomer is unchanged. The sulfuric acid solution of the ortho and meta isomers is treated with $BaCO_3$ and Na_2CO_3 to eliminate excess H_2SO_4 and form the sodium salts of the sulfonates. The resulting solution is subjected to evaporative crystallization, whereupon the o-xylene sodium sulfonate compound precipitates first. The separation comes from the fact that the order of preferential sulfonation and the order of hydrolyzing tendency of the sulfonic acids both are meta > ortho > para. Several patents suggesting commercial processes have been based upon this behavior (Meek, 1961), but there has been no large-scale installation, most likely because of the need for consumption of expensive reactant chemicals or for elaborate reprocessing to recover them.

A more successful approach to chemical separation of the isomers involves reversible chemical complexing (Fig. 14-2). All three isomers react rapidly and reversibly with a mixture of hydrogen fluoride, HF, and boron trifluoride, BF_3, to form complexes (Meek, 1961). The relative stabilities of the complexes favor the form with m-xylene. The xylene complexes with $HF-BF_3$ are soluble in excess HF, but the unreacted xylenes are not; this leads to an extraction process based on immiscible phases. Figure 14-7 shows a flow diagram of a process using this behavior (Davis, 1971). m-Xylene is preferentially extracted into $HF-BF_3$. A nearly complete separation of the m-xylene from the other isomers is obtained by countercurrent staging. Since the m-xylene is removed at this point, p-xylene can be recovered and separated from ethylbenzene and o-xylene in a series of distillation steps, thereby avoiding a low-temperature crystallization process. The m-xylene complex with $HF-BF_3$ can be decomposed upon heating; hence decomposition of a portion of the extract from the extraction column gives a quite pure m-xylene product. The $HF-BF_3$ mixture also serves as a low-temperature isomerization catalyst. The remaining extract passes through an isomerization reactor, following which the $HF-BF_3$ is removed by decomposition from the isomerized xylenes. The recycle isomerized xylene stream is smaller than in the crystallization-plus-isomerization process because no p-xylene is fed to the isomerization reactor.

Saito et al. (1971) describe another chemical approach, based upon preferential trans alkylation of m-xylene with t-butylbenzene, catalyzed by $AlCl_3$ and carried out in a distillation column. The trans alkylation reaction produces benzene and t-butyl-3,5-dimethylbenzene. Conversion is promoted by driving off benzene in the distillation, while regeneration is accomplished by adding benzene to reverse the reaction.

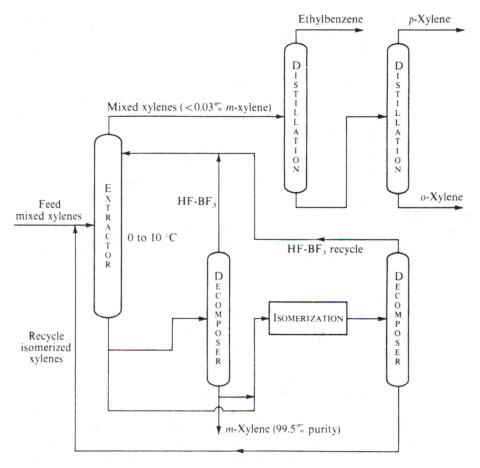

Figure 14-7 Japan Gas Chemical Co. process for xylene separation by HF–BF₃ extraction.

Concentration and Dehydration of Fruit Juices

Concentrated fruit juices are produced in very large quantity. In the United States the consumption of reconstituted juice concentrates is more than 4×10^9 kg/year equivalent of fresh juice (Tressler and Joslyn, 1971). This is greater than the production of p-xylene, which is one of the largest-volume petrochemicals.

There are two main advantages to concentration of fruit juices by removing between 60 and 99.5 percent of the water present: (1) a great economy in transportation and storage costs resulting from simple reduction in the volume and weight of the juice and (2) juice stability; a concentrated juice is more resistant to degradation of various kinds during storage than fresh juice under similar conditions.

A concentrated juice is reconstituted by the addition of cold water in the proper amount. Since the aim of juice concentration is to provide a reconstituted product that tastes and appears as much as possible like the original fresh juice, a juice-concentration process should remove water selectively. Ideally, components other than water should not be lost from the concentrate during processing, and no component should undergo chemical or biochemical change. This is a difficult goal to meet, in view of the fact that fruit juices are complex mixtures containing many substances.

Apple juice, for example, contains about 14 weight percent dissolved substances in the fresh juice (Tressler and Joslyn, 1971). The most prominent dissolved species are sugars; apple juice contains 4 to 8% levulose, 1 to 2% dextrose, and 2 to 4% sucrose. Also present in apple juice are malic acid and lesser amounts of other acids, along with tannins, pectins, enzymes, and other substances. The taste and aroma of a juice reflect the synergistic contributions of a vast number of volatile compounds present in the juice, which have been identified in the vapor given off by apple juice through flame-ionization gas chromatography, mass spectrometry, and other techniques (Flath et al., 1967).

Orange juice contains about 12 percent dissolved substances and about 0.5 percent suspended material; 5 to 10 percent sugars are present (Tressler and Joslyn, 1971). Sucrose is the most prominent sugar, levulose and dextrose being present to lesser extents. The most prominent acid is citric acid (about 1 percent). Numerous other nonvolatile components are present (pectins, glycosides, pentosans, proteins, etc.), along with a large number of volatile compounds. Table 14-3 lists some of the volatile components which have been identified in the equilibrium vapor over orange juice (Wolford et al., 1963). d-Limonene is the one compound which has been most directly related to characteristic orange aroma, although the other compounds marked with a dagger in Table 14-3 also have been shown to be prominent and important. Several hundred compounds have been identified in all.

The most common process for fruit-juice concentration is *evaporation*. Since the sugars and other heavier dissolved solids are all much less volatile than water, evaporation was a logical choice. It is a well-known and well-developed process and simple to carry out. Steam costs have always been reduced in practice through the use of multieffect evaporation. Despite the fact that evaporation is far and away the most common process, it has several problems:

1. Fruit juices have substantial thermal sensitivity and develop off-flavor and/or off-color when held at too high a temperature for too long a time. Ponting et al. (1964) indicate that most berry and fruit juices can be kept 2 or 3 h at 328 K without detectable flavor change. At higher temperatures the time is much less, typically under 1 min at 367 K and about 1 s at 389 K. Vitamin C in citrus juices is similarly heat-sensitive.
2. Again because of the thermal sensitivity of juices, there is a strong tendency toward fouling of heat-transfer surfaces (buildup of a semisolid layer next to the surface) in evaporators. This fouling reduces the heat-transfer coefficient across the evaporator surface and accentuates tendencies toward off-flavor because of the long residence time of the fouling layer.
3. The volatile flavor and aroma compounds escape readily from the juice during evaporation, causing a flat lifeless taste.

Approaches to dealing with these problems have followed two paths: improvement of evaporation processes and development of other kinds of separation processes.

Considering improvement of evaporation processes first, the most obvious approach toward overcoming the problem of too high a temperature for too long a

Table 14-3 Compounds present in the equilibrium vapor above Florida orange juice (data from Wolford et al., 1963)

Acetaldehyde	Ethyl n-caprylate	Methyl heptenol	2-Octenal
Acetone	Ethyl formate	Methyl isovalerate	n-Octyl butyrate
n-Amylol	Geranial	Methyl-n-methyl	n-Octyl isovalerate
Δ^3-Carene	Geraniol	anthranilate	α-Pinene†
trans-Carveol	n-Hexanal	β-Myrcene†	1-Propanol
l-Carvone†	2-Hexanal	Neral	α-Terpineol
Citronellol	n-Hexanol	Nerol	Terpinen-4-ol
p-Cymene	2-Hexenal	n-Nonanal	α-Terpinene
n-Decanal	3-Hexenol	1-Nonanol	γ-Terpinene
Ethanol	d-Limonene†	2-Nonanol	Terpinolene
Ethyl acetate	Linalool†	n-Octanal†	Terpinyl acetate
Ethyl butyrate	Methanol	n-Octanol†	n-Undecanal

† Proved to be closely associated with characteristic flavor.

time is *vacuum evaporation*. When the evaporation is carried out under reduced pressure, the boiling point of the juice occurs at a lower temperature and there is less thermal degradation. Another approach is to reduce the residence time of the juice in the evaporator as much as possible and to make the residence time of different elements of juice as uniform as possible. For this purpose a high heat-transfer surface-to-volume ratio is required, along with high heat-transfer coefficients and an avoidance of pockets or corners giving a long residence time for some of the juice. Turbulent flow in low-diameter tubes gives a relatively uniform velocity distribution and a high rate of heat transfer into the juice, keeping the residence time small (Eskew et al., 1951). In such an evaporator, condensing steam outside the tubes supplies the heat for evaporation. Another way to obtain rapid heating and minimum residence time is to preheat the juice by direct injection of steam (Brown et al., 1951). The steam for this purpose must be clean, however. Rapid heating in evaporators can give conditions approaching those which are needed in any event for pasteurization (Tressler and Joslyn, 1971; Brown et al., 1951).

The fouling problem can be minimized by clever evaporator design. A number of different approaches are discussed by Armerding (1966) and Morgan (1967). Carroll et al. (1966) have suggested radio-frequency heating as a means of avoiding heat-transfer surfaces altogether during the later phases of evaporation.

The third problem in evaporation, that of the loss of volatile flavor and aroma species, is the result of using a separation process that does not provide the desired division of the many components present in a fruit juice into the two products. For good product quality the volatile flavor and aroma species should remain with the juice-concentrate product, but instead they leave with the water vapor. Several approaches have been used for coping with this problem:

1. Adding fresh juice (called *cutback*) to the concentrate
2. Obtaining flavor material from peels, cores, etc., and adding it to the concentrate
3. Separating the volatile flavor and aroma compounds from the water vapor and returning them to the concentrate
4. Accomplishing the juice concentration by some process other than evaporation

The first of these approaches involves overconcentrating the juice before the cutback is added and necessarily gives a product which contains the volatile compounds perhaps to 10 percent, at most, of their natural level in the fresh juice. Nevertheless, a little bit of retained volatile flavor has a substantial effect on the attractiveness of the juice, and this approach has been successfully used. The second approach of obtaining flavoring material from peels, cores, etc., has been most successful for citrus juices; however, it is known that the flavoring material in peels differs in distinguishable respects from the natural juice flavor.

The volatile flavor components generally have a volatility greater than that of water (Bomben et al., 1973). Many of the compounds listed in Table 14-3 have boiling points higher than that of water, but they are sufficiently unlike water for their activity coefficients at high dilution in water solution to be very large. Thus for essentially all these compounds the product of the activity coefficient and the pure-component vapor pressure is substantially greater than the vapor pressure of water, and the relative volatility of the compound compared with that of water is much greater than unity. Consequently it is not surprising that distillation is the most common and best-developed method for separating the volatile flavor and aroma species from water vapor. Such distillation processes are called *essence-recovery* processes (Walker, 1961).

A flow diagram of an essence-recovery process (Bomben et al., 1966) is shown in Figure 14-8. Depending upon the degree of volatility of the most important flavor components, it may be possible to treat only the first portion of the vapor generated to recover the volatile compounds. In apple-juice processing, for example, the flavor species are sufficiently volatile to be located almost exclusively in the first 10 to 20 percent of the vapor generated. This vapor is fed to a distillation column with a long stripping section, which is present to provide a high recovery of the volatile compounds. The column is operated under vacuum to minimize thermal damage to the flavor compounds. The overhead liquid aroma-solution product contains only 0.5 to 1.0 percent of the water vapor which entered the column and may typically contain about half of the original amount of flavor species. Essence recovery is relatively successful, but it adds another separation process to the overall juice-concentration process and means an appreciable increase in steam requirements for a juice-concentration process (see Prob. 5-I). Also, in some cases, e.g., coffee extract, concentrated essences become chemically unstable, which can result in off-flavors.

We next turn to the question of alternative or supplementary processes to evaporation. One degree of freedom is the fraction of the total water removed. Evaporated concentrates are typically reduced in volume by a factor of 3 or 4, but any degree of concentration is, in principle, possible, ranging from less than this up to nearly complete dryness. A three- or four-fold concentrate must be kept at freezer temperatures for stability during storage. A greater extent of water removal would allow storage under less severe conditions. Efforts to market a liquid concentrate with a greater degree of water removal have been hampered by the difficulty of reconstitution resulting from the high viscosity of the product. Production and marketing of a fully dried natural juice product has been held back by the stickiness and hygroscopicity of the powder; however, dry juice powders are currently made on a small-scale and specialty basis.

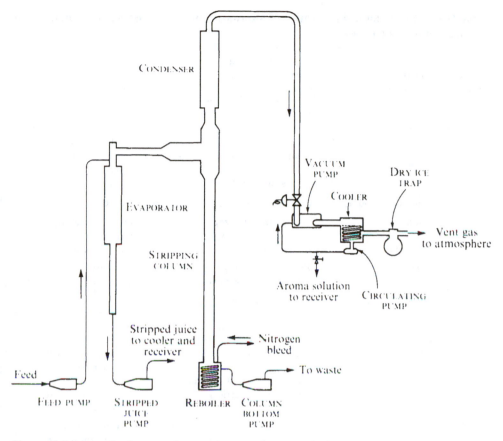

Figure 14-8 Schematic diagram of an essence-recovery process. (*Western Utilization Research and Development Division, Agricultural Research Service, USDA, Albany, California.*)

More possibilities for alternative water-removal processes can be generated by a form of morphological analysis (King, 1974a,b). Even though water is the major component in a fruit juice, it makes sense for a separation process to remove the water from the juice solutes. It is very unlikely that the alternative approach of removing everything else from the water could be sufficiently selective. If water is to be removed from the feed mixture, it is necessary that the water product be another phase, immiscible with the feed (equilibration processes) or that it be separated from the feed by a barrier (rate-governed processes). In either case, the chemical potential or activity of water in this product must be lower than that in the feed juice for transport of water into the second phase or across the barrier to take place. Different processes can be generated by considering different feed and product phases and different ways of creating the necessary chemical-potential difference.

Table 14-4 shows the processing alternatives which can be generated in this way. For equilibration processes, if the feed remains liquid and the water product is to be vapor, the necessary chemical-potential difference can be achieved by lowering the pressure of the water-vapor product, by increasing the temperature of the feed above

Table 14-4 Morphological generation of alternative processes for concentration and/or dehydration of fruit juices

Initial phase of water	Receiving phase for water	Means of creating chemical-potential difference	Example
		A. Equilibration processes	
Liquid	Vapor	Pressure (vacuum)	Flash evaporation
		Temperature (feed superheat)	Drying with superheated steam
		Composition (carrier)	Air drying
	Immiscible liquid	Composition (solvent)	Extraction
	Solid	Temperature (freeze)	Freeze concentration
		Composition (precipitate)	Clathration
		Composition (adsorb)	Solid desiccant
Solid	Vapor	Pressure (vacuum)	Ordinary freeze-drying
		Composition (carrier)	Carrier-gas freeze-drying
	Immiscible liquid	Composition (solvent)	
		B. Rate-governed processes	
Liquid	Vapor	Pressure (vacuum)	Pervaporation with compression
		Temperature (feed superheat)	
		Composition (carrier)	Pervaporation with carrier
	Liquid	Pressure (pressurize feed)	Reverse osmosis
		Composition (added solute)	Direct osmosis
		Composition (solvent)	Perstraction
		C. Mechanical processes	
Liquid		Screening	Pulp removal
		Density difference	Centrifugation before evaporation
Solid		Screening	Grinding and screening for frozen juices
		Density difference	Grinding and flotation in liquid of intermediate density, for frozen juices

Source: Adapted from King, 1974*a*, p. 21; used by permission.

the thermodynamic-saturation level, and/or by introducing a stripping gas, i.e., through pressure, temperature, or composition. Processes exemplifying these approaches are flash evaporation, drying using superheated steam as a vapor medium and heat supply, and hot-air drying. If the water-receiving phase is to be an immiscible liquid, it is difficult to create the chemical-potential difference through changes in temperature and pressure. One is then left with the use of a solvent to alter composition in the receiving phase, and this then leads to solvent extraction of water from the juice. If the water is to enter a solid phase, the most obvious approach is to lower the temperature and partially freeze the juice, but the morphological approach also suggests solidifying water through a composition effect, such as by adsorption of water onto a desiccant or by clathration. Achieving solidification by pressure change is difficult because the freezing point of water is relatively insensitive to pressure.

Alternatively, the feed can be converted into the solid phase by freezing the juice. With a vapor product, this leads to various forms of freeze-drying, where the water is removed by sublimation.

Similar alternatives exist for rate-governed processes. A process in which water (or some other substance) is vaporized across a selective membrane is known as pervaporation. One design complication in such processes is the need to supply the latent heat of vaporization to all portions of the membrane; this has held such processes back from commercial implementation (Michaels et al., 1967). Reverse osmosis is a rate-governed process with a liquid-water product, where the chemical-potential difference is created by raising the pressure of the feed stream enough to raise the chemical potential of water above that on the product-water side. The morphological analysis suggests that a membrane process with a liquid-water product can also be carried out by generating a temperature difference across the membrane or by adding a solute or miscible solvent on the product-water side in an amount great enough to diminish the chemical potential of water in that product below the chemical potential in the feed stream. The first of these possibilities is probably impractical because of the very large dissipation of heat, but the second alternative has received some attention. It is similar to dialysis and has also been called perstraction (Michaels et al., 1967). Since rate-governed processes with a solid feed seem impractical because of low transport rates in the solid phase, they have not been included in Table 14-4.

Juices like those from citrus fruits contain suspended matter. This leads to the possibility of separating pulp and/or "cloud" from serum first by a mechanical process and then being able to concentrate the serum under more severe conditions (Peleg and Mannheim, 1970). Alternatively, freezing gives a separation of ice crystals and residual amorphous concentrate on the microscale. Fine grinding can then lead to individual particles of ice and concentrate. Means of separating these from each other have been explored by Spiess et al. (1973).

Expanding the possibilities for rate-governed processes still further, Table 14-5 lists some of the potentially useful barriers. Especially interesting is the selective action which can be exerted by a dynamically formed surface layer. Particularly with carbohydrate solutions, such as juices, it has been found that achieving a surface layer of relatively low water content will serve to reduce greatly the diffusion of volatile flavor components, relative to diffusion of water, through that layer (Menting et al., 1970; Chandrasekaran and King, 1972). Thus once such a surface layer is formed, volatiles loss is greatly reduced, even if the subsurface material still has a high water content. Another example of using a different sort of selective membrane is osmotic dewatering of fruit pieces, where the fruit is placed in a solute-containing

Table 14-5 Barriers potentially useful for rate-governed dewatering processes

---

Filter media with extremely fine micropores:
  Ultrafiltration membranes
  Cellulose filters
  Irradiated polycarbonate filters

Membranes:
  Synthetic polymeric membranes (cellulose acetate, polyamide, etc.)
  Natural cell walls

Surface layers:
  Natural layers (skins, etc.)
  Dynamically formed surface layers of low water content
  Added surfactants forming a condensed surface phase

Source: Adapted from King, 1974a, p. 22; used by permission.

bath and water passes out selectively through natural cell walls (Farkas and Lazar, 1969; etc.).

It is instructive to compare the processes generated in Table 14-4 on the basis of selectivity of water removal. The vaporization equilibration processes suffer from the problem of volatiles loss. As already pointed out, this problem can be alleviated if a vaporization process can be carried out so that a layer of low water content forms rapidly at the surface of drying drops or particles. It turns out that freeze-drying has this property because of the prior concentration accomplished during the freezing step before drying (King, 1971). Also, volatiles retention in any evaporative drying process is promoted by lower water contents in the feed to the dryer; thus prior water removal by any aroma-retentive means is beneficial for lessening volatiles loss during a subsequent drying step. Among the nonvaporization processes, solvent extraction also runs the risk of volatiles loss, since the solvent will probably preferentially dissolve the organic volatile compounds. Here again, it has been found that rapid extraction of dispersed droplets will form a droplet surface layer of low water content, which is aroma-retentive (Kerkhof and Thijssen, 1974). However, like any mass-separating-agent process using a solvent, extraction runs the risk of contamination of the juice with residual solvent unless the solvent is food-compatible or can somehow be removed fully without detrimental solvent loss. Imposing a selective membrane can reduce solvent contamination and leads to perstraction as a rate-governed process. Membrane selectivities for water removal in reverse osmosis have been investigated by Merson and Morgan (1968) and by Feberwee and Evers (1970), using cellulose-acetate membranes. There is a substantial loss of low-molecular-weight polar organics, such as esters and aldehydes. For apple juice, where such compounds are predominant, the volatiles loss with the water permeate is marked, but for orange juice, which contains mainly terpenes (Table 14-3), the loss is much less.

Freezing gives the most selective removal of water at equilibrium. This is one of the benefits of the prior freezing step for freeze drying. It also leads to freeze-concentration (partial freezing, followed by settling, filtration, or centrifugation of ice crystals) as an attractive concentration process (Heiss and Schachinger, 1951; Muller, 1967). Although it is not yet used on a large scale for juices, freeze-concentration has found extensive use as a means of concentrating coffee extract before freeze-drying to give a highly aroma-retentive product. However, freeze-concentration must be carried out by indirect cooling; direct contact with a volatile refrigerant or cooling through vaporization of water (Prob. 13-E) would lead to a loss of volatiles.

Low operating temperatures also give an advantage to freeze-concentration, freeze-drying, other vacuum-drying processes, and membrane processes from the standpoint of minimizing thermal degradation of flavor, color, and nutrient content. However, at these lower temperatures concentrated juices become quite viscous. In reverse osmosis, this high viscosity results in low mass-transfer coefficients and there-by raises severe problems of concentration polarization, wherein the liquid adjacent to the membrane surface develops a much higher solute concentration than the bulk liquid. The high concentrate viscosity also makes it very difficult to wash residual concentrate from the ice crystals in freeze concentration. Compounding this problem is the fact that the ice crystals tend to be very small. A pulsed, pressurized wash column is one way of coping with this problem (Vorstmann and Thijssen, 1972). Even a small amount of entrained concentrate can be highly detrimental econo-mically because of the substantial product value. This point is illustrated in Examples 3-1 and 3-2.

Huang et al. (1965–1966) and Werezak (1969) have explored the use of hydrate formation, or *clathration*, to achieve formation of a solid phase at a higher tem-perature and hence with a lower solution viscosity. Methyl bromide, trichlorofluoromethane, 1,1-difluoroethane, ethylene oxide, and sulfur dioxide have been among the clathrating agents studied. Werezak (1969) has found that a clathra-tion process can form larger crystals than a freezing process in some instances, but the situation is usually the other way around, most likely because of slow mass transfer of the hydrating agent through the aqueous phase to the growing crystals, due to its low solubility. A clathration process involves addition of a mass separating agent, which must not be toxic and which must be removed as completely as possible from the juice concentrate product. It would be difficult to remove the clathrating agent from the concentrate without substantial loss of volatile flavor and aroma components.

Among processes for full dehydration of fruit juices, spray-drying has been plagued by the stickiness problem, volatiles loss, and thermal degradation. The two processes which have been used commercially and semicommercially to make an attractive product are freeze-drying and foam-mat drying (Ponting et al., 1964). Freeze-drying provides a porous product with good volatiles retention, which rehy-drates readily; however, careful packaging is required to avoid caking and/or dis-coloration, and for many juices freeze-drying must be carried out at very low temperatures to avoid product collapse (Bellows and King, 1973). Foam-mat drying is shown schematically in Fig. 14-9. A foaming agent is added, and the juice or juice concentrate is then blown with air or inert gas to give a stable foam, which is then dried to give a porous, easily rehydrated product. Rapid drying of thin foam films minimizes thermal degradation, but volatiles loss can be a problem.

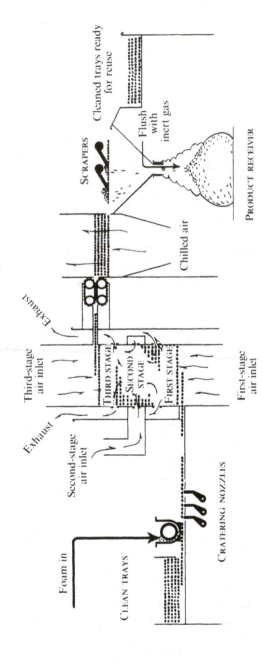

Figure 14-9 Moving-belt foam-mat drying process. (*Western Utilization Research and Development Division, Agricultural Research Service, USDA, Albany, California.*)

756

SOLVENT EXTRACTION

Solvent extraction illustrates several aspects of process selection which arise once the basic means of separation has been chosen, i.e., selection of an appropriate mass separating agent, selection of overall process configuration, and selection of equipment type.

Solvent Selection

Among the desirable features for an extraction solvent are the following:

1. It should have a high capacity for the species being separated into it. The higher the solvent capacity, the lower the solvent circulation rate required.
2. It should be selective, dissolving one or more of the components being separated to a large extent while not dissolving the other components to any large extent.
3. It should be chemically stable; i.e., it should not undergo irreversible reactions with components of the feed stream or during regeneration.
4. It should be regenerable, so that the extracted species can be separated from it readily and it can be reused again and again.
5. It should be inexpensive to keep the cost of maintaining solvent inventory and of replacing lost solvent low.
6. It should be nontoxic and noncorrosive and should not be a serious contaminant to the process streams being handled.
7. It should have a low enough viscosity to be pumped easily.
8. It should have a density different enough from that of the feed stream for the phases to counterflow and separate readily.
9. It should not form so stable an emulsion that the phases cannot be separated adequately.
10. It should allow formation of immiscible liquid phases, even at the highest solute concentrations which could be encountered.

In some cases a solvent mixture may be used to derive properties that cannot be achieved with pure solvents. Gerster (1966) discusses solvent selection in more detail.

Obviously no solvent will be best from all of these viewpoints, and the selection of a desirable solvent involves compromises between these various factors, e.g., between capacity and selectivity.

The separation factor for a liquid-liquid extraction process is given by the ratio of the activity coefficients of components i and j in liquid phases 1 and 2

$$\alpha_{ij} = \frac{\gamma_{i2}\gamma_{j1}}{\gamma_{i1}\gamma_{j2}} \tag{1-16}$$

This separation factor indicates the tendency for component i to be extracted more readily from phase 2 into phase 1 than component j is. If the solvent employed is not very soluble in the feed phase (denoted phase 2), the activity coefficients of components i and j in phase 2 will be nearly independent of the nature of the solvent. Consequently the selectivity between components exerted by the solvent will be determined by the ratio of the activity coefficients of the components in phase 1. This ratio can be called the selectivity S_{ij} of the solvent:

$$S_{ij} = \frac{\gamma_{j1}}{\gamma_{i1}} \tag{14-1}$$

The solubility of the preferentially extracted solute in the solvent phase, or the capacity of the solvent for the extracted solute, is also related to activity coefficients of the component being extracted:

$$\frac{x_{i1}}{x_{i2}} = \frac{\gamma_{i2}}{\gamma_{i1}} \tag{1-15}$$

Equation (1-15) gives the solubility of component i in phase 1 at equilibrium. Since the activity coefficient of the transferring solute in the feed phase (again denoted phase 2) is relatively independent of the nature of the solvent, the capacity of any solvent for the transferring solute will be related primarily to the activity coefficient of the solute in the solvent phase, the capacity of the solvent increasing as the activity coefficient of the solute in the solvent phase decreases.

Physical interactions The theory of regular solutions developed by Hildebrand (Hildebrand et al., 1970) leads to the following expression for activity coefficients in a liquid phase (Prausnitz, 1969), known as the *Scatchard-Hildebrand equation*:

$$\ln \gamma_i = \frac{V_i(\delta_i - \bar{\delta})^2}{RT} \tag{14-2}$$

where

$$\bar{\delta} = \frac{\sum\limits_{j=1}^{R} \delta_j V_j x_j}{\sum\limits_{j=1}^{R} V_j x_j} \tag{14-3}$$

In these equations V_i is the molal volume (or reciprocal molar density) of component i and is assumed to be the same as the partial molal volume of that component in solution; R is the gas constant, and T is the absolute temperature; δ_i, known as the *solubility parameter* of component i, is also the square root of the cohesive energy density of component i in the pure state. The cohesive energy density is a measure of the strength of intermolecular forces holding molecules together in the liquid state per unit volume of liquid and is given by the ratio of the latent energy of vaporization $\Delta E_v = \Delta H_v - P \Delta V_v$ of a pure component to the molal volume of that component:

$$\delta_i = \left[\frac{(\Delta E_v)_i}{V_i} \right]^{1/2} \tag{14-4}$$

x_j is the mole fraction of component j, and hence $V_j x_j / \sum\limits_{j=1}^{R} V_j x_j$ in Eq. (14-3) is the volume fraction of component j in a liquid mixture. Following Eq. (14-3), $\bar{\delta}$ in Eq. (14-2) is the volume average solubility parameter of all components present in the liquid phase in question. For a binary system of i and j, Eq. (14-2) for either component in a liquid phase becomes

$$\ln \gamma_i = \frac{V_i(x_j V_j)^2}{RT(x_i V_i + x_j V_j)^2} (\delta_i - \delta_j)^2 \tag{14-5}$$

and

$$\ln \gamma_j = \frac{V_j(x_i V_i)^2}{RT(x_i V_i + x_j V_j)^2} (\delta_i - \delta_j)^2 \tag{14-6}$$

Table 14-6 shows solubility parameters for various selected organic compounds. More extensive tabulations of solubility parameters are available (Gardon, 1966; Hildebrand et al., 1970). Solubility parameters are generally reported for 298 K, and can be calculated from measured volumes and latent heats of vaporization interpolated or extrapolated to 298 K. Since $P \Delta V_v$, where V_v is the volume difference between the gaseous and liquid states, is usually very nearly given by RT, ΔE_v can usually be computed as $\Delta H_v - RT$. The latent heat and molar volume can be estimated from various correlations (Reid et al., 1977) when measured values are not available. Lyckman et al. (1965) give correlations for predicting solubility parameters and molal volumes from the theory of corresponding states, and Rheineck and Lin (1968) suggest a group-contribution method for prediction of solubility parameters. Konstam and Feairheller (1970) also discuss calculation of solubility parameters for polar substances.

Table 14-6 Values of solubility parameters at 298 K† (data from Hildebrand et al., 1970, and Gardon, 1966)

	V, cm^3/mol	δ, (cal/cm^3)$^{1/2}$		V, cm^3/mol	δ, (cal/cm^3)$^{1/2}$
Water	18	23.2	Ethyl bromide	76	8.9
Ethylene glycol	56	15.7	Carbon tetrachloride	97	8.6
Phenol	88	14.5	Ethyl chloride	73	8.5
Methanol	40	14.5	Cyclohexane	109	8.2
Dimethyl sulfoxide	71	13.4	Cyclopentane	95	8.1
Nitromethane	54	12.6	Perfluorobenzene	115	8.1
Acetic acid	57	12.6	n-Hexadecane	295	8.0
Dimethyl formamide	77	12.1	Ethylene (169 K)	46	7.9
Acetonitrile	53	11.9	Methylcyclohexane	128	7.85
Furfural	83	10.9	CF$_4$ (145 K)	45	7.7
Aniline	91	10.8	n-Nonane	180	7.65
Benzaldehyde	101	10.8	Ethane (184 K)	55	7.6
Pyridine	81	10.7	Propylene (225 K)	69	7.6
Acrylonitrile	67	10.5	n-Octane	164	7.55
n-Butanol	92	10.4	Diethyl ether	105	7.5
Carbon disulfide	61	10.0	n-Heptane	147	7.45
Dioxane	86	10.0	n-Hexane	132	7.3
Acetone	74	9.9	cis-2-Butene	91	7.2
Nitrobenzene	108	9.9	Butadiene	88	7.1
Naphthalene	123	9.9	n-Pentane	116	7.05
1,2-Dichloroethane	79	9.8	trans-2-Butene	91	7.0
Chlorobenzene	102	9.5	Isooctane	166	6.85
Ethyl iodide	81	9.4	Methane (112 K)	38	6.8
Chloroform	81	9.3	2-Methylbutane	117	6.75
Styrene	115	9.3	Isobutene	94	6.7
Benzene	89	9.15	1-Butene	95	6.7
Ethyl acetate	99	9.1	Neopentane	122	6.25
o-Xylene	121	9.0	Perfluorocyclohexane	170	6.0
Toluene	107	8.9	Perfluoro-n-heptane	227	5.7

† Unless otherwise noted.

There have been a number of efforts to modify the solubility-parameter concept to take into account the different types of intermolecular forces (dipole-dipole, dipole-induced dipole, and dispersion forces; see Moore, 1963) as well as hydrogen bonding for the prediction of solubilities and activity coefficients (Prausnitz, 1969). In connection with the analysis of paint solvents Teas (1968) and others have suggested the use of triangular diagrams with axes corresponding to the ordinary solubility parameter, some measure of polarity, and some measure of hydrogen-bonding tendencies of any given substances. Prausnitz and coworkers (Prausnitz, 1969; Weimer and Prausnitz, 1965; Prausnitz et al., 1966) have developed an approach allowing for polarity and volume differences of molecules in predicting and analyzing activity coefficients through use of the Flory-Huggins parameter, the Wilson equation, and other concepts. Gardon (1966) has also suggested ways of allowing for polarity effects upon molecular interactions.

The development leading to the Scatchard-Hildebrand equation for predicting activity coefficients from solubility parameters assumes the molecules have similar sizes, undergo interaction through dispersion forces alone, and are not associated in solution (zero excess entropy of mixing). For the liquid mixtures encountered in extraction processes these assumptions often do not hold well, and Eq. (14-2) can be considered only a very rough first approximation; nonetheless, it has some use for screening extraction solvents and generalizing.

First of all, it is apparent from Eqs. (14-2), (14-5), and (14-6) that mixtures of components having nearly equal solubility parameters should exhibit activity coefficients near unity. Substances in solution with other components having a substantially different solubility parameter will have activity coefficients much greater than unity; if the solubility parameters are different enough, immiscibility may result. This thinking is in accord with the concept of similar molecules giving ideal solutions and dissimilar molecules giving strong positive deviations from ideality, as can be seen by judging the positions of different types of compounds in Table 14-6. Notice that polar molecules tend to have high solubility parameters while nonpolar molecules have low solubility parameters. As a rough approximation, substances must differ by about 3 $(cal/cm^3)^{1/2}$ or more in solubility parameter to generate two liquid phases. Thus paraffinic hydrocarbons are not miscible with aniline, furfural, dimethyl formamide, etc., but are miscible with most substances closer in solubility parameter.

Under special conditions even seemingly similar liquid phases can be made immiscible. One example of this is the separation of proteins, carbohydrates, and other biochemical substances by partitioning between two immiscible polymer-containing aqueous phases, e.g., a polyethylene glycol–water phase and a dextran-water phase (Albertsson, 1971). Organic solvents tend to denature proteins, and most proteins and carbohydrates are so hydrophilic that they are poorly extracted by organic solvents. Hence these two-aqueous-phase extractions can accomplish separations that cannot be made by conventional extraction.

The general relationship between solvent capacity and solvent selectivity for physically interacting solvent systems can be inferred from Eq. (14-2). If we suppose that species C is to be a solvent to extract B preferentially from solutions of A and B, the capacity of the solvent for B, given by Eq. (1-15), will decrease as the solubility parameter of C moves away from that of B. By Eq. (14-2) or (14-5), γ_B in the C-rich phase will increase as δ_C moves away from δ_B, and, since γ_B increases, Eq. (1-15) tells us that the solvent capacity for B decreases. On the other hand, the selectivity of extraction, given by Eq. (14-1), is related to solubility parameters by

$$\ln S_{BA} = \ln \gamma_A - \ln \gamma_B = \frac{V}{RT}[(\delta_A - \bar{\delta})^2 - (\delta_B - \bar{\delta})^2] \tag{14-7}$$

if V_A is assumed equal to V_B. Eq. (14-7) can be rearranged to

$$\ln S_{BA} = \frac{V}{RT}(\delta_A - \delta_B)(\delta_A + \delta_B - 2\bar{\delta}) \tag{14-8}$$

For C to be an effective extraction solvent we must have $\delta_A > \delta_B > \delta_C$ or $\delta_C > \delta_B > \delta_A$. As δ_C becomes more different from δ_A and δ_B, $\bar{\delta}$ must move away from δ_A and δ_B; the term in the right-hand-most parentheses of Eq. (14-8) will increase in absolute magnitude, and S_{BA} will become greater.

Thus for physically interacting solvents we have the interesting general observation that choosing a solvent with a solubility parameter more removed from the solubility parameters of the mixture being separated will enhance the solvent selectivity but reduce the solvent capacity. A compromise must be reached such that the solvent solubility parameter is far enough removed to give good selectivity (and to give immiscibility) but is not so different that the solvent has inadequate capacity.

Extractive distillation Regular-solution theory is somewhat more useful for analyzing the performance of solvents for extractive distillation, since in that case the solution nonideality is not strong enough to generate two liquid phases. For example, Gerster et al. (1960) measured selectivities and activity coefficients of 32 different solvents for effecting a separation of n-pentane and 1-pentene by extractive distillation. They found a strong correlation between the selectivity for the separation and the activity coefficient of n-pentane in the solvent. The selectivity increases with increasing activity coefficient, as predicted by regular-solution theory. For a given activity coefficient, hydrogen-bonding solvents gave somewhat less selectivity than non-hydrogen-bonding solvents.

Chemical complexing The regular-solution analysis illustrates why it is desirable to search for solvents which will chemically react, hydrogen-bond, or complex preferentially with the compound to be extracted. These effects are not accounted for in a physical-interaction analysis, and they have the desirable result of increasing the selectivity of the solvent while at the same time increasing its capacity. Thus a regenerable chemical base can be a more desirable solvent for removing a carboxylic acid from a hydrocarbon stream than a high-solubility-parameter physical solvent would be. Similarly, if acetone were to be removed from water by solvent extraction, chloroform would probably be preferable as a solvent to benzene (a compound with a solubility parameter close to that of chloroform) because chloroform hydrogen-bonds preferentially with acetone (see discussion preceding the extraction example in Chap. 7). On the other hand, chlorinated hydrocarbons are undesirable contaminants in effluent waters.

Candidate chemical-complexing mechanisms are outlined in Fig. 14-2.

An example Extraction of dilute acetic acid from aqueous streams is an important problem, in part because it is difficult to strip acetic acid from water. Dilute acetic acid solutions are found in many process effluents and are encountered in some proposed schemes for biological oxidation of solid waste material.

Since acetic acid is highly polar and water-loving, most conventional solvents which are immiscible with water give equilibrium distribution coefficients less than 1.0 for extraction of acetic acid from water (Treybal, 1973a). For example, ethyl acetate, which has often been used for extracting acetic acid at feed concentrations of 10 percent and greater, gives a distribution coefficient of about 0.90. Cyclohexanone and cyclohexanol, by virtue of their hydrogen-bonding abilities, give higher distribution coefficients, in the range of 1.2 to 1.3; however, cyclohexanol is highly viscous and cyclohexanone has a density very close to that of water. Therefore these solvents would probably be used only as components of solvent mixtures, along with diluents which improve the viscosity and/or density properties for extraction (Eaglesfield et al., 1953).

Solvents giving a higher equilibrium distribution coefficient for removal of acetic acid from water at low concentrations involve some additional chemical effect. As one example, Othmer (1978) has pointed out that acetic acid can be removed by extraction from aqueous effluents from chemical pulping processes in paper manufacture. These streams typically have high contents of salts and other dissolved solutes, such as sodium sulfate and lignosulfonates. Because of this high solute content it is possible to use acetone as a solvent, even though acetone is fully miscible with water in the absence of the other solutes. The resulting distribution coefficient for acetic acid is increased to the range 4.0 to 6.0. This is an example of a chemical salting-out effect, due to the presence of high concentrations of ionic salts.

Another chemical-complexing approach involves the use of regenerable organic bases, taking advantage of the acidity of acetic acid. Phosphoryl compounds, such as phosphates and phosphine oxides, are organic bases because of the directed nature of the $P \rightarrow O$ bond. Trioctyl phosphine oxide has been found to be an effective regenerable solvent for acetic acid extraction (Helsel, 1977), but it is comparatively expensive (about \$26 per kilogram). High-molecular-weight organic amines are also effective bases for acetic acid extraction and are about an order of magnitude less expensive (Wardell and King, 1978; Ricker et al., 1979a,b). Tertiary amines, such as tri(C_8 to C_{10})amines, are readily regenerable; secondary and primary amines give still higher distribution coefficients, but regeneration by distillation is hampered by irreversible formation of amides.

Amine and phosphine oxide solvents require that other substances be added to the solvent as diluents, both to dissolve the primary solvent and reduce the viscosity and/or density, if necessary, and also to provide a suitable solvating medium for the acid-base complex. Thus solvent mixtures of intermediate composition give much higher equilibrium distribution coefficients than either pure constituent in such cases. For the amine systems more polar compounds, such as alcohols and ketones, are the more effective diluents, from the standpoint of solvating the reaction complex. However, alcohols are subject to esterification with acetic acid during regeneration by distillation, and that reaction is difficult to reverse. For phosphine oxide solvents alcohol diluents diminish the solvent power because of preferential hydrogen bonding of the alcohol, rather than the acid, to the phosphoryl group, which is a strong hydrogen acceptor. A diluent such as a ketone, which is a hydrogen acceptor but not a hydrogen donor, is more effective (Ricker et al., 1979a).

Regeneration by back-extraction with an aqueous base, such as NaOH, is also a possibility, but in that case the chemical value of recovered acetic acid is diminished to that of sodium acetate.

Process Configuration

Once a separation method and a mass separating agent (if needed) have been chosen, there is still flexibility in picking the flow configuration of a separation process. As an example here, we shall use the basic idea of separating phenol from a dilute aqueous feed by extraction into an immiscible solvent, followed by phenol recovery and regeneration of the solvent by back-extraction into aqueous NaOH solution. Some of the schemes that can be used have been discussed by Boyadzhiev et al. (1977) and others and are shown in Fig. 14-10.

Scheme *a* is the straightforward approach of removing the phenol from the aqueous feed by countercurrent extraction with the solvent in a first column, followed by countercurrent back-extraction of the solvent with NaOH solution in a second column. The solvent circulation rate in such a process is limited by the maximum loading that can be achieved in the first column. For a strongly curved equilibrium relationship, a desirable alternative can be to withdraw a portion of the solvent part way along the regenerator and insert it part way along the first extraction column (scheme *b*). This enables the remaining solvent in the regenerator to be brought to a lower concentration of phenol (higher ratio of NaOH flow to solvent flow in the lower part of the regenerator), and this smaller but highly regenerated solvent stream can then be used to bring the aqueous effluent from the first column to a lower phenol content.

Scheme *c* involves recycle of individual solvent streams between isolated stages of the primary extractor and the regenerator (stage 1 paired with stage 1, stage 2 with stage 2, etc.) This leads to lower solvent flows in individual stages (Hartland, 1967). In scheme *d* the solvent is immobilized within a membrane, the feed flowing on one side and the NaOH regenerant on the other (Klein et al., 1973). This is a form of perstraction, mentioned earlier as an alternative for concentration of fruit juices. The system is now limited by the transport capacity of the membrane-solvent system.

In the *liquid-membrane process* [scheme *e*, see Cahn and Li (1974)] the NaOH solution is distributed as small droplets within larger drops of solvent, which rise through a downflowing continuous feed stream. This gives the benefits of the thin solvent membrane but in a form where a large interfacial area is more easily achieved than with a fixed-membrane device. The liquid-membrane process can also be viewed as an extraction process in which the solute capacity of a dispersed solvent has been increased by addition of islands of an irreversibly reactive material. The process does

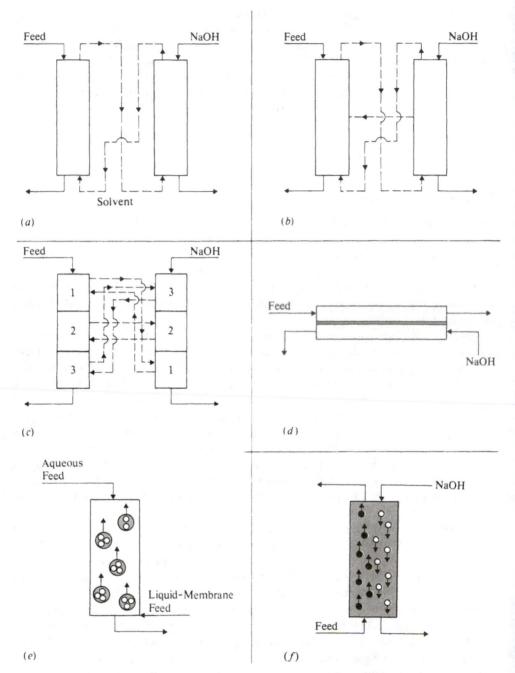

Figure 14-10 Alternative flow configurations for extraction of phenol from water, followed by regeneration with NaOH solution.

require stabilization of the solute-uptake droplets within the solvent drops, as well as facilities for separating both levels of liquid dispersion after the contacting.

Finally, scheme *f* is an approach where droplets of both the aqueous feed and the NaOH solute-uptake medium are dispersed in a continuous, nonflowing solvent phase (Boyadzhiev et al., 1977). Coalescence of the different kinds of drops is prevented by incorporation of appropriate surface-active agents. If the solvent has a density intermediate between that of the aqueous feed and that of the NaOH regenerant, countercurrent flow of the different kinds of drops can be achieved, in principle, as shown in Fig. 14-10*f*. This approach also presents design and operational problems.

Selection of Equipment

In Chap. 12 the relative merits of different sorts of tower internals for multistage gas-liquid contacting operations were considered in some detail (see Tables 12-1 and 12-2). In this section we explore ways of selecting an appropriate device for carrying out a liquid-liquid extraction process.

There are many different types of extraction equipment used in practice. Descriptions and comparison of these are given by Hanson (1968, 1971), Treybal (1963, 1973*b*), Akell (1966), Reman (1966), and Marello and Poffenberger (1950). Several of these devices are shown in Fig. 14-11. In summary, some of the different equipment types available are as follows.

1. *Spray column.* This is the simplest device to construct. The dispersed phase is sprayed as droplets into the continuous phase (the sprayer can be at the bottom of the column when the less dense phase is to be dispersed). The operation of these devices is hampered by a high degree of backmixing in the continuous phase.
2. *Packed column.* This is essentially a spray column with some form of divided packing inside it. The packing serves to reduce the backmixing in the continuous phase, but the backmixing is still important and hampers the action of a large number of transfer units.
3. *Plate columns.* Plate columns used for extraction are almost always perforated. The dispersed phase flows through the holes in the plates and collects on top of (for a heavy dispersed phase) or below (for a light dispersed phase) the next tray, which then redisperses the liquid. The discrete stages are effective for reducing backmixing.
4. *Pulsed column.* The contents of either a packed column or a plate column can be pulsed by applying intermittent surges of pump pressure to the column. This pulsing promotes mass-transfer rates within the column, both because of increased interfacial area (drop breakup) and increased mass-transfer coefficients. As a result, a pulsed column can give a specified separation in less tower height than an otherwise equivalent unpulsed column. The pulsed column provides some of the benefits of mechanical agitation without moving parts in the column. However, pulsing can increase axial dispersion.
5. *Baffle column.* This device (not shown in Fig. 14-11) is an open vertical column with various horizontal baffles built in at intervals along the height to reduce the extent of axial mixing. Common baffling devices are *disks and doughnuts*. The disks are solid horizontal circular plates, axially mounted and with a diameter less than that of the column. The doughnuts are horizontal annular rings attached to the walls of the column. The construction resembles that of the rotary disk contactor (RDC) column shown in Fig. 14-11, without the axial drive shaft.

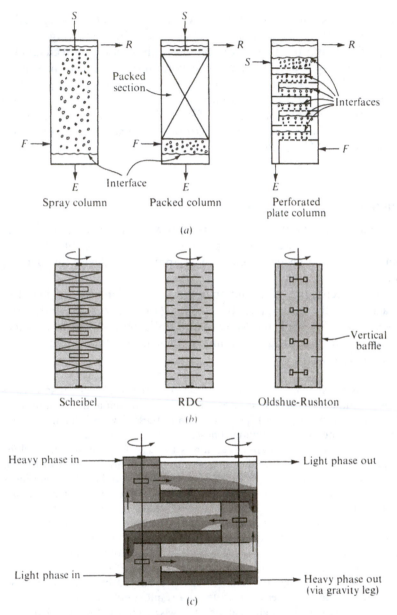

Figure 14-11 Varieties of extraction equipment: (*a*) unagitated column contactors; (*b*) mechanically agitated column types; (*c*) vertical type of mixer-settler.

6. *Mechanically agitated columns.* In these columns rotating agitators driven by a shaft extending axially along the column stir up the liquid phases, promoting drop breakup and mass transfer. Three varieties are shown in Fig. 14-11. In the *Scheibel column* regions agitated by axially mounted stirrers are separated vertically from each other by regions of wire mesh. The agitators promote dispersion and mass transfer. The mesh zones promote coalescence and phase separation to keep the light and heavy phases flowing in the desired

directions up and down the column. In the *RDC column* rapidly rotating horizontal disks serve to provide phase breakup and mass transfer through shear against the disks. Annular rings separate the rotating disk regions from each other to discourage backmixing effects. There is also an asymmetric rotating disk contactor (Hanson, 1968). The *Oldshue-Rushton* (Lightnin CMContactor) *column* uses turbine impellers, doughnuts, and vertical baffles to accomplish much the same result as the other devices in this category.

7. *Graesser raining-bucket contactor.* This is a unit quite different in concept, which is described by Hanson (1968). It consists of a large, slowly rotating, horizontal, cylindrical drum, inside of which are open " buckets" mounted on the cylinder wall. The two phases are stratified in the drum, filling it. The buckets catch quantities of either phase and transport them into the other phase, causing relatively large drops of each phase to fall or rise through the other. This gentle dispersion and the resultant easy settling are of use with systems which ordinarily do not settle easily because they tend to emulsify.

8. *Mixer-settler.* These devices provide separate compartments for mixing and for subsequent phase separation through settling. The mixing is usually accomplished by rotating mechanical agitators; however, one or both of the liquids may also be pumped through nozzles, orifices, etc., to cause the mixing (Treybal, 1973b). Mixer-settler devices generally give high mass-transfer efficiency, which makes reliable design possible using an equilibrium-stage analysis based solely on the equilibrium data of the system, no transfer-rate data really being required. Mixer-settler devices generally are more complex than other devices and occupy a relatively large volume. Figure 14-11 shows that it is possible to assemble mixer-settlers into a vertical staged configuration (Hanson, 1968).

9. *Centrifugal contactors.* These devices (not shown in Fig. 14-11) utilize centrifugal force to promote countercurrent flow of the phases past each other more rapidly than is possible through the action of gravity alone. The centrifugal force also promotes coalescence of droplets where that is difficult. Centrifugal extractors can provide several (but not many) equilibrium stages within a single device. A unique advantage of centrifugal extractors is the very short residence time of the phases in the device, a feature which is often attractive in the pharmaceutical industry. Different types of centrifugal contactor include the Podbielniak extractor, the Westfalia extractor, and the DeLaval extractor.

10. *Devices with two continuous liquid phases.* One of the newer types of extractor uses as internals a large number of long, continuous small-diameter fibers (Anon., 1974; Pan,

Table 14-7 Classification of extraction equipment

Countercurrent flow (if any) produced by	Gravity	Gravity	Gravity	Centrifugal force
Phase interdispersion produced by	Gravity	Pulsation	Mechanical agitation	Centrifugal force
Continuous-counterflow contacting devices	Spray column, packed column, baffle column, two-continuous-phase devices	Pulsed packed column	RDC contactor, Oldshue-Rushton column, Graesser raining-bucket contactor	Podbielniak extractor, Westfalia extractor, DeLaval extractor
Discrete-stage contacting devices (coalescence-redispersion cycle)	Perforated plate column	Pulsed plate column	Mixer-settler, Scheibel column, Treybal column	

Source: Adapted from Hanson, 1968, p. 83; used by permission.

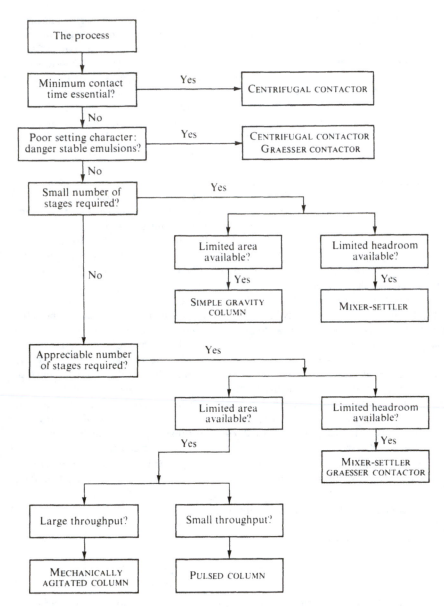

Figure 14-12 Selection guide for choosing extraction devices. (*Adapted from Hanson, 1968; p. 90; used by permission.*)

1974). One of the liquid phases wets the fibers preferentially and flows axially along them, while the other phase flows continuously in the interstices, in either cocurrent or counter-current flow. This flow scheme largely avoids the formation of droplets and is therefore effective for handling systems that are difficult to settle when a dispersion of droplets is formed. In one such device the fibers are about 50 μm in diameter and can be made of steel, glass, or any of various other materials that can be formed into fibers.

Table 14-8 Advantages and disadvantages of different extraction equipment (data from Akell, 1966)

Class of equipment	Advantages	Disadvantages
Mixer-settlers	Good contacting Handles wide flow ratio Low headroom High efficiency Many stages available Reliable scaleup	Large holdup High power costs High investment Large floor space Interstage pumping may be required
Continuous counterflow contactors (no mechanical drive)	Low initial cost Low operating cost Simplest construction	Limited throughput with small density difference Cannot handle high flow ratio High headroom Sometimes low efficiency Difficult scaleup
Continuous counterflow (mechanical agitation)	Good dispersion Reasonable cost Many stages possible Relatively easy scaleup	Limited throughput with small density difference Cannot handle emulsifying systems Cannot handle high flow ratio
Centrifugal extractors	Handles low density difference between phases Low holdup volume Short holdup time Low space requirements Small inventory of solvent	High initial costs High operating cost High maintenance cost Limited number of stages in single unit

Table 14-9 Order of preference for extraction contacting devices

Factor or condition	Preferred device(s)	Exceptions

Very low power input desired:
  One equilibrium stage
  Few equilibrium stages
  Many equilibrium stages
	
	Spray column
	Baffle column
	Perforated plate column, packed column	

Low to moderate power input desired, three or more stages:
  General and fouling service
  Nonfouling service requiring low residence time or small space
	
	Columns with rotating stirrers
	Centrifugal extractors, columns with rotating stirrers
		
		Strongly emulsifying systems

High power input	Centrifugal extractors	Use mixer-settlers for one to two stages
High phase ratio	Perforated plate column, mixer-settler	
Emulsifying conditions	Centrifugal extractors	Fouling systems
No design data on mass-transfer rates for system being considered	Mixer-settlers	
Radioactive systems	Pulsed extractors	

Source: Adapted from Oliver, 1966, p. 363; used by permission.

Table 14-7 classifies the various types of extractors by their distinguishing physical features.

Numerous authors have presented selection criteria for extraction equipment. A scheme of selection logic proposed by Hanson (1968) is shown in Fig. 14-12. Some of the reasons underlying the decision criteria indicated should be apparent from the foregoing summary of different equipment types. For comparison with Hanson's selection scheme and for augmentation of other factors not included in it, a list of advantages and disadvantages of different classes of equipment is shown in Table 14-8. Yet another selection list covering different devices is shown in Table 14-9. Selection criteria for extractors are also discussed by Reissinger and Schröter (1978).

SELECTION OF CONTROL SCHEMES

Any large-scale separation process requires a control scheme to assure relatively smooth operation in the face of upsets and to maintain product specifications. Analysis and selection of control systems is a complex field and largely beyond the scope of this book. However, it is true that the evaluation of control schemes can interact closely with process selection and evaluation. In the extreme, there are some separation processes which may seem attractive on the basis of steady-state analysis but which are not chosen for plant use because they are very difficult or impossible to control.

The number of control loops and the types of control loops which can be used with a separation process are determined by the same kind of thinking as enters into the application of the *description rule* (Chap. 2 and Appendix C). No more variables can be controlled than are necessary to specify the operation of the process fully. Installing a greater number of control loops will cause the operation of the process to cycle and probably become unstable, because an effort is being made to specify more independent variables than is possible. Installing fewer control loops than the number of specified variables will mean that the operation of the process cannot be well specified and that output variables will wander; also the process may not operate as smoothly as it would with a full control system. The installation and use of the control system may be looked upon as fixing the *operating* portion of those variables which are set by construction or controlled during operation by independent, external means.

In the use of the description rule for problem specification the variables chosen must be truly independent. No subset of specified variables should be uniquely related and determined by a subset of equations describing the system. The same restriction holds for the selection of control loops for a separation process: the controlled variables must in fact be independent of each other. Thus it is generally not workable to place *both* the products from a separation process on flow control, since these product flows are uniquely related by the overall material balance for the process. The feed flow rate will change from time to time (it will change somewhat even if it is under flow control itself), and it will therefore not be possible to maintain both product flows at the set-point values. The result will be oscillatory operation.

Control and dynamic behavior of distillation columns and other separation processes are reviewed by Buckley (1964) and Harriott (1964). The control of distillation columns is explored in more extensive detail by Rademaker et al. (1975) and Shinskey (1977). Some of the more practical aspects of distillation control are discussed by Lieberman (1977).

REFERENCES

Akell, R. B. (1966): *Chem. Eng. Prog.*, **62**(9):50.

Albertsson, P. A. (1971): "Partition of Cell Particles and Macromolecules," 2d ed., Wiley-Interscience, New York.

Anon. (1955): *Chem. Eng.*, December, p. 128.

———— (1963): *Chem. Eng.*, Aug. 5, p. 62.

———— (1968): *Chem. Eng.*, Feb. 26, p. 94.

———— (1971): *Chem. Eng.*, July 26, p. 68.

———— (1974): *Chem. Eng.*, Sept. 30, p. 54.

Armerding, G. F. (1966): in G. F. Stewart and E. Mrak (eds.), "Advances in Food Research," vol. 15, Academic, New York.

Bellows, R. J., and C. J. King (1973): *AIChE Symp. Ser.*, **69**(132):33.

Bird, R. B., W. E. Stewart, and E. N. Lightfoot (1960): "Transport Phenomena," table B-1, Wiley, New York.

Bomben, J. L., S. Bruin, H. A. C. Thijssen, and R. L. Merson (1973): in G. F. Stewart, E. Mrak and C. O. Chichester (eds.), "Advances in Food Research," vol. 20, Academic, New York.

————, J. A. Kitson, and A. I. Morgan, Jr. (1966): *Food Technol.*, **20**:1219.

Boyadzhiev, L., T. Sapundzhiev, and E. Bezenshek (1977): *Separ. Sci.*, **12**:541.

Brennan, P. J. (1966): *Chem. Eng.*, Jan. 17, p. 118.

Broughton, D. B. (1977): *Chem. Eng. Prog.*, **73**(10):49.

Brown, A. H., M. E. Lazar, T. Wasserman, G. S. Smith, and M. W. Cole (1951): *Ind. Eng. Chem.*, **43**:2949.

Buckley, P. S. (1964): "Techniques of Process Control," chaps. 27–30, Wiley, New York.

Cahn, R. P., and N. N. Li (1974): *Separ. Sci.*, **9**:505.

Carroll, D. E., A. Lopez, F. W. Cooler, and J. Eindhoven (1966): *Food Technol.*, **20**:823.

Chandrasekaran, S. K., and C. J. King (1972): *AIChE J.*, **18**:520.

Chem. Mark. Rep., Dec. 26, 1977.

Choo, C. Y. (1962): in J. J. McKetta (ed.), "Advances in Petroleum Chemistry and Refining," vol. 6, Interscience, New York.

Davis, J. C. (1971): *Chem. Eng.*, Aug. 9, p. 77.

Debreczeni, E. J. (1977): *Chem. Eng.*, June 6, p. 13

Eaglesfield, P., B. K. Kelly, and J. F. Short (1953): *Ind. Chem.*, April, p. 147; June, p. 243.

Egan, C. J., and R. V. Luthy (1955): *Ind. Eng. Chem.*, **47**:250.

Eskew, R. K., G. W. P. Phillips, R. P. Homiller, C. S. Redfield, and R. A. Davis (1951): *Ind. Eng. Chem.*, **43**:2949.

Farkas, D. F., and M. E. Lazar (1969): *Food Technol.*, **23**:688.

Feberwee, A., and G. H. Evers (1970): *Lebensm. Wiss. Technol.*, **3**(2):41.

Findlay, R. A., and J. A. Weedman (1958): in K. A. Kobe and J. J. McKetta (eds.), "Advances in Petroleum Chemistry and Refining," vol. 1, Interscience, New York.

Flath, R. A., D. R. Black, D. G. Guadagni, W. H. McFadden, and T. H. Schultz (1967): *J. Agr. Food Chem.*, **15**:29.

Gardon, J. L. (1966): *J. Paint Technol.*, **38**:43.

Gerster, J. A. (1966): *Chem. Eng. Prog.*, **62**(9):62.

————, J. A. Gorton, and R. B. Eklund (1960): *J. Chem. Eng. Data*, **5**:423.

Hanson, C. (1968): *Chem. Eng.*, Aug. 26, p. 76.

———— (ed.) (1971): "Recent Advances in Liquid-Liquid Extraction," Pergamon, Oxford.

Harriott, P. (1964): "Process Control," chap. 14, McGraw-Hill, New York.

Hartland, S. (1967): *Trans. Inst. Chem. Eng. Lond.*, **45**:T90.

Heiss, R., and L. Schachinger (1951): *Food Technol.*, **5**:211.

Helsel, R. W. (1977): *Chem. Eng. Prog.*, **73**(5):55.

Hildebrand, J. H., J. M. Prausnitz, and R. L. Scott (1970): " Regular and Related Solutions: The Solubility of Gases, Liquids, and Solids," Van Nostrand Reinhold, New York.
Huang, C. P., O. Fennema, and W. D. Powrie (1965–1966): *Cryobiology*, **2**:109, 240.
Keller, G. E., Union Carbide Corp., So. Charleston, W.Va. (1977): personal communication.
Kerkhof, P. J. A. M., and H. A. C. Thijssen (1974): *J. Food Technol.*, **9**:415.
King, C. J. (1971): " Freeze Drying of Foods," CRC Press, Cleveland.
——— (1974a): Understanding and Conceiving Chemical Processes, *AIChE Monogr. Ser.* 8.
——— (1974b): Novel Dehydration Techniques, in A. Spicer (ed.), "Advances in Dehydration and Preconcentration of Foods," Elsevier, New York.
Klein, E., J. K. Smith, R. E. C. Weaver, R. P. Wendt, and S. V. Desai (1973): *Separ. Sci.*, **8**:585.
Konstam, A. H., and W. R. Feairheller, Jr. (1970): *AIChE J.*, **16**:837.
Lieberman, N. (1977): *Chem. Eng.*, Sept. 12, p. 140.
Lyckman, E. W., C. A. Eckert, and J. M. Prausnitz (1965): *Chem. Eng. Sci.*, **20**:703.
Marello, V. S., and N. Poffenberger (1950): *Ind. Eng. Chem.*, **42**:1021.
McKay, D. L., G. H. Dale, and D. C. Tabler (1966): *Chem. Eng. Prog.*, **62**(11):104.
Meek, P. D. (1961): in J. J. McKetta (ed.), "Advances in Petroleum Chemistry and Refining," vol. 4, Interscience, New York.
Menting, L. C., B. Hoogstad, and H. A. C. Thijssen (1970): *J. Food Technol.*, **5**:127.
Merson, R. L., and A. I. Morgan, Jr. (1968): *Food Technol.*, **22**:631.
Michaels, A. S., H. J. Bixler, and P. N. Rigopoulos (1967): *Proc. 7th World Petrol. Cong., Mexico City.*
Mix, T. W., J. S. Dweck, M. Weinberg, and R. C. Armstrong (1978): *Chem. Eng. Prog.*, **74**(4):49.
Moore, W. J. (1963): " Physical Chemistry," 3d ed., chaps. 14 and 17, Prentice-Hall, Englewood Cliffs, N.J.
Moores, R. G., and A. Stefanucci (1964): Coffee, in " Kirk-Othmer Encyclopedia of Chemical Technology," 2d ed., vol. 5, p. 748, Wiley-Interscience, New York.
Morgan, A. I., Jr. (1967): *Food Technol.*, **21**:1353.
Muller, J. G. (1967): *Food Technol.*, **21**:49.
Oliver, E. D. (1966): " Diffusional Separation Processes: Theory, Design and Evaluation," chaps. 5 and 13, Wiley, New York.
Otani, S. (1973): *Chem. Eng.*, Sept. 17, p. 106.
Othmer, D. F. (1978): *Am. Chem. Soc. Natl. Meet., Anaheim*, March.
Pan, S. C. (1974): *Separ. Sci.*, **9**:227.
Peleg, M., and C. H. Mannheim (1970): *J. Food Sci.*, **35**:649.
Pitzer, K. S., and D. W. Scott (1943): *J. Am. Chem. Soc.*, **65**:803.
Ponting, J. D., W. L. Stanley, and M. J. Copley (1964): in W. B. van Arsdel and M. J. Copley (eds.), " Food Dehydration," vol. II, AVI, Westport, Conn.
Prausnitz, J. M. (1969): " Molecular Thermodynamics of Fluid-Phase Equilibria," chap. 7, Prentice-Hall, Englewood Cliffs, N.J.
———, C. A. Eckert, R. V. Orye, and J. P. O'Connell (1966): " Computer Calculations for Multicomponent Vapor-Liquid Equilibria," Prentice-Hall, Englewood Cliffs, N.J.
Prescott, J. H. (1968): *Chem. Eng.*, Oct. 7, p. 138.
Reid, R. C., J. M. Prausnitz, and T. K. Sherwood (1977): " The Properties of Gases and Liquids," 3d ed., McGraw-Hill, New York.
Rademaker, O., J. E. Rijnsdorp, and A. Maarleveld (1975): " Dynamics and Control of Continuous Distillation Units," Elsevier, New York.
Reissinger, K. H., and J. Schröter (1978): *Chem. Eng.*, Nov. 6, p. 109.
Reman, G. H. (1966): *Chem. Eng. Prog.*, **62**(9):56.
Rheineck, A. E., and K. F. Lin (1968): *J. Paint Technol.*, **40**:611.
Ricker, N. L., J. N. Michaels, and C. J. King (1979a): *J. Separ. Process Technol.*, **1**, No. 1.
———, E. F. Pittman, and C. J. King (1979b): *J. Separ. Process Technol.*, **1**, No. 1.
Rochelle, G. T. (1977): Process Synthesis and Innovation in Flue Gas Desulfurization, *Electr. Power Res. Inst. Rep.* EPRI-FP-463-SR, Palo Alto, Calif., July.
——— and C. J. King (1978): *Chem. Eng. Prog.*, **74**(2):65.
Saito, S., T. Michishita, and S. Maeda (1971): *J. Chem. Eng. Jap.*, **4**:37.
Schaeffer, W. D., and W. S. Dorsey (1962): in J. J. McKetta (ed.), "Advances in Petroleum Chemistry and Refining," vol. 6, Interscience, New York.

————, H. E. Rea, R. F. Deering, and W. W. Mayes (1963): *Proc. 6th World Petrol Congr., Frankfurt am Main*, vol. 4, p. 65.

Shinskey, F. G. (1977): " Distillation Control: For Productivity and Energy Conservation," McGraw-Hill, New York.

Souders, M. (1964): *Chem. Eng. Prog.*, **60**(2):75.

————, G. J. Pierotti, and C. L. Dunn (1970): *Chem. Eng. Prog. Symp. Ser.*, **66**(100):41.

Spiess, W. E. L., W. Wolf, W. Buttmi, and C. Jung (1973): *Chem. Ing. Techn.*, **45**:498.

Teas, J. P. (1968): *J. Paint Technol.*, **40**:19.

Tressler, D. K., and M. A. Joslyn (1971): " Fruit and Vegetable Juice Processing Technology," 2d ed., AVI, Westport, Conn.

Treybal, R. E. (1963): " Liquid Extraction," 2d ed., McGraw-Hill, New York.

———— (1973*a*): Liquid Extraction, in R. H. Perry and C. H. Chilton (eds.), " Chemical Engineers' Handbook," 5th ed., sec. 15, McGraw-Hill, New York.

———— (1973*b*): Liquid-Liquid Systems, in R. H. Perry and C. H. Chilton (eds.), " Chemical Engineers' Handbook," 5th ed., pp. 21-3 to 21-29, McGraw-Hill, New York.

Vorstmann, M. A. G., and H. A. C. Thijssen (1972): *Ingenieur*, **84**(45):CH65.

Walker, L. H. (1961): Volatile Flavor Recovery, chap. 12 in D. K. Tressler and M. A. Joslyn (eds.), " Fruit and Vegetable Juice Processing Technology," AVI, Westport, Conn.

Wardell, J. M., and C. J. King (1978): *J. Chem. Eng. Data*, **23**:144.

Weast, R. C. (ed.) (1968): " Handbook of Chemistry and Physics," 49th ed., CRC Press, Cleveland.

Weimer, R. F., and J. M. Prausnitz (1965): *Hydrocarbon Process.*, **44**(9):237.

Werezak, G. N. (1969): *Chem. Eng. Prog. Symp. Ser.*, **65**(91):6.

Whitmore, F. C. (1951): " Organic Chemistry," pp. 613-615, Van Nostrand, Princeton.

Wilkinson, T. K., and L. Berg (1964): *ACS Div. Petrol. Chem. Prepr.*, **9**:1, 13.

Wolford, R. W., J. A. Attaway, G. E. Alberding, and C. D. Atkins (1963): *J. Food Sci.*, **28**:320.

PROBLEMS

14-A$_1$ Suggest likely separation processes to be considered for separating an equimolal mixture of cyclohexane and benzene into relatively pure products on an industrial scale. If a mass separating agent is to be used, indicate what it should be.

14-B$_1$ Suggest two or more logical separation processes for the removal of 1 mol $\%$ benzene vapor from a waste nitrogen stream being discharged to the atmosphere. If a mass separating agent is to be employed, indicate a likely substance to use.

14-C$_1$ Suggest the most logical separation process for the separation of isopropanol from *n*-propanol on a large scale. If a mass separating agent is to be used, indicate what it should be.

14-D$_2$ Suggest one or more logical separation processes for the nearly complete removal of water present at saturation level in liquid benzene at ambient temperature. If a mass separating agent is to be used, indicate what it should be.

14-E$_2$ Environmental concerns require that concentrations of certain heavy metals in effluent waters be kept very low. Suppose that a plant has a water discharge of about 2.5 m^3/h which contains about 2 ppm cadmium. Indicate what separation processes could be useful for removal of this contaminant (*a*) if the cadmium is present as Cd^{2+} in solution and (*b*) if the cadmium is adsorbed on finely divided organic particles.

14-F$_2$ A company produces zirconium tetrachloride by chlorination of sands rich in zirconia. A substantial by-product is silicon tetrachloride, SiCl$_4$, for which a market exists at approximately 10 cents per pound. The company wants to install facilities for the recovery and purification of SiCl$_4$. The available feed stream is a liquid at atmospheric pressure and $-34°C$, containing 5000 kg/day SiCl$_4$ along with 7500 kg/day Cl$_2$. Titanium tetrachloride is present to approximately 0.3 mole percent. The product SiCl$_4$ should contain no more than 20 ppm Cl$_2$ and 5 ppm TiCl$_4$. The recycle chlorine to the chlorinator should be gaseous and should contain no more than 5 mol $\%$ SiCl$_4$. Give a flowsheet of an appropriate process for the purification of this SiCl$_4$-bearing stream. Show all vessels, heat exchangers, pumps, etc. Indicate approximate operating temperatures and pressures at pertinent points in the process. *Note:* Silicon tetrachloride decomposes when contacted with water.

14-G₂ As a new approach to the recovery of volatile flavor and aroma species during fruit-juice processing it is suggested that an immiscible liquid solvent be contacted with the fresh juice to extract the light organic volatile species. A suitable solvent might be a fluorocarbon, approved by the FDA. The fruit juice, once the volatiles had been extracted out, would be concentrated by evaporation of about 70 percent of the initial water present. The concentrate would then be contacted with the volatiles-laden solvent to pick the volatiles back up from the solvent. The solvent would then be recirculated. Assess the workability and desirability of such a process for volatiles retention.

14-H₂ Fermentation processes often produce a complex mixture of components, which require separation. Souders et al. (1970) discuss the separation of the fermentation broth in penicillin manufacture, using solvent extraction. Equilibrium distribution coefficients for the solvent considered are shown in Fig. 14-13 as a function of the pH of the aqueous phase (the broth). The particular shape of these curves results from the fact that all the components are weak acids, HA_i, for which there is an equilibrium distribution coefficient k_{1i} for the unionized form

$$k_{1i} = \frac{[HA_i]_o}{[HA_i]_w}$$

where subscripts o and w refer to the organic and aqueous phases, respectively. The degree of dissociation in the aqueous phase comes from an ionization constant k_{2i}

$$k_{2i} = \frac{[H^+]_w[A_i^-]_w}{[HA_i]_w}$$

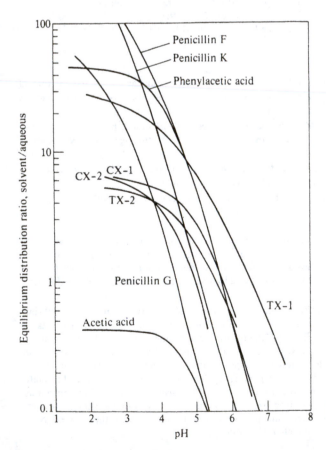

Figure 14-13 Equilibrium distribution ratios for various constituents in penicillin fermentation broths. (*From Souders et al., 1970; p. 41; used by permission.*)

Combining these expressions gives an overall distribution coefficient K_i

$$K_i = \frac{[HA_i]_o}{[HA_i]_w + [A_i^-]_w} = \frac{k_{1i}}{1 + k_{2i}/[H^+]}$$

At higher values of pH, $[H^+]$ is small and the second term in the denominator dominates. K_i is then directly proportional to $[H^+]$, and therefore log K_i decreases linearly with increasing pH, dropping one decade per pH unit. In the other extreme of low pH (high $[H^+]$) the first term in the denominator dominates, and K_i is effectively constant. The relative positions of the curves for different broth constituents in Fig. 14-13 are governed by the individual values of k_{1i} and k_{2i}.

Suppose that a broth concentrate has the following composition:

Component	Wt %	Component	Wt %
Penicillin F	12	CX-1	5
Penicillin G	30	CX-2	8
Penicillin K	30	Phenylacetic acid	2
TX-1	7	Acetic acid	1
TX-2	5		

Suggest a solvent-extraction scheme which will serve to remove the various other components to a large extent from penicillins G and K.

14-I₂ Explain the statement under Eq. (14-8): "For C to be an effective extraction solvent we must have $\delta_A > \delta_B > \delta_C$ or $\delta_C > \delta_B > \delta_A$."

14-J₃ Membrane permeation processes have been investigated in recent years as means of separating hydrocarbon liquid mixtures which are otherwise difficult to separate. For example, membranes have been found which, for a given fugacity-difference driving force, will pass benzene much more readily than cyclohexane. The permeate tending to pass through these polymeric membranes is enriched in benzene relative to the portion of the feed mixture which does not cross the membrane.

The design of a membrane separation device must somehow provide a fugacity difference of the preferentially passed component to cause it to migrate across the membrane from the feed side to the permeate side. Some difficulty arises in accomplishing this, since the permeate necessarily contains a greater proportion of that component. Thus, if the mixtures on both sides are binary and the pressures and temperatures are the same on both sides, the chemical potential of the preferentially passed component (benzene in the case cited) is greater on the permeate side than on the feed side and as a result that component will tend to cross the membrane in the reverse direction. What is needed is a method of increasing chemical potential in ways other than changing the relative proportions of components within a binary mixture.

Using as an example a case where the feed-side mixture contains benzene and cyclohexane in a 1 : 1 ratio and the permeate side contains these components in a 7 : 3 ratio, suggest *two* different practical ways in which the chemical potential difference can be changed to the desired direction. The feed is a liquid mixture of benzene and cyclohexane. The feed and permeate streams are to flow in thin channels along the membrane and on either side of it. Confirm the practicability of both your methods by appropriate calculations.

14-K₂ A possible flowsheet for the manufacture of decaffeinated instant coffee is shown in Fig. 14-14. Coffee beans, whole or cut, have caffeine extracted from them with an appropriate solvent. Residual solvent is removed, after which the beans are roasted and ground. Hot water is then used to extract coffee solution from the roasted grounds. This extract typically has a solute content in the range of 28 to 35 weight percent (Moores and Stefanucci, 1964).

Two routes are used commercially to convert extract into dry, instant-coffee particles. In the first, constituting about 70 percent of the market for instant coffee of all sorts, much of the water is removed by multieffect evaporation, after which the concentrated product is fed to a spray dryer, where the remaining water is removed through contact of droplets with hot air. In the second route, accounting for about 30 percent of the market for instant coffee, much of the water is removed from the extract by freeze concentration, and the resultant concentrate is freeze-dried.

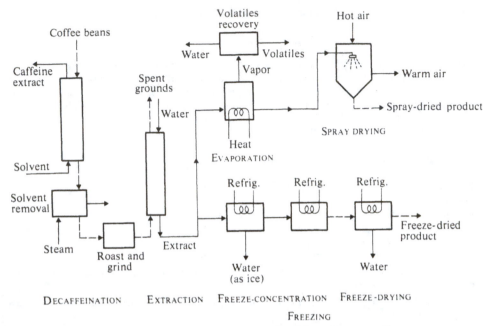

Figure 14-14 Processing routes for the manufacture of decaffeinated instant coffee.

The flavor and aroma of coffee are the result of numerous volatile organic compounds, which are easily lost during processing. Volatiles are recovered where possible, e.g., from the initial vapor formed upon evaporation. Better yet, processing steps are chosen and designed so as to minimize volatiles loss.

Caffeine has the structure

It is highly water-soluble but will also partition into some solvents. Chlorinated solvents, e.g., trichloro-ethylene, have classically been used for caffeine extraction, but there has been concern about the effects of residual quantities of these solvents left in the instant product. Recovered caffeine is used in the soft-drink industry.

(*a*) Why is it desirable to concentrate the extract first, by evaporation or freeze-concentration, before spray drying or freeze-drying?

(*b*) Suggest why evaporation is paired as a preconcentration process with spray drying, and freeze-concentration is paired with freeze-drying. Freeze-concentration is a more expensive process than ordinary evaporation.

(*c*) Hot water contacts the roast and ground coffee particles countercurrently to make the extract; this is typically done using the Shanks system of rotating fixed beds (Fig. 4-32). What is the probable main benefit achieved from the countercurrent flow?

(*d*) With the concern about chlorinated solvents, several alternate solvent possibilities for decaffeination have been explored. Assess (*i*) liquid carbon dioxide, (*ii*) water, and (*iii*) turpentine, listing desirable and undesirable features. Assume that caffeine can be removed efficiently in each case.

(*e*) In Fig. 14-14 decaffeination is accomplished by extraction of green coffee beans, before roasting. What would be the advantages and disadvantages of solvent decaffeinating (*i*) the extract, before concentration, (*ii*) the extract, after concentration, (*iii*) the final dried product, and (*iv*) roast and ground coffee, instead?

CONVERGENCE METHODS AND SELECTION OF COMPUTATION APPROACHES

A trial-and-error solution of an implicit equation involving a single variable consists of assuming values for the unknown variable until a value is found which satisfies the equation. An equation involving a single variable x can be written as

$$f(x) = 0 \qquad (A\text{-}1)$$

where $f(x)$ is the function resulting from putting all terms of the equation on the left-hand side. In a trial-and-error, or iterative, solution successive values of x are assumed according to a systematic plan until a value of x which causes $f(x)$ to be zero is found. Suitable systematic plans for this purpose are called *convergence methods*.

DESIRABLE CHARACTERISTICS

In devising or choosing a convergence method for a particular calculation, one should seek several desirable characteristics:

1. The convergence method should lead to the desired root of the equation. If the equation has multiple roots, the convergence method should lead reliably to the particular root in question.
2. The convergence method should be stable; it should approach the root asymptotically or in a well-damped oscillatory fashion, rather than developing large oscillations of successive values of the trial variable.
3. The convergence method should lead rapidly to the desired solution. Many iterations or many computations per iteration will require more computer time. This speed-of-convergence criterion is particularly important when the equation is involved in a subroutine which must be solved many times in the course of a main calculation.
4. Iteration should be avoided wherever possible. For example, it is usually better to solve a cubic equation by an algebraic approach than by an iterative solution.
5. If there is any doubt whether convergence has been achieved, it is desirable to surround the answer. i.e., come at it from both sides.

DIRECT SUBSTITUTION

A number of convergence methods have been developed for equations implicit in one variable and for simultaneous implicit equations involving more than one unknown variable (Beckett and Hurt, 1967; Lapidus, 1962; Southworth and DeLeeuw, 1965; Henley and Rosen, 1969;

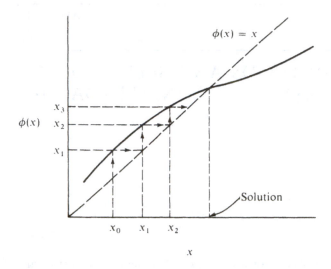

$$\phi(x) = x$$

$\phi(x)$ x_3

x_2

x_1

Solution

x_0 x_1 x_2

x

Figure A-1 Convergence by direct substitution.

etc.). One of the simplest is *direct substitution*, which can be used if the equation can be put in the form

$$\phi(x) = x \qquad (A-2)$$

where $\phi(x)$ is a function of x. This can be accomplished by adding x to both sides of Eq. (A-1).

In a direct-substitution approach one first assumes a value of x, which we shall call x_0. This x_0 is substituted into the left-hand side of Eq. (A-2) to give $\phi(x_0)$. A new *indicated* value of x can be obtained as $x_1 = \phi(x_0)$. This x_1, in turn, can be substituted into the left-hand side of Eq. (A-2) to give $\phi(x_1) = x_2$. This procedure is shown graphically in Fig. A-1. The solid curve represents $\phi(x)$, and the solution is that value of x for which the curve intersects the 45° diagonal.

For the situation shown in Fig. A-1 the direct-substitution procedure will achieve the converged solution for any starting value x_0 greater or less than the value of x corresponding to the solution. Such is not the case for the situation shown in Fig. A-2, however. In order for direct substitution to be convergent it is necessary that

$$\left| \frac{d\phi(x)}{dx} \right| < 1 \qquad (A-3)$$

at the solution. Multiple roots of Eq. (A-2) also can give trouble.

The direct-substitution procedure will converge faster in the vicinity of the solution to the extent that the derivative in Eq. (A-3) is small. Often the derivative is near unity, however, in which case *acceleration procedures* for direct substitution will be useful. One such acceleration procedure is the Wegstein method (Lapidus, 1962).

FIRST ORDER

Another procedure which is often employed as a convergence method is the *regula falsi*, or *false-position* approach, illustrated in Fig. A-3. Here we adopt Eq. (A-1)

$$f(x) = 0 \qquad (A-1)$$

and seek that value of x which makes the left-hand side of Eq. (A-1) (the solid curve in Fig. A-3) zero. This is accomplished by computing $f(x)$ for two initial values of x which we shall call x_0 and x_1. Preferably x_0 and x_1 should be selected so that $f(x_0)$ and $f(x_1)$ have

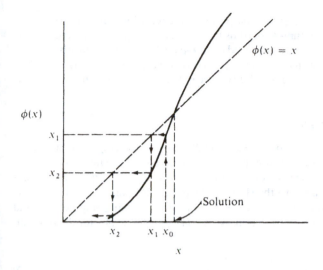

Figure **A-2** Divergent situation for direct substitution.

opposite signs. A linear interpolation is made between the points $[f(x_0), x_0]$ and $[f(x_1), x_1]$ to indicate x_2 at which $f(x_2)$ will be zero if the function is linear. Next a linear interpolation is made between the points corresponding to x_2 and either x_0 or x_1, whichever gave $f(x)$ of opposite sign from $f(x_2)$. The point for x_1 is used in the case shown in Fig. A-3. This initial point is used as a fixed pole for all succeeding iterations; in Fig. A-3 we next take a linear interpolation between the points for x_3 and x_1, then between the points for x_4 and x_1, etc. The trial value of x for the $(i + 1)$th iteration is computed by

$$x_{i+1} = x_i + (x_1 - x_i)\frac{f(x_i)}{f(x_i) - f(x_1)} \tag{A-4}$$

The *regula falsi* method is one of a general category, known as *secant methods*, which involves linear interpolation between past values of $f(x)$. These methods are also called *first*

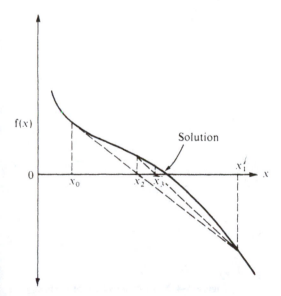

Figure **A-3** Regula falsi convergence.

order, since the error tends to decrease as the first power of the iteration number. First-order methods generally take a substantial number of trials to achieve convergence. A method that is somewhat more rapidly convergent than the fixed-pole *regula falsi* method involves linear interpolation between the two most recent points (Lapidus, 1962); however, this procedure can more readily run into oscillations and instability since it does not ensure that the answer is surrounded.

SECOND AND HIGHER ORDER

One of the most popular convergence procedures is the *Newton method* (Fig. A-4), a second-order scheme which tends to give an error diminishing as the square of the iteration number. Once again the solid curve represents $f(x)$. For an initial x_0, both $f(x_0)$ and $[df(x)/dx]_{x=x_0}$ are computed. The derivative corresponds to the slope of the dot-dash straight line in Fig. A-4. The intersection of this line with the abscissa gives x_1. At x_1 we once again compute $f(x_1)$ and $[df(x)/dx]_{x=x_1}$, and repeat the procedure to obtain x_2, etc. The trial value of x for the $(i + 1)$th iteration is computed as

$$x_{i+1} = x_i - \frac{f(x_i)}{[df(x)/dx]_{x=x_i}} \tag{A-5}$$

Even the Newton procedure does not guarantee convergence. For example, suppose that there were a maximum in the $f(x)$ curve between x_0 and the desired solution, as shown in Fig. A-5. In such a case the Newton method is divergent or reaches an undesired root.

Higher-order convergence methods also exist; there are third-order schemes involving calculations of both first and second derivatives, and fourth-order schemes which involve the first three derivatives. Usually fewer iterations are required the higher the order, since the error diminishes more rapidly from trial to trial. On the other hand, the higher-order methods require the evaluation of a number of derivatives at each point which is equal to the order minus 1. These derivatives must be obtained either through analytical expressions or through

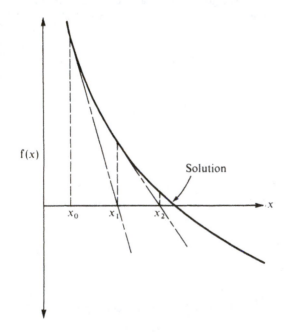

Figure A-4 Newton convergence scheme.

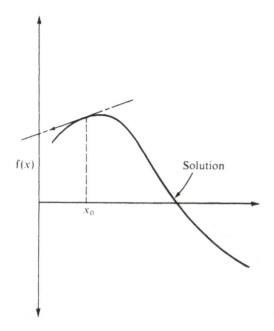

Figure A-5 Divergent situation for Newton convergence method.

the evaluation of $f(x)$ at incrementally different values of x. Either way, a higher-order method requires more computation per trial value of x. As a result, the choice between convergence methods of various orders is often not apparent a priori.

One effective higher-order version of the false-position method involves fitting three calculated points with a hyperbola (Hohmann and Lockhart, 1972).

INITIAL ESTIMATES AND TOLERANCE

In order to implement a convergence method for the computer it is necessary to provide some procedure for obtaining an *initial estimate* x_0 and to indicate the *tolerance*, which is the allowable error in $f(x)$ within which the calculation will be stopped. The initial estimate can be selected in one of two ways: one can specify a particular value for x_0 which is known to be in a region such that the convergence method will lead to the converged solution in a straightforward manner, or if the calculation is being repeated for a number of different values of other variables included in $f(x)$, one can use the last previous converged value of x as the first estimate for the next calculation.

The tolerance should be selected so that x will be found within the desired degree of precision but should not be low enough to require an unnecessarily large number of iterations. If there is a possibility that the specified tolerance is too large, it is useful to surround the answer by coming at it from both sides.

MULTIVARIABLE CONVERGENCE

Often a multivariable problem is encountered in which values of n variables are to be found so as to satisfy n independent, simultaneous, implicit equations. Two basic approaches can be used for such problems, *sequential* or *simultaneous*. A sequential convergence is illustrated in Fig. A-6. If two variables x and y in two equations are unknown, the approach is to assume a

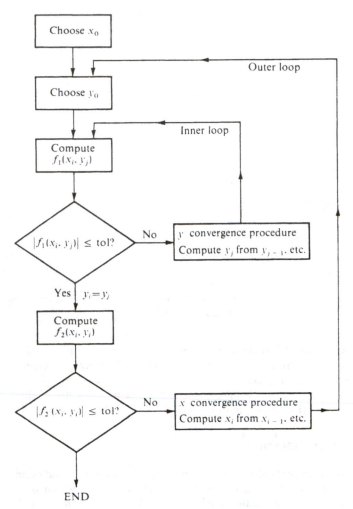

Figure A-6 Sequential convergence of two-unknown problem with two equations: $f_1(x, y) = 0$ and $f_2(x, y) = 0$.

value for x ($= x_0$) and proceed directly to a single-variable convergence loop that will find the converged value of y for $x = x_0$ from one of the two equations. The other equation is then used in an outer convergence loop to find a new value of x ($= x_1$). The inner loop is then entered once more† and produces a converged value of y for $x = x_1$. The outer loop then yields a value of x_2, and the calculation continues until the outer loop has also achieved convergence.

This concept of nesting loops with convergence of one variable at a time can be used for situations involving any number of unknown variables in an equivalent number of independent equations. As the number of variables increases, a very large number of trips through the inner loops will be required. At the possible sacrifice of stability in the calculation one can use instead a *simultaneous* approach in which all unknown variables are moved toward conver-

† It is reasonable to choose the initial estimate of y each time as the converged value of y from the previous trial.

gence together. There will be only one convergence loop in which the errors in all equations are used to give new values of all variables. The most popular simultaneous convergence method is the multivariate Newton approach, a generalization of the single-variable Newton method.

In the multivariate Newton method, corrections to each unknown variable are made by assuming that all partial derivatives are linear between the last calculated point and the converged solution. Therefore in a two-variable problem we choose x_{i+1} and y_{i+1} such that

$$-f_1(x_i, y_i) = \left[\frac{\partial f_1(x, y)}{\partial x}\right]_{x=x_i, y=y_i} (x_{i+1} - x_i) + \left[\frac{\partial f_1(x, y)}{\partial y}\right]_{x=x_i, y=y_i} (y_{i+1} - y_i) \qquad \text{(A-6)}$$

and

$$-f_2(x_i, y_i) = \left[\frac{\partial f_2(x, y)}{\partial x}\right]_{x=x_i, y=y_i} (x_{i+1} - x_i) + \left[\frac{\partial f_2(x, y)}{\partial y}\right]_{x=x_i, y=y_i} (y_{i+1} - y_i) \qquad \text{(A-7)}$$

In this way f_1 and f_2 should become zero. Solving for x_{i+1} and y_{i+1}, we have the two-variable analogs of Eq. (A-5)

$$x_{i+1} = x_i - \frac{[f_1(x_i, y_i)][\partial f_2(x, y)/\partial y] - [f_2(x_i, y_i)][\partial f_1(x, y)/\partial y]}{[\partial f_1(x, y)/\partial x][\partial f_2(x, y)/\partial y] - [\partial f_1(x, y)/\partial y][\partial f_2(x, y)/\partial x]} \qquad \text{(A-8)}$$

$$y_{i+1} = y_i - \frac{[f_1(x_i, y_i)][\partial f_2(x, y)/\partial x] - [f_2(x_i, y_i)][\partial f_1(x, y)/\partial x]}{[\partial f_1(x, y)/\partial y][\partial f_2(x, y)/\partial x] - [\partial f_1(x, y)/\partial x][\partial f_2(x, y)/\partial y]} \qquad \text{(A-9)}$$

All derivatives are evaluated at $x = x_i$ and $y = y_i$. Put in more compact determinant form, Eqs. (A-8) and (A-9) become

$$x_{i+1} - x_i = \frac{\begin{vmatrix} -f_1(x_i, y_i) & \partial f_1(x, y)/\partial y \\ -f_2(x_i, y_i) & \partial f_2(x, y)/\partial y \end{vmatrix}}{\begin{vmatrix} \partial f_1(x, y)/\partial x & \partial f_1(x, y)/\partial y \\ \partial f_2(x, y)/\partial x & \partial f_2(x, y)/\partial y \end{vmatrix}} \qquad \text{(A-10)}$$

following Cramer's rule for solving simultaneous linear equations. The corresponding expression for $y_{i+1} - y_i$ is obtained by interchanging y and x in Eq. (A-10).

In strictly analogous fashion we can obtain the convergence formula for each variable x_j in a multivariable situation where there are n unknown variables $x_1, x_2, \ldots, x_j, \ldots, x_n$ related by n equations of the form $f_1(x_1, x_2, \ldots, x_n) = 0$, $f_2(x_1, x_2, \ldots, x_n) = 0$, $\ldots, f_k(x_1, x_2, \ldots, x_n) = 0, \ldots, f_n(x_1, x_2, \ldots, x_j, \ldots, x_n) = 0$, as follows:

$$x_{j, i+1} - x_{j, i} = \frac{\begin{vmatrix} \partial f_1/\partial x_1 & \partial f_1/\partial x_2 & \cdots & -f_1 & \cdots & \partial f_1/\partial x_n \\ \partial f_2/\partial x_1 & \partial f_2/\partial x_2 & \cdots & -f_2 & \cdots & \partial f_2/\partial x_n \\ \hline \partial f_k/\partial x_1 & \partial f_k/\partial x_2 & \cdots & -f_k & \cdots & \partial f_k/\partial x_n \\ \hline \partial f_n/\partial x_1 & \partial f_n/\partial x_2 & \cdots & -f_n & \cdots & \partial f_n/\partial x_n \end{vmatrix}}{\begin{vmatrix} \partial f_1/\partial x_1 & \partial f_1/\partial x_2 & \cdots & \partial f_1/\partial x_j & \cdots & \partial f_1/\partial x_n \\ \partial f_2/\partial x_1 & \partial f_2/\partial x_2 & \cdots & \partial f_2/\partial x_j & \cdots & \partial f_2/\partial x_n \\ \hline \partial f_k/\partial x_1 & \partial f_k/\partial x_2 & \cdots & \partial f_k/\partial x_j & \cdots & \partial f_k/\partial x_n \\ \hline \partial f_n/\partial x_1 & \partial f_n/\partial x_2 & \cdots & \partial f_n/\partial x_j & \cdots & \partial f_n/\partial x_n \end{vmatrix}} \qquad \text{(A-11)}$$

The multivariate Newton convergence scheme generally gives rapid convergence when one is near the solution, but it may be divergent if some of the starting values are well removed from the solution. Often an effective procedure for a large multivariable problem is to combine the sequential and simultaneous approaches, taking a simultaneous solution for several of the variables as one loop in a nest of sequential loops for the other variables.

One disadvantage of the multivariate Newton method is that n^2 derivatives must be computed on each iteration. The amount of computation per iteration can often be substantially reduced with little loss in convergence speed or stability by using a *paired simultaneous approach*, also known as a *partitioned convergence scheme*. In such a method each variable is still corrected in each iteration in a single loop. In this case, however, the new value of each variable is determined from a single equation, instead of all equations being used to obtain new values of all variables, as in the multivariate Newton approach. Each variable is paired with a different equation in this modification, that is, $f_1(x, y)$ with x and $f_2(x, y)$ with y in a two-variable problem.

The paired simultaneous method works well if each equation is paired with that variable which has a dominant effect upon the equation, and which variable this is can often be determined from a physical analysis of the problem.

The sequential convergence scheme is also partitioned or paired, and again it is important to link each function with the independent variable which has the greater effect upon it. In Fig. A-6, f_1 has been paired with y and f_2 has been paired with x.

When there is no clear physical reasoning for pairing variables and equations in a certain way, it is probably best to use a full simultaneous approach.

CHOOSING $f(x)$

Often it will be possible through algebraic manipulation to put the function(s) which are to be reduced to zero into a number of different but equivalent forms. Certain of these forms will give more rapid convergence than others. The following guidelines are useful in selecting the best form for $f(x)$:

1. The range of allowable values of x should be bounded; i.e., solving for an unknown variable which varies between -1 and $+1$ is preferable to solving for one that can vary from $-\infty$ to $+\infty$.
2. The function $f(x)$ should have no spurious roots within the allowable range of x.
3. Maxima and minima and, to a lesser extent, second-order points of inflection in $f(x)$ hamper convergence.
4. To the extent that $f(x)$ is more nearly linear in x, convergence by almost any method will be more rapid.

REFERENCES

Beckett, R., and J. Hurt (1967): "Numerical Calculations and Algorithms," McGraw-Hill, New York.
Henley, E. J., and E. M. Rosen (1969): "Material and Energy Balance Computations," Wiley, New York.
Hohmann, E. C., and F. J. Lockhart (1972): *CHEMTECH*, **2**:614.
Lapidus, L. (1962): "Digital Computation for Chemical Engineers," McGraw-Hill, New York.
Southworth, R. W., and S. L. DeLeeuw (1965): "Digital Computation and Numerical Methods," McGraw-Hill, New York.

ANALYSIS AND OPTIMIZATION OF MULTIEFFECT EVAPORATION

In Chap. 4 it was shown that multieffect evaporation requires less steam to accomplish an evaporation than a single-effect evaporation. A three-effect evaporation process is shown in Fig. B-1. The feed is a salt solution entering the first effect. The steam to cause evaporation is fed at a high enough pressure and temperature to the coils of the first effect and causes evaporation of an amount of water from the salt solution equivalent in latent heat to the quantity of steam condensing. This evaporated water serves as condensing steam to cause evaporation from the salt solution in the second effect, and so on. In order for there to be a driving force for heat transfer in the desired direction across the evaporator coils each successive effect must operate at a lower pressure than the one before.

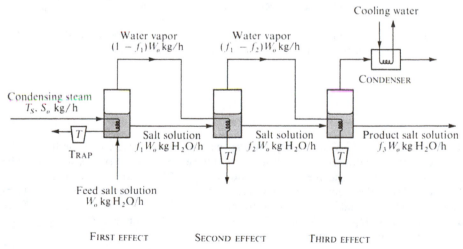

Figure B-1 Three-effect evaporation system.

SIMPLIFIED ANALYSIS

A simple analysis can be made of a multieffect evaporation system if we assume that latent-heat effects are completely dominant (no heat requirement for preheating the feed, etc.), that the elevation in boiling point of the salt solution due to dissolved salts is negligible, that the heat-transfer coefficient from condensing steam to boiling solution in each effect is constant at a value U, and that the latent heat of vaporization of water is independent of temperature and salt concentration.

Establishing notation for this analysis, we shall define the following variables (English units are given first, with SI units in parentheses):

U_i = heat-transfer coefficient in effect i, assumed constant and equal to U in simple analysis, Btu/h·ft²·°F (kJ/h·m²·°C)

a_i = heat-transfer area of coils in effect i, ft² (m²)

W_0 = amount of water in feed salt solution, lb/h (kg/h)

f_i = fraction of water in feed that remains in salt solution leaving effect i

N = number of effects

S_0 = steam condensation rate in coils of first effect, lb/h (kg/h)

T_s = saturation temperature of steam to first effect, °F (°C)

T_i = saturation temperature of vapor generated in effect i ($=$ boiling temperature of liquid in effect i if boiling-point elevation due to dissolved salts is neglected), °F (°C)

λ = latent heat of vaporization of water, Btu/lb (kJ/kg)

Two types of equations are required for this simplified analysis, enthalpy balances and heat-transfer rate equations. The enthalpy balances relate the amount of evaporation or condensation in one effect to the amount of evaporation or condensation in other effects:

$$S_0 = (1 - f_1)W_0 = (f_1 - f_2)W_0 = \cdots = (f_{N-1} - f_N)W_0 \qquad \text{(B-1)}$$

The amount of water evaporated in each effect is the same, since we have taken the latent heat of vaporization to be a constant and have neglected all sensible-heat effects. The heat release from condensation in the coils of each effect is $\lambda W_0(f_{i-2} - f_{i-1})$, and the heat consumption for boiling in that effect is $\lambda W_0(f_{i-1} - f_i)$.

The heat-transfer rate equations relate the rate of heat transfer across the coils of an effect to either the rate of condensation in the coils or the rate of boiling in the evaporation chamber:

$$\lambda(1 - f_1)W_0 = Ua_1(T_s - T_1)$$
$$\lambda(f_1 - f_2)W_0 = Ua_2(T_1 - T_2) \qquad \text{(B-2)}$$

$$\cdots\cdots\cdots\cdots\cdots\cdots\cdots\cdots\cdots\cdots\cdots\cdots$$

In a design problem we would typically specify the value of f_N, corresponding to the overall degree of concentration of the salt solution in the evaporator system; W_0 also would be specified, as would T_N, the temperature of condensation of the steam generated in the last effect, which is set by the available cooling water temperature for the final condenser. The term N also will be set, either independently or through an optimization (see following discussion). Equations (B-1) now represent N independent equations in N unknowns (S_0 and $f_1, f_2, \ldots, f_{N-1}$). Hence we can solve for these variables, finding that

$$f_{i-1} - f_i = \frac{1 - f_N}{N} \qquad \text{(B-3)}$$

which corresponds to $1/N$ times the total evaporation occurring in each effect, and

$$S_0 = \frac{1 - f_N}{N} W_0 \tag{B-4}$$

which indicates that the steam consumption rate is $1/N$ times the total evaporation.

We are now left with $2N - 1$ unknowns, as follows:

$$a_1, a_2, \ldots, a_N \qquad N \text{ unknowns}$$

$$T_1, T_2, \ldots, T_{N-1} \qquad N - 1 \text{ unknowns}$$

These equations are related by Eqs. (B-2), which are N independent equations. Hence $N - 1$ additional variables remain at our disposal to be specified. This gives us the opportunity to optimize the relative heat-transfer areas of the different effects of the evaporator system. Since the left-hand sides of Eqs. (B-2) are all equal, we can take advantage of the fact that

$$(T_s - T_1) + (T_1 - T_2) + \cdots + (T_{N-1} - T_N) = T_s - T_N \tag{B-5}$$

in the absence of boiling-point elevations due to dissolved solute. Equation (B-5) can be used to rearrange and add Eqs. (B-2), giving

$$\frac{1}{a_1} + \frac{1}{a_2} + \cdots + \frac{1}{a_N} = \frac{U(T_s - T_N)N}{W_0(1 - f_N)\lambda} \tag{B-6}$$

Following Eqs. (B-3) and (B-4), $W_0(1 - f_N)/N$ has been substituted into Eq. (B-6) for the constant left-hand sides of Eqs. (B-2). The right-hand side of Eq. (B-6) is composed of known quantities, and it remains to choose optimum values of the areas of the individual effects.

The installed cost of any effect of an evaporator system can generally be related to the heat-transfer area of the evaporator raised to a power m, which is usually less than unity (King, 1963; Badger and Standiford, 1958). Hence the total installed cost of the evaporator effects is given by

$$\text{Installed cost} = A(a_1^m + a_2^m + \cdots + a_N^m) \tag{B-7}$$

We would like to choose the areas of the effects so as to make Eq. (B-7) a minimum while satisfying the constraint expressed by Eq. (B-6). Inspection and common sense tell us that this will occur when all the areas are equal to each other, but it is also possible to prove that result formally. To do this we shall make use of the technique of *Lagrange multipliers* (Wilde and Beightler, 1967; Peters and Timmerhaus, 1968; etc.)

If a cost equation $F(x_1, x_2, \ldots, x_n) = 0$ is to be maximized or minimized, where $x_1, x_2, \ldots, x_n$ are independent variables, and if there is a constraint $\phi(x_1, x_2, \ldots, x_n) = 0$, which must be satisfied by the independent variables, the Lagrange multiplier technique is to write the cost equation as

$$G = F(x_1, x_2, \ldots, x_n) + \Lambda \phi(x_1, x_2, \ldots, x_n) = 0 \tag{B-8}$$

where Λ is an undefined parameter. Since the constraint must be satisfied at all points, the partial derivatives of the left-hand side of Eq. (B-8) with respect to $x_1, x_2, \ldots, x_n$ and Λ must all be equal to zero at the optimum. This provides $n + 1$ equations in $n + 1$ unknowns, which can be solved for the values of the independent variables at the optimum.

In the present case Eq. (B-8) becomes

$$G = A(a_1^m + a_2^m + \cdots + a_N^m) + \Lambda \left[\frac{1}{a_1} + \frac{1}{a_2} + \cdots + \frac{1}{a_N} - \frac{U(T_s - T_N)N}{W_0(1 - f_N)\lambda} \right] = 0 \tag{B-9}$$

Setting the partial derivatives equal to zero gives

$$\frac{\partial G}{\partial a_1} = mAa_1^{m-1} - \Lambda a_1^{-2} = 0$$

$$\frac{\partial G}{\partial a_2} = mAa_2^{m-1} - \Lambda a_2^{-2} = 0 \qquad \text{(B-10)}$$

$$\cdots\cdots\cdots\cdots\cdots\cdots\cdots\cdots\cdots$$

$$\frac{\partial G}{\partial a_N} = mAa_N^{m-1} - \Lambda a_N^{-2} = 0$$

The equation for $\partial G/\partial \Lambda$ is identical to Eq. (B-6). From Eqs. (B-10) $a_1 = a_2 = \cdots = a_N$ at the optimum.

Since the areas are equal, Eq. (B-6) becomes

$$a_i = \frac{W_0(1 - f_N)\lambda}{U(T_s - T_N)} \qquad \text{(B-11)}$$

Notice that the area per effect for this simple analysis is independent of the number of effects. This conclusion may be surprising at first, but it is the result of two compensating factors. As the number of effects increases, the amount to be evaporated in each effect decreases and the left-hand side of Eq. (B-2) decreases in inverse proportion to N. At the same time the temperature-difference driving force for heat transfer on the right-hand side of Eq. (B-2) also decreases in inverse proportion to N. Thus a_i is independent of N.

It is also interesting to note that once a multieffect evaporation system subject to this simple analysis has been built and is in operation with the areas of each effect now established, there are few independent operating variables left. For example, the water-vapor pressures or saturation temperatures in each effect are not independent, and will adjust as necessary to give equal rates of heat transfer across the coils of each effect [Eqs. (B-2)] so as to keep the enthalpy balance around each effect [Eqs. (B-1)] through the same amount of evaporation occurring in each effect. Similarly, the steam-condensation rate in the first effect cannot be adjusted independently and will level out to give the required amount of evaporation in the first effect, subject to the steady-state value of $T_s - T_1$.

OPTIMUM NUMBER OF EFFECTS

The determination of the optimum number of effects for a multieffect evaporation system is a classical optimization involving the balance between operating costs and capital equipment costs. The primary operating cost is for the steam consumption in the first effect. Through Eq. (B-4) this cost is given by

$$\text{Steam cost} = B\frac{(1 - f_N)W_0}{N} \qquad \text{(B-12)}$$

where B is the cost of steam per pound. When we combine Eqs. (B-7) and (B-11), the annual fixed charges for the evaporator equipment can be expressed as

$$\text{Fixed charges for evaporators} = C\left[\frac{W_0(1 - f_N)\lambda}{U(T_s - T_N)}\right]^m N \qquad \text{(B-13)}$$

where C is a constant equal to the product of A from Eq. (B-7) and the fraction of the installed equipment cost that makes up the annual fixed charges. The total annual cost is then

$$\text{Total cost} = B\frac{(1 - f_N)W_0}{N} + C\left[\frac{W_0(1 - f_N)\lambda}{U(T_s - T_N)}\right]^m N \qquad \text{(B-14)}$$

Equation (B-14) has the form of

$$\text{Cost} = \frac{\text{const}_1}{N} + (\text{const}_2)(N) \tag{B-15}$$

One of the interesting properties of an equation involving the sum of a term in N^{-1} and a term in N^{+1} is that the minimum cost will correspond to the value of N for which the two terms on the right-hand side are equal. The reader can prove this by simple differentiation. Thus the optimum number of effects is given by

$$N_{\text{opt}} = \left(\frac{\text{const}_1}{\text{const}_2}\right)^{1/2} = \left|\frac{B}{C}\left[\frac{U(T_s - T_N)}{\lambda}\right]^m [(1 - f_N)W_0]^{1-m}\right|^{1/2} \tag{B-16}$$

Since m, the cost-vs.-area exponent, is less than unity, the optimum number of effects will be larger for higher steam costs, lower evaporator costs, higher heat-transfer coefficients, higher steam-to-cooling-water temperature differences, lower latent heats of evaporation of the solution being concentrated, higher degrees of concentration of the solution, and higher feed rates.

MORE COMPLEX ANALYSES

Figure B-2 gives a flow diagram of a multieffect evaporator system for seawater conversion into fresh water, which was used in the U.S. Department of the Interior demonstration plant at Freeport, Texas, and is discussed by Standiford and Bjork (1960) and by King (1963). The scheme makes extensive use of additional heat exchangers which serve to preheat the seawater feed to the temperature of the first effect. The heat for the feed preheating is obtained from the sensible heat of the condensate leaving each of the effects and from portions of the overhead vapor from each effect which are drawn off and condensed. The system shown in Fig. B-2 uses *forward feed* of the brine from effect to effect, in the direction of decreasing evaporation temperatures and pressures. It is also possible to use *backward feed*, in which the feed seawater enters the last (lowest-pressure) effect and flows in the direction of increasing temperatures and pressures between effects. Such a scheme requires much less elaborate preheat equipment but does require pumps to transfer the brine between effects. In seawater conversion, a primary operating problem is the formation of calcium sulfate or other scales on

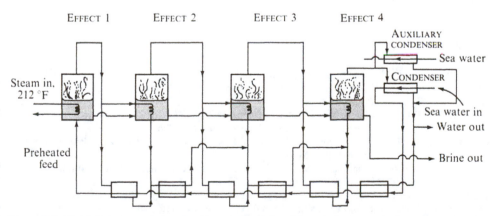

Figure B-2 Multieffect seawater-evaporation system using forward feed and preheat through vapor bleed and condensate heat exchangers. *(Adapted from King, 1963, p. 149; used by permission.)*

the heat-transfer surfaces within the effects. The tendency for calcium sulfate to precipitate is greatest where the brine concentration is highest or the temperature is highest, because of the inverse solubility curve of calcium sulfate with respect to temperature. Backward feed has the disadvantage of producing the highest temperatures and highest brine concentrations in the first effect together, whereas forward feed has the advantage of bringing the most dilute brine to the high temperatures of the first effect.

King (1963) has given the results of an optimization calculation to determine the optimum number of effects in the seawater conversion plant shown in Fig. B-2. The analysis allows for a number of complicating effects, e.g., variation of the heat-transfer coefficient with respect to temperature, boiling-point elevation due to the salt content of the brine (a function of concentration), installation costs for all the heat exchangers, and the need for purging a certain amount of the overhead vapor to accomplish full feed preheating, but retains the condition that the different effects all have the same heat-transfer area. Economic conditions have changed substantially since this analysis was made.

When the heat-transfer coefficient for the evaporators varies from effect to effect, the vapor-bleed preheat is used and/or the boiling-point elevation can vary from effect to effect, Eq. (B-6) is no longer valid, and other secondary cost terms must be considered in Eq. (B-7). As a result the equal-area-per-effect case is not necessarily still optimum. If the areas per effect are not held equal, an optimization problem with $N - 1$ independent variables results for any set number of effects. Itahara and Stiel (1968) have found that the technique of *dynamic programming* is well suited to this problem and have obtained a solution to the same problem solved for equal areas by King (1963), allowing the areas of the effects to be different.

The very large increases in steam (energy) costs occurring in recent years have served to increase the design optimum number of effects to values of the order of 17 and greater for seawater desalination. In part, the upper limit on the number of effects used is placed by the need to have a sufficient thermal driving force in each effect to give stable operation, and it has therefore become useful to investigate and develop economical evaporator designs which give stable operation at very low ΔT. Another new development is the combination of multieffect evaporation with multistage flashing (Prob. 4-G) for feed preheat (see, for example, Howe, 1974).

Dynamic programming has also been applied to optimization of solvent feed to each stage in crosscurrent multistage extraction (Rudd and Watson, 1968) and to the determination of the optimum pattern of reflux ratio vs. time in multistage batch distillation (Converse and Gross, 1963; Coward, 1967).

REFERENCES

Badger, W. L., and F. C. Standiford (1958): *Natl. Acad. Sci. Natl. Res. Counc. Publ.* 568, p. 103.
Converse, A. O., and G. D. Gross (1963): *Ind. Eng. Chem. Fundam.*, **2**:217.
Coward, I. (1967): *Chem. Eng. Sci.*, **22**:503.
Howe, E. D. (1974): "Fundamentals of Water Desalination," chap. 9, Dekker, New York.
Itahara, S., and L. I. Stiel (1968): *Ind. Eng. Chem. Process Des. Dev.*, **7**:6.
King, C. J. (1963): Fresh Water from Sea Water, in T. K. Sherwood, "A Course in Process Design," chap. 7, M.I.T. Press, Cambridge, Mass.
Peters, M. S., and K. D. Timmerhaus (1968): "Plant Design and Economics for Chemical Engineers," 2d ed., McGraw-Hill, New York.
Rudd, D. F., and C. C. Watson (1968): "Strategy of Process Engineering," Wiley, New York.
Standiford, F. C., and H. F. Bjork (1960): *ACS Adv. Chem. Ser.*, **27**:115.
Wilde, D. J., and C. S. Beightler (1967): "Foundations of Optimization," Prentice-Hall, Englewood Cliffs, N.J.

PROBLEM SPECIFICATION FOR DISTILLATION

THE DESCRIPTION RULE

As brought out in Chap. 2, the description rule can be used for identifying the number of variables which must be specified in a problem involving a separation process. For single-stage separation processes it may seem simpler to list all variables pertaining to the process and subtract the number of independent equations relating these variables in order to find the number of independent variables which must be specified. As processes and problems become more complex, however, the description rule presents a major saving of time over the method of counting variables and counting equations. This is particularly true for multistage separation processes (Hanson et al., 1962).

Consider the simple plate-distillation column of Fig. C-1 processing a feed of R components. The column is equipped with a series of stages above the stage where feed enters (the rectifying section) and a series of stages below the feed stage (the stripping section). The numbers of stages in each of these sections are denoted as n and m, respectively. We shall consider that these stages are *equilibrium* stages; i.e., the vapor and liquid leaving each stage are in equilibrium with each other.

A reboiler and a condenser are provided. The heat introduced through the reboiler has been denoted as Q_R and the heat removed in the condenser as Q_C. The condenser is a partial condenser.

The pressure in the column is governed by a pressure controller, which adjusts a valve on the overhead product (vapor) line to maintain a predetermined pressure. This fixed pressure is the *set point* of the pressure controller. In order to ensure that operation will occur at steady state, two level controllers have been provided. One of these adjusts the rate of reflux return (or a flow controller governing this rate) so as to hold a constant level (the set point) in the reflux accumulator drum. The other level controller adjusts the bottoms product rate so as to hold a constant level in the reboiler. The feed rate, cooling-water rate, and reboiler steam rate are manually set by means of valves, which are shown.

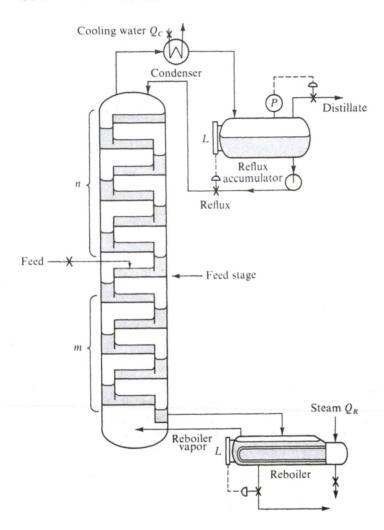

Figure C-1 Typical distillation column with partial condenser.

In order to apply the description rule to distillation we want to identify and count the variables set during *construction* and *operation* of the process: (1) It is apparent that we can arbitrarily set n and m at any values we please during construction of the column. If we pick specific numbers of equilibrium stages for each of these sections of stages, we have set two independent variables. (2) We can operate the column of stages at an arbitrarily chosen pressure by adjusting the set point on the pressure controller. The pressures we can use may be restricted to values between certain limits, but within these limits we are free to make any arbitrary choice, and hence the pressure constitutes another independent variable. (3) We can feed an arbitrarily chosen amount of each of the R components in the feed by altering the feed composition and adjusting the valve on the feed line. This sets R more independent variables. (4) We can arbitrarily set the enthalpy of the feed. This could be done, for example, by adjusting the temperature of the steam in a feed preheater. (5) We can arbitrarily introduce as much heat as we want into the reboiler by adjusting the steam valve or steam temperature. (6)

Again between limits, we can remove an arbitrary amount of heat from the condenser by adjusting the cooling-water flow rate. The set points for the liquid levels in the reboiler and reflux drum are not independent variables. These levels must be kept constant in order for there to be steady-state operation. The particular level in the reflux drum has no effect on the separation process, and the particular level in the reboiler can, at most, affect Q_R, which is already an independent variable.

If the variables set in construction and operation are noted down, the list is:

Amount of each component in the feed	R
Feed enthalpy	1
Pressure	1
Stages above feed entry n	1
Stages below feed entry m	1
Reboiler load Q_R	1
Condenser load Q_C	1
	$\overline{R + 6}$

These $R + 6$ variables completely describe the process, and if a value is set for each of them, the separation obtained under these values of the variables is completely determined and can be calculated.

While counting the number of independent variables by noting down those set by construction and operation is simple, as a practical matter the particular variables developed in such a list would seldom be set in the description of a given problem. Any or all of them could be replaced with other independent variables to which we are more interested in assigning values. In essentially every problem description, however, certain of the variables just listed will be set, namely, the variables describing the feed and the variable of pressure. If these are excluded from the variables to be further considered for setting or replacement, the remaining variables total four: n, m, Q_C, Q_R, independent of the total number of components. Thus, in describing any distillation problem concerning the column of Fig. C-1, after the feed and pressure have been set, four more independent variables must be set.

The variables which might be used to replace the four listed above could be (1) separation variables, (2) flows at some point or points in the process, and (3) temperatures at one or more points, or in general, any independent variable which characterizes the process. If the column already existed and we wanted to consider the possibility of using it for a new separation, a likely problem might be described by assigning values to the four variables

Stages above feed n
Stages below feed m
Recovery fraction of A in top product $(/_A)_D$
Concentration of A in top product $x_{A, D}$

A second common type of problem is the design of a new column. The separation to be accomplished is specified through two separation variables. A third variable set is usually a flow at some point, often the ratio of reflux to distillate. The fourth variable set is usually the location of the feed. Thus the problem could be described by the four variables

$(/_A)_D$
$(/_B)_D$
Reflux ratio (reflux flow divided by distillate flow)
Feed-stage location

where A and B are two components of the feed.

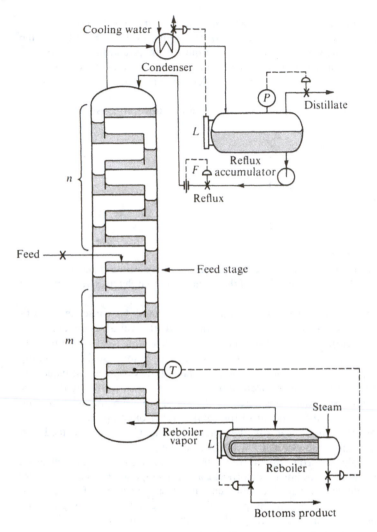

Figure C-2 Alternate control scheme for distillation column of Fig. C-1.

The *number* of independent variables which are set during construction and operation does not depend upon the type of controllers put on the tower. Figure C-2 shows the same tower as in Fig. C-1, but certain changes have been made in the control scheme. The level controller now governs the cooling-water flow rate, the reflux flow may be set by a valve or flow controller, and the reboiler steam rate is controlled by a signal from a thermocouple measuring the temperature of the second stage from the bottom. In this scheme the condenser must be overdesigned. Aside from pressure and feed variables, the following variables have now been set by construction and external means:

Stages above the feed n
Stages below the feed m
Temperature of second stage T_2
Reflux flow rate r

The number of independent variables has not changed. For example, Q_C and Q_R can no longer be independently set by adjusting valves, but T_2 can now be held at a determinable set point (within limits), and r can now be adjusted independently by means of the valve. There are still four additional independent variables. Other control schemes could be shown, all with the same result.

Our approach to the description rule so far has involved the assumption of equilibrium stages; yet if we build five plates in a distillation column we do not necessarily obtain the action of five equilibrium stages. The degree of equilibration of the vapor and liquid stage exit streams will depend upon such factors as the flow patterns on the plate, the intimacy of contact provided between vapor and liquid, etc. However, we are justified in saying that we have provided through construction the action of n *equilibrium* stages above the feed stage and the action of m *equilibrium* stages below the feed stage; n and m are numbers of equivalent equilibrium stages rather than the actual number of plates provided.

TOTAL CONDENSER VS. PARTIAL CONDENSER

If the column of Fig. C-1 is changed by using a total condenser at the top rather than a partial condenser, the column shown in Fig. C-3 results. If the variables defining the feed and the pressure are considered set, the remaining variables are found to be

Equilibrium stages above feed stage n
Equilibrium stages below feed stage m
Reboiler heat duty Q_R
Condenser heat duty Q_C
Reflux flow rate r

Here the remaining variables number five, compared with four for the same column using a partial condenser. In Fig. C-3 it is apparent that the liquid flow leaving the condenser can be split in any desired ratio by adjusting the valve in the reflux line. With a partial condenser, on the other hand, the ratio of distillate to reflux is set by the percent vapor in the total stream leaving the condenser. Thus in a problem description for a distillation column with a total condenser one more variable must be set independently than for a problem where a column has a partial condenser.

A certain amount of consideration reveals that the five variables for a column with a total condenser cannot all be replaced by separation variables or by other variables which influence the separation. This results from the fact that the amount of reflux and the amount of heat removed in the condenser are both controlling only one variable which affects the fractionation, namely, the *internal* liquid flow in the section of the column above the feed; r and Q_C are not independent of each other. One can increase the internal liquid flow either by increasing the reflux flow rate or by increasing the condenser duty while holding the rate of reflux return from the accumulator drum constant. In the latter case the reflux would become cooler and would produce more internal liquid flow when equilibrating with the vapor on the top stage. Hence, if one of these two variables were changed to change the fractionation, the other variable could be changed in reverse direction to return the fractionation to its original condition. This is not true of any other pair of variables we have listed for the case of a total condenser.

Five variables must be set to describe a problem for the column of Fig. C-3 nevertheless. Since all the five listed cannot be replaced with variables which independently affect the fractionation, it is necessary to set at least one variable associated with the condenser load or the reflux. Often this is done by simply specifying the temperature of the reflux, normally with

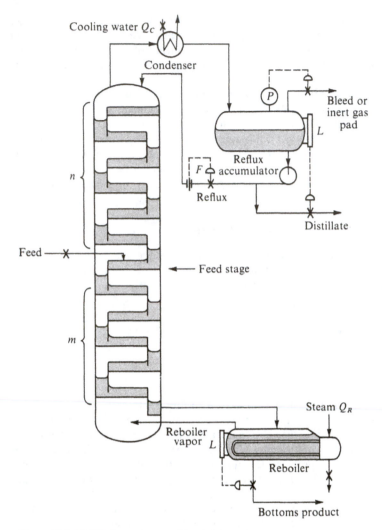

Figure C-3 Distillation column with a total condenser.

the statement that the reflux will be liquid at its saturation temperature or at some other set temperature.

RESTRICTIONS ON SUBSTITUTIONS AND RANGES OF VARIABLES

There are several other restrictions on the process of substituting variables. An obvious one, already mentioned, is that some prospective independent variables can be varied only *within limits*. For instance, in the column of Fig. C-1 with a partial condenser the product streams leave as thermodynamically saturated streams. As a result the overall enthalpy balance with a given feed will limit the extent to which Q_C and Q_R can change with respect to each other. Also the distillate rate cannot exceed the feed rate. The number of stages cannot be less than the minimum for the desired separation, nor can the reflux ratio or boil-up ratio be less than the minimum, etc.

In principle, more than two separation variables can be set in the problem description (Forsyth, 1970), but this is difficult since any separation variables beyond the first two will be bounded within a narrow range. For example, with a four-component feed one can readily set recovery fractions for two of the components, i.e., the keys, but setting a recovery fraction for a third component, e.g., a nonkey, can only be made within the narrow range of possible distributions for that component, given the set recovery fractions for the first two components and all combinations of reflux ratio and number of stages (see Distribution of Nonkey Components in Chap. 9).

If the feed rate is set, we cannot substitute both b and D as additional independent variables. Once F and b are specified, D is immediately fixed by overall mass balance. Any variables *uniquely related by a single equation or subset of equations* cannot be specified independently.

The feed rate or some capacity variable (a rate per unit time) must remain as an independent variable or else the list of independent variables will be reduced by 1. The quality of separation obtained is independent of the capacity if there are equilibrium stages. In the case of the column of Fig. C-1 we could specify the separation completely through the following list of variables, although the capacity would be indeterminate.

Feed composition z_i	$R - 1$
Feed specific enthalpy h_F/F	1
Pressure P	1
Stages above feed stage n	1
Stages below feed stage m	1
Reboiler duty per unit feed Q_R/F	1
Condenser duty per unit feed Q_C/F	1
	$\overline{R + 5}$

By eliminating all variables having to do with the actual capacity of the column for processing feed (number of moles processed per unit time) we have reduced the number of independent variables by 1 from $R + 6$ to $R + 5$. Note that there are only $R - 1$ feed *composition* variables since Σz_i must equal 1.0.

OTHER APPROACHES AND OTHER SEPARATIONS

The method of counting variables and counting equations has been applied to distillation by Gilliland and Reed (1942) and Kwauk (1956), the results giving the same number of independent variables as the description rule. The method of counting variables and equations is also covered in the first edition of this book, along with examples of applications to several other types of separations.

REFERENCES

Forsyth, J. S. (1970): *Ind. Eng. Chem. Fundam.*, **9**:507.
Gilliland, E. R., and C. E. Reed (1942): *Ind. Eng. Chem.*, **34**:551.
Hanson, D. N., J. H. Duffin, and G. F. Somerville (1962): "Computation of Multistage Separation Processes," chap. 1, Reinhold, New York.
Kwauk, M. (1956): *AIChE J.*, **2**:240.

D

OPTIMUM DESIGN OF DISTILLATION PROCESSES

In the design of a distillation column it is necessary to fix values of a complete set of independent variables. The feed variables are normally already known, and so, typically, it is necessary to pick near-optimum values of the reflux ratio, the column pressure, the column diameter, and the product purities. For any set of values of these additional independent variables it is then possible to determine the number of stages, etc., by the techniques outlined in this book. Depending upon the situation, the optimum value of one or several of these independent variables can be determined in the course of the design.

COST DETERMINATION

Costs associated with a distillation column itself are presented by Miller and Kapella (1977). Costs for bubble-cap columns and references for other sources of costs are given by Woods (1975). Costs of column auxiliaries (condensers, reboilers, etc.) and various other separation equipment are covered by Guthrie (1969). In all cases sources of costs should be updated by means of the *cost indexes* for plant, equipment, chemicals, construction, etc., reported biweekly in *Chemical Engineering*.

OPTIMUM REFLUX RATIO

Peters and Timmerhaus (1968) give an example of the determination of the optimum reflux ratio for a binary distillation with set feed conditions, a set pressure, and set product specifications. The specified conditions are:

Feed rate = 700 lb mol/h
Feed thermal condition = saturated liquid
Feed composition = 45 mol % benzene, 55 mol % toluene
Column pressure = 1 atm
Distillate composition = 92 mol % benzene
Bottoms composition = 5 mol % benzene
Average cooling-water temp in condenser = 90°F
Gain in cooling-water temp in condenser = 50°F
Steam to reboiler = saturated, at 60 lb/in² abs
Max allowable vapor velocity in tower = 2.5 ft/s
Stage efficiency = 70% (overall)

The column is to contain bubble-cap trays and will operate with a total condenser returning saturated liquid reflux. Constant heat-transfer coefficients are assumed for the reboiler (80 Btu/h·ft²·°F) and the condenser (100 Btu/h·ft²·°F).

Purchase and installation costs are considered for the column itself and for the reboiler and condenser. The unit is to operate 8500 h/year (97 percent time on stream), and the annual fixed charges for depreciation, maintenance, interest, etc., amount to 15 percent of the total cost for installed equipment counting piping, instrumentation, and insulation. The annual operating costs are for steam (50 cents per 1000 lb) and for cooling water (0.36 cent per 1000 lb). Other costs, such as labor, are presumed to be unaffected by the choice of reflux ratio in the column.

The remaining variable to be set in order to describe the column completely is the reflux ratio. This is chosen as the *optimum value*, defined as that value of reflux ratio which causes the total variable annual cost (annual fixed charges plus annual operating costs) to be a minimum. The results of computations of column size and of the various contributions to the total variable annual cost for different values of the reflux ratio are shown in Table D-1 and Fig. D-1.

Several trends in Table D-1 should be emphasized. As the reflux ratio increases above the minimum, the number of plates required in the column becomes less, since the operating lines

Table D-1. Individual costs contributing to total variable annual cost for benzene-toluene distillation example

| | | | Annual cost | | | | | |
| | Number of | | Fixed charges | | | Operating | | Total |
Reflux ratio	actual plates required	Column diameter, ft	Column	Condenser	Reboiler	Cooling water	Steam	variable annual cost
1.14	∞	6.7	$ ∞	$1870	$3960	$5780	$44,300	$ ∞
1.2	29	6.8	8930	1910	4040	5940	45,500	66,320
1.3	21	7.0	6620	1950	4130	6200	47,500	66,400
1.4	18	7.1	5920	2000	4240	6470	49,600	68,230
1.5	16	7.3	5490	2050	4340	6740	51,700	70,320
1.7	14	7.7	5290	2150	4540	7290	55,700	74,970
2.0	13	8.0	5210	2280	4800	8100	61,800	82,190

Source: Adapted from Peters and Timmerhaus, 1968, p. 316; used by permission.

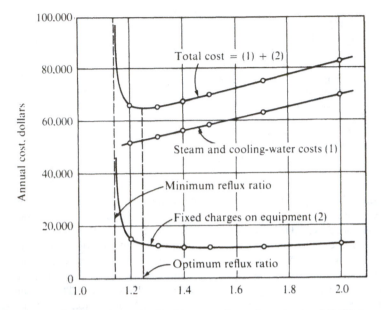

Reflux ratio, moles liquid returned to column/mole of distillate

Figure D-1 Total variable annual cost for benzene-toluene distillation as a function of reflux ratio. (*From Peters and Timmerhaus, 1968, p. 312; used by permission.*)

are moving away from the equilibrium curve. Column costs are directly proportional to the number of plates (Peters and Timmerhaus, 1968). The column diameter, on the other hand, increases since the reflux ratio and hence the vapor rate through the column are increasing. Despite the increase in column diameter, the annual fixed cost for the column goes down as the reflux increases because the saving in tower height more than offsets the increase in diameter. This will not continue to be the case as reflux increases, however. At very high reflux ratios the plate requirement approaches a constant value characteristic of the minimum stage requirement, while the diameter continues to increase; hence at some reflux ratio higher than those shown in Table D-1 the annual fixed charges for the column will begin to rise again.

The annual fixed charges for the reboiler and the condenser and the annual operating costs for steam and cooling water all rise in proportion to the vapor rate in the column (the fixed charges rise less rapidly because the installed costs of the heat exchangers are proportional to the exchanger duty to a power less than unity). Hence the optimization in this case reflects a balance between the annual fixed charges for the column, which decrease from infinity as reflux increases in this range, and the fixed charges and operating costs associated with the heat exchangers, which increase toward infinity as reflux increases. A minimum total annual cost exists at an intermediate reflux ratio.

In this case the minimum occurs at a reflux ratio of about 1.25. The minimum reflux ratio, as shown in Table D-1, is 1.14; hence the minimum total variable annual cost occurs at a reflux ratio 1.10 times the minimum.

It is very important, however, to notice the shape of the cost-vs.-reflux curve in Fig. D-1. The curve rises steeply and suddenly toward infinity at reflux ratios less than the optimum; in fact, the cost must become infinite at a reflux ratio just 10 percent less than the optimum. On the other hand, the curve rises much more slowly at reflux ratios above the optimum, and one could design at 1.20 to 1.30 times the minimum reflux and still have a total variable annual cost that was only 2 to 6 percent greater than that at the optimum reflux ratio.

Another point from Table D-1 should also be brought out. The single most important variable cost at the optimum reflux conditions is the operating cost for reboiler steam, which contributes some 70 percent of the total variable annual cost. This result is general for steam-driven water-cooled columns; the steam costs are usually an order of magnitude larger than the coolant costs. With refrigerated-overhead subambient-temperature columns, the refrigeration costs usually will be dominant.

Heaven (1969) used typical economic conditions for the 1960s to find the optimum reflux ratio for 70 different hydrocarbon distillations carried out at atmospheric pressure or above. Except for two towers with minimum reflux ratios under 0.2, he found the optimum reflux to be between 1.11 and 1.24 times the minimum in all cases. Brian (1972) reports a calculation of optimum reflux ratio for an atmospheric benzene-toluene distillation with a steam cost of 70 cents per 1000 lb and obtains an optimum 1.17 times the minimum. Fair and Bolles (1968) present calculated results for three cases, all giving an optimum reflux less than 1.1 times the minimum. Van Winkle and Todd (1971) evaluated a large number of cases and concluded that the optimum reflux ratio lay between 1.1 and 1.6 times the minimum, lower multiples of the minimum being favored by high relative volatilities and/or nonsevere separation specifications. Conversely, relative volatilities closer to unity and sharper separations led to higher ratios of optimum reflux ratio to minimum reflux ratio, within that range.

Costs of steam and other forms of energy have risen much more than materials costs in the years since these calculations were made. Typical steam costs for 1978 are in the range of $1.50 to $4 per 1000 lb. The percentage increase in steam costs being substantially greater than the percentage increase in materials cost means that optimum reflux ratios have become a still smaller multiple of the minimum reflux ratio. Tedder and Rudd (1978) considered an equimolar isobutane–n-butane distillation and found the optimum reboiler boil-up ratio to be 1.11 and 1.03 times the minimum for steam costs of $0.44 and $4.40 per 1000 lb, respectively. The corresponding reflux ratios should be very nearly the same as the boil-up ratios for this case.

It is safe to say that optimum reflux ratios in the late 1970s tend to be *less* than 1.10 times the minimum, on the basis of calculations like those leading to Table D-1 and Fig. D-1. However, under these circumstances precise knowledge of vapor-liquid equilibrium becomes very important because cost curves, for example, Fig. D-1, rise so sharply as the minimum reflux is approached. Changes in the vapor-liquid equilibrium data used change the minimum reflux ratio. Similarly, errors in the stage efficiency or changes in feed composition can change the reflux ratio needed to accomplish a given separation with a fixed number of stages. Consequently, when there is uncertainty in the vapor-liquid equilibrium data, the stage efficiency, and/or the feed composition, it is best to design for a reflux ratio somewhat higher than the economic optimum found by this sort of analysis. Optimum overdesign is discussed later in this appendix.

Higher energy costs, in particular the need for refrigeration overhead, lead to optimum reflux ratios closer to the minimum. On the other hand, more expensive materials of construction, more severe separations, greater rates of equipment write-off, and/or relative volatilities closer to 1 all lead to higher optimum reflux ratios.

OPTIMUM PRODUCT PURITIES AND RECOVERY FRACTIONS

Often the product purities to be achieved in a distillation column will be determined by the specifications imposed upon marketable material by the buyers. Thus, for example, in ethylene production the purity required in the product ethylene and the different allowable levels of various impurities in the product are set by the needs of the consumers of the ethylene. On the other hand, the recovery fraction of product material to be obtained is frequently subject to an economic optimization. Taking the production of ethylene as an example again, the final distillation separates ethylene from ethane (see, for example, Fig. 13-28). The ethylene is product, subject to imposed purity specifications, but the ethane is to be recycled for thermal cracking. The recovery fraction of ethylene in the overhead product is related to internal plant economics and reflects the increased value of ethylene in the product as opposed to the value of recycled ethylene.

Example D-1 Suppose that the benzene product purity in the foregoing benzene-toluene distillation example is held by consumer specification to 92 mole percent but that the recovery fraction of benzene overhead (and hence the bottoms purity) may vary subject to an optimization. The toluene product will be used for gasoline blending. The increased value of benzene in the product as opposed to benzene returned to fuel is 2 cents per gallon. Using the same economic factors as in the optimum-reflux-ratio example, find the optimum recovery fraction of benzene in the overhead product. Make simplifications where appropriate.

SOLUTION Because the recovery fraction of benzene in the distillate probably will be relatively high, we shall assume that the relative flows of the products remain very nearly the same as in the optimum-reflux-ratio example. The overhead composition remains the same, and hence the minimum reflux ratio remains the same. The optimization will reflect an economic balance between the value of recovered benzene, on the one hand, and the additional plates in the stripping section required to recover that benzene, on the other. Reboiler, condenser, steam, and cooling water costs will not vary.

As a base case we shall take the solution in Table D-1 for a reflux ratio of 1.3 (about 15 percent above the minimum). The column cost for the base case (5 mol $\%$ benzene in the bottoms) is $6620 per year for 21 plates, or $315 per plate. Since the overall stage efficiency is 70 percent, the annual cost per equilibrium stage is $315/0.70 = $450.

Recovered benzene is worth an additional 2 cents per gallon, or since the density is 0.879 (Perry and Chilton, 1973), and the molecular weight is 78,

$$\text{Value of recovered benzene} = \frac{(\$0.02/\text{gal})(78 \text{ lb/lb mol})}{(8.33)(0.879) \text{ lb/gal}} = \$0.21/\text{lb mol}$$

The base case bottoms flow rate is 378 lb mol/h. With 8500 operating hours per year, the value of each incremental mole percent benzene removed from the bottoms is

Value of each mol $\%$ benzene removed

$$= (\$0.21/\text{lb mol})(378 \text{ lb mol/h})(8500 \text{ h/y})(0.01 \text{ mol } \%/\text{mole fraction})$$

$$= \$6800/\text{y}$$

This calculation neglects the small changes in product flows as the bottoms composition changes.

The variable number of stages for recovering benzene will come at the low-benzene-mole-fraction end of the column, where the relative volatility is nearly constant. At the boiling point of toluene the relative volatility of benzene to toluene is 2.38 (Maxwell, 1950). Since the operating line and equilibrium curve are both nearly straight in this region, it is probably simplest to use the KSB equations [Eqs. (8-15) and (8-16)]. Some complication arises due to the fact that the base point for the stripping-section operating lines will shift from case to case, causing changes higher in the column; however, this will be a secondary effect. To allow for it we shall compute the stage requirement up to $x_B = 0.10$ for each case.

Since the overhead product rate is 322 lb mol/h, the reflux ratio of 1.3 corresponds to a vapor flow of 322 $(1 + 1.3) = 740.6$ lb mol/h and a liquid flow of 1118.6 lb mol/h below the feed. Hence the value of mV'/L in the zone of variable stages is

$$\frac{mV'}{L} = \frac{2.38(740.6)}{1118.5} = 1.54$$

We can use the solution of the KSB equation presented in Fig. 8-3 if we convert y to x, L to V, and m to $1/m$. Hence the vertical axis becomes

$$\frac{x_{B,\,out} - K y_{B,\,in}}{x_{B,\,in} - K y_{B,\,in}}$$

When we denote the bottoms composition by x_B and take a fixed upper mole fraction of 0.10, this group becomes

$$\frac{x_B - (1/2.38)x_B}{0.10 - (1/2.38)x_B} = \frac{1.38 x_B}{0.238 - x_B}$$

The parameter on Fig. 8-3 is now mV'/L', or 1.54.

Taking as a base case $x_B = 0.05$, we can compute the following table of additional equilibrium-stage requirements vs. bottoms composition:

x_B	$\dfrac{1.38 x_B}{0.238 - x_B}$	Number of equilibrium stages, N	Additional equilibrium stages	Additional plates	Additional benzene recovery	Total
					Variable annual costs	
0.05	0.369	1.1	0	$ 0	$ 0	$ 0
0.02	0.126	2.9	1.8	800	−20,400	−19,600
0.01	0.061	4.5	3.4	1500	−27,200	−25,700
0.005	0.021	6.8	5.7	2600	−30,600	−28,000
0.002	0.0085	8.6	7.5	3400	−32,640	−29,240
0.001	0.0042	10.2	9.1	4100	−33,320	−29,220
0.0005	0.0021	12.0	10.9	4900	−33,660	−28,760
0.0002	0.00085	13.9	12.8	5800	−33,864	−28,064

The optimum bottoms composition is about 0.0015 mole fraction benzene, corresponding to a benzene recovery fraction of about 0.9985 in the distillate product. Recovery of this additional benzene provides a savings of an additional $29,000+ per year. □

In a multitower sequence components of intermediate volatility are key components in more than one tower. The optimum distribution of recovery fractions for such components between towers, to make a given overall recovery fraction, has been explored by Gawin (1975).

OPTIMUM PRESSURE

Operation of a distillation column under vacuum requires extra equipment for producing and maintaining the vacuum. Also, operation under vacuum gives a higher volumetric vapor flow rate corresponding to any given molar vapor flow rate, and as a result vacuum columns have a relatively large diameter. As noted in Chap. 12, there is also a high tendency toward capacity limitation by entrainment in vacuum columns. These factors, taken together, imply that distillation towers probably will not be operated at vacuum pressures unless some aspect of the separation problem *requires* vacuum operation. For example, it may be necessary to operate under vacuum in order to achieve a low enough bottoms temperature to avoid thermal decomposition of the bottoms and/or to allow the use of readily obtainable heating media to accomplish the vaporization in the reboiler.

Operation at pressures substantially above atmospheric requires that the column shell be thicker to withstand the pressure. Also, it is a general characteristic of distillation systems that the relative volatility becomes closer to unity as the system pressure increases; consequently plate and reflux requirements for a given quality of separation increase as pressure increases. In nearly all cases these factors more than offset the savings in tower diameter which can accrue from the higher vapor density and lower volumetric vapor flow rate at higher pressure. Hence high-pressure operation is usually justified only in situations where the high-pressure operation is needed to allow condensation of the overhead stream with cooling water or where refrigeration is required for overhead condensation anyhow.

The foregoing analysis leads to the conclusion that the column pressure for distillation should be slightly above atmospheric as long as the condensation overhead can be accomplished with cooling water and the reboiling can be accomplished with ordinary heating media without thermal damage to the bottoms material. If high pressure (up to perhaps 250 lb/in^2 abs) is necessary to enable condensation of the overhead with cooling water, the column pressure should ordinarily be such as to give an average temperature difference driving force of 5 to 15°C in the overhead condenser. Heaven (1969) examined economically optimum column pressures for 70 hydrocarbon distillations requiring pressures in this range and found this criterion to be generally true.

If the column pressure required to accomplish overhead condensation with cooling water is above 250 lb/in^2 abs, it is worth considering the alternative of using a refrigerant on the overhead and running the column at a lower pressure. In this case an optimization calculation may be useful, the variable being the column pressure or the refrigerant temperature.

Griffin (1966) has given the results of a determination of the optimum pressure for an ethylene-ethane distillation, operated using the vapor recompression scheme of Fig. 13-16b. The conditions of the problem are shown in Table D-2.

The optimization calculation allows for variable operating costs for refrigeration and compressor power and for the fixed charges on the column, the compressors, and the various heat exchangers. The relative volatility of ethylene to ethane increases from 1.4 to 1.6 at a tower pressure of 250 lb/in^2 abs to 1.7 to 2.0 at a pressure of 80 lb/in^2 abs. The optimization represents a balance of the reflux and stages saving due to this higher relative volatility against the refrigeration and materials costs of low temperatures, along with several other factors. The annual cost figures reported include constant contributors to the cost, i.e., labor, as well as the variable costs.

The cost of the separation, expressed as cents per pound of ethylene produced, is shown as a function of pressure in Fig. D-2. Contributions to the purchased equipment costs and annual operating costs at various pressures are shown in Table D-3. The tower costs are computed by allowing different materials of construction for plates at different locations in the tower. Even so, there are discontinuities in the product cost. The discontinuity at about 160 to 175 lb/in^2 abs tower pressure is associated with the change in the material of construction for the reboiler from ordinary carbon steel to killed carbon steel as the temperature in the reboiler drops below $-20°F$ at pressures below 160 lb/in^2 abs, and with the change in the material for the overhead vapor compressor from killed carbon steel to $3\frac{1}{2}\%$ nickel steel as the overhead vapor drops below $-50°F$ at pressures below 175 lb/in^2 abs. There is also a discontinuity in the cost function at about 94 lb/in^2 abs, as the material for the reboiler goes from killed carbon steel to $3\frac{1}{2}\%$ nickel steel.

The purchased tower cost decreases with increasing pressure. This trend reflects the saving due to less expensive materials of construction as the tower goes from all $3\frac{1}{2}\%$ nickel steel at 80 lb/in^2 abs to 70 percent of the trays being ordinary carbon steel and the remainder being killed carbon steel at 250 lb/in^2 abs. This saving offsets the greater plate requirement caused by the lower relative volatility at higher pressures. The reboiler purchased cost be-

Table D-2. Conditions for optimum-pressure example

Feed:

Flow rate	41.500 lb/h
Composition	32.5 wt"$_0$ ethylene, 67.5 wt $^0_0$ ethane
Condition	Satd liquid at 290 lb/in² abs

Ethylene:

Product purity	98 wt $^0_0$
Product delivery pressure	500 lb/in² abs
Recovery fraction	0.97
Temperature difference across reboiler[†]	14°F
Compressor efficiencies[‡]	0.65

Materials of construction:

Above −20°F	Ordinary carbon steel
−20 to −50°F	Killed carbon steel
Below −50°F	Nickel steel, $3\frac{1}{2}$"$_0$

Levels of refrigeration available

	Temp. °F	Annual cost, per 10⁶ Btu/h
Propane	60	$ 3,700
	18	7,000
	0	8,600
	−34	10,600
Ethylene	−90	14,000
	−150	16,600

[†] Condensing ethylene to evaporating ethane.
[‡] (Isentropic work)/(actual work).
Source: Data from Griffin (1966).

comes less whenever an increase in pressure allows a less expensive material but rises with pressure for any given material of construction because the lower relative volatility at higher pressures increases the reflux and boilup requirements. The overhead-vapor compressor cost increases with increasing tower pressure because the overhead-vapor flow rate becomes greater at the higher reflux ratios, but there is a drop in compressor cost when the transition from $3\frac{1}{2}$"$_0$ nickel steel to killed carbon steel becomes possible at 175 lb/in² abs. The product compressor cost is less at higher tower pressures because a smaller compression ratio is required to bring the product up to 500 lb/in² abs.

The utilities costs in the second half of Table D-3 are composed of costs for refrigerant in the desuperheater, for power to drive the product compressor, and for steam to drive the overhead-vapor compressor. The other factors making up the annual operating costs vary in near proportion to the purchased equipment costs. Notice that the overhead-vapor compressor is the largest single purchased item of equipment in cost. The purchased cost of this compressor and the desuperheater along with the utilities cost for the compressor drive and refrigerant for the desuperheater (78 percent of the utilities costs between them) can be attributed to refrigeration for condensation of the overhead vapor. If the vapor-recompression system were not used, these units would be replaced by an expensive refrigeration system. Thus

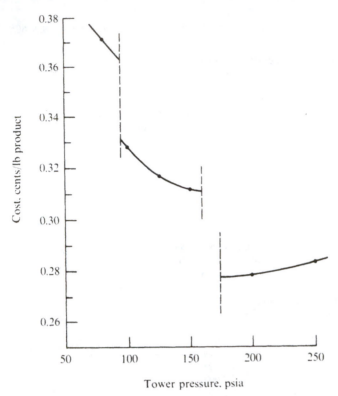

Figure D-2 Effect of distillation pressure on cost of ethylene recovery from an ethylene-ethane mixture. *(Adapted from Griffin, 1966, p. 16; used by permission.)*

Table D-3. Contributions to ethylene recovery costs (data from Griffin, 1966)

	Tower operating pressure, lb/in² abs					
	80	100	125	150	200	250
Purchased equipment costs:						
Distillation tower	$ 68.800	$ 65.000	$ 60.500	$ 54.800	$ 46.820	$ 46.000
Reboiler	72.500	50.000	51.500	56.200	43.000	47.500
Desuperheater (refrigerated cooler)	13.750	9.450	6.740	6.900	8.150	9.300
Vapor compressor (entire overhead vapor)	135.000	135.000	136.500	137.250	134.400	138.300
Product compressor (ethylene product)	30.650	27.300	23.700	19.100	9.240	0
Instruments	5.200	5.200	5.200	5.200	5.200	4.880
Total	$325.900	$291.950	$284.140	$279.450	$246.810	$245.980
Annual operating costs:						
Labor	$ 13.875	$ 13.875	$ 13.875	$ 13.875	$ 13.875	$ 13.875
Depreciation	76.406	66.858	64.629	63.843	55.807	56.087
Maintenance	38.203	33.429	32.315	31.922	27.904	28.044
Property taxes	11.450	10.028	9.700	9.576	8.370	8.420
Utilities	43.872	40.175	38.513	35.820	35.890	39.140
Return on investment	229.218	200.574	193.887	191.527	167.421	168.261
Total	$413.024	$364.939	$352.919	$346.563	$309.267	$313.827

this example bears out the earlier statement that refrigeration costs are usually dominant in columns operating with a refrigerated overhead.

From Fig. D-2 it would appear that the optimum pressure is just above 175 lb in² abs. In actuality it would probably be better to choose a higher pressure, such as 200 lb/in² abs, so that temperature fluctuations during operation will not impose a materials-damage problem in the killed-carbon-steel overhead-vapor compressor.

The optimum operating pressure in ethylene-ethane fractionators is also discussed by Davison and Hays (1958); the optimum pressure and reflux ratio for propylene-propane fractionators is discussed by Smy and Hay (1963); and the optimum pressure for distillation of isobutane from *n*-butane is discussed by Tedder and Rudd (1978).

OPTIMUM PHASE CONDITION OF FEED

Feed preheating, including partial or complete vaporization of the feed, reduces the required reboiler heating load but not in direct compensation since the vapor generated in a preheater is used only in the rectifying section. The extent to which feed preheating is advantageous, if at all, depends upon the relative costs of the heating media that could be used in the preheater and the reboiler. Tedder and Rudd (1978) have examined the optimum degree of feed preheating for an isobutane–*n*-butane distillation. Petterson and Wells (1977) also consider optimum levels of feed preheat.

OPTIMUM COLUMN DIAMETER

The column diameter for distillation is almost always set through a design heuristic rather than through an optimization calculation. In principle, it is possible to determine an optimum diameter by balancing savings due to a smaller diameter against the additional plate requirement resulting from a lower stage efficiency caused by entrainment and/or flooding. The result of such an optimization, however, would be a diameter giving a vapor rate where entrainment was relatively large or where the operation was quite close to flooding. Such a design would give a tower with poor operating flexibility. Because of unavoidable fluctuations in conditions during operation, the tower would have a tendency toward frequent floodings or gross losses of efficiency. So as to give operating flexibility to guard against this behavior, column diameters are usually selected to give a safe distance between the design conditions and the ultimate capacity limit. Common practice for plate columns is to set the column diameter to give a vapor velocity equal to 70 to 85 percent of that at the flooding or entrainment limit. A lower percentage is commonly used for packed columns.

OPTIMUM TEMPERATURE DIFFERENCES IN REBOILERS AND CONDENSERS

Reboiler temperatures should be kept low enough to avoid bottoms degradation and/or fouling. The general levels of reboiler and condenser temperatures reflect the pressure chosen for a distillation column. Common temperature differences used for heat exchange across reboilers and condensers (Frank, 1977) are:

	Temp, K
Condenser:	
Refrigeration	3–10
Cooling water	6–20
Pressurized fluid	10–20
Boiling water	20–40
Air	20–50
Reboiler:	
Process fluid	10–20
Steam	10–60
Hot oil	20–60

Source: Data from Frank, 1977.

OPTIMUM OVERDESIGN

The design of separation equipment is complicated by uncertainties in the phase equilibrium data and in the stage efficiencies. One approach to this difficulty is to adopt a conservative design, using the most pessimistic estimates of the equilibrium relationship and the stage efficiency. This usually results in a considerably overdesigned device, however, and a more common approach is to carry out the design for the best estimates of the equilibrium and efficiency and then apply an overdesign factor to the number of stages and/or the capacity parameters to allow for the uncertainty. For a distillation column the numbers of plates provided and/or the column diameter could be increased by whatever factor is chosen.

There have been some attempts to use probability analysis to determine what amount of overdesign is best. Villadsen (1968) applied such an analysis to find the amount of overdesign of distillation columns warranted by uncertainties in the stage efficiency. The approach is to assume that the stage efficiency may have any value between a given lower limit and a given upper limit once the column is built and that no one efficiency within this range has a greater probability of occurring than any other. The yearly cost of the separation (including both operating costs and fixed charges for equipment) is then related to the number of plates in the tower N, the overhead reflux ratio R, and the stage efficiency E. This annual cost is denoted by $U(N, R, E)$. There is an interrelationship between the reflux, the number of plates and the efficiency, however, such that the reflux should be that amount which is required to give the specified product purities with the prevailing values of N and E, assuming that N is still above N_{min}. Hence the cost can be considered to be a function of only two independent variables $U(N, E)$. The *expected cost* $\bar{U}(N)$ can now be defined as the sum of the costs for each possible value of the stage efficiency, weighted by the probability $p(E)$ of that stage efficiency's occurring:

$$\bar{U}(N) = \int_E U(N, E)p(E)\, dE \qquad (D-1)$$

The optimum number of plates to provide in the column would then be the value of N which makes $\bar{U}(N)$ in Eq. (D-1) a minimum.

Figure D-3 shows the results giving the optimum overdesign factor as a function of the range of the uncertainty in the stage efficiency. This result is relatively insensitive to the mean level of the efficiency, the relative volatility, the recovery fractions, and the percentage annual amortization of the equipment costs. The overdesign factor in Fig. D-3 is defined as the

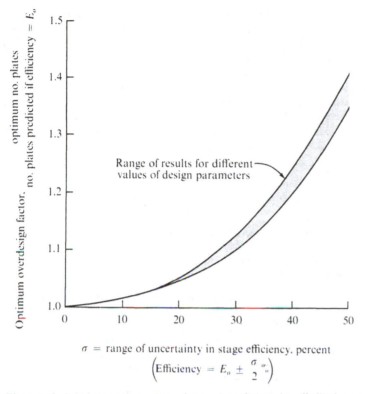

optimum no. plates

no. plates predicted if efficiency = E_o

Optimum overdesign factor, = $\dfrac{\text{optimum no. plates}}{\text{no. plates predicted if efficiency} = E_o}$

Range of results for different values of design parameters

σ = range of uncertainty in stage efficiency, percent

$$\left(\text{Efficiency} = E_o \pm \frac{\sigma}{2} {}_{\text{\tiny o}}\right)$$

Figure D-3 Optimal overdesign factor for number of plates in a distillation tower. (*After Villadsen, 1968.*)

number of plates determined as the optimum by minimizing Eq. (D-1), divided by the number of plates which would be indicated if the efficiency were known with certainty to be equal to the mean of the lower and upper limits on stage efficiency. This analysis as presented by Rudd and Watson (1968) implies that the diameter of the tower can be varied or, more realistically, that the tower capacity can be varied. Presumably a similar analysis could be used to obtain an optimum overdesign factor for the column diameter.

Lashmet and Szczepanski (1974) compared observed stage efficiencies for a large number of real distillation columns with the predictions of the AIChE method for stage efficiencies (Chap. 12), thereby obtaining an estimate of the uncertainty in predicting stage efficiencies. They used these results to determine the overdesign in number of plates required to give 90 percent confidence of achieving the desired separation with 1.3 times the minimum reflux ratio. Overdesign factors ranged from 1.07 to 1.16 for typical conditions.

In addition to uncertainties in the stage efficiency and the vapor-liquid equilibrium data there also will be uncertainties in the vapor-handling capacity of a column of given diameter, in the vapor-generation capacity of a reboiler of given size, and in the vapor-condensation capacity of a condenser of given size. Saletan (1969) indicated how these last three uncertainties can be combined with the stage-efficiency uncertainty to give the probability distribution of the feed-handling capacity of a column of given size which must make products of specified purity. The vapor-handling capacity of a distillation system can be limited by either the reboiler or the column diameter or the condenser, the one with the least vapor capacity providing the limit. Hence the probability P that the vapor-handling capacity of a distillation system is greater than some specified value is given by the product of three probabilities, $P = P_{reb} P_{col} P_{con}$. The terms P_{reb}, P_{col}, and P_{con} are the probabilities that the vapor-handling capacities of the reboiler, column, and condenser, respectively, are greater than the specified value. The probability distribution for the stage efficiency can be converted into a probability distribution for the reflux ratio required to accomplish the specified separation with the set number of plates. Once again, there may be a finite probability that the separation cannot be attained at all because of the minimum-stages limitation. The probability distribution of vapor-handling capacities for the distillation system can then be combined with the probability distribution of required reflux ratios to give the probability distribution for the feed-handling capacity of the distillation system. One might then ensure that there is an 80, 90, 95, or 98 percent probability that the distillation system can handle the desired feed capacity.

REFERENCES

Brian, P. L. T. (1972): "Staged Cascades in Chemical Processing," Prentice-Hall, Englewood Cliffs, N.J.
Davison, J. W., and G. E. Hays (1958): *Chem. Eng. Prog.*, **54**(12): 52.
Fair, J. R., and W. L. Bolles (1968): *Chem. Eng.*, Apr. 22, p. 156.
Frank, O. (1977): *Chem. Eng.*, Mar. 14, p. 111.
Gawin, A. F. (1975): M.S. thesis in chemical engineering, University of California, Berkeley.
Griffin, J. D. (1966): first-prize-winning solution for 1959, in "Student Contest Problems and First-Prize-Winning Solutions, 1959-65," American Institute of Chemical Engineers, New York.
Guthrie, K. M. (1969): *Chem. Eng.*, Mar. 24, p. 114.
Heaven, D. L. (1969): M.S. thesis in chemical engineering, University of California, Berkeley.
Lashmet, P. K., and S. Z. Szczepanski (1974): *Ind. Eng. Chem. Process Des. Dev.*, **13**:103.
Maxwell, J. B. (1950): "Data Book on Hydrocarbons," Van Nostrand, Princeton, N.J.
Miller, J. S., and W. A. Kapella (1977): *Chem. Eng.*, Apr. 11, p. 129.
Perry, R. H., and C. H. Chilton (1973): "Chemical Engineers' Handbook," 5th ed., McGraw-Hill, New York.
Peters, M. S., and K. D. Timmerhaus (1968): "Plant Design and Economics for Chemical Engineers," 2d ed., McGraw-Hill, New York.
Petterson, W. C., and T. A. Wells (1977): *Chem. Eng.*, Sept. 26, p. 79.
Rudd, D. F., and C. C. Watson (1968): "Strategy of Process Engineering," Wiley, New York.
Saletan, D. I. (1969): *Chem. Eng. Prog.*, **65**(5): 80.
Smy, K. G., and J. M. Hay (1963): *Can. J. Chem. Eng.*, **41**:39.
Tedder, D. W., and D. F. Rudd (1978): *AIChE J.*, **24**:303, 316.
Van Winkle, M., and W. G. Todd (1971): *Chem. Eng.*, Sept. 20, p. 136.
Villadsen, J. (1968): cited in D. F. Rudd and C. C. Watson, "Strategy of Process Engineering," Wiley, New York.
Woods, D. R. (1975): "Financial Decision Making in the Process Industry," p. 172, Prentice-Hall, Englewood Cliffs, N.J.

SOLVING BLOCK-TRIDIAGONAL SETS OF LINEAR EQUATIONS; BASIC DISTILLATION PROGRAM

In this appendix the form of block-tridiagonal matrices and the types of equations for which they are applicable are outlined. An efficient computer program (BAND) is given for solving these systems of equations. Finally, a simple distillation program, using the block-tridiagonal-matrix solution, is presented. The approach and programs are those developed by Newman (1967, 1968, 1973).

BLOCK-TRIDIAGONAL MATRICES

Block-tridiagonal matrices can be generated from sets of simultaneous linear difference equations in several unknowns written for successive positions. In order to become block-tridiagonal, the equations must involve unknowns at only the position in question and the two adjacent positions. This condition is met in countercurrent-staged and continuous-contactor separation processes.

If there is only one unknown variable to be evaluated at each position, only one equation is written for each position, relating the values of the unknowns at that position and the two adjacent positions. In that case the equations become a simple tridiagonal set, given by Eqs. (10-22). The solution can then be made efficiently by the Thomas method, outlined in Eqs. (10-23) to (10-30).

Generalization of the tridiagonal matrix to the case where there are n unknowns to be evaluated at each position (and n simultaneous equations at each position) leads to the block-tridiagonal matrix. Extension of the Thomas method from tridiagonal to block-tridiagonal matrices leads to the highly efficient BAND method presented here.

A block-tridiagonal system of equations takes the form

$$\sum_{k=1}^{n} A_{i,k}(j)C_k(j-1) + B_{i,k}(j)C_k(j) + D_{i,k}(j)C_k(j+1) = G_i(j) \qquad \text{(E-1)}$$

Here the n unknowns are $C_1, \ldots, C_k, \ldots, C_n$ at each position, $j = 1, 2, \ldots, j_{max}$. This amounts to nj_{max} total unknowns. The subscript i refers to the equation number, $i = 1, 2, \ldots, n$, again at each position j. The coefficients are $A_{i,k}(j)$, $B_{i,k}(j)$, and $D_{i,k}(j)$, as shown in Eq. (E-1), and are independent of the C values for a set of linear equations.

Written in block-tridiagonal form, Eq. (E-1) becomes

$$
\begin{bmatrix}
\mathbb{B}(1) & \mathbb{D}(1) & \mathbb{X} & & & \\
\mathbb{A}(2) & \mathbb{B}(2) & \mathbb{D}(2) & & & \\
\cdots\cdots & \cdots\cdots & \cdots\cdots & \cdots\cdots & \cdots\cdots & \cdots\cdots \\
 & & \mathbb{A}(j) & \mathbb{B}(j) & \mathbb{D}(j) & \\
\cdots\cdots & \cdots\cdots & \cdots\cdots & \cdots\cdots & \cdots\cdots & \cdots\cdots \\
 & & \mathbb{Y} & & \mathbb{A}(j_{max}) & \mathbb{B}(j_{max})
\end{bmatrix}
\begin{bmatrix}
C(1) \\ C(2) \\ \cdots \\ C(j) \\ \cdots \\ C(j_{max})
\end{bmatrix}
=
\begin{bmatrix}
G(1) \\ G(2) \\ \cdots \\ G(j) \\ \cdots \\ G(j_{max})
\end{bmatrix}
\tag{E-2}
$$

where the elements $\mathbb{A}$, $\mathbb{B}$, and $\mathbb{D}$ of the main matrix are themselves $n \times n$ matrices of coefficients, i.e.,

$$
\mathbb{B}(j) =
\begin{bmatrix}
B_{1,1} & B_{1,2} & \cdots & B_{1,n} \\
B_{2,1} & B_{2,2} & \cdots & B_{2,n} \\
\cdots & \cdots & \cdots & \cdots \\
B_{n,1} & B_{n,2} & \cdots & B_{n,n}
\end{bmatrix}
\tag{E-3}
$$

all evaluated at position j. The row subscripts refer to the equation number and the column subscripts to the unknown number.

$\mathbb{X}$ and $\mathbb{Y}$ are $n \times n$ matrices to be used in cases of certain boundary conditions (see below). The $C(j)$ and $G(j)$ elements in Eq. (E-2) are $1 \times n$ and $n \times 1$ matrices, respectively; i.e.,

$$
C(j) = [C_1 \quad C_2 \quad \cdots \quad C_n]
\tag{E-4}
$$

and

$$
G(j) =
\begin{bmatrix}
G_1 \\ G_2 \\ \vdots \\ G_n
\end{bmatrix}
\tag{E-5}
$$

again all evaluated at position j.

The symbols used here are a little different from those used in Chap. 10 for the Thomas method for $n = 1$ [Eq. (10-22)]; that is, D instead of C, C instead of I, G instead of D. This is done to be consistent with the notation used by Newman (1967, 1968, 1973).

Block-tridiagonal sets of difference equations arise whenever a staged process has flow linkages only between adjacent stages and when there is more than one unknown (mole fractions, total flows, temperature, etc.) at each position in independent sets of equations. They also arise for numerical solution of any coupled set of ordinary first- or second-order boundary-value differential equations, where numerical approximations of derivatives are made in the standard manner. For certain types of boundary conditions involving first derivatives Newman (1967, 1968, 1973) has shown that it is convenient to use the concept of an *image point*. This leads to additional terms in Eqs. (E-2), denoted by the $\mathbb{X}$ and $\mathbb{Y}$ entries. $\mathbb{X}$ is an $n \times n$ matrix of terms from the boundary conditions at one end, and $\mathbb{Y}$ is a similar matrix of terms from the boundary conditions at the other end.

The BAND method (below) solves Eqs. (E-2) under the presumption that the equations are linear, i.e., that the terms in the A, B, and D matrices are not themselves functions of C_1, $C_2, \ldots, C_n$. If the equations are, in fact, nonlinear, the BAND solution couples well with the full Newton multivariate (SC) convergence method, which involves successive linearization and solution of the linearized equations (Chap. 10 and Appendix A).

Within the field of countercurrent separation processes, the following classes of problems lead to block-tridiagonal matrices, solvable by the BAND method, coupled with Newton SC convergence.

1. Staged processes involving complex equilibria, that is. $K_j = f$ (all x_j and/or y_j, as well as T and P), and/or Murphree efficiencies not equal to unity (Chap. 10)
2. Continuous contactors, where complex equilibria and or variable mass-transfer coefficients occur (Chap. 11)
3. Complex staged or continuous-contactor processes involving axial dispersion, described by either the diffusion model or the stage or cell models of backmixing (Chap. 11)

Table E-1 lists the BAND program for solving block-tridiagonal sets of linear equations. Also included is an improved matrix-inversion routine (MATINV), itself a subroutine of BAND.

In BAND, A, B, C, D, G, X, and Y are taken directly from Eqs. (E-1) to (E-5). For A, B, D, X, and Y the first matrix index is the equation number i, and the second index is the unknown number k. For C the indices are k and j (position), and for G the index is i. The program is dimensioned for six unknowns (and therefore six simultaneous equations) at each position and 103 positions. These dimensions can readily be changed if desired. The E matrix $E(n, n + 1, j_{max})$ is used during the solution but is not input. The D matrix is made larger $[n \times (2n + 1)]$ than the D input $(n \times n)$ in order to provide working room during the solution. BAND is written to receive as input values of the A, B, D, and G matrices at each value of j, successively. The program transforms these values for storage, zeros the input matrices, and then receives values of the A, B, D, and G matrices for the next higher value of j. Upon reaching the specified j_{max} (denoted NJ), BAND then solves for all C_k at all j and returns these values in the C array. In order to use BAND, it is necessary to use or write a main program, which calls BAND at each j to supply values of A, B, D and G and which then receives the computed values of $C_{k,j}$ back after j reaches NJ.

BASIC DISTILLATION PROGRAM

Table E-2 gives a basic distillation program DIST, which uses BAND and MATINV as additional subroutines. The program calculates multicomponent distillation with varying molar overflow, using the Newton multivariate (SC) convergence scheme, BAND being used to solve linearized block-tridiagonal matrices during each iteration. The program is more pedagogical than broadly utilitarian, since as written it does not include provision for Murphree efficiencies other than unity and does not provide for nonideal phase-equilibrium data. The program is appropriate for student use to gain familiarity with the approach. The program can also be expanded in a straightforward fashion to incorporate more complex equilibrium data, since the equilibrium calculation is written as a separate function (EQUIL). This function can be made more complex in whatever way is desired. However, if the equilibrium relationships used cause K_i to depend upon other variables besides the component identity I and temperature T, it will be necessary to make those input variables to the new EQUIL function.

The program can accommodate several different types of problem specification, as noted below. The number of equilibrium stages must be a specified variable in all cases, as must be the locations of all feeds and sidestreams.

The following program description is paraphrased from Newman (1967).

The program is written to include as many as 40 stages (including reboiler and condenser) and as many as 10 components. For problems outside these limits, the dimensions can be changed appropriately. A total or a partial condenser can be used, and the possibility of a feed, a side draw of liquid, and a side draw of vapor on each stage has been included. A two-product condenser can be achieved by a liquid draw from the condenser.

Equilibrium ratios K_i in the form of power series in temperature or exponential functions can be used. These are put in a subroutine so that they can be changed without much trouble.

Table E-1 Subroutines BAND (Newman, 1967) and MATINV (Newman, 1978)† for solution of block-tridiagonal sets of linear equations

```
      SUBROUTINE BAND(J)
      DIMENSION C(6,103),G(6),A(6,6),B(6,6),D(6,13),E(6,7,103),X(6,6)
     1Y(6,6)
      COMMON A,B,C,D,G,X,Y,N,NJ
101   FORMAT (15H0DETERM=0 AT J=,I4)
      IF (J-2)  1,6,8
  1   NP1= N + 1
      DO 2 I=1,N
      D(I,2*N+1)= G(I)
      DO 2 L=1,N
      LPN= L + N
  2   D(I,LPN)= X(I,L)
      CALL MATINV(N,2*N+1,DETERM)
      IF (DETERM)  4,3,4
  3   PRINT 101, J
  4   DO 5 K=1,N
      E(K,NP1,1)= D(K,2*N+1)
      DO 5 L=1,N
      E(K,L,1)= - D(K,L)
      LPN= L + N
  5   X(K,L)= - D(K,LPN)
      RETURN
  6   DO 7 I=1,N
      DO 7 K=1,N
      DO 7 L=1,N
  7   D(I,K)= D(I,K) + A(I,L)*X(L,K)
  8   IF (J-NJ)  11,9,9
  9   DO 10 I=1,N
      DO 10 L=1,N
      G(I)= G(I) - Y(I,L)*E(L,NP1,J-2)
      DO 10 M=1,N
 10   A(I,L)= A(I,L) + Y(I,M)*E(M,L,J-2)
 11   DO 12 I=1,N
      D(I,NP1)= - G(I)
      DO 12 L=1,N
      D(I,NP1)= D(I,NP1) + A(I,L)*E(L,NP1,J-1)
      DO 12 K=1,N
 12   B(I,K)= B(I,K) + A(I,L)*E(L,K,J-1)
      CALL MATINV(N,NP1,DETERM)
      IF (DETERM)  14,13,14
 13   PRINT 101, J
 14   DO 15 K=1,N
      DO 15 M=1,NP1
 15   E(K,M,J)= - D(K,M)
      IF (J-NJ)  20,16,16
 16   DO 17 K=1,N
 17   C(K,J)= E(K,NP1,J)
      DO 18 JJ=2,NJ
      M= NJ - JJ + 1
      DO 18 K=1,N
      C(K,M)= E(K,NP1,M)
      DO 18 L=1,N
 18   C(K,M)= C(K,M) + E(K,L,M)*C(L,M+1)
      DO 19 L=1,N
      DO 19 K=1,N
 19   C(K,1)= C(K,1) + X(K,L)*C(L,3)
 20   RETURN
      END
```

† Courtesy of Professor Newman.

```
      SUBROUTINE MATINV(N,M,DETERM)
      DIMENSION A(6,6),B(6,6),C(6,401),D(6,13),ID(6)
      COMMON A,B,C,D
      DETERM=1.0
      DO 1 I=1,N
    1 ID(I)=0
      DO 18 NN=1,N
      BMAX=1.0
      DO 6 I=1,N
      IF(ID(I).NE.0) GOTO 6
      BNEXT=0.0
      BTRY=0.0
      DO 5 J=1,N
      IF(ID(J).NE.0) GOTO 5
      IF(ABS(B(I,J)).LE.BNEXT) GOTO 5
      BNEXT=ABS(B(I,J))
      IF(BNEXT.LE.BTRY) GOTO 5
      BNEXT=BTRY
      BTRY=ABS(B(I,J))
      JC=J
    5 CONTINUE
      IF(BNEXT.GE.BMAX*BTRY) GOTO 6
      BMAX=BNEXT/BTRY
      IROW=I
      JCOL=JC
    6 CONTINUE
      IF(ID(JC).EQ.0) GOTO 8
      DETERM=0.0
      RETURN
    8 ID(JCOL)=1
      IF(JCOL.EQ.IROW) GOTO 12
      DO 10 J=1,N
      SAVE=B(IROW,J)
      B(IROW,J)=B(JCOL,J)
   10 B(JCOL,J)=SAVE
      DO 11 K=1,M
      SAVE=D(IROW,K)
      D(IROW,K)=D(JCOL,K)
   11 D(JCOL,K)=SAVE
   12 F=1.0/B(JCOL,JCOL)
      DO 13 J=1,N
   13 B(JCOL,J)=B(JCOL,J)*F
      DO 14 K=1,M
   14 D(JCOL,K)=D(JCOL,K)*F
      DO 18 I=1,N
      IF(I.EQ.JCOL) GO TO 18
      F=B(I,JCOL)
      DO 16 J=1,N
   16 B(I,J)=B(I,J)-F*B(JCOL,J)
      DO 17 K=1,M
   17 D(I,K)=D(I,K)-F*D(JCOL,K)
   18 CONTINUE
      RETURN
      END
```

Table E-2 Program DIST for solution of distillation by successive linearization (Newman 1967, 1978)†

```
      PROGRAM DIST(INPUT,OUTPUT)
C     PROGRAM FOR FRACTIONATING COLUMN WITH SIDE-STREAM DRAWS
      DIMENSION A(13,13),B(13,13),C(13,40),D(13,27),G(13),X(13,13),Y(13,
     113),AK(10),BK(10),CK(10),DK(10),AHL(10),BHL(10),CHL(10),AHV(10),BH
     2V(10),CHV(10),F(40),HF(40),FX(10,40),IN(5),SPECS(4),SL(40),SV(40)
     3,ERR(40),SAVE(10),T(40),AL(40),V(40)
      COMMON A,B,C,D,G,X,Y,NP3,NS,AK,BK,CK,DK,KTYP,NC,JCOTYP,ITERAT,NP1,
     1NP2,QC,QR,SL,SV,ERR,T,AL,V,F,HF,FX,IN,SPECS
  101 FORMAT (74HICOMPONENTS STAGES    FEEDS    COTYP    KTYP    LIMIT    OR
     1AWS      INSTRUCTIONS,26X,7HPROBLEM/(13I8))
  102 FORMAT (118H0  I       AK(I)        BK(I)        CK(I)        DK(I)
     1 AHL(I)       BHL(I)       CHL(I)       AHV(I)       BHV(I)       CHV(I)/
     2(I3,7E12.4,3E11.4))
  103 FORMAT (60H0      STAGE           FEED          ENTHALPY          COMPO
     1NENTS/I8,E22.7,6E15.7/(8E15.7))
  104 FORMAT (40H0      STAGE    LIQUID DRAW      VAPOR DRAW/(I8,2E17.6))
  105 FORMAT (69H0      SUMERR          HETERR          SUBCOL           DTLIM
     1         SPECS/(8E15.7))
    1 FORMAT (9E8.4)
    2 FORMAT (13I4)
    3 READ 2, NC,NS,NF,JCOTYP,KTYP,LIM,NDRAW,(INT(I),I=1,5),NPROB
      IF (NC)  98,98,99
   98 STOP
   99 READ 1, (AK(I),BK(I),CK(I),DK(I),AHL(I),BHL(I),CHL(I),AHV(I),
     1BHV(I),CHV(I),I=1,NC)
      NP1= NC + 1
      NP2= NC + 2
      NP3= NC + 3
      READ 1, (AL(J),J=1,NS)
      READ 1, (T(J),J=1,NS)
      PRINT 101, NC,NS,NF,JCOTYP,KTYP,LIM,NDRAW,(INT(I),I=1,5),NPROB
      PRINT 102, (I,AK(I),BK(I),CK(I),DK(I),AHL(I),BHL(I),CHL(I),AHV(I),
     1BHV(I),CHV(I),I=1,NC)
      DO 4 J=1,NS
      SL(J)= 0.0
      SV(J)= 0.0
      F(J)= 0.0
      HF(J)= 0.0
      DO 4 I=1,NC
    4 FX(I,J)= 0.0
      DO 6 JF=1,NF
      READ 2, J
      READ 1, HF(J),(FX(I,J),I=1,NC)
      DO 5 I=1,NC
    5 F(J)= F(J) + FX(I,J)
    6 PRINT 103, J,F(J),HF(J),(FX(I,J),I=1,NC)
      IF (NDRAW) 9,9,7
    7 DO 8 J=1,NDRAW
      READ 2, JD
      READ 1, SL(JD),SV(JD)
    8 PRINT 104, JD,SL(JD),SV(JD)
    9 READ 1, SUMERR,HETERR,SUBCOL,DTLIM,(SPECS(I),I=1,4),CHECK
      PRINT 105, SUMERR,HETERR,SUBCOL,DTLIM,(SPECS(I),I=1,4)
      L= NS - 1
      DO 11 I=1,NC
   11 SAVE(I)= 0.0
      ITERAT= -1
      IF (ABSF(CHECK - 11.111) - 0.01) 12,12,3
   12 ITERAT= ITERAT + 1
```

Courtesy of Professor Newman.

```
      V(1)= AL(2) - AL(1) - SL(1) - SV(1) + F(1)
      DO 10 J=2,L
10    V(J)= AL(J+1) + V(J-1) - AL(J) - SL(J) - SV(J) + F(J)
      V(NS)= V(L) - AL(NS) - SL(NS) - SV(NS) + F(NS)
      DO 17 I=1,NC
      EQ= EQUIL(I,T(1))
      ERR(1)= 1.0/(AL(1)+SL(1) + (V(1)+SV(1))*EQ)
      C(I,1)= FX(I,1)*ERR(1)
      DO 16 J=2,NS
      EQB= EQ
      EQ= EQUIL(I,T(J))
      IF (J-NS)  15,13,13
13    IF (JCOTYP)  15,15,14
14    EQ= 1.0
15    ERR(J)=1.0/(AL(J)+SL(J)+(V(J)+SV(J))*EQ-ERR(J-1)*V(J-1)*EQB*AL(J))
16    C(I,J)= (FX(I,J) + V(J-1)*EQB*C(I,J-1))*ERR(J)
      DO 17 JD=1,L
      J= NS - JD
17    C(I,J)= C(I,J) + ERR(J)*C(I,J+1)*AL(J+1)
      DO 19 J=1,NS
      SUMX= 0.0
      DO 18 I=1,NC
18    SUMX= SUMX + C(I,J)
      DO 19 I=1,NC
19    C(I,J)= C(I,J)/SUMX
      IF (ITERAT-LIM)  20,20,22
20    DO 21 I=1,NC
      IF (ABSF(C(I,1)-SAVE(I))-SUMERR)  21,21,27
21    CONTINUE
22    HL= 0.0
      HV= 0.0
      HLU= 0.0
      TU= T(2)
      DO 23 I=1,NC
      HL= HL+C(I,1)*(AHL(I)+BHL(I)*T(1)+CHL(I)*T(1)**2)
      HV= HV+C(I,1)*EQUIL(I,T(1))*(AHV(I)+BHV(I)*T(1)+CHV(I)*T(1)**2)
23    HLU= HLU + C(I,2  )*(AHL(I)+BHL(I)*TU+CHL(I)*TU**2)
      QR= (V(1)+SV(1))*HV + (AL(1)+SL(1))*HL - HLU*AL(2) - HF(1)
      HL= 0.0
      HV= 0.0
      HVB= 0.0
      TB= T(L)
      DO 24 I=1,NC
      HL= HL+C(I,NS)*(AHL(I)+BHL(I)*T(NS)+CHL(I)*T(NS)**2)
      HV=HV+C(I,NS)*EQUIL(I,T(NS))*(AHV(I)+BHV(I)*T(NS)+CHV(I)*T(NS)**2)
24    HVB= HVB + C(I,L)*EQUIL(I,TB)*(AHV(I)+BHV(I)*TB+CHV(I)*TB**2)
      IF (JCOTYP)  26,26,25
25    HV= HL
26    QC= HVB*V(L)-HL*(AL(NS)+SL(NS))-HV*(V(NS)+SV(NS)) + HF(NS)
      CALL OUTPUT
      GO TO 3
27    DO 28 I=1,NC
28    SAVE(I)= C(I,1)
      J= 0
      DO 29 I=1,NP3
      DO 29 K=1,NP3
      Y(I,K)= 0.0
29    X(I,K)= 0.0
30    J= J + 1
```

```
      DO 31 I=1,NP3
      G(I)= 0.0
      DO 31 K=1,NP3
      A(I,K)= 0.0
      B(I,K)= 0.0
   31 D(I,K)= 0.0
      G(NP1)= 1.0
      DO 38 I=1,NC
      B(NP1,I)= 1.0
      EQ= EQUIL(I,T(J))
      B(I,NP1)= EQDT(I,T(J))*C(I,J)*(V(J)+SV(J))
      IF (J-NS)   32,33,33
   32 D(I,I)= - AL(J+1)
      D(I,NP2)= - C(I,J+1)
      GO TO 35
   33 IF (JCOTYP)   35,35,34
   34 EQ= 1.0
      B(I,NP1)= 0.0
      B(NP1,I)= EQUIL(I,T(J)+SUBCOL)
      B(NP1,NP1)= B(NP1,NP1) + C(I,J)*EQDT(I,T(J)+SUBCOL)
   35 IF (J-1) 37,37,36
   36 TB= T(J-1)
      EQB= EQUIL(I,TB)
      A(I,I)= - EQB*V(J-1)
      A(I,NP1)= - EQDT(I,TB)*C(I,J-1)*V(J-1)
      A(I,NP3)= - EQB*C(I,J-1)
   37 B(I,I)= AL(J) + SL(J) + EQ*(V(J)+SV(J))
      B(I,NP2)= C(I,J)
      B(I,NP3)= EQ*C(I,J)
   38 G(I)= FX(I,J)
      IF (ITERAT-IN(5)) 39,39,40
   39 NP3= NP3 - 2
      CALL BAND(J)
      NP3= NP3 + 2
      IF (J-NS) 30,65,65
   40 B(NP3,NP2)= 1.0
      B(NP3,NP3)= 1.0
      G(NP3)= F(J) - SL(J) - SV(J) - AL(J) - V(J)
      D(NP3,NP2)= - 1.0
      A(NP3,NP3)= - 1.0
      IF (J-1) 41,41,49
   41 B(NP2,NP2)= 1.0
      G(NP3)= G(NP3) + AL(J+1)
      CALL BAND(J)
      GO TO 30
   49 IF (J-NS)  50,52,52
   50 TU= T(J+1)
      G(NP3)= G(NP3) + AL(J+1) + V(J-1)
      DO 51 I=1,NC
      EQ= EQUIL(I,T(J))
      EQB= EQUIL(I,TB)
      HVIB= AHV(I) + BHV(I)*TB + CHV(I)*TB**2
      HVI= AHV(I) + BHV(I)*T(J) + CHV(I)*T(J)**2
      HLIU= AHL(I) + BHL(I)*TU + CHL(I)*TU**2
      HLI= AHL(I) + BHL(I)*T(J) + CHL(I)*T(J)**2
      B(NP2,NP2)= B(NP2,NP2) + HLI*C(I,J)
      B(NP2,NP3)= B(NP2,NP3) + HVI*EQ*C(I,J)
      D(NP2,NP2)= D(NP2,NP2) - HLIU*C(I,J+1)
      A(NP2,NP3)= A(NP2,NP3) - HVIB*EQB*C(I,J-1)
```

```
      A(NP2,I)= - HVIB*EQB*V(J-1)
      B(NP2,I)= HLI*(AL(J)+SL(J))  +  HVI*(V(J)+SV(J))*EQ
      D(NP2,I)= - HLIU*AL(J+1)
      A(NP2,NP1)= A(NP2,NP1)-V(J-1)*C(I,J-1)*(HVIB*EQDT(I,TB)+EQB*(BHV(I
     1)+2.0*CHV(I)*TB))
      B(NP2,NP1)= B(NP2,NP1)+C(I,J)*((AL(J)+SL(J))*(BHL(I)+2.0*CHL(I)*T(
     1J))+(V(J)+SV(J))*(HVI*EQDT(I,T(J))+EQ*(BHV(I)+2.0*CHV(I)*T(J))))
   51 D(NP2,NP1)= D(NP2,NP1)  - AL(J+1)*C(I,J+1)*(BHL(I)+2.0*CHL(I)*TU)
      G(NP2)= HF(J)
      CALL BAND(J)
      GO TO 30
   52 B(NP2,NP2)= 1.0
      G(NP3)= G(NP3) + V(J-1)
      CALL BAND(J)

      RAT=0.0
      DO 63 J=1,NS
      RATJ=ABS(C(NP2,J)/AL(J))
      RATVJ=ABS(C(NP3,J)/V(J))
      IF(RATVJ.GT.RAT) RAT=RATVJ
      IF(RATJ.GT.RAT)RAT=RATJ
   63 CONTINUE
      FAC=1.0
      IF(RAT.GT.0.40) FAC=0.4/RAT
      DO 64 J=1,NS
   64 AL(J)=AL(J)+FAC*C(NP2,J)
   65 DO 67 J=1,NS
      IF (ABSF(C(NP1,J))-DTLIM)  67,67,66
   66 C(NP1,J)= DTLIM*C(NP1,J)/ABSF(C(NP1,J))
   67 T(J)= T(J) + C(NP1,J)
      GO TO 12
      END
```

```
      SUBROUTINE OUTPUT
      DIMENSION A(13,13),B(13,13),C(13,40),D(13,27),G(13),X(13,13),Y(13,
     113),AK(10),BK(10),CK(10),DK(10), SL(40),SV(40),SUMX(40),T(40),AL(4
     20),V(40)
      COMMON A,B,C,D,G,X,Y,NP3,NS,AK,BK,CK,DK,KTYP,NC,JCOTYP,ITERAT,NP1,
     1NP2,QC,QR,SL,SV,SUMX,T,AL,V
110   FORMAT (1H1,14,11H ITERATIONS)
111   FORMAT (13,3E15.6,3E14.6)
112   FORMAT (118H0 J        T(J)             L(J)             V(J)
     1X(1,J)        X(2,J)        X(3,J)           X(4,J)           X(5,J) )
113   FORMAT (119H0 J      X(6,J)           X(7,J)           X(8,J)
     1X(9,J)        X(10,J)        X(11,J)          X(12,J)          X(13,J) )
114   FORMAT (105H0 J      X(14,J)         X(15,J)          X(16,J)
     1X(17,J)        X(18,J)        X(19,J)          X(20,J) )
115   FORMAT (17H0CONDENSER LOAD =,E14.6,19H,   REBOILER LOAD =,E14.6)
116   FORMAT (34H0TOP PRODUCT AMOUNTS BY COMPONENTS/E63.6,E15.6,3E14.6/
     1(E18.6,4E15.6,3E14.6))
117   FORMAT (15H0BOTTOM PRODUCT/E63.6,E15.6,3E14.6//(E18.6,4E15.6,3E14.6
     1))
118   FORMAT  (20H0VAPOR DRAW ON STAGE,14/E63.6,E15.6,3E14.6//(E18.6,4E
     115.6,3E14.6))
119   FORMAT (21H0LIQUID DRAW ON STAGE,14/E63.6,E15.6,3E14.6//(E18.6,4E
     115.6,3E14.6))
      PRINT 110, ITERAT
      PRINT 112
      IF (NC-5)  120,120,122
120   DO 121 J=1,NS
121   PRINT 111,  J,T(J),AL(J),V(J),(C(I,J),I=1,NC)
      GO TO 127
122   PRINT 111, (J,T(J),AL(J),V(J),(C(I,J),I=1,5),J=1,NS)
      PRINT 113
      IF (NC-13)  123,123,125
123   DO 124 J=1,NS
124   PRINT 111,  J,(C(I,J),I=6,NC)
      GO TO 127
125   PRINT 111,(J,(C(I,J),I=6,13),J=1,NS)
      PRINT 114
      DO 126 J=1,NS
126   PRINT 111,  J,(C(I,J),I=14,NC)
127   DO 128 I=1,NC
128   SUMX(I) = C(I,1)*AL(1)
      PRINT 117, (SUMX(I),I=1,NC)
      DO 130 I=1,NC
      EQ= EQUIL(I,T(NS))
      IF (JCOTYP)  130,130,129
129   EQ= 1.0
130   SUMX(I) = C(I,NS)*EQ*V(NS)
      PRINT 116, (SUMX(I),I=1,NC)
      DO 137 J=1,NS
      IF (SL(J))  133,133,131
131   DO 132 I=1,NC
132   SUMX(I)= C(I,J)*SL(J)
      PRINT 119, J,(SUMX(I),I=1,NC)
133   IF (SV(J))  137,137,134
134   DO 136 I=1,NC
      EQ= EQUIL(I,T(J))
      IF (J-NS)  136,138,138
138   IF (JCOTYP)  136,136,135
135   EQ= 1.0
```

```
136 SUMX(I) = C(I,J)*EQ*SV(J)
    PRINT 118, J,(SUMX(I),I=1,NC)
137 CONTINUE
    PRINT 115, QC,QR
    RETURN
    END


    FUNCTION EQUIL(I,T)
    DIMENSION A(13,13),B(13,13),C(13,40),D(13,27),G(13),X(13,13),Y(13,
   113),AK(10),BK(10),CK(10),DK(10)
    COMMON A,B,C,D,G,X,Y,NP3,NS,AK,BK,CK,DK,KTYP
    IF (KTYP-1) 1,1,2
  1 EQUIL= EXPF(AK(I)/(T+DK(I)) + BK(I)+CK(I)*(T+DK(I)))
    RETURN
  2 EQUIL= AK(I) + BK(I)*T + CK(I)*T**2 + DK(I)*T**3
    RETURN
    END


    FUNCTION EQDT(I,T)
    DIMENSION A(13,13),B(13,13),C(13,40),D(13,27),G(13),X(13,13),Y(13,
   113),AK(10),BK(10),CK(10),DK(10)
    COMMON A,B,C,D,G,X,Y,NP3,NS,AK,BK,CK,DK,KTYP
    IF (KTYP-1) 1,1,2
  1 EQDT= EXPF(AK(I)/(T+DK(I))+BK(I)+CK(I)*(T+DK(I)))*(CK(I)-AK(I)/(
   1T+DK(I))**2)
    RETURN
  2 EQDT= BK(I) + 2.0*CK(I)*T + 3.0*DK(I)*T**2
    RETURN
    END
```

The number of unknowns n on each stage is taken to be $n = NC + 3$, where NC is the number of components. The three additional unknowns are proposed *changes* in the temperature, liquid flow rate, and vapor flow rate.

The flow rate of the bottom product is controlled, i.e., left unchanged, by statement 41, and that of the reflux by statement 52. These represent the two remaining degrees of freedom after the number of stages, nature of feed, feed and side-draw locations, pressure, etc., have been specified. At the top or bottom of the column one might wish to control any of the following:

1. Bottom-product amount or top-product amount
2. Reflux or vapor flow from the reboiler
3. The mole fraction of a component in the top or bottom product
4. The flow rate of a component in the top or bottom product
5. The reboiler or condenser temperature
6. The heat load for the condenser or the reboiler

In Table E-3 Fortran statements are given for implementing the first five of these possibilities at both the top and the bottom of the column. Statement 41 and the following statement should be replaced by statements 41 to 43, and, for the top, statement 52 and that following should be replaced by statements 52 to 62. These added statements use IN(1) to IN(4) and SPECS(1) and SPECS(2) to decide which specification to use, which component to control, and what value to achieve.

These additional, more flexible specifications must be used with caution. First, they can contradict each other. One cannot specify both the top and bottom products independently, for example. Second, one must stay within the range of possible operating conditions of the column. For example, the reboiler temperature can be no higher than the boiling point of the heaviest component.

The input data are outlined below:

$$NC = \text{number of components}$$
$$NS = \text{number of stages, including reboiler and condenser}$$
$$NF = \text{number of feeds}$$
$$JCOTYP = \begin{cases} 0 & \text{for partial condenser} \\ 2 & \text{for total condenser} \end{cases}$$
$$KTYP = \begin{cases} 0 & \text{for exponential equilibrium ratios} \\ 2 & \text{for power-series equilibrium ratios} \end{cases}$$
$$LIM = \text{limit on total number of iterations}$$
$$NDRAW = \text{number of stages on which side draws occur}$$

IN(1) to IN(4). Used for alternate problem specifications; see above.

IN(5). Controls the number of times that temperature corrections are made without changes in the liquid and vapor flow rates (see the statement just before statement 39). A low value for IN(5) would typically be used. This serves to hold the total-flow-rate portions of the matrix of partial derivatives inactive for IN(5) iterations.

NPROB. Problem number for identification of output.

AK, BK, CK, DK. Parameters in the expressions for the equilibrium ratios, as follows:

Power-series expression:

$$K_i = AK_i + (BK_i)T + (CK_i)T^2 + (DK_i)T^3 \tag{E-6}$$

Table E-3 Alternate statements for DIST to allow changes in problem specification (Newman, 1967)

```
    41 IF (IN(1)) 47,47,42
    42 K= IN(1)
       GO TO (43,44,45,46),K
C      FIXED BOTTOM PRODUCT COMPOSITION, SPECS(1)= X(I)
    43 I= IN(3)
       B(NP2,I)= 1.0
       G(NP2)= SPECS(1)
       GO TO 48
C  FIXED BOTTOM PRODUCT COMPONENT AMOUNT, SPECS(1)= X(I)*AL(J)
    44 I= IN(3)
       B(NP2,I)= AL(I)
       B(NP2,NP2)= C(I,1)
       G(NP2)= SPECS(1)
       GO TO 48
C      FIXED REBOILER TEMPERATURE, SPECS(1)= T(J)
    45 B(NP2,NP1)= 1.0
       G(NP2)= SPECS(1) - T(1)
       GO TO 48
C      FIXED VAPOR FLOW FROM REBOILER, SPECS(1)= V(I)
    46 B(NP2,NP3)= 1.0
       G(NP2)= SPECS(1) - V(1)
       GO TO 48
C      FIXED BOTTOM PRODUCT AMOUNT
    47 B(NP2,NP2)= 1.0
    48 G(NP3)= G(NP3) + AL(J+1)

    52 IF (IN(2)) 61,61,53
    53 K= IN(2)
       GO TO (54,56,59,60),K
C      FIXED TOP PRODUCT COMPOSITION, SPECS(2)= Y(I)
    54 I= IN(4)
       B(NP2,I)= EQUIL(I,T(NS))
       B(NP2,NP1)= C(I,NS)*EQDT(I,T(NS))
       G(NP2)= SPECS(2)
       IF (JCOTYP) 62,62,55
    55 B(NP2,I)= 1.0
       B(NP2,NP1)= 0.0
       GO TO 62
C      FIXED TOP PRODUCT COMPONENT AMOUNT, SPECS(2)= Y(I)*V(J)
    56 I= IN(4)
       EQ= EQUIL(I,T(NS))
       B(NP2,NP1)= C(I,NS)*V(NS)*EQDT(I,T(NS))
       IF (JCOTYP) 58,58,57
    57 EQ= 1.0
       B(NP2,NP1)= 0.0
    58 B(NP2,NP3)= C(I,NS)*EQ
       B(NP2,I)= EQ*V(NS)
       G(NP2)= SPECS(2)
       GO TO 62
C      FIXED CONDENSER TEMPERATURE, SPECS(2)= T(J)
    59 B(NP2,NP1)= 1.0
       G(NP2)= SPECS(2) - T(NS)
       GO TO 62
C      FIXED TOP PRODUCT AMOUNT
    60 B(NP2,NP3)= 1.0
       GO TO 62
C      FIXED REFLUX
    61 B(NP2,NP2)= 1.0
    62 G(NP3)= G(NP3) + V(J-1)
```

Exponential expression:

$$K_i = \exp\left[\frac{AK_i}{T + DK_i} + BK_i + CK_i(T + DK_i)\right]$$ (E-7)

AHL, BHL, CHL. Parameters in a power-series expression for the enthalpy of a liquid stream. Per mole of mixture,

$$h = \sum_i x_i[AHL_i + (BHL_i)T + (CHL_i)T^2]$$ (E-8)

AHV, BHV, CHV. Parameters in a power-series expression for the enthalpy of a vapor stream. Per mole of mixture

$$H = \sum_i y_i[AHV_i + (BHV_i)T + (CHV_i)T^2]$$

AL. Initial estimates of the flow rates of the liquid streams leaving each stage (the last one being the reflux).
T. Estimated temperatures for each stage.
J. Feed stage.
HF. Total enthalpy of the feed.
FX. Feed rate for each component (J, HF, and FX are repeated for each feed stage).
JD. Number of stage with a side draw.
SL and SV. Molal flow rates for liquid and vapor sidestreams (JD, SL, and SV are repeated for each sidestream).
SUMERR. Used to check convergence. The mole fraction of each component in the reboiler must change by less than SUMERR between one iteration and the next in order to satisfy the convergence criterion.
SUBCOL. Subcooling of reflux for total condenser (degrees below bubble point).
DTLIM. Upper limit on the temperature correction, degrees.
SPECS(1) and SPECS(2). Used with alternate problem specifications (see above).
CHECK. 011111 + 1 in columns 65 to 72. This is used to make sure that the correct number of cards has been read.

Any number of problems can be run consecutively. In the output, J is the stage number, T is the temperature, AL is the liquid flow (or reflux for a total condenser), SUMX is the sum of the mole fractions, and X are component mole fractions. The component flow rates are then listed for the bottom product, the top product, and any sidestreams.

Fredenslund et al. (1977) give complete listings of distillation programs using the UNIFAC method to generate vapor-liquid equilibrium data and using the Newton multivariate SC method for convergence. Two programs are given, one allowing for variable molar overflow and the other postulating constant molal overflow.

REFERENCES

Fredenslund, A., J. Gmehling, and P. Rasmussen (1977): "Vapor-Liquid Equilibria Using UNIFAC," Elsevier, Amsterdam.
Newman, J. S. (1967): *Lawrence Berkeley Lab. Rep.* UCRL-17739, Berkeley, Calif.
——— (1968): *Ind. Eng. Chem. Fundam.*, 7:514.
——— (1973): "Electrochemical Systems," Prentice-Hall, Englewood Cliffs, N.J.
——— (1978): University of California, Berkeley, personal communication.

SUMMARY OF PHASE-EQUILIBRIUM AND ENTHALPY DATA

Type	Components	Location
	Phase equilibrium	
	General references	Pp. 42–43
Gas-liquid	H_2S, CO_2, C_2H_4, O_2, CO, N_2 in water	Fig. 6-7
	CO_2-potassium carbonate solution	Fig. 6-32
	CO_2-water; NH_3-water	Fig. 7-39
	H_2S-CO_2-monoethanolamine solution	Figs. 10-2 to 10-5
	CH_4-220-MW paraffinic oil	Fig. 13-6
Vapor-liquid	n-Butane, n-pentane, n-hexane	Fig. 2-2
	Ethanol-water	Fig. 2-21
	Hydrogen-methane	Fig. 2-23
	Acetone–acetic acid	Example 2-7
	Methanol-water	Prob. 5-B
	Ethylene-ethane	Prob. 5-M
	Acetone-water	Table 6-2
	Water vapor–NaOH solution	Prob. 6-E
	Ethanol-water-benzene	Fig. 7-29
	Methylcyclohexane-toluene-phenol	Fig. 7-31
	n-Heptane–toluene–methyl ethyl ketone	Fig. 7-33
	Alcohol mixtures	Probs. 2-R, 8-D
	Propylene-propane	Prob. 8-J
	Propyne-propylene-propane	Prob. 8-L
	Methanol-water-formaldehyde	Fig. 10-13

Liquid-liquid	Vinyl acetate–acetic acid–water	Fig. 1-21
	$Zr(NO_3)_4$–$NaNO_3$–HNO_3–H_2O–tributyl phosphate	Example 6-1
	Water–acetone–methyl isobutyl ketone	Example 6-6
	Methylcyclohexane–n-heptane–aniline	Prob. 6-F
	Water–phenol–isoamyl acetate	Prob. 6-K
Gas-solid	Water vapor–activated alumina	Fig. 3-17
	Water vapor–molecular sieve	Fig. 3-19
Liquid-solid	m-Cresol-p-cresol	Fig. 1-25
	Gold-platinum	Fig. 1-27
	p-Xylene-m-xylene	Fig. 14-4
	p-Xylene-m-xylene-o-xylene	Fig. 14-5

Enthalpy

n-Butane, n-pentane, n-hexane	Fig. 2-10
Ethanol-water	Figs. 2-20, 2-21
Ethanol-isopropanol-n-propanol	Prob. 2-P
Acetone-water	Table 6-2
Hydrogen (Mollier diagram)	Fig. 13-4
Methane (Mollier diagram)	Fig. 13-5

Other

Properties of xylene isomers	Table 14-2
Solubility parameters of various compounds	Table 14-6

NOMENCLATURE

Symbol	Definition	Dimensions[†]
a	Interfacial area per unit tower volume	L^{-1}
$a_1, \ldots, a_6$	Constants in Martin equations. defined in Eqs. (8-64) and (8-65)	
a_i	Heat-transfer area of coils in effect i (Appendix B)	L^2
a_p	Surface area of dry packing per unit packed volume (Chap. 12)	L^{-1}
A	Tower cross-sectional area; cross-sectional area of liquid in direction of flow (Chap. 12); area per membrane (dialysis)	L^2
A	Absorbent flow rate; airflow rate	mol/t
A	Constant defined by Eq. (8-71)	
A_p	Membrane area in stage p	L^2
A_p	Coefficients defined by Eq. (10-15)	
b	Bottoms flow rate in distillation column	mol/t
b	Intercept of straight line	
b'	Bottoms product in batch distillation	mol
B	Constant defined by Eq. (8-72)	
B	Available energy, $H - T_0 S$ (Chap. 13)	Q/mol
B_p	Coefficients defined by Eqs. (10-16) to (10-18)	
Bi	Biot number	
c	Molar density; concentration	mol/L^3
C	Number of components, in phase rule; constant defined by Eq. (8-73)	
$C(s, p)$	Binomial coefficient; number of combinations which can be made from s objects taken p at a time	
C_i	Concentration of component i	mol/L^3
C_{ij}	Concentration of component i in stream j. Eq. (1-22)	mol/L^3
$C_{j, p}$	Component mass-balance function. Eq. (10-12)	
C_p	Heat capacity	Q/MT
C_p	Coefficients defined by Eq. (10-1)	
d	Differential operator	
d	Distillate flow rate (liquid)	mol/t
d	Diameter	L
d_p	Drop diameter	L
D	Distillate flow rate (vapor)	mol/t
D_{AB}	Diffusivity for A in B	L^2/t
D_E	Effective diffusivity for mixing in direction of flow (Chap. 12)	L^2/t
D_G	Molecular diffusivity in gas phase	L^2/t
D_L	Molecular diffusivity in liquid phase	L^2/t
D_p	Coefficients defined by Eq. (10-20)	

[†] L = length Q = heat or energy
M = mass T = temperature
mol = moles t = time
P = pressure

Symbol	Definition	Dimensions†
$(DR)_i$	Distribution ratio for component i, defined by Eq. (9-26)	
e	Base of natural logarithms	
e	Moles of entrainment per unit time (Chap. 12)	mol/t
E	Extract flow rate	mol/t
E	Axial dispersion coefficient	L^2/t
$\mathbf{E}$	Eq. (10-1) (type of equation)	
E_1, E_2	Constants defined by Eq. (8-74)	
E_a	Murphree vapor efficiency: apparent value in the presence of entrainment	
E_i	Hausen stage efficiency, Eq. (12-38)	
E_{ip}	Holland vaporization efficiency for component i on stage p, Eq. (12-36)	
E_{MVi}	Murphree stage efficiency for component i based on stream V	
E_O	Overall stage efficiency	
E_{OG}	Point efficiency (Chap. 12)	
E_V	Energy dissipation rate	$Q/L^3 t$
$(\Delta E_V)_i$	Latent energy of vaporization of component i	Q/mol
erf (x)	Error function of x, defined by Eq. (8-51)	
f	Fraction of gas flowing to the left of location considered (Chaps. 3 and 12); fraction back mixing (Chap. 11); Fanning friction factor (Chap. 11)	
$f(x)$	Function of x	
f_i	Flow of component i in feed	mol/t
f_i	Fraction of water in feed which remains in solution leaving effect i (Appendix B); probability of component i going to next stage in any one transfer (countercurrent distribution)	
$f_{j,p}$	Flow of component j in feeds (less products) to stage p (Chap. 10)	mol/t
F	Feed flow rate	mol/t
F	Degrees of freedom, in phase rule	
F	$u_G \sqrt{\rho_G}$ (Chap. 12)	ft/s·(lb/ft³)$^{1/2}$
F'	Charge to a batch separation process	mol
F_i	Moles of component i in feed pulse (chromatography)	mol
Fo	Fourier number	
$g(x)$	Function of x	
g_p	Coefficients in Thomas method, Eqs. (10-27) and (10-28)	
G	Gas flow rate	mol/t; in Chap. 12, lb/h·ft²
G	Gibbs free energy	Q
G'	Flow rate of inert components of a gas	mol/t
G_M	Moles of gas per unit time per unit tower cross-sectional area (Chap. 12)	mol/tL^2
h	Heat-transfer coefficient	$Q/L^2 Tt$
h	Specific enthalpy of a liquid	Q/mol
h	Tower height	L
h^*	Specific enthalpy of feed at temperature corresponding to tower pressure bubble point of feed mixture (Chap. 5)	Q/mol
h_D	Liquid head corresponding to pressure drop due to gas flow through dispersing unit (Chap. 12)	L
h_{da}	Liquid head corresponding to frictional loss from flow through downcomer and under downcomer apron (Chap. 12)	L
h_{eq}	Molal enthalpy of the liquid which would be in equilibrium with the feed if the feed were a vapor at its column pressure dew point (Chap. 5)	Q/mol
h_F	Specific enthalpy of feed	Q/mol
h_i	Specific enthalpy of component i	Q/mol
h_l	Equivalent clear liquid height of liquid on a plate (Chap. 12)	L
h_L	Specific enthalpy of liquid phase	Q/mol

Symbol	Definition	Dimensions†
h_{ow}	Clear liquid crest over weir (Chap. 12)	L
h_t	Tray-to-tray pressure drop, expressed as liquid head (Chap. 12)	L
h_V	Specific enthalpy of vapor phase	Q/mol
h_w	Weir height (Chap. 12)	L
H	Specific enthalpy of a vapor	Q/mol
H	Henry's law constant	PL^3/mol
H	Total enthalpy of a stream	Q
H	Height of liquid in downcomer (Chap. 12)	L
H	Eqs. (10-3) (type of equation)	
H'	Flow rate of high-pressure product from gaseous diffusion stage	mol/t
H^*	Specific enthalpy of feed at temperature corresponding to tower pressure dew point of feed mixture (Chap. 5)	Q/mol
H'_0	Feed flow rate to gaseous diffusion stage	mol/t
ΔH_{abs}	Heat of absorption	Q/mol
H_{eq}	Molal enthalpy of the vapor which would be in equilibrium with the feed if the feed were a liquid at its column pressure bubble point (Chap. 5)	Q/mol
H_L	Enthalpy of a liquid product	Q
H_p	Enthalpy function for stage p, Eqs. (10-3)	
H_s	Length equivalent to an equilibrium stage (chromatography)	L
H_V	Enthalpy of a vapor product	Q
ΔH_v	Latent heat of vaporization	Q/mol
HETP	Height equivalent to a theoretical plate	L
$(HTU)_{OG}$	Height of an overall transfer unit, based on stream G	L
$(HTU)_G$	Height of an individual phase transfer unit, based on stream G	L
i	Square root of -1	
j_D	Mass transfer j factor, Eq. (11-44)	
j_H	Heat transfer j factor, Eq. (11-45)	
J	Molar flux	mol/$L^2 t$
k	Boltzmann's constant	Q/molecule $\cdot T$
k	Thermal conductivity	Q/LtT
k_c	Mass-transfer coefficient, based upon concentration driving force	L/t
k_G	Individual gas phase mass-transfer coefficient, based upon partial-pressure driving force	mol/tPL^2
k_L	Individual liquid phase mass-transfer coefficient, based upon concentration driving force	L/t
k_s	Rate proportionality constant for salt transport across a membrane, defined by Eq. (1-23)	L/t
k_w	Rate proportionality constant for water transport across a membrane, defined by Eq. (1-22)	mol/tPL^2
k_x	Individual liquid-phase mass-transfer coefficient, based upon mole-fraction driving force	mol/$L^2 t$
k_y	Individual gas-phase mass-transfer coefficient, based upon mole-fraction driving force	mol/$L^2 t$
K	Equilibrium constant for chemical reaction	
K_G	Overall mass-transfer coefficient, based on gas phase and partial-pressure driving force	mol/tPL^2
K_i	Equilibrium ratio of component i between phases; ratio of mole fractions at equilibrium	
K'_i	Equilibrium ratio of component i between phases; ratio of concentrations (mol/L^3) at equilibrium	
K_L	Overall mass-transfer coefficient, based on liquid phase and concentration driving force	L/t
$K_{j,p}$	Equilibrium ratio for component j on stage p (Chap. 10)	

Symbol	Definition	Dimensions†
K_v	Factor in Eq. (12-1)	
l_i	Flow rate of component i in liquid	mol/t
$l_{j,p}$	Flow rate of component j in liquid leaving stage p	mol/t
ln	Natural logarithm	
log	Base 10 logarithm	
L	Total liquid flow rate; liquid flow rate in rectifying section	mol/t; in Chap. 12, lb/h·ft^2
L	Liquid flow rate across a plate (Chap. 12)	gal/min·ft of flow path width
L	Characteristic length (Chap. 11)	L
L'	Liquid flow rate in stripping section; flow rate of inert components in liquid	mol/t
L'	Amount of liquid in the still pot in a Rayleigh distillation	mol
L'_0	Initial liquid charge to a Rayleigh distillation	mol
L''	Liquid flow rate in intermediate section of a multistage separation process	mol/t
L_F	Flow rate of liquid in feed	mol/t
$(\Delta L)_f$	Change in L at feed stage, $= L' - L$	mol/t
L_M	Liquid flow rate per unit cross-sectional area of tower	mol/tL^2
m	Slope of the equilibrium curve, $= dy/dx$; number of stages below feed	
M	Liquid holdup on a stage	mol
M	Eqs. (10-2) (type of equation)	
M_g	Moles of gas per unit column volume (chromatography)	mol/L^3
M_i	Molecular weight of component i	M/mol
M_{iF}	Amount or concentration of component i in feed (countercurrent distribution)	mol or mol/L^3
M_{ips}	Amount or concentration of component i in stage p after transfer step s	mol or mol/L^3
M_l	Moles of liquid per unit column volume (chromatography)	mol/L^3
M_v	Molecular weight of vapor	M/mol
n	Number of stages above feed	
N	Number of stages	
N_i	Flux of component i across interface or barrier	mol/tL^2
N_R	Number of stages in rectifying section	
N_S	Number of stages in stripping section	
$(\text{NTU})_{OG}$	Number of overall transfer units, based upon stream G	
$(\text{NTU})_G$	Number of individual phase transfer units, based upon stream G	
p	Stage number	
$p(E)$	Probability that stage efficiency is E (Appendix D)	
p_i	Partial pressure of component i	P
P	Total pressure	P
P	Number of phases in equilibrium, in phase rule	
ΔP	Pressure drop across membrane, Eq. (1-22)	P
P_1, P_2	Flow rates of products 1 and 2, respectively	mol/t
P_1, P_2	Total pressures of side 1 and 2, respectively, in gaseous diffusion process	P
P'_1, P'_2	Total products from batch separation process	mol
P_i^0	Vapor pressure of pure component i	P
$[P_i(p)]_s$	Probability that a molecule of component i has made p transfers forward from one stage to the next during s transfer steps (countercurrent distribution)	
Pe	Peclet number	
q	Heat flux	$Q/L^2 t$
q_p	Net amount of heat *added* to stage p	Q/t
Q	Heat flow	Q/t
ΔQ	Net flow of enthalpy in positive direction, Eq. (10-38)	Q/t
Q_C	Condenser duty	Q/t
Q_R	Reboiler duty	Q/t

Symbol	Definition	Dimensions†
r	Radius	L
r	Reflux flow rate	mol/t
r	Moles MEA/mol of inlet gas (Example 10-2)	
R	Gas constant	Q/mol·T or PL^3/mol·T
R	Raffinate flow rate	mol/t
R	Number of components	
R_A	High-flux parameter (Eq. 11-57)	
R_i	Ratio of effective velocity of component i along the column to the gas velocity (chromatography)	
Re	Reynolds number	
s	Surface renewal rate	1/t
s	Number of transfers (countercurrent distribution)	
S	Solids flow rate	Various
S	Solvent flow rate	mol/t
S	Entropy	Q/T
S	Eqs. (10-4) and (10-5) (type of equation)	
S'	Tray spacing (Chap. 12)	L
S_0	Steam-condensation rate in coils of first effect (Appendix B)	M/t
S_{ij}	Solvent selectivity for i over j, Eq. (14-1)	
$S_{L,p}$	Liquid sidestream flow rate (Chap. 10)	mol/t
S_R	Number of different column sequences possible for separating R products	
$S_{V,p}$	Vapor sidestream flow rate (Chap. 10)	mol/t
$S_{y,p}$	Summation of mole fraction function, Eqs. (10-4) and (10-5)	
Sc	Schmidt number, $\mu/\rho D$	
Sh	Sherwood number	
t	Time	t
t_i	Time elapsed between feed sample injection and emergence of peak for component i (chromatography)	t
t_L	Residence time of liquid on a plate in a distillation column (Chap. 12)	t
t_R	Residence time within separation device	t
T	Temperature	T
T_0	Ambient temperature (Chap. 13)	T
T_i	Saturation temperature of vapor generated in effect i (Appendix B)	T
T_S	Saturation temperature of steam to first effect (Appendix B)	T
u	Parameter in Thomas method, Eq. (10-24)	
u_G	Gas velocity (superficial) in chromatography	L/t
u_i	Velocity of peak maximum for component i along column in chromatography	L/t
U	Liquid holdup	mol
U	Average heat-transfer coefficient	Q/tTL^2
$U(N, R, E)$	Annual cost of distillation system for given values of number of stages N, reflux ratio R, and stage efficiency E (Appendix D)	Dollars
$\bar{U}(N)$	Expected cost of distillation system for a given number of stages N (Appendix D)	Dollars
U_{flood}	U_s at flooding (Chap. 12)	L/t
U_L	Interstitial velocity of fluid	L/t
U_s	Gas flow rate, volumetric flow per unit time and per unit active tray area (Chap. 12)	L/t
v	Velocity (Chap. 11)	L/t
v_i	Flow rate of component i in a vapor stream	mol/t
v_i	Dimensionless cumulative carrier gas flow, defined by Eq. (8-47)	
$v_{j,p}$	Flow of component j in vapor leaving stage p	mol/t
V	Total vapor flow rate; vapor flow rate in rectifying section	mol/t

Symbol	Definition	Dimensions[†]
V	Velocity	L/t
V	Cumulative volume of carrier gas passed through column since feed sample injection (chromatography); volume of a phase	L^3
$\bar{V}$	Partial molal volume	L^3/mol
V'	Vapor flow rate in stripping section	mol/t
V''	Vapor flow rate in intermediate section of a multistage separation process	mol/t
V_0	Initial vapor charge to a batch separation process	mol
V_F	Flow rate of vapor in feed	mol/t
$(\Delta V)_f$	Change in vapor flow rate at feed stage, $= V - V'$	mol/t
V_G	Volume of gas within each vessel in stage chromatography model	L^3
V_i	Molal volume of component i	L^3/mol
V_L	Volume of liquid within each vessel in stage chromatography model	L^3
V_p	Vapor flow leaving stage p (Chap. 10)	mol/t
V_p	Permeate volumetric flow from stage p	L^3/t
w_C	Weight of water in product concentrate (Example 3-1)	M
w_i	Weight fraction of component i	
w_i'	Weight fraction of component i on a solvent-free basis	
w_i'	Time lapse corresponding to peak width for component i (chromatography)	t
w_I	Weight of ice crystals (Example 3-1)	M
w_p	Coefficients in Thomas method; Eqs. (10-23) and (10-25)	
W	Weight	M
W	Water flow rate	mol/t
W	Weir height (Chap. 12)	Inches
W	Work (Chap. 13)	Q/mol
W_0	Amount of water in feed salt solution (Appendix B)	M/t
W_n	Net work consumption of process	Q/mol
W_n'	Net work consumption of process	Q/t
x	Solute mole fraction (usually liquid)	
x	Distance from leading edge (Chap. 11)	L
x'	Mole fraction on the basis of the keys alone (liquid) (Chap. 7)	
x_i	Mole fraction of component i (usually liquid)	
$x_{i,j}$	Liquid-phase mole fraction of component i in stream j	
x_L	Solute mole fraction in the liquid phase	
x_S	Solute mole fraction in the solid phase	
X, X_i	Mole ratio of component i in liquid phase (moles i/moles inert)	
y	Solute mole fraction in vapor or gas phase	
y'	Mole fraction on the basis of the keys alone, vapor phase (Chap. 7); vapor-phase mole fraction, including entrained liquid (Chap. 3)	
y_i	Mole fraction of component i, vapor phase	
$y_{i,j}$	Mole fraction of component i in stream j (vapor)	
Y, Y_i	Mole ratio of component i in vapor phase (mol i/mol inert)	
z	Distance in the direction of liquid flow across a distillation plate (Chap. 12); distance	L
z_i	Mole fraction of component i in total feed stream, even if mixed phases	
$z_{j,p}$	Mole fraction of component j in net feed to stage p (Chap. 10)	
Z_c	Tray holdup of liquid, in^3 of liquid/in^2 of active bubbling area; also known as clear liquid height (Chap. 12)	L
Z_l	Length of liquid flow path across plate (Chap. 12)	L

Symbolic

$(/i)_j$	Recovery fraction of component i in product j; amount	

Symbol	Definition	Dimensions†
	of component i in product j divided by amount of component i fed	
Greek		
α_i	Relative volatility of component i with respect to reference component	
α_{ij}	Separation factor: relative volatility of component i with respect to component j; K_i/K_j: equilibrium or inherent separation factor	
$\alpha_{ij}^\lor$	Actual separation factor between components i and j, based upon actual product compositions	
β	Constant defined in Eq. (8-78); constant in Eq. (9-27)	
β'	Aeration factor	
γ	Surface tension of liquid phase	dyn/cm
γ_i	Activity coefficient of component i	
γ_{ij}	Activity coefficient of component i in phase j	
δ	Film thickness	L
δ_i	Solubility parameter of component i, defined by Eq. (14-4)	$(Q/L^3)^{1/2}$
$\bar{\delta}$	Average solubility parameter of liquid mixture, defined by Eq. (14-3)	$(Q/L^3)^{1/2}$
$\delta_{s,p}$	Kronecker delta, $\delta = \begin{cases} 1 & \text{if } s = p \\ 0 & \text{if } s \neq p \end{cases}$	
Δ	Difference in a quantity	
Δ	Hydraulic gradient of liquid across a plate (Chap. 12)	L
Δ_{DB}	Temperature difference between dew point and bubble point	T
ϵ	Void fraction of a bed of solids	
ϵ	Lennard-Jones interaction potential	Q/molecule
ϵ_L	Fraction of vessel volume occupied by liquid	
η	Thermodynamic efficiency. $= W_{\min, r_0}/W_n$	
θ	Total exposure time	t
θ	Constant in Eq. (9-27); constant defined by Eq. (8-77); convergence factor, Eq. (10-32)	
θ_1, θ_2	Constants defined by Eqs. (8-75) and (8-76)	
θ_A	High-flux parameter, defined by Eq. (11-55)	
λ	Average latent heat of vaporization of mixture (Chap. 13)	Q/mol
λ_i	$K_i V/L$ or mG/L: stripping factor: reciprocal absorption factor	
μ	Liquid viscosity	M/Lt
ξ	Separation index	
π	3.14159	
$\Delta\pi$	Difference in osmotic pressure across membrane, Eq. (1-22)	P
ρ_G	Gas-phase density	M/L^3
ρ_L	Liquid-phase density	M/L^3
ρ_M	Molar density	mol/L^3
ρ_W	Molar density of water	mol/L^3
σ	Collision diameter	L
Σ	Summation	
ϕ	Parameter in Underwood equations, zone above feed	
ϕ_A	High-flux parameter, Eq. (11-56)	
ϕ'	Parameter in Underwood equations, zone below feed	
$\phi(x)$	A function of x (Appendix A)	
ϕ_{HK-}	Value of ϕ next below the α of the heavy key	
ϕ'_{LK+}	Value of ϕ' next above the α of the light key	
χ_A	Moles of liquid absorbent required to dissolve 1 mol of gas at the feed stage	mol/mol
ψ	Moles of entrained liquid/mole of gross downflowing liquid	mol/mol
$\psi(x)$	A function of x (Appendix A)	
Ω_D	Collision integral: function of kT/ϵ_{AB}	
Subscripts		
A, B, C, ...	Components	

Symbol	Definition	Dimensions†
a, A	Aqueous phase	
av	Average	
b	Bottoms	
d	Distillate (liquid)	
D	Distillate (vapor)	
diff	Difference point	
E	Extract	
E, eq	Equilibrium with other phase	
f	Final; film factor, Eqs. (11-65) and (11-66); feed stage	
F	Feed stream	
flood	Flooding	
G	Gas	
H	High-pressure side (rate-governed separation processes); high temperature	
HK	Heavy key	
HNK	Heavy nonkey	
i, j	Components	
i	Inlet; iteration number (Chap. 10)	
in	Inlet	
J	Defined by Eqs. (11-53) and (11-54); based upon ratio of J to driving force	
L	Lower phase (countercurrent distribution); liquid; low temperature; low-pressure side (rate-governed separation processes)	
lim	Limiting value in zone of constant mole fraction	
LK	Light key	
LM	Logarithmic mean	
LNK	Light nonkey	
min	Minimum	
mix	Mixing	
N	Based on ratio of N to driving force	
o	Outlet	
o, O	Organic phase	
op	Operating	
opt	Optimum	
out	Outlet	
p	Stage p	
r	Reflux; reference component	
R	Reboiler; raffinate	
S	Sidestream; steam; solvent; solids	
sep	Separation	
spec	Specified	
t	Top	
T_0	With heat sources and sinks at ambient temperature	
U	Upper phase (countercurrent distribution)	
w	Water	
x	x phase	
y	y phase	
0	Initial value	
1, 2	Ends of a column; products	
∞	Limiting value in zone of constant composition	

Superscripts

*	Equilibrium with prevailing value in other phase (equilibrium with exit value of other phase in Murphree efficiencies)	
*	Mole average (Chap. 11)	
V	Volume average (Chap. 11)	

INDEX

INDEX